# Introduction to Remote Sensing

FOURTH EDITION

# Introduction to Remote Sensing

## FOURTH EDITION

JAMES B. CAMPBELL

LONDON AND NEW YORK

First published in 1996
by Taylor & Francis
2 Park Square, Milton Park, Abingdon, Oxon OX14 4RN
Fourth edition 2006

*Taylor & Francis is an imprint of the Taylor & Francis Group, an informa business*



Typeset, printed and bound in the United States of America



*British Library Cataloguing in Publication Data*
A catalogue record for this book is available from the British Library

*Library of Congress Cataloging in Publication Data*
A catalogue record for this book is available from the Library of Congress

ISBN10: 0-415-41688-4
ISBN13: 978-0-415-41688-7

*To the memory of my parents,*

**JAMES BATCHELDER CAMPBELL**
**1911–1982**

**and**

**DOROTHY ANN TISON CAMPBELL**
**1917–1987**

# Contents

# Preface

Even a few decades ago, the diversity and capabilities of today's remote sensing systems would have been unimaginable, even for the most committed visionaries. Likewise, few could have anticipated the development of the analytical tools and techniques now available for examination of remotely sensed data, the widespread availability of remotely sensed imagery, or the multiplicity of its uses throughout society. These developments alone present a challenge for any text on this subject.

Further, the tremendous volume of relevant material, and the thorough access provided by the World Wide Web (WWW), can cause anyone to ponder the role of a university text—isn't its content already available to any reader? Despite the value of the WWW as a resource for any student of remote sensing, its complexity and dynamic character actually increases, rather than diminishes, the value of an introductory text. Because of the overwhelming volume of unstructured information on the WWW, students often require a guide to provide structure and context that enables them to select and assess the many sources at hand. In this context, then, this text forms a guide for the use of the many other sources available—sources that may be more comprehensive and up to date than the context of any text could be. Thus, I encourage students to use this volume in partnership with online materials—the text as a guide, and the online materials as a reference for additional depth.

Instructors should supplement the content of this volume with material of significance in their own programs. Supplementary materials will, of course, vary greatly from one institution to the next, depending on access to facilities and equipment, as well as on the varying expectations and interests of instructors, students, and curricula. It is assumed that the text will be used as the basis for readings and lectures, and that most courses will include at least brief laboratory exercises that permit students to examine more images than can be presented here. Because access to specific equipment and software varies so greatly, and because of the great variation in emphasis noted above, this book does not include laboratory exercises. But each chapter does conclude with a set of Review Questions and problems that can assist students in review and assessment of concepts and material.

For students who intend to specialize in remote sensing, this text offers not only an introduction but also a framework for subjects to be studied in greater detail. Students who do plan specialization in remote sensing should consult their instructors to plan a comprehensive course of study based on work in several disciplines as discussed in Chapter 1. This approach is reflected in the text itself: it introduces students to principal topics of significance for remote sensing, but recognizes that students will require additional depth in their chosen fields of study.

For those students who do not intend to pursue remote sensing beyond the introductory level, this book serves as an overview and introduction, so that they can understand remote sensing, its applications in varied disciplines, and its significance in today's world. For many, the primary emphasis will be study of those chapters and methods of greatest significance in their major field of study.

Instructors may benefit from a preview of some of the changes in the fourth edition. Material concerning lidar has been presented in its own chapter, in recognition of the increasing maturity of this technology and the accelerating pace of its applications within the remote sensing arena. Lidar's significance will continue to grow, and will likely continue to displace important applications now based on photogrammetric analysis of aerial photography.

Use of remote sensing to monitor phenology (seasonal changes in both agricultural and native plant cover) opens opportunities for improvements in analysis of agricultural landscapes, derivation of land-cover classes (based on differences in phenological cycles), and examination of land-use change.

Further, new material on remote sensing applications for plant pathology presented in Chapter 17 has significance for the growing awareness of the impact of introduced plant pathogens and their effect on native flora and on important agricultural crops. Remote sensing is an important tool (among others) in society's efforts to address these important challenges, so it is important to introduce to students to this topic, even at an introductory level.

Other developments present more problematic issues for the structure of this text. As these words are written, it is clear that there will be a programmatic gap in Landsat coverage—even if current problems are resolved, its remaining days are numbered. The nature of programs to acquire substitute imagery, and the design of a replacement system, are far too uncertain to discuss at this time. Likewise, although the sensors that form the core global remote sensing programs described in Chapter 21 will be in operation for many more years, there are no programs now under way to extend these systems, so they will expire in due course without a clear avenue to continue their missions. Therefore, although both of these topics will continue to be significant for the field of remote sensing, the specifics are rather murky, and will continue to remain so for some years to come. Instructors should not neglect to discuss these topics, but should monitor current developments to provide details as needed.

Other changes for the fourth edition include the revision of illustrations for Chapter 5 (image interpretation) to include more concrete examples focused on practical needs. Material in earth sciences (Chapter 18), land use and land cover (Chapter 20), and remote sensing and geographic information systems (Chapter 16) has been restructured to provide students with content of a more practical nature, with new illustrations.

To permit instructors to tailor assignments to meet specific structures for their own courses, content is organized at several levels. At the broadest level, a rough division into four parts offers a progression in the subjects presented, with only a few concessions to practicality (such as placing the "Image Interpretation" chapter under "Image Acquisition" rather than in its logical position in "Analysis"). Each part consists of two or more chapters organized as follows:

*Part I. Foundations*

1. History and Scope of Remote Sensing
2. Electromagnetic Radiation

*Part II. Image Acquisition*
3. Photographic Sensors
4. Digital Data
5. Image Interpretation
6. Land Observation Satellites
7. Active Microwave
8. Lidar
9. Thermal Radiation
10. Image Resolution

*Part III. Analysis*
11. Preprocessing
12. Image Classification
13. Field Data
14. Accuracy Assessment
15. Hyperspectral Remote Sensing

*Part IV. Applications*
16. Remote Sensing and Geographic Information Systems
17. Plant Sciences
18. Earth Sciences
19. Hydrospheric Sciences
20. Land Use and Land Cover
21. Global Remote Sensing

Conclusion: The Outlook for the Field of Remote Sensing

The 21 numbered chapters each constitute more or less independent units that can be selected as necessary to meet the specific needs of each instructor. Numbered sections within chapters form even smaller units that can be selected and combined with other material as desired by the instructor.

I gratefully acknowledge the contributions of those who assisted in identifying and acquiring images used in this book. Individuals and organizations in both private industry and governmental agencies have been generous with their advice and support. Daedalus Enterprises Incorporated, EROS Data Center, Digital Globe, Environmental Research Institute of Michigan, EOSAT, GeoSpectra Corporation, IDELIX Software, and SPOT Image Corporation are among the organizations that assisted my search for suitable images. For the fourth edition, I am grateful for the continued support of these organizations and for the assistance of the U.S. Geological Survey, RADARSAT International, Earth Satellite Corporation, and the Jet Propulsion Laboratory. Also, I gratefully recognize the assistance of EarthData, Emerge, and Orbital Imaging Corporation.

Much of what is good about this book is the result of the assistance of colleagues in many disciplines in universities, corporations, and research institutions who have contributed through their correspondence, criticisms, explanations, and discussions. Students in my classes have, through their questions, mistakes, and discussions, contributed greatly to my own learning, and therefore to this volume. Faculty who use this text at other universities have provided suggestions and responded to questionnaires designed by The Guilford Press.

At Guilford, Janet Crane, Peter Wissoker, and Kristal Hawkins guided preparation of the earlier editions. For these earlier editions I am grateful for the special contributions of Chris Hall, Soren Popescu, Russ Congalton, Bill Carstensen, Don Light, David Pitts, and Jim Merchant. Buella Prestrude and George Will assisted me with preparation of the illustrations.

Many individuals have supported preparation of this edition, although none are responsible for the errors and shortcomings that remain. The assistance of Chris North, Bill Carstensen, Steve Prisley, and Maggi Kelly is gratefully acknowledged, as are the suggestions and corrections offered by readers. Teaching assistants, including Dave Trible, Sara Hyland, and Sam Chambers, have contributed to development of materials used in this edition. Anonymous reviewers provided insightful and detailed comments and critiques that extended beyond the scope of the usual manuscript reviews. At The Guilford Press, Kristal Hawkins has guided the launch of this fourth edition. Seymour Weingarten, editor in chief, continued his support of this project through the course of its four editions.

Users of this text can inform the author of errors, suggestions, and other comments at:

*jayhawk@vt.edu*

James B. Campbell
*Blacksburg, Virginia*

# List of Tables

# List of Figures

# List of Plates

PART ONE

# FOUNDATIONS

CHAPTER ONE

# History and Scope of Remote Sensing

## 1.1. Introduction

A picture is worth a thousand words. Is this true, and, if so, why?

Pictures concisely convey information about positions, sizes, and interrelationships between objects. By their nature, they portray information about things that we can recognize as objects. These objects in turn can convey deep levels of meaning. Because humans possess a high level of proficiency in deriving information from such images, we experience little difficulty in interpreting even those scenes that are visually complex. We are so competent in such tasks that it is only when we attempt to replicate these capabilities using computer programs, for instance, that we realize how powerful our abilities are to derive this kind of intricate information. Each picture therefore can truthfully be said to distill the meaning of at least a thousand words.

This text is devoted to the analysis of a special class of pictures that employ an overhead perspective (e.g., maps, aerial photographs, and similar images), including many that are based upon radiation not visible to the human eye. These images have special properties that offer unique advantages for the study of the Earth's surface—we can see patterns instead of isolated points, and relationships between features that otherwise seem independent. They are especially powerful because they permit us to monitor changes over time; to measure sizes, areas, depths, and heights; and, in general, to acquire information that is very difficult to acquire by other means. However, our ability to extract this kind of information is not innate—we must work hard to develop the knowledge and skills that allow us to use images (Figure 1.1).

Specialized knowledge is important because remotely sensed images have qualities that differ from those we encounter in everyday experience:

- Image presentation
- Unfamiliar scales and resolutions
- Overhead views from aircraft or satellites
- Use of several regions of the electromagnetic spectrum

This text explores these and other elements of remote sensing, including some of its many practical applications. Our purpose in Chapter 1 is to briefly outline its content, origins, and scope as a foundation for the more specific chapters that follow.

**FIGURE 1.1.** Two examples of visual interpretation of images. Humans have an innate ability to derive meaning from the complex patterns of light and dark that form this image—we can interpret patterns of light and dark as people and objects. At another, higher, level of understanding, we learn to derive meaning beyond mere recognition of objects, to interpret the arrangement of figures, to notice subtle differences in posture, and to assign meaning not present in the arbitrary pattern of light and dark. Thus this picture tells a story: it conveys a meaning that can be received only by observers who can understand the significance of the figures, the statue, and their relationship.

## 1.2. Definitions

The field of remote sensing has been defined many times (Table 1.1). Examination of common elements in these varied definitions permits identification of the topic's most important themes. From a cursory look at these definitions, it is easy to identify a central concept: the gathering of information at a distance. This excessively broad definition, however, must be refined if it is to guide us in studying a body of knowledge that can be approached in a single course of study.

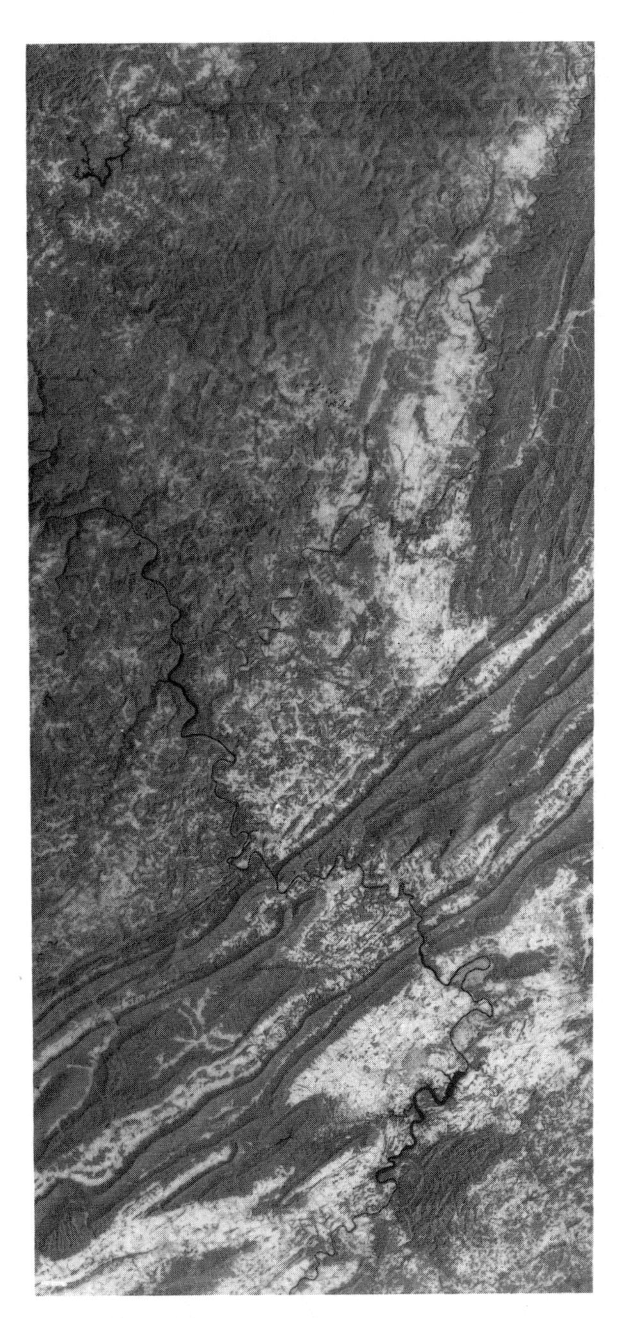

**FIGURE 1.1. (cont).** So it is also with this second image, a satellite image of southwestern Virginia. With only modest effort and experience, we can interpret these patterns of light and dark to recognize topography, drainage, rivers, and vegetation. But there is a deeper meaning here as well: the pattern of white tones tells a story about the interrelated human and natural patterns within this landscape, a story that can be understood by those prepared with the necessary knowledge and perspective. Because this image employs an unfamiliar perspective, and is derived from radiation outside the visible portion of the electromagnetic spectrum, and our everyday experience intuition are not adequate to interpret the meaning of the patterns recorded here, so it is necessary to consciously learn and apply acquired knowledge to understand the meaning of this pattern.

The kind of remote sensing to be discussed here is devoted to observation of the Earth's land and water surfaces by means of reflected or emitted electromagnetic energy. This more focused definition excludes applications that could be reasonably included in broader definitions, such as sensing the Earth's magnetic field or atmosphere, or the temperature of the human body. Also, we will focus upon instruments that present information in an image format (similar to Figure 1.1), so we must largely exclude instruments (e.g., certain lasers) that collect data at a distance, but do not portray the results in image format. Such exclusions can, of course, be considered "remote sensing" in its broad meaning, but they are omitted here as a matter of conve-

**TABLE 1.1. Remote Sensing: Some Definitions**

Remote sensing has been variously defined but basically it is the art or science of telling something about an object without touching it. (Fischer et al., 1976, p. 34)

Remote sensing is the acquisition of physical data of an object without touch or contact. (Lintz and Simonett, 1976, p. 1)

. . . Imagery is acquired with a sensor other than (or in addition to) a conventional camera through which a scene is recorded, such as by electronic scanning, using radiations outside the normal visual range of the film and camera—microwave, radar, thermal, infrared, ultraviolet, as well as multispectral, special techniques are applied to process and interpret remote sensing imagery for the purpose of producing conventional maps, thematic maps, resources surveys, etc., in the fields of agriculture, archaeology, forestry, geography, geology, and others. (American Society of Photogrammetry)

Remote sensing is the observation of a target by a device separated from it by some distance. (Barrett and Curtis, 1976, p. 3)

The term "remote sensing" in its broadest sense merely means "reconnaissance at a distance." (Colwell, 1966, p. 71)

Remote sensing, though not precisely defined, includes all methods of obtaining pictures or other forms of electromagnetic records of the Earth's surface from a distance, and the treatment and processing of the picture data. . . . Remote sensing then in the widest sense is concerned with detecting and recording electromagnetic radiation from the target areas in the field of view of the sensor instrument. This radiation may have originated directly from separate components of the target area; it may be solar energy reflected from them; or it may be reflections of energy transmitted to the target area from the sensor itself. (White, 1977, pp. 1–2)

"Remote sensing" is the term currently used by a number of scientists for the study of remote objects (earth, lunar, and planetary surfaces and atmospheres, stellar and galactic phenomena, etc.) from great distances. Broadly defined . . . , remote sensing denotes the joint effects of employing modern sensors, data-processing equipment, information theory and processing methodology, communications theory and devices, space and airborne vehicles, and large-systems theory and practice for the purposes of carrying out aerial or space surveys of the earth's surface. (National Academy of Sciences, 1970, p. 1)

Remote sensing is the science of deriving information about an object from measurements made at a distance from the object, i.e., without actually coming in contact with it. The quantity most frequently measured in present-day remote sensing systems is the electromagnetic energy emanating from objects of interest, and although there are other possibilities (e.g., seismic waves, sonic waves, and gravitational force), our attention . . . is focused upon systems which measure electromagnetic energy. (D. A. Landgrebe, quoted in Swain and Davis, 1978, p. 1)

nience. For our purposes, the definition can be based on modification of concepts given in Table 1.1:

> *Remote sensing is the practice of deriving information about the Earth's land and water surfaces using images acquired from an overhead perspective, by employing electromagnetic radiation in one or more regions of the electromagnetic spectrum, reflected or emitted from the Earth's surface.*

This definition serves as a concise expression of the scope of this volume. It is not, however, universally applicable, and is not intended to be so, because practical constraints limit the scope of this volume. So, although this text must omit many interesting topics (e.g., meteorological or extraterrestrial remote sensing), it can review knowledge areas and perspectives necessary for pursuit of topics that cannot be covered in full here.

## 1.3. Milestones in the History of Remote Sensing

The scope of the field of remote sensing can be elaborated by examining its history to trace the development of some of its central concepts. A few key events can be offered to trace the evolution of the field (Table 1.2). More complete accounts are given by Fischer (1975), Simonett (1983), and others.

Because the practice of remote sensing focuses upon the examination of images of the Earth's surface, its origins lie in the beginnings of the practice of photography. The first attempts to form images by photography date from the early 1800s, when a number of scientists, now largely forgotten, conducted experiments with photosensitive chemicals. In 1839 Louis Daguerre (1789–1851) publicly reported the results of his experiments with photographic chemicals; this date forms a convenient, although arbitrary, milestone for the birth of photography.

The use of photography to record an aerial view of the Earth's surface from a captive balloon dates from 1858. In succeeding years numerous improvements were made in photographic technology and in methods of acquiring photographs of the Earth from balloons and kites. These aerial images of the Earth are among the first to fit the definition of remote sensing given previously, but most must be regarded as curiosities rather than the basis for a systematic field of study.

**TABLE 1.2. Milestones in the History of Remote Sensing**

| | |
|---|---|
| 1800 | Discovery of infrared by Sir William Herschel |
| 1839 | Beginning of practice of photography |
| 1847 | Infrared spectrum shown by A. H. L. Fizeau and J. B. L. Foucault to share properties with visible light |
| 1850–1860 | Photography from balloons |
| 1873 | Theory of electromagnetic energy developed by James Clerk Maxwell |
| 1909 | Photography from airplanes |
| 1914–1918 | World War I: aerial reconnaissance |
| 1920–1930 | Development and initial applications of aerial photography and photogrammetry |
| 1929–1939 | Economic depression generates environmental crises that lead to governmental applications of aerial photography |
| 1930–1940 | Development of radars in Germany, United States, and United Kingdom |
| 1939–1945 | World War II: applications of nonvisible portions of electromagnetic spectrum; training of persons in acquisition and interpretation of airphotos |
| 1950–1960 | Military research and development |
| 1956 | Colwell's research on plant disease detection with infrared photography |
| 1960–1970 | First use of term "remote sensing" |
| | TIROS weather satellite |
| | Skylab remote sensing observations from space |
| 1972 | Launch of Landsat 1 |
| 1970–1980 | Rapid advances in digital image processing |
| 1980–1990 | Landsat 4: new generation of Landsat sensors |
| 1986 | SPOT French Earth observation satellite |
| 1980s | Development of hyperspectral sensors |
| 1990s | Global remote sensing systems, lidars |

The use of powered aircraft as platforms for aerial photography is the next milestone. In 1909 Wilbur Wright piloted the plane that acquired motion pictures of the Italian landscape near Centocelli; these are said to be the first aerial photographs taken from an airplane. The maneuverability of the airplane provided a capability for the control of speed, altitude, and direction required for systematic use of the airborne camera. Although there were many attempts to combine cameras with airplanes, the instruments of this era were clearly not tailored for use with each other (Figure 1.2).

World War I (1914–1918) marked the beginning of the acquisition of aerial photography on a routine basis. Although cameras used for aerial photography during this conflict was designed specifically for use with the airplane, the match between the two instruments was still rather rudimentary by the standards of later decades (Figure 1.3). The value of aerial photography for military reconnaissance and surveillance became increasingly clear as the war continued, and its applications became increasingly sophisticated. By the conclusion of the conflict, aerial photography's role in military operations was recognized, although training programs, organizational structures, and operational doctrine had not yet matured.

Numerous improvements followed from these beginnings. Camera designs were improved and tailored specifically for use in aircraft. The science of *photogrammetry*—the practice of making accurate measurements from photographs—was applied to aerial photography, with the development of instruments specifically designed for analysis of aerial photos. Although the fundamentals of photogrammetry had been defined much earlier, the field developed toward its modern form in the 1920s, with the application of accurate photogrammetric instruments. From these origins, another landmark was established: the more-or-less routine application of aerial photography in government programs, initially for topographic mapping, but later for soil survey, geologic mapping, forest surveys, and agricultural statistics.

**FIGURE 1.2.** Early aerial photography by the U.S. Navy, 1914. This photograph illustrates the difficulties encountered in early efforts to match the camera with the airplane: neither is well suited for use with the other. From U.S. Navy, National Archives and Records Administration, ARC 295605.

**FIGURE 1.3.** Aerial photography, World War I. By the time of World War I, attempts to match the camera and the airplane had progressed only to a modest extent, as illustrated by this example. This camera has been designed specifically for use in an aircraft, but the photographer must lean out of the open cockpit to use the camera. From U.S. Army, U.S. National Archives and Records Administration, ARC 530712.

During this period, the well-illustrated volume by Lee (1922), *The Face of the Earth as Seen from the Air,* surveyed the range of possible applications of aerial photography in a variety of disciplines from the perspective of those early days. Although the applications that Lee had the foresight to envision were achieved at a slow pace, the expression of governmental interest assured continuity in the scientific development of the acquisition and analysis of aerial photography, increased the number of photographs available, and trained many people in uses of aerial photography. Nonetheless, the acceptance of the use of aerial photography in most governmental and scientific activities developed slowly because of resistance among traditionalists, imperfections in equipment and technique, and genuine uncertainties regarding the proper role of aerial photography in scientific inquiry and practical applications.

The worldwide economic depression of 1929–1939 was not only an economic and financial crisis, but for many nations also an environmental crisis. National concerns about social and economic impacts of rural economic development, widespread soil erosion, reliability of water supplies, and similar issues lead to some of the first governmental applications of aerial surveys to record and monitor rural economic development. In the United States, the U.S. Department of Agriculture and the Tennessee Valley Authority led efforts to apply aerial photography to guide environmental planning and economic development. Such efforts formed an important contribution both to the institutionalization of the use of aerial photography in government and to the creation of a body of practical experience in applications of aerial photography (Figure 1.4).

These developments led to the eve of World War II (1939–1945), which is the next milestone in our history. During the war years, use of the electromagnetic spectrum was extended from almost exclusive emphasis upon the visible spectrum to other regions, most notably the infrared and microwave regions (far beyond the range of human vision). Knowledge of these regions of the spectrum had been developed in both basic and applied sciences during the pre-

**FIGURE 1.4.** Progress in applications of aerial photography, 1919–1939. During the interval between World War I and World War II (1919–1939) efforts were made to integrate the camera and the airplane and to institutionalize uses of aerial photography. By June 1943, the date of this photography, progress on both fronts was obvious. Here an employee of the U.S. Geological Survey uses a specialized instrument, the *Oblique Sketchmaster,* to match detail on an aerial photograph to a map. By the time of this photograph, aerial photography formed an integral component of U.S. Geological Survey operations. From U.S. Geological Survey and U.S. Library of Congress, fsa 8d38549.

ceding 150 years (Table 1.2). However, during the war years application and further development of this knowledge accelerated, as did dissemination of the means to apply it. Although research scientists had long understood the potential of the nonvisible spectrum, the equipment, materials, and experience necessary to apply it to practical problems were not at hand. Wartime research and operational experience provided both the theoretical and the practical knowledge required for everyday use of the nonvisible spectrum in remote sensing.

Furthermore, the wartime training and experience of large numbers of pilots, camera operators, and photointerpreters created a large pool of experienced personnel who were able to transfer their skills and experience into civilian occupations after the war. Many of these people assumed leadership positions in the efforts of business, scientific, and governmental programs to apply aerial photography and remote sensing to a broad range of problems.

The postwar era saw the continuation of trends set in motion by wartime research. On the one hand, established capabilities found their way into civilian applications. On the other hand, the beginnings of the Cold War between the Western democracies and the Soviet Union created the environment for further development of reconnaissance techniques (Figure 1.5), which were often closely guarded as defense secrets. But as newer, more sophisticated instruments were continuously developed, superseded technologies were released for wider, nondefense, applications in the civilian economy (Figure 1.6).

Among the most significant developments in the civilian sphere was the work of Robert Col-

**FIGURE 1.5.** A U.S. Air Force intelligence officer examines aerial photography, Korean conflict, July 1951. From US. Air Force, National Achieves and Records Administration, ARC 542288.

**FIGURE 1.6.** A photograph from the 1950s shows a forester examining aerial photography to delineate landscape units. By the 1950s, aerial photography and related forms of imagery had become integrated into day-to-day operations of a multitude of businesses and industries throughout the world. From Forest History Society, Durham, North Carolina. Reproduced by permission.

well (1956), who applied color infrared film (popularly known as "camouflage detection film," developed for use in World War II) to problems of identifying small-grain cereal crops and their diseases and other problems in the plant sciences. Although many of the basic principles of his research had been established earlier, his systematic investigation of their practical dimensions was one of the most significant steps in the development of the field of remote sensing. Even at this early date, Colwell delineated the outlines of modern remote sensing and anticipated many of the opportunities and difficulties of this field of inquiry.

The 1960s saw a series of important developments occur in rapid sequence. The first meteorological satellite (TIROS-1) was launched in April 1960. This satellite was designed for climatological and meteorological observations, but provided the foundation for later development of land observation satellites. During this period, some of the remote sensing instruments originally developed for military reconnaissance and classified as defense secrets were released for civilian use as more advanced designs became available for military application. These instruments extended the reach of aerial observation outside the visible spectrum into the infrared and microwave regions.

It was in this context that the term "remote sensing" was first used. Evelyn Pruitt, a scientist working for the U.S. Navy's Office of Naval Research, coined this term when she recognized that the term "aerial photography" no longer accurately described the many forms of imagery collected using radiation outside the visible region of the spectrum. Early in the 1960s the U.S. National Aeronautics and Space Administration (NASA) established a research program in remote sensing—a program that, during the next decade, was to support remote sensing research at institutions throughout the United States. During this same period, a committee of the U.S. National Academy of Sciences (NAS) studied opportunities for application of remote sensing in the field of agriculture and forestry. In 1970, NAS reported the results of their work in a document that outlined many of the opportunities offered by this emerging field of inquiry.

In 1972, the launch of Landsat 1, the first of many Earth-orbiting satellites designed for observation of the Earth's land areas, marked another milestone. Landsat provided, for the first time, systematic repetitive observation of the Earth's land areas. Each Landsat image depicted large areas of the Earth's surface in several regions of the electromagnetic spectrum, yet provided modest levels of detail sufficient for practical applications in many fields. Landsat's full significance may not yet be fully appreciated, yet it is possible to recognize three of its most important contributions. First, the routine availability of multispectral data for large regions of the Earth's surface greatly expanded the number of people who acquired experience with and interest in analysis of multispectral data. Multispectral data had been acquired previously, but their use were largely confined to specialized research laboratories. Landsat's data greatly expanded the population of scientists with interests in multispectral analysis.

Second, Landsat created an incentive for the rapid and broad expansion of uses of digital analyses for remote sensing. Before Landsat, image analyses were usually completed visually by examining prints and transparencies of aerial images (Figure 1.7). Analyses of digital images by computer were possible mainly in specialized research institutions; personal computers, and the variety of image analysis programs that we now take for granted, did not exist. Although Landsat data were initially used primarily as prints or transparencies, they were also provided in digital form. The routine availability of digital data in a standard format created the context that permitted the growth in popularity of digital analysis, and set the stage for the development of image analysis software that is now commonplace. During this era, photogrammetric processes

**FIGURE 1.7.** During the Cold War era, analysts continued to use traditional photointerpretation techniques during the development of digital systems. Here a U.S. Air Force analyst uses a light table for examination of imagery acquired by advanced sensor systems. From Staff Sergeant Brendan Kavanaugh, U.S. Air Force, 031008-F-000C-0004.

originally implemented using mechanical instruments were redefined as digital analyses, leading to improvements in image precision, and to streamlining the acquisition, processing, production, and distribution of remotely sensed data. Third, the Landsat program served as a model for the development of other land observation satellites designed and operated by diverse organizations throughout the world.

In the 1980s, scientists at the Jet Propulsion Laboratory (Pasadena, CA) began, with NASA's support, to develop instruments that could create images of the Earth at unprecedented levels of spectral detail. Whereas previous multispectral sensors collected data in a few rather broadly defined spectral regions, these new instruments could collect 200 or more very precisely defined spectral regions. These instruments created the field of *hyperspectral remote sensing,* which is still developing as a field of inquiry. Hyperspectral remote sensing will advance remote sensing's analytical powers to new levels, and will form the basis for a more thorough understanding of how to best develop future remote sensing capabilities.

By the 1990s, satellite systems had been designed specifically to collect remotely sensed data representing the entire Earth. Although Landsat had offered such a capability in principle, in practice effective global remote sensing requires sensors and processing techniques specifically designed to acquire broad-scale coverage. Such capabilities had existed on an ad hoc basis since the 1980s, primarily developed from the synoptic scope of meteorological satellites. By December 1999, NASA had launched Terra-1, the first satellite of a system specifically designed to acquire global coverage to monitor changes in the nature and extent of Earth's ecosystems. These data serve as a milestone marking the initiation of an era of broad-scale remote sensing of the Earth.

## 1.4. Overview of the Remote Sensing Process

Because remotely sensed images are the result of many interrelated processes, an isolated focus on any single component will produce a fragmented understanding. Therefore, our initial view of the field can benefit from a broad perspective that identifies the kinds of knowledge required for the practice of remote sensing (Figure 1.8).

Consider first the *physical objects,* consisting of buildings, vegetation, soil, water, and the like. These are the objects that applications scientists wish to examine. Knowledge of these physical objects resides within specific disciplines, such as geology, forestry, soil science, geography, and urban planning.

*Sensor data* are formed when an instrument (e.g., a camera or radar) views the physical objects by recording electromagnetic radiation emitted or reflected from the landscape. Chapters 3 through 10 are devoted to descriptions of the acquisition of image data using cameras and other sensors. Although the image domain can consist of pictorial images such as those familiar to us all (Figure 1.1), often images are most useful in their digital forms, which present information as numerical arrays that can be displayed and analyzed by computers. For many of us, sensor data often seem to be abstract and foreign because of their unfamiliar overhead perspective, unusual resolutions, and use of spectral regions outside the visible spectrum. As a result, effective use of sensor data requires analysis and interpretation to convert data to information that can be used to address practical problems, such as siting landfills or searching

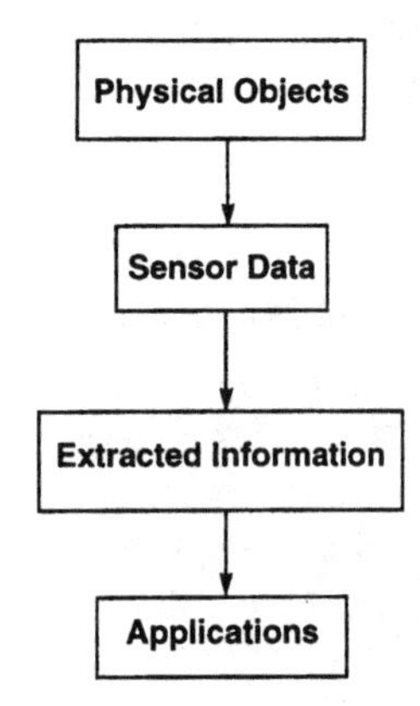


**FIGURE 1.8.** Schematic overview of knowledge used in remote sensing.

for mineral deposits. Chapters 11 to 15 describe methods used to interpret remotely sensed data.

These interpretations create extracted information, which consists of transformations of sensor data designed to reveal specific kinds of information (Figure 1.8). Actually, a more realistic view (Figure 1.9) illustrates that the same sensor data can be examined from alternative perspectives to yield different interpretations. Therefore, a single image can be interpreted to provide information about soils, land use, or hydrology, for example, depending on the specific image and the purpose of the analysis. In this text, Chapters 10 to 15 are devoted to image analysis, broadly defined: the extraction of information, specific to particular disciplines, from raw remotely sensed data.

Finally, we proceed to the *applications,* in which the analyzed remote sensing data can be combined with other data to address a specific practical problem, such as land-use planning, mineral exploration, or water-quality mapping. Applications are addressed in Chapters 16 to 21. When digital remote sensing data are combined with other geospatial data, applications are implemented in the context of geographic information systems (GIS). For example, remote sensing data may provide accurate land-use information that can be combined with soils, geologic, transportation, and other information to guide the siting of a new landfill. This process is largely beyond the scope of this text, although Chapter 16 introduces readers to GIS by outlining some of the ways that remotely sensed data can be used in GIS.

## 1.5. A Specific Example

Let's consider how a remotely sensed image is acquired by tracing the path of energy used to make an image. A cursory examination of an aerial image of a deciduous forest (Figure 1.10) can introduce some dimensions of the field of remote sensing. (All of the concepts mentioned here will be discussed in detail in subsequent chapters.)

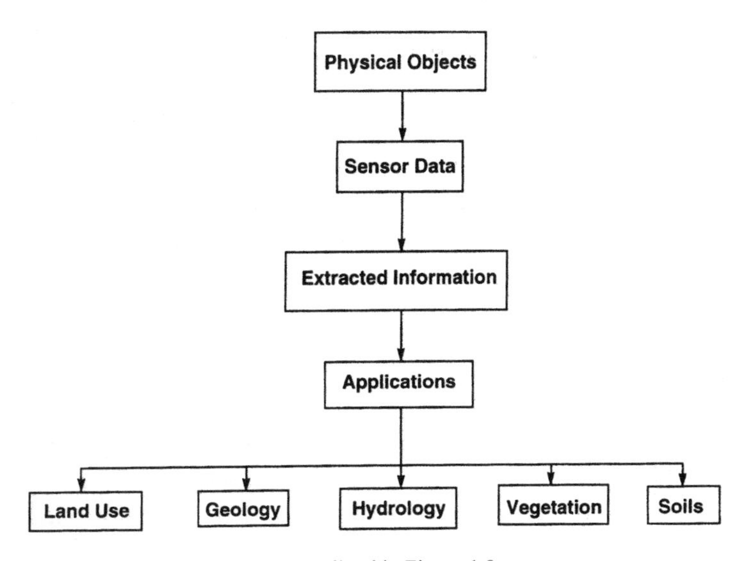


**FIGURE 1.9.** Expanded view of the process outlined in Figure 1.8.

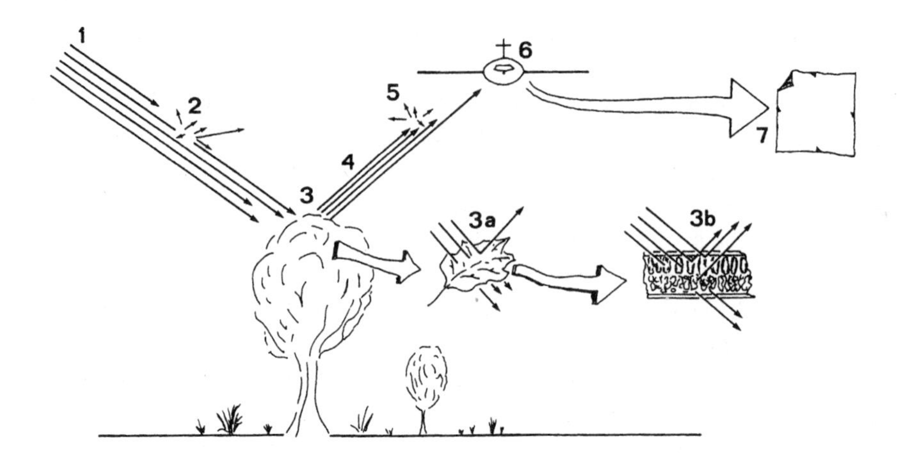


**FIGURE 1.10.** Idealized overview of acquisition of a remotely sensed image. This diagram depicts a schematic view of some of the elements that must be considered in the acquisition and interpretation of a remote sensing image. In this example, the image is derived from reflected solar radiation. Incoming solar radiation (1) is in part scattered or attenuated by the Earth's atmosphere (2). The remaining energy reaches the Earth's surface to interact with the landscape—in this example, a living tree (3). Much of the reflection of energy from the tree canopy is controlled by interactions with individual leaves (3a), which selectively absorb, transmit, and reflect energy according to its wavelength (3b), as discussed in later chapters. The reflected energy (4) is again subject to atmospheric attenuation (5) before it is recorded by an airborne or satellite sensor (6). The results of a multitude of such interactions are recorded as a photograph-like image (7), or as an array of quantitative values. The image is then available for interpretation and analysis to extract information concerning the landscape, based upon the interpreter's examination of the image and knowledge of items 1 through 7.

### *Physical Features*

Energy that reaches the Earth from the Sun is composed of many kinds of radiation, including that in the visible region of the spectrum (blue, green, and red light), as well as other radiation (such as the infrared spectrum) outside the range of human vision. The solar beam must pass through the Earth's atmosphere to reach the forest canopy; some of this energy is absorbed and scattered before it reaches the forest. Much scattering is selective by wavelength—for example, the atmosphere scatters blue light much more than it scatters green or red light. Of the remaining energy, some will reach the leaves of the tree canopy; of energy reaching the leaves, the infrared radiation (wavelengths just longer than those of visible radiation) will be reflected. The green region of the spectrum will be reflected by a different portion of the leaf, while blue and red radiation will be absorbed for use in photosynthesis. Energy that reaches the camera lens must again pass through the atmosphere, where it is again subject to attenuation.

### *Sensor Data*

Thus energy recorded by the camera lens is much different than the sunlight that entered the atmosphere. Some of the blue light was scattered by the atmosphere, and reaches the camera without being reflected from the Earth's surface. Blue, red, green, and infrared light have been reflected from the canopy, but reach the camera in proportions that differ from the proportions that were intercepted by the canopy. The film portrays these different kinds of radiation in different colors, which may not (as explained in Chapter 3) match those of the radiation they represent. This image represents a portion of the Earth's surface, but its meaning is hidden from casual observation.

### *Extracted Information*

Data within the image can be translated into information only through the process of *image interpretation,* or *image analysis.* An *image analyst,* an individual skilled in the examination of images, studies the image to identify, for example, timber types, and forest areas damaged by disease and insect activity. For some, much of this knowledge is based on everyday experience rather than formally structured learning. We do not need to "learn" that deciduous trees lose their leaves in the winter, or that wheat tends to be planted in fields of regular size and shape. This store of *implicit* knowledge is important because it forms such a large portion of what we know about a scene, and because we often are unaware of its role in image analysis. More *formal, explicit,* knowledge of the landscape is especially significant in two contexts. First, we have no means of acquiring an implicit understanding of how radiation behaves outside the visible spectrum. So knowledge that living vegetation is highly reflective in the near infrared portion of the spectrum must be acquired in a formalized learning situation. Second, our implicit knowledge has geographic limits; when we travel outside the range of our own experience, it is no longer valid in our assessment of the landscape. For example, there are large regions of the Earth where crops are not customarily planted in regular fields and where trees do not all shed their leaves during the same season. Therefore, analysts often must consciously acquire the information necessary to analyze images collected in areas outside the range of their direct experience.

### *Applications*

Applications of remotely sensed data to practical problems usually require links to other kinds of information, including, for example, topography, political boundaries, and pedologic, geologic, or hydrologic data. In recent years, such links have increasingly been made within the framework of GIS. GIS are devoted to the analysis of geospatial data, and especially to analysis of the interrelationships between different kinds of data, matched to each other within a specific geographic region. Although there is no sharp line between remote sensing and GIS, it is only partially incorrect to state that remote sensing is primarily a means of collecting data and that GIS are primarily a means of storing and analyzing data.

At local levels of government, remote sensing imagery records large-scale representations of topography and drainage and the basic infrastructure of highways, buildings, and utilities. At county and regional levels of detail, remotely sensed data provide a basis for outlining broad-scale patterns of development; for coordinating the relationships between transportation, residential, industrial, and recreational land uses; for siting landfills; and for planning future development. State governments require information from remotely sensed data to create broad-scale inventories of natural resources; to monitor environmental issues, including land reclamation and water quality; and to plan economic development.

At broader levels of examination, national governments apply the methods of remote sensing and image analysis for environmental monitoring (both domestically and internationally), management of federal lands, crop forecasting, disaster relief, and support activities to further national security and international relations. Remote sensing is also used to support activities of international scope, including analysis of broad-scale environmental issues, international development, disaster relief, aid for refugees, and investigation of environmental issues of global scope.

These activities are conducted or supported by industrial and commercial enterprises that design and manufacture equipment, instruments, and materials. Others operate systems to collect and distribute data to customers. And still others provide services to plan for effective acquisition of remotely sensed data, and to analyze and interpret data to provide information for customers. Finally, some industrial and commercial organizations conduct basic research to develop and refine instruments and analyses.

Within subject-area disciplines, remote sensing imagery is almost always combined with other kinds of data. Geologists and geophysicists use remotely sensed images to study lithologies, structures, surface processes, and geologic hazards. Hydrologists examine images that show land-cover patterns, soil moisture status, drainage systems, sediment content of lake and rivers, ocean currents, and other characteristics of water bodies. Geographers and planners examine imagery to study settlement patterns, inventory land resources, and track changes in human uses of the landscape. Foresters use remote sensing and GIS to map timber stands, estimate timber volume, monitor insect infestations, fight forest fires, and plan harvesting of timber. Agricultural scientists can examine the growth, maturing, and harvesting of crops, and monitor the progress of diseases, infestations, and droughts to forecast their impact on crop yields. Soil scientists use remotely sensed imagery to plot boundaries of soil units and to examine relationships between soil patterns and those of land use and vegetation. In brief, remotely sensed imagery has found applications in virtually all fields that require the analysis of distributions of natural or human resources on the Earth's surface.

## 1.6. Key Concepts of Remote Sensing

Scientists usually attempt to avoid the pitfalls of relying upon improvised approaches for conducting research. Improvised, or ad hoc, methods may repeat previously discovered methodological errors, lead to trivial results, promote inefficiency, or simply create inaccuracies. Instead, scientists prefer to base their methods upon systems of facts and methods found to yield consistent accurate results.

The practice of remote sensing is young enough that the basic facts and methods are not yet known completely. Scientists are still investigating to define many of the fundamental methods and concepts central to remote sensing. Nonetheless, it is useful to base our study of remote sensing upon a set of principles that seem to convey the essential dimensions of the practice of remote sensing. The principles outlined below are tentatively proposed as concepts that address ideas central to the practice of remote sensing, regardless of specific disciplinary applications.

### *Spectral Differentiation*

Remote sensing depends upon observed spectral differences in the energy reflected or emitted from features of interest. Expressed in everyday terms, one might say that we look for differences in the "colors" of objects. (This analogy may be useful even though remote sensing is often conducted outside the visible spectrum, where "colors," in the usual meaning of the word, do not exist.) This principle is the basis of *multispectral remote sensing,* the science of observing features at varied wavelengths in an effort to derive information about these features and their distributions. The term *spectral signature* has been used to refer to the spectral response of a feature, as observed over a range of wavelengths (Parker and Wolff, 1965). For the beginning student, this term can be misleading because it implies a distinctiveness and a consistency that seldom can be observed in nature. For example, consider the notion of a spectral signature for "corn." A cornfield has one spectral response when it is planted, another as the plants emerge, another as the plants mature, and yet another after harvest. Selection of a single time and place to designate a single spectral signature for "corn" is futile. Yet the term conveys a more general concept that *is* useful because at specific times and places the spectral response of cornfields may be the basis for reliable separation of corn from wheat, for example, even though neither crop forms universally applicable spectral signatures.

### *Radiometric Differentiation*

Examination of any image acquired by remote sensing ultimately depends upon detection of differences in the brightness of objects and their features. The scene itself must have sufficient contrast (at a specific spectral region) in brightnesses, and the remote sensing instrument must be capable of recording this contrast, before information can be derived from the image. As a result, the sensitivity of the instrument, and the existing contrast in the scene between objects and their backgrounds, are always issues of significance in remote sensing investigations.

### *Spatial Differentiation*

Every sensor is limited in respect to the size of the smallest area that can be separately recorded as an entity on an image. This minimum area determines the *spatial detail*—the fineness of the patterns—on the image. For some remote sensing systems, these smallest areal units—picture elements ("pixels")—are in fact discrete, distinct units, identifiable on the image (Figure 1.11). In other instances, the spatial detail (determined by the quality of a camera lens or the film that has been used) is less obvious, but is nonetheless present. Our ability to record spatial detail is influenced primarily by the choice of sensor and the altitude at which it is used to record images of the Earth. Note also that some landscapes vary greatly in their spatial complexity: some may be represented clearly at coarse levels of detail, whereas others are so complex that the finest level of detail is required to record their essential characteristics.

### *Geometric Transformation*

Every remotely sensed image represents a landscape in a specific geometric relationship determined by the design of the remote sensing instrument, specific operating conditions, terrain relief, and other factors. The ideal remote sensing instrument would be able to create an image with accurate, consistent geometric relationships between points on the ground and their corre-

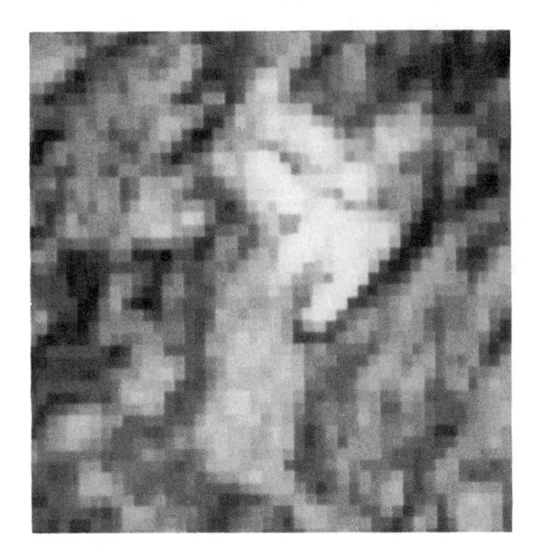

**FIGURE 1.11.** Digital remote sensing image. This image is a small section of Figure 1.1 shown in fine detail, so that individual picture elements ("pixels") are visible as separate cells of varied brightness in proportion to the brightness of the ground area at the time the image was acquired. In its "pictorial" form (Figure 1.1), the image can be examined by human interpreters; in its digital form (as an array of integer values), it can be easily analyzed by quantitative methods to extract information not evident to the human analyst. Any pictorial image can be represented in digital form (this image), and any digital image can be portrayed in pictorial form (Figure 1.1).

sponding representations on the image. Such an image could form the basis for accurate measurements of areas and distances. In reality, of course, each image includes positional errors caused by the perspective of the sensor optics, the motion of scanning optics, terrain relief, and Earth's curvature. Each source of error can vary in significance in specific instances, but the result is that geometric errors are inherent, not accidental, characteristics of remotely sensed images. In some instances we may be able to remove or reduce locational error, but it must always be taken into account before images are used as the basis for measurements of areas and distances.

### *Interchangeability of Pictorial and Digital Formats*

Most remote sensing systems generate photograph-like images of the Earth's surface. Any such image can be represented in digital form by systematically subdividing the image into tiny areas of equal size and shape, then representing the brightnesses of these areas by discrete values (Figure 1.11). Conversely, many remote sensing systems generate, as their first-generation output, digital arrays that represent brightnesses of areas of the Earth's surface in digital form. Digital images can be displayed as pictorial images by displaying each digital value as a brightness level scaled to the magnitude of the value. The two forms for remote sensing data —pictorial and digital—represent different methods of display and representation, but there is no real difference in the information conveyed by the two forms. Any image can be portrayed in either form (although sometimes with a loss of detail in converting from one form to another) according to the purposes of our investigation.

### *Remote Sensing Instrumentation Acts as a System*

The image analyst must always be conscious of the fact that the many components of the remote sensing process *act as a system,* and therefore cannot be isolated from one another. For example, upgrading the quality of a camera lens makes little sense unless we also use a film of sufficient quality to record the improvements produced by the superior lens, thus allowing the analyst to have the ability to derive improved information from the image.

Components of the system must be appropriate for the task at hand. This means that the interpreter must not only have intimate knowledge of the remote sensing system itself, but must also be deeply informed about the subject of the interpretation, to include the amount of detail required, to choose the appropriate time of year to acquire the data, to be aware of the best spectral regions to use, and so on. Like the physical components of the system, the interpreter's knowledge and experience also interact to form a whole.

### *Role of the Atmosphere*

All energy reaching the remote sensing instrument must pass through a portion of the Earth's atmosphere. For satellite remote sensing in the visible and near infrared ranges, energy received by the sensor must pass through a considerable depth of the earth's atmosphere. In doing so, the Sun's energy is altered in intensity and wavelength by particles and gases in the Earth's atmosphere. These changes appear on the image in ways that degrade image quality or influence the accuracy of interpretations.

## 1.7. Career Preparation and Professional Development

For the student, a course of study in remote sensing offers opportunities to enter a field of knowledge that can contribute to several dimensions of a university education and subsequent personal and professional development. Students enrolled in introductory remote sensing courses often view the topic as an important part of their occupational and professional preparation. It is certainly true that skills in remote sensing are valuable in the initial search for employment. But is equally important to acknowledge that this topic should form part of a comprehensive program of study that includes work in GIS and in-depth study of a specific discipline. A well-thought-out program appropriate for a student's specific interests and strengths should combine studies in several interrelated topics, such as:

- Geology, hydrology, geomorphology, soils
- Urban planning, transportation, urban geography
- Forestry, ecology, soils

Such programs are based on a foundation of supporting courses, including statistics, computer science, and the physical sciences.

Students should avoid studies that provide only narrowly based, technique-oriented content. Such highly focused studies, perhaps with specific equipment or software, may provide immediate skills for entry-level positions, but they leave the student unprepared to participate in the broader assignments required for effective performance and professional advancement. Employers report that they seek employees who:

- Have a good background in at least one traditional discipline.
- Are reliable, and able to follow instructions without detailed supervision.

- Can write and speak effectively.
- Work effectively in teams with others in other disciplines.
- Are familiar with common business practices.

Because this kind of preparation is seldom encompassed in a single academic unit within a university, students often have to apply their own initiative to identify the specific courses they will need to best develop these qualities. Table 1.3 shows a selection of sample job descriptions in remote sensing and related fields, as an indication of the kinds of knowledge and skills expected of employees in the geospatial information industry.

Possibly the most important but least visible contributions are those that lead to the development of concep tual thinking concerning the role of basic theory and method, integration of knowledge from several disciplines, and proficiency in identifying practical problems in a spatial context. Although skills with and knowledge of remote sensing are very important, it is usually a mistake to focus exclusively upon methodology and technique. At least two pitfalls are obvious. First, emphasis upon fact and technique without consideration of basic principles and theory provides a narrow, empirical foundation in a field that is characterized by diversity and rapid change. A student equipped with a narrow background is ill prepared to compete with those trained in other disciplines or to adjust to unexpected developments in science and technology. Thus any educational experience is best perceived not as a catalog of facts to be memorized, but as an experience in *how to learn* to equip oneself for independent learning later, outside the classroom. This task requires a familiarity with basic references, fundamental principles, and the content of related disciplines, as well as the core of facts that form the substance of a field of knowledge.

Second, many employers have little interest in hiring employees with shallow preparation in either their major discipline or in remote sensing. Lillesand (1982) reports that a panel of managers from diverse industries concerned with remote sensing recommended that prospective employees develop "an ability and desire to interact at a conceptual level with other specialists" (p. 290). Campbell (1978) quotes other supervisors who are also concerned that students receive a broad preparation in remote sensing and in their primary field of study:

> It is essential that the interpreter have a good general education in an area of expertise. For example, you can make a geologist into a good photo geologist, but you cannot make an image interpreter into a geologist.
>
> Often people lack any real philosophical understanding of why they are doing remote sensing, and lack the broad overview of the interrelationships of all earth science and earth-oriented disciplines (geography, geology, biology, hydrology, meteorology, etc.). This often creates delays in our work as people continue to work in small segments of the (real) world and don't see the interconnections with another's research. (p. 35)

These same individuals have recommended that those students who are interested in remote sensing should complete courses in computer science, physics, geology, geography, biology, engineering, mathematics, hydrology, business, statistics, and a wide variety of other disciplines. No student could possibly take all the recommended courses during a normal program of study, but it is clear that neither a haphazard selection of university courses or one that focused exclusively upon remote sensing courses would form a substantive background in remote sensing. In addition, many organizations have been forceful in stating that they desire employees who can write well, and several have expressed an interest in persons with expertise in remote

**TABLE 1.3.** Sample Job Descriptions Relating to Remote Sensing

**SURVEY ENGINEER**

XCELIMAGE is a spatial data, mapping, and geographic information systems (GISs) services company that provides its clients with customized products and services to support a wide range of land-use and natural resource management activities. The company collects geospatial data using a variety of airborne sensing technologies and turns that data into tools that can be used in GIS or design and engineering environments. With over 500 employees in offices nationwide, the XCELIMAGE group and affiliates represent one of the largest spatial data organizations in the world. XCELIMAGE is affiliated with six member companies and two affiliates. XCELIMAGE Aviation, located in Springfield, MA, has an immediate opening for a Survey Engineer.

XCELIMAGE Aviation supplies the aerial photography and remote sensing data from which terrain models, mapping, and GIS products are developed. XCELIMAGE Aviation operates aircraft equipped with analog, digital, and multispectral cameras; global positioning systems (GPS); a light detection and ranging system (LIDAR); a passive microwave radiometer; and thermal cameras. This position offers a good opportunity for advancement and a competitive salary and benefits package. Position requires a thorough knowledge of computer operation and applications including GIS software; a basic understanding of surveying, mapping theories, and techniques; a thorough knowledge of GPS concepts; exposure to softcopy techniques; and the ability to efficiently aid in successful implementation of new technologies and methods. This position includes involvement in all functions related to data collection, processing, analysis, and product development for aerial remote sensing clients; including support of new technology.

**ECOLOGIST/REMOTE SENSING SPECIALIST**

The U.S. National Survey Northern Plains Ecological Research Center is seeking an Ecologist/Remote Sensing Specialist to be a member of the Regional Gap Analysis Project team. The incumbent's primary responsibility will be mapping vegetation and land cover for our region from analysis of multitemporal Landsat Thematic Mapper imagery and environmental data in a geographic information system. To qualify for this position, applicants must possess (1) ability to conduct digital analysis of remotely sensed satellite imagery for vegetation and land cover mapping; (2) ability to perform complex combinations and sequences of methods to import, process, and analyze data in vector and raster formats in a geographic information system; and (3) knowledge of vegetation classification, inventory, and mapping.

**REMOTE SENSING/GIS AGRICULTURAL ANALYST**

Position located in Washington, DC. Requires US citizenship. Unofficial abstract of the "Crop Assessment Analyst" position: An interesting semianalytic/technical position is available with the Foreign Agricultural Service working as an international and domestic agriculture commodity forecaster. The position is responsible for monitoring agricultural areas of the world, performing analysis, and presenting current season production forecasts. Tools and data used in the position include imagery data (AVHRR, Landsat TM, SPOT), vegetation indexes, crop models, GIS software (ArcView, ArcInfo), image processing s/w (Erdas, PCI), agro-meteorological data models, web browsers, web page design s/w, graphic design s/w, spreadsheet s/w, GIS software, digital image processing, weather station data, climate data, historical agricultural production data, and assessing news stories. The main crops of concern are soybeans, canola, wheat, barley, corn, cotton, peanuts, and sorghum of the major export/import countries.

A background in agronomy, geographical spatial data, good computer skills, information management, and ag economics will prove beneficial in performing the work.

**REMOTE SENSING SPECIALIST POSITION, GEOSPATIAL AND INFORMATION TECHNOLOGIES INFORMATION RESOURCES UNIT**

The Remote Sensing Specialist is generally responsible for implementing remote sensing technology as a tool for natural resources management. Responsibilities include planning, coordinating, and managing a regional remote sensing program, working with resource specialists at the regional level, forests and districts, to meet information needs using remotely sensed data. Tasks include working with resource and GIS analysts to implement the recently procured national image processing software, keeping users informed about the system, and assisting with installation and training. The person in this position is also the primary contact in the region for national remote sensing issues, and maintains contact with other remote sensing professionals within the Forest Service, in other agencies, and in the private and academic sectors.

The position is located in the Geospatial and Information Technologies (GIT) group of the Information Resources (IRM) unit. IRM has regional responsibility for all aspects of information and systems management. The GIT group includes the remote sensing, aerial photography, photogrammetry, cartography, database, and GIS functions.

*Note.* Actual notices edited to remove information identifying specific firms. Although listed skills and abilities are typical, subject area specialties are not representative of the range of applications areas usually encountered.

sensing who have knowledge of a foreign language. The key point is that educational preparation in remote sensing should be closely coordinated with study in traditional academic disciplines, and should be supported by a program of courses carefully selected from offerings in related disciplines.

Students should consider joining a professional society devoted to the field of remote sensing. In the United States and Canada, the American Society for Photogrammetry and Remote Sensing (ASPRS; 5410 Grosvenor Lane, Suite 210, Bethesda, MD 20814-2160; 301-493-0290; *www.asprs.org*) is the principal professional organization in this field. ASPRS offers students discounts on membership dues, publications, and meeting registration, and conducts job fairs at its annual meetings. ASPRS is organized on a regional basis, so local chapters conduct their own activities, which are open to student participation. Other professional organizations often have interest groups devoted to applications of remote sensing within specific disciplines, usually with similar benefits for student members.

Students should also investigate local libraries to become familiar with professional journals in the field. The field's principal journals include:

Photogrammetric Engineering and Remote Sensing
Remote Sensing of Environment
International Journal of Remote Sensing
IEEE Transactions on Geoscience and Remote Sensing
Computers and Geosciences
GIScience & Remote Sensing

Although beginning students may not yet be prepared to read research articles in detail, those who make the effort to familiarize themselves with these journals will have prepared the way to take advantage of their content later. Students may find *Photogrammetric Engineering and Remote Sensing* particularly useful because it lists of job opportunities, announces scheduled meetings, and discusses new products.

The world of practical remote sensing has changed dramatically in recent decades. Especially since the early 1990s, commercial and industrial applications of remote sensing have expanded dramatically to penetrate well beyond the specialized applications of an earlier era, to extend, for example, into marketing, real estate, and agricultural enterprises. Aspects of remote sensing that formerly seemed to require a highly specialized knowledge became available to a much broader spectrum of users as data became less expensive and more widely available, and as manufacturers designed software for use by the nonspecialist.

These developments are leading to formation of a society in which remote sensing, GIS, global positioning systems (GPS), and related technological systems will become commonly used tools within the workplace, analogous in accessibility to spreadsheets, word processing, or cell phones. In such a setting, citizens, the nonspecialists, must be prepared to use spatial data effectively and appropriately, and to understand its strengths and limitations. Because of the need to produce a wide variety of ready-to-use products tailored for specific populations of users, the remote sensing community will continue to require specialists, especially those who can link a solid knowledge of remote sensing with subject-area knowledge (e.g., hydrology, planning, forestry). People who will work in this field will require skills and perspectives that differ greatly from those of previous graduates—even those who graduated just a few years ago.

## Review Questions

1. Aerial photography and other remotely sensed images have found rather slow acceptance into many, if not most, fields of study. Imagine that you are the director of a unit engaged in geological mapping in the early days of aerial photography (e.g., in the 1930s). Can you suggest reasons why you might be reluctant to devote your efforts and resources to use of aerial photography rather than to continue use of your usual procedures?

2. Satellite observation of the Earth provides many advantages over aircraft-borne sensors. Consider fields such as agronomy, forestry, or hydrology. For one such field of study, list as many of the advantages as you can. Can you suggest some disadvantages?

3. Much (but not all) information derived from remotely sensed data is derived from spectral information. To understand how spectral data may not always be as reliable as one might first think, briefly describe the spectral properties of a maple tree and a corn field. How might these properties change over the period of a year? Or a day?

4. All remotely sensed images observe the Earth from above. Can you list some *advantages* to the overhead view (as opposed to ground-level views) that make remote sensing images inherently advantageous for many purposes? List some disadvantages to the overhead view.

5. Remotely sensed images show the combined effects of many landscape elements, including vegetation, topography, illumination, soil, drainage, and others. In your view, is this diverse combination an advantage or a disadvantage? Explain.

6. List ways in which remotely sensed images differ from maps. Also list advantages and disadvantages of each. List some of the tasks for each which might be more useful.

7. Chapter 1 emphasizes how the field of remote sensing is formed by knowledge and perspectives from many different disciplines. Examine the undergraduate catalogue for your college or university and prepare a comprehensive program of study in remote sensing from courses listed. Identify gaps, courses or subjects that would be desirable but are not offered.

8. In your university library, find copies of *Photogrammetric Engineering and Remote Sensing, International Journal of Remote Sensing,* and *Remote Sensing of Environment,* some of the most important English-language journals reporting remote sensing research. Examine some of the articles in several issues of each journal. Although titles of some of these articles may now seem rather strange, as you progress through this course you will be able to judge the significance of most. Refer to these journals again as you complete the course.

9. Inspect library copies of some of the remote sensing texts listed in the references for Chapter 1. Examine the tables of contents, selected chapters, and lists of references. Many of these volumes may form useful references for future study or research in the field of remote sensing.

10. Examine issues of the journals mentioned in Question 8, noting the affiliations and institutions of authors of articles. Be sure to look at issues that date back for several years, so you can identify some of the institutions and agencies that have been making a continuing contribution to remote sensing research.

## References

Alföldi, T., P. Catt, and P. Stephens. 1993. Definitions of Remote Sensing. *Photogrammetric Engineering and Remote Sensing,* Vol. 59, pp. 611–613.

Arthus-Bertrand, Y. 1999. *Earth from Above.* New York: Abrams, 414 pp.

Avery, T. E., and G. L. Berlin. 1992. *Fundamentals of Remote Sensing and Airphoto Interpretation.* Upper Saddle River, NJ: Prentice-Hall, 472 pp.

Campbell, J. B. 1978. Employer Needs in Remote Sensing in Geography. *Remote Sensing Quarterly,* Vol. 5, No. 2, pp. 52–65.

Collier, P. 2002. The Impact on Topographic Mapping of Developments in Land and Air Survey 1900–1939. *Cartography and Geographic Information Science.* Vol. 29, pp. 155–174.

Colwell, R. N. 1956. Determining the Prevalence of Certain Cereal Crop Diseases by Means of Aerial Photography. *Hilgardia,* Vol. 26, No. 5, pp. 223–286.

Colwell, R. N. 1966. Uses and Limitations of Multispectral Remote Sensing. In *Proceedings of the Fourth Symposium on Remote Sensing of Environment.* Ann Arbor: Institute of Science and Technology, University of Michigan, pp. 71–100.

Colwell, R. N. (ed.). 1983. *Manual of Remote Sensing* (2nd ed.). Falls Church, VA: American Society of Photogrammetry, 2 vols., 2240 pp.

Curran, P. 1985. *Principles of Remote Sensing.* New York: Longman, 282 pp.

Curran, P. 1987. Commentary: On Defining Remote Sensing. Photogrammetric Engineering and Remote Sensing, Vol. 53, pp. 305–306.

Estes, J. E., J. R. Jensen, and D. S. Simonett. 1977. The Impact of Remote Sensing on United States' Geography: The Past in Perspective, Present Realities, Future Potentials. In *Proceedings of the Eleventh International Symposium on Remote Sensing of Environment.* Ann Arbor: Institute of Science and Technology, University of Michigan, pp. 101–121.

Estes, J. E., et al. 1993. The NCGIA Core Curriculum in Remote Sensing. *Photogrammetric Engineering and Remote Sensing,* Vol. 59, pp. 945–948.

Fischer, W. A. (ed). 1975. History of Remote Sensing. Chapter 2 in *Manual of Remote Sensing* (R. G. Reeves, ed.). Falls Church, VA: American Society of Photogrammetry, pp. 27–50.

Fischer, W. A., W. R. Hemphill, and A. Kover. 1976. Progress in Remote Sensing. *Photogrammetria,* Vol. 32, pp. 33–72.

Fussell, J. D. Rundquist, and J. A. Harrington. 1986. On Defining Remote Sensing. *Photogrammetric Engineering and Remote Sensing,* Vol. 52, pp. 1507–1511.

Hall, S. S. 1992. *Mapping the Next Millennium: The Discovery of New Geographies.* New York: Random House, 384 pp.

Hall, S. S. 1993. *Mapping the Next Millennium: How Computer-Driven Cartography Is Revolutionizing the Face of Science.* New York: Random House, 360 pp.

Jensen, J. R. 2000. *Remote Sensing of the Environment: An Earth Resource Perspective.* Upper Saddle River, NJ: Prentice-Hall, 544 pp.

Landgrebe, D. 1976. Computer-Based Remote Sensing Technology: A Look to the Future. Remote Sensing of Environment, Vol. 5, pp. 229–246.

Lee, W. T. 1922. *The Face of the Earth as Seen from the Air* (American Geographical Society Special Publication No. 4). New York: American Geographical Society.

Lillesand, T. M. 1982. Trends and Issues in Remote Sensing Education. *Photogrammetric Engineering and Remote Sensing,* Vol. 48, pp. 287–293.

Lillesand, T. M., R. W. Kiefer, and J. W. Chipman. 2004. *Remote Sensing and Image Interpretation.* New York: Wiley, 763 pp.

National Academy of Sciences. 1970. *Remote Sensing with Special Reference to Agriculture and Forestry.* Washington, DC: National Academy of Sciences, 424 pp.

Monmonier, M. 2002. Aerial Photography at the Agricultural Adjustment Administration: Acreage Controls, Conservation Benefits, and Overhead Surveillance in the 1930s. *Photogrammetric Engineering and Remote Sensing,* Vol. 68, pp. 1257–1261.

Parker, D. C., and M. F. Wolff. 1965. Remote Sensing. *International Science and Technology,* Vol. 43, pp. 20–31.

Ray, R. G. 1960. *Aerial Photographs in Geological Interpretation and Mapping* (U.S. Geological Survey Professional Paper 373). Washington, DC: U.S. Geological Survey, 230 pp.

Reeves, R. G. (ed.). 1975. *Manual of Remote Sensing.* Falls Church, VA: American Society of Photogrammetry, 2 vols., 2,144 pp.

Simonett, D. S. 1966. Present and Future Needs of Remote Sensing in Geography. In *Proceedings of the Fourth International Symposium on Remote Sensing of the Environment.* Ann Arbor: Institute of Science and Technology, University of Michigan, pp. 37–47.

Simonett, D. S. (ed.). 1983. Development and Principles of Remote Sensing. Chapter 1 in *Manual of Remote Sensing* (R. N. Colwell, ed.). Falls Church, VA: American Society of Photogrammetry, pp. 1–35.

CHAPTER TWO

# Electromagnetic Radiation

## 2.1. Introduction

With the exception of objects at absolute zero, all objects emit electromagnetic radiation. Objects also reflect radiation that has been emitted by other objects. By recording emitted or reflected radiation, and applying knowledge of its behavior as it passes through the Earth's atmosphere and interacts with objects, remote sensing analysts develop knowledge of the character of features such as vegetation, structures, soils, rock, or water bodies on the Earth's surface. Interpretation of remote sensing imagery depends on a sound understanding of electromagnetic radiation and its interaction with surfaces and the atmosphere. The discussion of electromagnetic radiation in this chapter builds a foundation to permit development in subsequent chapters of the many other important topics within the field of remote sensing.

The most familiar form of electromagnetic radiation is visible light, which forms a small (but very important) portion of the full electromagnetic spectrum. The large segments of this spectrum that lie outside the range of human vision require our special attention because they may behave in ways that are quite foreign to our everyday experience with visible radiation.

## 2.2. The Electromagnetic Spectrum

Electromagnetic energy is generated by several mechanisms, including changes in the energy levels of electrons, acceleration of electrical charges, decay of radioactive substances, and the thermal motion of atoms and molecules. Nuclear reactions within the Sun produce a full spectrum of electromagnetic radiation, which is transmitted through space without experiencing major changes. As this radiation approaches the Earth, it passes through the atmosphere before reaching the Earth's surface. Some is reflected upward from the Earth's surface; it is this radiation that forms the basis for photographs and similar images. Other solar radiation is absorbed at the surface of the Earth, where it is then reradiated as thermal energy. This thermal energy can also be used to form remotely sensed images, although they differ greatly from the aerial photographs formed from reflected energy. Finally, man-made radiation, such as that generated by imaging radars, is also used for remote sensing.

Electromagnetic radiation consists of an electrical field ($E$) that varies in magnitude in a direction perpendicular to the direction of propagation (Figure 2.1). In addition, a magnetic field ($H$) oriented at right angles to the electrical field is propagated in phase with the electrical field.

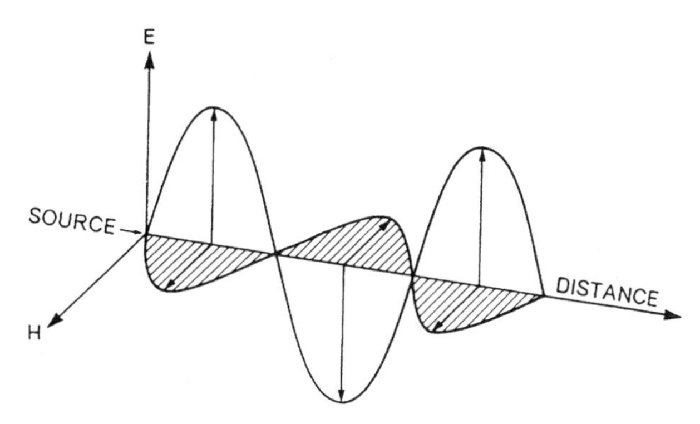


**FIGURE 2.1.** Electric ($E$) and magnetic ($H$) components of electromagnetic radiation. The electric and magnetic components are oriented at right angles to one another, and vary along an axis perpendicular to the axis of propagation.

Electromagnetic energy displays three properties (Figure 2.2):

1. *Wavelength* is the distance from one wave crest to the next. Some wavelengths can be measured in everyday units of length, but very short wavelengths have such small distances between wave crests that extremely short (and therefore less familiar) measurement units are required (Table 2.1).
2. *Frequency* is measured as the number of crests passing a fixed point in a given period of time. Frequency is often measured in *hertz,* units each equivalent to one cycle per second (Table 2.2), and multiples of the hertz.
3. *Amplitude* is equivalent to the height of each peak (see Figure 2.2). Amplitude is often measured as energy levels (formally known as *spectral irradiance*), expressed as watts per square meter per micrometer (i.e., as energy level per wavelength interval).

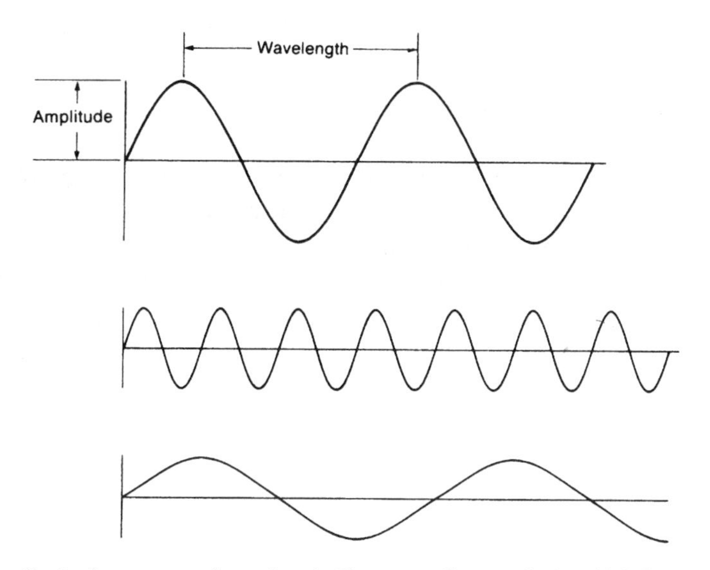


**FIGURE 2.2.** Amplitude, frequency, and wavelength. The center diagram displays high frequency, short wavelength; the bottom diagram shows low frequency, long wavelength.

**TABLE 2.1. Units of Length Used in Remote Sensing**

| Unit | Distance |
|---|---|
| Kilometer (km) | 1,000 m |
| Meter (m) | 1.0 m |
| Centimeter (cm) | 0.01 m $= 10^{-2}$ m |
| Millimeter (mm) | 0.001 m $= 10^{-3}$ m |
| Micrometer (μm)[a] | 0.000001 m $= 10^{-6}$ m |
| Nanometer (nm) | $10^{-9}$ m |
| Ångstrom unit (Å) | $10^{-10}$ m |

[a]Formerly called the "micron" (μ); the term "micrometer" is now used by agreement of the General Conference on Weights and Measures.

The speed of electromagnetic energy (c) is constant at 299,893 km per second. Frequency ($\nu$) and wavelength ($\lambda$) are related:

$$c = \lambda\nu \qquad \text{(Eq. 2.1)}$$

Therefore, characteristics of electromagnetic energy can be specified using either frequency or wavelength. Varied disciplines, and varied applications, follow different conventions for describing electromagnetic radiation, using either wavelength (measured in Angström units [Å], microns, micrometers, nanometers, millimeters, etc., as appropriate), or frequency (using hertz, kilohertz, megahertz, etc., as appropriate). Although there is no authoritative standard, a common practice in the field of remote sensing is to define regions of the spectrum on the basis of wavelength, often using micrometers (each equal to one one-millionth of a meter, symbolized as μm), millimeters (mm), and meters (m) as units of length. Departures from this practice are common; for example, electrical engineers who work with microwave radiation traditionally use frequency to designate subdivisions of the spectrum. This text usually employ wavelength designations. The student should, however, be prepared to encounter different usages in scientific journals and in references.

## 2.3. Major Divisions of the Electromagnetic Spectrum

Major divisions of the electromagnetic spectrum (Table 2.3) are, in essence, arbitrarily defined. In a full spectrum of solar energy there are no sharp breaks at the divisions indicated graphically in Figure 2.3. Subdivisions are established for convenience and by traditions within different disciplines, so do not be surprised to find different definitions in other sources or in references pertaining to other disciplines.

**TABLE 2.2. Frequencies Used in Remote Sensing**

| Unit | Frequency (cycles per second) |
|---|---|
| Hertz (Hz) | 1 |
| Kilohertz (kHz) | $10^3$ (= 1,000) |
| Megahertz (MHz) | $10^6$ (= 1,000,000) |
| Gigahertz (GHz) | $10^9$ (= 1,000,000,000) |

**TABLE 2.3. Principal Divisions of the Electromagnetic Spectrum**

| Division | Limits |
|---|---|
| Gamma rays | <0.03 nm |
| X-rays | 0.03–300 nm |
| Ultraviolet radiation | 0.30–0.38 μm |
| Visible light | 0.38–0.72 μm |
| Infrared radiation | |
| Near infrared | 0.72–1.30 μm |
| Mid-infrared | 1.30–3.00 μm |
| Far infrared | 7.0–1,000 μm (1 mm) |
| Microwave radiation | 1 mm–30 cm |
| Radio | ≥ 30 cm |

Two important categories are not shown in Table 2.3. The *optical spectrum,* from 0.30 to 15 μm, defines those wavelengths that can be reflected and refracted with lenses and mirrors. The *reflective spectrum* extends from about 0.38 to 3.0 μm; it defines that portion of the solar spectrum used directly for remote sensing.

### The Ultraviolet Spectrum

For practical purposes, radiation of significance for remote sensing can be said to begin with the ultraviolet region, a zone of short-wavelength radiation that lies between the X-ray region and the limit of human vision. Often the ultraviolet region is subdivided into the near ultraviolet (sometimes known as *UV-A;* 0.32–0.40 μm), the *far ultraviolet* (*UV-B;* 0.32–0.28 μm), and the *extreme ultraviolet* (*UV-C;* below 0.28 μm). The ultraviolet region was discovered in 1801 by the German scientist Johann Wilhelm Ritter (1776–1810). Literally, *ultraviolet* means "beyond the violet," designating it as the region just outside the violet region, the shortest wavelengths visible to humans. Near ultraviolet radiation is known for its ability to induce *fluorescence,* emission of visible radiation, in some materials; it has significance for a specialized form of remote sensing (see Section 2.6). However, ultraviolet radiation is easily scattered by the Earth's atmosphere, so it is not generally used for remote sensing of Earth materials.

### The Visible Spectrum

Although the visible spectrum constitutes a very small portion of the spectrum, it has obvious significance in remote sensing. Limits of the visible spectrum are defined by the sensitivity of

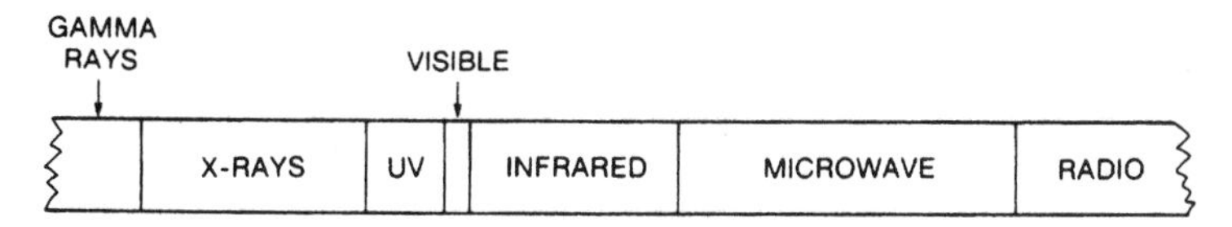


**FIGURE 2.3.** Major divisions of the electromagnetic spectrum. This diagram gives only a schematic representation—sizes of divisions are not shown in correct proportions. (See Table 2.3.)

the human visual system. Optical properties of visible radiation were first investigated by Isaac Newton (1641–1727), who during 1665 and 1666 conducted experiments that revealed that visible light can be divided (using prisms, or, in our time, diffraction gratings) into three segments. Today we know these segments as the *additive primaries,* defined approximately from 0.4 to 0.5 μm (blue), 0.5 to 0.6 μm (green), and 0.6 to 0.7 μm (red) (Figure 2.4). Primary colors are defined such that no single primary can be formed from a mixture of the other two, and that all other colors can be formed by mixing the three primaries in appropriate proportions. Equal proportions of the three additive primaries combine to form white light.

The color of an object is defined by the color of the light that it reflects (Figure 2.4). Thus a "blue" object is "blue" because it reflects blue light. Intermediate colors are formed when an object reflects two or more of the additive primaries, which combine to create the sensation of "yellow" (red and green), "purple" (red and blue), or other colors. The additive primaries are significant whenever we consider the colors of light, as, for example, in the exposure of photographic films.

In contrast, *representations* of colors in films, paintings, and similar images are formed by combinations of the three *subtractive primaries* that define the colors of pigments and dyes. Each of the three subtractive primaries absorbs a third of the visible spectrum (Figure 2.4). *Yellow* absorbs blue light (and reflects red and green); *cyan* (a greenish blue) absorbs red light (and reflects blue and green); and *magenta* (a bluish red) absorbs green light (and reflects red and blue light). A mixture of equal proportions of pigments of the three subtractive primaries yields black (complete absorption of the visible spectrum). The additive primaries are of interest in matters concerning radiant energy, whereas the subtractive primaries specify colors of the pigments and dyes used in reproducing colors on films, photographic prints, and other images.

### *The Infrared Spectrum*

Wavelengths longer than the red portion of the visible spectrum are designated as the infrared region, discovered in 1800 by the British astronomer William Herschel (1738–1822). This segment of the spectrum is very large relative to the visible region, as it extends from 0.72 to 15

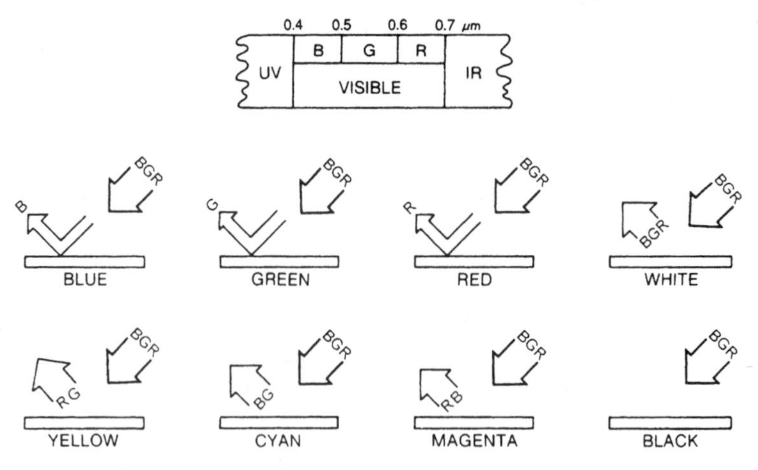


**FIGURE 2.4.** Colors.

μm—making it more than 40 times as wide as the visible light spectrum. Because of its broad range, it encompasses radiation with varied properties. Two important categories can be recognized here. The first consists of *near infrared* and *mid-infrared* radiation, defined as those regions of infrared spectrum closest to visible light. Radiation in the near infrared region behaves, with respect to optical systems, in a manner analogous to radiation in the visible spectrum. Therefore, remote sensing in the near infrared region can use films, filters, and cameras with designs similar to those intended for use with visible light.

The second category of infrared radiation consists of the *far infrared* region, consisting of wavelengths well beyond the visible, extending into regions that border the microwave region (Table 2.3). This radiation is fundamentally different from that in the visible and the near infrared regions. Whereas near infrared radiation is essentially solar radiation reflected from the Earth's surface, far infrared radiation is emitted by the Earth. In everyday language, the far infrared consists of "heat," or "thermal energy." Sometimes this portion of the spectrum is referred to as the *emitted infrared.*

### *Microwave Energy*

The longest wavelengths commonly used in remote sensing are those from about 1 mm to 1 μm in wavelength. The shortest wavelengths in this range have much in common with the thermal energy of the far infrared. The longer wavelengths of the microwave region merge into the radio wavelengths used for commercial broadcasts. Our knowledge of the microwave region originates from the work of the Scottish physicist James Clerk Maxwell (1831–1879) and the German physicist Heinrich Hertz (1857–1894).

## 2.4. Radiation Laws

The propagation of electromagnetic energy follows certain physical laws. In the interests of conciseness, some of these laws are outlined in abbreviated form because our interest here is the basic relationships they express rather than the formal derivations that are available to the student in more comprehensive sources.

Isaac Newton was among the first to recognize the dual nature of light (and by extension, all forms of electromagnetic radiation), which simultaneously displays behaviors associated with both discrete and continuous phenomena. Newton maintained that light is a stream of minuscule particles ("corpuscles") that travel in straight lines. This notion is consistent with the modern theories of Max Planck (1858–1947) and Albert Einstein (1879–1955). Planck discovered that electromagnetic energy is absorbed and emitted in discrete units now called *quanta,* or *photons.* The size of each unit is directly proportional to the frequency of the energy's radiation. Planck defined a constant ($h$) to relate frequency ($v$) to radiant energy ($Q$):

$$Q = hv \qquad \text{(Eq. 2.2)}$$

His model explains the *photoelectric effect,* the generation of electric currents by the exposure of certain substances to light, as the effect of the impact of these discrete units of energy (quanta) upon surfaces of certain metals, causing the emission of electrons.

Newton knew of other phenomena, such as the refraction of light by prisms, which are best explained by assuming that electromagnetic energy travels in a wave-like manner. James Clerk Maxwell was the first to formally define the wave model of electromagnetic radiation. His mathematical definitions of the behavior of electromagnetic energy are based upon the assumption from classical (mechanical) physics that light and other forms of electromagnetic energy propagate as a series of waves. The wave model best explains some aspects of the observed behavior of electromagnetic energy (e.g., refraction by lenses and prisms, and diffraction), whereas quantum theory provides explanations of other phenomena (notably, the photoelectric effect).

The rate at which photons (quanta) strike a surface is the *radiant flux* ($\phi_e$), measured in watts (W); this measure specifies energy delivered to a surface in a unit of time. We also need to specify a unit of area; the *irradiance* ($E_e$) is defined as radiant flux per unit area (usually measured as watts per square meter). Irradiance measures radiation that strikes a surface, whereas the term radiant exitance ($M_e$) defines the rate at which radiation is emitted from a unit area (also measured in watts per square meter).

All objects with temperatures above absolute zero have temperature and emit energy. The amount of energy and the wavelengths at which it is emitted depend upon the temperature of the object. As the temperature of an object increases, the total amount of energy emitted also increases, and the wavelength of maximum (peak) emission becomes shorter. These relationships can be expressed formally using the concept of the "blackbody." A *blackbody* is a hypothetical source of energy that behaves in an idealized manner. It absorbs all incident radiation; none is reflected. A blackbody emits energy with perfect efficiency; its effectiveness as a radiator of energy varies only as temperature varies.

The blackbody is a hypothetical entity because in nature all objects reflect at least a small proportion of the radiation that strikes them, and thus do not act as perfect reradiators of absorbed energy. Although truly perfect blackbodies cannot exist, their behavior can be approximated using laboratory instruments. Such instruments have formed the basis for the scientific research that has defined relationships between the temperatures of objects and the radiation they emit. *Kirchhoff's* law states that the ratio of emitted radiation to absorbed radiation flux is the same for all blackbodies at the same temperature. This law forms the basis for the definition of *emissive* ($\varepsilon$), the ratio between the emittance of a given object ($M$) and that of blackbody at the same temperature ($M_b$):

$$\varepsilon = M/M_b \qquad \text{(Eq. 2.3)}$$

The emissivity of a true blackbody is 1, and that of a perfect reflector (a *whitebody*) would be 0. Blackbodies and whitebodies are hypothetical concepts, approximated in the laboratory under contrived conditions. In nature, all objects have emissivities that fall between these extremes (*graybodies*). For these objects, emissivity is a useful measure of their effectiveness as radiators of electromagnetic energy. Those objects that tend to absorb high proportions of incident radiation and then to reradiate this energy will have high emissivities. Those that are less effective as absorbers and radiators of energy have low emissivities (i.e., they return much more of the energy that reaches them). (In Chapter 9, further discussion of emissivity explains that the emissivity of an object can vary with its temperature.)

The *Stefan–Boltzmann* law defines the relationship between the total emitted radiation ($W$) (often expressed in watts $\cdot$ $cm^{-2}$) and temperature ($T$) (absolute temperature, K):

$$W = \sigma T^4 \quad \text{(Eq. 2.4)}$$

Total radiation emitted from a blackbody is proportional to the fourth power of its absolute temperature. The constant ($\sigma$) is the Stefan–Boltzmann constant ($5.6697 \times 10^{-8}$) (watts $\cdot$ m$^{-2}$ $\cdot$ K$^{-4}$), which defines unit time and unit area. In essence, the Stefan–Boltzmann law states that hot blackbodies emit more energy per unit area than do cool blackbodies.

*Wien's displacement law* specifies the relationship between the wavelength of radiation emitted and the temperature of a blackbody:

$$\lambda = 2{,}897.8/T \quad \text{(Eq. 2.5)}$$

where is the wavelength at which radiance is at a maximum, and $T$ is the absolute temperature (K). As blackbodies become hotter, the wavelength of maximum emittance shifts to shorter wavelengths (Figure 2.5).

All three of these radiation laws are important for understanding electromagnetic radiation. They have special significance later in discussions of detection of radiation in the far infrared spectrum (Chapter 9).

## 2.5. Interactions with the Atmosphere

All radiation used for remote sensing must pass through the Earth's atmosphere. If the sensor is carried by a low-flying aircraft, effects of the atmosphere upon image quality may be negligible. In contrast, energy that reaches sensors carried by Earth satellites (Chapter 6) must pass through the *entire depth* of the Earth's atmosphere. Under these conditions, atmospheric effects may have substantial impact upon the quality of images and data that the sensors generate. Therefore, the practice of remote sensing requires knowledge of interactions of electromagnetic energy with the atmosphere.

In cities we often are acutely aware of the visual effects of dust, smoke, haze, and other atmospheric impurities due to their high concentrations. We easily appreciate their effects upon brightnesses and the colors we see. But even in clear air, visual effects of the atmosphere are

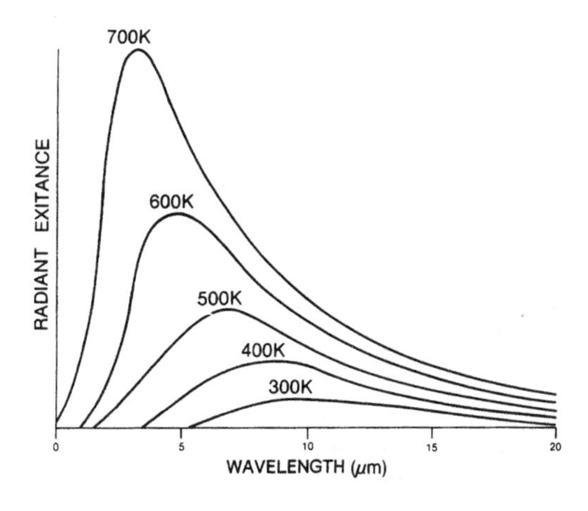


**FIGURE 2.5.** Wien's displacement law. For blackbodies at high temperatures, maximum radiation emission occurs at short wavelengths. Blackbodies at low temperatures emit maximum radiation at longer wavelengths.

numerous, although so commonplace that we may not recognize their significance. In both settings, as solar energy passes through the Earth's atmosphere, it is subject to modification by several physical processes, including (1) scattering, (2) absorption, and (3) refraction.

### *Scattering*

*Scattering* is the redirection of electromagnetic energy by particles suspended in the atmosphere or by large molecules of atmospheric gases (Figure 2.6). The amount of scattering that occurs depends upon the sizes of these particles, their abundance, the wavelength of the radiation, and the depth of the atmosphere through which the energy is traveling. The effect of scattering is to redirect radiation, so that a portion of the incoming solar beam is directed back toward space, as well as toward the Earth's surface.

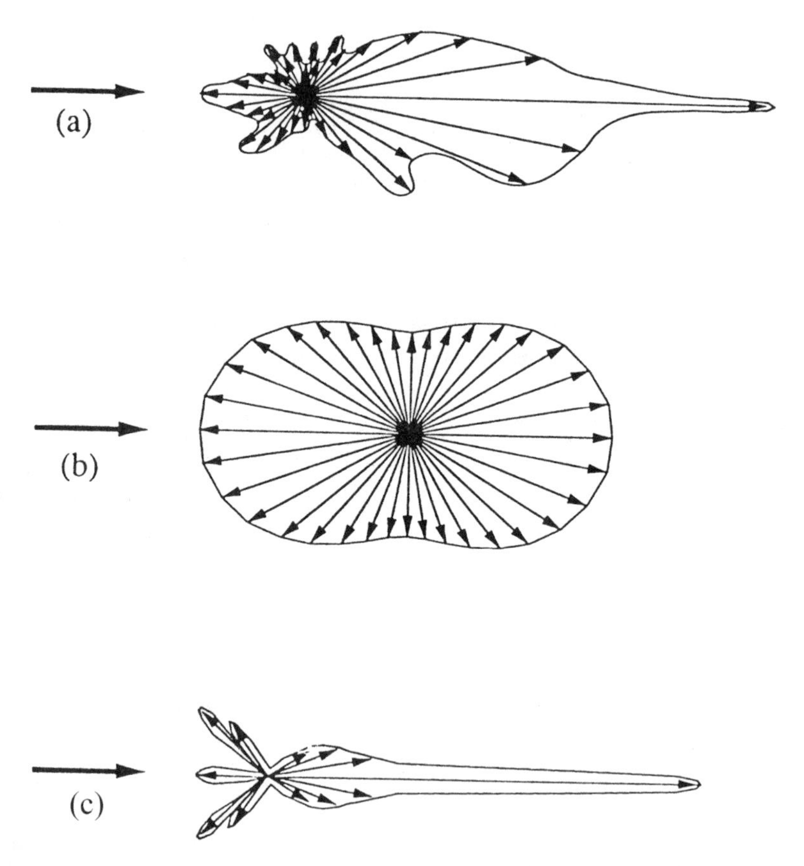


**FIGURE 2.6.** Scattering behaviors of three classes of atmospheric particles. (a) Atmospheric dust and smoke form rather large irregular particles that create a strong forward-scattering peak, with a smaller degree of backscattering. (b) Atmospheric molecules are more nearly symmetrical in shape, creating a pattern characterized by preferential forward- and backscattering, but without the pronounced peaks observed in the first example. (c) Large water droplets create a pronounced forward-scattering peak, with smaller backscattering peaks. From Lynch and Livingston (1995). Reprinted with the permission of Cambridge University Press.

A common form of scattering was discovered by the British scientist Lord J. W. S. Rayleigh (1824–1919) in the late 1890s. He demonstrated that a perfectly clean atmosphere, consisting only of atmospheric gases, causes scattering of light in a manner such that the amount of scattering increases greatly as wavelength becomes shorter. *Rayleigh scattering* occurs when atmospheric particles have diameters that are very small relative to the wavelength of the radiation. Typically, such particles could be very small specks of dust, or some of the larger molecules of atmospheric gases, such as nitrogen ($N_2$) and oxygen ($O_2$). These particles have diameters that are much smaller than the wavelength ($\lambda$) of visible and near infrared radiation (on the order of diameters less than $\lambda$).

Because Rayleigh scattering can occur in the absence of atmospheric impurities, it is sometimes referred to as *clear atmosphere scattering*. It is the dominant scattering process high in the atmosphere, up to altitudes of 9–10 km, the upper limit for atmospheric scattering. Rayleigh scattering is *wavelength-dependent,* meaning that the amount of scattering changes greatly as one examines different regions of the spectrum (Figure 2.7). Blue light is scattered about four times as much as is red light, and ultraviolet light is scattered almost 16 times as much as is red light. *Rayleigh's law* states that this form of scattering is in proportion to the inverse of the fourth power of the wavelength.

Rayleigh scattering is the cause both for the blue color of the sky and for the brilliant red and orange colors often seen at sunset. At midday, when the sun is high in the sky, the atmospheric path of the solar beam is relatively short and direct, so an observer at the Earth's surface sees mainly the blue light preferentially redirected by Rayleigh scatter. At sunset, observers on the Earth's surface see only those wavelengths that pass through the longer atmospheric path caused by the low solar elevation; because only the longer wavelengths penetrate this distance without attenuation by scattering, we see only the reddish component of the solar beam. Variations of concentrations of fine atmospheric dust or of tiny water droplets in the atmosphere may contribute to variations in atmospheric clarity, and therefore to variations in colors of sunsets.

Although Rayliegh scattering forms an important component of our understanding of atmospheric effects upon transmission of radiation in and near the visible spectrum, it applies only to a rather specific class of atmospheric interactions. In 1906 the German physicist Gustav Mie (1868–1957) published an analysis that describes atmospheric scattering involving a broader range of atmospheric particles. *Mie scattering* is caused by large atmospheric particles, includ-

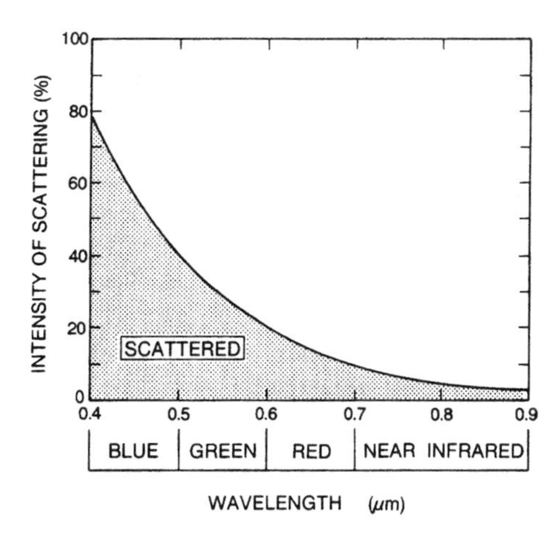


**FIGURE 2.7.** Rayleigh scattering. Scattering is much higher at shorter wavelengths.

ing dust, pollen, smoke, and water droplets. Such particles may seem to be very small by the standards of everyday experience, but they are many times larger than those responsible for Rayleigh scattering. Those particles that cause Mie scattering have diameters that are roughly equivalent to the wavelength of the scattered radiation. Mie scattering can influence a broad range of wavelengths in and near the visible spectrum; Mie's analysis accounts for variations in the size, shape, and composition of such particles. Mie scattering is wavelength-dependent, but not in the simple manner of Rayleigh scattering; it tends to be greatest in the lower atmosphere (0–5 km), where larger particles are abundant.

*Nonselective scattering* is caused by particles that are much larger than the wavelength of the scattered radiation. For radiation in and near the visible spectrum, such particles might be larger water droplets or large particles of airborne dust. "Nonselective" means that scattering is not wavelength-dependent, so we observe it as a whitish or grayish haze: all visible wavelengths are scattered equally.

### *Effects of Scattering*

Scattering causes the atmosphere to have a brightness of its own. In the visible portion of the spectrum, shadows are not jet-black (as they would be in the absence of scattering), but are merely dark; we can see objects in shadows because of light redirected by particles in the path of the solar beam. The effects of scattering are also easily observed in vistas of landscapes: colors and brightnesses of objects are altered as they are positioned at locations more distant from the observer. Landscape artists take advantage of this effect, called *atmospheric perspective,* to create the illusion of depth by painting more distant features in subdued colors and those in the foreground in brighter, more vivid colors.

For remote sensing, scattering has several important consequences. Because of the wavelength dependency of Rayleigh scattering, radiation in the blue and ultraviolet regions of the spectrum (which is most strongly affected by scattering) is usually not considered useful for remote sensing. Images that record these portions of the spectrum tend to record the brightness of the atmosphere rather than the brightness of the scene itself. For this reason, remote sensing instruments often exclude short-wave radiation (blue and ultraviolet wavelengths) by use of filters or by decreasing sensitivities of films to these wavelengths. (However, some specialized applications of remote sensing, not discussed here, do use ultraviolet radiation.) Scattering also directs energy from outside the sensor's field of view toward the sensor's aperture, thereby decreasing the spatial detail recorded by the sensor. Furthermore, scattering tends to make dark objects appear brighter than they would otherwise be and bright objects appear darker, thereby decreasing the contrast recorded by a sensor (Chapter 3). Because "good" images preserve the range of brightnesses present in a scene, scattering degrades the quality of an image.

Some of these effects are illustrated in Figure 2.8. Observed radiance at the sensor, $I$, is the sum of $I_S$, radiance reflected from the Earth's surface, conveying information about surface reflectance; $I_O$, radiation scattered from the solar beam directly to the sensor without reaching the Earth's surface, and $I_D$, diffuse radiation, directed first to the ground, then to the atmosphere, before reaching the sensor. Effects of these components are additive within a given spectral band (Kaufman, 1984):

$$I = I_S + I_O + I_D \quad \text{(Eq. 2.6)}$$

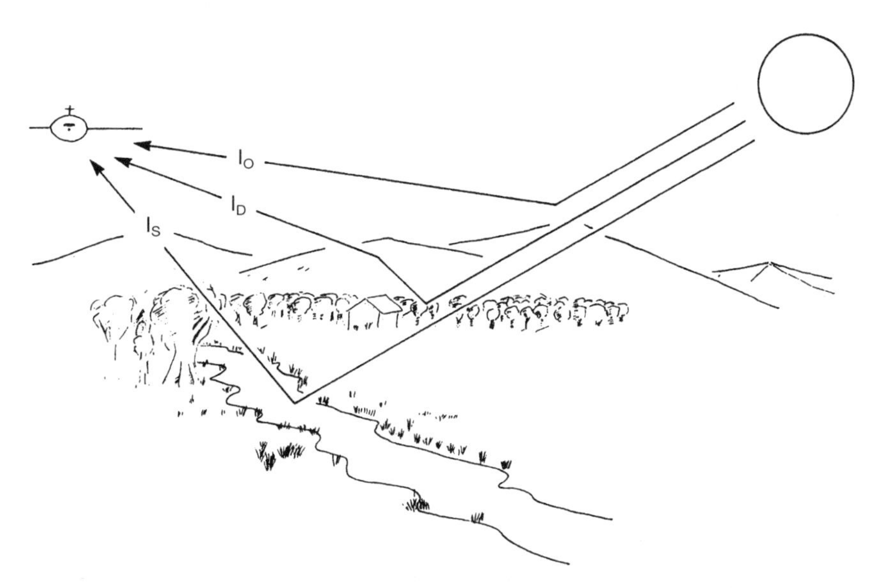


**FIGURE 2.8.** Principal components of observed brightness. $I_S$ represent radiation refected from the ground surface, $I_O$ is energy scattered by the atmosphere directly to the sensor, and $I_D$ represents diffuse light directed to the ground, then to the atmosphere, before reaching the sensor. This diagram describes behavior of radiation in and near the visible region of the spectrum. From Campbell and Ran (1993). Copyright 1993 by Elsevier Science Ltd.. Reproduced by permission.

$I_S$ varies with differing surface materials, topographic slopes and orientation, and angles of illumination and observation. $I_O$ is often assumed to be more or less constant over large areas, although most satellite images represent areas large enough to encompass atmospheric differences sufficient to create variations in $I_O$. Diffuse radiation, $I_D$, is expected to be small relative to other factors, but varies from one land surface type to another, so in practice would be difficult to estimate. We should note the special case presented by shadows, in which $I_S = 0$, because the surface receives no direct solar radiation. However, shadows have their own brightness, derived from $I_D$, and their own spectral patterns, derived from the influence of local land cover upon diffuse radiation. Remote sensing is devoted to the examination of $I_S$ at different wavelengths to derive information about the Earth's surface. Figure 2.9 illustrates how vary with wavelength for surfaces of differing brightness.

### *Refraction*

*Refraction* is the bending of light rays at the contact area between two media that transmit light. Familiar examples of refraction are the lenses of cameras or magnifying glasses (Chapter 3), which bend light rays to project or enlarge images, and the apparent displacement of objects submerged in clear water. Refraction also occurs in the atmosphere as light passes through atmospheric layers of varied clarity, humidity, and temperature. These variations influence the density of atmospheric layers, which in turn causes a bending of light rays as they pass from one layer to another. Everyday examples are the shimmering appearances on hot summer days of objects viewed in the distance as light passes through hot air near the surface of heated high-

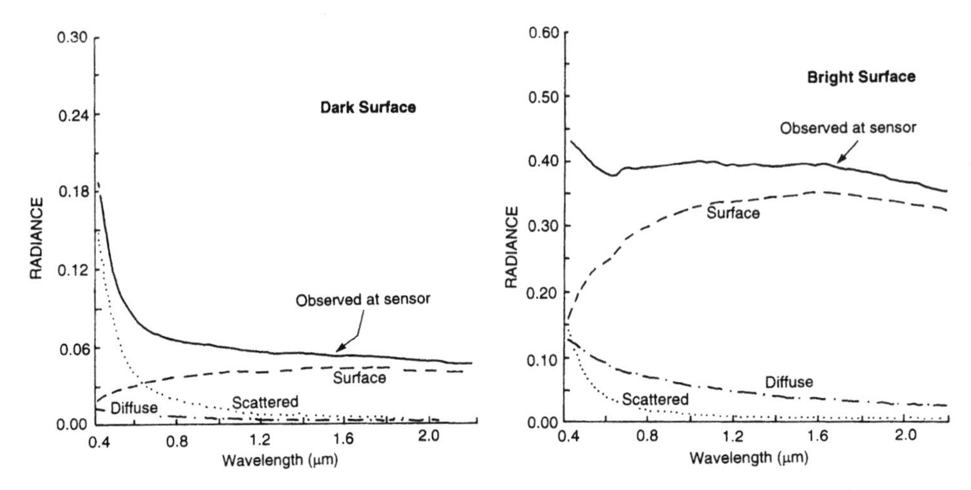


**FIGURE 2.9.** Changes in reflected, diffuse, scattered, and observed radiation over wavelength for dark (left) and bright (right) surfaces. The diagram shows the magnitude of the components illustrated in Figure 2.8. Atmosphere effects constitute a larger proportion of observed brightness for dark objects than for bright objects, especially at short wavelengths. Radiance has been normalized; note also the differences in scaling of the vertical axes for the two diagrams. Redrawn from Kaufman (1984). Reproduced by permission of the author and the Society of Photo-Optical Instrumentation Engineers.

ways, runways, and parking lots. The index of refraction ($n$) is defined as the ratio between the velocity of light in a vacuum ($c$) to its velocity in the medium ($c_n$)

$$n = c/c_n \quad \text{(Eq. 2.7)}$$

Assuming uniform media, as the light passes into a denser medium it is deflected toward the *surface normal,* a line perpendicular to the surface at the point when the light ray enters the denser medium, as represented by the solid line in Figure 2.10. The angle that defines the path of the refracted ray is given by *Snell's law:*

$$n \sin \theta = n' \sin \theta' \quad \text{(Eq. 2.8)}$$

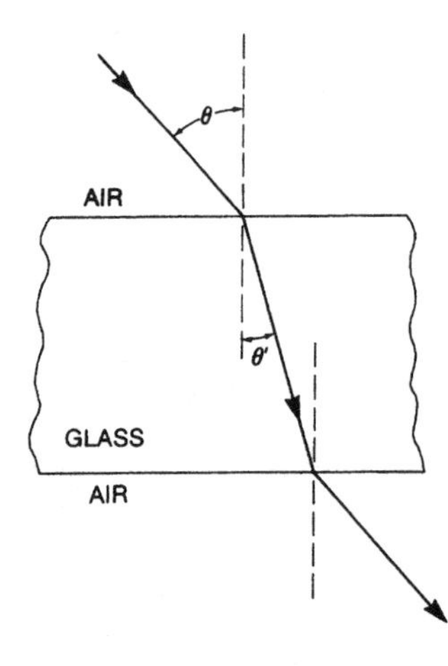


**FIGURE 2.10.** Refraction. This diagram represents the path of a ray of light as it passes from one medium (air) to another (glass), and again as it passes back to the first.

where $n$ and $n'$ are the indices of refraction of the first and second media, respectively, and $\theta$ and $\theta'$ are angles measured with respect to the surface normal, as defined in Figure 2.10.

### *Absorption*

*Absorption* of radiation occurs when the atmosphere prevents, or strongly attenuates, transmission of radiation or its energy through the atmosphere. (Energy acquired by the atmosphere is subsequently reradiated at longer wavelengths.) Three gases are responsible for most absorption of solar radiation. Ozone ($O_3$) is formed by the interaction of high-energy ultraviolet radiation with oxygen molecules ($O_2$) high in the atmosphere (maximum concentrations of ozone are found at altitudes of about 20–30 km in the stratosphere). Although naturally occurring concentrations of ozone are quite low (perhaps 0.07 parts per million at ground level, 0.1–0.2 parts per million in the stratosphere), ozone plays an important role in the Earth's energy balance. Absorption of the high-energy, short-wavelength portions of the ultraviolet spectrum (mainly less than 0.24 μm) prevents transmission of this radiation to the lower atmosphere.

Carbon dioxide ($CO_2$) also occurs in low concentrations (about 0.03% by volume of a dry atmosphere), mainly in the lower atmosphere. Aside from local variations caused by volcanic eruptions and mankind's activities, the distribution of $CO_2$ in the lower atmosphere is probably relatively uniform (although human activities that burn fossil fuels have apparently contributed to increases during the past 100 years or so). Carbon dioxide is important in remote sensing because it is effective in absorbing radiation in the mid- and far infrared regions of the spectrum. Its strongest absorption occurs in the region from about 13 to 17.5 μm, in the mid-infrared.

Finally, water vapor ($H_2O$) is commonly present in the lower atmosphere (below about 100 km) in amounts that vary from 0 to about 3% by volume. (Note the distinction between *water vapor,* discussed here, and droplets of *liquid* water, mentioned previously.) From everyday experience we know that the abundance of water vapor varies greatly from time to time and from place to place. Consequently, the role of atmospheric water vapor, unlike the roles of ozone and carbon dioxide, varies greatly with time and location. It may be almost insignificant in a desert setting or in a dry air mass, but may be highly significant in humid climates and in moist air masses. Furthermore, water vapor is several times more effective in absorbing radiation than are all other atmospheric gases combined. Two of the most important regions of absorption are in several bands between 5.5 and 7.0 μm, and above 27.0 μm; absorption in these regions can exceed 80% if the atmosphere contains appreciable amounts of water vapor.

### *Atmospheric Windows*

Thus the Earth's atmosphere is by no means completely transparent to electromagnetic radiation because these gases together form important barriers to transmission of electromagnetic radiation through the atmosphere. It selectively transmits energy of certain wavelengths; those wavelengths that are relatively easily transmitted through the atmosphere are referred to as *atmospheric windows* (Figure 2.11). Positions, extents, and effectiveness of atmospheric windows are determined by the absorption spectra of atmospheric gases. Atmospheric windows are of obvious significance for remote sensing because they define those wavelengths that can be

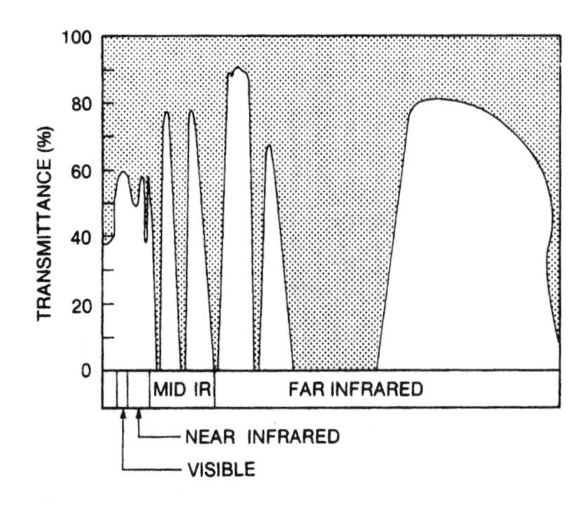


**FIGURE 2.11.** Atmospheric windows. This is a schematic representation that can depict only a few of the most important windows. The shaded region represents absorption of electromagnetic radiation.

used for forming images. Energy at other wavelengths, not within the windows, is severely attenuated by the atmosphere, and therefore cannot be effective for remote sensing. In the far infrared region, the two most important windows extend from 3.5 to 4.1 μm and from 10.5 to 12.5 μm. The latter is especially important because it corresponds approximately to wavelengths of peak emission from the Earth's surface. A few of the most important atmospheric windows are tabulated in Table 2.4; other smaller windows are not given here, but are listed in reference books.

### *Overview of Energy Interactions in the Atmosphere*

Remote sensing is conducted in the context of all the atmospheric processes discussed thus far, so it is useful to summarize some of the most important points by outlining a perspective that integrates much of the preceding material. Figure 2.12 is an idealized diagram of interactions of shortwave solar radiation with the atmosphere; values are based upon typical, or average, values derived from many places and many seasons, so they are by no means representative of values that might be observed at a particular time and place. This diagram represents only the behavior of "shortwave" radiation (defined loosely here to include radiation with wavelengths less than 4.0 μm). It is true that the Sun emits a broad spectrum of radiation, but the maximum intensity is

**TABLE 2.4. Major Atmospheric Windows**

| | |
|---|---|
| Ultraviolet and visible | 0.30–0.75 μm<br>0.77–0.91 μm |
| Near infrared | 1.55–1.75 μm<br>2.05–2.4 μm |
| Thermal infrared | 8.0–9.2 μm<br>10.2–12.4 μm |
| Microwave | 7.5–11.5 mm<br>20.0+ mm |

*Note.* Data selected from Fraser and Curran (1976, p. 35). Reproduced by permission of Addison-Wesley Publishing Co., Inc.

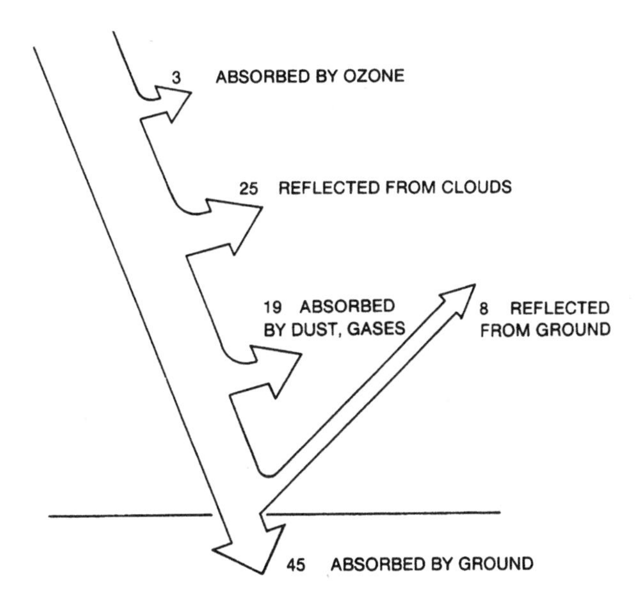


**FIGURE 2.12.** Incoming solar radiation. This diagram represents radiation at relatively short wavelengths, in and near the visible region. Values represent approximate magnitudes for the Earth as a whole—conditions at any specific place and time would differ from those given here.

emitted at approximately 0.5 μm within this region, and little solar radiation at longer wavelengths reaches the ground surface.

Of 100 units of shortwave radiation that reach the outer edge of the Earth's atmosphere, about three units are absorbed in the stratosphere as ultraviolet radiation interacts with oxygen ($O_2$) to form ozone ($O_3$). Of the remaining 97 units, about 25 are reflected from clouds, and about 19 are absorbed by dust and gases in the lower atmosphere. About eight units are reflected from the ground surface (this value varies greatly with different surface materials), and about 45 units ("about 50%") are ultimately absorbed at the Earth's surface. For remote sensing in the visible spectrum, it is the portion reflected from the Earth's surface that is of primary interest (see Figure 1.10), although knowledge of the quantity scattered is also important.

The 45 or so units that are absorbed are then reradiated by the Earth's surface. From Wien's displacement law (Eq. 2.5), we know that the Earth, being much cooler than the Sun, must emit radiation at much longer wavelengths than does the Sun. The Sun, at 6,000 K, has its maximum intensity at 0.5 μm (in the green portion of the visible spectrum); the Earth, at 300 K, emits with maximum intensity near 10 μm, in the far infrared spectrum.

Terrestrial radiation, with wavelengths longer than 10 μm, is represented in Figure 2.13. There is little, if any, overlap between the wavelengths of solar radiation, depicted in Figure 2.12, and the terrestrial radiation, shown in Figure 2.13. This diagram depicts the transfer of long-wave radiation (as defined above) from the ground surface to the atmosphere, in three separate categories.

About eight units are transferred from the ground surface to the atmosphere by "turbulent transfer" (heating of the lower atmosphere by the ground surface, which causes upward movement of air, then movement of cooler air to replace the original air). About 22 units are lost to the atmosphere by evaporation of moisture in the soil, water bodies, and vegetation (this energy

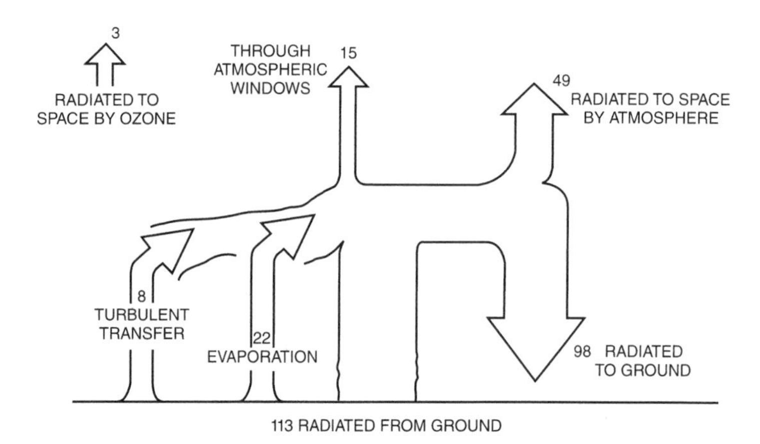


**FIGURE 2.13.** Outgoing terrestrial radiation. This diagram represents radiation at relatively long wavelengths—what we think of as sensible heat, or thermal radiation. Because the Earth's atmosphere absorbs much of the radiation emitted by the Earth, only those wavelengths that can pass through the atmospheric windows can be used for remote sensing.

is transferred as the latent heat of evaporation). Finally, about 113 units are radiated directly to the atmosphere.

Because atmospheric gases are very effective in absorbing this long-wave (far infrared) radiation, much of the energy that the Earth radiates is retained (temporarily) by the atmosphere. About 15 units pass directly through the atmosphere to space; this is energy emitted at wavelengths that correspond to atmospheric windows (chiefly 8–13 µm). Energy absorbed by the atmosphere is ultimately reradiated to space (49 units) and back to the Earth (98 units). For meteorology, it is these reradiated units that are of interest because they are the source of energy for heating of the Earth's atmosphere. For remote sensing, it is the 15 units that pass through the atmospheric windows that are of significance, as it is this radiation that conveys information concerning the radiometric properties of features on the Earth's surface.

## 2.6. Interactions with Surfaces

As electromagnetic energy reaches the Earth's surface, it must be reflected, absorbed, or transmitted. The proportions accounted for by each process depend upon the nature of the surface, the wavelength of the energy, and the angle of illumination.

### *Reflection*

*Reflection* occurs when a ray of light is redirected as it strikes a nontransparent surface. The nature of the reflection depends upon sizes of surface irregularities (roughness or smoothness) in relation to the wavelength of the radiation considered. If the surface is smooth relative to wavelength, *specular* reflection occurs (Figure 2.14a). Specular reflection redirects all, or almost all, of the incident radiation in a single direction. For such surfaces, the angle of incidence is equal to the angle of reflection (i.e., in Eq. 2.8, the two media are identical, so $n = n$,

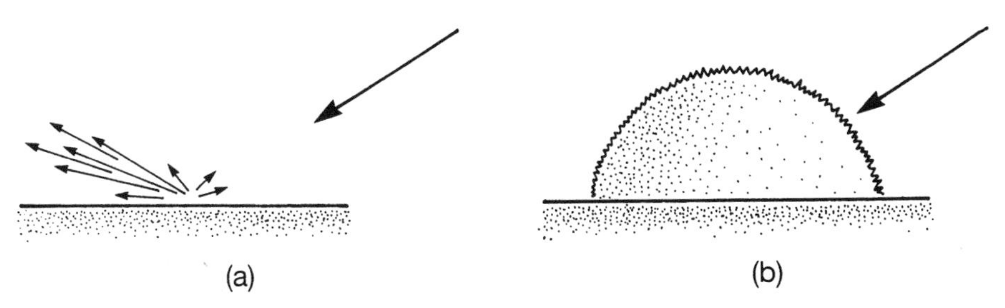


**FIGURE 2.14.** Specular (*a*) and diffuse (*b*) reflection. Specular reflection occurs when a smooth surface tends to direct incident radiation in a single direction. Diffuse reflection occurs when a rough surface tends to scatter energy more or less equally in all directions.

and therefore $\theta = \theta'$). For visible radiation, specular reflection can occur with surfaces such as a mirror, smooth metal, or a calm water body.

If a surface is rough relative to wavelength, it acts as a *diffuse,* or *isotropic,* reflector. Energy is scattered more or less equally in all directions. For visible radiation, many natural surfaces might behave as diffuse reflectors, including, for example, uniform grassy surfaces. A perfectly diffuse reflector (known as a *Lambertian surface*) would have equal brightnesses when observed from any angle (Figure 2.14b).

The idealized concept of a perfectly diffuse reflecting surface is derived from the work of Johann H. Lambert (1728–1777), who conducted many experiments designed to describe the behavior of light. One of Lambert's laws of illumination states that the perceived brightness (radiance) of a perfectly diffuse surface does not change with the angle of view. This is Lambert's cosine law, which states that the observed brightness ($I$) of such a surface is proportional to the cosine of the incidence angle ($\theta$), where $I$ is the brightness of the incident radiation as observed at zero incidence:

$$I' = I/\cos\theta \qquad \text{(Eq. 2.9)}$$

This relationship is often combined with the equally important inverse square law, which states that observed brightness decreases according to the square of the distance from the observer to the source:

$$I' = (I/D^2)(\cos\theta) \qquad \text{(Eq. 2.10)}$$

Both the cosine law and the inverse square law are depicted in Figure 2.15.

### *Bidirectional Reflectance Distribution Function*

Because of its simplicity and directness, the concept of a Lambertian surface is frequently used as an approximation of the optical behavior of objects observed in remote sensing. However, the Lambertian model does not hold precisely for many, if not most, natural surfaces. Actual surfaces exhibit complex patterns of reflection determined by details of surface geometry (e.g., the sizes, shapes, and orientations of plant leaves). Some surfaces may approximate Lambertian

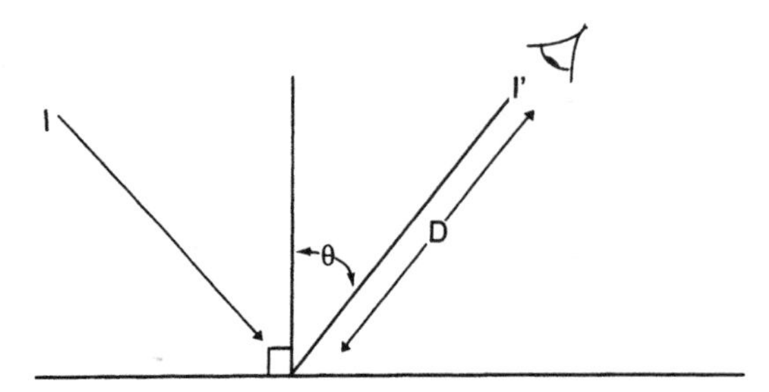


FIGURE 2.15. Inverse square law and Lambert's cosine law.

behavior at some incidence angles, but exhibit clearly non-Lambertian properties at other angles.

Reflection characteristics of a surface are described by the *bidirectional reflectance distribution function* (BRDF). The BRDF is a mathematical description of the optical behavior of a surface with respect to angles of illumination and observation, given that it has been illuminated with a parallel beam of light at a specified azimuth and elevation. (The function is "bidirectional" in the sense that it accounts both for the angle of illumination and the angle of observation.) The BRDF for a Lambertian surface has the shape depicted in Figure 2.14b, with even brightnesses as the surface is observed from any angle. Actual surfaces have more complex behavior. Description of BRDFs for actual, rather than idealized, surfaces permits assessment of the degrees to which they approach the ideals of specular and diffuse surfaces (Figure 2.16).

### Transmission

*Transmission* of radiation occurs when radiation passes through a substance without significant attenuation (Figure 2.17). From a given thickness, or depth, of a substance, the ability of a medium to transmit energy is measured as the transmittance ($t$):

$$t = \frac{\text{Transmitted radiation}}{\text{Incident radiation}} \qquad \text{(Eq. 2.11)}$$

In the field of remote sensing, the transmittance of films and filters is often important. With respect to naturally occurring materials, we often think only of water bodies as capable of transmitting significant amounts of radiation. However, the transmittance of many materials varies greatly with wavelengths, so our direct observations in the visible spectrum do not transfer to other parts of the spectrum. For example, plant leaves are generally opaque to visible radiation but transmit significant amounts of radiation in the infrared.

### Fluorescence

*Fluorescence* occurs when an object illuminated with radiation of one wavelength emits radiation at a different wavelength. The most familiar examples are some sulfide minerals, which emit visible radiation when illuminated with ultraviolet radiation. Other objects also fluoresce, although observation of fluorescence requires very accurate and detailed measurements, not now routinely available for most applications. Figure 2.18 illustrates the fluorescence of healthy

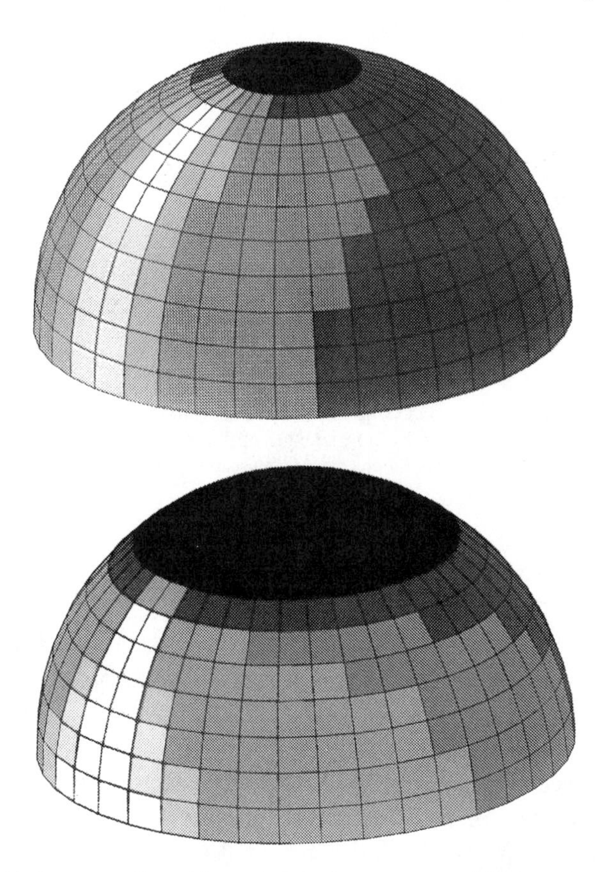

FIGURE 2.16. BRDFs for two surfaces. The varied shading represents differing intensities of observed radiation. (Calculated by Pierre Villeneuve.)

and senescent leaves, using one axis to describe the spectral distribution of the illumination and the other to show the spectra of the emitted energy. These contrasting surfaces illustrate the effectiveness of fluorescence in revealing differences between healthy and stressed leaves.

### *Spectral Properties of Objects*

Remote sensing consists of the study of radiation emitted and reflected from features at the Earth's surface. In the instance of emitted (far infrared) radiation, the object itself is the imme-

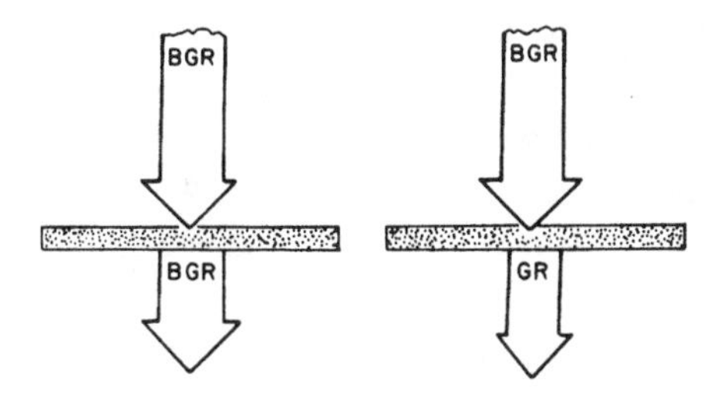


FIGURE 2.17. Transmission. Incident radiation passes through an object without significant attenuation (left), or may be selectively transmitted (right). The object on the right would act as a yellow ("minus blue") filter, as it would transmit all visible radiation except for blue light.

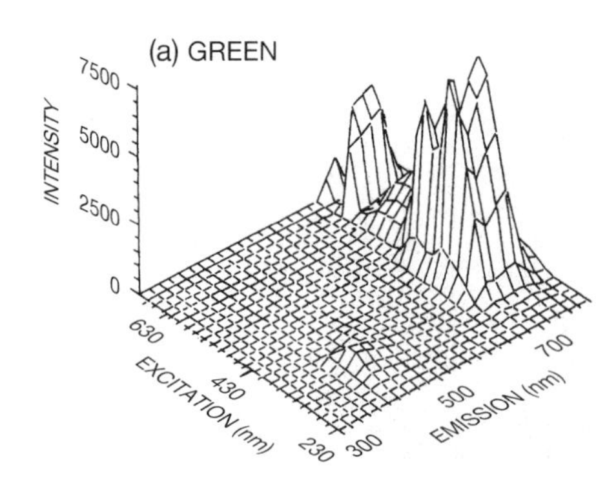


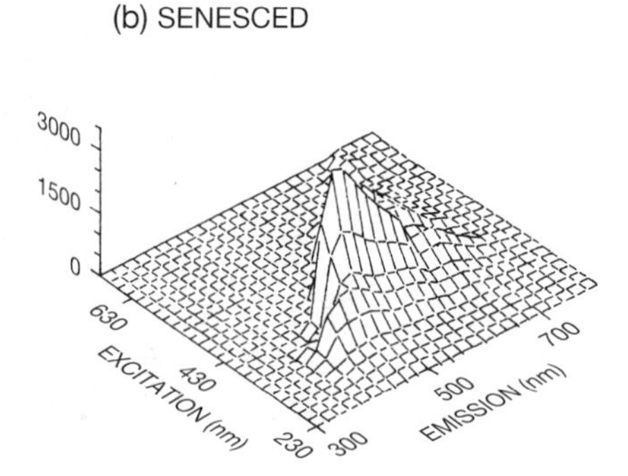


**FIGURE 2.18.** Fluorescence. Exitation and emission are shown along the two horizontal axes (with wavelengths given in nanometers). The vertical axes show strength of fluorescence, with the two examples illustrating the contrast in fluorescence between healthy green (*a*) and senesced (*b*) leaves. From Rinker (1994).

diate source of radiation. For reflected radiation, the source may be the sun, the atmosphere (by means of scattering of solar radiation), or man-made radiation (chiefly imaging radars).

A fundamental premise in remote sensing is that we can learn about objects and features on the Earth's surface by studying the radiation reflected and/or emitted by these features. Using cameras and other remote sensing instruments, we can observe the brightnesses of objects over a range of wavelengths, so that there are numerous points of comparison between brightnesses of separate objects. A set of such observations or measurements constitute a spectral response pattern, sometimes called the *spectral signature* of an object (Figure 2.19). In the ideal, detailed knowledge of a spectral response pattern might permit identification of features of interest, such as separate kinds of crops, forests, or minerals. This idea has been expressed as follows:

> Everything in nature has its own unique distribution of reflected, emitted, and absorbed radiation. These spectral characteristics can—if ingeniously exploited—be used to distinguish one thing from another or to obtain information about shape, size, and other physical and chemical properties. (Parker and Wolff, 1965, p. 21)

This statement expresses the fundamental concept of the spectral signature, the notion that features display unique spectral responses that would permit clear identification, from spectral information alone, of individual crops, soils, and so on, from remotely sensed images. In practice, it is now recognized that spectra of features change both over time (e.g., as a cornfield grows during a season) and over distance (e.g., as proportions of specific tree species in a forest change from place to place).

Nonetheless, the study of the spectral properties of objects forms an important part of remote sensing. Some research has been focused upon examination of spectral properties of different classes of features. Thus, although it may be difficult to define unique signatures for specific kinds of vegetation, we can recognize distinctive spectral patterns for vegetated and nonvegetated areas, and for certain classes of vegetation, and we can sometimes detect the existence of diseased or stressed vegetation. In other instances, we may be able to define spectral patterns that are useful within restricted geographic and temporal limits as a means of studying the distributions of certain plant and soil characteristics. Chapter 15 describes how very detailed spectral measurements permit application of some aspects of the concept of the spectral signature.

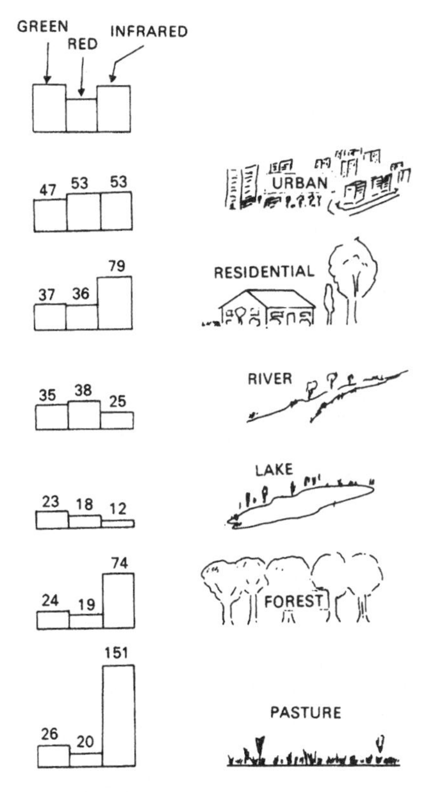


**FIGURE 2.19.** Spectral signatures.

## 2.7. Summary: Three Models for Remote Sensing

Remote sensing typically takes one of three basic forms depending on the wavelengths of energy detected and on the purposes of the study. In the simplest form, one records the reflection of solar radiation from the Earth's surface (Figure 2.20). This is the kind of remote sensing most nearly similar to everyday experience. For example, film in a camera records radiation from the Sun after it is reflected from the objects of interest, regardless of whether one uses a simple handheld camera to photograph a family scene or a complex aerial camera to photograph a large area of the Earth's surface. This form of remote sensing mainly uses energy in the visible and near infrared portions of the spectrum. Key variables include atmospheric clarity, spectral properties of objects, angle and intensity of the solar beam, choices of films and filters, and others explained in Chapter 3.

A second strategy for remote sensing is to record radiation *emitted* from (rather than *reflected* from) the Earth's surface. Because emitted energy is strongest in the far infrared spectrum, this kind of remote sensing requires special instruments designed to record these wavelengths. (There is no direct analogue to everyday experience for this kind of remote sensing.) Emitted energy from the Earth's surface is mainly derived from shortwave energy from the Sun that has been absorbed, then reradiated at longer wavelengths (Figure 2.21).

Emitted radiation from the Earth's surface reveals information concerning thermal properties

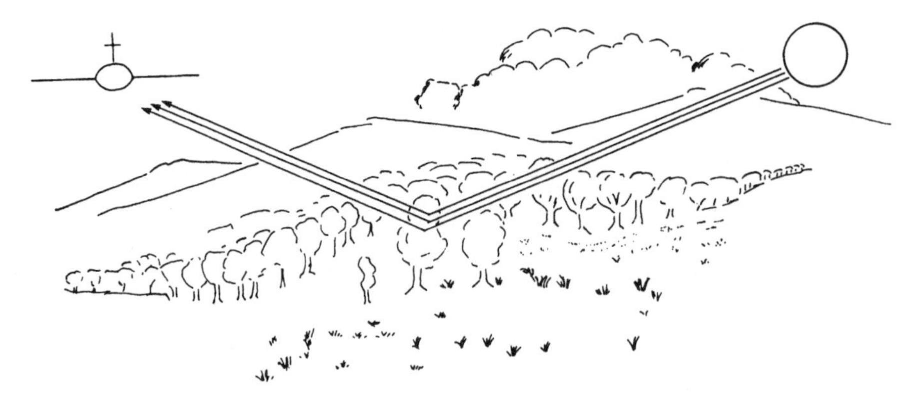

**FIGURE 2.20.** Remote sensing using reflected solar radiation. The sensor detects solar radiation that has been reflected from features at the Earth's surface. (See Figure 2.12.)

of materials, which can be interpreted to suggest patterns of moisture, vegetation, surface materials, and man-made structures. Other sources of emitted radiation (of secondary significance here, but often of primary significance elsewhere) include geothermal energy and heat from steam pipes, power plants, buildings, and forest fires. This example also represents "passive" remote sensing, because it employs instruments designed to sense energy emitted by the Earth, not energy generated by a sensor.

Finally, sensors belonging to a third class of remote sensing instruments generate their own energy, then record the reflection of that energy from the Earth's surface (Figure 2.22). These are "active" sensors—"active" in the sense that they provide their own energy, so they are independent of solar and terrestrial radiation. As an everyday analogy, a camera with a flash attachment can be considered to be an active sensor. In practice, active sensors are best represented by imaging radars and lidars (Chapters 7 and 8), which transmit energy toward the Earth's surface from an aircraft or satellite, then receive the reflected energy to form an image. Because they sense energy provided directly by the sensor itself, such instruments have the capability to operate at night and during cloudy weather.

**FIGURE 2.21.** Remote sensing using emitted terrestrial radiation. The sensor records solar radiation that has been absorbed by the Earth, and then reemitted as thermal infrared radiation. (See Figures 2.12 and 2.13.)

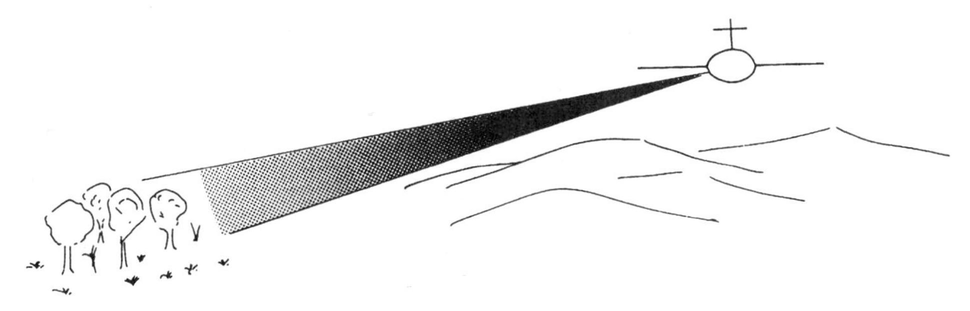

**FIGURE 2.22.** Active remote sensing. The sensor illuminates the terrain with its own energy, then records the reflected energy as it has been altered by the Earth's surface.

## Review Questions

1. Using books provided by your instructor or available through your library, examine reproductions of landscape paintings to identify artistic use of atmospheric perspective. Perhaps some of your own photographs of landscapes illustrate the optical effects of atmospheric haze. Look for examples of atmospheric effects upon color, brightness, contrast, and spatial detail.

2. Some streetlights are deliberately manufactured to provide illumination with a reddish color. From material presented in this chapter, can you suggest why?

3. Although this chapter has largely dismissed ultraviolet radiation as an important aspect of remote sensing, there may well be instances where it might be effective, despite problems associated with its use. Under what conditions might it prove practical to use ultraviolet radiation for remote sensing?

4. The human visual system is most nearly similar to which of the models for remote sensing described in the last sections of this chapter?

5. Can you identify analogues from the animal kingdom for each of the models for remote sensing discussed in Section 2.7?

6. Examine Figures 2.12 and 2.13. Discuss how the values in this figure might change in different environments, including (a) desert, (b) the arctic, and (c) an equatorial climate. How might these differences influence our ability to conduct remote sensing in each region?

7. Special signatures can be illustrated using values indicating the brightness in several spectral regions.

| | UV | Blue | Green | Red | IR |
|---|---|---|---|---|---|
| Forest | 28 | 29 | 36 | 27 | 56 |
| Water | 22 | 23 | 19 | 13 | 8 |
| Corn | 53 | 58 | 59 | 60 | 71 |
| Pasture | 40 | 39 | 42 | 32 | 62 |

Assume for now that these are "pure" signatures, not influenced by effects of the atmosphere. Can all categories be reliably separated, based upon these spectral values? Which bands are most useful for distinguishing between these classes?

8. Describe ideal atmospheric conditions for remote sensing.
9. Can you identify some advantages that active sensors have relative to passive sensors? Do they have disadvantages?
10. List ways that the spectral signature of a field of corn might vary throughout a year. How do such variations influence the concept of a spectral signature?

## References

Bohren, C. F. 1987. *Clouds in a Glass of Beer: Simple Experiments in Atmospheric Physics.* New York: Wiley, 195 pp.

Campbell, J. B., and L. Ran. 1993. CHROM: A C Program to Evaluate the Application of the Dark Object Subtraction Technique to Digital Remote Sensing Data. *Computers and Geosciences,* Vol. 19, pp. 1475–1499.

Chahine, M. T. 1983. Interaction Mechanisms within the Atmosphere. Chapter 5 in *Manual of Remote Sensing* (R. N. Colwell, ed.). Falls Church, VA: American Society of Photogrammetry, pp. 165–230.

Chameides, W. L., and D. D. Davis. 1982. Chemistry in the Troposphere. *Chemical and Engineering News,* Vol. 60, pp. 39–52.

Estes, J. E. 1978. The Electromagnetic Spectrum and Its Use in Remote Sensing. Chapter 2 in *Introduction to Remote Sensing of Environment* (B. F. Richason, ed.). Dubuque, IA: Kendall-Hunt, pp. 15–39.

Fraser, R. S., and R. J. Curran. 1976. Effects of the Atmosphere on Remote Sensing. Chapter 2 in *Remote Sensing of Environment* (C. C. Lintz and D. S. Simonett, eds.). Reading, MA: Addison-Wesley, pp. 34–84.

Goetz, A. F. H., J. B. Wellman, and W. L. Barnes. 1985. Optical Remote Sensing of the Earth. *Proceedings of the IEEE,* Vol. 73, pp. 950–969.

Kaufman, Y. J. 1984. Atmospheric Effects on Remote Sensing of Surface Reflectance. Special issue: *Remote Sensing* (P. N. Slater, ed.). *Proceedings, SPIE,* Vol. 475, pp. 20–33.

Kaufman, Y. J. 1989. The Atmospheric Effect on Remote Sensing and Its Correction. Chapter 9 in *Theory and Applications of Optical Remote Sensing* (Ghassam Asrar, ed.). New York: Wiley, pp. 336–428.

Lynch, D. K., and W. Livingston. 1995. *Color and Light in Nature.* New York: Cambridge University Press, 254 pp.

Minnaert, M. 1954. *The Nature of Light and Color* (revision by H. M. Kremer-Priest; translation by K. E. Brian Jay). New York: Dover, 362 pp.

Rees, W. G. 1990. *Physical Principles of Remote Sensing.* New York: Cambridge University Press, 247 pp.

Rinker, J. N. 1994. ISSSR Tutorial I: Introduction to Remote Sensing. In *Proceedings of the International Symposium on Spectral Sensing Research '94.* Alexandria, VA: U.S. Army Topographic Engineering Center, pp. 5–43.

Slater, P. N. 1980. *Remote Sensing: Optics and Optical Systems.* Reading, MA: Addison-Wesley, 575 pp.

Stimson, A. 1974. *Photometry and Radiometry for Engineers.* New York: Wiley, 446 pp.

Swain, P. H., and S. M. Davis. 1978. *Remote Sensing: The Quantitative Approach.* New York: McGraw-Hill, 396 pp.

Turner, R. E., W. A. Malila, and R. F. Nalepka. 1971. Importance of Atmospheric Scattering in Remote Sensing. In *Proceedings of the 7th International Symposium on Remote Sensing of Environment.* Ann Arbor: Willow Run Laboratories, University of Michigan, pp. 1651–1697.

PART TWO

# IMAGE ACQUISITION

CHAPTER THREE

# Photographic Sensors

## 3.1. Introduction

The word *photography* was coined in France in the mid-1800s, when it was fashionable to use Greek and Latin words to name new scientific discoveries. The word *photography* means "to write with light"—a literal description of what the newly invented camera could do. Today our meaning is often expanded to include radiation just outside the visible spectrum, in the ultraviolet and near infrared regions.

Although the status of photography is challenged by continuing innovations in digital imaging technology, photography remains the most practical, inexpensive, and widely used means of remote sensing. Further, the basic optical principals used for photography are also employed in optical systems involving nonphotographic sensors, and we often use photographic film to record images generated by nonphotographic sensors. Therefore, knowledge of photography is vital for understanding the field of remote sensing.

The formation of images by refraction of light is a surprisingly old practice. In antiquity, Greek and Arab scholars knew that images could be formed as light passed through a pinhole opening in a dark enclosure. Refraction of light at the tiny opening bends light rays to form an inverted image in a manner analogous to the effect of a simple lens. In medieval Europe, a device known as the *camera obscura* ("dark chamber") employed this principle to project an image onto a screen as an aid for artists, who could then trace the outline of the image as the foundation for more elaborate drawings or paintings. During the Renaissance, the addition of a simple convex lens improved the camera obscura, although there was still no convenient means of recording the image formed on the screen. Later, with the development of photographic emulsions (described below) as a means of making a detailed record of the image, the camera obscura began its evolution toward the everyday cameras that we know today, which in turn are models for the more complex cameras used for aerial survey.

Despite the current availability of more sophisticated imaging systems, aerial photography still remains the most accessible and versatile form of remote sensing imagery. Routine use of aerial photography has incalculable value throughout the world as a major source of information concerning the landscape and as the primary means of producing modern topographic maps. Its economic contributions to surveying the Earth and effective planning are considerable. Aerial photography will remain as a primary source of remote sensing imagery for many years to come.

## 3.2. The Aerial Camera

In their most basic elements, aerial cameras are similar to the simple handheld cameras we all have used. Both share the four main components of all cameras: (1) a lens to focus light on the film, (2) a light-sensitive film to record the image, (3) a shutter that controls entry of light into the camera, and (4) the camera body, a light-tight enclosure that holds the film, lens, and shutter in their correct positions.

In addition, aerial cameras include three other elements not usually encountered in our personal experiences with photography: the film magazine, the drive mechanism, and the lens cone (Figure 3.1).

### *The Lens*

The *lens* gathers reflected light and focuses it on the film. In its simplest form, a lens is a glass disk carefully ground into a shape with nonparallel curved surfaces (Figure 3.2). The change in optical densities as light rays pass from the atmosphere to the lens and back to the atmosphere causes refraction of light rays; the sizes, shapes, arrangements, and compositions of lenses are carefully designed to control this bending of light rays to maintain color balance and to minimize optical distortions. Optical characteristics of lenses are determined largely by the refractive index of the glass (Chapter 2) and the degree of curvature present in the lens surface. The quality of a lens is determined by the quality of its glass, the precision with which that glass is shaped, and the accuracy with which it is positioned within a camera. Imperfections in lens shape contribute to *spherical aberration,* a source of error that distorts images and causes loss of image clarity. For modern aerial photography, spherical aberration is usually not a severe problem because most modern aerial cameras use lenses of very high quality.

Figure 3.2 shows the simplest of all lenses: a simple positive lens. Such a lens is formed from a glass disk with equal curvature on both sides; light rays are refracted at both edges to form an

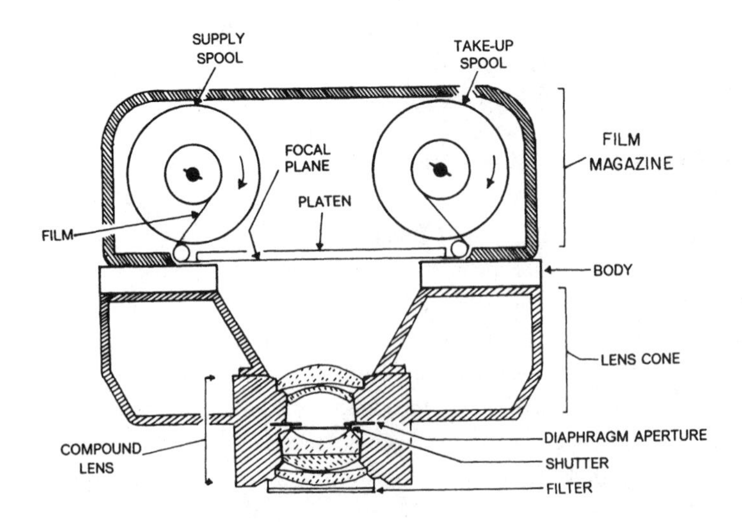


**FIGURE 3.1.** Schematic diagram of an aerial camera, cross-sectional view. Labeled items are discussed in text.

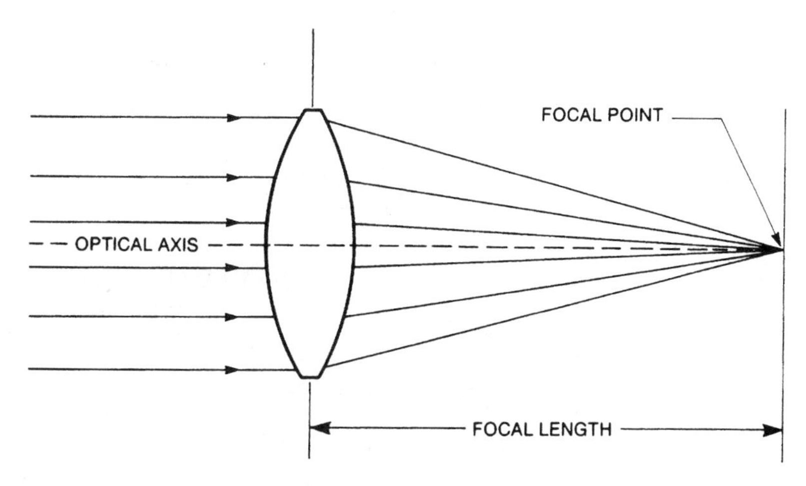


**FIGURE 3.2.** Simple lens.

image. Most aerial cameras use *compound lenses,* formed from many separate lenses of varied sizes, shapes, and properties. These many components are designed to correct for the errors that may be present in any single component, so the whole unit is much more accurate than any single element. For present purposes, consideration of a simple lens will be sufficient to define the most important features of lenses, even though a simple lens differs from those actually used in modern aerial cameras.

The *optical axis* joins the centers of curvature of the two sides of the lens. Although refraction occurs throughout a lens, a plane passing through the center of the lens, known as the *image principal plane,* is considered to be the center of refraction within the lens (Figure 3.3). The image principal plane intersects the optical axis at the *nodal point.*

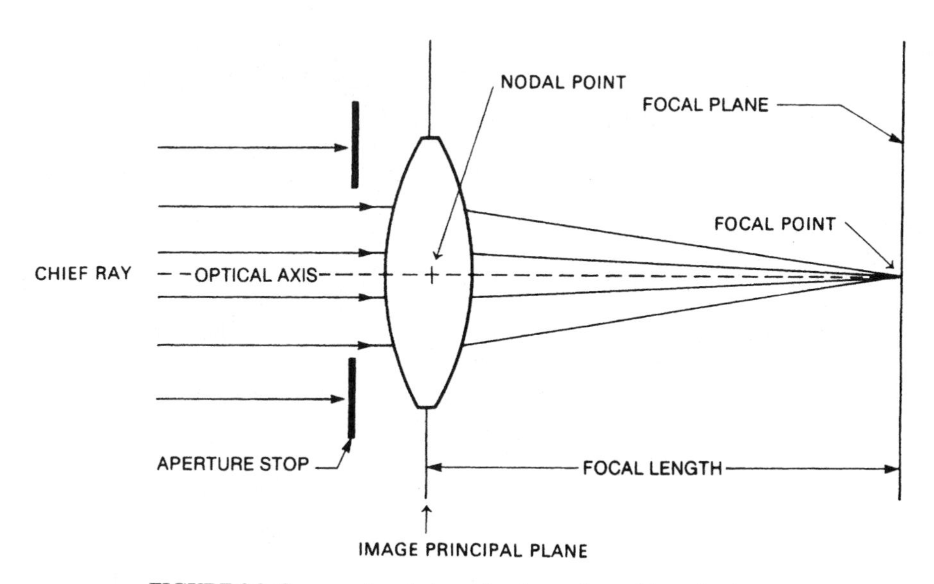


**FIGURE 3.3.** Cross-sectional view of an image formed by a simple lens.

Parallel light rays reflected from an object at a great distance (at an "infinite" distance) pass through the lens and are brought to focus at the principal *focal point,* the point at which the lens forms an image of the distant object. The chief ray passes through the nodal point without changing direction; all other rays are bent by the lens. A plane passing through the focal point parallel to the image principal plane is known as the *focal plane.* For handheld cameras, the distance from the lens to the object is important because the image is brought into focus at distances that increase as the object is positioned closer to the lens. For such cameras, it is important to use lenses that can be adjusted to bring each object to a correct focus as the distance from the camera to the object changes. For aerial cameras, the scene to be photographed is always at such large distances that the focus can be fixed at infinity, with no need to change the focus of the lens.

For a simple positive lens, the *focal length* is defined as the distance from the center of the lens to the focal point, usually measured in inches or millimeters. (For a compound lens, the definition is more complex.) For a given lens, the focal length is not identical for all wavelengths. Blue light is brought to a focal point at a shorter distance than are red or infrared wavelengths (Figure 3.4). This effect is the source of chromatic aberration. Unless corrected by lens design, chromatic aberration would cause the individual colors of an image to be out of focus in the photograph. Chromatic aberration is corrected in high-quality aerial cameras to assure that the radiation used to form the image is brought to a common focal point.

The field of view of a lens can be controlled by a *field stop,* a mask positioned just in front of the focal plane. An *aperture stop* is usually positioned near the center of a compound lens; it consists of a mask with a circular opening of adjustable diameter (Figure 3.5). An aperture stop can control the intensity of light at the focal plane, but does not influence the field of view or the size of the image. Manipulation of the aperture stop controls only the brightness of the image without changing its size. Usually aperture size is measured as the diameter of the adjustable opening that admits light to the camera. Relative aperture is defined as

$$f = \text{Focal length/aperture size} \qquad \text{(Eq. 3.1)}$$

where focal length and aperture are measured in the same units of length, and $f$ is the *f number,* the relative aperture. A large $f$ number means that the aperture opening is small relative to focal length; a small $f$ number means that the opening is large relative to focal length.

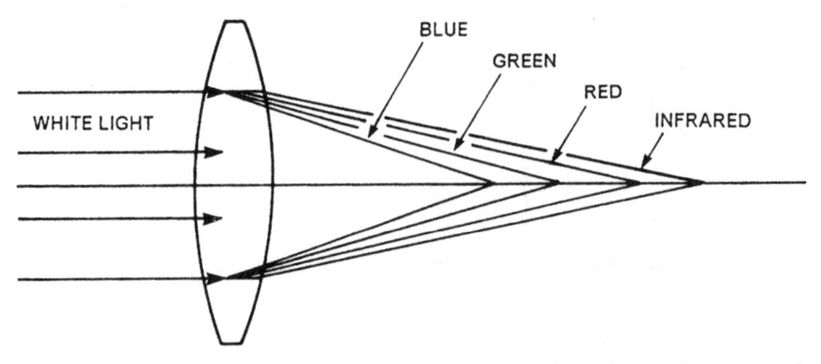


**FIGURE 3.4.** Chromatic aberration. Energy of differing wavelengths is brought to a focus at varying distances from the lens. More complex lenses are corrected to bring all wavelengths to a common focal point.

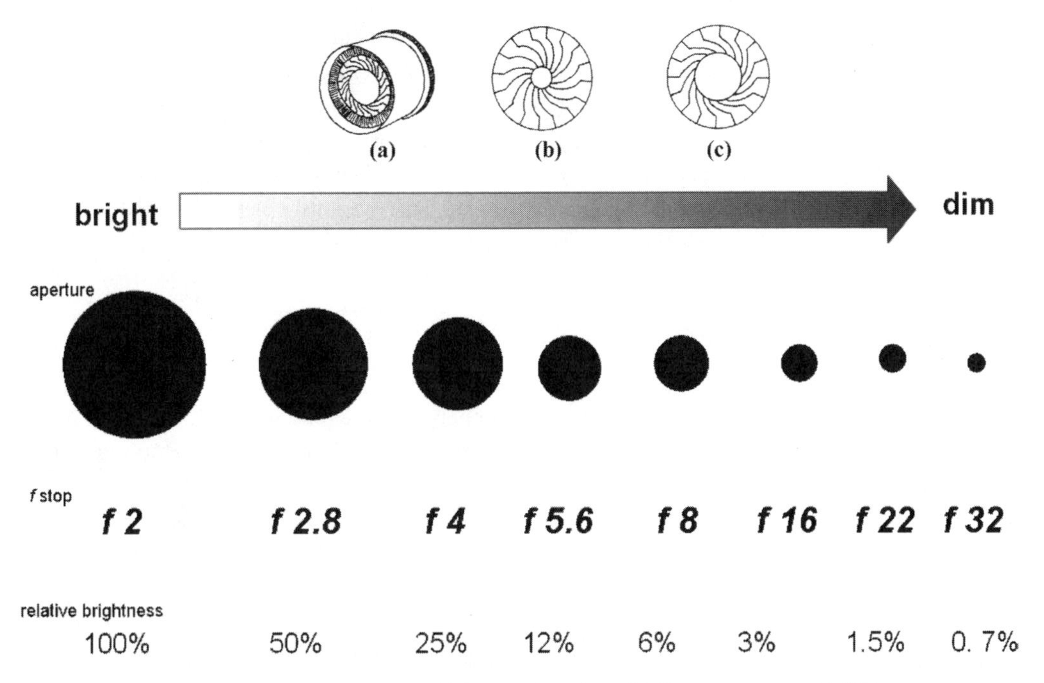


**FIGURE 3.5.** Diaphragm aperture stop. (a) Perspective view. (b) Narrow aperture. (c) Wide aperture.

Why use *f* numbers rather than direct measurements of aperture? One reason is that standardization of aperture with respect to focal length permits specification of aperture sizes using a value that is independent of camera size. Specification of an aperture as "23 mm" has no practical meaning unless we also know the size (focal length) of the camera. Specification of an aperture as "*f* 4" has meaning for cameras of all sizes; we know that it is one-fourth of the focal length for any size camera.

The standard sequence of apertures is: *f* 1, *f* 1.4, *f* 2, *f* 2.8, *f* 4, *f* 5.6, *f* 8, *f* 11, *f* 16, *f* 22, *f* 32, *f* 64, . . . This sequence is designed to change the amount of light by a factor of 2 as the *f*-stop is changed by one position. For example, a change from *f* 2 to *f* 2.8 halves the amount of light entering the camera; a change from *f* 11 to *f* 8 doubles the amount of light. A given lens, of course, is capable of using only a portion of the range of apertures mentioned above.

Lenses for aerial cameras typically have wide fields of view. As a result, light reaching the focal plane from the edges of the field of view is typically dimmer than light reflected from an object positioned near the center of the field of view. This effect creates a dark rim around the center of the aerial photograph—this effect is known as *vignetting.* It is possible to employ an *antivignetting filter,* darker at the center and clearer at the periphery, which can be partially effective in evening brightnesses across the photograph.

### *The Shutter*

The *shutter* controls the length of time that the film is exposed to light. The simplest shutters are often metal blades positioned between elements of the lens, forming "intralens," or "between-the-lens," shutters. An alternative form of shutter is the focal plane shutter, consisting of a metal or fabric curtain positioned just in front of the film, near the focal plane. The curtain is constructed with a number of slits; the choice of shutter speed by the operator selects the opening that produces the desired exposure. Although some aerial cameras use focal plane shutters, the between-the-lens shutter is preferred for most aerial cameras. The between-the-lens shutter subjects the entire negative to illumination simultaneously, and presents a clearly defined perspective that permits use of the image negative as the basis for precise measurements.

### *The Film Magazine*

The *film magazine* (Figure 3.1) is a light-tight container that holds the supply of film. The magazine usually includes a supply spool, holding perhaps several hundred feet of unexposed aerial film, and a take-up spool to accept exposed film.

### *The Lens Cone*

The *lens cone* (Figure 3.1) supports the lens and filters and holds them in their correct positions in relation to the film. The lens cone is usually detachable to permit the use of different lenses with the same camera body. The camera manufacturer carefully aligns the lens with the other components of the camera to assure geometric accuracy of photographs. Common focal lengths for typical aerial cameras are 150 mm (about 6 in.), 300 mm (about 12 in.), and 450 mm (about 18 in.). Slater (1975) lists characteristics (including focal lengths and apertures) for a number of specific models of aerial cameras.

### *The Drive Mechanism*

The *drive mechanism* advances the film after each exposure, using electric motors activated in coordination with the shutter and the motion of the plane. At the time of exposure, it is important that the film lie flat in the camera's focal plane. This function is performed by the *platen,* which for simple handheld cameras is a small, spring-mounted, metal plate positioned to hold the film flat at the instant of exposure. Because of the difficulty of holding large sheets of film flat, aerial cameras use special platens. A *vacuum platen* consists of a flat plate positioned at the focal plane; a vacuum pump draws air through small holes in the plate to hold the film flat and stationary during exposure. The vacuum sucks the film flat against the platen to prevent bending of the film or formation of bubbles of air as the film is positioned in the focal plane. The vacuum is released after exposure to allow the film to advance for the next exposure, then is applied again as the next frame is ready for exposure.

High-quality aerial cameras usually include a capability known as *image motion compensation* (*or forward motion compensation*), achieved by a mechanism that moves the film platen (or other components of the camera's optical system) during exposure at a speed and in a direction that compensates for the apparent motion of the image in the focal plane. As outlined in Section 3.4, high-resolution aerial films will have slower film speeds (i.e., will be less sensitive to light), so they will require slower shutter speeds, thus subjecting the image to blur when the aircraft is operated at relatively low altitudes. Image motion compensation permits the photographer to use a slower speed film than otherwise would be practical, and therefore to acquire higher spatial resolution images at lower altitudes (where image motion is fastest) or at lower light levels than would otherwise be feasible.

## 3.3. Kinds of Aerial Cameras

Most civilian aerial photography has been acquired using metric cameras (sometimes called "cartographic cameras") (Figure 3.1). These are aerial cameras designed to provide high-quality images with a minimum of optical and geometric error. Metric cameras used for professional work have been calibrated at special laboratories operated by the manufacturer or by governmental agencies. During calibration, each camera is used to photograph a target image having features positioned with great accuracy. Then precise measurements are made of focal length, flatness of the focal plane, and other variables. Such precise knowledge of the internal geometry of a camera permits photogrammetrists to make accurate measurements from photographs.

Other kinds of aerial cameras are less frequently used for routine photography, but may have uses for special applications. *Reconnaissance cameras* have been designed chiefly for military use. For such applications, geometric accuracy may be less important than the ability to take photographs at high air speed, at low altitude, or under unfavorable light conditions. As a result, photographs from reconnaissance cameras do not have the geometric accuracy expected from those taken by metric cameras.

*Strip cameras* acquire images by moving film past a fixed slit that serves as a form of shutter (Figure 3.6). The speed of film movement as it passes the slit is coordinated with the speed and altitude of the aircraft to provide proper exposure. The resulting image is one long continuous

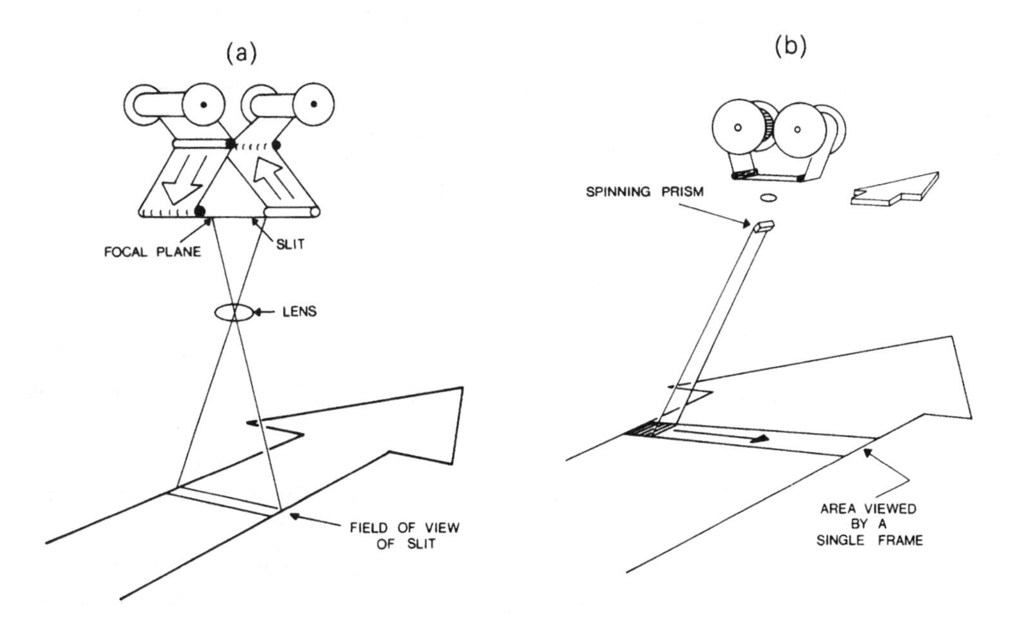


**FIGURE 3.6.** (a) Strip camera and (b) panoramic camera.

strip of imagery without the individual frames formed by conventional cameras. Strip cameras are capable of acquiring high-quality images from planes flying at high speed and low altitudes—optical conditions that are so extreme that conventional cameras often cannot provide the fast shutter speeds necessary to acquire sharp images.

*Panoramic cameras* (Figure 3.6) are designed to record a very wide field of view. Usually, a lens with a narrow field of view scans across a wide strip of land; its side-to-side motion forms the image as the aircraft moves forward. Photographs from panoramic cameras show a long narrow strip of terrain that extends perpendicular to the flight track from horizon to horizon. Because of the forward motion of the aircraft during the side-to-side scan of the lens, panoramic photographs have serious geometric distortions that require correction before they can be used as the basis for measurements. Panoramic aerial photographs are useful because of the large areas they represent, but only the central portions are suitable for detailed interpretation because of the large variations in scale and detail present near the outside edges of the images.

## 3.4. Black-and-White Aerial Films

Today the field of remote sensing encompasses a wide variety of sensors, both photographic and nonphotographic. Yet we have only one practical medium or recording images on paper or film: the photographic emulsion. Therefore, even if we do not use a camera to record an image, we must still use photographic film to prepare film or paper copies of that image. Knowledge of the qualities and limitations of photographic film, therefore, is central to understanding the field of remote sensing.

In the late 1700s and early 1800s a number of amateur scientists experimented with light-sensitive (*photosensitive*) chemicals. For example, silver nitrate ($AgNO_3$), familiar to high school chemistry students, darkens when exposed to sunlight; therefore, a glass or metal plate coated with silver nitrate formed the basis for recording a crude image. Those areas on the plate where the light is brightest became dark; those where the light is dim remained light in tone.

Joseph Nicephone Niepce (1765–1833), a French chemist, is one of the many who experimented with such chemicals. He is often assigned credit for devising the first negative image (1826). Niepce worked with Louis Daguerre (1789–1851), a French scientist and artist, to design a silver-coated metal plate treated with iodine vapor. Their invention was the first practical means of recording projected images. Their experiments were conducted over many years; by tradition, the year that Daguerre ceded rights to their invention to the French Academy of Sciences (1839) is given as the birth date of photography.

*Daguerrotypes,* an early name for photographic images made using Daguerre's method, were used for many years, with many modifications. Many features of early photography differ greatly from modern equipment and practice, and were clearly impractical for routine aerial photography. In the 1800s equipment was large, heavy, and cumbersome. Exposure times were long, cameras required bright light, and images were recorded on metal or glass plates, which were heavy, fragile, and awkward to use. Nonetheless, many aerial photographs were taken in the early days of photography, mainly by using balloons or large kites as a means of elevating the camera. Of course, such photographs were primarily curiosities rather than scientific tools because of the difficulty of controlling the orientation of the camera. Furthermore, each photographer tended to have tailor-made equipment; photographers often prepared their own chemicals and used individually formulated emulsions. The lack of standardization of equipment, materials, and practice meant that even the fundamentals of photographic practice were as much an art as a science.

The more compact photographic equipment required for modern aerial photography was made possible by developments started by George Eastman (1854–1932), who invented roll film and improved and standardized methods of photographic processing. His invention of the Kodak camera in 1888, and formation of the Eastman Kodak Company in 1892, popularized the practice of photography by mass production of standardized photographic products. Widening the scope of photography greatly increased the number of people knowledgeable about photography, standardized photographic practice, and decreased the cost of photographic materials. In brief, his work created the environment in which modern aerial survey could develop and grow into its present form.

Initially, photographic films were sensitive primarily to portions of the visible spectrum, and could portray only those brightnesses in a single broad region of the spectrum. In contrast, modern photographic films can be designed to be sensitive to nonvisible portions of the spectrum, and can represent reflectances in much more specific spectral regions. Therefore, the photographer has a choice of films that can extend his or her reach beyond the visible spectrum.

### *Major Components*

Aerial films have essentially the same structure as photographic films used in handheld cameras. The film base, or support, is usually a thin (40–100 μm), flexible, transparent material that holds a light-sensitive coating. In the early days of photography, the support was often formed from metal or glass plates, but today such materials are inconvenient for everyday use. Modern films have bases of polyester film. These materials are useful because they can be fabricated into thin, lightweight, flexible strips that are strong enough to withstand the forceful motions of winding and unwinding as film is moved within the camera.

The base must be able to resist changes in size caused by variation in temperature and humidity. Photogrammetrists measure distances on images so precisely that even small differences in image size due to shrinking or expanding of the base can introduce significant errors. Thus, glass plates are still used for images that are to be used with some photogrammetric instruments because they are insensitive to variations in temperature and humidity.

The base is coated with a light-sensitive coating, the *photographic emulsion* (Figure 3.7). Photosensitive coatings used in the early stages of photography were formed from silver nitrate (metallic silver dissolved in nitric acid). When a surface coated with silver nitrate is exposed to light, the silver nitrate darkens as the action of light changes it to metallic silver. The darkening

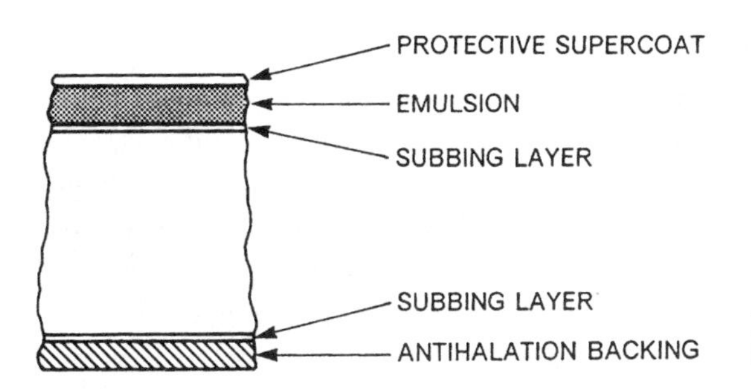


**FIGURE 3.7.** Schematic cross-sectional view of black-and-white photographic film.

effect increases as the light becomes more intense or as the length of exposure is increased. This effect provided a crude means of recording the image of a scene, but a number of practical problems (including the long exposures required to darken the coating) provided incentives to develop improved photosensitive coatings.

Modern emulsions consist of extremely small crystals of silver halide (typically, silver bromide [95%] and silver iodide [5%]) suspended in a gelatin matrix (possibly 5 m thick). These crystals form the light-sensitive portion of modern films, just as silver nitrate was the light-sensitive agent in the early days of photography. Although the gelatin that holds the grains is ostensibly a mundane substance, it possesses several important characteristics. Silver halide crystals are insoluble and have other physical characteristics that prevent them from adhering directly to the base. Gelatin holds the crystals in suspension, permitting the manufacturer to spread them evenly on the base. Furthermore, gelatin is transparent, porous (to allow photographic chemicals to contact the crystals), and absorbs halogen gases released when light strikes the emulsion.

Physical characteristics of the silver halide crystals assume some importance. They are extremely small, irregular in shape, with many sharp edges—shapes that favor the interception of photons that pass into the emulsion. The finer the size of the grains, the finer the detail that can be recorded. Coarser grains can record less detail, but produce a film with greater sensitivity to light. Thus the spatial resolution of a film is inversely related to its speed: as we increase the size of the crystals to improve the film's sensitivity to light, the finest level of detail that the emulsion can record becomes coarser. Alternatively, if we design a film with very fine grains to record fine detail, the emulsion exhibits decreased sensitivity—we must have brighter light or use longer exposures.

Recent research by the Eastman Kodak Company has produced film emulsions with grains that are flat in shape. These new grains have essentially the same volume as those in older emulsions, but their surface area is greatly increased. The flat grains, oriented parallel to the film surface, expose large surface areas to the light, thereby increasing the speed of the film without decreasing the resolution of the film. At present that type of film has been developed and marketed for popular photography (color prints only) rather than for aerial survey.

The film emulsion is coated with a thin layer of clear gelatin, the *protective supercoat,* designed to shield the emulsion from scratches during handling. Despite the presence of the supercoat, the emulsion is still vulnerable to damage from dust and from moisture and oil from handling with bare hands. As a result, analysts and interpreters should always wear cotton gloves when handling film or should protect film with transparent plastic sleeves.

Below the emulsion is a *subbing layer,* designed to ensure that the emulsion adheres to the base. On the reverse side of the base is an *antihalation* backing. This backing absorbs light that passes through the emulsion and the base, to prevent reflection back to the emulsion. In the absence of such a backing, the images of bright objects will be surrounded by halos caused by these reflections. The backing also acts as an anticurl agent, to counteract the curling effect of the emulsion that coats the upper side of the film.

When the shutter opens, it allows light to enter and strike the emulsion. The silver halide crystals are so small that even a small area of the film contains many thousands of them. When light strikes a crystal, it changes a very small portion of the crystal (perhaps only a single molecule) to metallic silver. The more intense the light striking a portion of the film, the greater the number of crystals affected. Thus the pattern of crystals influenced by light forms a record of patterns of light reflected from the scene. If it were possible to examine the exposed film without again subjecting it to the effects of light, it would appear no different than before exposure because of the extremely subtle effect of light upon the emulsion. At this point the image is recorded only as a *latent image*; processing is required to reveal this image.

*Development* is the process of bathing the exposed film in an alkaline chemical, the *developer,* that reduces the silver halide grains that have been exposed to light (Figure 3.8). Crystals in the latent image that were altered minimally are now completely changed to metallic silver in a process that in effect amplifies the pattern recorded by the latent image. In the latent image only a tiny portion of each grain has been altered by the effect of light; after development, each grain exposed to light is changed entirely to metallic silver. The developer acts most rapidly on those grains that have been exposed to light, so those areas that were exposed to the most intense light have the greatest density of metallic silver in the final image. Application of an acidic *stop bath* allows exact control of the time the film is in contact with the developer by counteracting the chemical effect of the alkaline developer. Next a *fixer* is applied to dissolve, then remove, unexposed silver halide grains. If the fixer were not used, these unexposed grains would darken when the film was next exposed to daylight.

After development and fixing, the resulting image is a negative representation of the scene, because those areas that were brightest in the scene are represented by the greatest concentra-

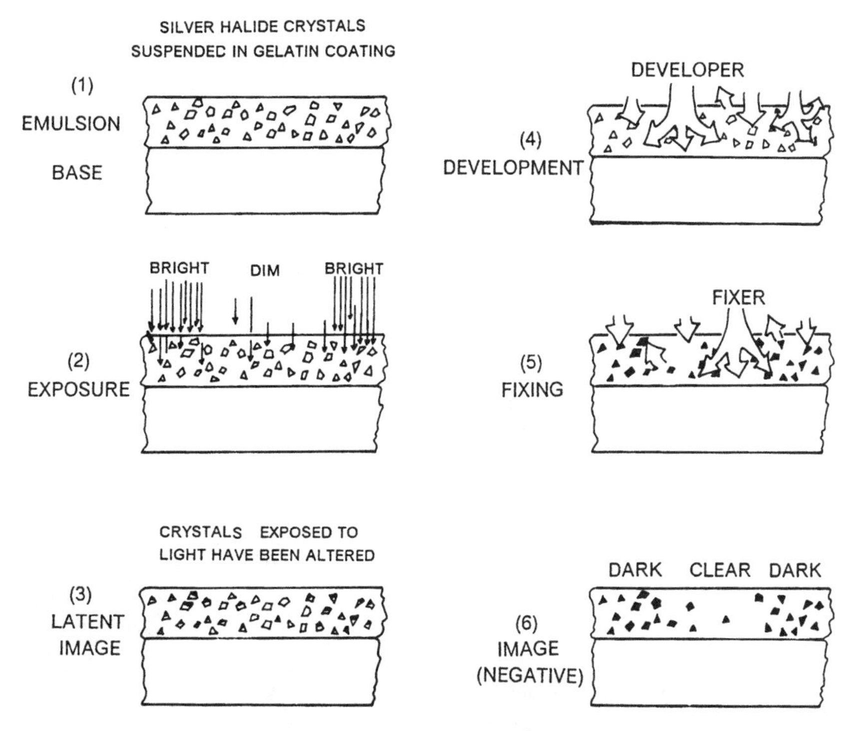


**FIGURE 3.8.** Schematic representation of processing of black-and-white photographic film. (1) Basic structure of the film as seen in cross section; photosensitive chemicals are suspended in a gelatin coating on the film base. (2) During exposure, light strikes the emulsion at varied intensities, depending upon the brightness levels in the scene. (3) Light creates a chemical reaction in the photosensitive chemicals that changes only a few molecules of each grain, creating the latent image. (4) During development, the emulsion is bathed in an alkaline chemical that changes to metallic silver all grains modified in Step 3; not shown here is the addition of an acidic chemical, the stop bath, that stops the action of the developer. (5) During fixing, unexposed grains are removed from the emulsion, leaving only those that had been exposed to light in Step 2. (6) The final image is a negative; those areas exposed to the most intense light in Step 2 are darkest; those exposed to dim light are clear.

tions of metallic silver, which appears dark on the processed image (Figure 3.9). Thus, in the negative, brightnesses are reversed from their original values in the scene.

*Film speed* is a measure of the sensitivity of an emulsion to light. A fast film requires relatively low intensity of light for proper exposure; a slow film requires more light, meaning that the aperture must be opened wider or that a longer exposure time must be used. As mentioned previously, film speed is directly related to grain size, and is inversely related to the ability of the film to record fine detail. Amateur photographers are familiar with the Deutsches Institut für Normung (DIN) and American Standards Association (ASA) ratings for assessing the speeds of films for handheld cameras. The analogous scales for aerial films include the *aerial film speed* (AFS) and the *aerial exposure index* (AEI).

*Contrast* indicates the range of gray tones recorded by a film. *High contrast* means that the film records the scene largely in blacks and whites, with few intermediate gray tones. *Low contrast* indicates a representation largely in grays, with few really dark or really bright tones. Often the interpreter needs information about the intermediate brightnesses in a film, so for aerial photography low-contrast representation may be desirable. Fine-grained emulsions tend to have low contrast, so slower films tend to have higher spatial resolution and lower contrast than do the coarser grained fast films. Emulsions on photographic papers typically have higher contrast than emulsions on films, so interpreters often prefer to use film transparencies if they are available.

*Spectral sensitivity* records the spectral region to which a film is sensitive (Figure 3.10). The spectral sensitivity curve for Kodak Tri-X Aerographic Film 2403 shows typical features of black-and-white films (Figure 3.10a). It is sensitive throughout the visible spectrum, but is also sensitive to ultraviolet radiation. Because of the scattering of these shorter (ultraviolet and blue) wavelengths, filters are often used with black-and-white aerial films to screen out blue light (Figure 3.11). This film presents a black-and-white representation of a scene (Figure 3.12a) that is essentially in accord with our view of the scene as we see it directly with our own eyes. This

**FIGURE 3.9.** Black-and-white negative image (left), and the corresponding positive print (right).

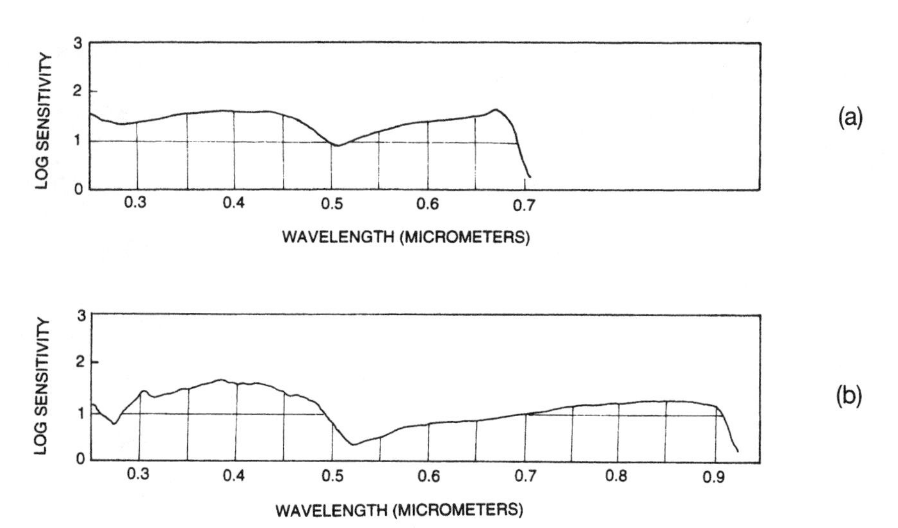


**FIGURE 3.10.** Spectral sensitivities of two photographic films. (a) Black-and-white panchromatic film (Kodak TRI-X Aerographic Film 2403). (b) Black-and-white infrared film (Kodak Infrared Aerographic Film 2424). Copyright Eastman Kodak Company. Permission has been granted to reproduce this material from *KODAK Data for Aerial Photography* (Code: M-29), courtesy of Silver Pixel Press, official licensee and publisher of Kodak books.

is because the Tri-X film is an example of a *panchromatic* film, an emulsion that is sensitive to radiation throughout the visible spectrum, much the same as the human visual system is sensitive throughout the visible spectrum. The term *orthochromatic* designates films with preferred sensitivity in the blue and green, usually with peak sensitivity in the green.

Figure 3.10 also shows the spectral sensitivity curve for Kodak Infrared Aerographic Film

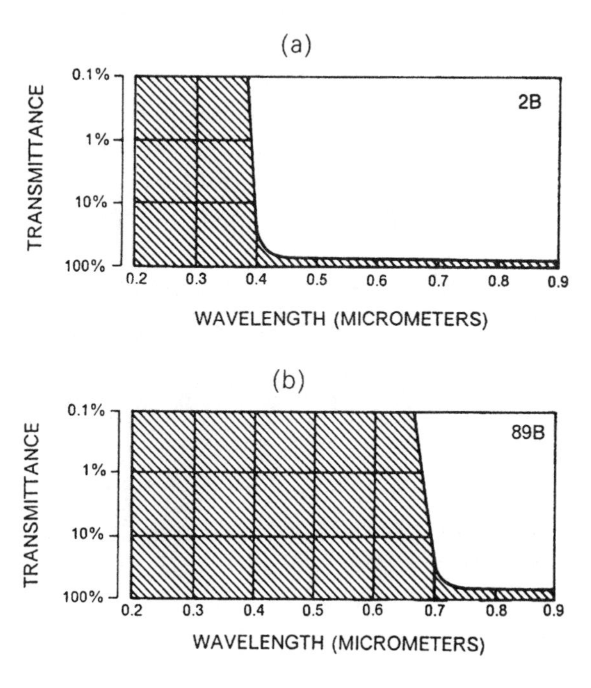


**FIGURE 3.11.** Transmission curves for two filters. (*a*) Pale yellow filter (Kodak filter 2B) to prevent ultraviolet light from reaching the film; it is frequently used with panchromatic film. (*b*) Kodak 89B filter used to exclude visible light, used with black-and-white infrared film. (Shaded portions of the diagrams signify that the filter is blocking transmission of radiation at specified wavelengths.) Copyright Eastman Kodak Company. Permission has been granted to reproduce this material from *KODAK Photographic Filters Handbook* (Code: B-3), courtesy of Silver Pixel Press, official licensee and publisher of Kodak books.

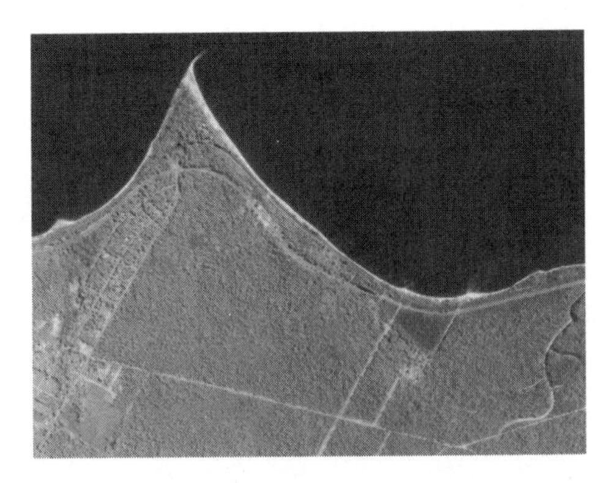

**FIGURE 3.12.** Aerial photographs. Left: Panchromatic film. Right: Black and white infrared film. Both show the coastline near Pensacola, Florida, 1965 (right) and 1974 (left). Credit: U.S. Geological Survey.

2424, a black-and-white infrared film. Note that its sensitivity extends well beyond the visible into the infrared portion of the spectrum. Usually it is desirable to exclude visible radiation, so this film is often used with a deep red filter that blocks visible radiation, but allows infrared radiation to pass (Figure 3.11). An image recorded by black-and-white infrared film (Figure 3.12b) is quite different from its representation in the visible spectrum. For example, living vegetation is many times brighter in the near infrared portion of the spectrum than it is in the visible portion, so vegetated areas appear bright white on the black-and-white infrared image.

### *The Characteristic Curve*

If we examine a negative after development and fixing, we find a pattern of dark and light related to the patterns of metallic silver formed in the processed film. Where the original scene was bright, the negative now has large amounts of silver, which create the dark areas. Where the original scene was dark, the film is clear, due to the absence of metallic silver.

We can see intermediate shades of brightness because the crystals in the emulsion are much smaller that the human eye can resolve. The areas that we perceive as shades of gray are actually variations in the abundance of the tiny grains of silver in the processed film. Thus each crystal is either present (black) or absent (clear), with shades of gray formed by variations in the abundance of crystals, which occur in proportion to the brightness of the original scene.

If we shine a light of intensity $I_0$ through a very small area of the negative (perhaps a fraction of a mm in diameter), the brightness of the light measured on the other side ($I$) is a measure of darkness of that region of the film (Figure 3.13). $I$, of course, is less than $I_0$. The ratio

$$I_0/I \qquad \text{(Eq. 3.2)}$$

is defined as the *opacity*. The darkness of the film is of interest because it is related to the brightness of the original scene. By convention, the darkness of the film is expressed as density, defined as the $\log_{10}$ of opacity (Table 3.1).

The effect of light upon the emulsion of a film is determined by the product of intensity ($i$)

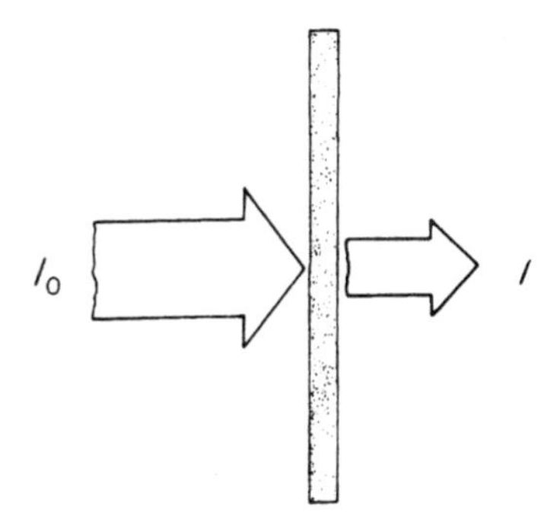


**FIGURE 3.13.** Measurement of opacity.

("brightness," equivalent to irradiance) and time ($t$). For a given density, the product $i \times t$ has a constant value $D$, the exposure:

$$E = i \times t \qquad \text{(Eq. 3.3)}$$

In everyday photography we use this relationship whenever we compensate for use of a fast shutter speed by opening the aperture to allow more light to enter. The *characteristic* curve expresses the relationship between brightness in the scene and density on the film (Figure 3.14). Or, more formally, it is a plot of the relationship between the density of a negative and the $\log_{10}$ of exposure. A specific emulsion will be characterized by a set of characteristic curves that differ in slope as development times vary. For present purposes, it is sufficient to consider only a single curve as an example.

The S-shape is typical of the characteristic curve. The lower part (the toe) is curved, the central part forms a straight line (the straight-line segment), and the upper portion (the shoulder) is curved. For scientific purposes, the straight-line segment is of interest; for a range of exposures, there is a consistent relationship between exposure and density. If exposure is increased by a certain amount, then the relationship depicted by the characteristic curve permits prediction of the corresponding increase in image density. The predictability of this relationship permits scientists to define consistent relationships between the darkness in a photograph and the brightness in the original scene. Given suitable controls, an image scientist can use measurements

**TABLE 3.1. Transmission, Opacity, and Density**

| Percentage of light transmitted | Opacity | Density ($\log_{10}$ opacity) | Example |
|---|---|---|---|
| 100 | 100/100 = 1 | 0 | Clear |
| 80 | 100/80 = 1.25 | 0.09 | |
| 50 | 100/50 = 2 | 0.3 | |
| 20 | 100/20 = 5 | 0.7 | |
| 10 | 100/10 = 10 | 1.0 | |
| 5 | 100/5 = 20 | 1.3 | |
| 1 | 100/1 = 100 | 2. | Dark sunglasses |
| 0.1 | 100/0.1 = 1,000 | 3. | |
| 0.01 | 100/0.01 = 10,000 | 4. | Very dark |
| 0.0 | 100/0.0 = ∞ | — | Opaque |

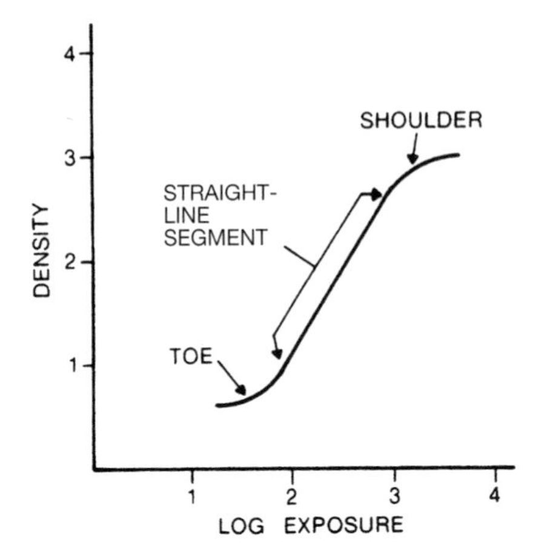


**FIGURE 3.14.** Characteristic curve of a photographic emulsion.

from the photograph to learn about brightnesses in the scene. Note that the lowest point of the curve does not reach the bottom of the graph; this difference is the *fog level*—a very small amount of density caused by the development process itself, unrelated to exposure. At the other extreme, the shoulder of the characteristic curve, the curved shape indicates that further increases in exposure will not produce a corresponding increase in density. This effect is referred to as *saturation*—the exposure is so extreme that the density will no longer have a predictable association with exposure.

Examine Figure 3.15; here two characteristic curves illustrate how differences in the slope of the straight-line segment translate a given *brightness range* (BR) into different *density ranges* (DR) in the final processed images. If the slope of the straight-line segment is steep, then a small range in exposure translates into a big range in density. Such an image has high contrast, meaning that the image displays a large range in brightness. If the slope is shallow (Figure 3.15b), a given range in scene brightness is translated into a smaller range in image brightness—a low-contrast image.

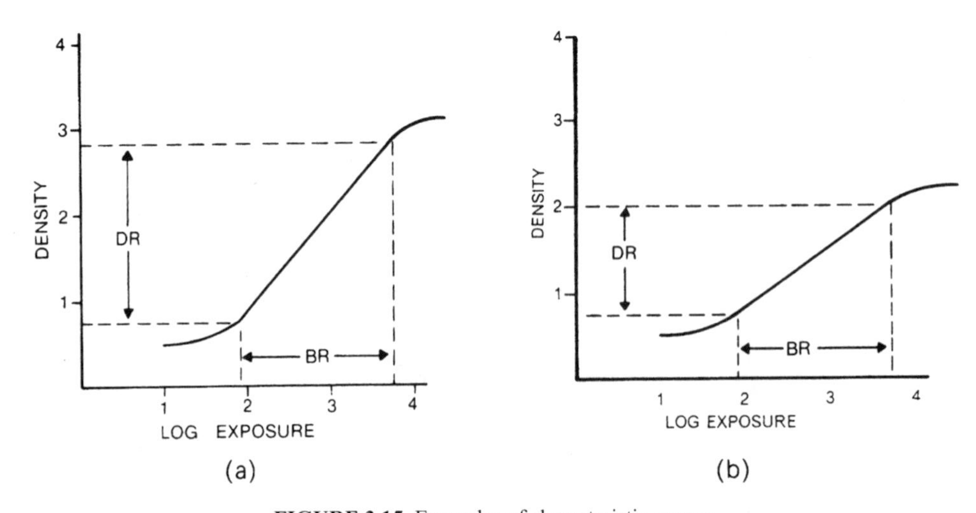


**FIGURE 3.15.** Examples of characteristic curves.

Note that the toe and the shoulder of the curve do not depict consistent relationships between exposure and density. This means that for very high and very low exposures, the film will not produce the predictable densities that it will for exposures in the straight-line segment. For these exposures, the film no longer forms a reliable scientific tool for portraying scene brightness because we cannot measure a density and then relate the density to the brightness of the original scene. Often, professional photographers use the toe or the shoulder to create artistic effects, but scientists always avoid use of image measurements that may be based on very high or very low image densities—they have an unknown relationship to brightnesses in the scene they portray.

Because photographic films are used to record images acquired by nonphotographic sensors (as described in subsequent chapters), knowledge of the characteristic curve is especially important in the field of remote sensing. Such sensors often record a large range of brightnesses—a range so large that it may exceed the capability of the film to record it. If so, the photographic record of the image will inevitably be inaccurate as a record of scene brightness. The very dark areas, the very bright areas, or perhaps both will be represented in nonlinear portions of the characteristic curve, and the image will show only a small portion of the brightnesses present in the scene. It is partially for this reason that digital image data, which can represent very large ranges of brightnesses, often offer advantages in comparison with photographic images for some kinds of analyses.

## 3.5. Color Reversal Films

Many of the color films used in remote sensing are reversal films, similar to those used in handheld cameras for color slides. Their basic elements are similar to those of black-and-white photographic film except that they are coated with three separate emulsions, each sensitive to one of the three additive primaries (Figure 3.16). The protective supercoat, the backing, and the subbing layer are present, and between the several emulsions are spacer layers of gelatin to prevent mixing of adjacent emulsions. The layer between the uppermost blue-sensitive emulsion and the middle (green-sensitive) emulsion is treated to act as a yellow filter to prevent blue light from passing through the upper layers to expose the lower emulsions. This filter is necessary because of the difficulty of manufacturing emulsions sensitive to red and green light without also sensitizing them to blue light.

Upon exposure, blue light exposes the blue layer, passes through the blue layer, but is pre-

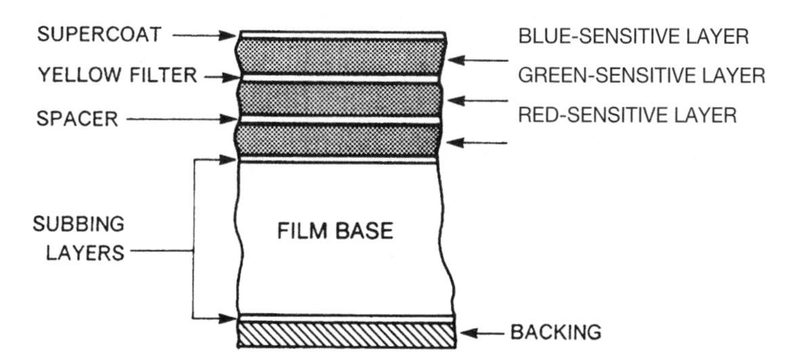


**FIGURE 3.16.** Idealized cross-sectional diagram of color reversal film.

vented from exposing the other two layers by the yellow filter (Figure 3.17). Green light passes through the blue layer and exposes the green-sensitive emulsion. Red light passes through the upper emulsions to expose only the lower, red-sensitive layer.

After processing, all areas not exposed to blue light on the blue-sensitive emulsion are represented by a yellow dye, while those areas exposed to blue are left clear. Areas exposed by green light on the green-sensitive emulsion are left clear; other areas are shown in magenta dye. Areas on the red-sensitive layer not exposed to red light are represented by cyan dye; images of red objects are clear on this emulsion. Thus each emulsion is sensitive to one of the additive primaries; after processing, each emulsion contains one of the subtractive primaries. The strategy of dying all areas not exposed in each separate emulsion differs fundamentally from the process used for the black-and-white films described earlier. Here there is no negative image; the dyes in the film processed combine to form a positive image in which brightness in the image corresponds (approximately) to brightness in the original scene.

When the processed film is viewed as a transparency against a light source, the magenta and cyan dyes present in those areas exposed to blue light combine to form a blue color. Likewise yellow and cyan combine to represent green, and yellow and magenta combine to form red (Figure 3.17 and Plate 1). This process is the same as that used for production of the 35-mm color slides that are familiar to many readers, so it should be easy to visualize the result of this process, even though the explanation shown in Figure 3.17 may seem rather abstract.

## 3.6. Color Infrared Films

*Color infrared* (CIR) *films* are based upon the same principles as color reversal films except for differences in the sensitivity of the emulsions and conventions in representation of colors. This film is used with a yellow filter to prevent blue light from entering the camera. The blue-sensi-

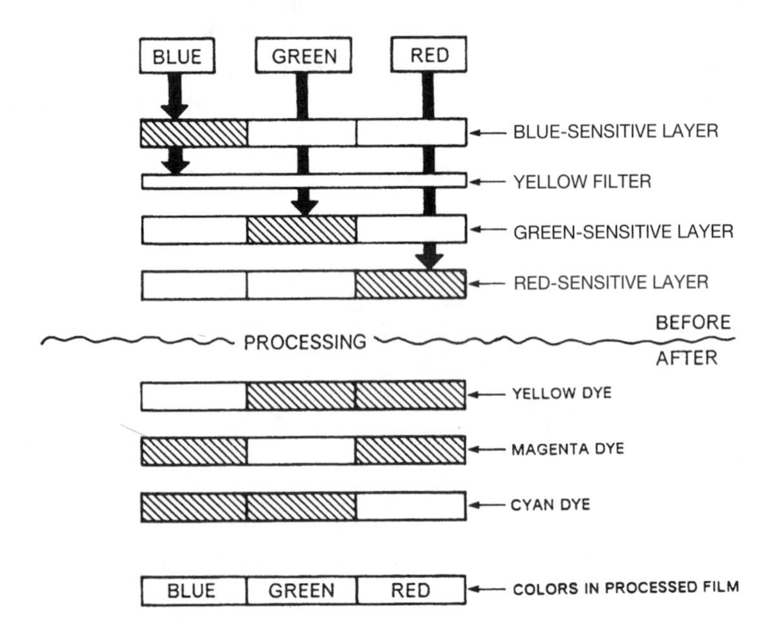


**FIGURE 3.17.** Color representation in color reversal film.

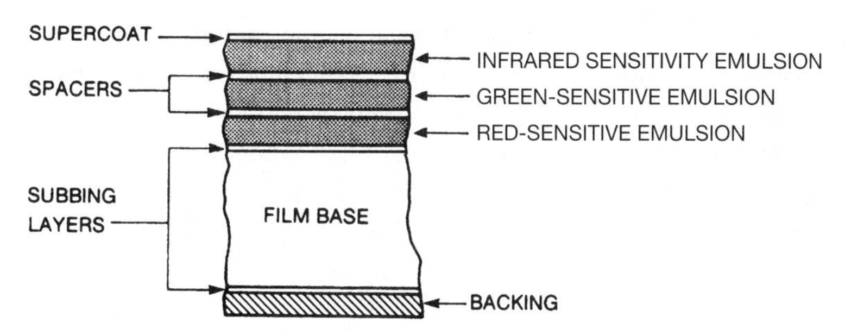


**FIGURE 3.18.** Idealized cross-sectional diagram of color infrared film.

tive layer is replaced by an emulsion sensitive to a portion of the near infrared region (Figure 3.18). After developing, representation of colors in the scene is shifted one position in the spectrum, so that green in the scene appears as blue on the image, red appears as green, and objects reflecting strongly in the near infrared are depicted in red. The comparison with normal color films can be represented schematically as follows:

| | | | | |
|---|---|---|---|---|
| Object in the scene reflects: | Blue | Green | Red | Infrared |
| Color reversal film represents the object as: | Blue | Green | Red | ***** |
| Color infrared film represents the object as: | **** | Blue | Green | Red |

Most objects, of course, reflect in several portions of the spectrum, so the CIR image shows a variety of colors derived from the varied reflectances in the scene. CIR film is designed in a manner analogous to, but not identical to, that of color reversal film (Figure 3.19). A yellow filter over the camera lens excludes all blue light. Green light exposes the green-sensitive layer,

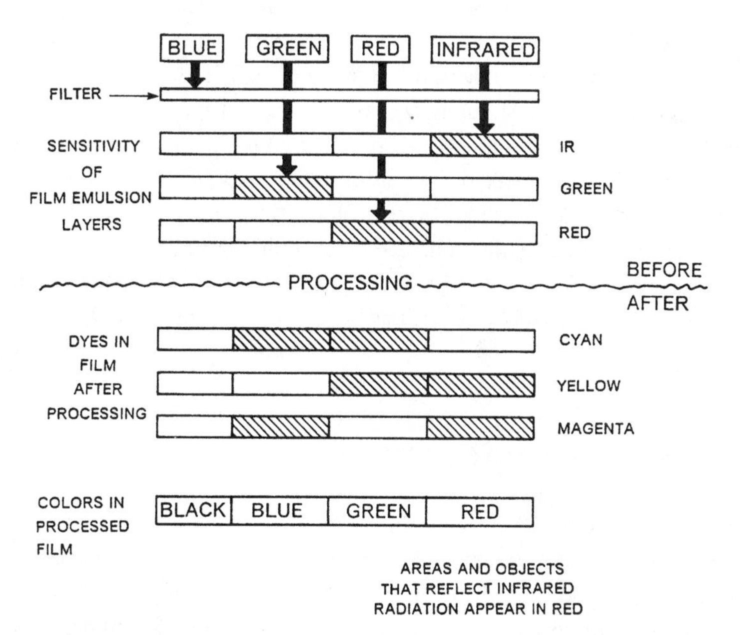


**FIGURE 3.19.** Color representation in color infrared film.

red light exposes the red-sensitive layer, and IR radiation exposes the IR-sensitive layer. After processing, areas not exposed to the green light are colored in yellow dye. Areas not exposed to red light are colored magenta. And areas not exposed to IR radiation are represented as cyan. In the final transparency, cyan and magenta combine to represent green areas as blue, cyan and yellow combine to represent red areas as green, and yellow and magenta combine to form the red color that represents objects that reflect strongly in the near infrared (Plate 2).

## 3.7. Film Format and Annotation

The term *format* designates the size of the image acquired by a camera. Mapping cameras generally produce a square image, with a size controlled by the width of the film. Common formats are 23 cm × 23 cm (approximately 9 in. × 9 in.) and 5.7 cm × 5.7 cm (approximately 2.5 in. × 2.5 in). Film 35 mm in width (24 mm × 36 mm image format) has also been used for specialized purposes, although generally it is not practical for routine applications of remote sensing. The 23-cm format is a standard for most cartographic cameras; paper prints made directly from the negative (*contact prints*) are among the most frequently used form of aerial photography. Although enlargements of the original negative can be made to any size desired, the 23-cm format is convenient for storage and handling; moreover, because the prints are made directly from the negative without enlargement, the image retains maximum detail and sharpness.

In some instances, paper prints are not made, and the film is examined simply as a strip of film—a *positive transparency*—wound on large spools and viewed against an illuminated background (Chapter 5). Because emulsions of transparencies typically represent a greater range of image tones than do paper prints, positive transparencies are often preferred for detailed interpretations, and especially for color and color infrared films.

Most aerial photographs carry some form of *annotation,* markings that identify the photographs and provide details concerning their acquisition. Typically, aerial photographs are annotated at the forward edge of the photograph (Figure 3.20). Annotation consists of a series of letters and numerals that can vary in meaning from one aerial survey firm to the next, but usually they include the *date* of the photography, a series of letters and numbers that *identify each project,* and the *film roll* number. Usually the last three digits specify the *frame number,* which shows the sequence in which photographs were taken. Other annotations, such as *image scale,* may also be included. In some instances the information may be recorded directly by the camera itself as the image is acquired; in other instances it may be added later as the film is processed and prepared for dissemination. The most sophisticated cameras may record on each frame information such as date, project identifier, focal length, time, image of a bubble level to indicate degree of tilt, and locational coordinates provided by onboard global positioning systems.

## 3.8. Geometry of the Vertical Aerial Photograph

Aerial photographs can be classified according to the orientation of the camera in relation to the ground at the time of exposure (Figure 3.21). *Oblique* aerial photographs have been acquired by cameras oriented toward the side of the aircraft. *High oblique* photographs (Figure 3.21a and Figure 3.22) show the horizon; *low oblique* photographs (Figure 3.21b) are acquired with the

**FIGURE 3.20.** Vertical aerial photograph.

camera aimed more directly toward the ground surface and do not show the horizon. Oblique photographs have the advantage of showing very large areas in a single image. Often those features in the foreground are easily recognized, as the view in an oblique photograph may resemble that from a tall building or mountain top. However, oblique photographs are not widely used for analytic purposes, primarily because the drastic changes in scale that occur from foreground to background prevent convenient measurement of distances, areas, and elevations.

*Vertical* photographs are acquired by a camera aimed directly at the ground surface from above (Figures 3.20 and 3.21c). Although objects and features are often difficult to recognize from their representations on vertical photographs, the map-like view of the Earth, and the predictable geometric properties of vertical photographs provide practical advantages. It should be noted that few, if any, aerial photographs are truly vertical; most have some small degree of tilt due to aircraft motion and other factors. The term *vertical photograph* is commonly used to designate aerial photographs that are within a few degrees of a corresponding (hypothetical) truly vertical photograph.

Because the geometric properties of vertical and nearly vertical aerial photographs are well understood and can be applied to many practical problems, they form the basis for making accurate measurements using aerial photographs. The science of making accurate measurements from aerial photographs (or from any photograph) is known as *photogrammetry.* The following paragraphs outline some of the most basic elements of introductory photogrammetry; the reader should consult a photogrammetry text (e.g., Wolf, 1983) for complete discussion of this subject. Aerial cameras are manufactured to include adjustable index marks attached rigidly to the cam-

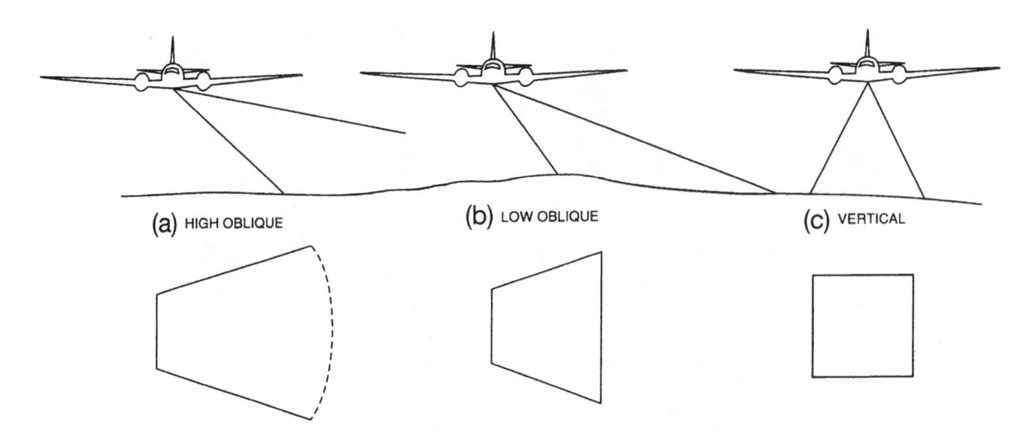


**FIGURE 3.21.** Oblique and vertical aerial photographs.

era so that the positions of the index marks are recorded on the photograph during exposure. These *fiducial marks* (usually four or eight in number) appear as silhouettes at the edges and/or corners of the photograph (Figures 3.20 and 3.23; in Figure 3.20 they appear at the corners). Lines that connect opposite pairs of fiducial marks intersect to identify the *principal point,* the optical center of the image. The *ground nadir* is defined as the point on the ground vertically beneath the center of the camera lens at the time the photograph was taken (Figure 3.24). The *photographic nadir* is defined by the intersection with the photograph of the vertical line that intersects the ground nadir and the center of the lens (i.e., the image of the ground nadir).

The *isocenter* can be defined informally as the focus of tilt. Imagine a truly vertical photograph that was taken at the same instant as the real, almost vertical, image. The almost vertical image would intersect with the (hypothetical) perfect image along a line that would form a "hinge"; the isocenter is a point on this hinge. On a truly vertical photograph, the isocenter, the principal point, and the photographic nadir coincide. The most important positional, or geometric, errors in the vertical aerial photograph can be summarized as follows:

1. *Optical distortions* are errors caused by an inferior camera lens, camera malfunction, or similar problems. These distortions are probably of minor significance in most modern photography flown by professional aerial survey firms.

2. *Tilt* is caused by displacement of the focal plane from a truly horizontal position by aircraft motion (Figure 3.24). The focus of tilt, the isocenter, is located at or near the principal point. Image areas on the upper side of the tilt are displaced further away from the ground than is the isocenter; these areas are therefore depicted at scales smaller than the nominal scale. Image areas on the lower side of the tilt are displaced down; these areas are depicted at scales larger than the nominal scale. Therefore, because all photographs have some degree of tilt, measurements confined to one portion of the image run the risk of including systematic error caused by tilt (i.e., measurements may be consistently too large or too small). To avoid this effect, it is a good practice to select distances used for scale measurements (Chapter 5) as lines that pass close to the principal point; then errors caused by the upward tilt compensate for errors caused by the downward tilt. The resulting value for image scale is not, of course, precisely accurate for either portion of the image, but it will not include the large errors that can arise in areas located further from the principal point.

**FIGURE 3.22.** High oblique aerial photograph. From author's photographs.

3. Because of routine use of high-quality cameras and careful inspection of photography to monitor photo quality, today the most important source of positional error in vertical aerial photography is probably *relief displacement* (Figure 3.25). Objects positioned directly beneath the center of the camera lens will be photographed so that only the top of the object is visible (e.g., object A in Figure 3.25). All other objects are positioned such that both their tops and their sides are visible from the position of the lens. That is, these objects appear to lean outward from the central perspective of the camera lens (e.g., see objects in Figure 3.25). Correct planimetric positioning of these features would represent only the top view, yet the photograph shows both the top and sides of the object. For tall features, it is intuitively clear that the base and the top cannot both be in their correct planimetric positions.

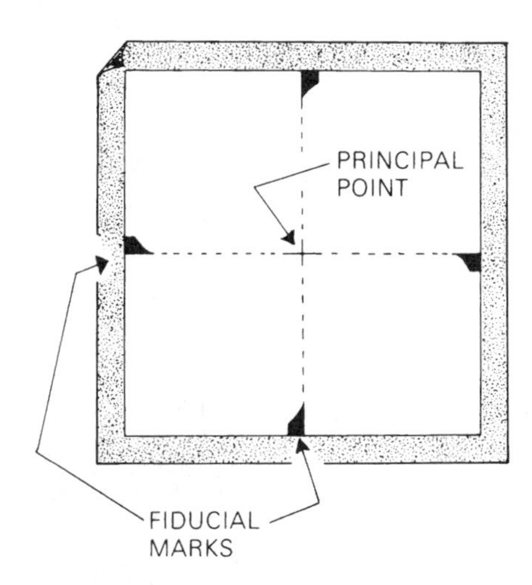


**FIGURE 3.23.** Fiducial marks and principal point.

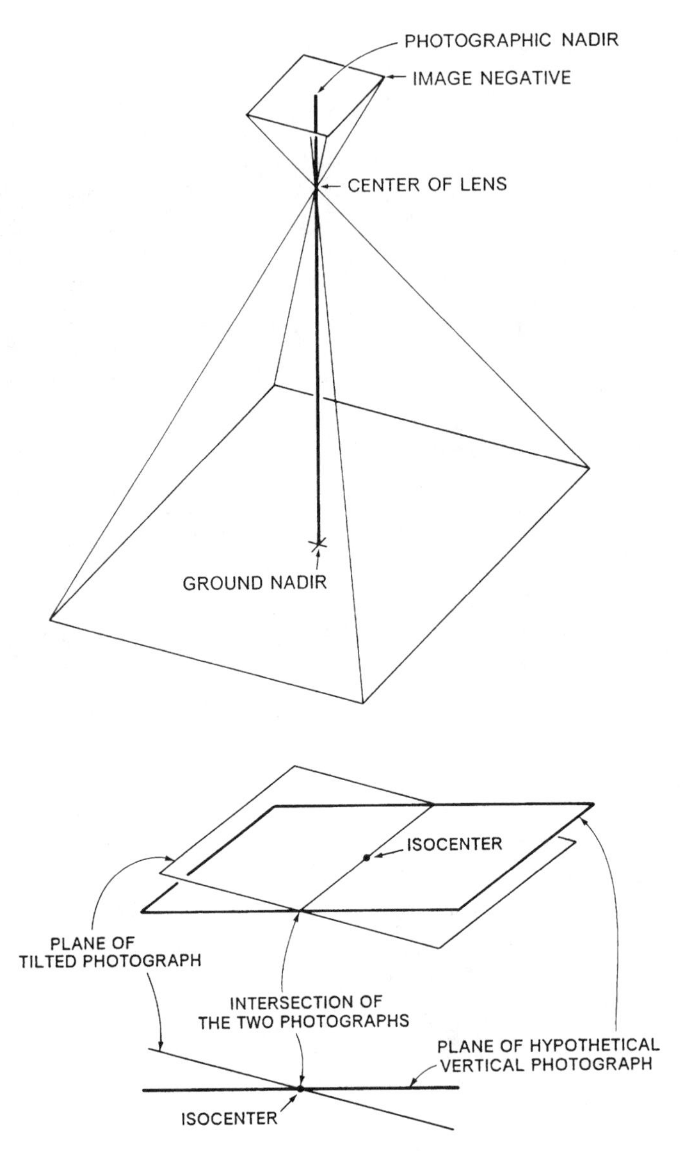


**FIGURE 3.24.** Schematic representation of terms to describe geometry of vertical aerial photographs.

This difference in apparent location is due to the height (*relief*) of the object and forms an important source of positional error in vertical aerial photographs. The direction of relief displacement is always radial from the nadir; the amount of displacement depends upon (1) the height of the object and (2) the distance of the object from the nadir. Relief displacement increases with increasing heights of features and with increasing distances from the nadir. (It also depends upon focal length and flight altitude, but these may be regarded as constant for a few sequential photographs.)

Relief displacement forms the basis of measurements of heights of objects, but its greatest significance is its role as a source of positional error. Uneven terrain can create significant relief displacement, so all measurements made directly from uncorrected aerial photographs are suspect.

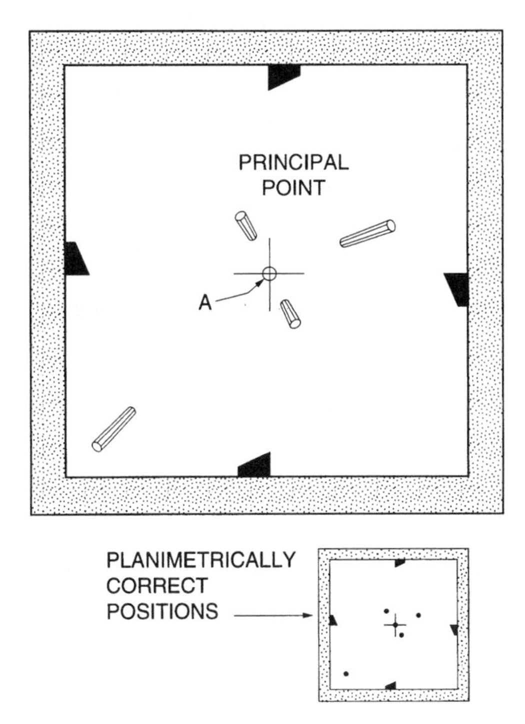


**FIGURE 3.25.** Relief displacement. The diagram depicts a vertical aerial photograph of a level terrain with five towers of equal height located at different positions with respect to the principal point. Images of the tops of towers are displaced away from the principal point along lines that radiate from the nadir, as discussed in the text.

## 3.9. Coverage by Multiple Photographs

Pilots normally acquire vertical aerial photographs by flying a series of parallel flight lines that together build up complete coverage of a specific region. Each flight line consists of individual frames, usually numbered in sequence (Figure 3.26). Often the camera operator can view the area to be photographed through a viewfinder attached to the camera. The operator can manually trigger the shutter as aircraft motion brings predesignated landmarks into the field of view or can set controls to automatically acquire photographs at intervals tailored to provide the desired coverage.

Individual frames form ordered strips, as shown in Figure 3.26a. If the plane's course is deflected by a crosswind, the positions of ground areas shown by successive photographs form the pattern shown in Figure 3.26b, known as *drift. Crab* (Figure 3.26c) is caused by correction of the flight path to compensate for drift without a change in the orientation of the camera.

Usually flight plans call for a certain amount of *forward overlap* (Figure 3.27), duplicate coverage by successive frames in a flight line, usually by about 50—60% of each frame. If forward overlap is 50% or more, then the image of the principal point of one photograph is visible on the next photograph in the flight line. These are known as *conjugate principal points* (Figure 3.27). When it is necessary to photograph large areas, coverage is built up by means of several parallel strips of photography; each strip is called a *flight line.* Sidelap between adjacent flight lines may vary from about 5 to 15%, in an effort to prevent gaps in coverage of adjacent flight lines.

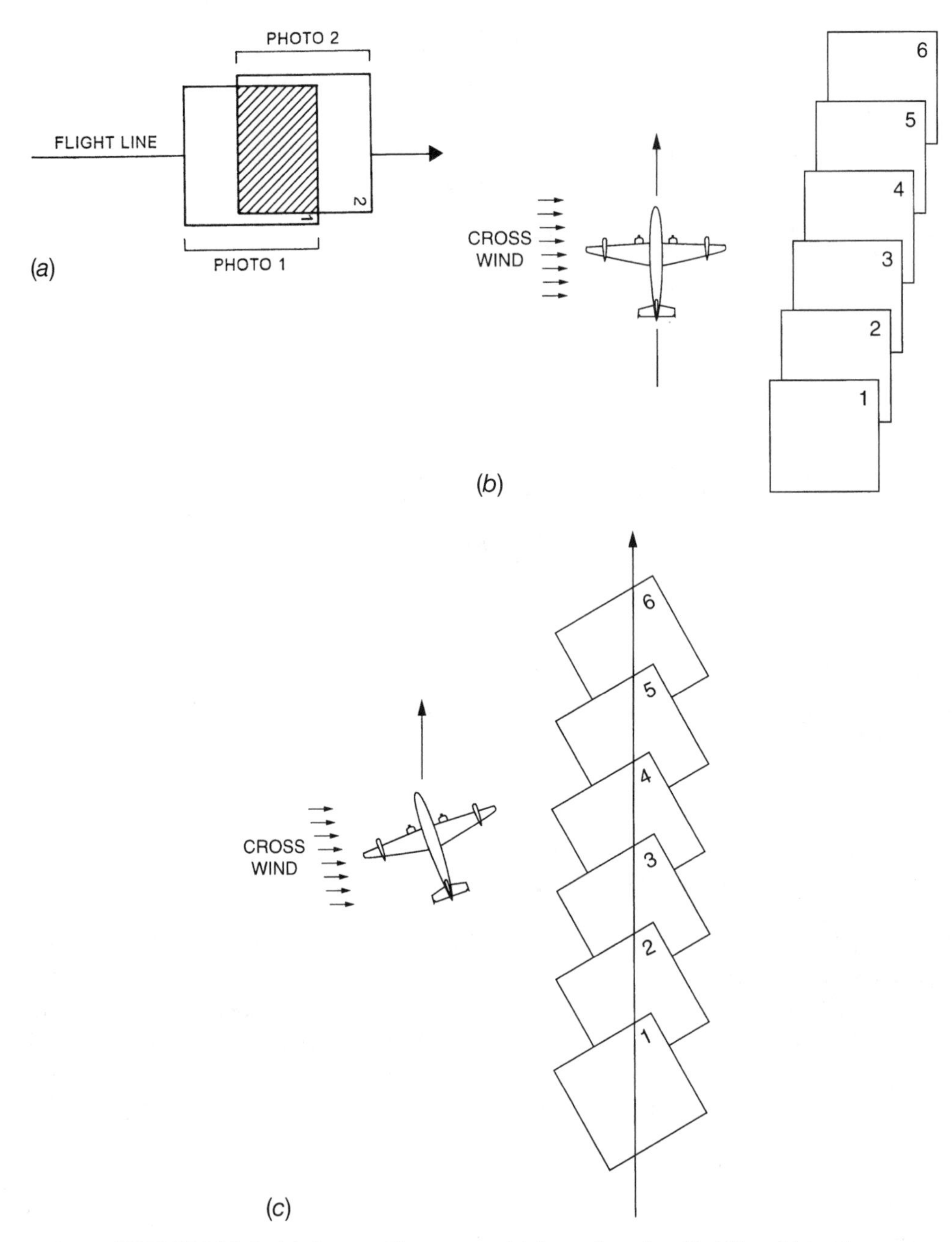


**FIGURE 3.26.** Aerial photographic coverage: (*a*) forward overlap, (*b*) drift, and (*c*) crab.

However, as pilots and other crew members collect complete photographic coverage of a region, there may still be gaps (known as *holidays*) in coverage due to equipment malfunction, navigation errors, cloud cover, or other problems. Sometimes photography flown later to cover holidays differs noticeably from adjacent images with respect to sun angle, vegetative cover, and other qualities. For planning flight lines, the number of photographs required for each line can be estimated using the relationship

$$\text{Number of photos} = \frac{\text{Length of flight line}}{(gd \text{ of photo}) \times (1 - \text{overlap})} \quad \text{(Eq. 3.4)}$$

where gd is the ground distance represented on a single frame, measured in the same units as the length of the planned flight line. For example, if a flight line is planned to be 33 mi. in length, each photograph is planned to represent 3.4 mi. on a side, and forward overlap is to be 0.60, then 33/ [3.4 × (1 – .60)] = 33/(1.36) = 24.26, or about 25 photographs are required. (Chapter 5 shows how to calculate the coverage of a photograph for a given negative size, focal length, and flying altitude.)

### *Stereoscopic Parallax*

If we have two photographs of the same area taken from different perspectives (i.e., from different camera positions), we observe a displacement of images of objects from one image to the other. You can observe this effect now by simple observation of nearby objects. Look up from this book at the objects near you. Close one eye, then open it and close the other. As you do this, you observe a change in the appearance of objects from one eye to the next. Nearby objects are slightly different in appearance because one eye tends to see, for example, only the front of an object, whereas the other, because of its position (about 2.5 in.) from the other, sees the front and some of the side of the same object. This difference in appearances of objects due to change in perspective is known as *stereoscopic parallax.* The amount of parallax decreases as objects increase in distance from the observer (Figure 3.28). If you repeat the experiment looking out the window at a landscape, you can confirm this effect by noting that distant objects display little or no observable parallax.

Stereoscopic parallax can therefore be used as a basis for measuring distance or height. Overlapping aerial photographs record parallax due to the shift in position of the camera as aircraft motion carries the camera forward between successive exposures. If forward overlap is 50% or more, then the entire ground area shown on a given frame can be viewed in stereo using three adjacent frames (a *stereo triplet*). Forward overlap of 50–60% is common. This amount of overlap doubles the number of photographs required, but assures that the entire area can be viewed in stereo because each point on the ground will appear on two successive photographs in a flight line.

Displacement due to stereo parallax is always parallel to the flight line. Tops of tall objects,

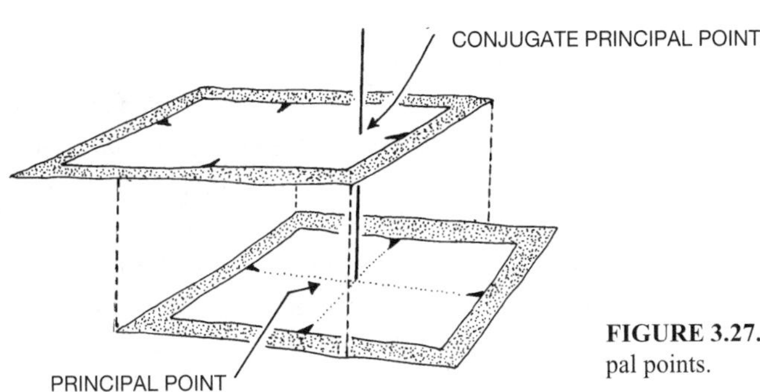


**FIGURE 3.27.** Forward overlap and conjugate principal points.

**FIGURE 3.28.** Stereoscopic parallax. These two photographs of the same scene were taken from slightly different positions. Note the differences in the appearances of objects due to the difference in perspective; note also that the differences are greatest for objects nearest the camera and least for objects in the distance. From author's photographs.

nearer to the camera, show more displacement than do shorter objects, which are more distant from the camera. Measurement of parallax therefore provides a means of estimating heights of objects. Manual measurement of parallax can be accomplished as follows. Tape photographs of a stereo pair to a work table so the axis of the flight line is oriented from right to left (Figure 3.29). For demonstration purposes, distances can be measured with an engineer's scale.

1. Measure the distance between two principal points ($X$).
2. Measure the distance between separate images of the base of the object as represented on the two images ($Y$). Subtract this distance from that found in (1) to get $P$.
3. Measure top-to-top distances ($B$), and base-to-base ($A$) distances, then subtract to find $dp$.

In practice, parallax measurements can be made more conveniently using a *parallax wedge,* or *parallax bar* (Chapter 5), devices that permit accurate measurement of small amounts of parallax.

### *Mosaics*

A series of vertical aerial photographs that show adjacent regions on the ground can be joined together to form a mosaic. Aerial mosaics belong to one of two classes. *Uncontrolled mosaics* are formed by placing the photographs together in a manner that provides continuous coverage of an area, without concern for preservation of consistent scale and positional relationships. The most rudimentary form of uncontrolled mosaic is formed simply by placing the photographs in their correct sequence, but with only rough alignment at their edges (Figure 3.30). Then another photograph is taken of these individual photographs, forming a single image that gives an overview of the landscape within a given region. An uncontrolled mosaic of this kind is often used as an aerial index, a key to the locations and coverages of the individual photographs. An

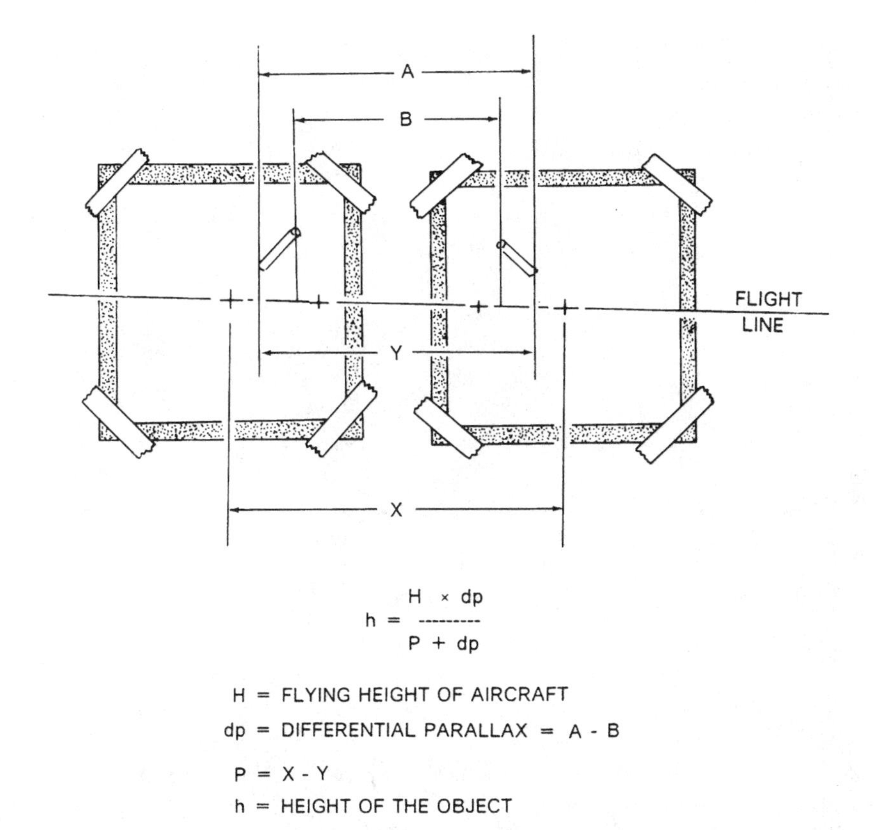


**FIGURE 3.29.** Measurement of stereoscopic parallax.

aerial index provides a means of identifying those particular photographs that might be needed for a specific purpose, and does give a small-scale overview of a particular region. Uncontrolled mosaics cannot be used for measurements of distance or area because no effort has been made to ensure that images are positioned in their correct relative positions. Other kinds of uncontrolled mosaics can be made using only the center sections of individual photographs, with efforts to improve matching of detail at edges of images. Such mosaics are more attractive, but still lack the geometric accuracy necessary to provide a basis for accurate measurement.

*Controlled mosaics* are formed from individual photographs assembled in a manner that preserves correct positional relationships between the features they represent. Often the most accurate region of each photograph, that near the principal point, is cut out and used for the mosaic. Locational control must, of course, be provided by ground survey or by information from accurate maps. Preparation of controlled mosaics can be expensive, but may have sufficient accuracy to serve as a map substitute in some instances.

### *Orthophotos and Orthophotomaps*

Aerial photographs are not planimetric maps because they have geometric errors, most notably the effects of relief displacement, in the representations of the features they show. That is,

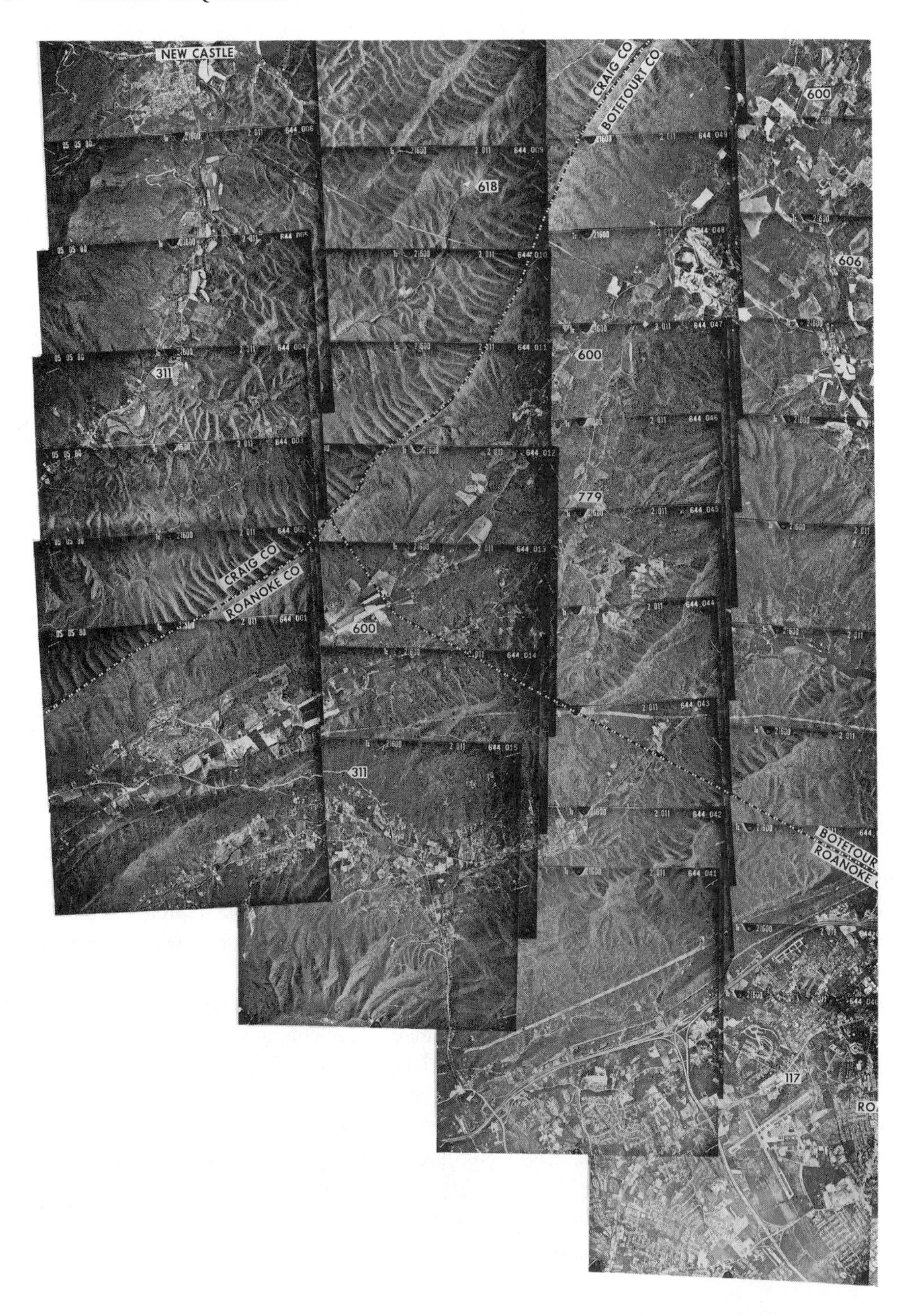


**FIGURE 3.30.** Uncontrolled mosaic. From Virginia Department of Transportation. Reproduced by permission.

objects are not represented in their correct relative positions, and as a result the images cannot be used as the basis for accurate measurements.

Stereoscopic photographs can be used to generate a corrected form of an aerial photograph known as an *orthophoto* that shows photographic detail without the errors caused by tilt and relief displacement. An instrument known as an orthophotoscope can, for a given instant, project a corrected version of a very small portion of an image. An orthophotoscope is an optical–mechanical instrument that, instead of exposing an entire image from a central perspective (i.e., through a single lens), exposes each small section individually in a manner that corrects for the elevation of that small section. The result is an image that has orthographic properties rather than those of the central perspective of the original aerial photograph. The orthophotoscope is capable of scanning an entire image piece by piece to generate a corrected version of that image. The projection orientation is adjusted to correct for tilt, and the instrument continuously varies the projection distance to correct for relief displacement. Thus, as the instrument scans an image, the operator views the ground surface in stereo and the new image is formed as a geometrically correct version of the original image. The result is a photo image that shows the same detail as the original aerial photograph but without the geometric errors introduced by tilt and relief displacement. *Orthophotomaps* therefore can be used for most purposes as maps because they show correct planimetric position and preserve consistent scale throughout the image. Orthophotographs form the basis for orthophotomaps, which are orthophotographs presented in map format, with annotations, scale, and geographic coordinates.

Orthophotomaps are valuable because they show the fine detail of an aerial photograph without the geometric errors that are normally present, and they can be compiled much more quickly and cheaply than the usual topographic maps. Therefore, they can be very useful as map substitutes in instances where topographic maps are not available; or as map supplements when maps are available, but the analyst requires the finer detail, and more recent information provided by an image.

### *Digital Orthophoto Quadrangles*

*Digital orthophoto quadrangles* (DOQs) are orthophotos prepared in a digital format, designed to correspond to the 7.5-minute quadrangles of the U.S. Geological Survey (USGS). DOQs are presented either as black-and-white or as color images that have been processed to attain the geometric properties of a planimetric map (Figure 3.31).

DOQs are prepared from National Aerial Photography Program (NAPP) photography (high-altitude photography described in Section 3.13), at 1:40,000 scale, supplemented by other aerial photography as needed. The rectification process is based upon the use of DEMs to represent variations in terrain elevation. The final product is presented to correspond to the matching USGS 7.5-minute quadrangle, with a supplementary border of imagery representing 50—300 m beyond the limits of the quadrangle, to facilitate matching and mosaicking with adjacent sheets. DOQs provide image detail equivalent to 2 m or so for DOQs presented in the quadrangle format, and finer detail for quarter-quad DOQs. The USGS has responsibility for leading the U.S. federal government's effort to prepare and disseminate digital cartographic data. For more information on DOQs, visit the USGS website at http://edc.usgs.gov/glis/hyper/guide/usgs_doq.

**FIGURE 3.31.** Digital orthophoto. From USGS.

## 3.10. Photogrammetry

*Photogrammetry* is the science of making accurate measurements from photographs. Photogrammetry applies the principles of optics and knowledge of the interior geometry of the camera and its orientation to reconstruct the dimensions and positions of objects represented within photographs. Therefore, its practice requires detailed knowledge of specific cameras and the circumstances under which they were used and accurate measurements of features within photographs. Photographs used for photogrammetry have traditionally been prepared on glass plates or other dimensionally stable materials (i.e., materials that do not change in size as temperature and humidity change).

Photogrammetry can be applied to any photograph, provided the necessary information is at hand. However, by far the most frequent application of photogrammetry is the analysis of stereo aerial photography to derive estimates of topographic elevation for topographic mapping. With the aid of accurate locational information describing key features within a scene (*ground control*), photogrammetrists estimate topographic relief by estimating stereo parallax for any array of points within a region. Although stereo parallax can be measured manually, it is far more practical to employ specialized instruments designed for stereoscopic analysis.

Such instruments, known as *analytical stereoplotters,* first designed in the 1920s, reconstruct the orientations of photographs at the time they were taken. Operators then can view the image in stereo; by maintaining constant parallax visually, they can trace lines of uniform elevation. The quality of information derived from such instruments depends upon the quality of the photography, the accuracy of the data, and the operator's skill in setting up the stereo model and tracing lines of uniform parallax. As the design of instruments improved, it eventually became possible to automatically match corresponding points on stereo pairs and thereby identify lines of uniform parallax with limited assistance from the operator.

## 3.11. Digital Photography

Photographs can be electronically scanned to record the patterns of blacks, grays, and whites as digital values, each representing the brightness of a specific point within the image. Although these values can be displayed in the form of a conventional photograph, the digital format offers advantages of compact storage and the power of numerical representation (see Chapters 4 and 11). Further, it is possible to manufacture cameras that replace the film in the focal plane with an array of light-sensitive detectors (Chapter 4) that directly record images in digital form, thereby bypassing the scanning step. For example, Plate 2 shows a high-resolution digital CIR image acquired using a proprietary imaging system employed by Emerge, Inc., a subsidiary of Litton TASC, Inc. Because this imaging system collects navigational and positional data that can be used with elevation data of the region imaged, it can produce imagery of high positional accuracy.

## 3.12. Softcopy Photogrammetry

With further advances in instrumentation, it became possible to extend automation of the photogrammetric process so as to conduct the analysis completely within the digital domain. Satellite images or aerial photographs can be acquired digitally (or conventional photographs can be scanned to create digital products). With the use of global positioning systems (GPS; Chapter 16) to acquire accurate positional information and the use of data recorded from the aircraft's navigational system to record the orientations of photographs, it then became feasible to reconstruct the geometry of the image using those data gathered as the image was acquired.

This process forms the basis for softcopy photogrammetry, so named because it does not require the physical (hardcopy) form of the photograph necessary for traditional photogrammetry. Instead the digital (softcopy) version of the image is used as input for a series of mathematical models that reconstruct the orientation of each image to create planimetrically correct representations. This process requires specialized computer software installed in workstations, which analyze digital data specifically acquired for the purpose of photogrammetric analysis. Softcopy photogrammetry offers advantages of speed and accuracy, and also creates output data that are easily integrated into other production and analytical systems, including GIS (Chapter 16).

## 3.13. Sources of Aerial Photography

Aerial photography can be acquired by (1) the user or (2) purchased from organizations that serve as repositories for imagery flown by others (*archival imagery*). In the first instance, aerial photography is produced upon request by firms that specialize in taking high-quality aerial photography. Such firms are listed in the business sections of most metropolitan phone directories. Customers may be individuals, governmental agencies, or other businesses that use aerial photography. Such photography is, of course, customized to meet the specific needs of customers with respect to date, scale, film, and coverage. As a result, costs may be prohibitive for many noncommercial uses.

Thus, for financial reasons, many users of aerial photography turn to archival photography as a means of acquiring the images they need. Although such photographs may not exactly meet users' requirements with respect to scale or date, low costs and ease of access may compensate for any shortcomings. For some tasks that require reconstruction of conditions at earlier dates (such as the Environmental Protection Agency's search for abandoned toxic waste dumps), archival images may be the only source of information.

It is feasible to take "do-it-yourself" aerial photography. Many small cameras are suitable for aerial photography. Often the costs of local air charter services for an hour or so of flight time are relatively low. Small-format cameras, such as the usual 35-mm cameras, can be used for aerial photography if the photographer avoids the effects of aircraft vibration (Do not rest the camera against the aircraft!). If the altitude is low and the atmosphere is clear, ordinary films can produce satisfactory results. Be sure to use a high-wing aircraft to ensure that the photographer will have a clear view of the landscape. If the camera can accommodate filters, it is possible to use other films (such as infrared or color infrared) similar to those described above. Some experimentation may be necessary for the first-time user to obtain proper exposures, but most people can learn rather quickly to take satisfactory photographs. Usually the best lighting is when the camera is aimed away from the sun. Photographs acquired in this manner (Figure 3.22) may be useful for illustrative purposes, although for scientific or professional work the large-format, high-quality work of the fully equipped air survey firm is probably necessary.

### *EROS Data Center*

The EROS Data Center (EDC), in Sioux Falls, South Dakota, is operated by the USGS as a repository for aerial photographs and satellite images acquired by NASA, the USGS, and other agencies. A computerized database at EDC provides an indexing system for information pertaining to aerial photographs and satellite images. For more information contact:

Customer Services
U.S. Geological Survey
Earth Resources Observation & Science (EROS)
47914 252nd Street
Sioux Falls, SD 57198-0001
Tel: 800-252-4547 or 605-594-6151
Fax: 605-594-6589
E-mail: custserv@usgs.gov
EDC home page: *http://edc.usgs.gov/*

### *Earth Science Information Centers*

The Earth Science Information Centers (ESIC; *http://geography.usgs.gov/esic/esic_index.html*) are operated by the USGS as a central source for information pertaining to maps and aerial photographs. ESIC has a special interest in information pertaining to federal programs and agencies, but also collects data pertaining to maps and photographs held by state and local governments. The ESIC headquarters is located at Reston, Virginia, but ESIC also maintains seven other offices throughout the United States; other federal agencies have affiliated offices. ESIC can provide information to the public concerning the availability of maps and remotely sensed images. The following sections describe two programs administered by ESIC that can provide access to archival aerial photography.

#### *Aerial Photography Summary Record System*

The Aerial Photography Summary Record System (APSRS) is maintained by USGS as a computer-based information system for recording detailed information pertaining to aerial photography held by numerous federal, state, and private organizations. Prior to the establishment of centralized records in 1975, citizens desiring comprehensive information on coverage by aerial photographs were required to query holdings by numerous agencies, which each followed different conventions for reporting coverage. After 1975, users could obtain information on the integrated holdings of numerous agencies, reported in a standard format.

Those who request information from APSRS receive information sorted by date; describing image scale, cloud cover, and camera focal length; and identifying the organization holding the photography. Coverage is indexed by USGS 7.5-minute quadrangles, which are listed by the latitude and longitude of the southeastern corner of each quadrangle. Users first identify the quadrangle that covers their area of interest, and then use the latitude and longitude to search for coverage of the region. For each photographic mission, APSRS provides a listing giving the date of coverage, amount of cloud coverage, scale, film type and format, focal length, and other qualities. Also listed is the agency that holds the photography; these listing are keyed to a directory that lists addresses so that the user can order copies of the photographs. Listings also include the Federal Information Processing Standards (FIPS) code for each area, which permits cross-reference to political and census units. Further infromation is available at *http://erg.usgs.gov/isb/pubs/factsheets/fs22096.html.*

#### *National Aerial Photography Program*

The National Aerial Photography Program (NAPP) acquires aerial photography for the coterminous United States, according to a systematic plan that ensures uniform standards. NAPP was initiated in 1987 by the USGS as a replacement for the National High Altitude Aerial Photography Program (NHAP), begun in 1980 to consolidate the many federal programs that use aerial photography. The USGS manages the NAPP, but it is funded by the federal agencies that are the primary users of its photography. Program oversight is provided by a committee of representatives from the USGS, the Bureau of Land Management, the National Agricultural Statistics Service, the National Resources Conservation Service (NRCS; previously known as the Soil Conservation Service), the Farm Services Agency (previously known as the Agricultural Stabilization and Conservation Service), the U.S. Forest Service, and the Tennessee Valley Authority. Light (1993) and Plasker and TeSelle (1988) provide further details.

Under NHAP, photography was acquired under a plan to obtain complete coverage of the coterminous 48 states, then to update coverage as necessary to keep pace with requirements for current photography. Current plans call for updates at intervals of 5 years, although the actual schedules are determined in coordination with budgetary constraints. NHAP flight lines were oriented north–south, centered on each of four quadrants systematically positioned within USGS 7.5-minute quadrangles, with full stereoscopic coverage at 60% forward overlap and sidelap of at least 27%. Two camera systems were used to acquire simultaneous coverage: Black-and-white coverage was acquired at scales of about 1:80,000 using cameras with focal lengths of 6 in. Color infrared coverage was acquired at scales of about 1:58,000, using a focal length of 8.25 in. Plate 3 shows a high-altitude CIR image illustrating the broad-scale coverage provided by this format.

Dates of NHAP photography varied according to geographic region. Flights were timed to provide optimum atmospheric conditions for photography and to meet specifications for sun angle, snow cover, and shadowing, with preference for autumn and winter seasons to provide images that show the landscape without the cover of deciduous vegetation.

Specifications for NAPP photographs differ from those of NHAP. NAPP photographs are acquired at 20,000-ft. altitude using a 6-in. focal length lens. Flight lines are centered on quarter quads (1:24,000-scale USGS quadrangles). NAPP photographs are planned for 1:40,000, black-and-white or color infrared film, depending on specific requirement for each area.

Photographs are available to all who may have an interest in their use. Their detail and quality permit use for land-cover surveys; assessment of agricultural, mineral, and forest resources; as well as examination of patterns of soil erosion and water quality. Further infromation is available at *http://edc.usgs.gov/guides/napp.html.* Two other sources of aerial photography include the USDA Aerial Photography Field Office (*http://www.apfo.usda.gov/*) and the U.S. National Archives and Records Administration (*http://www.nara.gov/nara/nn/nns/nnscord.html* ).

## 3.14. Summary

Aerial photography is a simple, reliable, and inexpensive means of acquiring remotely sensed images. It has been used to make images from very low altitudes and from Earth-orbiting satellites, so it can be said to be one of the most flexible strategies for remote sensing. Aerial photography is useful mainly in the visible and near infrared portions of the spectrum, but its principles are important throughout the field of remote sensing. For example, lenses are used in many nonphotographic sensors and photographic films are used to record images acquired by a variety of instruments.

Aerial photographs form the primary source of information for compilation of many maps, especially large-scale topographic maps. Vertical aerial photographs are valuable as map substitutes or as map supplements. Geometric errors in the representation of location prevent direct use of aerial photographs as the basis for measurement of distance or area. But, because these errors are known and are well understood, it is possible for photogrammetrists to use photographs as the basis for reconstruction of correct positional relationships and the derivation of accurate measurements. Aerial photographs record complex detail of the varied patterns that constitute any landscape. Each image interpreter must develop the skills and knowledge necessary to resolve these patterns by disciplined examination of aerial images.

## Review Questions

1. List several reasons why time of day might be very important in flight planning for aerial photography.

2. Outline advantages and disadvantages of high-altitude photography. Explain why routine high-altitude aerial photography was not practical before infrared films were available.

3. List several problems that you would encounter in acquiring and interpreting large-scale aerial photography of a mountainous region.

4. Speculate upon the likely progress of aerial photography since 1890 if George Eastman had not been successful in popularizing the practice of photography to the general public.

5. Should an aerial photograph be considered a "map"? Explain.

6. Assume you have recently accepted a position as an employee of an aerial survey company; your responsibilities include preparation of flight plans for the company's customers. What are the factors that you must consider as you plan each mission?

7. If color films and color infrared films are now available, why are black-and-white films still widely used?

8. Suggest circumstances in which oblique aerial photography might be more useful than vertical aerial photography. Identify situations in which oblique aerial photography would clearly not be useful.

9. It might seem that large-scale aerial photographs would always be more useful than small-scale aerial photographs, yet larger scale images are not always the most useful. What are the disadvantages to the use of large-scale images?

10. A particular object will not always appear the same when photographed by an aerial camera. List some of the factors that can cause the appearance of an object to change from one photograph to the next.

## References

American Society of Photogrammetry and Remote Sensing. 1985. Interview: Frederick J. Doyle and Gottfried Konecny. *Photogrammetric Engineering and Remote Sensing,* Vol. 5, pp. 1160–1169.

Anon. 1968. *Applied Infrared Photography* (Kodak Technical Publication M-28). Rochester, NY: Eastman Kodak Co., 88 pp.

Anon. 1970. *Kodak Filters for Scientific and Technical Uses* (Kodak Publication B-3. Rochester, NY: Eastman Kodak Co., 90 pp.

Anon. 1974. *Kodak Infrared Films* (Kodak Publication N-17). Rochester, NY: Eastman Kodak Co., 16 pp.

Anon. 1976. *Kodak Data for Aerial Photography* (Kodak Publication M-29). Rochester, NY: Eastman Kodak Co., 92 pp.

Doyle, F. J. 1985. The Large Format Camera on Shuttle Mission 41-G. *Photogrammetric Engineering and Remote Sensing,* Vol. 51, pp. 200–201.

Estes, J. E. 1974. Imaging with Photographic and Non-Photographic Sensor Systems. Chapter 2 in *Remote Sensing: Techniques for Environmental Analysis* (J. E. Estes and L. Senger, eds.). Santa Barbara, CA: Hamilton, pp. 15–50.

Langford, M. J. 1965. *Basic Photography: A Primer for Professionals.* New York: Focal Press, 376 pp.

Light, D. L. 1993. The National Aerial Photography Program as a Geographic Information System Resource. *Photogrammetric Engineering and Remote Sensing,* Vol. 59, pp. 61–65.

Miller, S. B., U. V. Helavea, and K. D. Helavea. 1992. Softcopy Photogrammetric Workstations. *Photogrammetric Engineering and Remote Sensing,* Vol. 58, pp. 77–83.

Plasker, J. R., and G. W. TeSelle. 1988. Present Status and Future Applications of the National Aerial Photography Program. In *Proceedings of the ACSM/ASPRS Convention.* Bethesda, MD: American Society for Photogrammetry and Remote Sensing, pp. 86–92.

Silva, L. F. 1978. Radiation and Instrumentation in Remote Sensing. Chapter 2 in *Remote Sensing: The Quantitative Approach* (P. H. Swain and S. M. Davis, eds.). New York: McGraw-Hill, pp. 21–135.

Skalet, C. D., G. Y. G. Lee, and L. J. Ladner. 1992. Implementation of Softcopy Photogrammetric Workstations at the U.S. Geological Survey. *Photogrammetric Engineering and Remote Sensing,* Vol. 58, pp. 57–63.

Slater, P. N. 1975. Photographic Systems for Remote Sensing. Chapter 6 in *Manual of Remote Sensing* (R. G. Reeves, ed.). Bethesda, MD: American Society for Photogrammetry and Remote Sensing, pp. 235–323.

Stimson, A. 1974. *Photometry and Radiometry for Engineers.* New York: Wiley, 446 pp.

Wolf, P. R. 1983. *Elements of Photogrammetry.* New York: McGraw-Hill, 628 pp.

## YOUR OWN INFRARED PHOTOGRAPHS

Anyone with even modest experience with amateur photography can take infrared photographs using commonly available materials. A 35-mm camera, with some of the usual filters, will be satisfactory. Infrared films can be purchased at camera stores (but are unlikely to be available at stores that do not specialize in photographic supplies). Infrared films are essentially similar to the usual films, but should be used promptly, as the emulsions deteriorate much more rapidly than do those of normal films. To maximize life of the film, it should be stored under refrigeration according to manufacturer's instructions.

Black-and-white infrared films should be used with a deep red filter to exclude most of the visible spectrum. Black-and-white infrared film can be developed using normal processing for black-and-white emulsions, as specified by the manufacturer.

Color infrared (CIR) films are also available in 35-mm format. They should be used with a yellow filter, as specified by the manufacturer. Processing of CIR film will require the services of a photographic laboratory that specializes in customized work rather than the laboratories that handle only the more popular films. Before purchasing the film, it is best to inquire concerning the availability and costs of processing.

Results are best with bright illumination. The photographer should take special care to face away from the sun while taking photographs. Because of differences in the reflectances of objects in the visible and the near infrared spectrums, the photographer should anticipate the nature of the scene as it will appear on the infrared film. Artistic photographers have sometimes used these differences to create special effects. The camera lens will bring infrared radiation to a focal point that differs from that for visible radiation, so infrared images may be slightly out of focus if the normal focus is used. Some lenses have special markings to show the correct focus for infrared films.

## YOUR OWN 3D PHOTOGRAPHS

You can take your own stereo photographs using a handheld camera simply by taking a pair of overlapping photographs. Two photographs of the same scene, taken from slightly different positions, create a stereo effect in the same manner that overlapping aerial photographs provide a three-dimensional view of the terrain.

This effect can be accomplished by aiming the camera to frame the desired scene, taking the first photograph, then moving the camera laterally a short distance, then taking a second photograph that overlaps the field of view of the first. The lateral displacement need only be a few inches (equivalent to the distance between the pupils of a person's eyes), but a displacement of a few feet will often provide a modest

**FIGURE 3.32.** Black-and-white infrared photograph (top), with a normal black-and-white photograph of the same scene shown for comparison (bottom). From author's photographs.

**FIGURE 3.33.** Stereo photographs. From author's photographs.

exaggeration of depth that can be useful in distinguishing depth. However, if the displacement is too great, the eye cannot fuse the two images to simulate the effect of depth.

Prints of the two photographs can then be mounted side by side to form a stereo pair that can be viewed with a stereoscope, just as a pair of aerial photos can be viewed in stereo. Stereo images can provide three-dimensional ground views that illustrate conditions encountered within different regions delineated on aerial photographs. Section 5.13 provides more information about viewing of stereo photographs.

CHAPTER FOUR

# Digital Data

## 4.1. Introduction

Thus far our discussion has focused on examination of remotely sensed images as photographs or photograph-like images. Such images can also be represented in digital form, in which the pattern of image brightness (often in several spectral channels) constitutes an array of numbers recorded in digital mode. When an image is represented as numbers, brightness can be added, subtracted, multiplied, divided, and in general subjected to statistical manipulation that is not possible if an image is presented only as a photograph. Thus the digital format greatly increases our ability to display, examine, and analyze remotely sensed data.

Although digital analysis of remotely sensed data dates from the early days of remote sensing, the launch of the first Landsat Earth observation satellite in 1972 began an era of increasing interest in machine processing. Previously, digital remote sensing data could be analyzed only at specialized remote sensing laboratories. The expensive equipment and highly trained personnel necessary to conduct routine machine analysis of data were not widely available, in part because of limited availability of digital remote sensing data and a lack of appreciation of their qualities.

After the Landsat 1 Multispectral Scanner Subsystem (to be introduced in Chapter 6) began to generate a steady stream of digital data, analysts realized that the usual visual examination of Landsat images would not permit full exploitation of the information they conveyed. In time, routine availability of digital data increased interest among businesses and institutions, computers and peripheral equipment became less expensive, more personnel acquired the necessary training and experience, software became more widely available, and managers developed knowledge of the capabilities of digital analysis for remote sensing applications. Today digital analysis has a significance that far exceeds that of purely visual interpretation of remotely sensed data.

## 4.2. Electronic Imagery

Digital data can be created by a family of instruments that can systematically scan portions of the Earth's surface, recording photons reflected or emitted from individual patches of ground, known as pixels (from "picture elements"). A digital image is composed of many thousands of pixels, usually each too small to be individually resolved by the human eye. Each pixel represents the brightness of a small region on the Earth's surface, recorded digitally as a numeric value, usually with separate values for each of several regions of the electromagnetic spectrum (Figure 4.1).

Digital images can be generated by two kinds of instruments, each described in later chapters. *Optical-mechanical scanners* physically move mirrors or lenses to systematically aim the field of view over the Earth's surface (Figure 4.2). As the instrument scans the Earth's surface, it generates an electrical current that varies in intensity as the land surface varies in brightness.

Sensors sensitive in several regions of the spectrum use filters to separate energy into several spectral regions, each represented by a separate electrical current. Each electrical signal must be subdivided into distinct units to create the discrete values necessary for digital analysis. This conversion from the continuously varying analog signal to the discrete digital values is accomplished by sampling the current at a uniform interval (analog-to-digital, or *A-to-D, conversion*) (Figure 4.3). Because all signal values within this interval are represented as a single average, all variation within this interval is lost. Thus the choice of sampling interval establishes one dimension to the resolution of the sensor (Chapter 10).

A second basic design for sensors uses *charge-coupled devices* (CCDs) (Figure 4.4). A CCD is formed from light-sensitive material embedded in a silicon chip. The *potential well* receives photons from the scene, usually through an optical system designed to collect, filter, and focus radiation. The sensitive components of CCDs can be manufactured to be very small, perhaps as small as 1μm in diameter, and sensitive to visible and near infrared radiation. These elements can be connected using microcircuitry to form *arrays;* detectors arranged in a single line form a *linear array,* detectors arranged in several rows and columns form *two-dimensional arrays.* Individual detectors are so small that a linear array smaller than 2 cm in length might have as many as 1,000 separate detectors. Two-dimensional arrays of 800 × 800 have been used, and larger arrays can be formed by mosaicking several smaller arrays together.

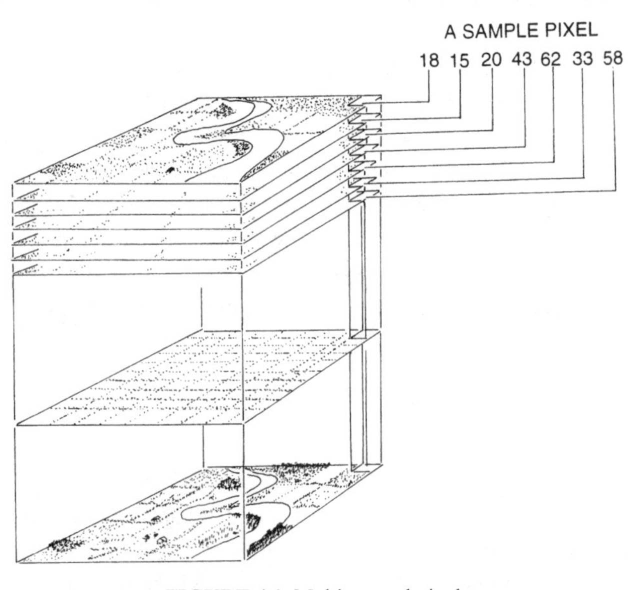


**FIGURE 4.1.** Multispectral pixels.

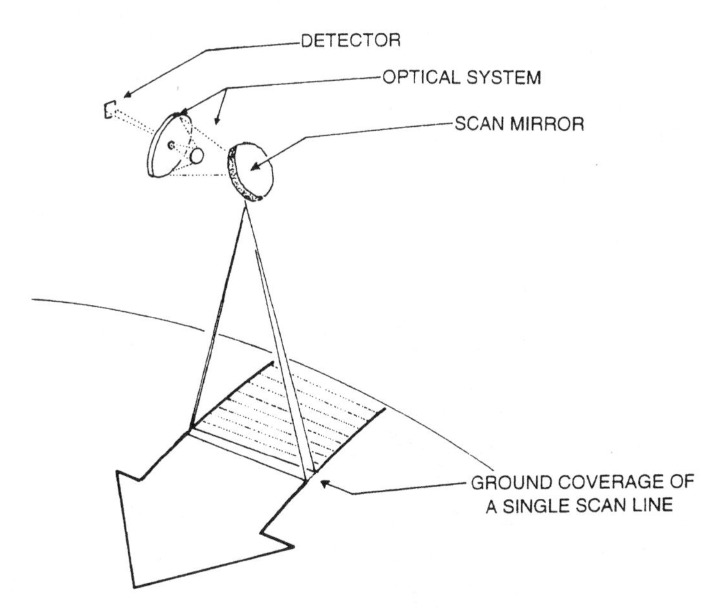


**FIGURE 4.2.** Optical-mechanical scanner.

Each detector collects photons that strike its surface and accumulates a charge proportional to the intensity of the radiation it receives. At a specified interval, charges accumulated at each detector pass through a *transfer gate,* which controls the flow of information from the detectors. Microcircuits connect detectors within an array to form *shift registers.* Shift registers permit charges received at each detector to be passed to adjacent elements (in a manner analogous to a bucket brigade), temporarily recording the information until it is convenient to transfer it to another portion of the instrument. Through this process, information read from the shift register is read sequentially, much in the same manner that a mechanical scanner collects a line of data through its side-to-side motion.

A CCD therefore scans electronically, without the mechanical motion necessary for the optical scanners described above. CCDs are compact; moreover, relative to other sensors, they are more efficient in detecting photons, so CCDs are especially effective when intensities are dim. Further, they tend to respond linearly to brightness, so they produce images that have more con-

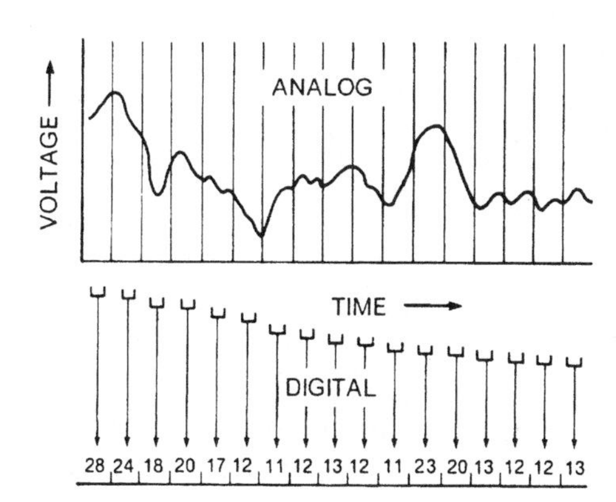


**FIGURE 4.3.** Analog-to-digital conversion.

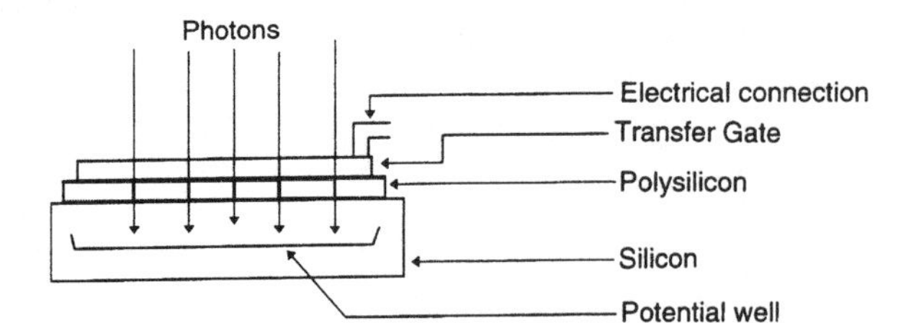


**FIGURE 4.4.** Charge-coupled device (CCD).

sistent relationships to scene brightness than is possible to produce using photographic processes (Chapter 3).

An alternative imaging technology, *complementary metal-oxide semiconductors* (CMOS), is often used in camcorders and related consumer products. Recent improvements in CMOS-based instruments provide fine detail at a low cost and with a low power requirement. CCDs expose all pixels at the same instant, then read these values as the next image is acquired. In contrast, CMOS instruments expose a line, then expose the next line in the image while data for the previous line is transferred. Therefore, pixels within a CMOS image are not exposed at the same instant. This property of CMOS technology, plus the advantage of the low noise that characterizes CCD imagery, favors use of CCDs for digital remote sensing instruments. Nonetheless, continued technological advances in CMOS technology, and continued decreases in cost, are likely to lead to its increasing use in imaging systems.

Optical sensors often use prisms and filters to separate light into separate spectral regions. Electronic sensors usually use diffraction gratings, considered more efficient because of their effectiveness, small size, and lightweight. *Diffraction gratings* are closely spaced transmitting slits cut into a flat surface (a *transmission grating*) or grooves cut into a polished surface (a *reflection grating*). Effective transmission gratings must be very accurately and consistently spaced, and must have very sharp edges. Light from a scene is passed though a *colluminating lens,* designed to produce a beam of parallel rays of light that is oriented to strike the diffraction grating at an angle (Figure 4.5).

Light striking a diffraction grating experiences both destructive and constructive interference as wavefronts interact with the grating. *Destructive interference* causes some wavelengths to be suppressed, whereas *constructive interference* causes others to be reinforced. Because the grating is oriented at an angle with respect to the beam of light, different wavelengths are diffracted

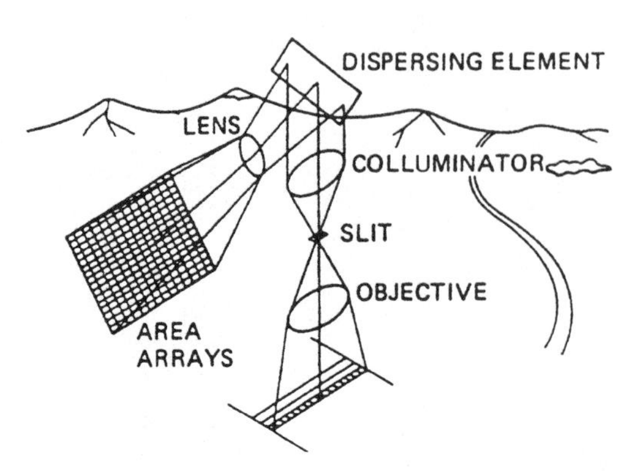


**FIGURE 4.5.** Diffraction grating and colluminating lens. (NASA diagram.)

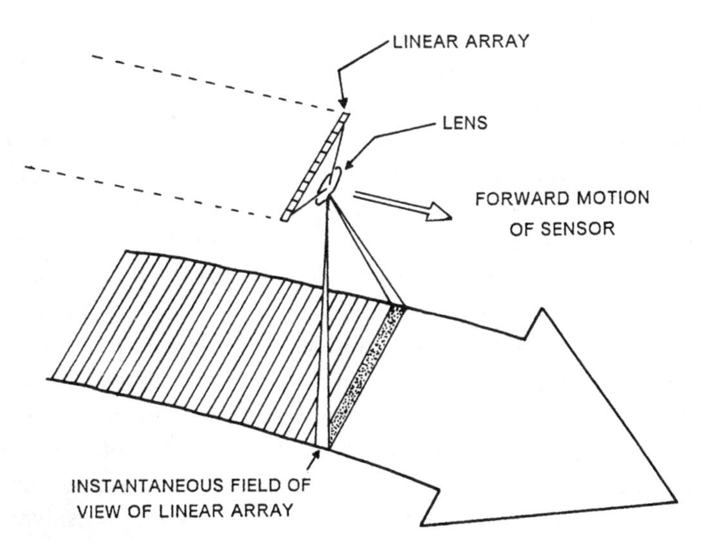


**FIGURE 4.6.** CCD used in pushbroom scanning.

at different angles, and the radiation can be separated spectrally. This light can then illuminate detectors to achieve the desired spectral sensitivity.

CCDs can be positioned in the focal plane of a sensor such that they view a thin rectangular strip oriented at right angles to the flight path (Figure 4.6). The forward motion of the aircraft or satellite moves the field of view forward along the flight path, building up coverage. By analogy, this means of generating an image is known as *pushbroom scanning.* (In contrast, mechanical scanning can be visualized by analogy to a whisk broom, which creates an image using the side-to-side motion of the scanner.)

The *instantaneous field of view* (IFOV) of any sensor refers to the hypothetical area viewed by the instrument if it were possible to suspend the motion of the aircraft and the scanning of the sensor for an instant (Figure 4.7). The IFOV therefore defines the smallest area viewed by the sensor and establishes a lower limit for the level of spatial detail that can be represented in a

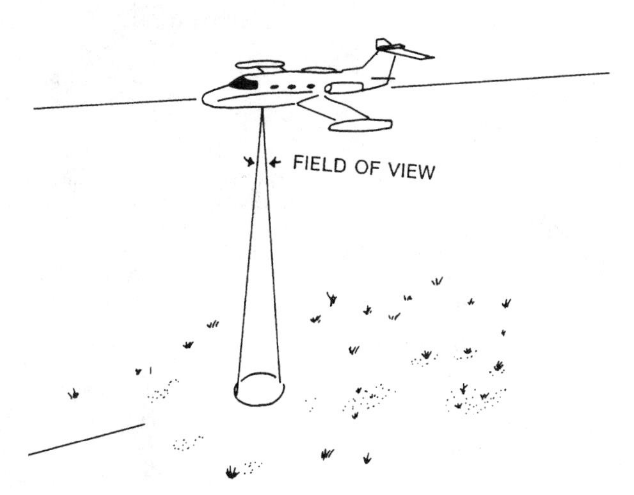


**FIGURE 4.7.** Instantaneous field of view (IFOV).

digital image. Although data in the final image can be aggregated so that an image pixel represents a ground area larger than the IFOV, it is not possible for pixels to carry information about ground areas smaller than the IFOV.

Electronic sensors must be operated within the limits of their design capabilities. Altitudes and speeds of aircraft and satellites must be selected to match the sensitivities of the sensors, such that detectors view a given ground area long enough to accumulate enough photons to generate strong signals (this interval is known as dwell time). At the lower end of an instrument's sensitivity is the *dark current* signal (Figure 4.8). At low levels of brightness, a CCD can record a small level of brightness even when there is none in the scene. At an instrument's upper threshold, bright targets *saturate* the sensor's response—the instrument fails to record further increases in target brightness. Between these limits, sensors are designed to generate signals that have predictable relationships with scene brightness; these relationships are established by careful calibration of each individual sensor. These characteristics of electronic sensors are analogous to the characteristic curve previously described for photographic films (Chapter 3). In other words, they define the upper and lower limits of the system's sensitivity to brightness, and the range of brightnesses over which a system can generate measurements with consistent relationships to scene brightnesses.

The range of brightnesses that can be accurately recorded is known as the sensor's *dynamic range.* Most electronic sensors have rather large dynamic ranges compared to those of photographic films, computer displays, or the human visual system. Therefore, photographic representations of electronic imagery tend to lose information at the upper and/or lower ranges of brightness. Because visual interpretation forms such an important dimension of our understanding of images, the way that photographic films (Chapter 3), image displays (discussed below) and image-enhancement methods (Section 5.15) handle this problem forms an important dimension of the field of image analysis.

Each sensor creates responses unrelated to target brightness. This "noise" is the result, in part, of the accumulated electronic errors from various components of the sensor. (In this context "noise" refers specifically to noise generated by the sensor, although the noise that the ana-

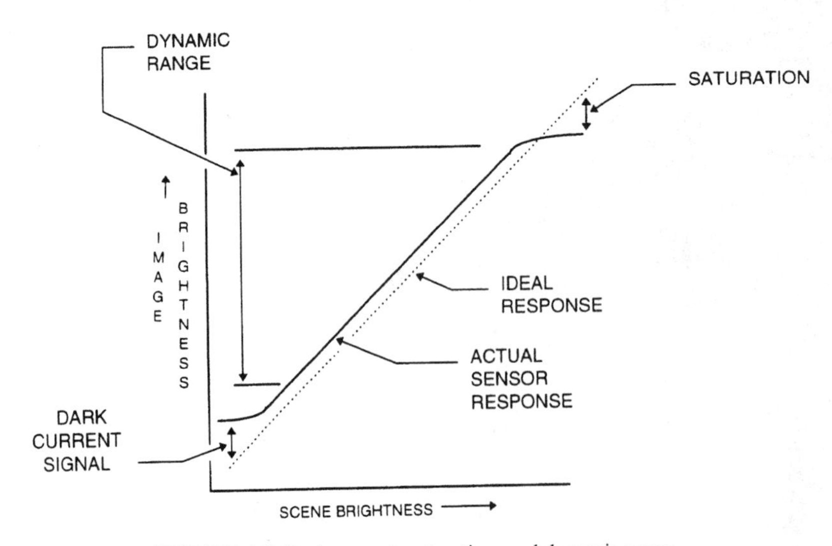


**FIGURE 4.8.** Dark current, saturation, and dynamic range.

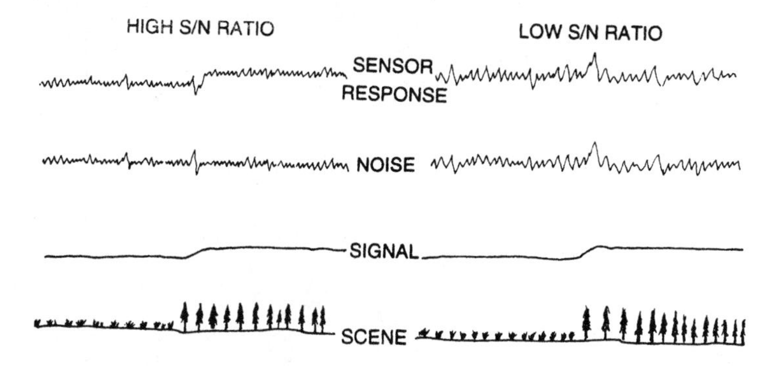


**FIGURE 4.9.** Signal-to-noise (S/N) ratio. At the bottom, a hypothetical scene is composed of two cover types. The signal records this region, with only a small difference in brightness between the two classes. Atmospheric effects, sensor error, and other factors contribute to noise, which is added to the signal. The sensor then records a combination of signal and noise. When noise is small relative to the signal (left: high S/N ratio), the sensor conveys the difference between the two regions. When the signal is small relative to noise (right: low S/N ratio), the sensor cannot portray the difference in brightness between the two regions.

lyst receives originates not only in the sensor but also in the atmosphere, the interpretation process, etc.) For effective use, instruments must be designed such that their noise levels are small relative to the signal (brightness of the target). This is measured as the signal-to-noise ratio (S/N or SNR) (Figure 4.9). Analysts desire signals to be large relative to noise, so the SNR should be large not only for bright targets, when the signal is large, but over the entire dynamic range of the instrument, especially at the lower levels of sensitivity, when the signal is small relative to noise. Engineers who design sensors must balance the radiometric sensitivity of the instrument with pixel size, dynamic range, operational altitude, and other factors to maintain acceptable SNRs.

## 4.3. Spectral Sensitivity

Because the various filters and diffraction gratings that instruments use to define the spectral limits (i.e., the "colors" that they record) do not define discrete limits, spectral sensitivity varies across a specific defined interval. For example, an instrument designed to record radiation in the green region of the spectrum will not exhibit equal sensitivity across the green region, but will exhibit greater sensitivity near the center of the region than at the transitions to the red and blue regions on either side (Figure 4.10). Defining the spectral sensitivity to be the extreme limits of the energy received would not be satisfactory because it is clear that the energy received at the extremes is so low that the effective sensitivity of the instrument is defined by a much narrower wavelength interval.

As a result, the spectral sensitivity of an instrument is often specified using the definition of *full width, half maximum* (FWHM), the spectral interval measured at the level at which the instrument's response reaches one-half of its maximum value (Figure 4.10). Thus FWHM forms a definition of *spectral resolution,* the narrowest spectral interval that can be resolved by an instrument. (Even though the instrument is sensitive to radiation at the extreme limits, beyond the limits of FWHM, the response is so weak and unreliable at these limits that FWHM forms a measure of functional sensitivity.) Figure 4.10 also illustrates the definition of the spec-

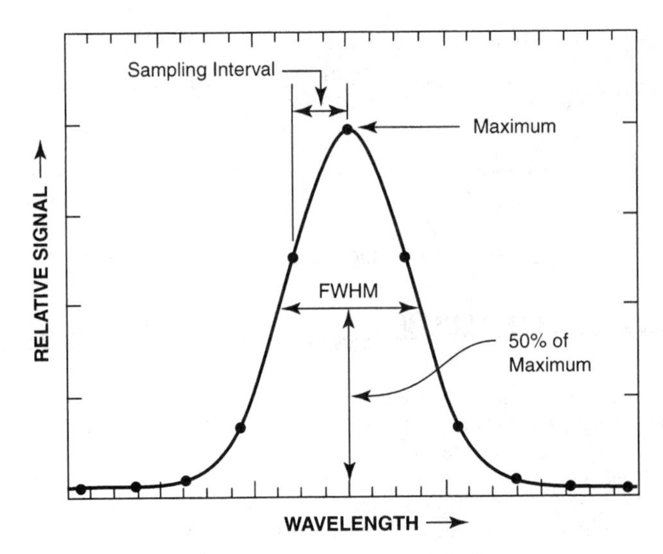


**FIGURE 4.10.** Full width, half maximum.

tral *sampling interval* (known also as *spectral bandwidth*), which specifies the spectral interval used to record brightness in relation to wavelength.

## 4.4. Digital Data

Output from electronic sensors reaches the analyst as a set of numeric values. Each digital value is recorded as a series of binary values known as *bits.* Each bit records an exponent of a power of 2, with the value of the exponent determined by the position of the bit in the sequence. As an example, consider a system designed to record 7 bits for each digital value. This means that seven binary places are available to record the brightness sensed for each band of the sensor. The seven values record, in sequence, successive powers of 2. A "1" signifies that a specific power of 2 (determined by its position within the sequence) is to be evoked; a "0" indicates a value of zero for that position. Thus the 7-bit binary number "1111111" signifies $2^6 + 2^5 + 2^4 + 2^3 + 2^2 + 2^1 + 2^0 = 64 + 32 + 16 + 8 + 4 + 2 + 1 = 127$. And "1001011" records $2^6 + 0^5 + 0^4 + 2^3 + 0^2 + 2^1 + 2^0 = 64 + 0 + 0 + 8 + 0 + 2 + 1 = 75$. Figure 4.11 shows different examples. Eight bits constitute a *byte,* intended to store a single character. Larger amounts of memory can be indicated in terms of *kilobytes* (KB) 1,024 ($2^{10}$) bytes; *megabytes* (MB) 1,048,576 ($2^{20}$) bytes; and *gigabytes* (GB), 1,073,741,824 ($2^{30}$) bytes (Table 4.1).

In this manner, discrete digital values for each pixel are recorded in a form suitable for storage on tapes or disks and for analysis by digital computer. These values, as read from tape or disk, are popularly known as "digital numbers" (DNs), "brightness values" (BVs), or "digital counts," in part as a means of signifying that these values do not record true radiances from the scene.

The number of brightness values within a digital image is determined by the number of bits available. The 7-bit example given above permits a maximum range of 128 possible values (0–127) for each pixel. A decrease to 6 bits would decrease the range of brightness values to 64 (0–63); an increase to 8 bits would extend the range to 256 (0–255). Thus the number of bits

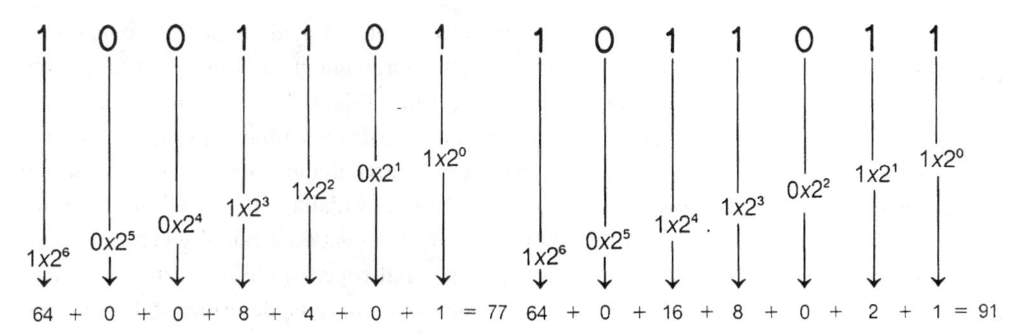


**FIGURE 4.11.** Digital representation of values in 7 bits.

determines the radiometric resolution (Chapter 10) of a digital image. The number of bits available is determined by the design of the system, especially the sensitivity of the sensor (adding too many extra bits would simply record system noise rather than provide additional information about scene brightness) and its capabilities for recording and transmitting data (each added bit increases transmission requirements). If we assume that transmission and storage resources are fixed, then increasing the number of bits for each pixel means that we will have fewer pixels per image and that pixel sizes must be larger. Thus design of remote sensing systems requires trade-offs between image coverage and radiometric, spectral, and spatial resolutions.

### *Radiances*

The brightness of radiation reflected from the Earth's surface is measured as brightness (watts) per wavelength interval (micrometer) per angular unit (steradian); thus the measured brightness is defined with respect to wavelength (i.e., "color"), spatial area (angle), and intensity (brightness). Radiances record actual brightnesses, measured in physical units, given as real values (to include decimal fractions).

Use of digital numbers (DNs) facilitates the design of instruments, data communications, and the visual display of image data. For visual comparison of different scenes, or analyses that examine relative brightnesses, use of DNs is satisfactory. However, because a DN from one scene does not represent the same brightness as the same DN from another scene, DNs are not comparable from scene to scene if an analysis must examine actual scene brightnesses for purposes that require use of original physical units. Such applications include comparisons of scenes of the same area acquired at different times, or matching adjacent scene to make a mosaic.

**TABLE 4.1. Terminology for Computer Storage**

| | | |
|---|---|---|
| Bit | A binary digit (0 or 1) | |
| Byte | 8 bits, 1 character | |
| Kilobyte (K or KB) | 1,024 bytes | ($2^{10}$ bytes) |
| Megabyte (MB) | 1,048,576 bytes | ($2^{20}$ bytes) |
| Gigabyte (GB) | 1,073,741,824 bytes | ($2^{30}$ bytes) |
| Terabyte (TB) | 1,099,511,627,776 bytes | ($2^{40}$ bytes) |

For such purposes, it is necessary to convert the DNs to the original radiances. This process requires knowledge of calibration data specific to each instrument. To ensure that a given sensor provides an accurate measure of brightness, it must be calibrated against targets of known brightness. The sensitivities of electronic sensors tend to drift over time, so to maintain their accuracy it is necessary to recalibrate them periodically. Although those sensors used in aircraft can be recalibrated regularly, those used in satellites are not available after launch for the same kind of recalibration. Typically, such sensors are designed so that they can observe calibration targets onboard the satellite or they are calibrated using landscapes of uniform brightness (e.g., in desert regions). Nonetheless, calibration errors, such as those to be described in Chapter 11, sometimes remain.

## 4.5. Data Formats

Digital image analysis is usually conducted using raster data structures in which each image is treated as an array of values (Figure 4.12). Additional spectral channels form additional arrays that register to one another. Each pixel is treated as a separate unit, which can always be located within the image by its row and column coordinates. In most remote sensing analysis, coordinates originate in the upper left-hand corner of an image, and are referred to as rows and columns, or as lines and pixels, to measure position down and to the right, respectively.

Raster data structures offer advantages for manipulation of pixel values by image-processing systems, as it is easy to find and locate pixels and their values. The disadvantages are usually apparent only when we need to represent not the individual pixels, but areas of pixels as discrete patches or regions. Then the alternative structure, vector format, becomes more attractive. Vector format (discussed in Chapter 16) uses polygonal patches and their boundaries as the fundamental units for analysis and manipulation. The vector format is not appropriate for digital analysis of remotely sensed data, although sometimes we may wish to display the results of our analysis using a vector format. Almost always equipment and software for digital processing of remotely sensed data must be tailored for raster format.

Digital remote sensing data are typically organized according to one of three alternative strategies for storing image data. Consider an image consisting of four spectral channels, which together can be visualized as four superimposed images, with corresponding pixels in one band registering exactly to those in the other bands.

One of the earliest formats for digital data was *band interleaved by pixel* (BIP). Data are

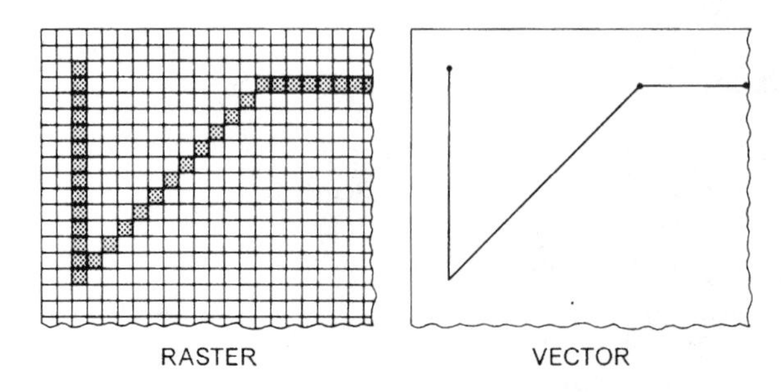


**FIGURE 4.12.** Raster and vector formats.

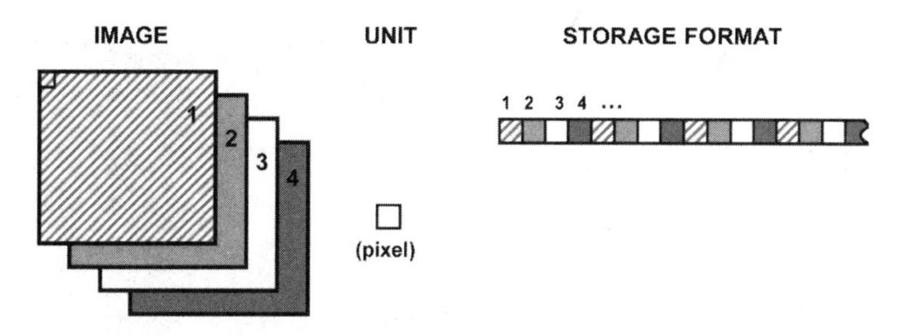


FIGURE 4.13. Band interleaved by pixel (BIP) format.

organized in sequence values for line 1, pixel 1, band 1; then for line 1, pixel 1, band 2; then for line 1, pixel 1, band 3; and finally for line 1, pixel 1, band 4. Next are the four bands for line 1, pixel 2, and so on (Figure 4.13). Thus values for all four bands are written before values for the next pixel are represented. Any given pixel, once located within the data, is found with values for all four bands written in sequence one directly after the other. This arrangement may be advantageous in some situations, but for most applications it is awkward to sort through the entire sequence of data (which typically are very large in number) in order to sort the four bands into their respective images.

The *band interleaved by line* (BIL) format treats each line of data as a separate unit (Figure 4.14). In sequence the analyst finds line 1 for band 1, line 1 for band 2, line 1 for band 3, line 1 for band 4, line 2 for band 1, line 2 for band 2, and so on. Each line is represented in all four bands before the next line is encountered.

A third convention for recording remotely sensed data is the *band sequential* (BSQ) format (Figure 4.15). All data for band 1 are written in sequence, followed by all data for band 2, and so on. Each band is treated as a separate unit. For many applications, this format is the most practical, as it presents data in the format that most closely resembles the data structure used for the display and analysis. However, if areas smaller than the entire scene are to be examined, the analyst must read all four images before the subarea can be identified and extracted.

Actual data formats used to distribute digital remote sensing data are usually variations on these basic alternatives. Exact details of data formats are specific to particular organizations and to particular forms of data, so whenever an analyst acquires data, he or she must make sure to acquire detailed information regarding the data format. Although organizations attempt to standardize formats for specific kinds of data, it is also true that data formats change as new mass

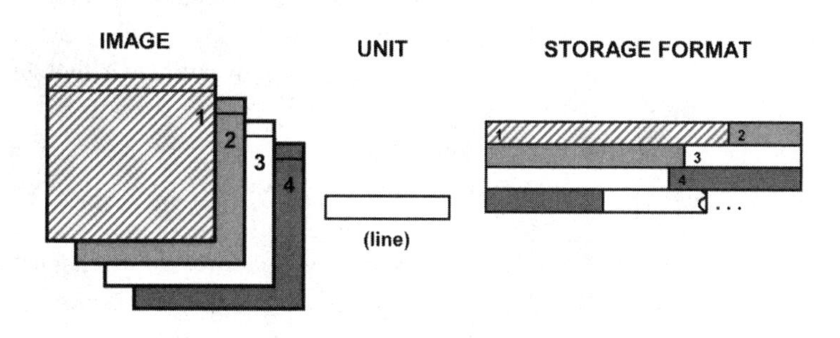


FIGURE 4.14. Band interleaved by line (BIL) format.

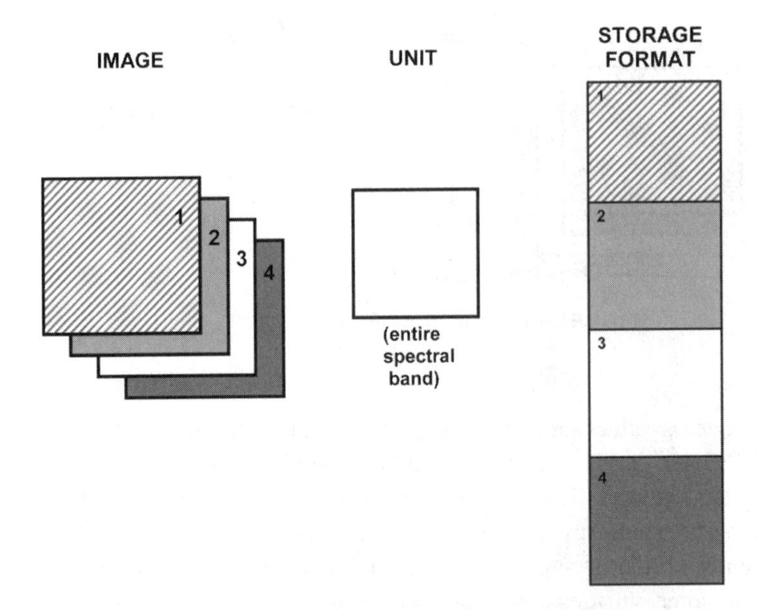


**FIGURE 4.15.** Band sequential (BSQ) format.

storage media come into widespread use and as user communities employ new kinds of hardware or software.

The "best" data format depends upon immediate context and often upon the specific software and equipment available. If all bands for an entire image must be used, then the BSQ and BIL formats are useful because they are convenient for reconstructing the entire scene in all four bands. If the analyst knows beforehand the exact position on the image of the subarea that is to be studied, then the BIP format is useful because values for all bands are found together and it is not necessary to read through the entire data set to find a specific region. In general, however, the analyst must be prepared to read the data in the format in which they are received, and to convert them into the format most convenient for use at a specific laboratory.

Other formats are less common in everyday applications, but are important for applications requiring use of long sequences of multispectal images. *Hierarchical data format* (HDF) is a specialized data structure developed and promoted by the National Center for Supercomputing Applications (*http://hdf.ncsa.uiuc.edu/index.html*), designed specifically to promote effective management of scientific data. Whereas the formats discussed thus far organize data conveyed by a specific image, HDF and related structures provide frameworks for organizing collections of images. For example, conventional data formats become awkward when it is necessary to portray three-dimensional data structures as they might vary over time. Although such structures might typically portray complex atmospheric data as it varies hourly daily, seasonally, or yearly, they also lend themselves to recording large sequences of multispectral images. HDF therefore enables effective analysis and visualization of such large, multifaceted data structures.

A related but distinctly different format, *Network common data form* (NetCDF), also provides structures tailored for handling dynamic, array-oriented data. It is specifically designed to be compatible with a wide variety of computer platforms, so that it can facilitate sharing of data over the World Wide Web. NetCDF was created to work with the Unidata system (*http://www.unidata.ucar.edu/software/netcdf/*), designed for rapid transmission of meteorological data to a wide range of users.

Although HDF and NetCDF structures are unlikely to be encountered in usual remote sensing applications, they are becoming more common in advanced applications requiring the handling of very large sequences of images, such as those encountered in geophysics, meteorology, and environmental modeling. Often, remotely sensed data contribute to such studies.

*Data compression* reduces the amount of digital data required to store or transmit information by exploiting the redundancies within a data set. If data arrays contain values that are repeated in sequence, then compression algorithms can exploit that repetition to reduce the size of the array, while retaining the ability to restore the array to its original form. When the complete array is needed for analysis, then the original version can be restored by *decompression.* Because remotely sensed images require large amounts of storage, and usually are characterized by modest levels of redundancies, data compression is an important tool for effective storage and transmission of digital remote sensing data. Compression and decompression are accomplished, for example, by executing computer programs that receive compressed data as input and produce a decompressed version as output.

The *compression ratio* compares the size of the original image to the size of the compressed image. A ratio of 2:1 indicates that the compressed image is one-half the size of the original. *Lossless compression* techniques restore compressed data to their exact original form; *lossy* techniques degrade the reconstructed image, although in some applications the visual impact of a lossy technique may be imperceptible. For digital satellite data, lossless compression techniques can achieve ratios from 1.04:1 to 1.9 to 1. For digitized cartographic data, ratios of 24:1 using lossy techniques have been reported to exhibit good quality.

It is beyond the scope of this discussion to describe the numerous techniques and algorithms available for image compression. Probably the most well-known compression standard is the JPEG (Joint Photographic Experts Group) format, a lossy technique that applies the discrete cosine transform (DCT) as a compression–decompression algorithm. Although the JPEG algorithm has been widely accepted as a useful technique for compression of continuous-tone photographs, it is not likely to be ideal for remote sensing images, nor for geospatial data in general. The recent modification to the JPEG format, JPEG2000 (*http://www.jpeg.org/*), provides a high compression rate with high fidelity (Liu et al., 2005). Depending upon its application, JPEG2000 can be either lossy or nonlossy.

Generally stated, lossy compression techniques should not be applied to data intended for analysis or as archival copies. Lossy compression may be appropriate for images to present visual records of results of an analytical process, provided they do not form input to other analyses.

## 4.6. Equipment for Digital Analysis

Digital analysis requires specialized equipment tailored for the storage, display, and analysis of the large amounts of data that are necessary to record remotely sensed images. Remote sensing analysis has specific needs for input, storage, analysis, and display of data that are often not fully met by standard equipment and programs.

### *Computers*

Although there was a time when some computers were specifically designed to analyze and display remotely sensed data, these computers were too expensive and inflexible for widespread use. Therefore, although remote sensing analysis may sometimes require extra memory, disk storage, or peripheral devices, today the vast majority of remote sensing analysis uses general-purpose computers with specialized remote sensing software.

The heart of any digital computer is its *central processing unit* (CPU), which consists of three components (Figure 4.16). The *control unit* manages the flow of data and instructions to and from the CPU and to and from input and output devices. Once CPUs were massive networks of vacuum tubes, wires, and switches that filled several rooms. Today the electrical pathways for a CPU are tiny circuits printed on silicon chips perhaps half the size of a credit card. Successive generations of chips have become increasingly more powerful and efficient, so the once arcane designations assigned by manufacturers to chips have become common shorthand for designating the power of a computer. At the time of this writing, the Intel® Pentium® IV processor is the current standard for consumer computers and for many business applications.

Within the CPU, the *arithmetic/logic unit* (ALU) manipulates input data to produce new (output) values that form results of a particular operation. Any computer program, when resolved into its simplest components, consists of a sequence of fundamental operations such as addition, multiplication, and comparisons of values. These operations are often performed by specially designed silicon chips that form the ALU. Whereas most chips simply store bits that record data without altering the values, those in the ALU are designed to perform arithmetic or logical operations, so that the data they return to the CPU are altered from the data that they receive. The ALU receives data from the control unit, performs the required operations, then returns the results to the control unit. Although each operation of the ALU is itself very simple, a computer's power is achieved by its ability to perform millions of such simple operations quickly, with accuracy, and with precision.

The third component of the CPU is its *primary storage,* the CPU's internal memory. Primary storage is known as *random access memory* (RAM), meaning that the CPU has direct access to each location in primary memory without sequential searching through preceding addresses and without mechanical movement (which are required with other forms of memory). Every value is stored in binary form, as described previously. The capabilities of main memory are a func-

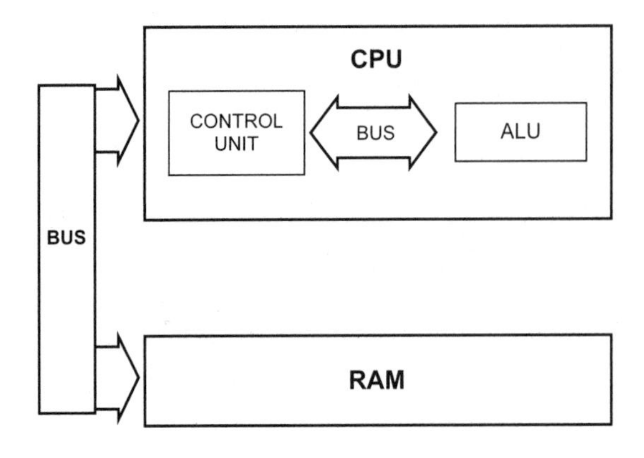


**FIGURE 4.16.** Schematic representation of the CPU.

tion of the number of bits used to record each value in memory, which is a function of the design of the chips that form the computer's main memory. Computers used for image processing now use 32-bit processors, meaning that each value stored in main memory is recorded with the precision of 32 bits, as explained previously. Because the CPU must load a program and data into primary storage, the amount of primary storage determines many of the capabilities of the computer. Small primary storage means that the computer can run rather short programs on small data sets; large primary storage means that the computer can run longer programs on larger data sets. For image-processing work, 1 GB of memory is now common.

The CPU, RAM, and ALU are connected by *data buses,* electrical pathways that transfer data from one portion of the computer to another. A bus consists of wires, connectors, and controllers; a bus must provide one wire for each bit required to specify the address of a location in memory, plus additional pathways to distinguish among the alternative data transfer options possible. For PCs, the design of the bus, often known as the *expansion bus,* determines the extent to which a computer's memory, speed, and peripheral support can be expanded by adding or upgrading components. The capacity of the bus to transmit data determines the speed at which it can operate; current computers transmit data at 850 MHz or more, with speeds of 1.5 GHz representing an upper limit at the time of this writing.

Some advanced research applications in remote sensing are based upon the use of supercomputers. *Supercomputers* are highly specialized computing systems, often constructed by networking many smaller computers in a way that allows computing tasks to be subdivided and allocated to the individual CPUs. Whereas our usual computers are tailored to efficiently perform a wide variety of logical operations, a supercomputer is designed to perform many executions of a few operations. Because remote sensing analysis often requires analysis of very large image arrays using many repetitions of the same calculations, supercomputers offer many opportunities for advanced research in remote sensing analysis.

### *Mass Storage*

Because remotely sensed data typically require very large storage capacities, image-processing systems use substantial amounts of disk storage. Disks can be manufactured from either rigid or flexible materials that are coated with a magnetized substance that is sensitized to record the bits that represent each digital value. The computer's disk drive rapidly spins the disk, allowing the computer to read data from both sides. Rigid (*hard*) disks provide rapid access to the very large amounts of storage (perhaps 50–100 gigabytes [GB]) required for image processing. Flexible (*floppy*) disks have far smaller storage capacities (1.44 MB) and provide slower access to the data; they were once used to convey data between smaller computers, but now have been replaced by systems that provide much larger capacities. These include a variety of removable disk drives that can store 100–250 MB (e.g., Iomega® Zip® drives) to 2 GB (e.g., Iomega® JAZ® drives) of data in a compact, transportable form. Even more convenient are the several versions of the miniaturized USB keys and jump drive storage devices, available now with capacities large enough to transport some remote sensing data sets.

*CD-ROM* (compact disc, read-only memory) storage is the oldest and most reliable form of optical storage. Each CD-ROM consists of an internal disc coated with a very thin metal (usually aluminum) surface, then encased in a plastic shell. Data are recorded as microscopic indentations on the silver surface of the internal disc. CD-ROMs must be read using special drives, available as standard equipment on personal computers. The CD-ROM drive reads the disc by directing a laser beam at the disc's surface. The device examines the way in which the laser is reflected from the disc and reads the pattern of indentations formed when the disc was created. Because each disc can hold 680 MB, it forms a compact, durable means of storing remote sensing data.

Efforts to create more flexible mass storage have resulted in the *CD-R* and the *CD-RW* formats. CD-R (compact disc-recordable) are CDs that can be written by a user one time, but not altered once written. In contrast, a CD-RW (compact disc-rewritable) drive can rewrite the same disk several times. Disks written using CD-R and CD-RW devices can be read by CD-ROM drives. CD-RW and CD-R media are inexpensive—their principal disadvantage is that the ability to rewrite data is limited. CD-R disks are not rewritable, and until recently CD-RW discs had to be reformatted to recover the space taken by deleted files; CD-RW rewriteability is imperfect because successive rewrites reduce the CD-RW disc's capacity. In recent years CD-ROM discs and CD-ROM drives have become so inexpensive that they can be used as a routine means of distributing image data.

*Digital video disc* (DVD) is another form of optical disc storage technology, analogous to CD technology, but faster, and providing larger capacities. DVD is best known for its applications to cinema-like video data, but its large capacity and compact size offer obvious advantages for storage of other digital data, including digital remote sensing imagery. (It is important to note that although the discs used for video and computer applications may be physically identical, the application formats for DVDs used for video storage will differ from those used for data storage.) Capacities of DVDs depend upon the format employed. A single-sided DVD might hold 4.7 billion bytes. (Note that the convention in the realm of computer science is to specify capacity in multiples of 1,024 ($2^{10}$) bytes (Section 4.4), whereas the commercial world of DVDs measures capacity in multiples of 1,000 bytes, following usual conventions of SI units. Therefore, it is important to carefully evaluate the true capacity of a DVD before relying on it to store a specific set of data because for large capacities the difference in the two conventions can be significant.) Although DVDs are now mainly used for storage of consumer video data, we can expect increases in the use of DVDs as a medium for storing large amounts of remotely sensed data.

### *Image Display*

For remote sensing computing, the image display is especially important because the analyst must be able to examine images and to inspect results of analyses, which often are themselves images. At the simplest level, an image display can be thought of as a high-quality television screen, although those tailored specifically for image processing have image-display processors, which are special computers designed to very rapidly receive digital data from the main computer, then display them as brightnesses on the screen. The capabilities of an image display are determined by several factors. First is the size of the image it can display, usually specified by the number of rows and columns it can show at any one time. The usual desktop computer display might show 1,600 rows and 1,280 columns. A smaller display with respectable capabilities could show a 1,024 × 768 image; others of much more limited capabilities could show only smaller sizes, perhaps 800 × 600.

Second, a display has a given *radiometric resolution* (Chapter 10), that is, for each pixel, it has a capability to show a range of brightnesses. One-bit resolution would give the capability to represent either black or white—certainly not enough detail to be useful for most purposes. In practice, 6 bits (64 brightness levels) are minimally necessary for gray-scale images to appear "natural," and high-quality displays typically display 8 bits (256 brightness levels) or more.

A third factor controls the *rendition of color* in the displayed image. The method of depicting color is closely related to the design of the image display and the display processor. Image-display data are held in the *frame buffer,* a large segment of computer memory dedicated to handling data for display. The frame buffer provides one or more bits to record the brightness of each pixel to be shown on the screen (the "bit plane"); thus the displayed image is generated, bit by bit, in the frame buffer. The more bits that have been designed in the frame buffer for each pixel, the greater the range of brightnesses that can be shown for that pixel, as explained above. Rendition of color is referred to as *color depth,* measured in *bits per pixel* (bpp). Higher bpp provides more ability to portray a greater range of colors.

For actual display on the screen, the digital value for each pixel is converted into an electrical signal that controls the brightness of the pixel on the screen. This requires a digital-to-analog (D-to-A) converter that translates discrete digital values into continuous electrical signals (the opposite function of the A-to-D converter mentioned previously). Currently there are three strategies for displaying images, each outlined here in a an abbreviated form.

The *cathode ray tube* (CRT) dates from the early 1940s, when it formed the basis for the first television displays. A CRT is formed from a large glass tube, wide at one end (the "screen"), and narrow at the other. The inside of the wide end is coated with phosphor atoms. An electron gun positioned at the narrow end directs a stream of electrons against the inside of the wide end of the tube. As the electrons strike the phosphor coating, it glows, creating an image as the intensity of the electron beam varies according to the strength of the video signal. Electromagnets positioned on four sides of the narrow portion of the tube control the scan of the electron stream across the face of the tube left to right, top to bottom.

As each small region of the screen is illuminated from the inside by the stream of electrons, it glows. Because the gun directs the stream of electrons systematically, and very rapidly, it creates the images we can see on a computer display or television screen. Because the electron gun scans very rapidly (30–70 times each second), it can return to refresh, or update, the brightness at each pixel before the phosphor coating fades. In this manner, the image appears to the eye to be a continuous image.

The video signal determines the brightness and colors of the regions (which can be thought of as pixels), thereby forming an image. CRTs produce very clear images, but because an increase in the size of the screen requires a commensurate increase in the depth of the tube (so the gun can illuminate the entire width of the screen), the size and weight of a CRT display creates a major inconvenience.

An alternative display technology was developed in the early 1970s, but became available for computer display only much later. *Liquid crystal displays* (LCDs) depend upon *liquid crystals,* substances that are intermediate between solid and liquid phases. The state assumed at a specific time depends upon temperature so an electrical current can change the orientation of the molecules within a liquid crystal.

LCD displays use two sheets of polarizing materials that enclose a liquid crystal solution between them. When the video signal sends an electrical current to the display, the crystals align to block the passage of light between them. In effect, the liquid crystal at each pixel acts like a shutter, either blocking or transmitting light, thereby forming an image. LCDs are used in a variety of consumer products, but especially for the flat-panel displays in portable computers and for the compact displays now used for desktop computers.

A third display technology, *plasma display,* is not yet widely available. Each pixel is represented by three tiny fluorescent lights. Fluorescent lights are small sealed glass tubes containing an internal phosphor coating, an inert gas, mercury, and two electrodes. As an electrical current flows across the electrodes, it vaporizes some of the mercury. The electrical current also raises the energy levels of some of the mercury atoms; when they return to their original state, the emit photons in the ultraviolet potion of the spectrum. The ultraviolet light strikes the phosphor coating on the tube, creating visible light used to make an image. Variations in the coatings can create different colors. The positions of the tiny fluorescent lights can be accessed as intersections in a raster grid, so the tube required for the CRT is not necessary for a plasma display, and the screen can be much more compact. Plasma display are suitable for large, relatively compact, image displays, but are expensive, so are not now preferred for analytical use.

Table 4.2 lists a selection of current graphics standards. Image display is important because remote sensing instruments typically record a much wider range of values than can be accurately displayed by film products (Chapter 3) or by any single representation on an image-display device. As a result, displays that provide the analyst with the means to conveniently change image scale, color assignments, and viewing area, provide a capability to explore visually the varied dimensions of an image. Such capabilities are not available with film images, where considerable effort is required to study varied representations of a digital image.

### *Film Recorders and Color Printers*

*Film recorders* are instruments designed to write image information directly to photographic paper or film. Usually the film is mounted on a cylindrical drum; during operation, the drum spins rapidly. A beam of light is varied in intensity in proportion to image brightness. As the drum spins, the light slowly moves the length of the image, exposing the entire image line by line. If a color image is to be made, usually it is necessary to make three separate exposures, one for each primary color. Many color inkjet and color laser printers and plotters (even some that are relatively inexpensive) now can provide paper copies of images of good quality that are satisfactory for many uses in the practice of remote sensing.

**TABLE 4.2. Some Graphics Standards for Computer Displays**

| Name and designation | Color depth (bpp) | Screen size |
|---|---|---|
| Color graphics adapter (CGA) | 2 | 320 × 200 |
| Extended graphics adapter (EGA) | 4 | 640 × 350 |
| Video graphics array (VGA) | 4 | 640 × 450 |
| Extended graphics array (XGA) | 32 | 1,024 × 768 |
| Widescreen ultra extended graphics array (WUXGA) | 32 | 1,920 × 1,600 |
| Widescreen quad extended graphics array (WQEGA) | 32 | 2,560 × 1,600 |

*Note:* This list is not complete, but simply illustrates some of the graphics standards that have been used over the years.

### *Advanced Image Display*

Current remotely sensed imagery is often characterized by the large arrays and the fine detail of the information they portray. Conventional systems permit the users to examine regions in fine detail only by zooming in to display the region of interest at the cost of losing the broader context. Or users can discard the finer detail and examine broad regions at coarse detail. Although this trade-off sometimes causes little or no inconvenience, in other situations the sacrifice of one quality for the other means that some of the most valuable qualities of the data are discarded.

*"Fisheye,"* or *"focus + context,"* displays enable the analyst to view selected detail without discarding the surrounding context. Fisheye displays use existing display hardware but include a simulated magnifier that can roam over the image to enlarge selected regions within the context of the coarser resolution display (Figure 4.17). Software "lenses" locally magnify a subset of the image, while maintaining the visual context of the unmagnified image. Such capabilities can be linked to analytical software to enable the analyst to annotate, measure, and delineate regions to improve the functionality of existing software. Similar effects can be achieved in a different manner by linking two windows within the display, one for detail, one for the broader context, and providing the analyst with the capability to alter the size of the window. This approach is known as "multiple linked views" or "overview + detail" display.

*Multiple-monitor systems* (tiled displays) are formed as arrays of flat-panel monitors (or rear-projection displays) that can display very large images at high levels of spatial detail (Figure 4.18). The highest quality tiled displays can project 50 million pixels–within the near future, gigapixel displays (1 billion pixels) will likely be attempted. Multiple-monitor systems enable the computer's operating system to use two or more display devices to create a single large image. Rear-projection systems (sometimes referred to as "power walls") do not have the seams

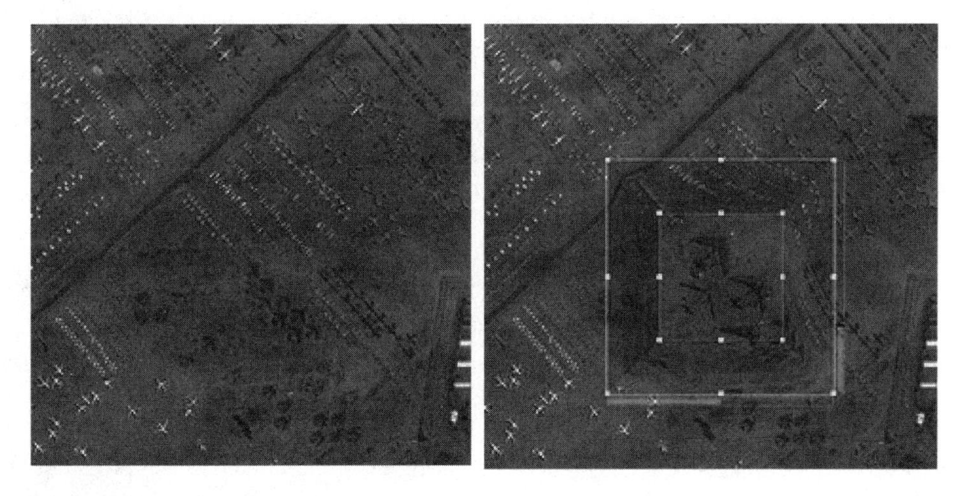

**FIGURE 4.17.** Example of "fisheye"-type image display. Left: Image of aircraft parked in a storage area for out-of-service aircraft. Right: Same image as viewed with a fisheye-type display that magnifies the central region of the image, while preserving the context of the surrounding image. Credit: Pliable Display Technology: IDELIX Software Inc, Vancouver, B.C. (http://www.idelix.com/imageintel.shtml); Satellite Image credit: Digital Globe, Longmont, CO (http://www.digitalglobe.com/).

**FIGURE 4.18.** Example of tiled image display. From Chris North, Virginia Tech Center for Human–Computer Interaction.

between tiles that characterize the LCD tiled systems, so they have greater visual continuity, but tiled displays are less expensive and provide better visual quality.

Multiple-monitor systems became practical as the costs of RAM and LCD displays decreased enough to enable economical development of composite displays. Tiled displays form a mosaic of screens, supported by operating systems configured to support multiple displays. Special mounts hold the multiple monitors provided by commercial vendors, and specialized software systems permit integrated use of the several monitors. Analysts can display images across the seams formed by the edges of the displays, and zoom, create multiple windows, execute multiple applications, and arrange windows as necessary for specific tasks.

Obvious applications include display of remotely sensed and GIS images that require the analyst to simultaneously exploit fine detail and broad areal coverage. Analysts sometimes desire to run two or more analytical programs simultaneously, with independent displays of the images generated by each analysis. Both fisheye and multiple-monitor systems are emerging technologies in the sense that while they have both proven to be successful, they are still under investigation to explore how they can be most effectively used in specific applications.

## 4.7. Image-Processing Software

Digital remote sensing data can be interpreted by computer programs that manipulate the data recorded in pixels to yield information about specific subjects, as described in subsequent chapters. This kind of analysis is known as *image processing,* a term that encompasses a very wide range of techniques. Image processing requires a system of specialized computer programs tailored to the manipulation of digital image data. Although such programs vary greatly in purpose and in detail, it is possible to identify the major components likely to be found in most image-processing systems.

A separate specific portion of the system is designed to read image data, usually from CD-

ROM or other storage media, and to reorganize the data into the form to be used by the program. For example, many image-processing programs manipulate the data in BSQ format. Thus the first step may be to read BIL or BIP data, and then reformat the data into the BSQ format required for the analytical components of the system. Another portion of the system may permit the analyst to subdivide the image into subimages; to merge, superimpose, or mosaic separate images; and in general to prepare the data for analysis, as described later in Chapter 11. The heart of the system consists of a suite of programs that analyze, classify (Chapter 12), and manipulate data to produce output images and the statistics and data that may accompany them. Finally, a section of the image-processing system must prepare data for display and output, either to the display processor or to the line printer. In addition, the program requires "housekeeping" subprograms that monitor movement and labeling of files from one portion of the program to another, generate error messages, and provide online documentation and assistance to the analysts.

Widely used image-processing systems run on PCs, Macintoshes, or workstations. More elaborate systems can be supported by peripheral equipment, including extra mass storage, digitizers, scanners, color printers, disk drives, and related equipment. Almost all such systems are directed by menus and graphic user interfaces that permit the analyst to select options from a list.

Although there are many good image-processing systems available, some of the most commonly used are:

- **ER Mapper** (Earth Resources Mapping, 4370 13400 Sabre Springs Parkway, Suite 150, San Diego, CA 92128; 858-391-5638; *http://www.ermapper.com/*)
- **EASI/PACE** (PCI Geomatics Headquarters, 50 West Wilmot Street, Richmond Hill, Ontario, Canada, L4B 1M5; 905-764-0614; *http://www.pci.on.ca/*)
- **ENVI** (Research Systems Inc., 4990 Pearl East Circle, Boulder, CO 80301; 303-786-9900; *http://www.rsinc.com/envi/*)
- **ERDAS Imagine** (Leica Geosystems Geospatial Imaging, LLC, Worldwide Headquarters, 5051 Peachtree Corners Circle, Norcross, GA 30092-2500; 770-776-3400; *http://gis.leica-geosystems.com*)
- **GRASS GIS** (Center for Applied Geographic and Spatial Research, Baylor University, P.O. Box 97351, Waco, TX 76798-7351; 254-710-6814; *http://www.baylor.edu/grass/*)
- **IDRISI** (The IDRISI Project, Clark Labs, Clark University, 950 Main Street, Worcester, MA 01610-1477; 508-793-7526; *http://www.idrisi.clarku.edu/*)

(Lemmens, 2004, provides a point-by-point comparison of image-processing systems designed for remote sensing applications.) IDRISI was designed for use on PCs; the others listed here were designed for use on workstations, but often have PC versions available. The specific systems listed here are general-purpose image-processing systems; others have been designed specifically to address requirements for specific kinds of analysis (e.g., geology, hydrology). Some of the general-purpose systems have added optional modules that focus on more specific topics. Further details of image analysis systems are given by user manuals or help files for specific systems. Authors of image-processing systems typically upgrade their systems to add new or improved capabilities, accommodate new equipment, or address additional application areas.

Several image-processing systems are available to the public either without cost or at minimal cost. For students, some of these systems offer respectable capabilities for illustrating the basic capabilities for image processing. A partial list would include:

- MultiSpec: *http://dynamo.ecn.purdue.edu/~biehl/MultiSpec*
- TNTlite: *http://www.microimages.com*
- MicroMSI: *http://www.nga.mil/portal/site/nga01/index.jsp?epi-content=GENERIC&itemID=cdc86591e1b3af00VgnVCMServer23727a95RCRD&beanID=1629630080&viewID=Article*

## 4.8. The Internet

Digital data once were available only through physical transfer of disks or tapes, usually an inconvenient, awkward process. The availability of the Internet has facilitated not only the acquisition of digital data, but also the equally troublesome task of searching indexes and archives to identify the appropriate coverage. The *Internet* is a network of computers connected by the world's telecommunications infrastructure of fiber optic cables, phone lines, microwave relays, and satellite links. This hardware infrastructure is supported by software that employs common conventions for the electronic transmission of data. Thus the Internet consists of *servers,* computers that run programs designed to share their data with other computers by permitting remote access, and *clients,* computers that run programs designed to access servers.

A large part of the power of the Internet derives from the simultaneous use of some computers as both clients and servers by the establishment of temporary links between computers at distant locations holding related data. Internet users, therefore, can access vast amounts of data residing at varied locations as if the data were immediately at hand. Further, because different users can define their own pathways through the network, it can be tailored with the specific requirements of varied users. Because servers are maintained by a vast and growing number of individuals, businesses, universities, and other institutions, the amount and kinds of information available provides an unprecedented resource for students and researchers in the field of remote sensing.

To use these networks users require access to a computer, either an IBM-compatible PC or a Macintosh, equipped with communications software, and a modem or Ethernet card, and access to communications such as a phone line or fiber optic cable. Finally, Internet users require accounts or user identification, such as those issued by university computing centers or commercial networks, known as *Internet service providers* (ISPs).

### *File Transfer Protocol and Telnet*

*File transfer protocol* (FTP) is a program for transferring computer files between two computers connected by the Internet. FTP permits transfer of either text or binary files. Remotely sensed images are almost always stored as binary files, although they may be accompanied by text files that convey important information about the binary images.

FTP usually requires that a user know a user ID and password for the other ("target") computer, as well as those for the user's own computer. FTP is a powerful tool for practitioners of remote sensing because of the convenience it offers in transferring large images from one computer to another—even those separated by large distances. Equally valuable, however, is the access that FTP provides to archives of free, public-domain software and imagery. Some computers permit outside users to log on without a user ID or password using a convention called *anonymous FTP*. When prompted for a user ID, the analyst enters "anonymous"; when prompted for the password, the user enters his or her own Internet email address.

Whereas FTP permits users to transfer data to or from a remote computer, TELNET permits users to log on to other computers as if they were its owners. Use of TELNET therefore requires knowledge of the user ID and password and the permission of the owner. Exceptions include some public computers, such as those of some libraries or governmental agencies, which may publish the user ID and password, or may configure the account so that a password is not required. (With the development of the World Wide Web, as described below, FTP and TELNET have often been incorporated into web browsers so that they function behind the scenes, without requiring users to explicitly evoke these functions.)

### *The World Wide Web*

The World Wide Web (WWW) is a system for organizing information on the Internet. Internet browsers. Although many browsers are available, the best-known example is probably Internet Explorer (Microsoft Corporation, Redmond, Washington), permit users to access information on the WWW using the computer mouse to point and click to select topics or icons presented on the screen.

The WWW was proposed in 1989 at the European Laboratory for Particle Physics (CERN), in Geneva, Switzerland, the home of the WWW. Key features of the WWW are rules that standardize communications between computers so that users can easily access information at varied locations around the world without explicitly changing from one computer to the other. Users access the WWW using a specialized program called a browser, which permits convenient access to remote sites and rapid transfer of text and graphics over the Internet.

Users of these programs view images, transmitted from the remote computer, which resemble the cover of a magazine, with images and lists of topics. This image forms the home page for a specific computer, or set of information on a specific computer. The home page is both an index and an advertisement for the information available at that site. This information is a form of *hypertext* in which highlighted words or phrases are linked to additional text that explains or elaborates on the topic identified by the highlighted expression. (Data prepared for access in this form are prepared using *hypertext markup language* [HTML].) By employing the mouse to select highlighted items, the user can access the more detailed information. These cross-references may identify information resident on computers at completely different locations, perhaps separated by thousands of miles (although necessarily linked by the Internet). The browser enables the user to access information at diverse sites, but with a speed and convenience that presents the impression of having the information immediately at hand locally. These browsers extend the reach of FTP and related systems by permitting users to receive graphic information (maps, images, photographs, and diagrams) in addition to text and data. Each web page includes *metatags*—HTML that reports information about the content of the web page, its creator, and its administration.

### *Search Engines*

Search engines are specialized Internet programs that can search registries for millions of Internet sites to locate sites of interest to specific users based upon keywords supplied by the user. Because of the immense size of the Internet, its frequent changes, and the absence of a central index, search engines have become an essential component of the WWW. Services such as Google™, Lycos®, or Excite compete for advertising customers by providing the fastest, most efficient, searches based upon the effectiveness of their search software, their capacity to store large indexes, and their ability to update them frequently.

Search engines can employ any of three alternative designs. Some are based upon use of *spiders,* programs that can reach out to web sites, read their metatags, and report a summary back to the search engine, which records the results in a massive database, which is indexed for rapid access. Other search engines depend upon submissions by web page creators, and still others use a combination of the two strategies. A user who queries a search engine is searching the index that a search engine has created. Search engines that report inactive links have not updated their indexes recently, so they include inactive sites in their search results. Typically, each search engine uses its own proprietary software to form the index, to search the index, and to optimize each search for speed and completeness.

### *Image Formats*

The abundance of images on the WWW has increased the use of image formats employed to transmit digital imagery. A *tagged image file format* (TIFF) image consists of (1) an image file header; (2) an image file directory (IFD)—information describing the image, and pointers to the actual image data; and (3) the image data, represented as a sequence of 8-bit bytes, organized in a raster structure, in a specified number of rows and columns. TIFF images can be represented in any of four forms: (1) *bi-level* (black and white) (2) *grayscale,* (3) *palette-color,* or (4) *RGB full-color.* The TIFF structure is complex, but it permits specification of a range of image characteristics, including the actual physical size of the image, its resolution, and the software application that generated the image.

Although the TIFF structure has become a widely used raster file format, it has been limited in applications relating to cartography and remote sensing imagery because it does not provide an effective means of conveying information relating to geographic location. The *GeoTIFF* format uses a portion of the usual TIFF identifying information to record geographic location, thereby enabling geographic applications of the image data. GeoTIFF supports projections such as UTM, U.S. State Plane and national grids, and basic cartographic projections, such as the Universal Transverse Mercator and Lambert Conformal Conic.

The *graphics interchange format* (GIF) was designed to facilitate convenient exchange of digital graphics data, especially using the WWW. A GIF image is composed of 2, 4, 8, 16, 32, 64, 128, or 256 colors, stored in a color palette within the image file. Each color in the palette is described as a mixture of the red, green, and blue (RGB) primaries, each ranging between 0 and 255. Therefore, a GIF image has access to over 16.8 million colors, although only a maximum of 256 colors can be used for a specific GIF image.

These formats have facilitated the transfer of images over the WWW and in digital documents. Although they are convenient for representing remotely sensed images, it should be emphasized that they do not convey the full range of remotely sensed data that constitute each image, but only RGB (or black-and-white) representations of these images. Therefore, the data they convey permits visual representation of an image, but does not support the analytical examination that is essential for the practice of remote sensing.

A large number of public-domain data sets are available over the WWW. WWW addresses are known as *uniform resource locators* (URLs), with the distinctive format of those given in Table 4.3, a short list of URLs of interest to students of remote sensing. Any list of Internet addresses is quickly outdated as new sites are added and older addresses are changed or removed altogether. Most users prepare their own lists of useful URLs, which they regularly update as they find new addresses on the Internet itself. Internet users should note some of its unwritten practices. Not all sites are open to remote users throughout the day and some sites restrict access during peak demand times, so it may be necessary to access some sites during evening or weekend hours. Before downloading data, be sure to verify that you have sufficient

**TABLE 4.3. Some WWW Addresses of Interest to Remote Sensing Students**

| | |
|---|---|
| http://gis.leica-geosystems.com/default.aspx | Leica Geosystems |
| http://www.jpl.nasa.gov/ | Jet Propulsion Laboratory |
| http://earth.google.com/ | Google Earth |
| http://www.noaa.gov/ | National Oceanic and Atmospheric Administration |
| http://www.esa.int/esaCP/index.html | European Space Agency |
| http://www.ccrs.nrcan.gc.ca/ | Canada Centre for Remote Sensing |
| http://daac.gsfc.nasa.gov/ | NASA DAAC (Distributed Active Achieve Centers) |
| http://www.ulrmc.org.ua/ | Ukrainian Land and Resources Management Centre |
| http://www.calmit.unl.edu/ | Center for Advanced Land Management Information Technologies |
| http://www.usgs.gov/ | U.S. Geological Survey |
| http://mapping.usgs.gov/ | USGS EROS Data Center |
| http://www.cast.uark.edu/ | The Center for Advanced Spatial Technologies, University of Arkansas |
| http://www.sandia.gov/RADAR/sar.html | Sandia National Laboratories, Synthetic Aperture Radar |
| http://nssdc.gsfc.nasa.gov/ | National Space Science Data Center |
| http://gs.mdacorporation.com/ | RadarSat International |
| http://www.ngdc.noaa.gov/ | NOAA's National Geophysical Data Center (NGDC) |
| http://www.lib.berkeley.edu/EART/aerial.html | Earth Sciences and Map Library, University of California, Berkeley |
| http://modis.gsfc.nasa.gov/ | MODIS (Moderate Resolution Imaging Spectroradiometer) |
| http://www.remotesensing.org/tiki-index.php | Open source software for remote sensing |
| http://www.asprs.org/ | American Society for Photogrammetry & Remote Sensing |
| http://www.grss-ieee.org/ | IEEE Geoscience and Remote Sensing Society |
| http://www.aag.org/ | Association of American Geographers |
| http://www.isprs.org/ | International Society for Photogrammetry and Remote Sensing |
| http://www.ga.gov.au/acres/ | Australian Centre for Remote Sensing |

*Note.* Check with current sources for up-to-date addresses, as entries on this list are likely to change over time. This list can only be a selection of the many possible entries, so omission of other addresses does not imply that they are not of interest. Abbreviations and acronyms used here for conciseness are explained in later chapters or are evident upon accessing the address.

disk space to receive the requested data, and remember that image files are often very large and require long intervals to download (perhaps hours, if your access is via modem). Executable files (programs) can transmit computer viruses to your computers, so executable files should be screened by a reliable virus-scanning program.

Once data are in hand, either from the Internet or from other sources, analysts must bring the data into a format that can be used by the specific image-processing system in use at a particular laboratory. This step requires knowledge of the file structure (number and sequence of files and header records), format of image data (BIP, BIL, BSQ, etc.), dimensions of images, number of bands, and number of bits assigned to each pixel. Such information is usually provided in the text files that accompany each image or in paper printouts that are provided with tapes or CD-ROMs. Given these data, utility programs within image-processing systems can translate data structures into formats that can be utilized by the image-processing system. Once data have entered the system, users typically are unaware of the file structure employed internally by a given system to organize the bits on the tape or CD-ROM into an image that can be displayed and analyzed.

### *Image Viewers and Online Digital Image Archives*

*Image viewers* (or, sometimes, *map viewers*) are programs designed to provide basic capabilities to view and navigate through digital maps and images. Some image viewers are available commercially; others are available online at minimal cost or as freeware. They provide a convenient means of examining digital maps, GIS data, and aerial imagery. Although most image viewers do not offer analytical capabilities, they do permit users to examine a wide range of spatial data by searching, roaming, magnifying, and applying a variety of projection and coordinate systems. For example, GIS Viewer 4.0 (University of California, Berkeley; *http://elib.cs.berkeley.edu/gis/*) provides an illustration of the basic functions of an image viewer. Image viewers are closely connected to the idea of *digital imagery archives* or *libraries,* which provide collections of digital imagery in standardized formats, such that viewers can easily retrieve and navigate through the collection. Microsoft's® Terraserver (*http://terraserver.homeadvisor.msn.com*) is one of the most comprehensive online archives of digital imagery. It includes digital aerial photographs and maps of large portions of the United States, with an ability to search by place-names and to roam across a landscape. Such systems may well form prototypes for design of more sophisticated image archive systems. *Google Earth* (*http://earth.google.com/earth.html*) provides a comprehensive coverage of images, with the ability to roam and change the perspective of view.

## 4.10. Summary

Image-processing hardware is a means to accomplish an end: the display and analysis of remotely sensed data. In this context, the details of the hardware may seem rather mundane and peripheral to our primary concerns. Yet it is important to recognize that the role of the equipment is much more important, and much more subtle, than we may first appreciate. Characteristics of display and analysis hardware determine in part how we perceive the data, and therefore how we use them. This equipment, and related software, is, in effect, a filter through which we visualize data. We can never view image data without using this filter, because there are so many data, and so many details, that we can never see them directly. So we depend upon image analysis hardware and software to assist us in viewing and understanding the data. The best equipment permits us flexibility as we choose between different levels of radiometric, spatial, and spectral detail, and allows us convenience as we select alternative ways of examining images.

## Review Questions

1. It may be useful to practice conversion of some values from digital to binary form as confirmation that you understand the concepts. Convert the following digital numbers to eight-bit binary values:

   a. 100 c. 24 e. 2 g. 256
   b. 15 d. 31 f. 111 h. 123

2. Convert the following values from binary to digital form:

   a. 10110 c. 10111 e. 0011011
   b. 11100 d. 1110111 f. 1101101

3. Consider the implications of selecting the appropriate number of bits for recording remotely sensed data. One might be tempted to say, "Use a large number of bits to be sure that all values are recorded precisely." What would be the disadvantage of using, for example, seven bits to record data that are accurate only to five bits?

4. Describe in a flow chart or diagram the steps required to read data in a BIP format, then organize them in a BSQ sequence.

5. State the minimum number of bits required to precisely represent the following values:

   a. 1,786 d. 32,000
   b. 32 e. 17
   c. 689 f. 3
   g. 29

6. One of the primary effects of the routine availability of Landsat digital data in the 1970s was the increased availability of image-processing-systems, and especially the development of image-processing systems for microcomputers. One limitation of such systems is that they typically work with subsets of satellite images. Examine a small-scale map of your city or county to develop a plan for selecting a subset that will be satisfactory for studying the region.

7. Would it possible to evaluate a sensor's SNR by visual examination of an image? How?

8. Much of remote sensing analysis depends upon effective display of images. Can you think of ways that contemporary computer displays limit the ability of the analyst to perceive remotely sensed data?

9. Increasingly, remotely sensed images are collected and/or stored in digital formats. Most of the reasons for use of digital formats for recording remotely sensed data are obvious. Can you think

of some disadvantages, or difficulties, to widespread adoption of digital storage of remotely sensed data?

10. This chapter lists only a few of the many image-processing software systems tailored for analysis of remotely sensed data. Can you identify some of the reasons that there should be so many different products intended to achieve the same purpose?

## References

Ball, R., and C. North. 2005. Effects of Tiled High-Resolution Display on Basic Visualization and Navigation Tasks. In *Proceedings, Extended Abstracts of ACM Conference on Human Factors in Computing Systems* (CHI 2005) Portland, Oregon: Association for Computing Machinery, pp. 1196–1199.

Bracken, P. A. 1983. Remote Sensing Software Systems. Chapter 19 in *Manual of Remote Sensing* (R. N. Colwell, ed.). Falls Church, VA: American Society of Photogrammetry, pp. 807–839.

Davidson, D. B., and E. Chen. 1995. A Brief Introduction to the Internet. *Computers and Geosciences,* Vol. 21, pp. 731–735.

Holkenbrink, P. F. 1978. *Manual on Characteristics of Landsat Computer-Compatible Tapes.* Washington, DC: U.S. Government Printing Office, for EROS Data Center Digital Image Processing System, U.S. Geological Survey. (Stock No. 024-001-03116-7, with change 1 August 1979)

Hopper, G. M., and S. L. Mandell. 1984. *Understanding Computers.* New York: West, 490 pp.

Hutchinson, S., and S. C. Sawyer. 1992. *Microcomputers: The User Perspective.* Boston: Irwin, 769 pp.

Jensen, J. R. 1986. *Introductory Digital Image Processing: A Remote Sensing Perspective.* Englewood Cliffs, NJ: Prentice-Hall, 316 pp.

Lemmens, M. 2004. Remote Sensing Processing Software. *GIM International.* Vol 18, pp. 53–57.

Liu, J.-K., H. Wu, and T. Shih. 2005. Effects of JPEG2000 on the Information and Geometry Content of Aerial Photo Compression. *Photogrammetric Engineering and Remote Sensing,* Vol. 71, pp. 157–167.

Merchant, J. W. 1995. A Guide to GIS, Remote Sensing and Other Useful Internet Addresses. *RSSG Newsletter* (Association of American Geographers, Remote Sensing Specialty Group), Vol. 16, No. 2, pp. 13–19.

Nichols, D. 1983. Digital Hardware. Chapter 20 in *Manual of Remote Sensing* (R. N. Colwell, ed.). Falls Church, VA: American Society of Photogrammetry, pp. 841–871.

Rogers, D. F. 1985. *Procedural Elements for Computer Graphics.* New York: McGraw-Hill, 433 pp.

Root, R. R. 1995. Introduction to E-mail and Other Internet Services. *Photogrammetric Engineering and Remote Sensing,* Vol. 61, pp. 875–880.

Thomas, B. J. 1995. *The Internet for Scientists and Engineers: Online Tools and Resources.* Bellingham, WA: SPIE—The International Society for Optical Engineering, 450 pp.

Thomas, V. L. 1975. *Generation and Physical Characteristics of the LANDSAT 1 and 2 MSS Computer Compatible Tapes* (X-563-75-223). Greenbelt, MD: Goddard Space Flight Center.

Woronow, A., and S. Dare. 1995. On the Internet with a PC. *Computers and Geosciences,* Vol. 21, pp. 753–757.

CHAPTER FIVE

# Image Interpretation

## 5.1. Introduction

Earlier chapters have defined our interest in remote sensing as focused primarily on images of the Earth's surface—map-like representations of the Earth's surface based on the reflection of electromagnetic energy from vegetation, soil, water, rocks, and man-made structures. From such images we learn much that cannot be derived from other sources.

Yet such information is not presented to us directly: the information we seek is encoded in the varied tones and textures we see on each image. To translate images into information, we must apply a specialized knowledge, *image interpretation,* which we can apply to derive useful information from the raw uninterpreted images we receive from remote sensing systems. Proficiency in image interpretation is formed from three separate kinds of knowledge, of which only one—the final one listed here—falls within the scope of this text.

### *Subject*

Knowledge of the subject of our interpretation—the kind of information that motivates us to examine the image—is the heart of the interpretation. Accurate interpretation requires familiarity with the subject of the interpretation. For example, interpretation of geological information requires education and experience in the field of geology. Yet narrow specializations are a handicap because each image records a complex mixture of many kinds of information, requiring application of broad knowledge that crosses traditional boundaries between disciplines. For example, accurate interpretation of geological information may require knowledge of botany and the plant sciences as a means of understanding how vegetation patterns on an image reflect geologic patterns that may not be directly visible. As a result, image interpreters should be equipped with a broad range of knowledge pertaining to the subjects at hand and their interrelationships.

### *Geographic Region*

Knowledge of the specific geographic region depicted on an image can be equally significant. Every locality has unique characteristics that influence the patterns recorded on an image. Often the interpreter may have direct experience within the area depicted on the image that can be applied to the interpretation. In unfamiliar regions the interpreter may find it necessary to make a field reconnaissance or to use maps and books that describe analogous regions with similar climate, topography, or kinds of land use.

### *Remote Sensing System*

Finally, knowledge of the *remote sensing system* is obviously essential. The interpreter must understand how each image is formed and how each sensor portrays landscape features. Different instruments use separate portions of the electromagnetic spectrum, operate at different resolutions, and use different methods of recording images. The image interpreter must know how each of these variables influences the image to be interpreted, and how to evaluate their effects on his or her ability to derive useful information from the imagery. This chapter outlines how the image interpreter derives useful information from the complex patterns of tone and texture on each image.

## 5.2. The Context for Image Interpretation

Human beings are well prepared to examine images. Our visual system and our life experiences equip us to discern subtle distinctions in brightness and darkness, to distinguish between various image textures, to perceive depth, and to recognize complex shapes and features. Even in early childhood we apply such skills routinely in everyday experience so that few of us encounter difficulties as we examine, for example, family snapshots or photographs in newspapers. Yet image analyses require a conscious, explicit effort not only to learn about the subject matter, geographic setting, and imaging systems (as mentioned above) in unfamiliar contexts, but also to further develop our innate abilities for image analysis.

Three issues distinguish interpretation of remotely sensed imagery from interpretation conducted in everyday experience. First, remotely sensed images usually portray an *overhead view,* an unfamiliar perspective. Training, study, and experience are required to develop the ability to recognize objects and features from this perspective. Second, many remote sensing images use *radiation outside the visible* portion of the spectrum—in fact, use of such radiation is an important advantage that we exploit as often as possible. Even the most familiar features may appear quite different in nonvisible portions of the spectrum than they do in the familiar world of visible radiation. Third, remote sensing images often portray the Earth's surface at *unfamiliar scales and resolutions.* Commonplace objects and features may assume strange shapes and appearances as scale and resolution change from those to which we are accustomed.

This chapter outlines the art of image interpretation as applied to aerial photography. Students cannot expect to become proficient in image analysis simply by reading about image interpretation. Experience is the only sure preparation for skillful interpretation. Nonetheless, this chapter can highlight some of the issues that form the foundations for proficiency in image analysis.

In order to discuss this subject at an early point in the text, we must confine the discussion to interpretation of aerial photography, the only form of remote sensing imagery discussed thus far. But the principles, procedures, and equipment described here are equally applicable to other kinds of imagery acquired by the sensors described in later chapters. Manual image interpretation is discussed in detail by Paine and Kiser (2003), Avery and Berlin (1992), and Philipson (1996); older references that may also be useful are the text by Lueder (1959) and the *Manual of Photographic Interpretation* (Colwell, 1960).

## 5.3. Image Interpretation Tasks

The image interpreter must routinely conduct several kinds of tasks, many of which may be completed together in an integrated process. Nonetheless, for purposes of clarification, it is important to distinguish between these separate functions (Figure 5.1).

### *Classification*

*Classification* is the assignment of objects, features, or areas to classes based on their appearance on the imagery. Often a distinction is made between three levels of confidence and precision. *Detection* is the determination of the presence or absence of a feature. *Recognition* implies a higher level of knowledge about a feature or object, such that the object can be assigned an identity in a general class or category. Finally, *identification* means that the identity of an object or feature can be specified with enough confidence and detail to place it in a very specific class. Often an interpreter may qualify his or her confidence in an interpretation by specifying the identification as "possible" or "probable."

### *Enumeration*

*Enumeration* is the task of listing or counting discrete items visible on an image. For example, housing units can be classified as "detached single-family home," "multifamily complex,"

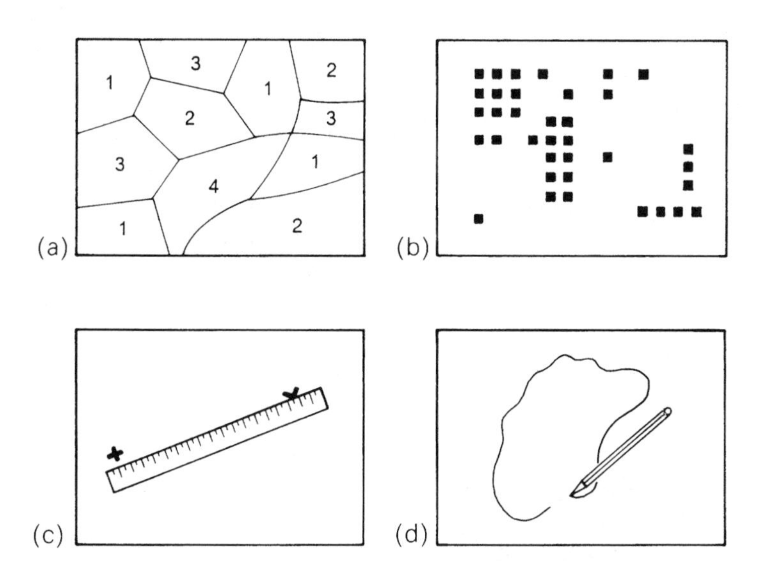


**FIGURE 5.1.** Image interpretation skills. (*a*) Classification. (*b*) Enumeration. (*c*) Mensuration. (*d*) Delineation.

"mobile home," and "multistory residential," and then reported as numbers present within a defined area. Clearly, the ability to conduct such an enumeration depends upon an ability to accurately identify and classify items as discussed above.

### *Measurement*

*Measurement,* or *mensuration,* is an important function in many image interpretation problems. Two kinds of measurement are important. First is the measurement of distance and height, and, by extension, of volumes and areas as well. The practice of making such measurements forms the subject of *photogrammetry* (Chapter 3), which applies knowledge of image geometry to the derivation of accurate distances. Although strictly speaking photogrammetry applies only to measurements from photographs, by extension it has analogs for the derivation of measurements from other kinds of remotely sensed images.

A second form of measurement is quantitative assessment of image brightness. The science of *photometry* is devoted to measurement of the intensity of light, and includes estimation of scene brightness by examination of image tone with the use of special instruments known as *densitometers,* described below. If the measured radiation extends outside the visible spectrum, the term *radiometry* applies. Both photometry and radiometry employ similar instruments and principles, so they are closely related.

### *Delineation*

Finally, the interpreter must often *delineate,* or outline, regions as they are observed on remotely sensed images. The interpreter must be able to separate distinct areal units that are characterized by specific tones and textures, and to identify edges or boundaries between separate areas. Typical examples include delineation of separate classes of forest or of land use—both of which occur only as areal entities (rather than as discrete objects). Typical problems include (1) selection of appropriate levels of generalization (e.g., when boundaries are intricate, or when many tiny but distinct parcels are present); and (2) placement of boundaries when there is a gradation (rather than a sharp edge) between two units.

The image analyst may simultaneously apply several of these skills in examining an image. Recognition, delineation, and mensuration may all be required as the interpreter examines an image. Yet specific interpretation problems may emphasize specialized skills. Military photointerpretation often depends upon accurate recognition and enumeration of specific items of equipment, whereas land-use inventory emphasizes delineation, although other skills are obviously important. Image analysts therefore need to develop proficiency in all of these skills.

## 5.4. Elements of Image Interpretation

By tradition, image interpreters are said to employ some combination of the eight *elements of image interpretation,* which describe characteristics of objects and features as they appear on remotely sensed images. Image interpreters quite clearly use these characteristics together in very complex, but poorly understood, processes as they examine images. Nonetheless, it is convenient to list them separately as a way of emphasizing their significance.

### *Image Tone*

For black-and-white images, *image tone* denotes the lightness or darkness of a region within an image (Figure 5.2). Tone may be characterized as "light," "medium gray," "dark gray," "dark," and so on, as the image assumes varied shades of white, gray, or black. For color or CIR imagery, *image tone* refers simply to "color," described informally perhaps in such terms as "dark green," "light blue," or "pale pink." Image tone refers ultimately to the brightness of an area of ground as portrayed by the film in a given spectral region (or in three spectral regions, for color or CIR film).

Image tone can also be influenced by the intensity and angle of illumination and by the processing of the film. Within a single aerial photograph, vignetting (Section 3.2) may create noticeable differences in image tone due solely to the position of an area within a frame of photography: the image becomes darker near the edges. Thus the interpreter must employ caution in relying solely on image tone for an interpretation, as it can be influenced by factors other than the absolute brightness of the Earth's surface. Analysts should also remember that very dark or very bright regions on an image may be exposed in the nonlinear portion of the characteristic curve (Chapter 3), so they may not be represented in their correct relative brightnesses. Also, nonphotographic sensors may record such a wide range of brightness values that they cannot all be accurately represented on photographic film—in such instances digital analyses (Chapter 4) may be more accurate.

Experiments conducted by Cihlar and Protz (1972) have shown that interpreters tend to be consistent in their interpretation of tones on black-and-white imagery, but less so in interpretation of color imagery. Interpreters' assessment of image tone is much less sensitive to subtle differences in tone than are measurements by instruments (as might be expected). For the range of tones used in the experiments, human interpreters' assessment of tone expressed a linear relationship with corresponding measurements made by instruments. Cihlar and Protz's results imply that a human interpreter can provide reliable estimates of relative differences in tone, but not be capable of accurate description of absolute image brightness.

### *Image Texture*

*Image texture* refers to the apparent roughness or smoothness of an image region. Usually texture is caused by the pattern of highlighted and shadowed areas created when an irregular surface is illuminated from an oblique angle. Contrasting examples (Figure 5.3) include the rough textures of a mature forest and the smooth textures of a mature wheat field. The human interpreter is very good at distinguishing subtle differences in image texture, so it is a valuable aid to interpretation—certainly it is equal in importance to image tone in many circumstances.

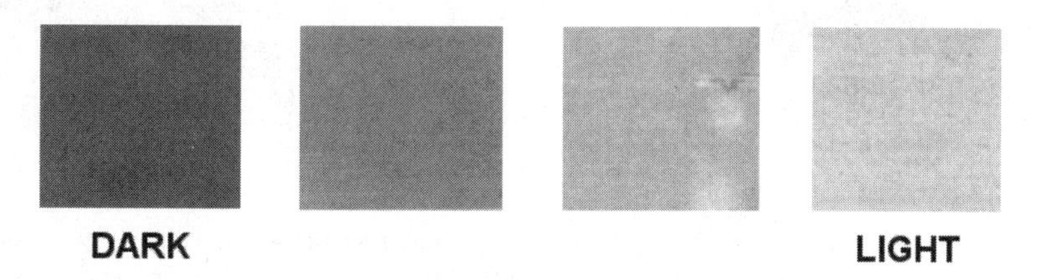


**FIGURE 5.2.** Varied image tones, dark to light (left to right). From USDA.

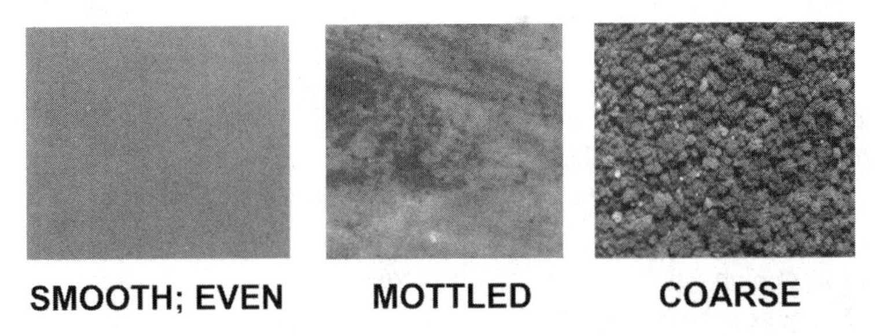


**FIGURE 5.3.** Varied image textures, with descriptive terms. From USDA.

Image texture depends not only upon the surface itself, but also upon the angle of illumination, so it can vary as lighting varies. Also, good rendition of texture depends upon favorable image contrast, so images of poor or marginal quality may lack the distinct textural differences so valuable to the interpreter.

### *Shadow*

*Shadow* is an especially important clue in the interpretation of objects. A building or vehicle, illuminated at an angle, casts a shadow that may reveal characteristics of its size or shape that would not be obvious from the overhead view alone (Figure 5.4). Because military photointerpreters often are primarily interested in identification of individual items of equipment, they have developed methods to use shadows to distinguish subtle differences that might not otherwise be visible. By extension, we can emphasize this role of shadow in interpretation of any man-made landscape in which identification of separate kinds of structures or objects is significant.

Shadow is of great significance also in interpretation of natural phenomena, even though its role may not be as obvious. For example, Figure 5.5 depicts an open field in which scattered shrubs and bushes are separated by areas of open land. Without shadows, the individual plants

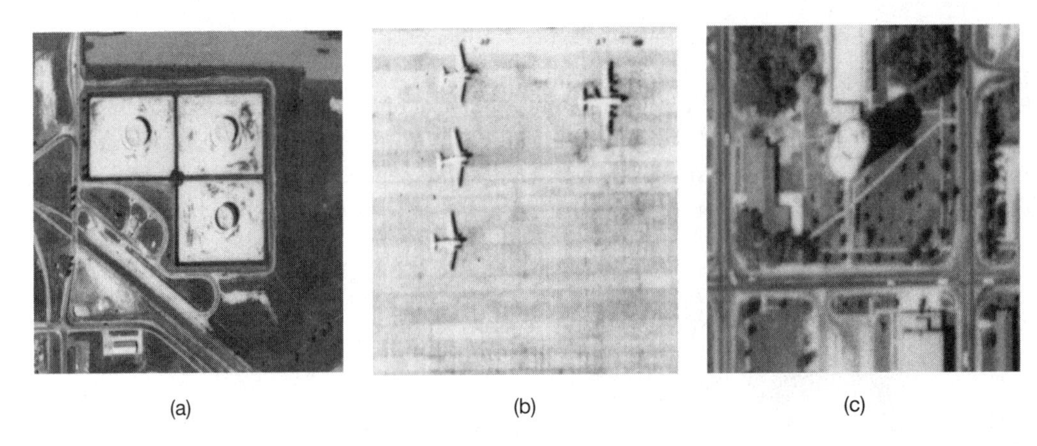


**FIGURE 5.4.** Examples of significance of shadow in image interpretation, as illustrated by (*a*) fuel storage tanks, (*b*) military aircraft on a runway, and (*c*) a water tower. From USDA.

**FIGURE 5.5.** Significance of shadow for image interpretation, as illustrated by the characteristic pattern caused by shadows of shrubs cast on open field, and shadows at the edge of a forest enhances the boundary between two different land covers. From USDA.

might be too small (as seen from above) and so nearly similar in tone to their background to be visible. Yet their shadows are large enough, and dark enough, to create the streaked pattern on the imagery typical of this kind of land. A second example is also visible in Figure 5.5: at the edges between the trees in the hedgerows and the adjacent open land, trees cast shadows that form a dark strip that enhances the boundary between the zones as seen on the imagery.

### *Pattern*

*Pattern* refers to the arrangement of individual objects into distinctive recurring forms that facilitate their recognition on aerial imagery (Figure 5.6). Pattern on an image usually follows from a functional relationship between the individual features that compose the pattern. Thus the buildings in an industrial plant may have a distinctive pattern due to their organization to permit economical flow of materials through the plant from receiving raw material to shipping of the finished product. The distinctive spacing of trees in an orchard arises from careful planti-

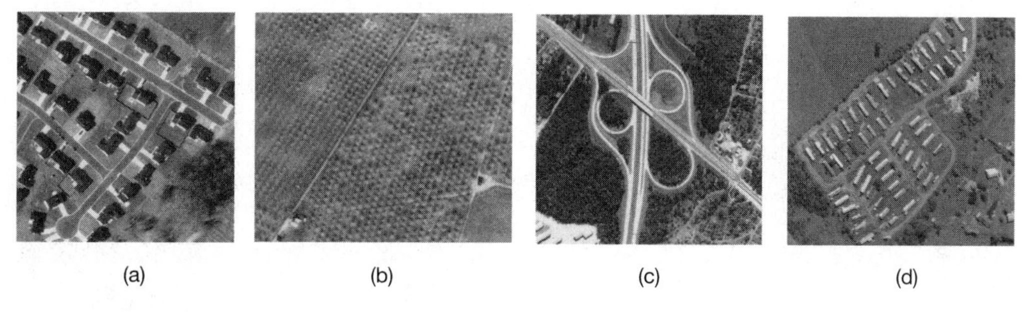


**FIGURE 5.6.** Significance of distinctive image pattern, as illustrated by (*a*) structures in a suburban residential neighborhood, (*b*) an orchard, (*c*) a highway interchange, and (*d*) a rural trailer park. From: VDOT (*a, d*); USDA (*b*); USGS (*c*).

ng of trees at intervals that prevent competition between individual trees and permits convenient movement of equipment through the orchard.

### *Association*

*Association* specifies the occurrence of certain objects or features, usually without the strict spatial arrangement implied by pattern. In the context of military photointerpretation, association of specific items has great significance, as, for example, when the identification of a specific class of equipment implies that other, more important, items are likely to be found nearby.

### *Shape*

*Shapes* of features are obvious clues to their identities (Figure 5.7). For example, individual structures and vehicles have characteristics shapes, which, if visible in sufficient detail, provide the basis for identification. Features in nature often have such distinctive shapes that shape alone might be sufficient to provide clear identification. For example, ponds, lakes, and rivers occur in specific shapes unlike others found in nature. Often specific agricultural crops tend to be planted in fields that have characteristic shapes (perhaps related to the constraints of the kinds of equipment used or the method of irrigation that the farmer employs).

### *Size*

*Size* is important in two ways. First, the relative size of an object or feature in relation to other objects on the image provides the interpreter with an intuitive notion of its scale and resolution, even though no measurements or calculations may have been made. This intuition is achieved via recognition of familiar objects (dwellings, highways, rivers, etc.), and extrapolation to relate the sizes of these known features to estimate the sizes and identities of those objects that are not so easily recognized. This is probably the most direct and important function of size.

Second, absolute measurements can be equally valuable as interpretation aids. Measure-

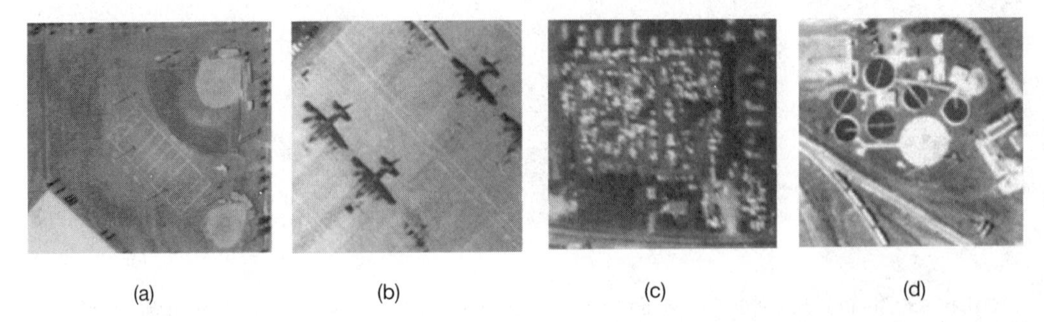


**FIGURE 5.7.** Significance of shape for image interpretation, as illustrated by (*a*) athletic fields, (*b*) aircraft parked on a runway, (*c*) automobiles in a salvage yard, and (*d*) a water treatment plant. From USDA.

ments of the size of an object can confirm its identification based upon other factors, especially if its dimensions are so distinctive that they form definitive criteria for specific items or classes of items. Furthermore, absolute measurements permit derivation of quantitative information, including lengths, volumes, or (sometimes) even rates of movement (e.g., of vehicles or ocean waves as they are shown in successive photographs).

### *Site*

*Site* refers to topographic position. For example, sewage treatment facilities are positioned at low topographic sites near streams or rivers to collect waste flowing through the system from higher locations. Orchards may be positioned at characteristic topographic sites—often on hill-sides (to avoid cold air drainage to low-lying areas) or near large water bodies (to exploit cooler spring temperatures near large lakes to prevent early blossoming).

## 5.5. Image Interpretation Strategies

An *image interpretation strategy* can be defined as a disciplined procedure that enables the interpreter to relate geographic patterns on the ground to their appearance on the image. Campbell (1978) defined five categories of image interpretation strategies, discussed below.

### *Field Observations*

*Field observations* are required when the image and its relationship to ground conditions are so imperfectly understood that the interpreter is forced to go to the field to make identification. In effect, the analyst is unable to interpret the image from his or her own knowledge and experience, and therefore must gather field observations to ascertain the relationship between the landscape and its appearance on the image. Field observations are, of course, a routine dimension to any interpretation as a check on accuracy or a means of familiarization with a specific region. Here their use as an interpretation strategy refers to the fact that when they are required for the interpretation, their use reflects a rudimentary understanding of the manner in which a landscape is depicted on a specific image.

### *Direct Recognition*

*Direct recognition* is the application of an interpreter's experience, skill, and judgment to associate the image patterns with informational classes. The process is essentially a qualitative subjective analysis of the image using the elements of image interpretation as visual and logical clues. In everyday experience direct recognition is applied in an intuitive manner; for image analysis, it must be a disciplined process, involving a very careful systematic examination of the image.

### *Interpretation by Inference*

*Interpretation by inference* is the use of a visible distribution to map one that is not itself visible on the image. The visible distribution acts as a surrogate, or proxy (i.e., a substitute), for the mapped distribution. For example, soils are defined by vertical profiles that cannot be directly observed by remotely sensed imagery. But soil distributions are sometimes very closely related to patterns of landforms and vegetation that are recorded on the image. Thus they can be surrogates for the soil pattern; the interpreter infers the invisible soil distribution from other patterns that are visible. Application of this strategy requires a complete knowledge of the link between the proxy and the mapped distribution; attempts to apply imperfectly defined proxies produce inaccurate interpretations.

### *Probabilistic Interpretation*

*Probabilistic interpretations* are efforts to narrow the range of possible interpretations by formally integrating nonimage information into the classification process, often by means of quantitative classification algorithms. For example, knowledge of the crop calendar can restrict the likely choices when identifying the crops of a specific region. For example, if it is known that winter wheat is harvested in June, one's choice of possible crops for interpretation of an image from the month of August can be restricted to eliminate wheat as a likely choice, and thereby avoid a potential classification error. Often such knowledge can be expressed as a statement of probability. Possibly certain classes might favor specific topographic sites, but occur over a range of sites, so a decision rule might express this knowledge as a 0.90 probability of finding the class on a well-drained site, but only a 0.05 probability of finding it on a poorly drained site. Several such statements systematically incorporated into the decision-making process can improve classification accuracy.

### *Deterministic Interpretation*

A fifth strategy for image interpretation is deterministic interpretation, the most rigorous and precise approach to image interpretation. *Deterministic interpretations* are based on quantitatively expressed relationships that tie image characteristics to ground conditions. In contrast with the other methods, most information is derived from the image itself. Photogrammetric analysis of stereo pairs for terrain information is a good example. Suppose a scene is imaged from two separate positions along a flight path. The photogrammetrist can measure the apparent displacement. Based upon his or her knowledge of the geometry of the photographic system, he or she can reconstruct a topographic model of the landscape. The result is therefore the derivation of precise information about the landscape using only the image itself and a knowledge of its geometric relationship with the landscape. Relative to the other methods, very little nonimage information is required.

Image interpreters, of course, may apply a mixture of several strategies in a given situation. For example, interpretation of soil patterns may require direct recognition to identify specific classes of vegetation, then application of interpretation by inference to relate the vegetation pattern to the underlying soil pattern.

## 5.6. Collateral Information

*Collateral,* or ancillary, *information* refers to nonimage information used to assist in the interpretation of an image. Actually, all image interpretations use collateral information in the form of the implicit, often intuitive, knowledge that every interpreter brings to an interpretation in the form of everyday experience and formal training. In its narrower meaning, it refers instead to the explicit, conscious effort to employ maps, statistics, and similar material to aid in analysis of an image. In the context of image interpretation, use of collateral information is permissible, and certainly desirable, provided two conditions are satisfied. First, the use of such information is to be explicitly acknowledged in the written report; and second, the information must not be focused upon a single portion of the image or map to the extent that it produces uneven detail or accuracy in the final map. For example, it would be inappropriate for an interpreter to focus upon acquiring detailed knowledge of tobacco farming in an area of mixed agriculture if he or she then produced highly detailed, accurate delineations of tobacco fields, but mapped other fields at lesser detail or accuracy.

Collateral information can consist of information from books, maps, statistical tables, field observations, or other sources. Written material may pertain to the specific geographic area under examination, or, if such material is unavailable, it may be appropriate to search for information pertaining to analogous areas—similar geographic regions (possibly quite distant from the area of interest) characterized by comparable ecology, soils, landforms, climate, or vegetation.

## 5.7. Imagery Interpretability Rating Scales

Remote sensing imagery can vary greatly in quality due both to environmental and to technical conditions influencing acquisition of the data. In the United States, some governmental agencies use rating scales to evaluate the suitability of imagery for specific purposes. The National Imagery Interpretability Rating Scale (NIIRS) has been developed for single-channel and panchromatic imagery, and the Multispectral Imagery Interpretability Rating Scale (MS IIRS) (Erdman et al., 1994) has been developed for multispectral imagery. Such scales are based on evaluations using a large number of experienced interpreters to independently evaluate images of varied natural and man-made features, as recorded by images of varying characteristics. They provide a guide for evaluation if a specific form of imagery is likely to be satisfactory for specific purposes.

## 5.8. Image Interpretation Keys

Image interpretation keys are valuable aids for summarizing complex information portrayed as images. They have been widely used for image interpretation (e.g., Coiner and Morain, 1972). Such keys serve either or both of two purposes: (1) they are a means of training inexperienced personnel in the interpretation of complex or unfamiliar topics, and (2) they are a reference aid for experienced interpreters to organize information and examples pertaining to specific topics.

An *image interpretation* key is simply reference material designed to permit rapid and accurate identification of objects or features represented on aerial images. A key usually consists of two parts: (1) a collection of annotated or captioned images or stereograms, and (2) a graphic or word description, possibly including sketches or diagrams. These materials are organized in a systematic manner that permits retrieval of desired images by, for example, date, season, region, or subject.

Keys of various forms have been used for many years in the biological sciences, especially botany and zoology. These disciplines rely upon complex taxonomic systems that are so extensive that even experts cannot master the entire body of knowledge. The key therefore is a means of organizing the essential characteristics of a topic in an orderly manner. It must be noted that scientific keys of all forms require a basic familiarity with the subject matter. A key is not a substitute for experience and knowledge, but a means of systematically ordering information so that an informed user can learn it quickly.

Keys were first routinely applied to aerial images during World War II, when it was necessary to train large numbers of inexperienced photointerpreters in the identification of equipment of foreign manufacture and in the analysis of regions far removed from the experience of most interpreters. The interpretation key formed an effective way of organizing and presenting the expert knowledge of a few individuals. After the war ended, interpretation keys were applied to many other subjects, including agriculture, forestry, soils, and landforms. Their use has since been extended from aerial photography to other forms of remotely sensed imagery. Today interpretation keys are still used for instruction and training, but they may have somewhat wider use as reference aids. Also, it is true that construction of a key tends to sharpen one's interpretation skills and encourages the interpreter to think more clearly about the interpretation process.

Keys designed solely for use by experts are referred to as *technical keys*. *Nontechnical keys* are those designed for use by those with a lower level of expertise. Often it is more useful to classify keys by their formats and organizations. *Essay keys* consist of extensive written descriptions, usually with annotated images as illustrations. A *file key* is essentially a personal image file with notes; its completeness reflects the interests and knowledge of the compiler. Its content and organization suit the needs of the compiler, so it may not be organized in a manner suitable for use by others.

## 5.9. Interpretive Overlays

Often in resource-oriented interpretations it is necessary to search for complex associations of several related factors that together define the distribution or pattern of interest. For example, often soil patterns may be revealed by distinctive relationships between separate patterns of vegetation, slope, and drainage. The *interpretive overlays* approach to image interpretation is a way of deriving information from complex interrelationships between separate distributions recorded on remotely sensed images. The correspondence between several separate patterns may reveal other patterns not directly visible on the image (Figure 5.8).

The method is applied by means of a series of individual overlays for each image to be examined. The first overlay might show the major classes of vegetation, perhaps consisting of dense forest, open forest, grassland, and wetlands. A second overlay maps slope classes, including perhaps level, gently sloping, and steep slopes. Another shows the drainage pattern, and still others might show land use and geology. Thus, for each image, the interpreter may have as many as five or six overlays, each depicting a separate pattern. By superimposing these overlays, the interpreter can derive information presented by the coincidence of several patterns.

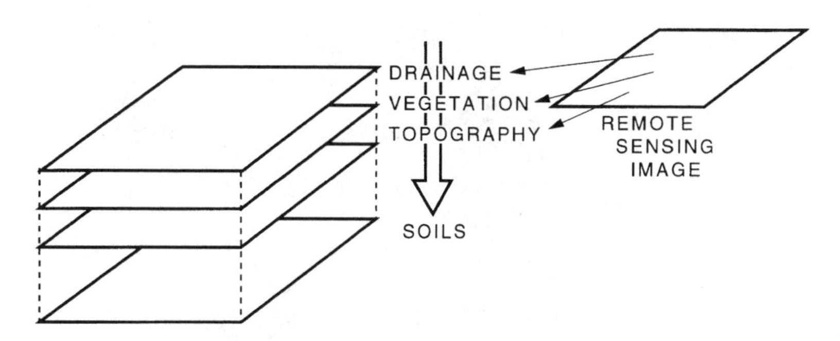


**FIGURE 5.8.** Interpretive overlays permit the analyst to extract and then combine information from several themes.

From his or her knowledge of the local terrain, the interpreter may know that certain soil conditions can be expected where the steep slopes and the dense forest are found together, and that others are expected where the dense forest matches to the gentle slopes. From the information presented by several patterns, the interpreter can resolve information not conveyed by any single pattern.

## 5.10. Photomorphic Regions

Another approach to interpretation of complex patterns is the search for *photomorphic regions,* regions of uniform appearance on the image. The interpreter does not attempt to resolve the individual components within the landscape (as he or she does when using interpretive overlays), but instead looks for their combined influence on image pattern (Figure 5.9). For this reason, application of photomorphic regions may be most valid for small-scale imagery where the coarse resolution tends to average together the separate components of the landscape. "Photomorphic regions," then, are simply "image regions" of relatively uniform tone and texture.

In the first step, the interpreter delineates regions of uniform image appearance, using tone, texture, shadow, and the other elements of image interpretation as a means of separating regions. In some instances, interpreters have used densitometers to quantitatively measure image tone and variation in image tone as an aid to more subjective interpretation techniques (Nunnally, 1969).

In the second step, the interpreter matches photomorphic regions to useful classes of interest to the interpreter. For example, the interpreter must determine if specific photomorphic regions match to vegetation classes. This step obviously requires field observations or collateral information because regions cannot be identified by image information alone. As the interpretation is refined, the analyst may combine some photomorphic regions or subdivide others.

Delineation of photomorphic regions is a powerful interpretation tool, but one that must be applied with caution. Photomorphic regions do not always correspond neatly to the categories of interest to the interpreter. The appearance of one region may be dominated by factors related to geology and topography, whereas that of another region on the same image may be controlled by the vegetation pattern. And the image appearance of a third region may be the result of the interaction of several other factors.

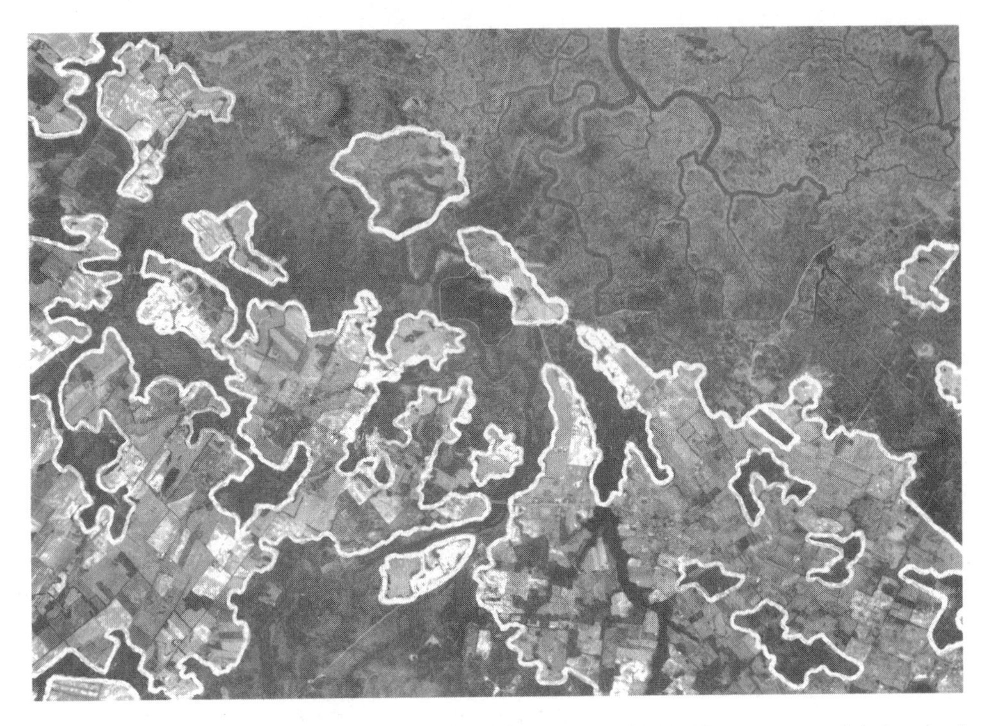

**FIGURE 5.9.** Photomorphic regions outline broad-scale regions of uniform appearance on aerial photographs.

## 5.11. Significance of Context

In Chapter 1 the discussion of Figure 1.1 introduced the significance of *context* in deriving meaning from an image. That is, a purely visual understanding of an image does not necessarily lead to an understanding of its underlying meaning. This topic deserves further exploration in the context of image interpretation.

Most of us are familiar with the kind of visual illusion illustrated in Figure 5.10, which illustrates the *Rubin illusion,* in which the viewer sees either a white vase against a black background or two faces in silhouette facing each other against a white background. The success of the illusion depends upon its ability to confuse the viewer's capacity to assess the *figure–ground relationship*. To make visual sense of an image, our visual system must decide which part of a scene is the *figure* (the feature of interest) and which the *ground* (the background that simply outlines the figure).

Normally, our visual system expects the background to constitute the larger proportion of a scene. The Rubin illusion, like most visual illusions, is effective because it is contrived to isolate the viewer's perception of the scene—in this instance, by designing the illustration so that figure and ground constitute equal proportions of the scene. The viewer's visual system cannot resolve the ambiguity, so the viewer experiences difficulty in interpreting the meaning of the scene.

Although such contrived images are not encountered in day-to-day practice, the principles that they illustrate apply to situations that are frequently encountered. For example, *relief inversion* occurs when aerial images of shadowed terrain are oriented in a manner that confuses our intuitive expectations. Normally, we expect to see terrain illuminated from the upper right (Fig-

**FIGURE 5.10.** Rubin face/vase illusion.

ure 5.11, left); most observers see such images in their correct relief. If the image is oriented so the illumination appears to originate from the lower right, most observers tend to perceive the relief as inverted (Figure 5.11, right). Experimentation with conditions that favor this effect confirms the belief that, like most illusions, relief inversion is perceived only when the context has confined the viewer's perspective to present an ambiguous visual situation.

Image analysts encounter many situations in which visual ambiguities can invite misleading or erroneous interpretations. When NASA analysts examined the 1976 Viking Orbiter images of the Cydonia region of Mars, they noticed, with some amusement, the superficial resemblance of an imaged feature to a humanoid face (Figure 5.12, left). Once the images were released to the public, however, the issue became a minor sensation, as many believed the images conveyed clear evidence of intelligent design, and openly speculated about the origin and meaning of a feature that appeared to have such visually obvious significance. In 1998 the same region of Mars was imaged again, this time by the Mars Global Surveyor, a spacecraft with instruments that provided images with much higher spatial resolution (Figure 5.12, right). These images reveal that the region in question does not in fact offer a striking resemblance to a face.

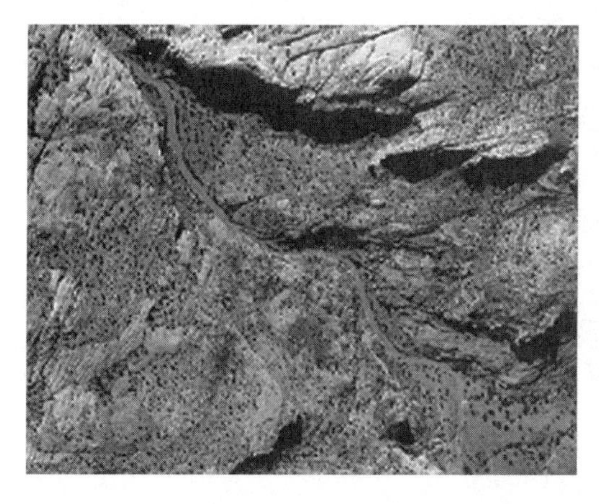

**FIGURE 5.11.** Photographs of landscapes with pronounced shadowing are usually perceived in correct relief when shadows fall toward the observer. Left: When shadows fall toward the observer, relief is correctly perceived. Right: When the image is rotated so that shadows fall in the opposite direction, away from the observer, topographic relief appears to be reversed. From USGS.

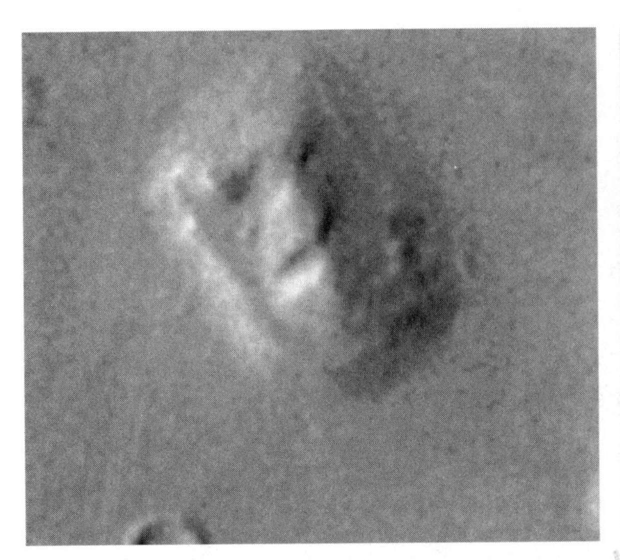

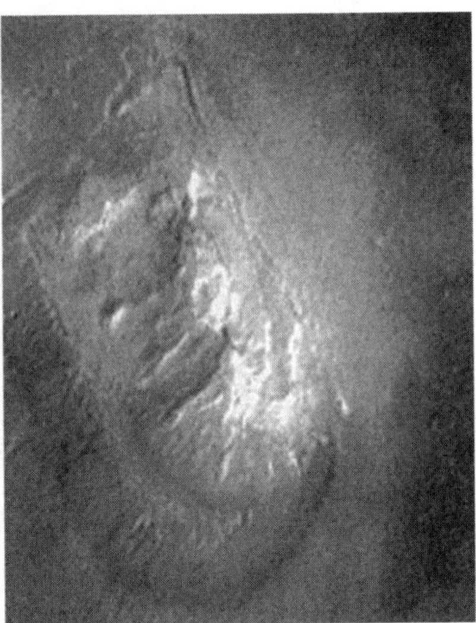

**FIGURE 5.12.** Mars face illusion. Left: This 1976 Viking Orbiter image of the Cydonia region of Mars was considered by some to present a strong resemblance to a human face, causing speculation that the feature was created by intelligent beings. From NASA. Right: In 1998 the Mars Global Surveyor re-imaged the Cydonia region at much higher spatial resolution. With greater spatial detail available to the eye, the previous features are seen to be interesting, but without any convincing resemblance to an artificial structure. From NASA.

Photointerpreters should remember that that the human visual system has a powerful drive to impose its own interpretation upon the neurological signals it receives from the eye, and can easily create plausible interpretations of images when the evidence is uncertain, confused, or absent. Image analysts must strive always to establish *several independent lines of evidence and reasoning* to set the context that establishes the meaning of an image. When several lines of evidence and reasoning converge, then an interpretation can carry authority and credibility. When multiple lines of evidence and reasoning do not converge, or are absent, then the interpretation must be regarded with caution and suspicion.

Image interpretation's successes illustrate the significance of establishing the proper context for understanding the meaning of an image. The use of photointerpretation to monitor the development and deployment of the German V-1 and V-2 missiles during World War II (Babbington-Smith, 1957; Irving, 1964), and to discuss the deployment of Soviet missiles in Cuba during the 1962 Cuban Missile Crisis (Brugioni, 1991) was successful because it provided information that could be examined and evaluated in a broader context. Image interpretation proved to be less successful in February 2003 when U.S. Secretary of State Colin Powell presented images to the United Nations to document the case for an active threat from weapons of mass destruction in Iraq; later it became quite clear that there was insufficient information at hand to establish the proper meaning of those images.

## 5.12. Image Interpretation Equipment

Image interpretation can often be conducted with relatively simple, inexpensive equipment, although some tasks may require expensive items. Typically, an image interpretation laboratory is equipped for storage and handling of images both as paper prints and as film transparencies. Paper prints are most frequently 9 in. × 9 in. contact prints, often stored in sequence in a standard file cabinet. Larger prints and indices must be stored flat in a map cabinet. Transparencies are available as individual 9 in. × 9 in. frames; often they are stored as long rolls of film wound on spools and sealed in canisters to protect them from dust and moisture.

### *Light Tables*

A *light table* is a translucent surface illuminated from behind to permit convenient viewing of film transparencies. In its simplest form, the light table is a box-like frame with a frosted glass surface (Figure 5.13). The viewing area can range in size from desk-size, to briefcase-size. If roll film is to be used, light tables must be equipped with special brackets to hold the film spools and rollers at the edges to permit the film to move freely without damage. More elaborate models have dimmer switches to control the intensity of the lighting, high-quality lamps to control the spectral properties of the illumination, and sometimes power drives to wind and unwind long spools of film.

### *Measurement of Length*

Ordinary household rulers are not satisfactory for image interpretation. Analysts should use an engineer's scale or a ruler with accurate graduations. Both SI units (to at least 1 mm) and English units (to at least 1/20 in.) are desirable. It is convenient for both measurement and calculation if English units are subdivided into decimal divisions.

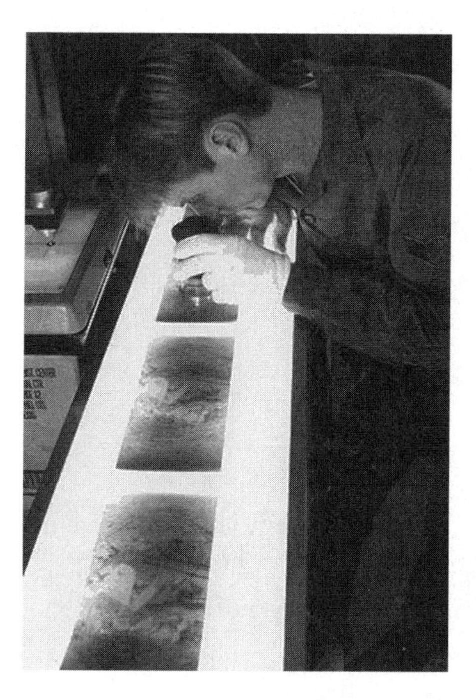

**FIGURE 5.13.** A light table provides a translucent surface that enables the analyst to examine image transparencies in detail. From SSGT Scott Stewart, Department of Defense, DFST8712359.

### *Measurement of Area*

Areas on maps or remote sensing imagery can be measured using any of several techniques. At one time, the *dot grid* was the standard technique for measuring areas. An array of equally spaced dots, printed on a transparent overlay, was superimposed over the area to be measured. The analyst could count the number of dots within a delineated area, and then apply a formula that related the density of dots on the grid, the number of dots counted, and the scale of the image to estimate the area in question. If applied properly, the dot grid was simple, inexpensive, and reasonably accurate under most circumstances. However, this method is time-consuming and impractical for complex maps.

The *polar planimeter* (Figure 5.14) is a compact instrument with a moveable arm that can be used to trace the outline of an area; a dial at the base of the arm records the area outlined by the perimeter, usually as the map area (e.g., in square centimeters or square inches), which can then be converted to ground area. Planimeters are simple and reliable. However, accuracy requires the averaging of results from several repetitions, so this method is cumbersome when complex maps must be analyzed.

Today most analysts measure areas using some form of electronic digitizer (Chapter 16), which records an electronic version of the outline traced by the analyst. From this electronic record, a microprocessor can compute areas and apply corrections for image scale. A special, less expensive, version of the electronic digitizer is the *electronic planimeter.* The analyst can trace outlines with a movable arm. Coordinates, areas, and distances can be read from the display, or transferred to a computer if desired. A keyboard permits the analyst to enter commands to the microprocessor and to enter identifying codes as areas are digitized.

### *Stereoscopes*

*Stereoscopes* are devices that facilitate stereoscopic viewing of aerial photographs (to be discussed further in Section 5.13). The simplest and most common is the *pocket stereoscope* (Fig-

**FIGURE 5.14.** A planimeter is a mechanical device used for measuring areas on maps and aerial photographs. The dot grid is a simple means for estimating area. From Forest History Society, Durham, NC.

ure 5.15). Its compact size and inexpensive cost make it one of the most widely used remote sensing instruments. The pocket stereoscope consists of a body holding two low-power lenses attached to a set of collapsible legs that can be folded so that the entire instrument can be stored in a space a bit larger than a deck of playing cards. The body is usually formed from two separate pieces, each holding one of the two lenses, which can be adjusted to control the spacing between the two lenses to accommodate the individual user.

Other kinds of stereoscopes include the *mirror stereoscope* (Figure 5.16), which permits stereoscopic viewing of large areas, usually at low magnification, and the *binocular stereoscope* (Figure 5.17), designed primarily for viewing transparencies on light tables. Often the binocular stereoscope has adjustable magnification that enables enlargement of portions of the image up to 20 or 40 times.

The stereoscope is only one of several devices designed to present separate images intended for each eye to create the stereo effect. Its way of doing this is known as the *optical separation* technique. Left and right images are presented side by side and an optical device is used to separate the analyst's view of the left and right images. The *red/blue anaglyph* presents images intended for each eye in separate colors, reds for the right eye, blues for the left eye, and shades of magenta for those portions of the image common to both eyes. The analyst views the image using special glasses with a red lens for the left eye and a blue lens for the left eye. The colored lenses cause the image intended for the other eye to blend into the background; the image intended for its own eye will appear as black. The anaglyph has been widely used for novelties, less often as an analytical device.

The use of *polarized lenses* for stereovision is based on the projection of images for each eye through separate polarizing filters (e.g., horizontal for the left eye, vertical for the right eye). The combined image must be viewed through special glasses that use orthogonal polarizations for the left and right lenses. This technique is one of the most effective means of stereoviewing for instructional and analytical applications. A proprietary variation of this technique (CrystalEyes®) displays left- and right-eye views of a digital image in sequential refresh scans on a monitor, then uses synchronized polarized shutter glasses to channel the correct image to the correct eye. This technique forms the basis for stereographic images in many virtual reality dis-

**FIGURE 5.15.** A USGS geologist uses a pocket stereoscope to examine vertical aerial photography, 1957. From USGS Photographic Library: photograph by E. F. Patterson, No. 22.

**FIGURE 5.16.** Image interpretation equipment, Korean conflict, March 1952. A U.S. Air Force image interpreter uses a tube magnifier to examine an aerial photograph in detail. A mirror stereoscope is visible in the foreground. From U.S. Air Force, U.S. National Achieves and Records Administration, ARC 542277.

play environments. There are many other techniques that are effective to varying degrees for stereovision (including random dot stereograms, Magic Eye® images, and others)—most are less effective for scientific and analytical applications than stereoscopes and polarized lenses.

### *Magnification*

Image analysts almost always wish to examine images using magnification, although the exact form depends upon individual preference and the nature of the task at hand. A simple handheld reading glass is satisfactory in many circumstances. Tube magnifiers (Figures 5.16 and 5.18) are low-power lenses (2× to about 8×), often mounted in a transparent tube-like stand. The base may include a reticule calibrated in units as small as 0.001 ft. or 0.1 mm, to permit accurate measurement of objects depicted on small-scale images. Sometimes it is necessary to use the much more expensive *binocular microscopes* (similar to the instrument depicted in Figure 5.17) for examination of film transparencies; such instruments may have magnification adjustable to as much as 40×, which will approach or exceed the limits of resolution for most images.

**FIGURE 5.17.** Binocular stereoscopes permit stereoscopic viewing of images at high magnification. From U.S. Air Force.

**FIGURE 5.18.** An image interpreter uses a magnifier to examine detail recorded on an aerial image. From SSGT Scott Stewart, Department of Defense, DFST8712359.

### *Densitometry*

*Densitometry* is the science of making accurate measurements of film density. In the context of remote sensing, the objective is often to reconstruct estimates of brightness in the original scene or sometimes merely to estimate relative brightnesses on the film (Chapter 3). A *densitometer* (Figure 5.19) is an instrument that measures image density by directing a light of known brightness through a small portion of the image, then measuring its brightness as altered by the film. Typically, the light beam might pass through an opening perhaps 1 mm in diameter; use of smaller openings (measured sometimes in micrometers) is known as *microdensitometry.* Such instruments find densities for selected regions within an image; an interpreter might use a densitometer to make quantitative measurements of image tone. For color or CIR images, filters are used to make three measurements, one for each of the three additive primaries.

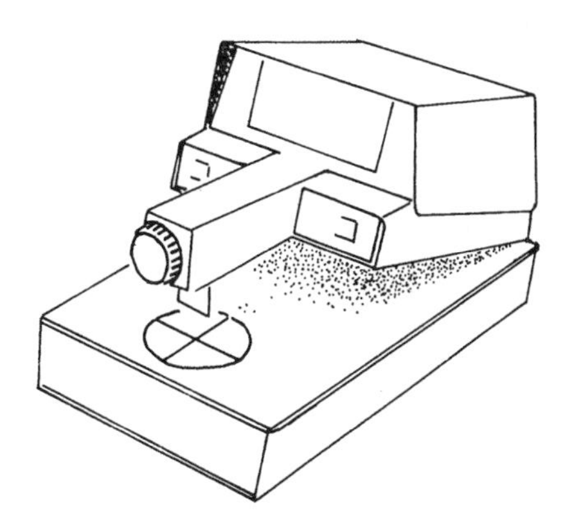

**FIGURE 5.19.** Densitometer, an instrument used to acquire measurements of image density at a specific point.

In principal, densitometric measurements can be used to estimate brightnesses in the original scene. However, several factors intervene to make such estimates difficult. The densitometer must be carefully calibrated; areas of known brightness must be represented on the film and subjected to the same processing as the image to be examined. Measurements of densities that fall in the nonlinear portion of the characteristic curve cannot, of course, be related to the brightness of the original scene. For these reasons and others, it is difficult to make reliable estimates of scene brightness by densitometry.

### *Image Scanning*

Paper or film images can be scanned for digital representation, as outlined in Chapter 4. Among the most accurate instruments for image scanning are *scanning densitometers* (Figure 5.20), designed to accurately and precisely measure image density by systematically scanning across an image, creating an array of digital values to represent the image pattern. Such instruments have been designed to very precisely measure position and image density, possibly using resolutions of a few micrometers or so. Although instruments designed for such precise tolerances are expensive, many other less precise, but still serviceable, scanners can sometimes be used for tasks that do not require high accuracy. Desktop scanners, although designed for office use, can be used to scan maps and images for visual analysis (Coburn et al., 2001; Carstensen and Campbell, 1991).

### *Parallax Bar*

The *parallax bar* (also known as a *stereometer bar* or a *height finder*) (Figure 5.21) is an instrument designed for use with a stereoscope; it permits estimation of topographic elevation or of the heights of features from stereo aerial photographs. The parallax bar attaches to the base of the stereoscope; the bar holds two plastic tabs, one under each lens of the stereoscope. Both tabs are marked with a small black dot, but one tab is fixed in position, whereas the other can be moved from side to side along a scale that measures its movement left to right parallel to the bar. The operator aligns the photographs for stereoscopic viewing after marking their principal points, their conjugate principal points, and the flight line. The stereoscope is positioned for stereoscopic viewing as normal (see Section 5.13). Then the interpreter views the scene stereoscopically, positioning the dot to appear to float above the terrain surface; when it is positioned

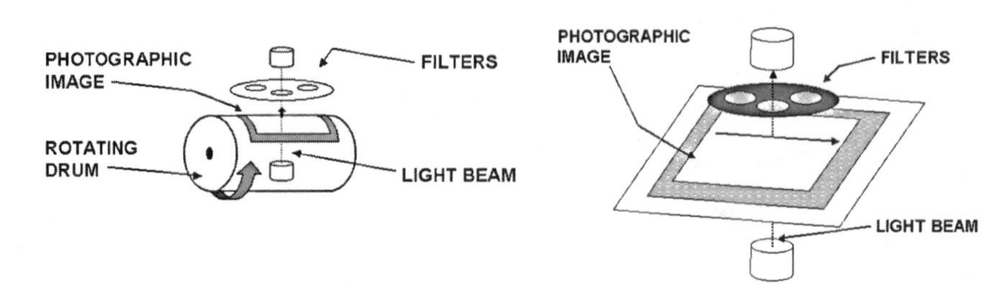


**FIGURE 5.20.** Scanning densitometers, instruments used to systematically measure density for an entire image. Left: Rotating drum densitometer. Right: Flat-bed scanning densitometer.

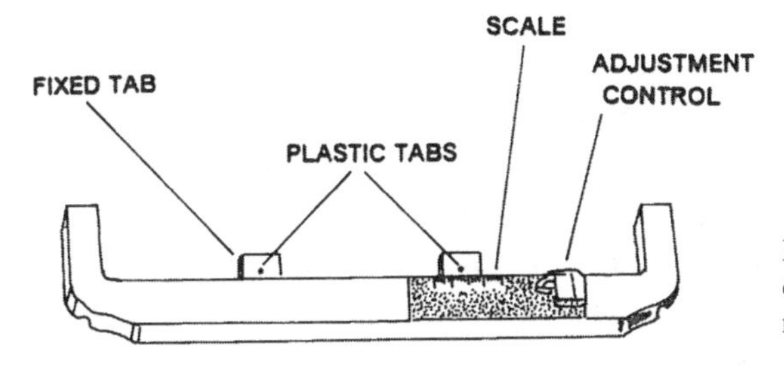


**FIGURE 5.21.** A parallax bar attaches to the legs of the pocket stereoscope; it can be used to estimate heights from stereoscopic photographs.

so it appears to rest on the terrain surface, a reading of the scale gives the parallax measurement for that point. A parallax factor is found by using a set of tables. Readings for several points, when combined with simple calculations, permit determination of elevation differences between the individual points (e.g., in feet or meters). If the photographs show a point of known elevation, then these points can be assigned elevations with reference to a datum (e.g., mean sea level); otherwise they provide only relative heights.

### *Data Transfer*

Analysts often need to transfer information from one map or image to another to assure accurate placement of features, to update superseded information, or to bring several kinds of information into a common format. Traditionally, these operations have been accomplished by optical projection of maps or images onto a working surface where they could be traced onto an overlay that registers to another image. Manipulation of the optical system permitted the operator to change scale and to selectively enlarge or reduce portions of the image to correct for tilt and other geometric errors. As digital analysis has become more important, such devices are designed to digitize imagery and match them to other data, with computational adjustments for positional errors (Figure 5.22).

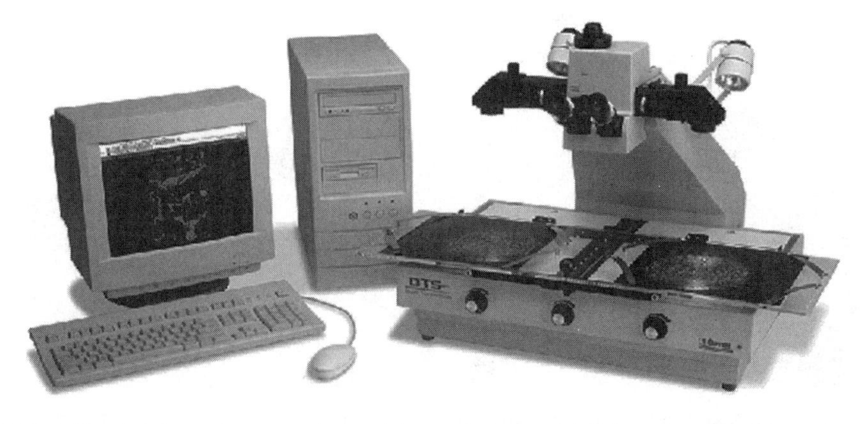

**FIGURE 5.22.** The digital transfer scope, designed to transfer image detail from the original image to a planimetrically correct map. From Thales Optem, Inc. Reproduced by permission.

### *Digital Photointerpretation*

Increasing use of digital photography and softcopy photogrammetry (Section 3.12) has blurred a previously distinct separation between manual and digital photointerpretation. Analyses that previously were conducted by visual examination of photographic prints or transparencies can now be completed by examination of digital images viewed on computer screens. Analysts record the results of their interpretations as on-screen annotations, using the mouse and cursor to outline and label images. Figure 5.23 illustrates a digital record of interpreted boundaries recorded by on-screen digitization (left), and the outlines shown without the image backdrop (center). The right-hand image shows an enlargement of a portion of the labeled region, illustrating the raster structure of the image and the boundaries.

Some systems employ photogrammetric software to project image detail in its correct planimetric location, without the positional or scale errors that might be present in the original imagery. Further, the digital format enables the analyst to easily manipulate image contrast (Section 5.15) to improve interpretability of image detail. Digital photogrammetric workstations (Figure 5.24), often based on the usual PC or UNIX operating systems, can accept scanned film imagery, airborne digital imagery, or digital satellite data. The full range of photogrammetric processes can be implemented digitally, including triangulation, compilation of digital terrain models (DTMs), feature digitization, construction of orthophotos, mosaics, and fly-throughs. Analysts can digitize features on-screen ("heads-up" digitization), using the computer mouse, to record and label features in digital format.

## 5.13. Use of the Pocket Stereoscope

*Stereoscopy* (Section 3.9) is the ability to derive height information from two images of the same scene. Stereovision contributes a valuable dimension to information derived from aerial photography. Although full development of its concepts and techniques are encompassed by the field of photogrammetry (Wolfe, 1974), here we can introduce its application through the use of the pocket stereoscope. This simple inexpensive instrument is an important image interpretation aid that can be employed in a wide variety of situations. Moreover, discussion of its use introduces concepts that underlie more advanced instruments.

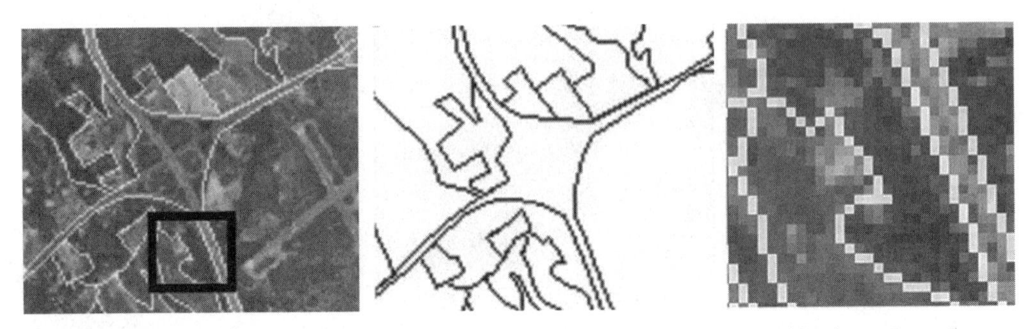

**FIGURE 5.23.** A digital record of image interpretation, showing the outlines as traced by the analyst using on-screen digitization (left), the outlines without the image backdrop (center), and a detail of the raster structure of the digitized outlines (right).

**FIGURE 5.24.** Digital photogrammetric workstation. The operator is depicted using polarizing stereovision glasses.

Although, at first glance, the stereoscope appears designed to magnify images, magnification is really an incidental feature of the instrument. In fact, the purpose of the stereoscope is to assist the analyst in maintaining parallel lines of sight. Stereoscopic vision is based upon the ability of our visual system to detect *stereoscopic parallax,* the difference in the appearance of objects due to differing perspectives. So, when we view a scene using only the right eye, we see a slightly different view than we do using only the left eye—this difference is stereoscopic parallax. Because stereoscopic parallax is greater for nearby objects than it is for more distant objects, our visual system can use this information to make accurate judgments about distance.

Aerial photographs are often taken in sequences designed to provide overlapping views of the same terrain—that is, they provide two separate perspectives of the same landscape, just as our eyes provide two separate images of a scene. So, we can use a stereo pair of aerial photographs to simulate a stereoscopic view of the terrain, provided we can maintain parallel lines of sight, just as we would normally do in viewing a distant object (Figure 5.25a). Parallel lines of sight assure that the right and the left eyes each see independent views of the same scene, to provide the parallax needed for the stereoscopic illusion. However, when we view objects that are nearby, our visual system instinctively recognizes that the objects are close, so our lines of sight converge (Figure 5.25b), depriving our visual system of the two independent views needed for stereoscopic vision. Therefore, the purpose of the stereoscope is to assist us in maintaining the parallel lines of sight that enable the stereoscopic effect (Figure 5.25c).

Although many students will require the assistance of an instructor as they learn to use the stereoscope, the following paragraphs may provide some assistance for beginners. First, stereo photographs must be aligned so that the flight line passes left to right (as shown in Figure 5.26). Check the photo numbers to be sure that the photographs have been selected from adjacent positions on the flight line. Usually (but not always) the numbers and annotations on photos are placed on the *leading edge* of the image, the edge of the image nearest the front of the aircraft at the time the image was taken. Therefore, these numbers should usually be oriented in sequence from left to right, as shown in Figure 5.26. If the overlap between adjacent photos does not correspond to the natural positions of objects on the ground, then the photographs are incorrectly oriented.

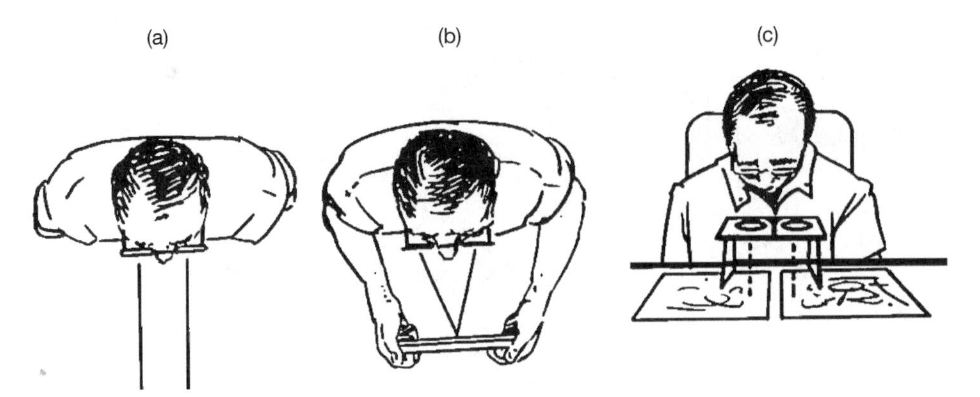


FIGURE 5.25. The role of the stereoscope in stereoscopic vision. (*a*) To acquire the two independent views of the same scene required for stereoscopic vision, we must maintain parallel lines of sight. (*b*) Normally, when we view nearby objects, our lines of sight converge, preventing us from acquiring the stereo effect. (*c*) The stereoscope is an aid to assist in maintaining parallel lines of sight even when the photographs are only a few inches away from the viewer.

Next, the interpreter should identify a distinctive feature on the image within the zone of stereoscopic overlap. The photos should then be positioned so that the duplicate images of this feature (one on each image) are approximately 64 mm (2.5 in.) apart. This distance represents the distance between the two pupils of a person of average size (referred to as the *interpupillary distance*), but for many it may be a bit too large or too small, so the spacing of photographs may require adjustment as the interpreter follows the procedure outlined here. The pocket stereoscope should be opened so that its legs are locked in place to position the lens at their correct height above the photographs. The two segments of the body of the stereoscope should be

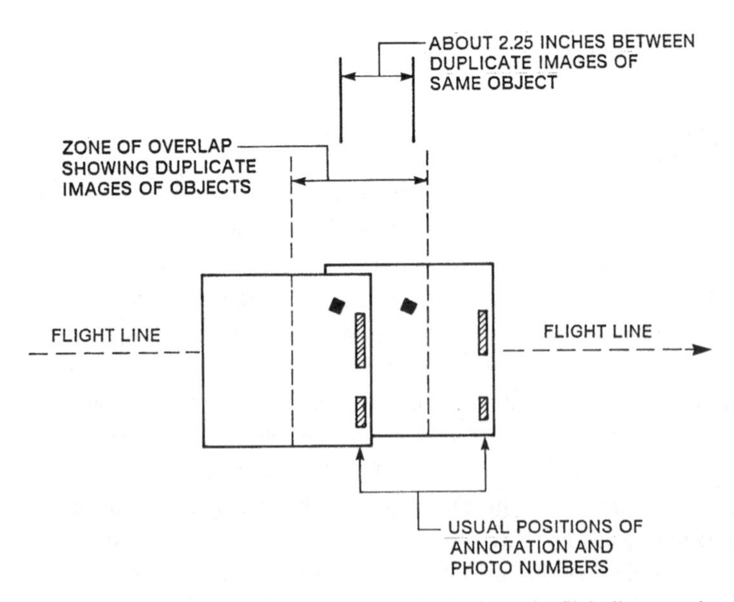


FIGURE 5.26. Positioning aerial photographs for stereoscopic viewing. The flight line must be oriented laterally in front of the viewer.

adjusted so that the centers of the eyepieces are about 64 mm (2.5 in) apart (or a slightly larger or smaller distance, as mentioned above).

Then the stereoscope should be positioned so that the centers of the lenses are positioned above the duplicate images of the distinctive feature selected previously. Looking through the two lenses, the analyst sees two images of this feature; if the images are properly positioned, the two images will appear to "float" or "drift." The analyst can, with some effort, control the apparent positions of the two images so that they fuse into a single image; as this occurs, the two images should merge into a single image that is then visible in three dimensions. Usually aerial photos show exaggerated heights, due to the large separation (relative to distance to the ground) between successive photographs as they were taken along the flight line. Although exaggerated heights can prevent convenient stereo viewing in regions of high relief, it can be useful in interpretations of subtle terrain features that might not otherwise be noticeable.

The student who has successfully used the stereoscope to examine a section of the photo should then practice moving the stereoscope over the image to view the entire region within the zone of overlap. As long as the axis of the stereoscope is oriented parallel to the flight line, it is possible to retain stereo vision while moving the stereoscope. If the stereoscope is twisted with respect to the flight line, the interpreter loses stereo vision. By lifting the edge of one of the photographs, it is possible to view the image regions near the edges of the photos. Although the stereoscope is a valuable instrument for examining terrain, drainage, and vegetation patterns, it does not provide the detailed measurements within the realm of photogrammetry and more sophisticated instruments.

## 5.14. Image Scale Calculations

Scale is a property of all images. Knowledge of image scale is essential for making measurements from images and for understanding the geometric errors present in all remotely sensed images. *Scale* is an expression of the relationship between the *image distance* between two points and the *actual distance* between the two corresponding points on the ground. This relationship can be expressed in several ways.

The *word statement* sets a unit distance on the map or photograph equal to the correct corresponding distance on the ground—for example, "One inch equals one mile," or just as correctly "One centimeter equals five kilometers." The first unit in the statement in the expression specifies the map distance, the second, the corresponding ground distance. A second method of specifying scale is the *bar scale,* which simply labels a line with subdivisions that show ground distances. The third method, the *representative fraction* (RF), is more widely used and often is the preferred method of reporting image scale. The representative fraction is the ratio between image distance and ground distance. It usually takes the form "1:50,000" or "1/50,000," with the numerator set equal to 1 and the denominator equal to the corresponding ground distance.

The representative fraction has meaning in any unit of length as long as both the numerator and the denominator are expressed in the same units. Thus "1:50,000" can mean "1 in. on the image equals 50,000 in. on the ground" or "1 cm on the image equals 50,000 cm on the ground." A frequent source of confusion is converting the denominator into the larger units that we find more convenient to use for measuring large ground distance. With metric units, the conversion is usually simple; in the example given above, it is easy to see that 50,000 cm is equal to 0.50 km, and that 1 cm on the map represents 0.5 km on the ground. With English units, the same process is not quite so easy. It is necessary to convert inches to miles to derive "1 in. equals 0.79 mi." from 1:50,000. For this reason, it is useful to know that 1 mi. equals 63,360 in. Thus, 50,000 in. is equal to 50,000/63,360 = 0.79 mi.

A typical scale problem requires estimation of the scale of an individual photograph. One method is to use the focal length and altitude method (Figure 5.27):

$$\text{RF} = \frac{\text{Focal length}}{\text{Altitude}} \qquad \text{(Eq. 5.1)}$$

Both values must be expressed in the same units. Thus, if a camera with a 6-in. focal length is flown at 10,000 ft., the scale is 0.5/10,000 = 1:20,000. (Altitude always specifies the flying height above the terrain, *not* above sea level.) Because a given flying altitude is seldom the exact altitude at the time the photography was taken, and because of the several sources that contribute to scale variations within a given photograph (Chapter 3), we must always regard the results of such calculations as an approximation of the scale of any specific portion of the image. Often such values are referred to as the "nominal" scale of an image, meaning that it is recognized that the stated scale is an approximation and that image scale will vary within any given photograph.

A second method is the use of a *known ground distance*. We identify two points on the aerial photograph that are also represented on a map. For example, in Figure 5.28, the image distance between points *A* and *B* is measured to be approximately 2.2 in. (5.6 cm). From the map, the same distance is determined to correspond to a ground distance of 115,000 in. (about 1.82 mi.). Thus the scale is found to be:

$$\text{RF} = \frac{\text{Image distance}}{\text{Ground distance}} = \frac{2.2\text{ in.}}{1.82\text{ mi.}} = \frac{2.2\text{ in.}}{115{,}000\text{ in.}} = \frac{1}{52{,}273} \qquad \text{(Eq. 5.2)}$$

In instances when accurate maps of the area represented on the photograph may not be available, the interpreter may not know focal length and altitude. Then an approximation of image scale can be made if it is possible to identify an object or feature of known dimensions. Such

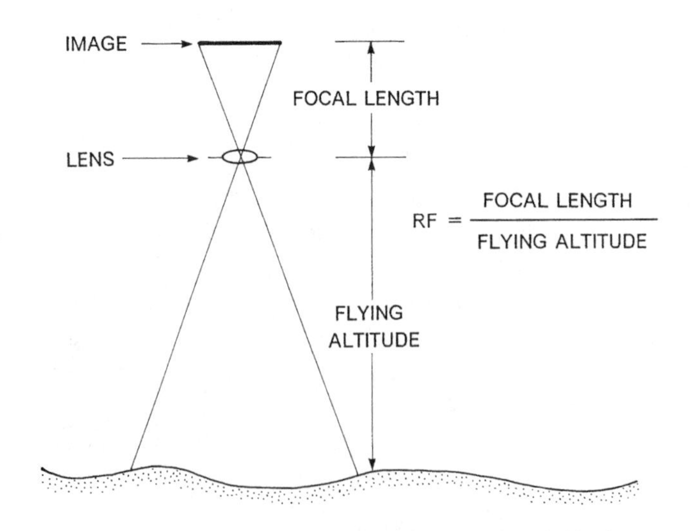


**FIGURE 5.27.** Estimating image scale by focal length and altitude.

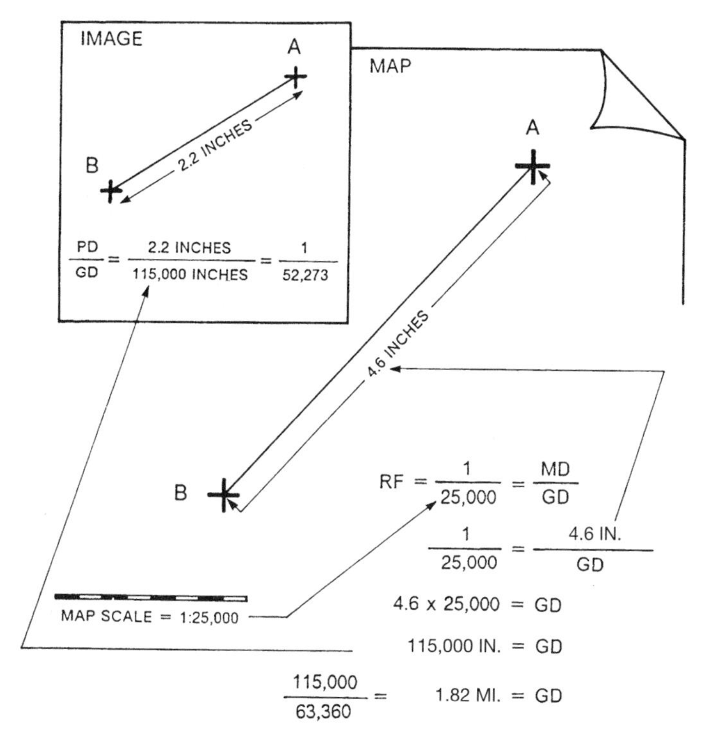


**FIGURE 5.28.** Measurement of image scale using a map to derive ground distance.

features might include a football field or baseball diamond; measurement of a distance from these features as they are shown on the image provides the "image distance" value needed to use the relationship given above. The "ground distance" is derived from our knowledge of the length of a football field or the distance between bases on a baseball diamond. Some photointerpretation manuals provide tables of standard dimensions of features commonly observed on aerial images including sizes of athletic fields (soccer, field hockey, etc.), lengths of railroad boxcars, distances between telephone poles, and so on as a means of using the known ground distance method.

A second kind of scale problem is the use of a known scale to measure a distance on the photograph. Such a distance might separate two objects on the photograph but not be represented on the map, or the size of a feature has changed since the map was compiled. For example, we know that image scale is 1:15,000. A pond not shown on the map is measured on the image as 0.12 in. in width. Therefore, we can estimate the actual width of the pond to be:

$$\frac{1}{15{,}000} = \frac{\text{Image distance}}{\text{Ground distance}} \qquad \text{(Eq. 5.3)}$$

$$\frac{0.12 \text{ in.}}{\text{Unknown GD}} = \frac{\text{Image distance}}{\text{Ground distance}}$$

$$\text{GD} = 0.12 \times 15{,}000 \text{ in.}$$

$$\text{GD} = 1{,}8000 \text{ in., or } 150 \text{ ft.}$$

This example can illustrate two other points. First, because image scale varies throughout the image, we cannot be absolutely confident that our distance for the width of the pond is accurate; it is simply an estimate, unless we have high confidence in our measurements and in the image scale at this portion of the photo. Second, measurements of short image distances are likely to have errors due simply to our inability to make accurate measurements of very short distances (e.g., the 0.12-in. distance measured above). As distances become shorter, our errors constitute a greater proportion of the estimated length. Thus an error of 0.005 in. is 0.08% of a distance of 6 in., but 4% of a distance of 0.12 in. mentioned above. Thus the interpreter should exercise a healthy skepticism regarding measurements made from images unless he or she has taken great care to ensure maximum accuracy and consistency.

## 5.15. Interpretation of Digital Imagery

Image interpretation was once practiced entirely within the realm of photographic prints and transparencies, using the equipment and techniques outlined in the preceding sections. As digital analyses have increased in significance, so has the importance of interpretation of imagery presented on computer displays. Although such interpretations are based on the same principles outlined above for traditional imagery, digital data have their own characteristics that require special treatment in the context of visual interpretation.

### *Image Enhancement*

*Image enhancement* is the process of improving the visual appearance of digital images. Image enhancement has increasing significance in remote sensing because of the growing importance of digital analyses. Although some aspects of digital analysis may seem to reduce or replace traditional image interpretation, many of these procedures require analysts to examine images on computer displays, doing tasks that require many of the skills outlined in earlier sections of this chapter.

Most image enhancement techniques are designed to improve the visual appearance of an image, often as evaluated by narrowly defined criteria. Therefore, it is important to remember that enhancement is often an arbitrary exercise—what is successful for one image or purpose may be unsuitable for another image or for another purpose. In addition, image enhancement is conducted without regard for the integrity of the original data: the original brightness values will be altered in the process of improving their visual qualities and they will lose their relationships to the original brightnesses on the ground. Therefore, enhanced images should not be used as input for additional analytical techniques; rather, any further analysis should use the original values as input.

### *Contrast Enhancement*

*Contrast* refers to the range of brightness values present on an image. Contrast enhancement is required because, as outlined in Chapter 4, digital data usually have brightness ranges that do not match the capabilities of the human visual system, nor those of photographic films (Chapter

3). Therefore, for analysts to view the full range of information conveyed by digital images, it is usually necessary to rescale image brightnesses to ranges that can be accommodated by human vision, photographic films, and computer displays.

For example, if the maximum possible range of values is 0 to 255 (i.e., 8 bits), but the display can show only the range from 0 to 63 (6 bits), then the image will have poor contrast, and important detail may be lost in the values that cannot be shown on the display (Figure 5.29a). Contrast enhancement alters each pixel value in the old image to produce a new set of values that exploits the full range of 256 brightness values (Figure 5.29b).

Figure 5.30 illustrates the practical effect of image enhancement. Before enhancement (left), detail is lost in the darker regions of the image. After enhancement has stretched the histogram of brightness values to take advantage of the capabilities of the display system, the detail is more clearly visible to the eye.

Many alternative approaches have been proposed to improve the quality of the displayed image. The appropriate choice of technique depends upon the image, the previous experience of the user, and the specific problem at hand. The following paragraphs illustrate a few of the simpler and more widely used techniques.

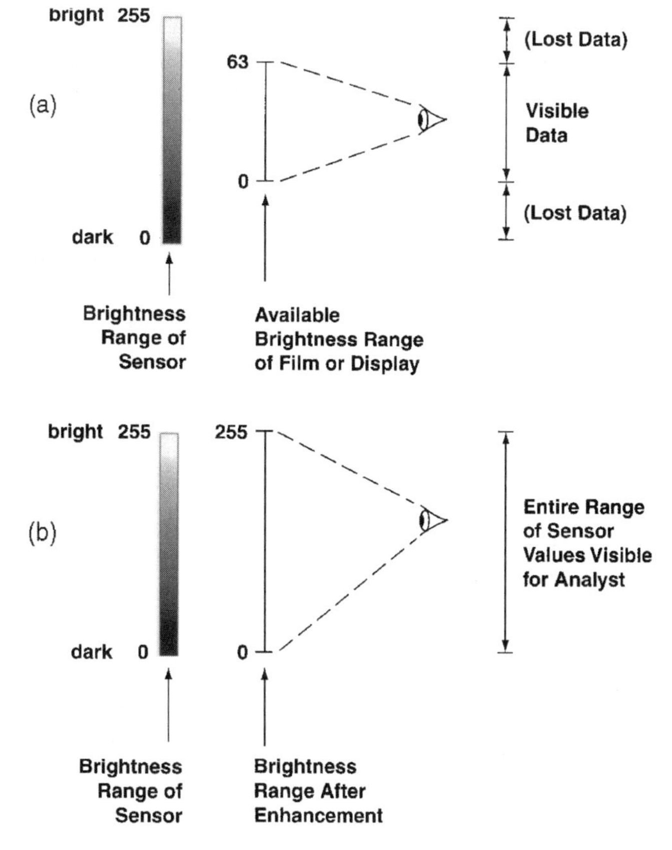


**FIGURE 5.29.** Schematic representation of the loss of visual information in display of digital imagery. (*a*) Often the brightness range of digital imagery exceeds the ability of the image display to represent it to the human visual system. (*b*) Image enhancement rescales the digital values to more nearly match the capabilities of the display system.

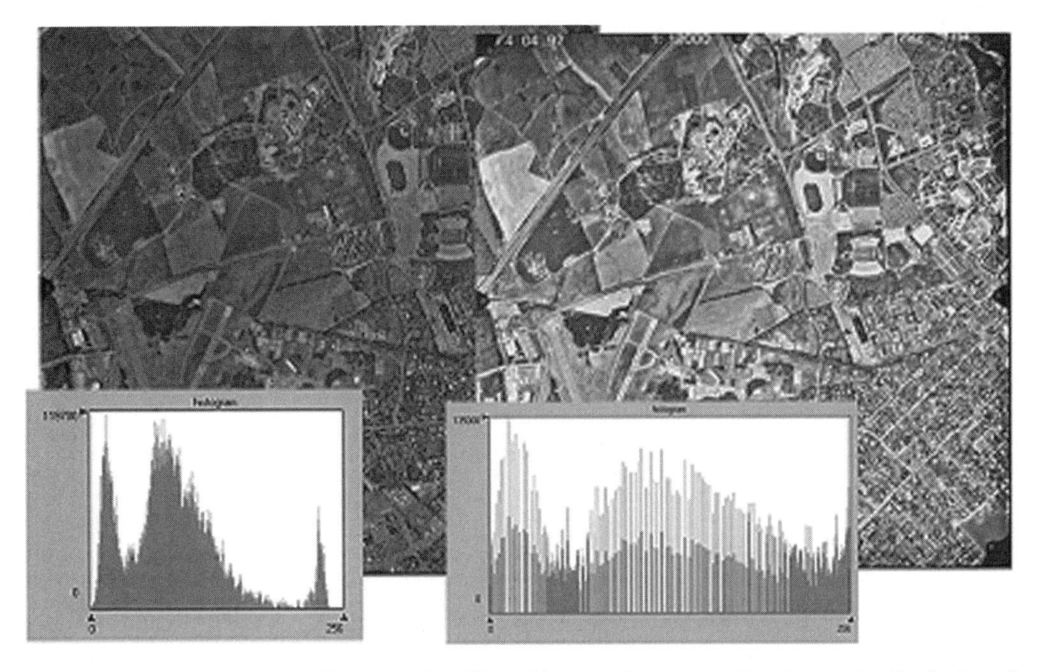

**FIGURE 5.30.** Pair of images illustrating the effect of image enhancement. By altering the distribution of brightness values, as discussed in the text, the analyst is able to view detail formerly hidden by the ineffective distribution of image brightnesses.

### *Linear Stretch*

*Linear stretch* converts the original digital values into a new distribution, using new minimum and maximum values specified. The algorithm then matches the old minimum to the new minimum and the old maximum to the new maximum. All of the old intermediate values are scaled proportionately between the new minimum and maximum values (Figure 5.31). *Piecewise linear stretch* means that the original brightness range was divided into segments before each segment was stretched individually. This variation permits the analyst to emphasize cer-

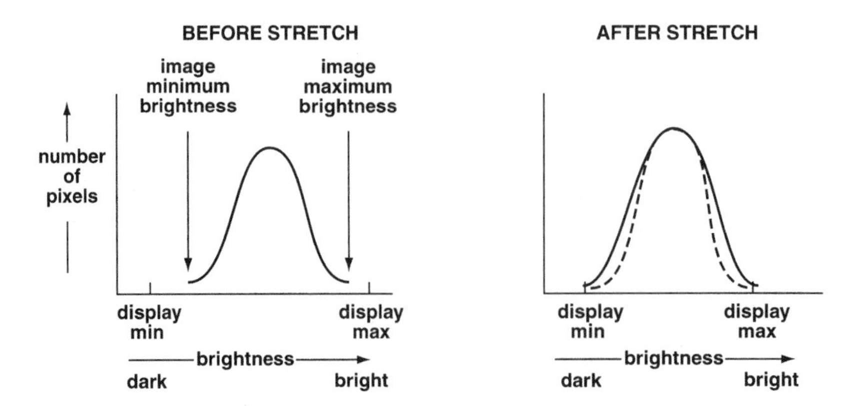


**FIGURE 5.31.** Linear stretch spreads the brightness values over a broader range, allowing the eye to see detail formerly concealed in the extremely dark or bright tones.

tain segments of the brightness range that might have more significance for a specific application.

### *Histogram Equalization*

*Histogram equalization* reassigns digital values in the original image such that brightnesses in the output image are equally distributed among the range of output values (Figure 5.32). Unlike contrast stretching, histogram equalization is achieved by applying a nonlinear function to reassign the brightnesses in the input image such that the output image approximates a uniform distribution of intensities. The histogram peaks are broadened and the valleys are made shallower. Histogram equalization has been widely used for image comparison processes (because it is effective in enhancing image detail) and for adjustment of artifacts introduced by digitizers or other instruments.

### *Density Slicing*

*Density slicing* is accomplished by arbitrarily dividing the range of brightnesses in a single band into intervals, then assigning each interval to a color (Figure 5.33 and Plates 4 and 17). Density slicing may have the effect of emphasizing certain features that may be represented in vivid colors, but, of course, does not convey any more information than the single image used as the source.

### *Edge Enhancement*

*Edge enhancement* is an effort to reinforce the visual boundaries between regions of contrasting brightness. Typically, the human interpreter prefers sharp edges between adjacent parcels, whereas the presence of noise, coarse resolution, and other factors often tends to blur or weaken the distinctiveness of these edges. Edge enhancement is in effect the strengthening of local con-

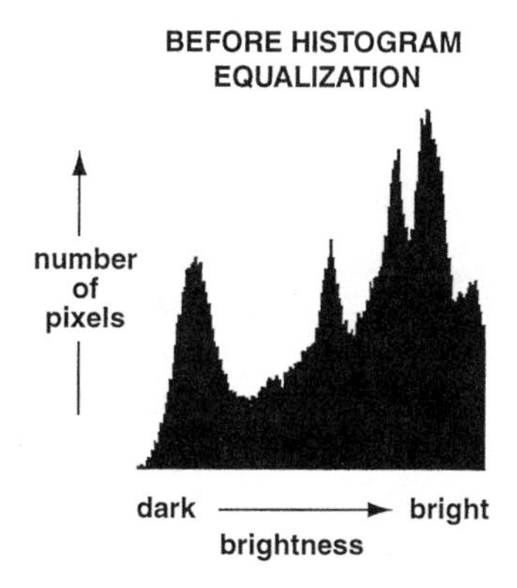


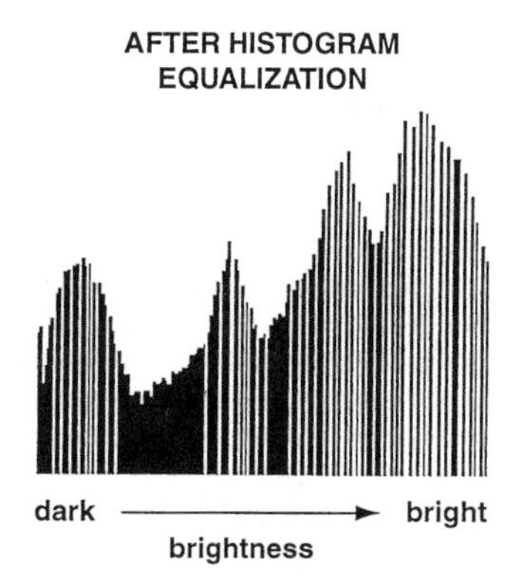


**FIGURE 5.32.** Histogram equalization spreads the range of brightness values, but preserves the peaks and valleys in the histogram.

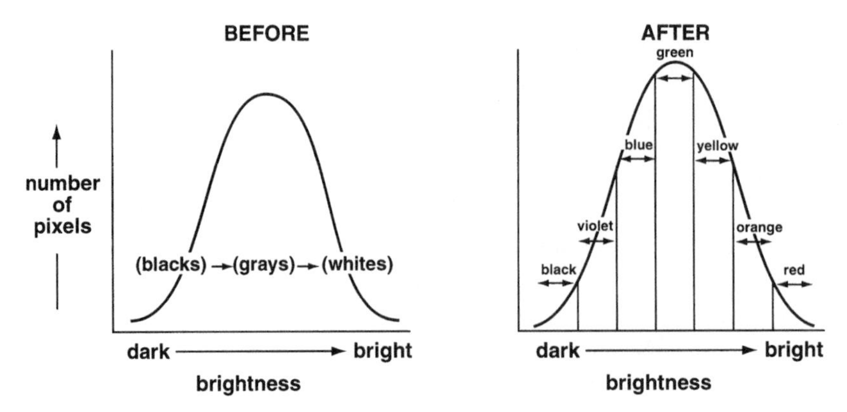


**FIGURE 5.33.** Density slicing assigns colors to specific intervals of brightness values. See also Plates 4 and 17.

trast—enhancement of contrast within a local region. A typical edge enhancement algorithm consists of a window that is systematically moved through the image, centered successively on each pixel. At each position, it is then possible to calculate a local average of values within the window; the central value can be compared to the averages of the adjacent pixels. If the value exceeds a specified difference from this average, the value can be altered to accentuate the difference in brightness between the two regions.

Rohde et al. (1978) describe an edge enhancement procedure used at the EROS Data Center (see Section 3.13). A new (*output*) digital value is calculated using the original (*input*) value and the local average of five adjacent pixels. A constant can be applied to alter the effect of the enhancement as necessary in specific situations. The output value is the difference between twice the input value and the local average, thereby increasing the brightness of those pixels that are already brighter than the local average, and to decrease the brightnesses of pixels that are already darker than the local average. Thus the effect is to accentuate differences in brightnesses, especially at places ("edges") where a given value differs greatly from the local average (Figure 5.34).

## 5.16. Summary

Despite the increasing significance of digital analysis in all aspects of remote sensing, image interpretation still forms a key component in the way that humans understand images. Analysts must evaluate imagery, either as paper prints or as displays on a computer monitor, using the skills outlined in this chapter. The fundamentals of manual image interpretation were developed for application to aerial photographs at an early date in the history of aerial survey, although it was not until the 1940s and 1950s that they were formalized in their present form. Since then, these techniques have been applied, without substantial modification, to other kinds of remote sensing imagery. As a result, we have a long record of experience in their application and comprehensive knowledge of their advantages and limitations.

Interesting questions remain. In what ways might image interpretation skills be modified in the context of interpretation using computer monitors? What new skills might be necessary? How have analysts already adjusted to new conditions? How might equipment and software be improved to facilitate interpretation in this new context?

Without sharpening

With sharpening

**FIGURE 5.34.** Edge enhancement and image sharpening. A sample image is shown with and without enhancement to accentuate edges. In this example, the effect of sharpening is especially noticeable at edges of some of the larger shadows.

## Review Questions

1. A vertical aerial photograph was acquired using a camera with a 9-in. focal length at an altitude of 15,000 ft. Calculate the nominal scale of the photograph.
2. A vertical aerial photograph shows two object to be separated by 6¾ in. The corresponding ground distance is 9½ mi. Calculate the nominal scale of the photograph.
3. A vertical aerial photograph shows two features to be separated by 4.5 in. A map at 1:24,000 shows the same two features to be separated by 9.3 in. Calculate the scale of the photograph.
4. Calculate the area represented by a 9 in. × 9 in. vertical aerial photograph acquired at an altitude of 10,000 ft. using a camera with a 6-in. focal length.
5. A vertical aerial photograph shows two features to be separated by 2.3 in. A map at 1:50,000 shows the same two features to be separated by 1.2 in. Calculate the scale of the photograph.
6. Calculate the area represented by a 9 in. × 9 in. vertical aerial photograph acquired at an altitude of 20,000 ft. using a camera with a 9-in. focal length.
7. You plan to acquire coverage of a region using a camera with 6-in. focal length and a 9 in. × 9 in. format. You require an image scale of 4 in. equal to 1 mi., 60% forward overlap, and sidelap of 10%. Your region is square in shape, measuring 15.5 mi. on a side. How many photographs are required? At what altitude must the aircraft fly to acquire these photos?
8. You have a flight line of 9 in. × 9 in. vertical aerial photographs taken by camera with a 9-in.

focal length at an altitude of 12,000 ft. above the terrain. Forward overlap is 60%. Calculate the distance (in miles) between ground nadirs of successive photographs.

9. You require complete stereographic coverage of your study area, which is a rectangle measuring 1.5 mi. × 8 mi. How many 9 in. × 9 in. vertical aerial photographs at 1:10,000 are required?

10. You have very little information available to estimate the scale of a vertical aerial photograph, but you are able to recognize a baseball diamond among features in an athletic complex. Using a tube magnifier, you measure the distance between first and second base and find it to be 0.006 ft. What is your estimate of the scale of the photo?

## References

Avery, T. E., and G. L. Berlin. 1992. *Fundamentals of Remote Sensing and Airphoto Interpretation.* New York: Macmillan, 472 pp.

Babington Smith, C. 1985. *Air Spy: The Story of Photo Intelligence in World War II.* Falls Church, VA: American Society for Photogrammetry and Remote Sensing, 256 pp. (Originally published 1957)

Brugioni, D. 1991. *Eyeball to Eyeball: The Inside History of the Cuban Missile Crisis.* New York: Random House, 622 pp.

Brugioni, D. 1996. The Art and Science of Photoreconnaissance. *Scientific American,* Vol. 274, pp. 78–85.

Campbell, J. B. 1978. A Geographical Analysis of Image Interpretation Methods. *Professional Geographer,* Vol. 30, pp. 264–269.

Carstensen, L. W., and J. B. Campbell. 1991. Desktop Scanning for Cartographic Digitization and Spatial Analysis. *Photogrammetric Engineering and Remote Sensing,* Vol. 57, pp. 1437–1446.

Chavez, P. S., G. L. Berlin, and W. B. Mitchell. 1977. Computer Enhancement Techniques of Landsat MSS Digital Images for Land Use/Land Cover Assessment. In *Proceedings of the Sixth Annual Remote Sensing of Earth Resources Conference,* Houston, TX: NASA, pp. 259–276.

Cihlar, J., and R. Protz. 1972. Perception of Tone Differences from Film Transparencies. *Photogrammetria,* Vol. 8, pp. 131–140.

Coburn, C., A. Roberts, and K. Bach. 2001. Spectral and Spatial Artifacts from the Use of Desktop Scanners for Remote Sensing. *International Journal of Remote Sensing,* Vol. 22, pp. 3863–3870.

Coiner, J. C., and S. A. Morain. 1972. *SLAR Image Interpretation Keys for Geographic Analysis* (Technical Report 177-19). Lawrence, KS: Center for Research, Inc., 110 pp.

Colwell, R. N. (ed.). 1960. *Manual of Photographic Interpretation.* Falls Church, VA: American Society of Photogrammetry, 868 pp.

Departments of the Army, Navy, and Air Force. 1967. *Image Interpretation Handbook* (TM 30-245; NAVAIR 10-35-685; AFM 200-50). Washinton, DC: U.S. Government Printing Office.

Erdman, C., K. Riehl, L. Mayer, J. Leachtenauer, E. Mohr, J. Odenweller, R. Simmons, and D. Hothem. 1994. Quantifying Multispectral Imagery Interpretability. In *International Symposium on Spectra Sensing Research,* Vol. 1, pp. 468–476. Alexandria, VA: U.S. Corps of Engineers.

Estes, J. E., E. J. Hajic, L. R. Tinney, et al. 1983. Fundamentals of Image Analysis: Analysis of Visible and Thermal Infrared Data. Chapter 24 in *Manual of Remote Sensing* (R. N. Colwell, ed.). Falls Church, VA: American Society of Photogrammetry, pp. 987–1124.

Haack, B., and S. Jampoler. 1995. Colour Composite Comparisons for Agricultural Assessments. *International Journal of Remote Sensing,* Vol. 16, pp. 1589–1598.

Irving, D. 1964. *The Mare's Nest: The War Against Hitler's Secret Vengence Weapons.* London: Granada, 270 pp.

Lueder, D. R. 1959. *Aerial Photographic Interpretation: Principles and Applications.* New York: McGraw-Hill, 462 pp.

Nunnally, N. R. 1969. Integrated Landscape Amalysis with Radar Imagery. *Remote Sensing of Environment,* Vol. 1, pp.1–6.

Paine, D. P., and J. D. Kiser. 2003. *Aerial Photography and Image Interpretation.* New York: Wiley, 648 pp.

Philipson, W. R. (ed.). 1996. *Manual of Photographic Interpretation* (2nd ed.). Bethesda, MD: American Society for Photogrammetry and Remote Sensing, 689 pp.

Rohde, W. G., J. K. Lo, and R. A. Pohl. 1978. EROS Data Center Landsat Digital Enhancement Techniques and Imagery Availability, 1977. *Canadian Journal of Remote Sensing,* Vol. 4, pp. 63–76.

Taranick, J. V. 1978. *Principles of Computer Processing of Landsat Data for Geological Applications* (USGS Open File Report 78-117). Reston, VA: U.S. Geological Survey, 50 pp.

Wolfe, P. R. 1974. *Elements of Photogrammetry.* New York: McGraw-Hill, 562 pp.

CHAPTER SIX

# Land Observation Satellites

## 6.1. Satellite Remote Sensing

Today many corporations and national governments operate satellite remote sensing systems specifically designed for observation of the Earth's surface to collect information concerning topics such as crops, forests, water bodies, land use, cities, and minerals. Satellite sensors offer several advantages over aerial platforms: they can provide a synoptic view (observation of large areas in a single image), fine detail, and systematic, repetitive coverage. Such capabilities are well suited to creating and maintaining a worldwide cartographic infrastructure and to monitoring changes in the many broad-scale environmental issues that the world faces today, to list two of many pressing concerns.

Because of the large number of satellite observation systems in use and the rapid changes in their design, this chapter cannot list or describe all systems currently in use or planned. It can, however, provide readers with the basic framework they need to understand key aspects of Earth observation satellites in general as a means of preparing readers to acquire knowledge of specific satellite systems as it becomes available. Therefore, this chapter outlines essential characteristics of the most important systems—past, present, and future—as a guide for understanding other systems not specifically discussed here.

Today's land observation satellites have evolved from earlier systems. The first Earth observation satellite, the Television and Infrared Observation Satellite (TIROS), was launched in April 1960 as the first of a series of experimental weather satellites designed to monitor cloud patterns. TIROS was the prototype for the operational programs that now provide meteorological data for daily weather forecasts throughout the world. Successors to the original TIROS vehicle have seen long service in several programs designed to acquire meteorological data. Although data from meteorological satellites have been used to study land resources (as discussed in Chapters 17, 20, and 21), this chapter focuses on satellite systems specifically tailored for observation of land resources, mainly via passive sensing of radiation in the visible and infrared regions of the spectrum.

## 6.2. Landsat Origins

Early meteorological sensors had limited capabilities for land resources observation. Although their sensors were valuable for observing cloud patterns, most had coarse spatial resolution, so they could provide only rudimentary detail concerning land resources. Landsat ("land satellite") was designed in the 1960s and launched in 1972 as the first satellite tailored specifically for broad-scale observation of the Earth's land areas; it was created to accomplish for land resource studies what meteorological satellites had accomplished for meteorology and climatology. Today the Landsat system is important both in its own right—as a remote sensing system that has contributed greatly to earth resources studies—and as an introduction to similar land observation satellites operated by other organizations.

Landsat was proposed by scientists and administrators in the U.S. government who envisioned application of the principles of remote sensing to broad-scale, repetitive surveys of the Earth's land areas. (Initially, Landsat was known as the "Earth Resources Technology Satellite," "ERTS" for short.) The first Landsat sensors recorded energy in the visible and near infrared spectrums. Although these regions of the spectrum had long been used for aircraft photography, it was by no means certain that they would also prove practical for observation of Earth resources from satellite altitudes. Scientists and engineers were not completely confident that the sensors would work as planned, that they would prove to be reliable, that detail would be satisfactory, or that a sufficient proportion of scenes would be free of cloud cover. Although many of these problems were encountered, the feasibility of the basic concept was demonstrated, and Landsat became the model for similar systems now operated by other organizations.

The Landsat system consists of spacecraft-borne sensors that observe the Earth and then transmit information by microwave signals to ground stations that receive and process data for dissemination to a community of data users. Early Landsat vehicles carried two sensor systems: the return beam videcon (RBV) and the multispectral scanner subsystem (MSS) (Table 6.1). The RBV was a camera-like instrument designed to provide, relative to the MSS, high spatial resolution and geometric accuracy, but lower spectral and radiometric detail. That is, the positions of features would be accurately represented, but without fine detail concerning their colors and brightnesses. In contrast, the MSS was designed to provide finer detail concerning spectral characteristics of the Earth, but less positional accuracy. Because technical difficulties restricted RBV operation, the MSS soon became the primary Landsats sensor. A second generation of Landsat vehicles (Landsats 4 and 5) added the thematic mapper (TM), a more sophisticated version of the MSS. The current family of Landsats carry ETM+, an advanced version TM.

**TABLE 6.1. Landsat Missions**

| Satellite | Launched | End of service[a] | Principal sensors[b] |
|---|---|---|---|
| Landsat 1 | 23 July 1972 | 6 January 1978 | MSS, RBV |
| Landsat 2 | 22 January 1975 | 25 January 1982 | MSS, RBV |
| Landsat 3 | 5 March 1978 | 3 March 1983 | MSS, RBV |
| Landsat 4 | 16 July 1982 | [c] | TM, MSS |
| Landsat 5 | 1 March 1984 | | TM, MSS |
| Landsat 6 | 5 October 1993 | Lost at launch | ETM |
| Landsat 7 | 15 April 1999 | [d] | ETM+ |

See *http://geo.arc.nasa.gov/sge/landsat/lpchron.html* for a complete chronology.
[a]Satellite systems typically operate on an intermittent or stand-by basis for considerable periods prior to formal retirement from service.
[b]Sensors are discussed in the text. MSS, multispectral scanner subsystem; RBV, return beam vidicon; TM, thematic mapper; ETM, enhanced TM; ETM+, enhanced TM plus.
[c]Transmission of TM data failed in August 1993.
[d]Malfunction of TM Scan Line Corrector has limited quality of imagery since May 2003.

## 6.3. Satellite Orbits

Satellites are placed into orbits tailored to match the objectives of each satellite mission and the capabilities of the sensors they carry. For simplicity, this section describes *normal* orbits, based on the assumption that the Earth's gravitational field is spherical, although in fact satellites actually follow *perturbed* orbits, due in part to distortion of the Earth's gravititational field by the Earth's *oblate* shape (flattened at the poles, and bulging at the equator), and in part due to lunar and solar gravity, tides, solar wind, and other influences.

A normal orbit forms an ellipse with the center of the Earth at one focus, characterized by an *apogee* (*A*; the point farthest from the Earth), a *perigee* (*P*; the point closest to the Earth), an *ascending node* (*AN*; the point where the satellite crosses the equator moving south to north), and a *descending node* (*DN*; the point where the satellite crosses the equator passing north to south). For graphical simplicity, the *inclination* (*i*) is shown in Figure 6.1 as the angle that a satellite track forms with respect to the equator at the descending node (Figure 6.1). (More precisely, the inclination should be defined as the angle between the Earth's axis at the North Pole and a line drawn perpendicular to the plane of the satellite orbit, viewed such that the satellite follows a counterclockwise trajectory.)

The time required for a satellite to complete one orbit (its period) increases with altitude. At an altitude of about 36,000 km, a satellite has the same period as the Earth's surface, so (if positioned in the equatorial plane) it remains stationary with respect to the Earth's surface—it is in a *geostationary* orbit. Geostationary orbits are ideal for meteorological or communications satellites designed to maintain a constant position with respect to a specific region on the Earth's surface. Earth observation satellites, however, are usually designed to satisfy other objectives.

Ideally, all remotely sensed images acquired by satellite would be acquired under conditions of uniform illumination, so that brightnesses of features within each scene would reliably indicate conditions on the ground rather than changes in the conditions of observation. In reality, brightnesses recorded by satellite images are not directly indicative of ground conditions because differences in latitude, time of day, and season lead to variations in the nature and intensity of light that illuminates each scene.

*Sun-synchronous* orbits are designed to reduce one important source of variation in illumina-

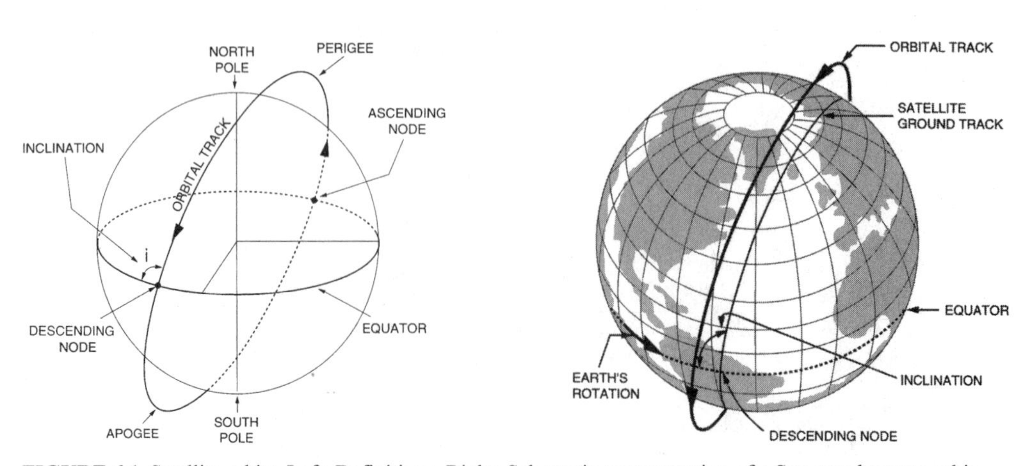


**FIGURE 6.1.** Satellite orbits. Left: Definitions. Right: Schematic representation of a Sun-synchronous orbit.

tion: that caused by differences in time of day, which arises from the fact that the spherical Earth rotates inside the solar beam. The *hour angle* ($h$) describes the difference in longitude between a point of interest and that of the direct solar beam (Figure 6.2). The value of $h$ can be found using the formula

$$h = [(GMT - 12.0) \times 15] - \text{longitude} \qquad \text{(Eq. 6.1)}$$

where GMT is Greenwich mean time and *longitude* is the longitude of the point in question. Because $h$ varies with longitude, to maintain uniform local sun angle, it is necessary to design satellite orbits that acquire each scene at the same local sun time. Careful selection of orbital height, eccentricity, and inclination can take advantage of the gravitational effect of the Earth's equatorial bulge to cause the plane of the satellite's orbit to rotate with respect to the Earth to match the seasonal motion of the solar beam. That is, the nodes of the satellite's orbit will move eastward about 1 degree each day, so that over a year's time the orbit will move through the complete 360° cycle. A satellite placed in a Sun-synchronous orbit will observe each part of the Earth within its view at the same local sun time each day (i.e., constantly), thereby removing time of day as a source of variation in illumination.

Although the optimum local sun time varies with the objectives of each project, most Earth observation satellites are placed in orbits designed to acquire imagery between 9:30 and 10:30 A.M. local sun time—a time that provides an optimum trade-off between ideal illumination for some applications and time of minimum cloud cover in tropical regions. Although it may initially seem that the characteristics of satellite orbits should be very stable and precisely known, in fact they are subject to numerous effects that disturb actual orbits and cause them to deviate from their idealized forms. Uncertainties in orbital path, timing, and orientation can lead to significant errors in estimation of satellite position, in aiming the sensor, and other variables that determine the geometric accuracy of an image. For example, a hypothetical satellite in an equatorial orbit that has a pointing (orientation) error of 1 degree will not be aimed at the intended point on the Earth's surface. The pointing error can be estimated as (altitude) × sin (angle). For an altitude of 800 km, the 1-degree pointing error could create a 14-km positional error in the

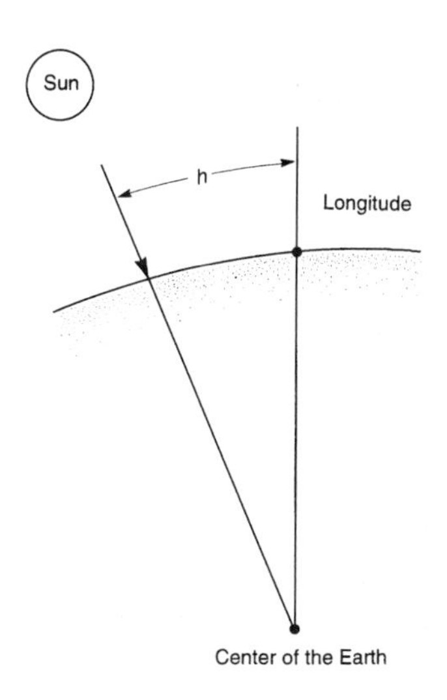


**FIGURE 6.2.** Hour angle. The hour angle ($h$) measures the difference in longitude between a point of interest ($A$) of known longitude and the longitude of the direct solar beam.

location of a point on an image. Although 1 degree might seem to be a small error, the 14 km result is obviously far too large for practical purposes, so it is clear that pointing errors for remote sensing satellites must be much smaller than 1 degree. (This example does not include contributions from other effects, such as uncertainties in knowledge of the orbital path, or in time, which determines the satellite's position along the orbital path.)

## 6.4. The Landsat System

Although the first generation of Landsat sensors is no longer in service, those satellites acquired a large library of images that are available as a baseline reference of environmental conditions for land areas throughout the world. Therefore, knowledge of these early Landsat images is important both as an introduction to later satellite systems and as a basis for work with the historical archives of images from Landsats 1, 2, and 3.

From 1972 to 1983, various combinations of Landsats 1, 2, and 3 orbited the Earth in sun-synchronous orbits every 103 minutes—14 times each day. After 252 orbits—completed every 18 days—Landsat passed over the same place on the Earth to produce repetitive coverage (Figure 6.3). When two satellites were both in service, their orbits were tailored to provide repetitive coverage every 9 days. Sensors were activated to acquire images only at scheduled times, so this capability was not always used. In addition, equipment malfunctions and cloud cover sometimes prevented acquisition of planned coverage.

As a function of the Earth's rotation on its axis from west to east, each successive north-to-south pass of the Landsat platform was offset to the west by 2,875 km (1,786 mi.) at the equator (Figure 6.4). Because the westward longitudinal shift of adjacent orbital tracks at the equator was approximately 159 km (99 mi.), gaps between tracks were incrementally filled during the 18-day cycle. Thus, on day 2, orbit number 1 was displaced 159 km to the west of the path of orbit number 1 on day 1. On the 18th day, orbit number 1 was identical to that of orbit number 1, day 1. The first orbit on the 19th day coincided with that of orbit number 1 on day 2, and so

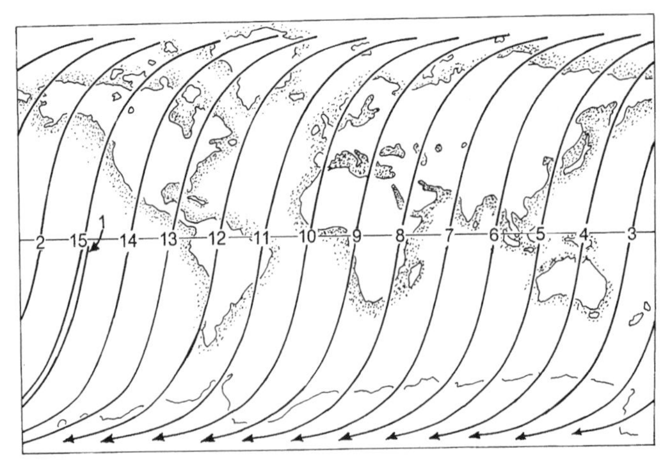


**FIGURE 6.3.** Coverage cycle, Landsats 1, 2, and 3. Each numbered line designates a northeast-to-southwest pass of the satellite. In a single 24-hour interval the satellite completes 14 orbits; the first pass on the next day (orbit 15) is immediately adjacent to pass 1 on the preceding day.

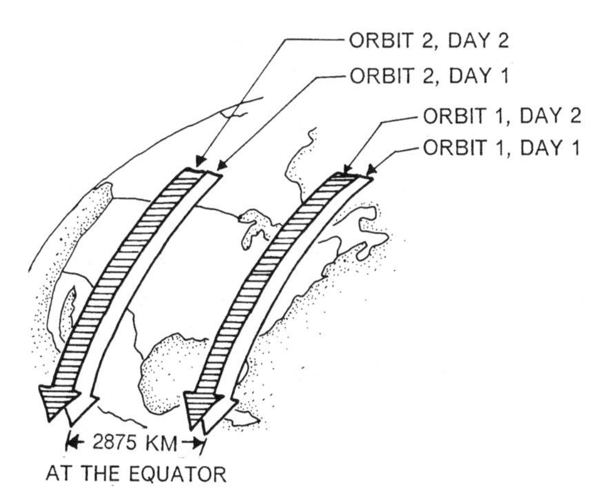


**FIGURE 6.4.** Incremental increases in Landsat 1 coverage. On successive days, orbital tracks begin to fill in the gaps left by the displacement of orbits during the preceding day. After 18 days, progressive accumulation of coverage fills in all gaps left by coverage acquired on day 1.

on. Therefore, the entire surface of the Earth between 81°N and 81°S latitude was subject to coverage by Landsat sensors once every 18 days (every 9 days, if two satellites were in service).

### *Support Subsystems*

Although our interest here is primarily focused upon the sensors that these satellites carried, it is important to briefly mention the *support subsystems,* units that are necessary to maintain proper operation of the sensors. Although this section refers specifically to the Landsat system, all Earth observation satellites require similar support systems. The *attitude control subsystem* (ACS) maintained orientation of the satellite with respect to the Earth's surface and with respect to the orbital path. The *orbit adjust subsystem* (OAS) maintained the orbital path within specified parameters after the initial orbit was attained. The OAS also made adjustments throughout the life of the satellite to maintain the planned repeatable coverage of imagery. The *power subsystem* supplied electrical power required to operate all satellite systems by means of two solar array panels and eight batteries. The batteries were charged by energy provided by the solar panels while the satellite was on the sunlit side of the Earth, then provided power when the satellite was in the Earth's shadow. The *thermal control subsystem* controlled the temperatures of satellite components by means of heaters, passive radiators (to dissipate excess heat), and insulation. The *communications and data-handling subsystem* provided microwave communications with ground stations for transmitting data from the sensors, commands to satellite subsystems, and information regarding satellite status and location.

Data from sensors were transmitted, in digital form, by microwave signal to ground stations equipped to receive and process data. Direct transmission from the satellite to the ground station as the sensor acquired data was possible only when the satellite had direct line-of-sight view of the ground antenna (a radius of about 1,800 km from the ground stations). In North America, stations at Greenbelt, Maryland; Fairbanks, Alaska; Goldstone, California; and Prince Albert, Saskatchewan, Canada, provided this capability for most of the United States and Canada. Elsewhere a network of ground stations has been established over a period of years through agreements with other nations. Areas outside the receiving range of a ground station could be imaged only by use of the two tape recorders onboard each of the early Landsats. Each tape recorder could record about 30 min. of data; then, as the satellite moved within range of a ground station, the recorders could transmit the stored data to a receiving station. Thus these satellites had, within the limits of their orbits, a capability for worldwide coverage. Unfortunately, the tape recorders proved to be one of the most unreliable elements of the Landsat system, and when they failed the system was unable to image areas beyond the range of the ground stations. These unobservable areas became smaller as more ground stations were established, but there were always some areas that Landsat could not observe. Later satellites were able to avoid this problem by use of communications relay satellites.

### *Return Beam Vidicon*

The return beam vidicon (RBV) camera system generated high-resolution television-like images of the Earth's surface. Its significance for our discussion is that it was intended to apply the remote sensing technology of the day for a spacecraft to use from orbital rather than from aircraft altitudes. Thus it provided three spectral channels, in the green, red, and near infrared, to replicate the information conveyed by color infrared film. The RBV was designed to provide a camera-like perspective, using a shutter, and a electronic record of the image projected into the focal plane, to provide an image that could be used for application of photogrammetry, in much the same way that photogrammetry was used to analyze aerial photographs acquired at aircraft altitudes.

On Landsats 1 and 2, the RBV system consisted of three independent cameras that operated simultaneously, with each sensing a different segment of the spectrum (Table 6.2). All three instruments were aimed at the same region beneath the satellite, so the images they acquired registered to one another to form a three-band multispectral representation of a 185 km × 170 km ground area, known as a *Landsat scene* (Figure 6.5). This area matched the area represented by the corresponding MSS scene. The RBV shutter was designed to open briefly, in the manner of a camera shutter, to simultaneously view the entire scene. Technical difficulties prevented routine use of the RBV, so attention was then directed to imagery from the other sensor on the Landsat, the MSS, then considered to be an experimental system of yet untested capabilities, but a system that proved to be highly successful, and over the next decades became the model for collection of imagery from space.

### *Multispectral Scanner Subsystem*

As a result of malfunctions in the RBV sensors early in the missions of Landsats 1 and 2, the multispectral scanner subsystem (MSS) became the primary Landsat sensor. Whereas the RBV was designed to capture images with known geometric properties, the MSS was tailored to provide multispectral data without as much concern for positional accuracy. In general, MSS imagery and data were found to be of good quality—indeed, much better than many expected—and clearly demonstrated the merits of satellite observation for acquiring Earth resources data. The economical routine availability of MSS digital data has formed the foundation for a size-

**TABLE 6.2. Landsats 1–5 Sensors**

| Sensor | Band | Spectral sensitivity |
|---|---|---|
| | Landsats 1 and 2 | |
| RBV | 1 | 0.475–0.575 µm (green) |
| RBV | 2 | 0.58–0.68 µm (red) |
| RBV | 3 | 0.69–0.83 µm (near infrared) |
| MSS | 4 | 0.5–0.6 µm (green) |
| MSS | 5 | 0.6–0.7 µm (red) |
| MSS | 6 | 0.7–0.8 µm (near infrared) |
| MSS | 7 | 0.8–1.1 µm (near infrared) |
| | Landsat 3 | |
| RBV | | 0.5–0.75 µm (panchromatic response) |
| MSS | 4 | 0.5–0.6 µm (green) |
| MSS | 5 | 0.6–0.7 µm (red) |
| MSS | 6 | 0.7–0.8 µm (near infrared) |
| MSS | 7 | 0.8–1.1 µm (near infrared) |
| MSS | 8 | 10.4–12.6 µm (far infrared) |
| | Landsats 4 and 5[a] | |
| TM | 1 | 0.45–0.52 µm (blue-green) |
| TM | 2 | 0.52–0.60 µm (green) |
| TM | 3 | 0.63–0.69 µm (red) |
| TM | 4 | 0.76–0.90 µm (near infrared) |
| TM | 5 | 1.55–1.75 µm (mid infrared) |
| TM | 6 | 10.4–12.5 µm (far infrared) |
| TM | 7 | 2.08–2.35 µm (mid infrared) |
| MSS | 1 | 0.5–0.6 µm (green) |
| MSS | 2 | 0.6–0.7 µm (red) |
| MSS | 3 | 0.7–0.8 µm (near infrared) |
| MSS | 4 | 0.8–1.1 µm (near infrared) |

[a]On Landsats 4 and 5 MSS bands were renumbered although the spectral definitions remained the same.

able increase in the number and sophistication of digital image-processing capabilities available to the remote sensing community. A version of the MSS was placed on Landsats 4 and 5; later systems have been designed with an eye toward maintaining the continuity of MSS data.

The MSS (Figure 6.6) is a scanning instrument utilizing a flat oscillating mirror to scan from west to east to produce a ground swath of 185 km (100 nautical mi.) perpendicular to the orbital track. The satellite motion along the orbital path provides the along-track dimension to the image. Solar radiation reflected from the Earth's surface is directed by the mirror to a telescope-like instrument that focuses the energy onto fiber optic bundles located in the focal plane of the telescope. The fiber optic bundles then transmit energy to detectors sensitive to four spectral regions (Table 6.2).

Each west-to-east scan of the mirror covers a strip of ground approximately 185 km long in the east–west dimension and 474 m wide in the north–south dimension. The 474-m distance corresponds to the forward motion of the satellite during the interval required for the west-to-east movement of the mirror and its inactive east-to-west retrace to return to its starting position. The mirror returns to start another active scan just as the satellite is in position to record

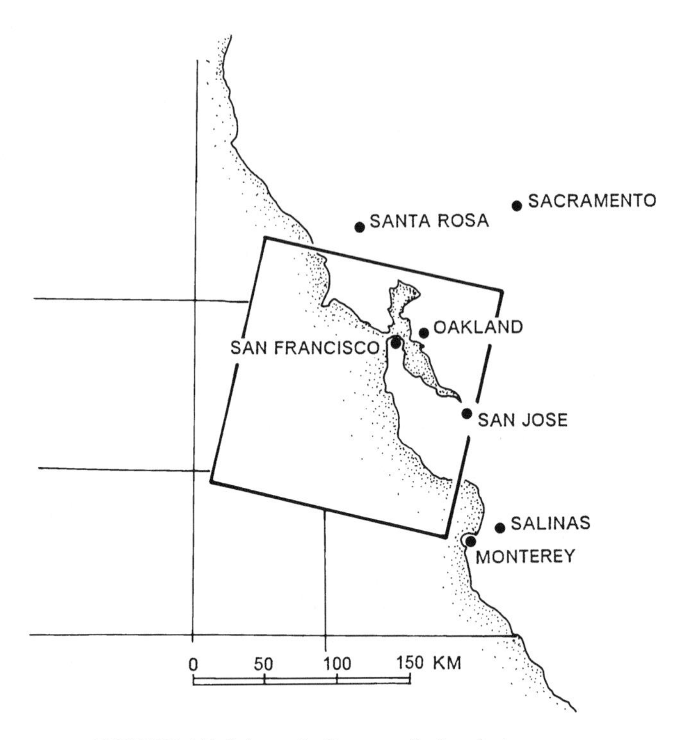


**FIGURE 6.5.** Schematic diagram of a Landsat scene.

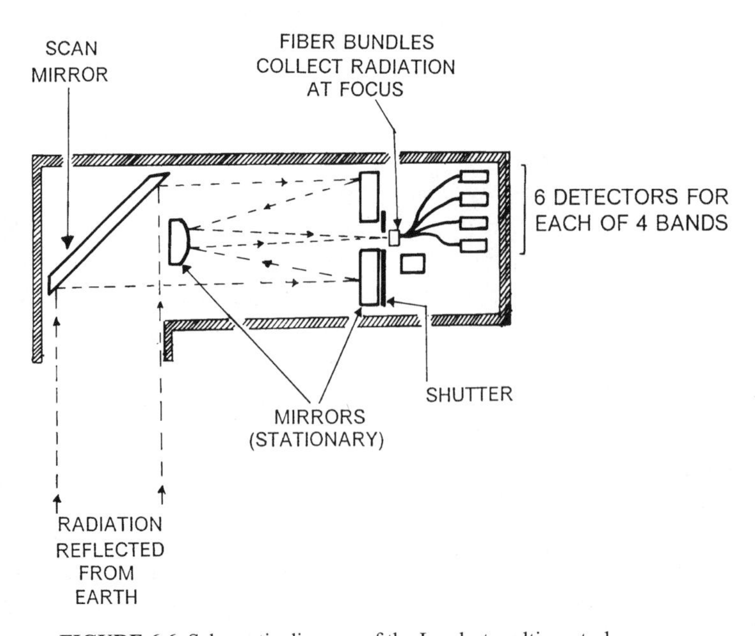


**FIGURE 6.6.** Schematic diagram of the Landsat multispectral scanner.

another line of data at a ground position immediately adjacent to the preceding scan line. Each motion of the mirror corresponds to six lines of data on the image because the fiber optics split the energy from the mirror into six contiguous segments.

The *instantaneous field of view* (IFOV) of a scanning instrument can be informally defined as the ground area viewed by the sensor at a given instant in time. The nominal IFOV for the MSS is 79 m × 79 m. Slater (1980) provides a more detailed examination of MSS geometry that shows the IFOV to be approximately 76-m square (about 0.58 ha, or 1.4 acres), although differences exist between Landsats 1, 2, and 3, and within orbital paths as satellite altitude varies. The brightness from each IFOV is displayed on the image as a pixel (picture element) formatted to correspond to a ground area said to be approximately 79 m × 57 m in size (about 0.45 ha, or 1.1 acres). In everyday terms, the ground area corresponding to a MSS pixel can be said to be somewhat less than that of a U.S. football field (Figure 6.7d).

For the MSS instruments onboard Landsat 1 and 2, the four spectral channels were located in the green, red, and infrared portions of the spectrum:

- *Band 1:* 0.5–0.6 µm (green)
- *Band 2:* 0.6–0.7 µm (red)
- *Band 3:* 0.7–0.8 µm (near infrared)
- *Band 4:* 0.8–1.1 µm (near infrared)

(Originally, the four MSS bands were designated as Bands 4, 5, 6, and 7. To maintain continuity of band designations, later bands designations were changed, even though the spectral designations remained consistent.)

The Landsat 3 MSS included an additional band in the far infrared from 10.4 to 12.6 µm. Because this band included only two detectors, energy from each scan of the mirror was subdivided into only two segments, each 234 m wide. Therefore, the IFOV for the thermal band was 234 × 234 m, much coarser than MSS images for the other bands.

Although the MSS has now been replaced by subsequent systems, it is significant because it introduces, in rather basic form, concepts used for later systems that have become too complex to be discussed in detail in an introductory text. Further, it has an important historic significance, as the techniques developed by scientists to interpret MSS imagery form the origins of practice of digital image processing, which is now the principal means of examining satellite imagery, and a significant portion of imagery from other sources.

## 6.5. Multispectral Scanner Subsystem Images

### *The Multispectral Scanner Subsystem Scene*

The *MSS scene* is defined as an image representing a ground area approximately 185 km in the east–west (across-track) direction, and 178 km in the north–south (along-track) direction (Figure 6.8). The across-track dimension is defined by the side-to-side motion of the MSS; the along-track dimension is defined by the forward motion of the satellite along its orbital path. If the MSS were operated continuously for an entire descending pass, it would provide a continuous strip of imagery representing an area 185-km wide. The 178-km north–south dimension simply divides this strip into segments of convenient size.

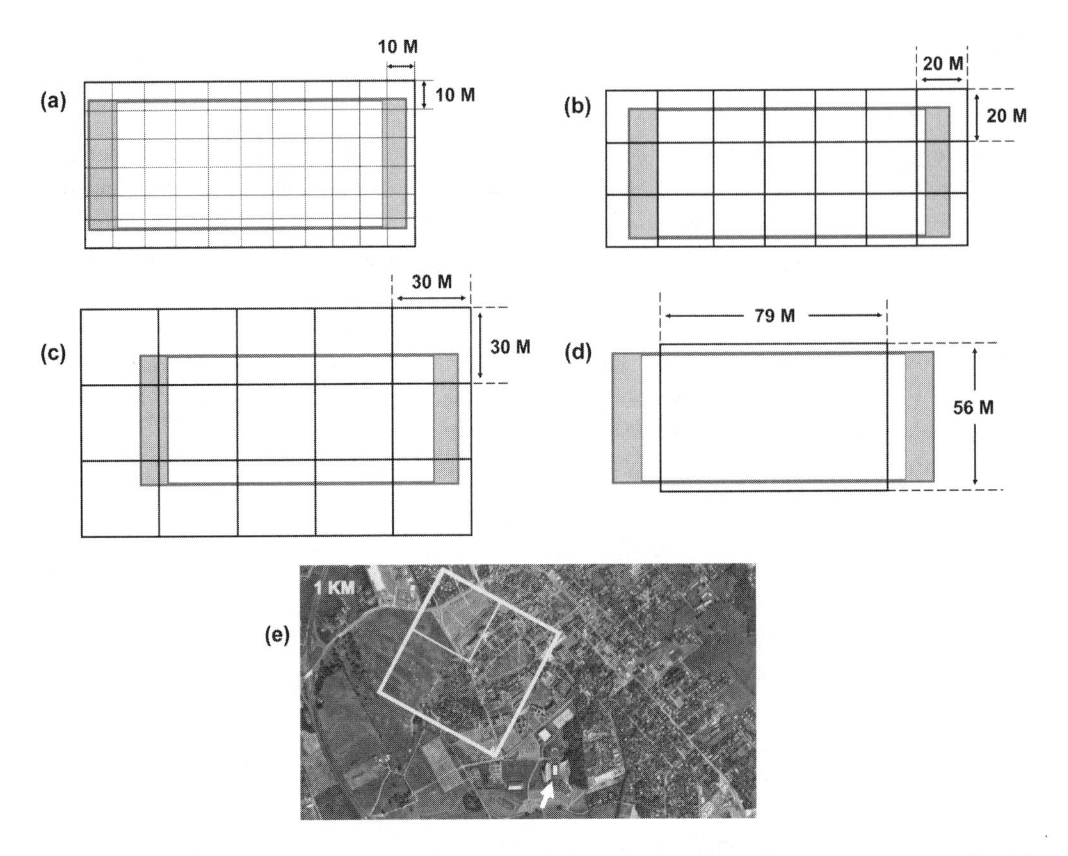


**FIGURE 6.7.** Schematic representations of spatial resolutions of selected satellite imaging systems. Spatial detail of satellite imagery varies greatly. This diagram attempts to depict the relative sizes of pixels with reference to the dimensions of a U.S. football field as an everyday reference. (a) SPOT HRV, panchromatic mode; (b) SPOT HRV, multispectral mode: (c) Landsat TM; (d) Landsat MSS imagery (for the MSS, the orientation of the pixel is rotated 90° to match to the usual rendition of a football field; on a MSS image, the narrow ends of MSS pixels are oriented approximately north and south); (e) broad-scale remote sensing satellites depict much coarser spatial resolution, often on the order of a kilometer or so, indicated by the large square, or a quarter of a kilometer, depicted by the smaller square, against the background of an aerial photograph of a small town. For comparison, note in the lower center of image (e) that the white rectangle (indicated by the white arrow) represents a football field of the size used as a reference in images (a) through (d). Examples (a) through (d) represent resolutions of Landsat-class systems, as discussed in the text. Example (e) represents detail portrayed by broad-scale observation systems. The fine-resolution systems discussed in the text are not represented in this illustration—such systems might represent the football field with hundreds to thousands of pixels.

The MSS scene, then, is an array of pixel values (in each of four bands) consisting of about 2,400 scan lines, each composed of 3,240 pixels (Figure 6.8). Although centerpoints of scenes acquired at the same location at different times are intended to register with each other, there is often in fact a noticeable shift from one date to another (known as the *temporal registration problem*) in ground locations of center points due to uncorrected drift in the orbit.

There is a small overlap (about 5%, or 9 km) between scenes to the north and south of a given scene. This overlap is generated by repeating the last few lines from the preceding image, not by stereoscopic viewing of the Earth. Overlap with scenes to the east and west depends upon latitude; sidelap will be a minimum of 14% (26 km) at the equator and increases with latitude to 57% at 60°, then to 85% at 80°N and S latitude. Because this overlap is created by viewing the

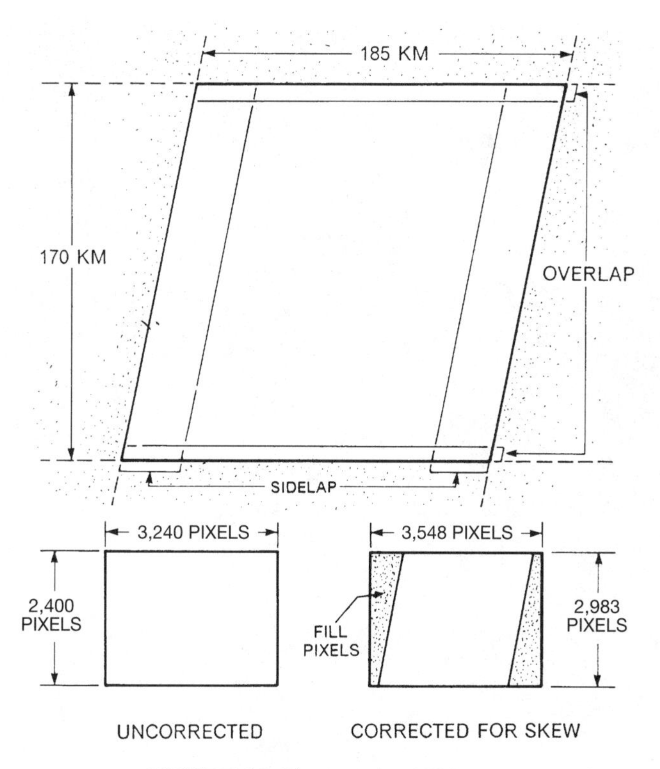


**FIGURE 6.8.** Diagram of an MSS scene.

same area of the Earth from different perspectives, the area within the overlap can be viewed in stereo. At high latitudes, this area can constitute an appreciable portion of a scene.

### *Image Format*

MSS data are available in several image formats that have been subjected to different forms of processing to adjust for geometric and radiometric errors. The following section describes some of the basic forms for MSS data, which provide a general model for other kinds of satellite imagery, although specifics vary with each kind of data.

In its initial form a digital satellite image consists of a rectangular array of pixels in each of four bands (Figure 6.8). In this format, however, no compensation has been made for the combined effects of spacecraft movement and rotation of the Earth as the sensor acquires the image (this kind of error is known as skew). When these effects are removed, the image assumes the shape of a parallelogram (Figure 6.8). For convenience in recording data, *fill* pixels are added to preserve the correct shape of the image. Fill pixels, of course, convey no information, as they are simply assigned values of zero as necessary to attain the desired shape.

Each of the spectral channels of a multispectral image forms a separate image, each emphasizing landscape features that reflect specific portions of the spectrum (Figures 6.9 and 6.10). These separate images, then, record in black-and-white form the spectral reflectance in the green, red, and infrared portions of the spectrum, for example. The green, red, and infrared

**FIGURE 6.9.** Landsat MSS band 2. Image of New Orleans, Louisiana, 16 September 1982 (scene ID 40062-15591-2). Image reproduced by permission of EOSAT. (MSS band 2 was formerly designated as band 5; see Table 6.2.)

bands can be combined into a single color image (Plate 6), known as a *false-color composite.* The near infrared band is projected onto color film through a red filter, the red band 2 through a green filter, and the green band through a blue filter. The result is a false-color rendition that uses the same assignment of colors used in conventional color infrared aerial photography (Chapter 3). Strong reflectance in the green portion of the spectrum is represented as blue on the color composite, red as green, and infrared as red. Thus living vegetation appears as bright red, turbid water as a blue color, and urban areas as a gray or sometimes pinkish gray.

On photographic prints of MSS images, the annotation block at the lower edge gives essential information concerning the identification, date, location, and characteristics of the image. During the interval since the launch of Landsat 1, the content and form of the annotation block have been changed several times, but some of the basic information can be shown in a simplified form to illustrate key items (Figure 6.11). The *date* has obvious meaning. The *format center and ground nadir* give, in degrees and minutes of latitude and longitude, the ground location of the center point of the image. The spectral band is given in the form "MSS 1" (meaning multispectral scanner, band 1). *Sun angle* and *sun elevation* designate, in degrees, the solar eleva-

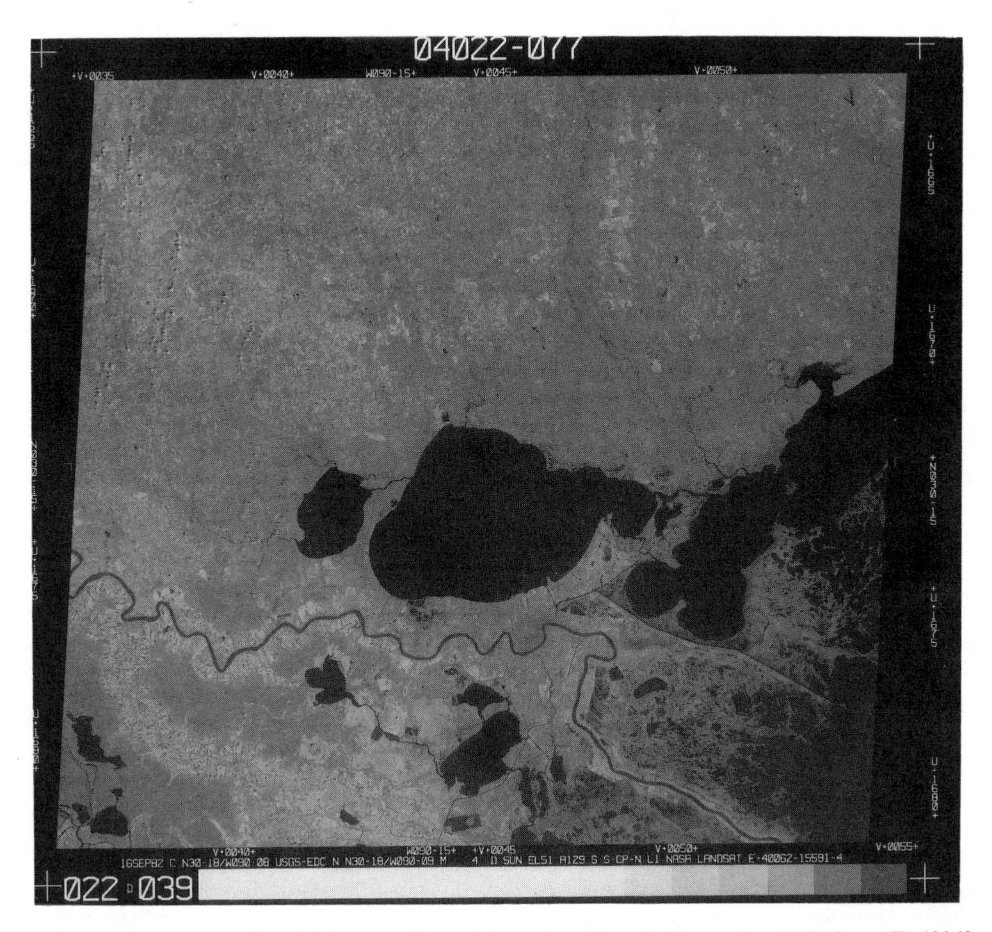


**FIGURE 6.10.** Landsat MSS band 4. Image of New Orleans, Louisiana, 16 September 1982 (Scene ID 40062-15591-7). Image reproduced by permission of EOSAT. (MSS band 4 was formerly designated as band 7; see Table 6.2.)

tion (above the horizon) and the azimuth of the solar beam from true north at the center of the image. Of the remaining items on the annotation block, the most important for most users is the *scene ID,* a unique number that specifies the scene and band. The scene ID uniquely specifies any MSS scene, so it is especially useful as a means of cataloging and indexing MSS images as explained below. MSS imagery can be purchased from the Space Imaging Corporation and the U.S. Geological Survey's EROS Data Center (see *http://geo.arc.nasa.gov/sge/landsat/daccess.html*).

### *Worldwide Reference System*

The *worldwide reference system* (WRS), a concise designation of nominal center points of Landsat scenes, is used to index Landsat scenes by location. The reference system is based upon a coordinate system in which there are 233 north–south paths corresponding to orbital tracks of the satellite, and 119 rows representing latitudinal center lines of Landsat scenes. The combina-

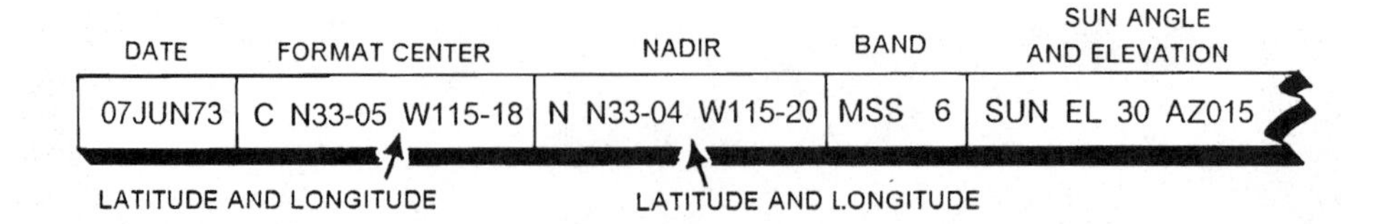


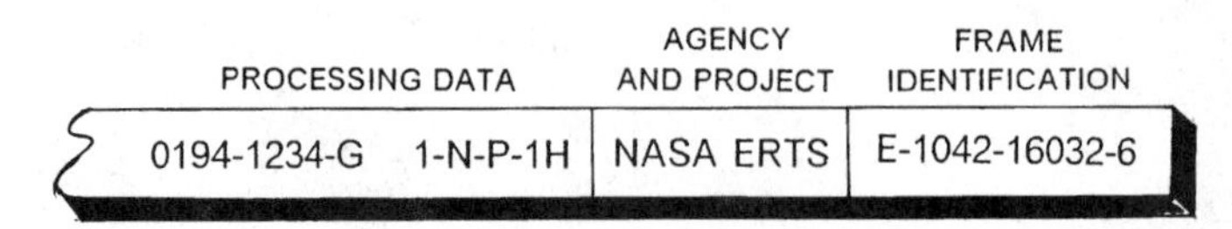


**FIGURE 6.11.** Landsat annotation block. The annotation block has been changed several times, but all show essentially the same information.

tion of a path number and a row number uniquely identifies a nominal scene center (Figure 6.12). Because of the drift of satellite orbits over time, actual scene centers may not match exactly to the path–row locations, but the method does provide a convenient and effective means of indexing locations of Landsat scenes. To outline the concept in its most basic form, this discussion has presented the Landsat 1–3 WRS; note that Landsats 4 through 7 each have WRS systems analogous to those described here, but they differ with respect to detail. Therefore, investigate the specifics of the WRS system you expect to use.

## 6.6. Landsat Thematic Mapper

Even before Landsat 1 was launched, it was recognized that existing technology could improve the design of the MSS. Efforts were made to incorporate improvements into a new instrument

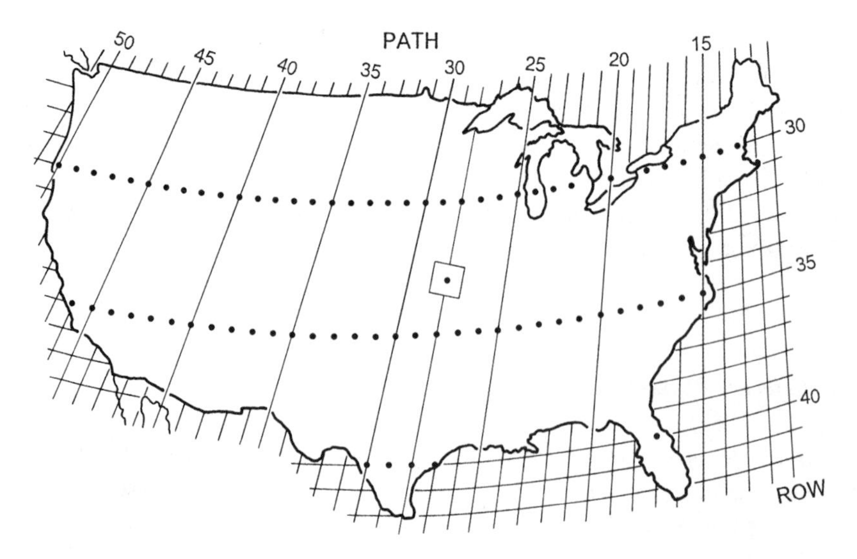


**FIGURE 6.12.** WRS path–row coordinates for the United States.

modeled on the basic design of the MSS. Landsats 4 and 5 carried a replacement for MSS known as the *thematic mapper* (TM), which can be considered as an upgraded MSS. In these satellites, both a TM and a MSS were carried on an improved platform that maintained a high degree of stability in orientation as a means of improving geometric qualities of the imagery.

TM was based upon the same principles as the MSS, but had a more complex design. It provides finer spatial resolution, improved geometric fidelity, greater radiometric detail, and more detailed spectral information in more precisely defined spectral regions. Improved satellite stability and an orbit adjust subsystem were designed to improve positional and geometric accuracy. The objectives of the second generation of Landsat instruments were to assess the performance of the TM, to provide ongoing availability of MSS data, and to continue foreign data reception.

Despite the historical relationship between the MSS and the TM, the two sensors are distinct. whereas the MSS has four broadly defined spectral regions, the TM records seven spectral bands (Table 6.3). TM band designations do not follow the sequence of the spectral definitions (the band with the longest wavelengths is band 6, rather than band 7) because band 7 was added so late in the design process that it was not feasible to relabel bands to follow a logical sequence.

TM spectral bands were tailored to record radiation of interest to specific scientific investigations rather than the more arbitrary definitions used for the MSS. Spatial resolution is about 30 m (about 0.09 ha, or 0.22 acre) (Figure 6.7c), compared to the 76-m IFOV of the MSS. (TM band 6 had a coarser spatial resolution of about 120 m.) The finer spatial resolution provided a noticeable increase (relative to the MSS) in spatial detail recorded by each TM image (Figure

**TABLE 6.3. Summary of TM Sensor Characteristics**

| Band | Resolution | Spectral definition | Some applications[a] |
|---|---|---|---|
| 1 | 30 m | Blue-green, 0.45–0.52 μm | Penetration of clear water; bathymetry; mapping of coastal waters; chlorophyll absorption; distinction between coniferous and deciduous vegetation |
| 2 | 30 m | Green, 0.52–0.60 μm | Records green radiation reflected from healthy vegetation; assesses plant vigor; reflectance from turbid water |
| 3 | 30 m | Red, 0.63–0.69 μm | Chlorophyll absorption important for plant-type discrimination |
| 4 | 30 m | Near infrared, 0.76–0.90 μm | Indicator of plant cell structure; biomass; plant vigor; complete absorption by water facilitates delineation of shorelines |
| 5 | 30 m | Mid-infrared, 1.55–1.75 μm | Indicative of vegetation moisture content; soil moisture mapping; differentiating snow from clouds; penetration of thin clouds |
| 6 | 120 m | Far infrared, 10.4–12.5 μm | Vegetation stress analysis; soil moisture discrimination; thermal mapping; relative brightness temperature; soil moisture; plant heat stress |
| 7 | 30 m | Mid-infrared, 2.08–2.35 μm | Discrimination of rock types; alteration zones for hydrothermal mapping; hydroxyl ion absorption |

[a]Sample applications listed here; these are not the only applications.

6.7). Digital values are quantized at 8 bits (256 brightness levels), which provide (relative to the MSS) a much larger range of brightness values. These kinds of changes produced images with much finer detail than those of the MSS (Figures 6.13 and 6.14).

Each scan of the TM mirror acquired 16 lines of data. Unlike the MSS, the TM scan acquires data as it moves in both the east–west and the west–east directions. This feature permitted engineers to design a slower speed of mirror movement, thereby improving the length of time the detectors can respond to brightness in the scene. However, this design required additional processing to reconfigure image positions of pixels to form a geometrically accurate image. TM detectors are positioned in an array in the focal plane (it does not use the fiber optics employed in the MSS to ensure perfect spatial registration of the MSS bands); as a result, there may be a slight misregistration of TM bands.

TM imagery is analogous to MSS imagery with respect to areal coverage and organization of data into several sets of multispectral digital values that overlay to form an image (Plate 5). In

**FIGURE 6.13.** TM band 3 image of New Orleans, Louisiana, 16 September 1982 (Scene ID 40062-15591-4).

**FIGURE 6.14.** TM band 4 image of New Orleans, Louisiana, 16 September 1982 (Scene ID 40062-15591-5).

comparison with MSS images, TM imagery has a much finer spatial and radiometric resolution, so TM images show relatively fine detail of patterns on the Earth's surface.

Use of seven rather than four spectral bands, and use of a smaller pixel size within the same image area, means that TM images consist of many more data values than do MSS images. As a result, each analyst must determine those TM bands that are most likely to provide the required information. Because the "best" combinations of TM bands vary according to the purpose of each study, season, geographic region, and other factors, a single selection of bands is unlikely to be equally effective in all circumstances.

A few combinations of TM bands appear to be effective for general purpose use. Use of TM bands 2, 3, and 4 creates an image that is analogous to the usual false-color aerial photograph. TM bands 1 (blue-green), 2 (green), and 3 (red) form a natural-color composite, approximately equivalent to a color aerial photograph in its rendition of colors. Experiments with other combinations have shown that Bands 2, 4, and 5; 2, 4, and 7; and 2, 4, and 5 are also effective for visual interpretation. Of course, there are many other combinations of the seven TM bands that may be useful in specific circumstances. TM imagery can be purchased from the USGS's EROS Data Center (see *http://edc.usgs.gov*).

### *Orbit and Ground Coverage: Landsats 4 and 5*

Landsats 4 and 5 were placed into orbits resembling those of earlier Landsats. Sun-synchronous orbits brought the satellites over the equator at about 9:45 A.M., thereby maintaining approximate continuity of solar illumination with imagery from Landsats 1, 2, and 3. Data were collected as the satellite passed northeast to southwest on the sunlit side of the earth. The image swath remained at 185 km. In these respects, coverage was compatible with that of the first generation of Landsat systems.

However, there were important differences. The finer spatial resolution of the TM was achieved in part by a lower orbital altitude, which required several changes in the coverage cycle. Earlier Landsats produced adjacent image swaths on successive days. However, Landsats 4 and 5 acquired coverage of adjacent swaths at intervals of 7 days. Landsat 4 completed a coverage cycle in 16 days. Successive passes of the satellite were separated at the equator by 2,752 km; gaps between successive passes were filled in over an interval of 16 days. Adjacent passes were spaced at 172 km. At the equator, adjacent passes overlapped by about 7.6%; overlap increased as latitude increased.

A complete coverage cycle was achieved in 16 days—233 orbits. Because this pattern differs from that at earlier Landsat systems, Landsats 4 and later required a new WRS indexing system for labeling paths and rows (Figure 6.15). Row designations remained the same as before, but a new system of numbering paths was required. In all there are 233 paths and 248 rows, with row 60 positioned at the equator.

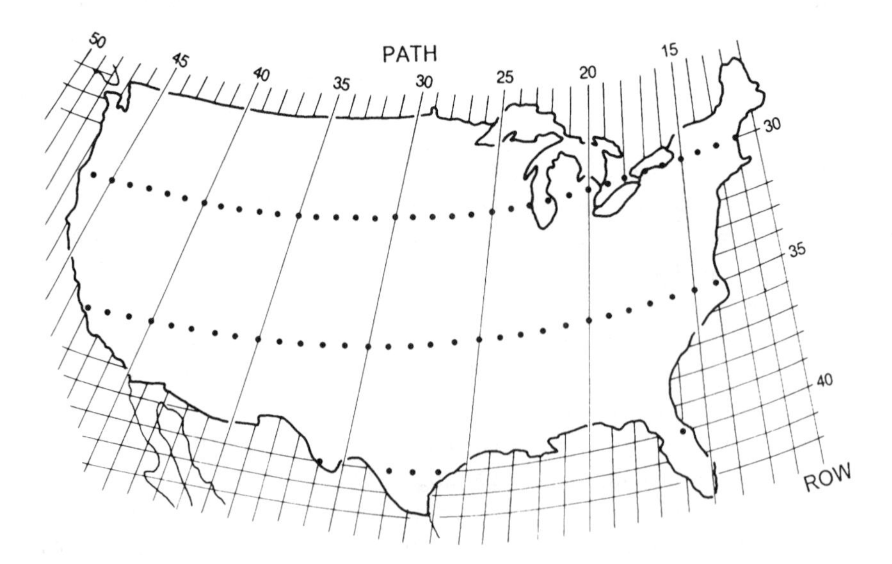


**FIGURE 6.15.** WRS path–row coordinates for Landsats 4 and 5.

## 6.7. Administration of the Landsat Program

Landsat was originally operated by NASA as a part of its mission to develop and demonstrate applications of new technology related to aerospace engineering. Although NASA had primary responsibility for Landsat, other federal agencies, including the USGS and the National Oceanographic and Atmospheric Administration (NOAA), contributed to the program. These federal agencies operated the Landsat system for many years, but Landsat was officially considered to be "experimental" because of NASA's mission to develop new technology (rather than assume responsibility for routine operations) and because other federal agencies were unwilling to assume responsibility for a program of such cost and complexity.

Over a period of many years, successive presidential administrations and numerous sessions of the U.S. Congress debated to define a model for operation of U.S. land remote sensing policy. In essence, the debate has centered on the merits of public operation (by agencies of the federal government) versus operation by a private corporation. Those favoring private operation emphasize the prospects for more efficient operation of the system and more aggressive pursuit of new applications and new technologies. Those who favored continued government operation stressed the need to maintain long-term continuity of data flow and data format, the significance of maintaining the scientific integrity of the data, the importance of providing for public access to data, and the significance of a public archive of data as a historical record.

Congress has passed two acts of significance to U.S. land remote sensing satellites. In June 1984 the U.S. Congress passed the Land Remote Sensing Act of 1984, which established a process by which the Landsat system passed from public to private operation. The structure specified by this act was not successful. In October 1992 Congress passed the Land Remote Sensing Policy Act of 1992. This act and related agreements among federal agencies are described by Sheffner (1994), Salomonson et al. (1996) and Lauer et al. (1997). Its provides measures to maintain continuity in the collection of data, establishment of the basis for a pricing policy for image data, and an outline of the nature of additional instruments to be designed for future satellites. The act outlines four options for continued development of the system: (1) funding and management by an international consortium, (2) funding and management by the U.S. government, joint operation by the U.S. government and private industry, or (4) funding and management through the private sector. NASA and the USGS have a mandate to investigate how these options might be implemented. Congress made clear its preference that private industry have a strong role in future development and operation of U.S. satellite remote sensing systems. At the time of this writing, there is no clear structure in place to provide for continued U.S. leadership for Landsat-class land remote sensing satellite observation.

## 6.8. Current Satellite Systems

This outline of early Landsat systems sets the stage for examination of the more advanced systems now in use or scheduled for deployment in the near future. In recent years the number of such systems has increased so rapidly that it is impractical to list them, let alone describe all of their characteristics. Understanding this wide range of satellite systems can be easier if we consider them as members of three families of satellites.

The first group consists of *Landsat-like systems,* designed for acquisition of rather broad geographic coverage at moderate levels of detail. Data from these systems have been used for an amazingly broad range of applications, which can be generally described as focused on survey and monitoring of land and water resources. A second group consists of those satellite observation systems designed to acquire *very-broad-scale images at coarse resolutions,* intended in part to acquire images that can be aggregated to provide continental or global coverage. Such images enable scientists to monitor broad-scale environmental dynamics. Finally, a third family of satellite systems provides *fine detail for small regions,* to acquire imagery that might assist in urban planning, or design of highway or pipeline routes, for example. Although this categorization is imperfect, it does provide a framework that helps us understand the capabilities of a very large number of satellite systems now available.

## *Landsat-Class Systems*

The success of Landsat in the early 1970s stimulated interests in other nations. The Landsat model has been the template for the land remote sensing systems proposed and in use by many other nations. Lauer, Morain, and Salomonson (1997) have listed 33 separate Landsat or Landsat-like systems launched or planned for launch between 1972 and 2007. This number attests to the value of observational systems designed to acquire regional overviews at moderate-to-coarse levels of spatial detail. Landsat has been the model for these systems with respect to all their essential characteristics: technological design, data management, and overall purpose. For example, Landsat established the value of digital data for the general community of remote sensing practitioners, set the expectation that imagery would be available for all customers, and established a model for organizing, archiving, and cataloging imagery. Related systems have followed Landsat's lead in almost every respect.

### *Landsat 7 and the Enhanced Thematic Mapper Plus*

Landsat's current flagship sensor, the enhanced thematic mapper plus (ETM+), was placed in orbit in April 1999, with the successful launch of Landsat 7. This system was designed by NASA and is operated by NOAA and USGS. ETM+ is designed to extend the capabilities of previous TMs by adding modest improvements to the TM's original design. In the visible, near infrared, and mid-infrared, its spectral channels duplicate those of earlier TMs (Table 6.4). The thermal channel has 60-m resolution, improved from the 120-m resolution of earlier TMs. The ETM+ also has a 15-m panchromatic channel. Swath width remains at 185 km, and the system is characterized by improvements in accuracy of calibration, data transmission, and other characteristics.

ETM+ extends the continuity of earlier Landsat data (from both MSS and TM) by maintaining consistent spectral definitions, resolutions, and scene characteristics, while taking advantage of improved technology, calibration, and efficiency of data transmission. Details concerning Landsat 7 and ETM+ are available at the Landsat Data Access web page (*http://geo.arc.nasa.gov/sge/landsat/daccess.html*) and at the Landsat 7 web page (*http://landsat.gsfc.nasa.gov/*).

In May 2003 the Landsat 7 TM experienced an anomaly in the functioning of the scan-line corrector (SLC), a portion of the sensor that assures that scan lines are oriented properly in the

**TABLE 6.4. Spectral Characteristics for ETM+**

| Band | Spectral range | Ground resolution |
|---|---|---|
| 1 | 0.450–0.515 μm | 30 m |
| 2 | 0.525–0.605 μm | 30 m |
| 3 | 0.630–0.690 μm | 30 m |
| 4 | 0.75–0.90 μm | 30 m |
| 5 | 1.55–1.75 μm | 30 m |
| 6 | 10.4–12.5 μm | 60 m |
| 7 | 2.09–2.35 μm | 30 m |
| Pan | 0.52–0.90 μm | 15 m |

*Note.* See Table 6.3.

processed imagery (Section 6.6). As a result, the images display wedge-shaped gaps near the edges, such that only the central two-thirds of the image presents a continuous view of the scene. Although images can be processed to address cosmetic dimension of the problem, the instrument is restricted in its ability to fulfill its mission as long as this problem persists. As a result, the aging Landsat 5 TM remains as the only fully functional Landsat system at the time of this writing. The system will be operated in a manner that will maximize its capabilities and minimize the impact of the SLC anomaly. Because there is no program underway to replace the Landsat system in a timely manner, it seems likely that there will be a gap in Landsat converge as the Landsat 5 TM approaches the end of its service. Although systems operated by other nations and other U.S. systems can provide useful imagery in the interim, it seems clear that there will be a loss in the continuity and compatibility of the image achieve.

Landsat 7 ETM+ data are distributed to the public from the USGS's EROS Data Center (EDC), the primary Landsat 7 receiving station in the United States:

Customer Services
U.S. Geological Survey
EROS Data Center
Sioux Falls, SD 57198-0001
Tel: 605-594-6151
Fax: 605-594-6589
E-mail: custserv@usgs.gov

U.S. Landsat remote sensing satellites are operated by and their images are marketed by Space Imaging Corporation:

Space Imaging
12076 Grant Street
Thornton, CO 80241
Tel: 303-254-2000;
toll Free US: 800-425-9037
Fax: 1-303-254-2215
E-mail: info@spaceimaging.com

## *SPOT*

### *SPOTs 1, 2, and 3*

SPOT—Le Système pour l'Observation de la Terre (Earth Observation System)—began operations in 1986, with the launch of SPOT 1. SPOT was conceived and designed by the French Centre National d'Etudes Spatiales (CNES) in Paris, with the cooperation of other European organizations. SPOT 1, launched in February 1986, was followed by SPOT 2 (January 1990), SPOT 3 (September 1993), SPOT 4 (March 1998), and SPOT 5 (May 2002), the first three of which carry sensors identical to those of SPOT 1. SPOT 5 carries sensors modified from those used on the earlier systems.

The SPOT system is designed to provide data for land-use studies, assessment of renewable resources, exploration of geologic resources, and cartographic work at scales of 1:50,000–1:100,000. Design requirements included provision for complete world coverage, rapid dissemination of data, stereo capability, high spatial resolution, and sensitivity in spectral regions responsive to reflectance from vegetation.

The SPOT *bus* is the basic satellite vehicle, designed to be compatible with a variety of sensors (Figure 6.16). The bus provides basic functions related to orbit control and stabilization, reception of commands, telemetry, monitoring of sensor status, and the like. The bus, with its sensors, is placed in a Sun-synchronous orbit at about 832 km, with a 10:30 A.M. equatorial crossing time. For vertical observation, successive passes occur at 26-day intervals, but because of the ability of SPOT sensors to be pointed off-nadir, successive imagery can be acquired, on the average, at 2½ day intervals. (The exact interval for repeat coverage varies with latitude.)

The SPOT payload consists of two identical sensing instruments, a telemetry transmitter, and magnetic tape recorders. The two sensors are known as HRV ("high resolution visible") instruments. HRV sensors use pushbroom scanning, based upon CCDs (discussed in Chapter 4), that simultaneously images an entire line of data in the cross-track axis (Figure 4.6). SPOT linear arrays consist of 6,000 detectors for each scan line in the focal plane; the array is scanned electronically to record brightness values (8 bits; 256 brightness values) in each line. Radiation from the ground is reflected to the two arrays by means of a moveable plane mirror. An innova-

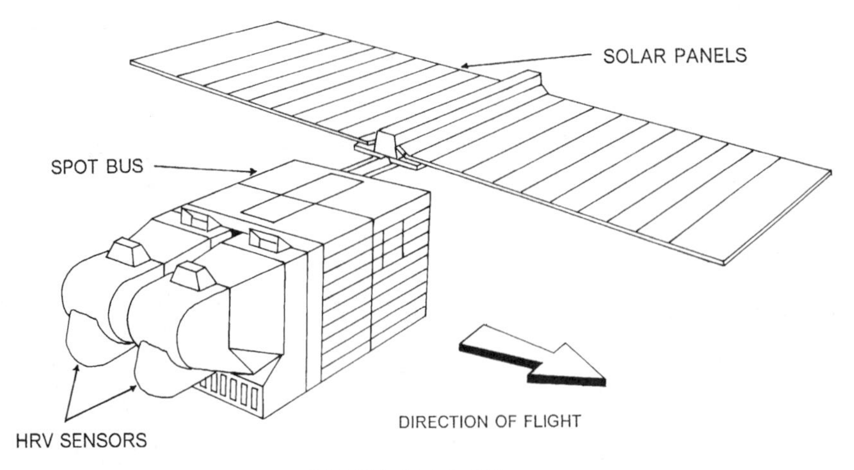


**FIGURE 6.16.** SPOT bus.

tive feature of the SPOT satellite is the ability to control the orientation of the mirror by commands from the ground—a capability that enables the satellite to acquire oblique images, as described below.

The HRV can be operated in either of two modes. In *panchromatic* (PN) mode, the sensor is sensitive across a broad spectral band from 0.51 to 0.73 μm. It images a 60-km swath with 6,000 pixels per line, for a spatial resolution of 10 m (Figure 6.7a). In this mode the HRV instrument provides fine spatial detail, but records a rather broad spectral region.

In the panchromatic mode, the HRV instrument provides coarse spectral resolution but fine (10 m) spatial resolution.

In the other mode, the *multispectral* (XS) configuration, the HRV instrument senses three spectral regions (Plate 6):

- *Band 1:* 0.50–0.59 μm (green)
- *Band 2:* 0.61–0.68 μm (red; chlorophyll absorption)
- *Band 3:* 0.79–0.89 μm (near infrared; atmospheric penetration)

In this mode the sensor images a strip 60 km in width using 3,000 samples for each line, at a spatial resolution of about 20 m (Figure 6.7b). Thus, in the XS mode, the sensor records fine spectral resolution but coarse spatial resolution. The three images from the XS mode can be used to form false-color composites, in the manner of CIR, MSS, and TM images. In some instances, it is possible to "sharpen" the lower spatial detail of multispectral images by superimposing them on the fine spatial detail of high-resolution panchromatic imagery of the same area.

With respect to sensor geometry, each of the HRV instruments can be positioned in either of two configurations (Figure 6.17). For nadir viewing (Figure 6.17a), both sensors are oriented in a manner that provides coverage of adjacent ground segments. Because the two 60-km swaths overlap by 3 km, the total image swath is 117 km. At the equator, centers of adjacent satellite

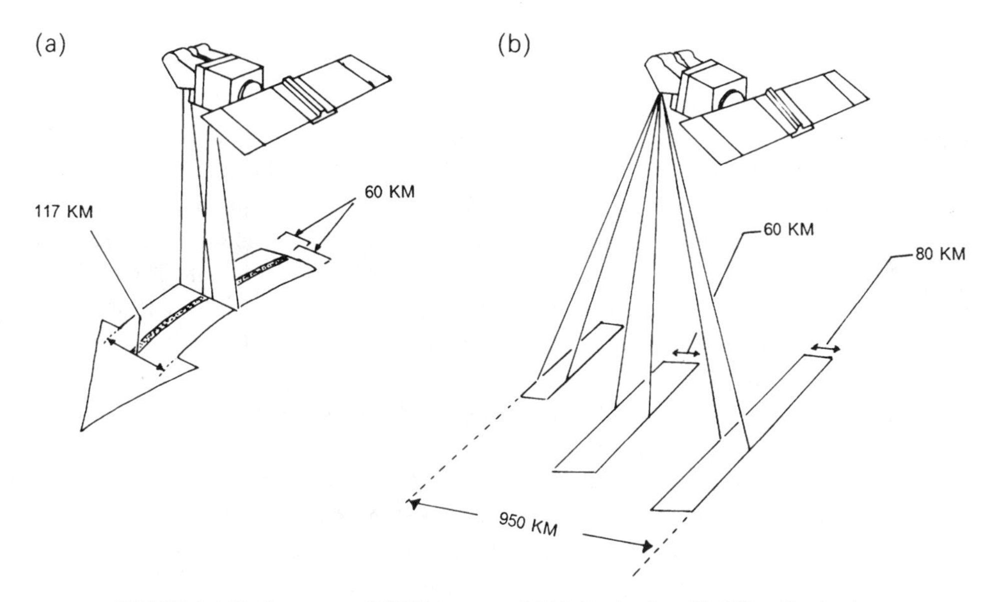


**FIGURE 6.17.** Geometry of SPOT imagery: (a) Nadir viewing. (b) Off-nadir viewing.

tracks are separated by a maximum of only 108 km, so in this mode the satellite can acquire complete coverage of the Earth's surface.

An *off-nadir viewing* capability is possible by pointing the HRV field of view as much as 27° relative to the vertical in 45 steps of 0.6° each (Figure 6.17b) in a plane perpendicular to the orbital path. Off-nadir viewing is possible because the sensor observes the Earth through a pointable mirror that can be controlled by command from the ground. (Note that although mirror orientation can be changed upon command, it is not a scanning mirror, as used by the MSS and the TM.) With this capability, the sensors can observe any area within a 950-km swath centered on the satellite track. The pointable mirror can position any given off-nadir scene center at 10-km increments within this swath.

When SPOT uses off-nadir viewing, the swath width of individual images varies from 60 to 80 km, depending upon viewing angle. Alternatively, the same region can be viewed from separate positions (from different satellite passes) to acquire stereo coverage. (Such stereo coverage depends, of course, upon cloud-free weather during both passes.) The twin sensors are not required to operate in the identical configuration; that is, one HRV can operate in the vertical mode while the other images obliquely. Using its off-nadir viewing capability, SPOT can acquire repeat coverage at intervals of 1 to 5 days, depending upon latitude. Reception of SPOT data is possible at a network of ground stations positioned throughout the world.

CNES processes SPOT data in collaboration with the Institut Geographique National (IGN); image archives are maintained by the Centre de Rectification des Images Spatiales (CRIS), also operated by CNES and IGN. Four levels of processing are listed below in ascending order of precision:

- *Level 1:* Basic geometric and radiometric adjustments
  - *Level 1a:* Sensor normalization
  - *Level 1b:* Level 1a processing, with the addition of simple geometric corrections
- *Level 2:* Use of GCPs to correct image geometry; no correction for relief displacement
- *Level 3:* Further corrections using digital elevation models.

### *SPOTs 4 and 5*

A principal feature of the SPOT 4 mission is the *high-resolution visible and infrared* (HRVIR) instrument, a modification of the HRV used for SPOTs 1, 2, and 3. HRVIR resembles the HRV, with the addition of a mid-infrared band (1.58–1.75 μm), designed to provide capabilities for geological reconnaissance, for vegetation surveys, and for survey of snow cover. Whereas the HRV's 10-m resolution band covers a panchromatic range of 0.51–0.73 μm, the HRVIR's 10-m band is positioned to provide spectral coverage identical to band 2 (0.61–0.68 μm). In addition, the 10-m band is registered to match data in band 2, facilitating use of the two levels of resolution in the same analysis.

SPOT 4 carries two identical HRVIR instruments, each with the ability to point 27° to either side of the ground track, providing a capability to acquire data within a 460-km swath for repeat coverage or stereo. HRVIR's monospectral (M) mode, at 10-m resolution, matches HRVIR band 2's spectral range. In multispectral (X) mode, the HRVIR acquires four bands of data (1, 2, 3, and mid-infrared) at 20-m resolution. Both M and X data are compressed onboard the satellite, then decompressed on the ground to provide 8-bit resolution.

SPOT 5 carries an upgraded version of the HRVIR, which acquires data at either 2.5-m or

5-m resolution and provide a capability for along-track stereo imagery. Stereo imagery at 5-m resolution is intended to provide data for compilation of large-scale topographic maps and data. The new instrument, the *high-resolution geometrical* (HRG), has the flexibility to acquire data using the same bands and resolutions as the SPOT 4 HRVIR, thereby providing continuity with earlier systems.

In the United States, SPOT imagery can be purchased from:

SPOT Image Company
14595 Avion Parkway, Suite 500
Chantilly, VA 20151
Tel: 703-715-3100
Fax: 703-715-3120
E-mail: *sales@spot.com*
Web site: *http://www.spotimage.fr/home/*

### *India Remote Sensing*

After operating two coarse-resolution remote sensing satellites in the 1970s and 1980s, India began to develop multispectral remote sensing programs in the style of the Landsat system. During the early 1990s, two India remote sensing (IRS) satellites were in service. IRS-1A, launched in 1988, and IRS-1B, launched in 1991, carried the LISS-I and LISS-II pushbroom sensors (Tables 6.5 and 6.6). These instruments collect data in four bands: blue (0.45–0.52 μm), green (0.52–0.59 μm), red (0.62–0.68 μm), and near infrared (0.77–0.86 μm), creating images of 2,400 lines in each band. LISS-I provides resolution of 72.5 m in a 148-km swath, and LISS-

**TABLE 6.5. Spectral Characteristics for LISS-I and LISS-II Sensors (IRS-1A and IRS-1B)**

| | | Resolution | |
|---|---|---|---|
| Band | Spectral limits | LISS-I | LISS-II |
| 1 | Blue-green 0.45–0.52 μm | 72.5 m | 36.25 m |
| 2 | Green 0.52–0.59 μm | 72.5 m | 36.25 m |
| 3 | Red 0.62–0.68 μm | 72.5 m | 36.25 m |
| 4 | Near infrared 0.77–0.86 μm | 72.5 m | 36.25 m |

**TABLE 6.6. Spectral Characteristics of a LISS-III Sensor (IRS-1C and IRS-1D)**

| Band | Spectral limits | Resolution |
|---|---|---|
| 1[a] | Blue — | |
| 2 | Green 0.52–0.59 μm | 23 m |
| 3 | Red 0.62–0.68 μm | 23 m |
| 4 | Near infrared 0.77–0.86 μm | 23 m |
| 5 | Midinfrared 1.55–1.70 μm | 70 m |

[a]Band 1 is not included in this instrument, although the numbering system from earlier satellites is maintained to provide continuity.

II has 36.25-m resolution. Two LISS-II cameras acquire data from 74-km-wide swaths positioned within the field of view of LISS-I (Figure 6.18), so that four LISS-II images cover the area imaged by LISS-I, with an overlap of 1.5 km in the cross-track direction and of about 12.76 km in the along-track direction. Repeat coverage is 22 days at the equator, with more frequent revisit capabilities at higher latitudes.

Image data can be acquired within range of receiving stations in Shadnagar, India, and Norman, Oklahoma. In the United States, IRS imagery can be purchased from:

Space Imaging
12076 Grant Street
Thornton, CO 80241
Tel: 303-254-2000; toll free US: 800-232-9037
Fax: 1-303-254-2215
E-mail: info@spaceimaging.com
url: *http://www.spaceimaging.com/*

The LISS-III instrument was designed for the IRS-1C and IRS-1D missions, launched in 1995 and 1997, respectively. These systems use tape recorders, permitting acquisition of data outside the range of the receiving stations mentioned above. They acquire data in four bands: green (0.52–0.59 μm), red (0.62–0.68 μm), near infrared (0.77–0.86 μm), and short-wave infrared (1.55–1.70 μm). LISS-III provides 23-m resolution for all bands, except for the short-

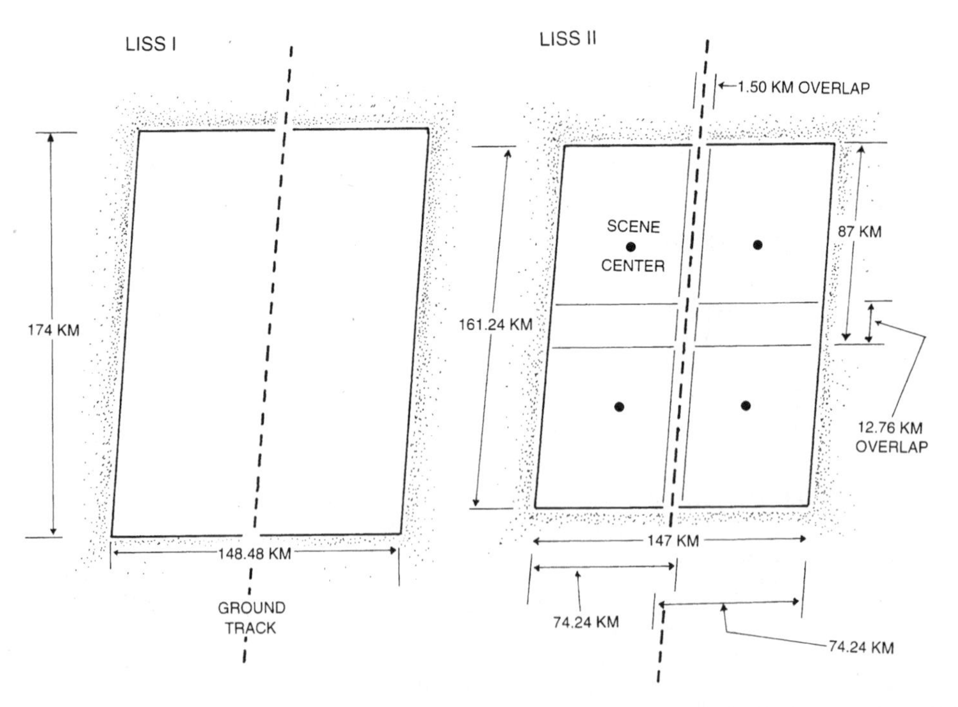


**FIGURE 6.18.** Coverage diagram for the IRS LISS-I and LISS-II sensors. Subscenes gathered by the LISS-II sensors (right) combine to cover the area imaged by the coarser resolution of the LISS-I instrument (left).

wave infrared, which will have 70-m resolution (Table 6.5). Swath width is 142 km for bands 2, 3, and 4, and 148 km in band 5. The satellite provides a capability for 24-day repeat coverage at the equator.

### *Broad-Scale Coverage*

This class of systems includes satellite systems that provide coarse levels of detail for very large regions. Images collected over a period of several weeks can be used to generate composites that represent large areas of the Earth without the cloud cover that would be present in any single scene. These images have opened a new perspective for remote sensing by allowing scientists to examine topics that require examination at continental or global scales that previously were outside the scope of direct observation.

#### *AVHRR*

AVHRR (advanced very-high-resolution radiometer) is a scanning system carried on NOAA's polar orbiting environmental satellites. The first version was placed in service in 1978, primarily as a meteorological satellite to provide synoptic views of weather systems and to assess the thermal balance of land surfaces in varied climate zones. The satellite makes 14 passes each day, viewing a 2,399-km swath. Unlike many other systems described here, the system provides complete coverage of the Earth from pole to pole.

At nadir, the system provides an IFOV of about 1.1 km (Figure 6.7e). Data presented at that level of detail are referred to as *local area coverage* (LAC). In addition, a *global areal coverage* (GAC) data set is generated by onboard averaging of the full-resolution data. GAC data are formed by the selection of every third line of data in the full-resolution data set; for each of these lines, four out of every five pixels are used to compute an average value that forms a single pixel in the GAC data set. This generalized GAC coverage provides pixels of about 4-km × 4-km resolution at nadir. A third AVHRR image product, *HRPT* (high-resolution picture transmission) data, is created by direct transmission of full-resolution data to a ground receiving station as the scanner collects the data. Like LAC data, the resolution is 1.1 km. (LAC data are stored onboard for later transmission when the satellite is within line of sight of a ground receiving station; HRPT data are directly transmitted to ground receiving stations.)

For recent versions of the AVHRR scanner, the spectral channels are:

- *Band 1:* 0.58–0.68 µm (red; matches TM Band 3)
- *Band 2:* 0.725–1.10 µm (near infrared; matches TM Band 4)
- *Band 3:* 3.55–3.93 µm (mid-infrared)
- *Band 4:* 10.3–11.3 µm (thermal infrared)
- *Band 5:* 11.5–12.5 µm (thermal infrared)

Data from several passes are collected to create georeferenced composites that show large regions of the Earth without cloud cover. As will be discussed in Chapter 17, scientists interested in broad-scale environmental issues can use such data to examine patterns of vegetation, climate, and temperature. Further information describing AVHRR is available at *http://www.ngdc.noaa.gov/seg/globsys/avhrr.shtml.*

### *SeaWiFS*

*SeaWiFS* (sea-viewing wide field of view sensor) (Orbview 2), launched in August 1997, is designed to observe the Earth's oceans. The instrument was designed by NASA–Goddard Space Flight Center, and was built and is operated by ORBIMAGE. ORBIMAGE is an affiliate of Orbital Sciences Corporation (Dulles, VA), which retains rights for commercial applications, but sells data to NASA in support of scientific and research applications. Under this data-sharing agreement, most data are held privately for a 2-week interval to preserve their commercial value, then released for more general distribution to the scientific community.

SeaWiFS's primary mission is to observe ocean color. The color of the ocean surface is sensitive to changes in abundance and types of marine phytoplankton, which indicate the rate that marine organisms convert solar energy into biomass (i.e., basic production of food for other marine organisms). Estimation of this rate leads to estimates of primary production of marine environments, an important consideration in the study of global marine systems. SeaWiFS data also contribute to studies of meteorology, climatology, and oceanography (e.g., currents, upwelling, sediment transport, shoals), and, over land bodies, to studies of broad-scale vegetation patterns.

To accomplish these objectives, SeaWiFS provides broad-scale imagery at coarse spatial resolution, but at fine spectral and radiometric resolution (Plate 7). That is, it is designed to survey very broad regions, but make fine distinctions of brightness and color within those regions. It provides imagery at a spatial resolution of 1.1 km, with a swath width of 2,800 km. The satellite is positioned in a sun-synchronous orbit (with a noon equatorial crossing) that can observe a large proportion of the Earth's oceans every 48 hours.

SeaWiFS uses eight spectral channels:

- *Band 1:* 0.402–0.422 µm (blue; yellow pigment/phytoplankton)
- *Band 2:* 0.433–0.453 µm (blue; chlorophyll)
- *Band 3:* 0.480–0.500 µm (blue-green; chlorophyll)
- *Band 4:* 0.500–0.520 µm (green; chlorophyll)
- *Band 5:* 0.545–0.565 µm (red; yellow pigment/phytoplankton)
- *Band 6:* 0.660–0.680 µm (red; chlorophyll)
- *Band 7:* 0.745–0.785 µm (near infrared; land–water contact, atmospheric correction, vegetation)
- *Band 8:* 0.845–0.885 µm (near infrared; land–water contact, atmospheric correction, vegetation

For more information, see *http://seawifs.gsfc.nasa.gov/SEAWIFS.html.*

### *VEGETATION*

SPOT 4 and SPOT 5 each carry an auxiliary sensor, the *VEGETATION* (VGT) instrument, a joint project of several European nations. VGT is a wide-angle radiometer designed for high radiometric sensitivity and broad areal coverage to detect changes in spectral responses of vegetated surfaces. The swath width is 2,200 km, with repeat coverage on successive days at latitudes above 35°, and, at the equator, coverage for 3 out of every 4 days.

The VGT instrument has a CCD linear array sensitive in four spectral bands designed to be compatible with the SPOT 4 HRVIR mentioned above:

- *Band 1:* 0.45–0.50 µm (blue)
- *Band 2:* 0.61–0.68 µm (red)
- *Band 3:* 0.79–0.89 µm (near infrared)
- *Band 4:* 1.58–1.75 µm (mid-infrared)

VEGETATION concurrently provides data in two modes. In direct (regional observation) mode, VGT provides a resolution of 1 km at nadir. In recording (worldwide observation) mode, each pixel corresponds to four of the 1-km pixels, aggregated by onboard processing. In worldwide observation mode, VGT can acquire data within the region between 60°N and 40°S latitude. Data from HRVIR and VGT can be added or mixed, using onboard processing, as required to meet requirements for specific projects.

The system provides two products. VGT-P offers full radiometric fidelity, with minimal processing, for researchers who require data as close as possible to the original values. VGT-S data have been processed to provide geometric corrections composites generated to filter out cloud cover. These images are designed for scientists who require data prepared to fit directly with applications projects.

### *Fine-Resolution Satellite Systems*

The third class of land observation satellites consist of those systems designed to provide very detailed coverage of very small regions. Landsat, SPOT, and other systems established the technical and commercial value of the land observations satellite concept in the 1970s and 1980s, but they also revealed the high costs and technical challenges of designing and operating a system intended to provide general-purpose data for a broad community of users who may have diverse requirements. During the late 1980s and early 1990s, interest increased among those who believed it would be feasible to operate smaller, special-purpose, Earth observations satellites focused on the requirements of very specific groups of users.

For example, the broad areal and multispectral coverage of Landsat and SPOT data might be sacrificed for much finer resolution in a single panchromatic band focused on a small area selected by a specific customer. In addition, it might be possible to link imaging technology with GPS (Chapter 13) and data relay systems to integrate image data with observations acquired directly in the field. This kind of approach to satellite remote sensing generated a multitude of proposals for small commercial satellite systems tailored for specific markets and purposes. At the time of this writing, it is not yet clear which of these concepts will prove to be commercially feasible, but even if only a few of the ideas that have been discussed are eventually implemented, the nature of satellite remote sensing will be greatly changed.

The ancestors of these systems are the military reconnaissance satellites employed for strategic surveillance during the Cold War era, designed to provide detailed images of very specific regions (see the Vignette, CORONA, at the end of this chapter). At the end of the Cold War, some of the older defense-related images were released to the public as a historical archive. In 1994 governmental restrictions on the commercial applications of high-resolution satellite imagery were relaxed. These changes in policy, linked with advances in technology, have opened opportunities for deployment of satellites designed to provide high-resolution data for rather small areas. These characteristics separate them clearly from the design of other systems described here, which have been specifically tailored to produce coarse-resolution data for very large regions.

Today, several commercial firms have exploited this opportunity by designing systems to provide spatial detail in the range of 4 m or so, to submeter levels. Routine availability of imagery at this level of detail opens a new arena for applications of satellite data, especially in fields such has urban planning, transportation analysis, utility management, and precision agriculture that require levels of detail that are not available from broader scale satellite systems. Imagery from these systems, used in combination with GPS (Chapter 13) and GIS (Chapter 16), will clearly have profound impacts on the ways that spatial data are used throughout the economies of developed nations. They will provide fine-detail digital imagery in a format compatible with existing software systems that will promote routine applications of remotely sensed imagery in areas such as real estate, marketing, urban mapping, transportation and pipeline planning, local land-use surveys, and precision agriculture and forestry—to mention only some of the more obvious opportunities.

Imagery at fine levels of detail seems likely to generate new users of the technology, users with interests that differ noticeably from those who use the broad-scale systems. Images will cover much more specific geographic regions, so they will be of interest to a much more distinct pool of customers, who will differ in numbers and interests from users of Landsat data, for example. Users of fine-detail imagery are likely to have an interest in observation of specific regions in fine detail rather than large regions in coarse detail.

### *IKONOS*

The *IKONOS* satellite system (named from the Greek word for "image"), launched in September 1999, is operated by ORBIMAGE, Dulles, Virginia, under the brand name GeoEye: (*http://www.geoeye.com*). In panchromatic mode, IKONOS provides spatial resolution at 1 m, in the spectral range 0.45–0.90 µm (Figures 6.19 and 6.20). In multispectral mode, it provides imagery at 4 m spatial resolution in four spectral bands (Plate 8):

- *Band 1:* 0.45–0.52 µm (blue)
- *Band 2:* 0.52–0.60 µm (green)
- *Band 3:* 0.63–0.69 µm (red)
- *Band 4:* 0.76–0.90 µm (near infrared)

The image swath is 11 km at nadir; imagery is acquired from a sun-synchronous orbit, with a 10:30 A.M. equatorial crossing. The revisit interval varies with latitude; at 40°, repeat coverage can be acquired at about 3 days in the multispectral mode and at about 11–12 days in the panchromatic mode.

### *OrbView 3*

*OrbView 3,* operated by ORBIMAGE (Dulles, VA; *http://www.orbimage.com*) provides 1-m resolution in a panchromatic band and 4-m resolution in four multispectral bands, at a swath width of 8 km.:

- *Band 1:* 0.45–0.52 µm
- *Band 2:* 0.52–0.60 µm
- *Band 3:* 0.625–0.695 µm
- *Band 4:* 0.76–0.90 µm

**FIGURE 6.19.** IKONOS panchromatic scene of Washington, DC, 30 September 1999. From Space Imaging, Inc.

### *QuickBird*

In October 2001, DigitalGlobe, Inc. (Longmont, CO) launched *Quickbird,* a satellite tailored to acquire fine-detail imagery using a panchromatic band with detail at 61-cm resolution and four multispectral bands with 2.44-m detail:

- *Band 1:* 0.45–0.52. μm (blue)
- *Band 2:* 0.52–0.60 μm (green)
- *Band 3:* 0.63–0.69 μm (red)
- *Band 4:* 0.76–0.890 μm (near infrared)
- *Band 5:* 0.76–0.890 μm (panchromatic)

Quickbird acquires data at a swath width of 16.5 km. For more information, see *http://www.digitalglobe.com*.

DigitalGlobe is now designing *WorldView I,* scheduled for launch in 2006, and *WorldView II,* planned for 2008. WorldView I will acquire panchromatic imagery at 0.5-m resolution, with

**FIGURE 6.20.** Detail of Figure 6.19 showing the Jefferson Memorial. From Space Imaging, Inc.

high positional accuracy, and a stereo capability. WorldView II is planned to acquire panchromatic imagery at 0.5-m resolution, and multispectal imagery, in eight spectral channels, at 1.8 m.

## 6.9. Computer Searches

Because of the unprecedented amount of data generated by Earth observation satellite systems, from the very beginning computerized databases formed the principal means of indexing satellite imagery. Each satellite system mentioned above is supported by an electronic catalogue that characterizes each scene by area, date, quality, cloud cover, and other qualities. Often users can examine such achieves through the World Wide Web.

One example is Earth Explorer, offered by the USGS as an index for USGS and Landsat imagery and cartographic data: *http://earthexplorer.usgs.gov*. Users can specify a geographic region by using geographic coordinates, by outlining a region on a map, or by entering a placename . Subsequent screens allow users to select specific kinds of data, specify desirable dates of coverage, and indicate the minimum quality of coverage. The result is a computer listing that provides a tabulation of coverage meeting the constraints specified by the user.

Another example is the USGS Global Visualization Viewer (GLOVIS; *http://glovis.usgs.gov/*). The GLOVIS screen allows the user to roam from one Landsat path/row to another, then to scroll through alternative dates for the path/row selected. The user can set constraints on date, cloud cover, and scene quality to display only those scenes likely to be of interest to the analyst. Other organizations offer variations on the same basic strategy. Some commercial organizations require users to register before they can examine their full achieves. Once a user has selected scenes that meet requirements for specific projects, the scene IDs can

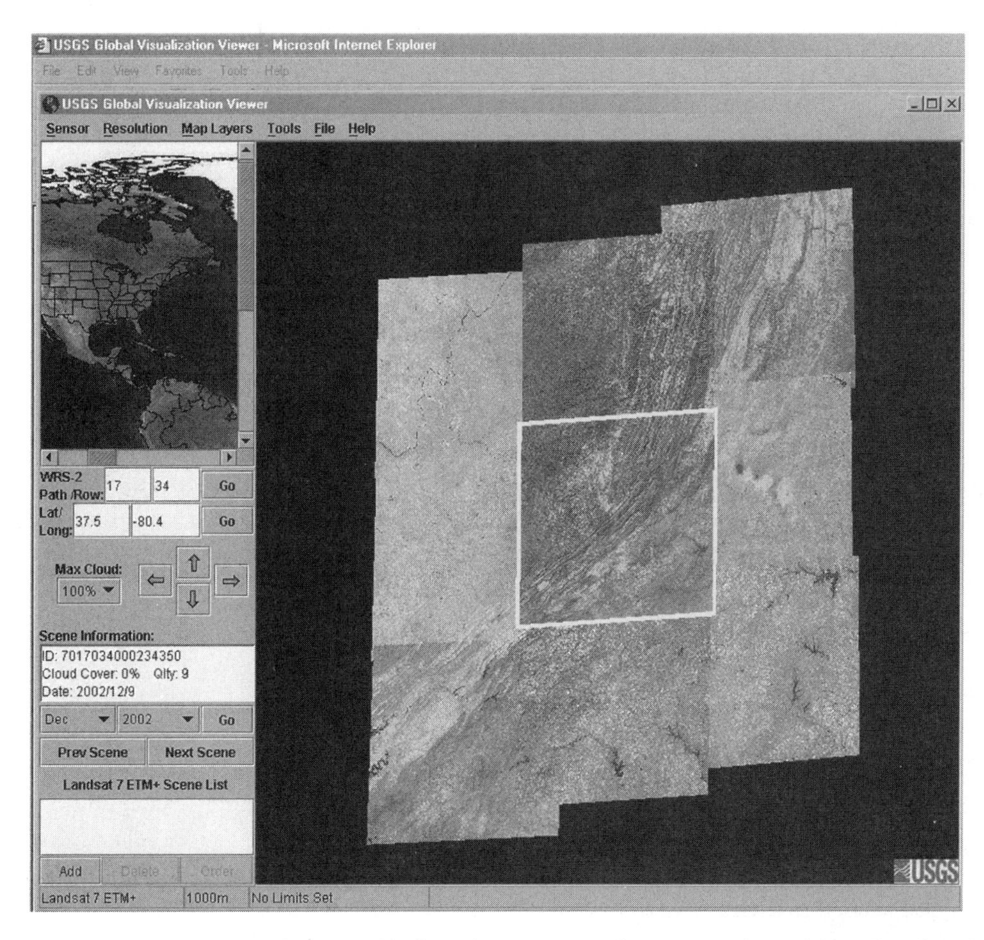


**FIGURE 6.21.** GLOVIS example. GLOVIS allows the user to roam from one Landsat path/row to another (the central scene in the viewer), then to scroll through alternative dates for the path/row selected. The user can set constraints on date, cloud cover, scene quality, and so on, to display only those scenes likely to be of interest to the analyst.

be used to order data from the organization holding the data. If the user requires imagery to be acquired specifically on a given date, an additional change is required to plan for the dedicated acquisition, which, of course, can be acquired subject to prevailing weather for the area in question.

Descriptive records describing coverages of satellite data are examples of a class of data known as *metadata*. Metadata consist of descriptive summaries of other data sets (in this instance, the satellite images themselves). As larger volumes of remotely sensed data are accumulated, the ability to search, compare, and examine metadata has become increasingly significant (Mather and Newman, 1995). In the context of remote sensing, metadata usually consists of text describing images: dates, spectral regions, quality ratings, cloud cover, geographic coverage, and so on. An increasing number of efforts are underway to develop computer programs and communication systems that will link databases together to permit searches of multiple archives.

### *AmericaView*

AmericaView, Inc. is a nonprofit organization funded chiefly through the USGA to promote uses of satellite imagery and to distribute imagery to users. Its activities consists of a coalition of state-based organizations, such as OhioView, VirginiaView, WisconsinView, AlaskaView, and other "stateviews," each at which is composed of a coalition of universities, businesses, state agencies, and nonprofit organizations, and each of which is led by a university within that state. Stateviews distribute image data to their members and to the public, conduct research to apply imagery to local and state problems, conduct workshops and similar educational activities, support research, and in general promote the use of satellite imagery within the widest possible range of potential users.

America View (*http://www.americaview.org* and *http://americaview.usgs.gov/*) began in 1998 as a program within Ohio that has grown to encompass 12 universities. By 2000, Congress directed the USGS to extend the Ohio model nationwide, creating the AmericaView framework described above. At the present time, AmericaView has 18 full members and six affiliate members; its goal is to extend to all 50 states, and U.S. territories.

Each stateview has its own web site, with a guide to its members, activities, and data resources. These web sites can guide users to the data services provided by each stateview.

## 6.10. Summary

Satellite observation of the earth has greatly altered the field of remote sensing. Since the launch of Landsat 1, a larger and more diverse collection of scientists than ever before has conducted remote sensing research and used its applications. Public knowledge of, and interest in, remote sensing has increased. Digital data for satellite images has contributed greatly to the growth of image processing, pattern recognition, and image analysis (Chapters 11–15). Satellite observation systems have increased international cooperation through joint construction and operation of ground receiving stations and though collaboration in the training of scientists.

The history of the U.S. Landsat system, as well as the histories of comparable systems of other nations, illustrates the continuing difficulties experienced in defining structures for financing the development, operation, and distribution costs required for satellite imaging systems. The large initial investments and high continuing operational costs (quite unlike those of aircraft systems) resemble those of a public utility. Yet there is no comparable shared understanding that such systems generate widespread public benefits, so users must contribute substantially to supporting their costs. But if such systems were to pass their full costs on to customers, the prices of imagery would be too high for the development of new applications and the growth of the market. In the United States, as well as in other nations, there has been increasing interest in development of policies that encourage governmental agencies, military services, and private corporations to share costs of development and operation of satellite systems. This approach (*dual use*) may represent an improvement over previous efforts to establish long-term support of satellite remote sensing, but it is now too early to determine if it constitutes a real solution to the long-term problem.

Another concern that will attract more and more public attention is the issue of personal privacy. As systems provide more-or-less routine availability of imagery with submeter resolution, governmental agencies and private corporations will have direct access to information that could provide very detailed information about specific individuals and their property. Although this imagery would not necessarily provide information not already available through publicly available aerial photography, the ease of access and the standardized format open new avenues for use of such information. The real concern should focus not so much on the imagery itself, but upon the effects of combining such imagery with other data from marketing information, census information, and the like. The combination of these several forms of information, each in itself rather benign, could develop capabilities that many people would consider to be objectionable or even dangerous.

A third issue concerns reliable public access to fine-resolution imagery. For example, news organizations might wish to maintain their own surveillance systems to provide independent sources of information regarding military, political, and economic developments throughout the world. Will governments, as investors and/or operators of satellite imaging systems, feel that they are in a position to restrict access when questions of national security are at stake?

This chapter forms an important part of the foundation necessary to develop topics presented in subsequent chapters. The specific systems described here are significant in their own right, but they also provide the foundation for understanding other satellite systems that operate in the microwave (Chapter 7) and far infrared (Chapter 9) regions of the spectrum. Finally, it can be noted that the discussion thus far has emphasized acquisition of satellite data. Little has been said about analysis of these data and their applications to specific fields of study. Both topics will be covered in subsequent chapters (Chapters 11–15, and 16–21, respectively).

## Review Questions

1. Outline the procedure for identifying and ordering SPOT, IRS, or Landsat images for a study area near your home. For each step, identify the *information* necessary to complete that step and proceed to the next. Can you anticipate some of the difficulties you might encounter?

2. In some instances it may be necessary to form a mosaic of several satellite scenes by matching several images together at the edges. List some of the problems you expect to encounter as you prepare such a mosaic.

3. What are some of the advantages (relative to use of aerial photography) of using satellite imagery? Can you identify disadvantages?

4. Manufacture, launch, and operation of Earth observation satellites is a very expensive undertaking—so large that it requires the resources of a national government to support the many activities necessary to continue operation. Many people question whether it is necessary to spend government funds for Earth resource observation satellites and have other ideas for use of these funds. What arguments can you give to justify the costs of such programs?

5. Why are orbits of land observation satellites so low relative to those of communications satellites?

6. Would it be feasible to design an Earth observation satellite with a sun-synchronous orbit to provide coverage of the poles? Explain.

7. Discuss problems that would arise as engineers attempt to design multispectral satellite sensors with smaller and smaller pixels. How might some of these problems be avoided?

8. Can you suggest some of the factors that might be considered as scientists select the observation time (local sun time) for a sun-synchronous Earth observation satellite?

9. Earth observation satellites do not continuously acquire imagery, but only those individual scenes as instructed by mission control. List factors that might be considered in planning scenes to be acquired during a given week. Design a strategy for acquiring satellite images worldwide, specifying rules for deciding which scenes are to be given priority.

10. Using information given in the text, calculate the number of pixels for a single band of an MSS scene, of a TM scene, and of a SPOT HRV image. (For SPOT, assume the image is acquired at nadir.) Recompute the numbers to include all bands available for each sensor.

11. Estimate the number of aerial photographs at a scale of 1:15,470 that would be required to show the land area represented on a single SPOT HRV scene. Assume end lap of 20% and side lap of 10%.

12. Explain why a satellite image and an aerial mosaic of the same ground area are not equally useful, even though image scale might be the same.

13. How many pixels are required to represent a complete SPOT scene (one band only)? A single band of a TM scene?

14. Prepare a template showing (at the correct scale) the dimensions of MSS, TM, and SPOT pixels for an aerial photograph of a nearby area, or other images provided by your instructor. Position the template at various sites throughout the aerial photograph, and assess the effectiveness of the sensors in recording various components of the landscape, including forested land, agricultural land, urban land, and so on. (If pixels are composed of only a single category or feature, they tend to be recorded more effectively than if pixels are composed of two or more classes.)

15. On a small-scale map (such as a road map or similar map provided by your instructor) plot at the correct scale the outlines of an MSS scene centered on a nearby city. How many different countries are covered by this area?

## References

Arnaud, M. 1995. The SPOT Programme. Chapter 2 in *TERRA 2: Understanding the Terrestrial Environment* (P. M. Mather, ed.). New York: Wiley, pp. 29–39.

Baker, J. C., K. M. O'Connell, and R. A. Williamson (eds.). 2001. *Commercial Observation Satellites at the Leading Edge of Global Transparency.* Bethesda, MD: American Society for Photogrammetry and Remote Sensing, 668 pp.

Begni, G. 1982. Selection of the Optimum Spectral Bands for the SPOT Satellite. *Photogrammetric Engineering and Remote Sensing,* Vol. 48, pp. 1613–1620.

Chevrel, M., M. Courtois, and G. Weill. 1981. The SPOT Satellite Remote Sensing Mission. *Photogrammetric Engineering and Remote Sensing,* Vol. 47, pp. 1163–1171.

Curran, P. J. 1985. *Principals of Remote Sensing.* New York: Longman, 282 pp.

General Electric Company. No date. *Data Users Handbook.* Philadelphia: Space Division, General Electric Company.

General Electric Company. No date. *Landsat 3 Reference Manual.* Philadelphia: Space Division, General Electric Company.

Lauer, D. T., S. A. Morain, and V. V. Salomonson. 1997. The Landsat Program: Its Origins, Evolution, and Impacts. *Photogrammetric Engineering and Remote Sensing,* Vol. 63, pp. 831–838.

Mack, P. 1990. *Viewing the Earth: The Social Constitution of the Landsat Satellite System.* Cambridge, MA: MIT Press, 270 pp.

Mather, P. M., and I. A. Newman. 1995. U.K. Global Change Federal Metadata Network. Chapter 9 in *TERRA-2: Understanding the Terrestrial Environment* (P. M. Mather, ed.). New York: Wiley, pp. 103–111.

McClain, E. P. 1980. *Environmental Satellites.* Entry in *McGraw-Hill Encyclopedia of Environmental Science.* New York: McGraw-Hill.

Morain, S. A., and A. M. Budge. 1995. *Earth Observing Platforms and Sensors CD-ROM.* Bethesda, MD: American Society for Photogrammetry and Remote Sensing.

Salomonson, V. V. (ed.). 1997. The 25th Anniversary of Landsat-1. (Special Issue.) *Photogrammetric Engineering and Remote Sensing,* Vol. 63, No. 7.

Salomonson, V. V., J. R. Irons, and D. L. Williams. 1996. The Future of Landsat: Implications for Commercial Development. In *American Institute for Physics, Proceedings 325* (M. El-Genk and R. P. Whitten, eds.). College Park, MD: American Institute of Physics, pp. 353–359.

Sheffield, C. 1981. *Earth Watch: A Survey of the World from Space.* New York: Macmillan, 160 pp.

Sheffield, C. 1983. *Man on Earth: How Civilization and Technology Changed the Face of the World—A Survey from Space.* New York: Macmillan, 166 pp.

Sheffner, E. J. 1994. The Landsat Program: Recent History and Prospects. *Photogrammetric Engineering and Remote Sensing,* Vol. 60, pp. 735–744.

Short, N. M. 1976. *Mission to Earth: Landsat Views the World.* Washington, DC: NASA, 459 pp.

Slater, P. N. 1979. A Re-examination of the Landsat MSS. *Photogrammetric Engineering and Remote Sensing,* Vol. 45, pp. 1479–1485.

Slater, P. N. 1980. *Remote Sensing: Optics and Optical Systems. Reading,* MA: Addison-Wesley, 575 pp.

Taranick, J. V. 1978. *Characteristics of the Landsat Multispectral Data System.* (USGS Open File Report 78-187). Reston, VA: U.S. Geological Survey, 76 pp.

## CORONA

CORONA is the project designation for the satellite reconnaissance system operated by the United States during the interval 1960–1972. CORONA gathered photographic imagery that provided strategic intelligence on the activities of Soviet industry and strategic forces. For many years this imagery and details of the CORONA system were closely guarded as national security secrets. In 1995, when the Soviet threat was no longer present, and CORONA had been replaced by more advanced systems, President Clinton announced that CORONA imagery would be declassified and released to the public. MacDonald (1995) and Ruffner (1995) provide detailed descriptions of the system. Curran (1985) provides an overview based on open sources available prior to the declassification.

### *Historical Context*

When Dwight Eisenhower became president of the United States in 1953, he was distressed to learn of the rudimentary character of the nation's intelligence estimates concerning the military stature of the Soviet Union. Within the Soviet sphere, closed societies and rigid security systems denied Western nations the information they required to prepare reliable estimates of Soviet and allied nations' military capabilities.

As a result, Eisenhower feared the United States faced the danger of developing policies based on speculation or political dogma rather than reliable estimates of the actual situation.

His concern led to a priority program to build the U-2 system, a high-altitude reconnaissance aircraft designed to carry a camera system of unprecedented capabilities. The first flight occurred in the summer of 1956. Although U-2 flights imaged only a small portion of the Soviet Union, they provided information that greatly improved U.S. estimates of Soviet capabilities. The U-2 program was intended only as a stop-gap measure—it was anticipated that the Soviet Union would develop countermeasures that would prevent long-term use of the system. In fact, the flights continued for about 4 years, ending in 1960 when a U-2 plane was shot down near Sverdlovsk, in the Soviet Union. This event ended use of the U-2 over the Soviet Union, although it continued to be a valuable military asset for observing other regions of the world. Later it was employed in the civilian sphere for collecting imagery for environmental analyses.

At the time of the U-2 incident, work was already underway to design a satellite system that would provide photographic imagery from orbital altitudes, thereby avoiding risks to pilots and the controversy arising from U.S. overflights of another nation's territory. Although CORONA had been conceived beforehand, the Soviet Union's launch of the first artificial Earth-orbiting satellite in October 1957 increased the urgency of the effort.

The CORONA satellite was designed and constructed as a joint effort of the U.S. Air Force and the Central Intelligence Agency (CIA), in collaboration with private contractors, who designed and manufactured launch systems, satellites, cameras, and films. Virtually every element of the system extended technical capabilities beyond their known limits. Each major component encountered difficulties and failures that were, in time, overcome, to eventually produce an effective, reliable system. During the interval June 1959–December 1960, the system experienced a succession of failures, first with the launch system, then with the satellite, next with the recovery system, and finally with the cameras and films. Each problem was identified and solved within a remarkably short time, so that CORONA was in effect operational by August 1959, only 3 months after the end of the U-2 flights over the Soviet Union. Today CORONA's capabilities may seem commonplace, as we are familiar with artificial satellites, satellite imagery, and high-resolution films. However, at the time, CORONA extended reconnaissance capabilities far beyond what even technical experts thought might be feasible.

Although the Soviet Union knew in a general way that a reconnaissance satellite effort was under development, details of the system, the launch schedule, the nature of the imagery, and the ability of interpreters to derive information from the imagery were closely guarded secrets. Secrecy denied Soviets knowledge of the capabilities of the system, which might have permitted development of effective measures for deception. In time, the existence of the program became more widely known in the United States, although details of the system's capabilities were still kept secret.

CORONA is the name given to the satellite system, whereas the camera systems carried KEYHOLE (KH) designations that were familiar to those who used the imagery. The designations KH-1 through KH-5 refer to different models of the cameras used for the CORONA program. Most of the imagery was acquired with the KH-4 and KH-5 systems (including KH-4A and KH-4B).

#### *Satellite and Orbit*

The reconnaissance satellites were placed into near-polar orbits, with, for example, an inclination of 77°, apogee of 502 mi., and perigee of 116 mi. Initially missions would last only 1 day; by the end of the program, missions extended for 16 days. Unlike the other satellite systems described here, images were returned to earth by a capsule (the satellite recovery vehicle) ejected from the principal satellite at the conclusion of each mission. The recovery vehicle was designed to withstand the heat of reentry into the earth's atmosphere and to deploy a parachute at an altitude of about 60,000 ft. The capsule was then recovered in the air by specially designed aircraft. Recoveries were planned for the Pacific Ocean near Hawaii. Capsules were designed to sink if not recovered within a few days to prevent recovery by other nations in the event of malfunction. Several capsules were in fact lost when this system failed. Later models of the satellite were designed with two capsules, which extended the length of CORONA missions to as long as 16 days.

### *Cameras*

Camera designs varied as problems were solved and new capabilities were added. MacDonald (1995) and Ruffner (1995) provide details of the evolution of the camera systems; here the description focuses on the main feature of the later models that acquired much of the imagery in the CORONA archive. The basic camera design, the KH-3, manufactured by Itek Corporation, was a vertically oriented panoramic camera with a 24-in. focal length. The camera's 70° panoramic view was acquired by the mechanical motion of the system at right angles to the line of flight. Image motion compensation (Chapter 3) was employed to correct effects of satellite motion relative to the Earth's surface during exposure.

The KH-4 camera acquired most of the CORONA imagery (Figure 6.22). The KH-4B (sometimes known as the MURAL system) consisted of two KH-3 cameras oriented to observe the same area from different perspectives, thereby providing stereo capability (Figure 6.23). One pointed 15° forward along the flight path, the other was aimed 15° aft. A small-scale index image provided the context for proper orientation of the panoramic imagery, stellar cameras viewed the pattern of stars, and horizon cameras viewed the Earth's horizon to assure correct orientation of the spacecraft and the panoramic cameras. A related system, named ARGON, acquired photographic imagery to compile accurate maps of the Soviet Union, using cameras with 3-in. and 1.5-in. focal lengths.

### *Imagery*

The CORONA archive consists of about 866,000 images, totaling about 400 mi. of film, acquired during the interval August 1960–May 1972. Although film format varied, most imagery was recorded on 70-mm film in strips about 25–30 in. long. Actual image width is typically either 4.4 in. or 2.5 in. wide. Almost all CORONA imagery was recorded on black-and-white panchromatic film, although a small portion used a color infrared emulsion, and some was in a natural color emulsion. The earliest imagery is said to have a resolution of about 40 ft.; by the end of the program in 1972, the resolution was as fine as 6 ft., although detail varied greatly depending upon atmospheric effects, illumination, and the nature of the target (Figure 6.24).

The imagery is indexed, to show coverage and dates. About 50% is said to be obscured by clouds, although the index does not record which scenes are cloud-covered. Most of the early coverage was directed at the Soviet Union (Figure 6.25), although later coverage shows other areas, often chosen because they were objects of current or potential international concern.

### *National Photographic Interpretation Center*

Interpretation of CORONA imagery was largely centralized at the National Photographic Interpretation Center (NPIC), maintained by the CIA in a building in the navy yard in southeastern Washington, D.C.

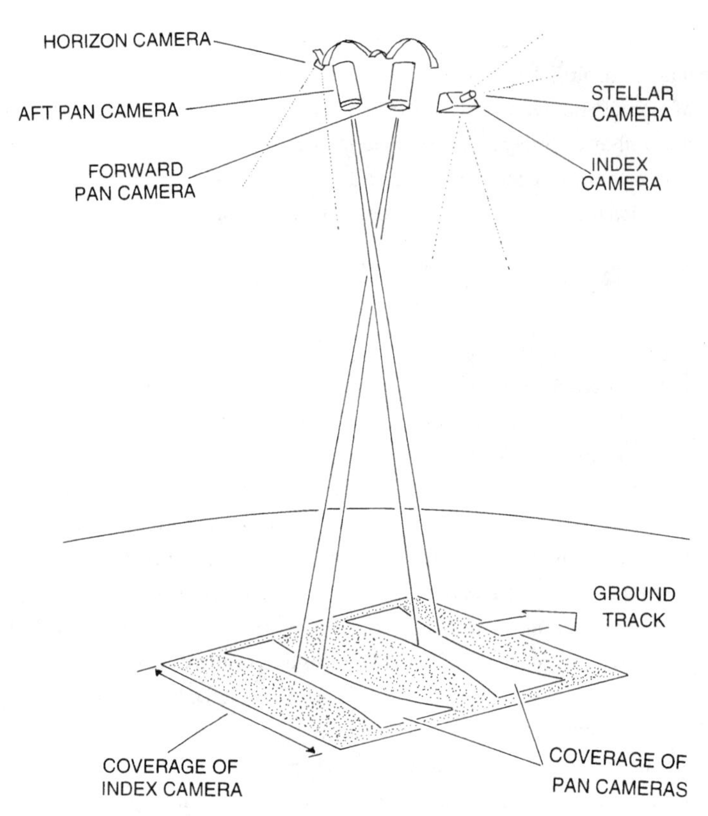


**FIGURE 6.22.** Coverage of a KH-4B camera system. The two stereo panoramic cameras point fore and aft along the ground track, so a given area can be photographed first by the aft camera (pointing forward), then by the forward camera (pointing aft) (see Figure 6.23). Each image from the panoramic camera represents an area about 134.8 mi. (216.8 km) in length and about 9.9 mi. (15.9 km) wide at the image's greatest width. The index camera provides a broad-scale overview of the coverage of the region; the stellar and horizon cameras provide information to maintain satellite stability and orientation. Based upon MacDonald (1995).

Here teams of photointerpreters from the CIA, armed forces, and allied governments were organized to examine imagery to derive intelligence pertaining to the strategic capabilities of the Soviet Union and its satellite nations. Film from each mission was immediately interpreted to provide immediate reports on topics of current significance.

Then imagery was reexamined to provide more detailed analyses of less urgent developments. In time, the system accumulated an archive that provided a retrospective record of the development of installations, testing of weapons systems, and operations of units in the field. By this means, image analysts could trace the development of individual military and industrial installations, recognize unusual activities, and develop an understanding of the strategic infrastructure of the Soviet Union. Although the priority focus was on understanding the strategic capabilities of Soviet military forces, CORONA imagery was also used to examine agricultural and industrial production, environmental problems, mineral resources, population patterns, and other facets of the Soviet Union's economic infrastructure.

Reports from NPIC photointerpreters formed only one element of the information considered by analysts as they prepared intelligence estimates; they also studied other sources, such as electronic signals, reports from observers, and press reports. However, CORONA photographs had a central role in this analysis because of their ready availability, reliability, and timeliness.

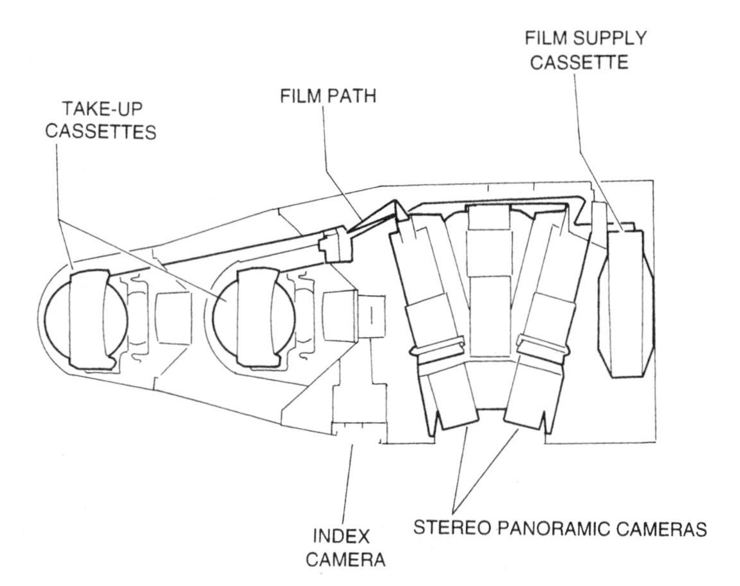


**FIGURE 6.23.** Line drawing of major components of the KH-4B camera (based upon MacDonald, 1995). The two panoramic cameras are pointed forward and aft along the ground track to permit acquisition of stereo coverage as shown in Figure 6.22. Each of the two take-up cassettes could be ejected sequentially to return exposed film to Earth. The index camera, as shown in Figure 6.22, provided a broad-scale overview to assist in establishing the context for coverage from the panoramic cameras.

**FIGURE 6.24.** Severodvinsk Shipyard (on the White Sea coastline, near Archangel, Russia, formerly USSR) as imaged by KH-4B camera, 10 February 1969. This image is much enlarged from the original image. From CIA.

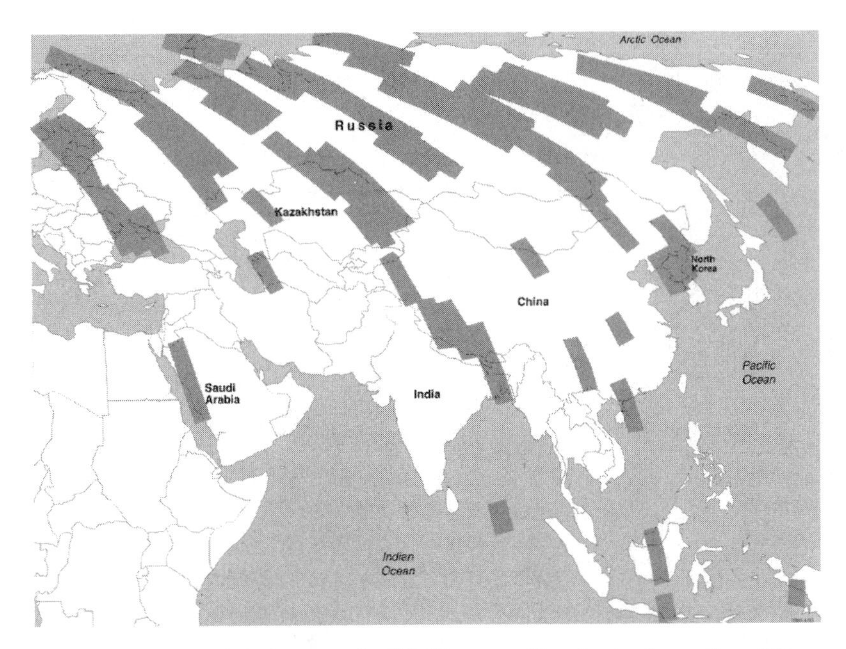


**FIGURE 6.25.** Typical coverage of a KH-4B camera, Eurasian landmass. From CIA.

### *Image Availability*

Until 1995, the CIA maintained the CORONA archive. It was released to the National Archives and Records Administration and to the USGS after declassification. The National Archive retains the original negatives, and provides copies for purchase by the public through the services of contractors. The USGS stores duplicate negatives to provide imagery to the public, and maintains a digital index as part of the Earth Explorer online product search index: http://earthexplorer.usgs.gov. Other organizations may decide to purchase imagery for resale to the public, possibly in digitized form, or to sell selected scenes of special interest.

### *Uses*

CORONA provided the first satellite imagery of the Earth's surface, and therefore extends the historical record of satellite imagery into the late 1950s, about 10 years before Landsat. Therefore, it may be able to assist in assessment of environmental change, trends in human use of the landscape, and similar phenomena (Plate 9). Further, CORONA imagery records many of the pivotal events of the Cold War, and therefore may form an important source for historians who wish to examine and reevaluate these events more closely than was possible before imagery was released. Ruffner (1995) provides examples that show archaeological and geological applications of CORONA imagery. There may well be other applications not yet identified.

However, today's interpreters of this imagery may face some difficulties. Many of the original interpretations depended not only upon the imagery itself, but also upon the skill and experience of the interpreter (who often were specialized in identification of specific kinds of equipment), access to collateral information, and availability of specialized equipment. Further, many critical interpretations depended upon the experience and expertise of interpreters with long experience in analysis of images of specific weapons systems—experience that is not readily available to today's analysis. Therefore, it may be difficult to reconstruct or reevaluate interpretations of an earlier era.

## References

Brugioni, D. A. 1991. *Eyeball to Eyeball: The Inside Story of the Cuban Missile Crisis.* New York: Random House, 622 pp.

Brugioni, D. 1996. The Art and Science of Photoreconnaissance. *Scientific American,* Vol. 274, pp. 78–85.

Day, D. A., J. M. Logsdon, and B. Latell (eds.). 1998. *Eye in the Sky: The Story of the Corona Spy Satellites.* Washington, DC: Smithsonian Institution Press, 303 pp.

MacDonald, R. A. 1995. CORONA: Success for Space Reconnaissance, a Look into the Cold War, and a Revolution for Intelligence. *Photogrammetric Engineering and Remote Sensing,* Vol. 61, pp. 689–719.

Ruffner, K. C. (ed.). 1995. *CORONA: America's First Satellite Program.* Washington, DC: Center for the Study of Intelligence, 360 pp.

CHAPTER SEVEN

# Active Microwave

## 7.1. Introduction

This chapter describes active microwave sensor systems. Active microwave sensors are one example of an *active sensor,* a sensor that broadcasts a directed pattern of energy to illuminate a portion of the Earth's surface, then receives the portion scattered back to the instrument. This energy forms the basis for the imagery we interpret. Because passive sensors (e.g., photography) are sensitive to variations in solar illumination, their use is constrained by time of day and weather conditions. In contrast, active sensors generate their own energy, so their use is subject to fewer constraints, and they can be used under a wider range of operational conditions. Further, because active sensors use energy generated by the sensor itself, its properties are known in detail. Therefore, it is possible to compare transmitted energy with received energy to judge with more precision than is possible with passive sensors the characteristics of the surfaces that have scattered the energy.

## 7.2. Active Microwave

The microwave region of the electromagnetic spectrum extends from wavelengths of about 1 mm to about 1 m. This region is, of course, far removed from those in and near the visible spectrum, where our direct sensory experience can assist in our interpretation of images and data. Thus formal understanding of the concepts of remote sensing are vital to understanding imagery acquired in the microwave region. As a result, the study of microwave imagery is often a difficult subject for beginning students, and requires more attention than is usually necessary for study of other regions of the spectrum. The family of sensors discussed here (Figure 7.1) are all active microwave sensors (imaging radars carried by either aircraft or satellites). These are *active* sensors because they illuminate the ground with their own energy, then record a portion of that energy reflected back to themselves. *Passive* microwave sensors, instruments sensitive to microwave energy emitted from the Earth's surface, are discussed in Chapter 9.

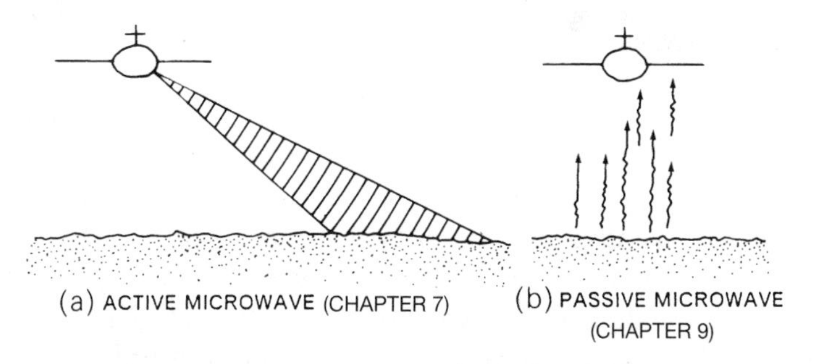


**FIGURE 7.1.** Active and passive microwave remote sensing. (*a*) Active microwave sensing, using energy generated by the sensor, as described in this chapter. (*b*) Passive microwave sensing, which detects energy emitted by the Earth's surface, described in Chapter 9.

## *Active Microwave Sensors*

Active microwave sensors are *radar* devices, instruments that transmit a microwave signal, then receive its reflection as the basis for forming images of the Earth's surface. The rudimentary components of an imaging radar system include a transmitter, a receiver, an antenna array, and a recorder (Figure 7.2). A *transmitter* is designed to transmit repetitive pulses of microwave energy at a given frequency. A *receiver* accepts the reflected signal as received by the antenna, then filters and amplifies it as required. An *antenna array* transmits a narrow beam of microwave energy. Such an array is composed of *waveguides,* devices that control the propagation of an electromagnetic wave, such that waves follow a path defined by the physical structure of the guide. (A simple waveguide might be formed from a hollow metal tube.) Usually the same antenna is used both to transmit the radar signal and to receive its echo from the terrain. Finally, a *recorder* records and/or displays the signal as an image. Numerous refinements and variations of these basic components are possible; a few are described below in greater detail.

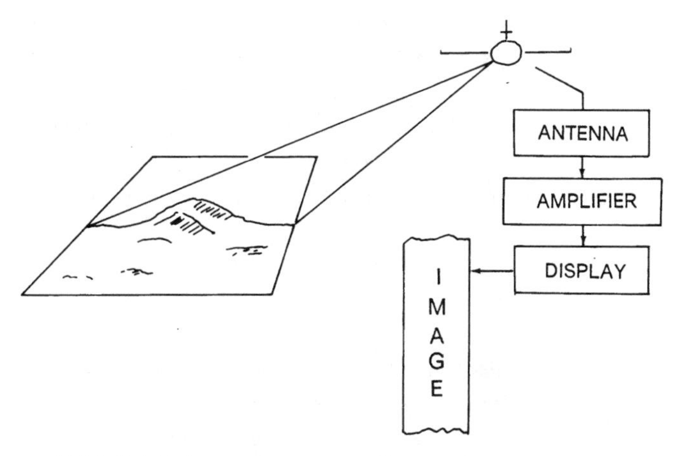


**FIGURE 7.2.** Schematic overview of an imaging radar system.

## *Side-Looking Airborne Radar*

*Radar* is an acronym for "radio detection and ranging." The "ranging" capability is achieved by measuring the time delay between the time a signal is transmitted toward the terrain and the time its echo is received. Through its ranging capability, possible only with active sensors, radar can accurately measure the distance from the antenna to features on the ground. A second unique capability, also a result of radar's status as an active sensor, is its ability to detect frequency and polarization shifts. Because the sensor transmits a signal of known wavelength, it is possible to compare the received signal with the transmitted signal. From such comparisons imaging radars detect changes in frequency that give them capabilities not possible with other sensors.

*Side-looking airborne radar* (SLAR) imagery is acquired by an antenna array aimed to the side of the aircraft, so that it forms an image of a strip of land parallel to, and at some distance from, the ground track of the aircraft (Figure 7.2). The resulting image geometry differs greatly from that of other remotely sensed images. These qualities establish radar imagery as a distinctive form of remote sensing imagery. One of SLAR's most unique and useful characteristics is its ability to function during inclement weather. SLAR is often said to possess an "all-weather" capability, meaning that it can acquire imagery in all but the most severe weather conditions. The microwave energy used for SLAR imagery is characterized by wavelengths long enough to escape interference from clouds and light rain (although not necessarily from heavy rainstorms). Because SLAR systems are independent of solar illumination, missions using SLAR can be scheduled at night or during early morning or evening hours when solar illumination might be unsatisfactory for acquiring aerial photography. This advantage is especially important for imaging radars carried by the Earth-orbiting satellites described later in this chapter.

Radar images typically provide crisp, clear representations of topography and drainage (Figure 7.3). Despite the presence of the geometric errors described below, radar images typically provide good positional accuracy; in some areas they have provided the data for small-scale base maps depicting major drainage and terrain features. Analysts have registered TM and SPOT data to radar images to form composites of the two quite different kinds of information. Some of the most successful operational applications of SLAR imagery have occurred in tropical climates, where persistent cloud cover has prevented acquisition of aerial photography, and where shortages of accurate maps create a context in which radar images can provide cartographic information superior to that on existing conventional maps. Another important characteristic of SLAR imagery is its synoptic view of the landscape. SLAR's ability to clearly represent the major topographic and drainage features within relatively large regions at moderate image scales makes it a valuable addition to our repertoire of remote sensing imagery. Furthermore, because it acquires images in the microwave spectrum, SLAR may show detail and information that differ greatly from that of sensors operating in the visible and near infrared spectra.

## *Origins and History*

The foundations for imaging radars were laid by scientists who first investigated the nature and properties of microwave and radio energy. James Clerk Maxwell (1831–1879) first defined the essential characteristics of electromagnetic radiation; his mathematical descriptions of the properties of magnetic and electrical fields prepared the way for further theoretical and practical

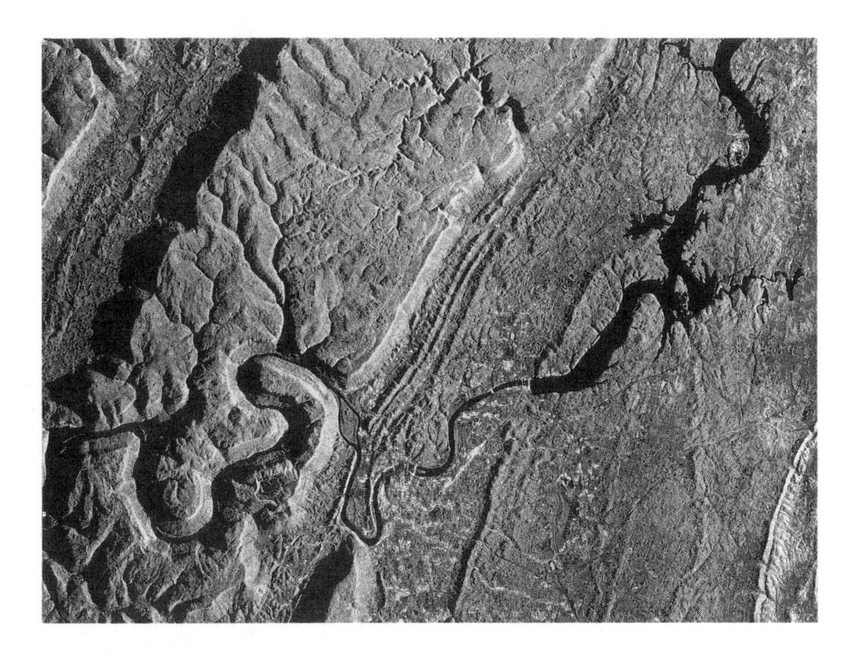

**FIGURE 7.3.** Radar image of a region near Chattanooga, TN, September 1985 (X-band, HH polarization). This image has been processed to produce pixels of about 11.5 m in size. From USGS.

work. In Germany, Heinrich R. Hertz (1857–1894) confirmed much of Maxwell's work, and further studied properties and propagation of electromagnetic energy in microwave and radio portions of the spectrum. The hertz, the unit for designation of frequencies (Chapter 2), is named in his honor. Hertz was among the first to demonstrate the reflection of radio waves from metallic surfaces, and thereby begin research that led to development of modern radios and radars.

In Italy, Guilielmo M. Marconi (1874–1937) continued the work of Hertz and other scientists, in part by devising a practical antenna suitable for transmitting and receiving radio signals. In 1895 he demonstrated the practicability of the wireless telegraph. After numerous experiments over shorter distances, in 1901 he demonstrated the feasibility of long-range communications by sending signals across the Atlantic; in 1909 he shared the Nobel Prize in Physics. Later he proposed that ships could be detected by using the reflection of radio waves, but there is no evidence that his suggestion influenced the work of other scientists.

The formal beginnings of radar date from 1922, when A. H. Taylor and L. C. Young, civilian scientists working for the U.S. Navy, were conducting experiments with high-frequency radio transmissions. Their equipment was positioned near the Anacostia River near Washington, D.C., with the transmitter on one bank of the river and the receiver on the other. They observed that the passage of a river steamer between the transmitter and the receiver interrupted the signal in a manner that clearly revealed the potential of radio signals as a means for detecting the presence of large objects (Figure 7.4). Taylor and Young recognized the significance of their discovery for marine navigation in darkness and inclement weather, and, in a military context, its potential for detection of enemy vessels. Initial efforts to implement this idea depended upon the placement of transmitters and receivers at separate locations, so that a continuous microwave signal was reflected from an object, then recorded by a receiver placed some dis-

**FIGURE 7.4.** Beginnings of radar: schematic diagram of the situation that led to the experiments by Young and Taylor.

tance away. Designs evolved so that a single instrument contained both the transmitter and the receiver at a single location, integrated in a manner that permitted use of a pulsed signal that could be reflected from the target back to the same antenna that transmitted the signal. Such instruments were first devised during the years 1933–1935 more or less simultaneously in the United States (by Young and Taylor), Great Britain, and Germany. A British inventor, Sir Robert Watson-Watt, is sometimes given credit as the inventor of the first radar system, although Young and Taylor have been credited by other historians.

Subsequent improvements were based mainly upon refinements in the electronics required to produce high-power transmissions over narrow wavelength intervals, to carefully time short pulses of energy, and to amplify the reflected signal. These and other developments led to rapid evolution of radar systems in the years prior to World War II. Due to the profound military significance of radar technology, World War II prompted radar's sophistication. The postwar era perfected imaging radars. Experience with conventional radars during World War II revealed that radar reflection from ground surfaces (*ground clutter*) varied greatly according to terrain, season, settlement patterns, and so on, and from ocean surfaces was modified by winds and waves. Ground clutter was undesirable for radars designed to detect aircraft and ships, but later systems were designed specifically to record the differing patterns of reflection from ground and ocean surfaces. The side-looking characteristic of SLAR was desirable because of the experiences of reconnaissance pilots during the war, who often were required to fly low-level missions along predictable flight paths in lightly armed aircraft to acquire the aerial photography required for battlefield intelligence. The side-looking capability provided a means for acquiring information at a distance using an aircraft flying over friendly territory.

Later, the situation posed by the emergence of the Cold War in Europe, with clearly defined frontiers, and strategic requirements for information within otherwise closed borders, also provided an incentive for development of sensors with SLAR's capabilities. The experience of the war years also must have provided an intense interest in development of SLAR's all-weather capabilities, because Allied intelligence efforts were several times severely restricted by the absence of aerial reconnaissance during inclement weather. Thus the development of imaging radars is linked to military and strategic reconnaissance, even though many current applications focus upon civilian requirements.

## 7.3. Geometry of the Radar Image

The basics of the geometry of a SLAR image are illustrated in Figure 7.5. Here the aircraft is viewed head-on, with the radar beam represented in vertical cross section as the fan-shaped fig-

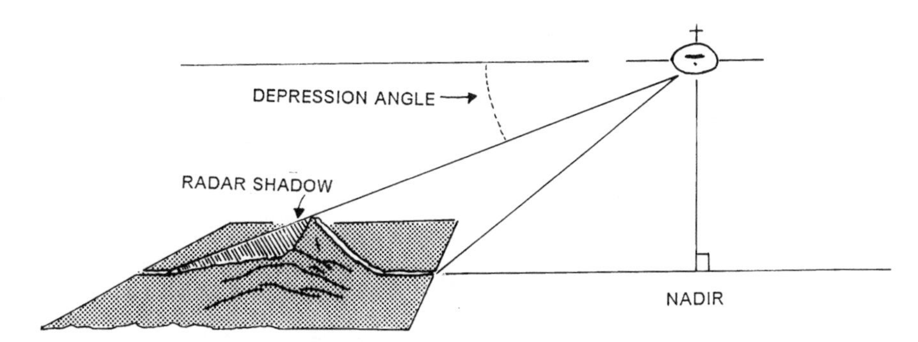


**FIGURE 7.5.** Geometry of an imaging radar system. The radar beam illuminates a strip of ground parallel to the flight path of the aircraft; the reflection and scattering of the microwave signal from the ground forms the basis for the image.

ure at the side of the aircraft. The upper edge of the beam forms an angle with a horizontal line extended from the aircraft; this angle is designated as the *depression angle* of the far edge of the image. Upper and lower edges of the beam, as they intersect with the ground surface, define the edges of the radar image; the forward motion of the aircraft (toward the reader, out of the plane of the illustration) forms what is usually the "long" dimension of the strip of radar imagery.

The smallest depression angle forms the *far-range* side of the image. The *near-range* region is the edge nearest to the aircraft. Intermediate regions between the two edges are sometimes referred to as *midrange* portions of the image. Steep terrain may hide areas of the imaged region from illumination by the radar beam, causing *radar shadow*. Note that radar shadow depends upon topographic relief and the direction of the flight path in relation to topography. Within an image, radar shadow depends also upon depression angle, so that (given equivalent topographic relief) radar shadow will be more severe in the far-range portion of the image (where depression angles are smallest), or for those radar systems that use shallow depression angles. (A specific radar system is usually characterized by a fixed range of depression angles.)

Radar systems measure distance to a target by timing the delay between a transmitted signal and its return to the antenna. Because the speed of electromagnetic energy is a known constant, the measure of time translates directly to a measure of distance from the antenna. Microwave energy travels in a straight path from the aircraft to the ground—a path that defines the *slant-range* distance, as if one were to stretch a length of string from the aircraft to a specific point on the ground as a measure of distance. Image interpreters prefer images to be presented in *ground-range* format, with distances portrayed in their correct relative positions on the Earth's surface. Because radars collect all information in the slant-range domain, radar images inherently contain geometric artifacts, even though the image display may ostensibly appear to match a ground-range presentation.

One such error is *radar layover* (Figure 7.6). At near range, the top of a tall object is closer to the antenna than its base. As a result, the echo from the top of the object reaches the antenna before the echo from the base. Because radar measures all distances with respect to time elapsed between transmission of a signal and the reception of its echo, the top of the object appears (i.e., in the slant-range domain) to be closer to the antenna than does its base. Indeed, it is closer, if only the slant-range domain is considered. However, in the ground-range domain (the context for correct positional representation and for accurate measurement), both the top and the base of the object occupy the same geographic position. In the slant-range domain of the radar image,

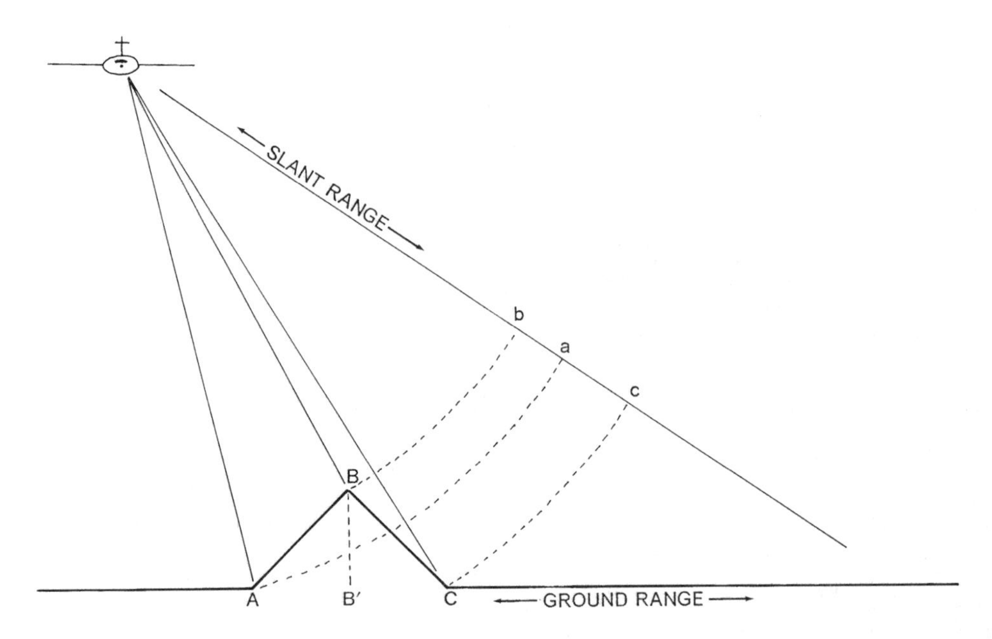


**FIGURE 7.6.** Radar layover. In the ground-range domain *AB* and *BC* are equal. Because the radar can measure only slant-range distances, *AB* and *BC* are projected onto the slant-range domain, represented by the line *bac*. The three points are not shown in their correct relationship because the slant-range distance from the antenna to the points does not match to their ground-range distances. Point *B* is closer to the antenna than is point *A,* so it is depicted on the image as closer to the edge of the image.

they occupy different image positions—a geometric error analogous perhaps to relief displacement in the context of aerial photography.

Radar layover is depicted in Figure 7.6. Here the topographic feature *ABC* is shown with *AB* = *BC* in the ground-range representation. However, because the radar can position *A, B,* and *C* only by the time delay with relation to the antenna, it must perceive the relationships between *A, B,* and *C* as shown in the slant range (image plane). Here A and B are reversed from their ground-range relationships, so that *ABC* is now *bac,* due to the fact that the echo from *B* must be received before the echo from *A*.

A second form of geometric error, *radar foreshortening,* occurs in terrain of modest-to-high relief depicted in the mid- to far-range portion of an image (Figure 7.7). Here the slant-range representation depicts *ABC* in their correct relationships abc, but the distances between them are not accurately shown. Whereas *AB* = *BC* in the ground-range domain, $ab < bc$ when they are projected into the slant range. Radar foreshortening tends to cause images of a given terrain feature to appear to have steeper slopes than they do in nature on the near-range side of the image and to have shallower slopes than they do in nature on the far-range side of the feature (Figure 7.8). Thus a terrain feature with equal fore and back slopes may be imaged to have shorter, steeper, and brighter slopes than it would in a correct representation, and the image of the back slope would appear to be longer, shallower, and darker than it would in a correct representation. Because depression angle varies with position on the image, the amount of radar foreshortening in the image of a terrain feature depends not only upon the steepness of its slopes, but also upon its position on the radar image. As a result, apparent terrain slope and shape on radar images are

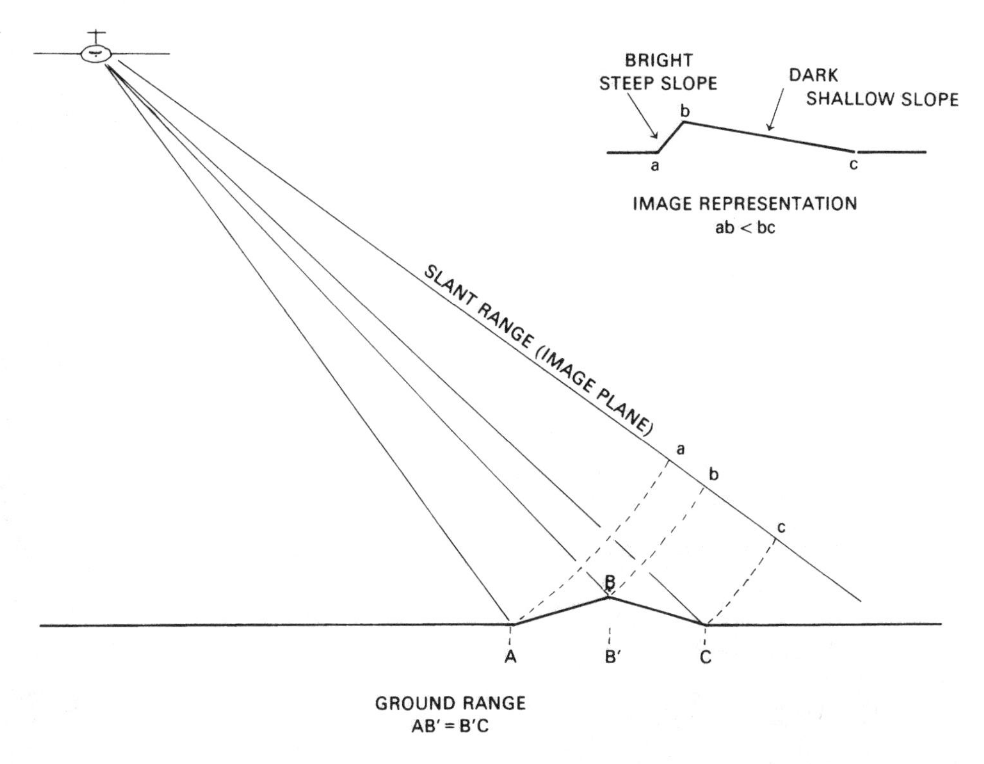


**FIGURE 7.7.** Radar foreshortening. Projection of *A, B,* and C into the slant-range domain distorts the representations of *AB* and *BC,* so that *ab* appears shorter, steeper, and brighter than it should be in a faithful rendition, and *bc* appears longer, shallower in slope, and darker than it should be.

not necessarily accurate representations of their correct character in nature. Thus care should be taken when interpreting these features.

## 7.4. Wavelength

Imaging radars normally operate within a small range of wavelengths with the rather broad interval defined at the beginning of this chapter. Table 7.1 lists primary subdivisions of the active microwave region, as commonly defined in the United States. These divisions and their designations have an arbitrary, illogical flavor that is the consequence of their origin during the development of military radars, when it was important to conceal the use of specific frequencies for given purposes. To preserve military security, the designations were designed as much to confuse unauthorized parties as to provide convenience for authorized personnel. Eventually these designations became established in everyday usage, and they continue to be used even though there is no longer a requirement for military secrecy.

Although experimental radars can often change frequency, or sometimes even use several frequencies (for a kind of "multispectral radar"), operational systems are generally designed to use a single wavelength band. Airborne imaging radars have frequently used C-, K-, and X-bands. As described elsewhere in this chapter, imaging radars used for satellite observations often use L-band frequencies.

**TABLE 7.1. Radar Frequency Designations**

| Band | Wavelengths |
|---|---|
| P-band | 107–77 cm |
| UHF | 100–30 cm |
| L-band | 30–15 cm |
| S-band | 15–7.5 cm |
| C-band | 7.5–3.75 cm |
| X-band | 3.75–2.40 cm |
| Ku-band | 2.40–1.67 cm |
| K-band | 1.67–1.18 cm |
| Ka-band | 1.18–0.75 cm |

**FIGURE 7.8.** Image illustrating radar foreshortening, Death Valley, CA. The unnatural appearance of the steep terrain illustrates the effect of radar foreshortening, when a radar system observes high, steep topography at steep depression angles. The radar observes this terrain from the right—radar foreshortening creates the compressed appearance of the mountainous terrain in this scene. From NASA-JPL; SIR-C/X-SAR, P43883, October 1994.

The choice of a specific microwave band has several implications for the nature of the radar image. For real aperture imaging radars, spatial resolution improves as wavelength becomes shorter with respect to antenna length (i.e., for a given antenna length, resolution is finer with use of a shorter wavelength). Penetration of the signal into the soil is in part a function of wavelength—for given moisture conditions, penetration is greatest at longer wavelengths. The longer wavelengths of microwave radiation (e.g., relative to visible radiation) mean that imaging radars are insensitive to the usual problems of atmospheric attenuation—usually only very heavy rain will interfere with transmission of microwave energy.

## 7.5. Penetration of the Radar Signal

In principle, radar signals are capable of penetrating what would normally be considered solid features, including vegetative cover and the soil surface. In practice, it is very difficult to assess the existence or amount of radar penetration in the interpretation of specific images. Penetration is assessed by specifying the *skin depth,* the depth to which the strength of a signal is reduced to $1/e$ of its surface magnitude, or about 37%. Separate features are subject to differing degrees of penetration; specification of the skin depth, measured in standard units of length, provides a means of designating variations in the ability of radar signals to penetrate various substances.

In the absence of moisture, skin depth increases with increasing wavelength. Thus optimum conditions for observing high penetration would be in arid regions, using long-wavelength radar systems. Penetration is also related to surface roughness and to incidence angle; penetration is greater at steeper angles and decreases as incidence angle increases. We should therefore expect maximum penetration at the near-range edge of the image and minimum penetration at the far-range portion of the image.

The difficulties encountered in the interpretation of an image that might record penetration of the radar signal would probably prevent practical use of any information that might be conveyed to the interpreter. There is no clearly defined means by which an interpreter might be able to recognize the existence of penetration or to separate its effects from the many other variables that contribute to radar backscatter. For radar systems operating near the X-band and the K-band, empirical evidence suggests that the radar signal is generally scattered from the first surface it strikes, probably foliage in most instances. Even at the L-band, which is theoretically capable of a much higher degree of penetration, the signal is apparently scattered from the surface foliage in densely vegetated regions, as reported by Sabins (1983) after examining L-band imagery of forested regions in Indonesia.

## 7.6. Polarization

The *polarization* of a radar signal denotes the orientation of the field of electromagnetic energy emitted and received by the antenna. Radar systems can be configured to transmit either horizontally or vertically polarized energy and to receive either horizontally or vertically polarized energy as it is scattered from the ground. Unless otherwise specified, an imaging radar usually transmits horizontally polarized energy and receives a horizontally polarized echo from the terrain. However, some radars are designed to transmit horizontally polarized signals, but to separately receive the horizontally and vertically polarized reflections from the landscape. Such systems produce two images of the same landscape (Figure 7.9). One is the image formed by the transmission of a horizontally polarized signal and the reception of a horizontally polarized return signal. This is often referred to as the *HH image* or the *like-polarized* mode. A second image is formed by the transmission of a horizontally polarized signal and the reception of the vertically polarized return; this is called the *HV image* or the *cross-polarized* mode.

By comparing the two images, the interpreter can identify features and areas that represent regions on the landscape that tend to depolarize the signal. Such areas will reflect the incident horizontally polarized signal back to the antenna as vertically polarized energy—that is, they change the polarization of the incident microwave energy. Such areas can be identified as bright regions on the HV image and as dark or dark gray regions on the corresponding HH image. Their appearance on the HV image is much brighter due to the effect of depolarization—the polarization of the energy that would have contributed to the brightness of the HH image has been changed, so it creates instead a bright area on the HV image. Comparison of the two images therefore permits detection of those areas that are good depolarizers.

This same information can be restated in a different way. A surface that is an ineffective depolarizer will tend to scatter energy in the same polarization in which it was transmitted; such areas will appear bright on the HH image and dark on the HV image. In contrast, a surface that is a "good" depolarizer will tend to scatter energy in a polarization different from that of the incident signal; such areas will appear dark on the HH image and bright on the HV image.

Causes of depolarization are related to physical and electrical properties of the ground surface. A rough surface (with respect to the wavelength of the signal) may depolarize the signal.

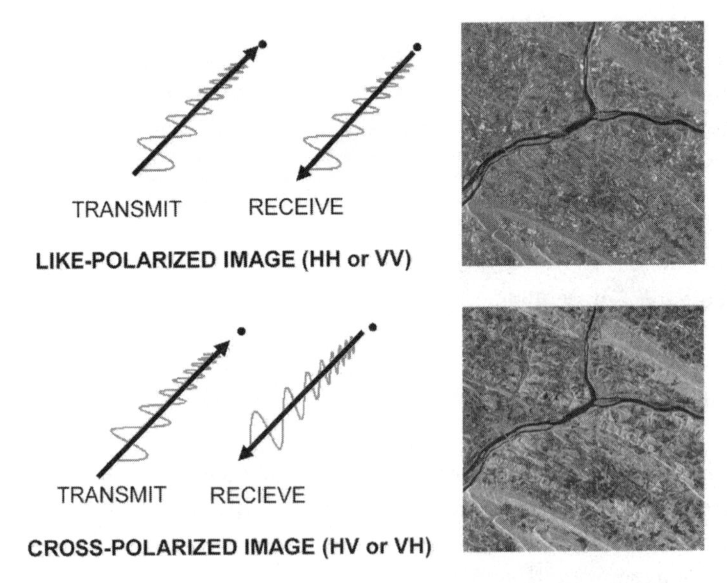


**FIGURE 7.9.** Radar polarization. Many imaging radars can transmit and receive signals in both horizontally and vertically polarized modes. By comparing the like-polarized (top) and cross-polarized images (bottom), analysts can learn about characteristics of the terrain surface. From NASA-JPL; SIR-C/X-SAR, P45541, October 1994.

Another cause of depolarization is volume scattering from an inhomogeneous medium; such scatter might occur if the radar signal is capable of penetrating beneath the soil surface (as might conceivably be possible in some desert areas where vegetation is sparse and the soil is dry enough for significant penetration to occur), where it might encounter subsurface inhomogeneities, such as buried rocks or indurated horizons.

## 7.7. Look Direction and Look Angle

### *Look Direction*

*Look direction,* the direction at which the radar signal strikes the landscape, is important in both natural and man-made landscapes. In natural landscapes, look direction is especially important when terrain features display a preferential alignment. Look directions perpendicular to topographic alignment will tend to maximize radar shadow, whereas look directions parallel to topographic orientation will tend to minimize radar shadow. In regions of small or modest topographic relief, radar shadow may be desirable as a means of enhancing microtopography or revealing the fundamental structure of the regional terrain. The extent of radar shadow depends not only upon local relief, but also upon orientations of features relative to the flight path; those features positioned in the near-range portion (other factors being equal) will have the smallest shadows, whereas those at the far-range edge of the image will cast larger shadows (Figure 7.10). In areas of high relief, radar shadow is usually undesirable, as it masks large areas from observation.

In landscapes that have been heavily altered by human activities, the orientation of structures and land-use patterns are often a significant influence on the character of the radar return, and therefore upon the manner in which given landscapes appear on radar imagery. For instance, if an urban area is viewed at a look direction that maximizes the scattering of the radar signal from structures aligned along a specific axis, it will have an appearance quite different from that of an image acquired at a look direction that tends to minimize reflection from such features.

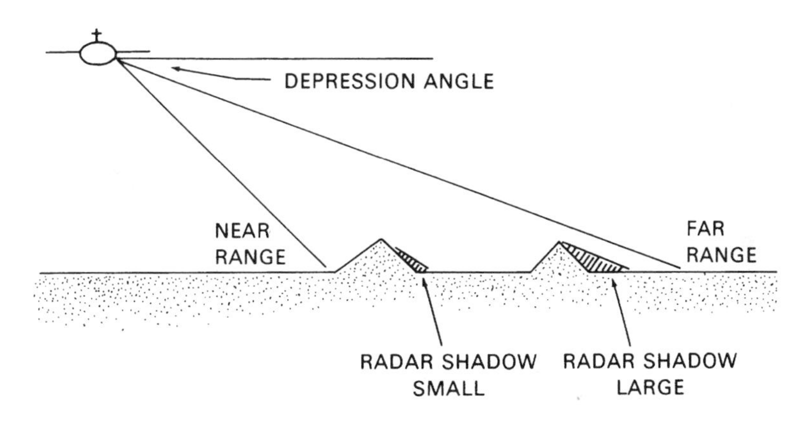


**FIGURE 7.10.** Radar shadow. Radar shadow increases as terrain relief increases and depression angle decreases.

### *Look Angle*

*Look angle,* the depression angle of the radar, varies across an image, from relatively steep at the near-range side of the image to relatively shallow at the far-range side (Figure 7.11). The exact values of the look angle vary with the design of specific radar systems, but some broad generalizations are possible concerning the effects of varied look angles. First, the basic geometry of a radar image ensures that the resolution of the image must vary with look angle; at steeper depression angles, a radar signal illuminates a smaller area than does the same signal at shallow depression angles. Therefore, the spatial resolution, at least in the across-track direction, varies with respect to depression angle.

It has been shown that the sensitivity of the signal to ground moisture is increased as depression angle becomes steeper. Furthermore, the slant-range geometry of a radar image means that all landscapes are viewed (by the radar) at oblique angles. As a result, the image tends to record reflections from the sides of features. The obliqueness, and therefore the degree to which we view sides, rather than tops, of features varies with look angle. In some landscapes, the oblique view may be very different from the overhead view to which we are accustomed in the use of other remotely sensed imagery (Figure 7.12). Such variations in viewing angle may contribute to variations in the appearance on radar imagery of otherwise similar landscapes.

## 7.8. Real Aperture Systems

*Real aperture* SLAR systems (sometimes referred to as *brute force* systems), one of the two strategies for acquiring radar imagery, are the oldest, simplest, and least expensive of imaging

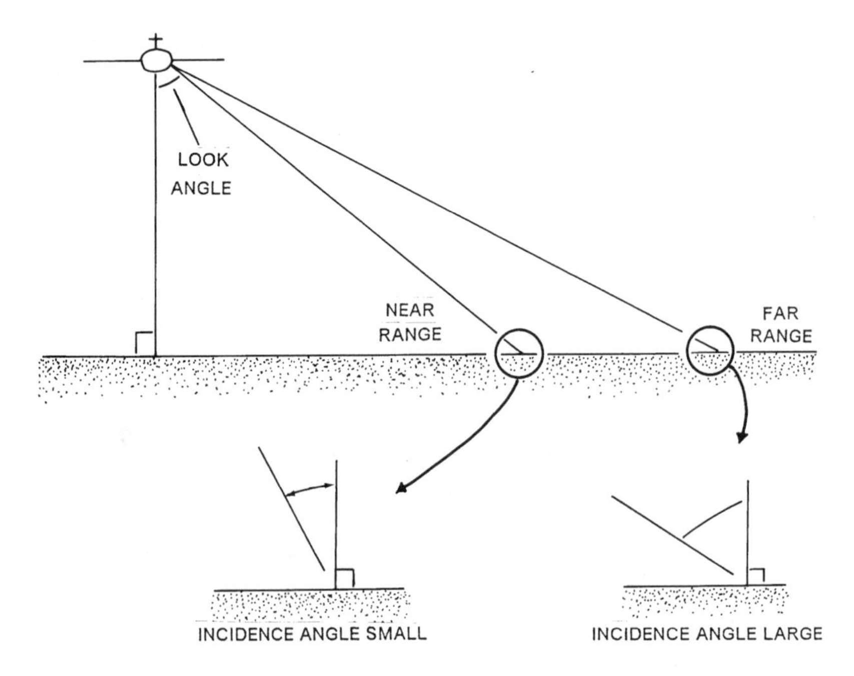


**FIGURE 7.11.** Look angle and incidence angle.

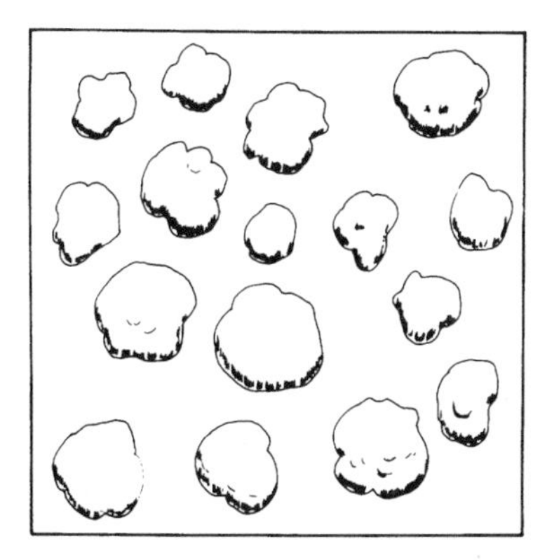

**FIGURE 7.12.** Perspective as related to depression angle. The side-looking nature of imaging radars means that radars tend to view the landscape from an oblique angle, which means that the image tends to portray the sides of objects rather than the overhead view that we are accustomed to in aerial photography.

radar systems. They follow the general model described earlier for the basic configuration of a SLAR system (Figure 7.2). The transmitter generates a signal at a specified wavelength and of a specified duration. The antenna directs this signal toward the ground, and then receives its reflection. The reflected signal is amplified, filtered, then displayed on a cathode ray tube, where a moving film records the radar image line by line as it is formed by the forward motion of the aircraft.

The resolution of such systems is controlled by several variables. One objective is to focus the transmitted signal to illuminate as small an area as possible on the ground, as it is the size of this area that determines the spatial detail recorded on the image. If the area illuminated is large, then reflections from diverse features may be averaged together to form a single graytone value on the image, and their distinctiveness is lost. If the area is small, individual features are recorded as separate features on the image, and their identities are preserved.

The size of the area illuminated is controlled by several variables. One is antenna length in relation to wavelength. A long antenna length permits the system to focus energy on a small ground area. Thus real aperture systems require long antennas in order to achieve fine detail; limits on the ability of an aircraft to carry long antennas forms a practical limit to the resolution of radar images. This restriction on antenna length forms a barrier for use of real aperture systems on spacecraft: small antennae would provide very coarse resolution from spacecraft altitudes, but practical limitations prevent use of large antennae.

The area illuminated by a real aperture SLAR system can be considered analogous to the spot illuminated by a flashlight aimed at the floor. As the flashlight is aimed straight down, the spot of light it creates is small and nearly circular in shape. As the flashlight beam is pointed toward the floor at increasingly further distances, the spot of light becomes larger, dimmer, and assumes a more irregular shape. By analogy, the near-range portions of a radar image will have finer resolution than the far-range portions. Thus antenna length in relation to wavelength determines the angular resolution of a real aperture system—the ability of the system to separate two objects in the along-track dimension of the image (Figure 7.13).

The relationship between resolution and antenna length and wavelength is given by the equation

$$\beta = \lambda / A \qquad \text{(Eq. 7.1)}$$

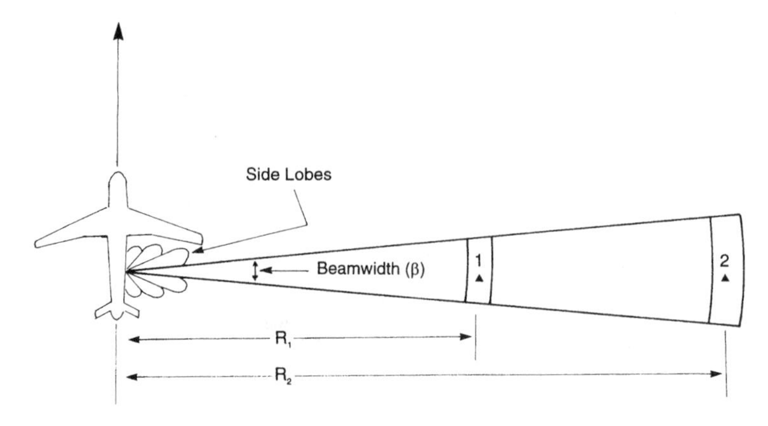


**FIGURE 7.13.** Azimuth resolution. For real aperture radar, the ability of the system to acquire fine detail in the along-track axis derives from its ability to focus the radar beam to illuminate a small area. A long antenna, relative to wavelength, permits the system to focus energy on a small strip of ground, improving detail recorded in the along-track dimension of the image. The *beam width* ($\beta$) measures this quality of an imaging radar. Beamwith, in relation to range ($R$) determines detail—region 1 at range $R_1$ will be imaged in greater detail than region 2 at greater range $R_2$. Also illustrated here are *side lobes,* smaller beams of microwave energy created because the antenna cannot be perfectly effective in transmitting a single beam of energy.

where $\beta$ is the beamwidth, $\lambda$ is the wavelength, and $A$ is the antenna length. Real aperture systems can be designed to attain finer along-track resolution by increasing the length of the antenna and by decreasing wavelength. Therefore, these qualities establish the design limits on the resolution of real aperture systems—antenna length is constrained by practical limits of aeronautical design and operation.

Radar systems have another, unique, means of defining spatial resolution. The length of the radar pulse determines the ability of the system to resolve the distinction between two objects in the cross-track axis of the image (Figure 7.14). Long pulses strike two nearby features at the same time, thereby recording the two objects as a single reflection and as a single feature on the

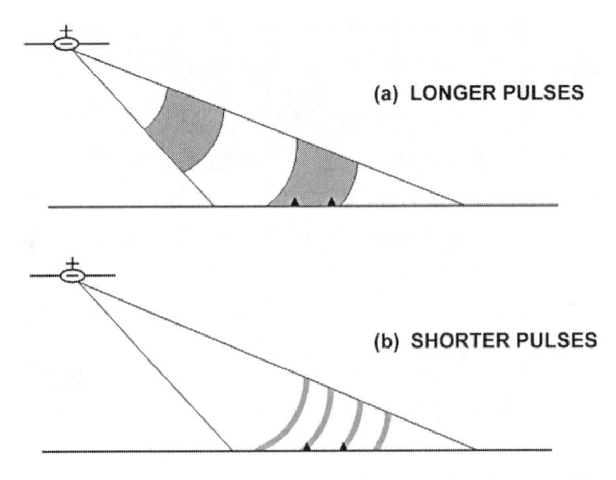


**FIGURE 7.14.** Effect of pulse length. (*a*) Longer pulse length means that the two objects shown here are illuminated by a single burst of energy, creating a single echo that cannot reveal the presence of two separate objects. (*b*) Shorter pulse length illuminates the two objects with separate pulses, creating separate echoes for each object. Pulse length determines resolution in the cross-track dimension of the image.

image. Shorter pulses are each reflected separately from adjacent features and can record the distinctive identities and locations of the two objects.

## 7.9. Synthetic Aperture Systems

The alternative design is the *synthetic aperture radar* (SAR), which is based upon principles and technology differing greatly from those of real aperture radars. Although SAR systems have greater complexity, and are more expensive to manufacture and operate than are real aperture systems, they can overcome some of the limitations inherent to real aperture systems, and therefore can be applied in a wider variety of applications, including observation from Earth-Orbiting satellites.

Consider a SAR that images the landscape as depicted in Figure 7.15. At *1,* the aircraft is positioned so that a specific region of the landscape is just barely outside the region illuminated by the SAR. At *2,* it is fully within the area of illumination. At *3,* it just at the trailing edge of the illuminated area. Finally, at *4,* the aircraft moves so that the region falls just outside the area illuminated by the radar beam. A SAR operates on the principal that objects within a scene are illuminated by the radar over an interval of time, as the aircraft moves along its flight path. A SAR system receives the signal scattered from the landscape during this interval, and saves the complete history of reflections from each object. Knowledge of this history permits later reconstruction of the reflected signals as though they were received by a single antenna occupying

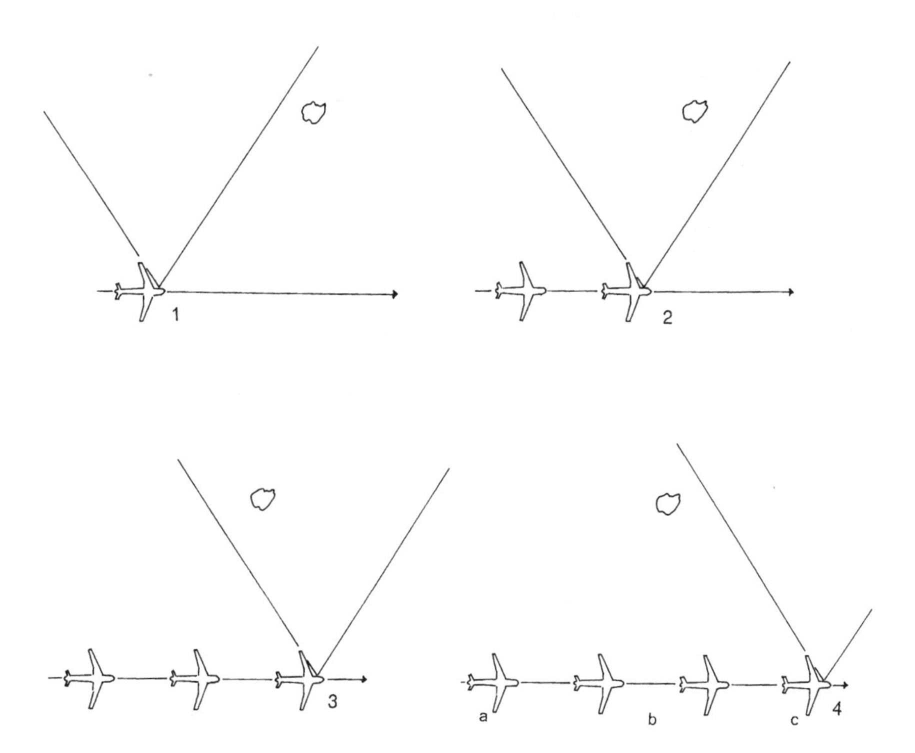


**FIGURE 7.15.** Synthetic aperture imaging radar. Synthetic aperture systems accumulate a history of backscattered signals from the landscape as the antenna moves along path *abc*.

physical space *abc,* even though they were in fact received by a much shorter antenna that was moved in a path along distance *1234.* (Thus the term "synthetic aperture" denotes the artificial length of the antenna, in contrast to the "real" aperture based upon the actual physical length of the antenna used with real aperture systems.)

In order to implement this strategy, it is necessary to define a practical means of assigning separate components of the reflected signal to their correct positions as the spatial representation of the landscape is recreated on the image. This process is, of course, extraordinarily complicated if each such assignment must be considered an individual problem in unraveling the complex history of the radar signal at each of a multitude of antenna positions.

Fortunately, this problem can be solved in a practical manner because of the systematic changes in frequency experienced by the radar signal as it is scattered from the landscape. Objects within the landscape experience different frequency shifts in relation to their distances from the aircraft track. At a given instant, objects at the leading edge of the beam reflect a pulse with an increase in frequency (relative to the transmitted frequency) due to their position ahead of the aircraft, and those at the trailing edge of the antenna experience a decrease in frequency (Figure 7.16). This is the *Doppler effect,* often explained by analogy to the change in pitch of a train whistle heard by a stationary observer as a train passes by at high speed. As the train approaches, the pitch appears higher than that of a stationary whistle due to the increase in frequency of sound waves. As the train passes the observer, then recedes into the distance, the pitch appears lower, due to the decrease in frequency. Radar, as an active remote sensing system, is operated with full knowledge of the frequency of the transmitted signal. As a result, it is possible to compare the frequencies of transmitted and reflected signals to determine the nature and amount of frequency shift. Knowledge of frequency shift permits the system to assign reflections to their correct positions on the image and to synthesize the effect of a long antenna.

## 7.10. Interpreting Brightness Values

Each radar image is composed of many image elements of varying brightness (Figure 7.17). Variations in image brightness correspond, at least in part, to place-to-place changes within the landscape; through knowledge of this correspondence, the image interpreter has a basis for making predictions, or inferences, concerning landscape properties.

Unlike passive remote sensing systems, active systems illuminate the land with radiation of

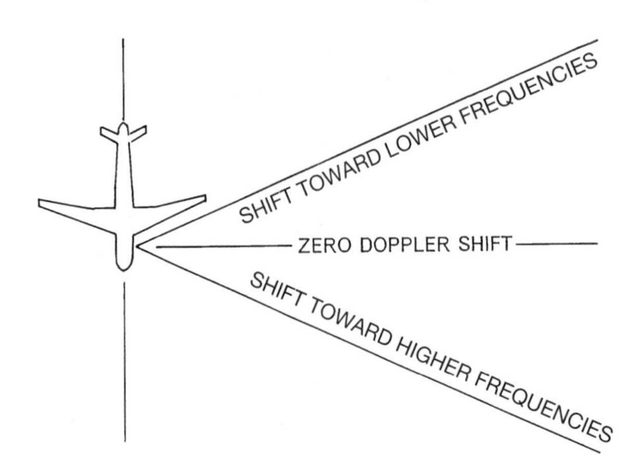


**FIGURE 7.16.** Frequency shifts experienced by features within the field of view of the radar system.

**FIGURE 7.17.** Two examples of radar images illustrating their ability to convey detailed information about quite different landscapes. Left: agricultural fields of the Maricopa Agricultural Experiment Station, Phoenix, AZ (Ku-band, spatial resolution, about 1 m); varied tones and textures distinguish separate crops and growth states. Right: structures near athletic fields, University of New Mexico, Albuquerque (Ku-band, spatial resolution about 1 m). Varied brightnesses and tones convey information about diffuse surfaces, specular reflection, and corner reflectors, which each carry specific meaning within the context of different landscapes. From Sandia National Laboratories. Reproduced by permission.

known and carefully controlled properties. Therefore, in principle, the interpreter should have a firm foundation for deciphering the meaning of the image because the only "unknowns" of the many variables that influence image appearance are the ground conditions, the object of study.

However, in practice, the interpreter of a radar image faces many difficult obstacles in making a rigorous interpretation of a radar image. First, most imaging radars are uncalibrated in the sense that image brightness values on an image cannot be quantitatively matched to backscattering values in the landscape. Typically, returned signals from a terrain span a very broad range of magnitudes from very low to very high; the ranges in values are often so large that they exceed the ability of photographic emulsions to accurately portray the actual range of values (Chapter 3). Very dark and very bright landscape elements may be portrayed in the nonlinear portions of the characteristic curve, and therefore lack consistent relationships with their representations on the image. Furthermore, the features that compose even the simplest landscapes have complex shapes and arrangements, and are formed from diverse materials of contrasting electrical properties.

As a result, there are often few detailed models of the kinds of backscattering that should in principle be expected from separate classes of surface materials. Direct experience and intuition are not always reliable guides to interpretation of images acquired outside the visible spectrum. In addition, many SLAR images observe the landscape at very shallow depression angles. Since interpreters gain experience from their observations at ground level, or from studying overhead aerial views, they may find the oblique radar view from only a few degrees above the horizon difficult to interpret.

### *The Radar Equation*

The fundamental variables influencing the brightness of a region on a radar image are formally given by the *radar equation:*

$$P_r = \frac{\sigma G^2 P_t \lambda^2}{(4\pi)^3 R^4} \qquad \text{(Eq. 7.2)}$$

Here $P_r$ designates the power returned to the antenna from the ground surface; $R$ specifies the range to the target from the antenna; $P_t$ is the transmitted power; $\lambda$ is the wavelength of the energy; and $G$ is the antenna gain (a measure of the system's ability to focus the transmitted energy). All of these variables are determined by the design of the radar system, and are therefore known or controlled quantities. The one variable in the equation not thus far identified is $\sigma$, the backscattering coefficient; $\sigma$ is, of course, not controlled by the radar system, but by the specific characteristics of the terrain surface represented by a specific region on the image. Whereas s is often an incidental factor for the radar engineer, it is the primary focus of study for the image interpreter, as it is this quantity that carries information about the landscape.

The value of $\sigma$ conveys information concerning the amount of energy scattered from a specific region on the landscape as measured by $\sigma^{\circ}$, the *radar cross section.* It specifies the corresponding area of an isotropic scatterer that would return the same power as does the observed signal. The backscattering coefficient ($\sigma^{\circ}$) expresses the observed scattering from a large surface area as a dimensionless ratio between two areal surfaces; it measures the average radar cross section per unit area. The backscattering coefficient $\sigma^{\circ}$ varies over such wide values that it must be expressed as a ratio rather than as an absolute value.

Ideally, radar images should be interpreted with the objective of relating observed ?o (varied brightnesses) to properties within the landscape. It is known that backscattering is related to specific *system variables,* including wavelength, polarization, and azimuth, in relation to landscape orientation and depression angle. In addition, *landscape parameters* are important, including surface roughness, soil moisture, vegetative cover, and microtopography. Because so many of these characteristics are interrelated, making detailed interpretations of individual variables is usually very difficult, in part due to the extreme complexity of landscapes, which normally are intricate compositions of diverse natural and man-made features. Often many of the most useful landscape interpretations of radar images have attempted to recognize integrated units defined by assemblages of several variables rather than to separate individual components. The notion of "spectral signatures" is very difficult to apply in the context of radar imagery because of the high degree of variation in image tone as incidence angle and look direction change.

### *Moisture*

Moisture in the landscape influences the backscattering coefficient through changes in the dielectric constant of landscape materials. (The *dielectric constant* is a measure of the ability of a substance to conduct electrical energy—an important variable determining the response of a substance that is illuminated with microwave energy.) Although natural soils and minerals vary in their ability to conduct electrical energy, these properties are difficult to exploit as the basis for remote sensing because the differences between dielectric properties of separate rocks and minerals in the landscape are overshadowed by the effects of even very small amounts of moisture, which greatly change the dielectric constant. As a result, the radar signal is sensitive to the presence of moisture both in the soil and in vegetative tissue; this sensitivity appears to be greatest at steep depression angles.

The presence of moisture also influences effective skin depth; as the moisture content of surface soil increases, the signal tends to scatter from the surface. As moisture content decreases, skin depth increases, and the signal may be scattered from a greater thickness of soil.

### *Roughness*

A radar signal that strikes a surface will be reflected in a manner that depends both upon characteristics of the surface and properties of the radar wave, as determined by the radar system and the conditions under which it is operated. The *incidence angle* ($\theta$) is defined as the angle between the axis of the incident radar signal and a perpendicular to the surface that the signal strikes (Figure 7.18). If the surface is homogeneous with respect to its electrical properties, and "smooth" with respect to the wavelength of the signal, then the reflected signal will be reflected at an angle equal to the incidence angle, with most of the energy directed in a single direction (i.e., specular reflection).

For "rough" surfaces, reflection will not depend as much upon incidence angle, and the signal will be scattered more or less equally in all directions (i.e., diffuse, or isotropic, scattering). For radar systems, the notion of a rough surface is defined in a manner considerably more complex than that familiar from everyday experience, as roughness depends not only upon the physical configuration of the surface, but also upon the wavelength of the signal and its incidence angle (Table 7.2). Consider the physical configuration of the surface to be expressed by the standard deviation of the heights of individual facets (Figure 7.19). Although definitions of surface roughness vary, one common definition defines a rough surface as one in which the standard deviation of surface height ($S_h$) exceeds one-eighth of the wavelength ($\lambda$) divided by the cosine of the incidence angle ($\cos \theta$):

$$S_h > \lambda/(8 \cos \theta) \qquad \text{(Eq. 7.3)}$$

where $h$ is the average height of the irregularities. In practice, this definition means that a given surface appears rougher as wavelengths become shorter. Also, for a given wavelength, surfaces will act as smooth scatterers as incidence angle becomes greater (i.e., equal terrain slopes will appear as smooth surfaces as depression angle becomes smaller, as occurs in the far-range portions of radar images).

### *Corner Reflectors*

The return of the radar signal to the antenna can be influenced not only by moisture and roughness, but also by the broader geometric configuration of targets. Objects that have complex geo-

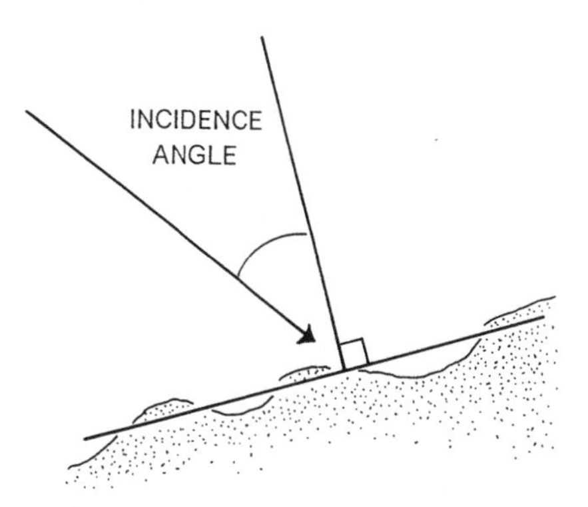


**FIGURE 7.18.** Incidence angle.

**TABLE 7.2. Surface Roughness Defined for Several Wavelengths**

| Roughness category | K-band ($\lambda$ = 0.86 cm) | X-band ($\lambda$ = 3 cm) | L-band ($\lambda$ = 25 cm) |
|---|---|---|---|
| Smooth | $h < 0.05$ cm | $h < 0.17$ cm | $h < 1.41$ cm |
| Intermediate | $h = 0.05–0.28$ cm | $h = 0.17–0.96$ cm | $h = 1.41–8.04$ cm |
| Rough | $h > 0.28$ cm | $h > 0.96$ cm | $h > 8.04$ cm |

*Note.* Data from Jet Propulsion Laboratory (1982).

metric shapes, such as those encountered in an urban landscape, can create radar returns that are much brighter than would be expected based upon size alone. This effect is caused by the complex reflection of the radar signal directly back to the antenna in a manner analogous to a ball that bounces from the corner of a pool table directly back to the player. This behavior is caused by objects classified as *corner reflectors,* which often are in fact corner-shaped features (such as the corners of buildings and the alleyways between them in a dense urban landscape), but are also formed by other objects of complex shape. Corner reflectors are common in urban areas due to the abundance of concrete, masonry, and metal surfaces constructed in complex angular shapes (Figure 7.20).

Corner reflectors can also be found in rural areas, formed sometimes by natural surfaces, but more commonly by metallic roofs of farm buildings, agricultural equipment, and items such as powerline pylons and guardrails along divided highways.

Corner reflectors are important in interpretation of the radar image. They form a characteristic feature of the radar signatures of urban regions, and identify other features such as powerlines, highways, and railroads (Figure 7.21). It is important to remember that the image of a corner reflector is not shown in proportion to its actual size: the returned energy forms a star-like burst of brightness that is proportionately much larger than the size of the object that caused it. Thus, corner reflectors can convey important information, but they do not appear on the image in their correct relative sizes.

## 7.11. Satellite Imaging Radars

Scientists working with radar remote sensing have been interested for years in the possibility of observing the Earth by means of imaging radars carried by Earth satellites. Whereas real aper-

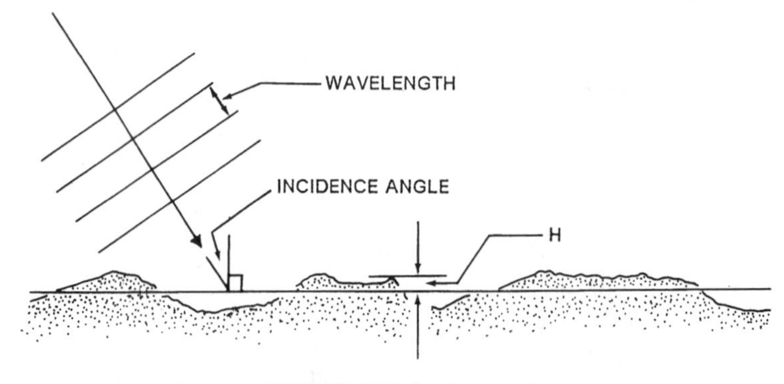


**FIGURE 7.19.** Surface roughness.

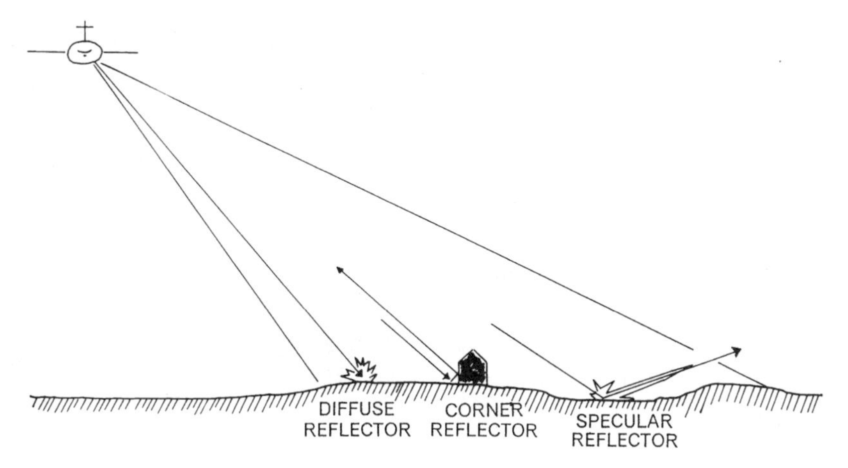


**FIGURE 7.20.** Three classes of features important for interpretation of radar imagery. See Figure 2.21 for examples.

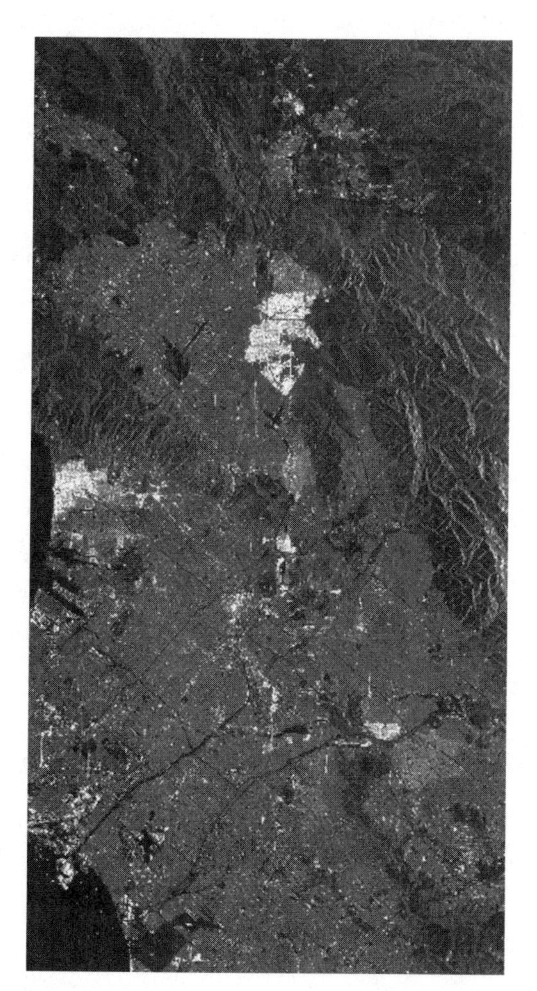

**FIGURE 7.21.** SAR image of Los Angeles, CA, acquired at L-band. This image illustrates the classes of features represented on radar imagery, including diffuse, specular, and corner reflectors. This image depicts the internal structure of the built-up urban region, including transportation, roads, and highways. Radar foreshortening is visible in its representation of the mountainous topography. From NASA-JPL; SIR-C/X-SAR, PIA1738, October 1994.

ture systems cannot be operated at satellite altitudes without unacceptably coarse spatial resolution (or use of impractically large antennas), the synthetic aperture principle permits compact radar systems to acquire imagery of fine spatial detail at high altitudes. This capability, combined with the ability of imaging radars to acquire imagery in darkness, through cloud cover, and during inclement weather, provides the opportunity for development of a powerful remote sensing capability, with potential to observe large areas of the Earth's ocean and land areas that might otherwise be unobservable because of remoteness and atmospheric conditions.

Several satellite missions have used imaging radars; all have been experimental programs to further develop the concepts and experience necessary for longer term efforts. These satellite imaging radars have been designed primarily to acquire a library of imagery representative of diverse terrains and to promote more experience and knowledge of the interpretation and analysis of radar images.

### *Seasat Synthetic Aperture Radar*

Seasat (Figure 7.22) was specifically tailored to observe the Earth's oceans by means of several sensors designed to monitor winds, waves, temperature, and topography. Many of the sensors detected active and passive microwave radiation, although one radiometer operated in the visible and near infrared spectrum. (Seasat carried three microwave radiometers, an imaging radar, and a radiometer that operated both in the visible and in the infrared. Radiometers are described in Chapter 9.) Some sensors were capable of observing 95% of the Earth's oceans every 36 hours. Specifications for the Seasat SAR are as follows:

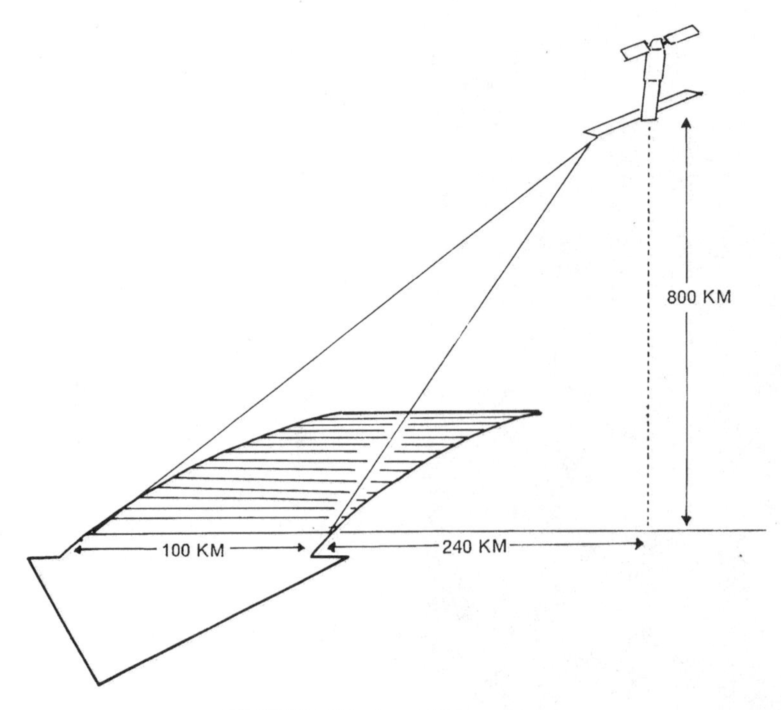


**FIGURE 7.22.** Seasat SAR geometry.

- *Launch:* 28 June 1978
- *Electrical system failure:* 10 October 1978
- *Orbit:* Nearly circular at 108° inclination; 14 orbits each day
- *Frequency:* L-band (1.275 GHz)
- *Wavelength:* 23 cm
- *Look direction:* Looks to starboard side of track
- *Swath width:* 100 km, centered 20° off-nadir
- *Ground resolution:* 25 m × 25 m
- *Polarization:* HH

Our primary interest here is the synthetic aperture radar (SAR), designed to observe ocean waves, sea ice, and coastlines. The satellite orbit was designed to provide optimum coverage of oceans, but its track did cross land areas, thereby offering the opportunity to acquire radar imagery of the Earth's surface from satellite altitudes. The SAR, of course, had the capability for operation during both daylight and darkness and during inclement weather, so (unlike most other satellite sensors) the Seasat SAR could acquire data on both ascending and descending passes.

The high transmission rates required to convey data of such fine resolution meant that Seasat SAR data could not be recorded onboard for later transmission to ground stations; instead the data was immediately transmitted to a ground station. Therefore, data could be acquired only when the satellite was within line of sight of one of the five ground stations equipped to receive Seasat data; these were located in California, Alaska, Florida, Newfoundland, and England.

The SAR was first turned on in early July 1978 (Day 10); it remained in operation for 98 days. During this time, it acquired some 500 passes of data: the longest SAR track covered about 4,000 km in ground distance. Data are available in digital form, as either digitally or optically processed imagery. All land areas covered are in the northern hemisphere, including portions of North America and western Europe.

Seasat SAR data have been used for important oceanographic studies. In addition, Seasat data have provided a foundation for the study of radar applications within the earth sciences and for the examination of settlement and land-use patterns. Because Seasat data were acquired on both ascending and descending passes, the same ground features can be observed from differing look angles under conditions that hold most other factors constant, or almost constant.

### *Shuttle Imaging Radar-A*

The Shuttle Imaging Radar (SIR) is a synthetic aperture imaging radar carried by the Shuttle Transportation System. (NASA's space shuttle orbiter has the ability to carry a variety of scientific experiments; the SIR is one of several.) SIR-A, the first scientific payload carried on board the shuttle, was operated for about 54 hours during the flight of Columbia in November 1981. Although the mission was reduced in length from original plans, SIR-A was able to acquire images of almost 4,000,000 mi$^2$ of the Earth's surface. Additional details of SIR-A are given below:

- *Launch:* 12 November 1981
- *Land:* 14 November 1981

- *Altitude:* 259 km
- *Frequency:* 1.278 GHz (L-band)
- *Wavelength:* 23.5 cm
- *Depression angle:* 40°
- *Swath width:* 50 km
- *Polarization:* HH
- *Ground resolution:* about 40 m × 40 m

SIR-A's geometric qualities were fixed, with no provision for changing depression angle. All data were recorded on magnetic tape and signal film onboard the shuttle; after the shuttle landed, it was physically carried to ground facilities for processing into image products. First-generation film images were typically 12 cm (5 in.) in width at a scale of about 1:5,250,000 (Figure 7.23). About 1,400 ft. of film were processed; most has been judged to be of high quality. Portions of all continents (except Antarctica) were imaged, to provide images of a wide variety of environments differing with respect to climate, vegetation, geology, land use, and other qualities..

### *Shuttle Imaging Radar-B*

The Shuttle Imaging Radar-B (SIR-B) was the second imaging radar experiment for the space shuttle. It was similar in design to SIR-A, except that it provided greater flexibility in acquiring imagery at varied depression angles, as shown below:

*Launched:* 5 October 1984
*Landed:* 11 October 1984

- *Orbital altitude:* 225 km
- *Orbital inclination:* 57°

**FIGURE 7.23.** SIR-A image, southwestern Virginia. From National Space Science Data Center.

- *Frequency:* 1.28 GHz
- *Wavelength:* 23 cm
- *Ground resolution:* about 25 m × 17 m (at 60° depression), or about 25 m × 58 m (at 15° depression)
- *Swath width:* 40–50 km

SIR-B used horizontal polarization at L-band (23 cm). Azimuth resolution was about 25 m; range resolution varied from about 17 m at an angle of 60° to 58 m at an angle of 15°. The shuttle orbited at 225 km; the orbital track drifted eastward 86 km each day (at 45° latitude). In contrast to the fixed geometric configuration of the Seasat SAR and SIR-A, the radar pallet onboard the shuttle permitted control of the angle of observation, which thereby enabled operation of the SIR in several modes. A given image swath of 40–50 km could be imaged repeatedly, at differing depression angles, by changing the orientation of the antenna as spacecraft position changed with each orbit. This capability provided the ability to acquire stereo imagery.

Also, the antenna could be oriented to image successive swaths and thereby build up coverage of a larger region than could be covered in any single pass. A mosaic composed of such images would be acquired at a consistent range of depression angles, although angles would, of course, vary within individual images.

The shuttle orbit also provided the opportunity to examine a single study area at several look directions. Varied illumination directions and angles permit imaging of a single geographic area at varied depression angles and azimuths, and thereby enable (in concept) derivation of information concerning the influence of surface roughness and moisture content upon the radar image, and information concerning the interactions of these variables with look direction and look angle. Thus an important function of SIR-B was its ability to advance understanding of the radar image itself and its role in remote sensing of the Earth's landscapes.

The significance of SIR imagery arises in part from the repetitive nature of the orbital coverage, with an accompanying opportunity to examine temporal changes in such phenomena as soil moisture, crop growth, land use, and the like. Because of the all-weather capability of SAR, the opportunities to examine temporal variation with radar imagery may be even more significant than it was with Landsat imagery. Second, the flexibility of the shuttle platform and the SIR experiment permit examination of a wide range of configurations for satellite radars, over a broad range of geographic regions, by many subject areas in the earth sciences. Therefore, a primary role for SIR-B was to investigate specific configurations and applications for the design of other satellite systems tailored to acquire radar imagery of the Earth's surface.

### *Shuttle Imaging Radar-C/X-Synthetic Aperture Radar System*

In August 1994 the space shuttle conducted a third imaging radar experiment. The Shuttle Imaging Radar-C (SIR-C) is an SAR operating at both L-band (23 cm) and C-band (6 cm), with the capability for transmitting and receiving both horizontally and vertically polarized radiation. SIR-C, designed and manufactured by the Jet Propulsion Laboratory (Pasadena, CA) and Ball Communications Systems Division, is one of the largest and most complex items ever built for flight on the shuttle. In addition to the use of two microwave frequencies and its dual polarization, the antenna has the ability to electronically aim the radar beam, to supplement the capability of the shuttle to aim the antenna by maneuvering the spacecraft. Data from SIR-C can be recorded onboard using tape storage, or it can be transmitted by microwave to the Tracking and Data Releay Satellite System (TDRSS) link to ground stations (Chapter 6).

The X-synthetic aperture radar (X-SAR) was designed and built in Europe as a joint German–Italian project for flight on the shuttle, to be used independently or in coordination with SIR-C. X-SAR was an L-band SAR (VV polarization) with the ability to create highly focused radar beams. It was mounted in a manner that permits it to be aligned with the L- and C-band beams of the SIR-C.

Together, the two systems had the ability to gather data at three frequencies and two polarizations. The SIR-C/X-SAR experiment was coordinated with field experiments that collected ground data at specific sites devoted to studies of ecology, geology, hydrology, agriculture, oceanography, and other topics. These instruments continue the effort to develop a clearer understanding of the capabilities of SAR data to monitor key processes at the Earth's surface (Plates 10 and 11).

X-SAR characteristics can be briefly summarized:

- *Frequencies:*
  - X-band (3 cm)
  - C-band (6 cm)
  - L-band (23 cm)
- *Ground swath:* 15–90 km, depending upon orientation of the antenna
- *Resolution:* 10–200 m

### *European Resource Satellite Synthetic Aperture Radar*

The European Space Agency (ESA), a joint organization of several European nations, designed a remote sensing satellite with several sensors configured to conduct both basic and applied research. Here our primary interest is the SAR for European Resource Satellite-1 (ERS-1) (launched in 1991, in service till 2000) and ERS-2 (launched in 1995). One of the satellite's primary missions is to use several of its sensors to derive wind and wave information from 5 km × 5 km SAR scenes positioned within the SAR's 80-km swath width. Nonetheless, the satellite has acquired a library of images of varied land and maritime scenes (Figure 7.24).

Because of the SAR's extremely high data rate, SAR data cannot be stored onboard, so it must be acquired within range of a ground station. Although the primary ground control and receiving stations are located in Europe, receiving stations throughout the world are equipped to receive ERS data. ERS has a sun-synchronous, nearly polar orbit that crosses the equator at about 10:30 A.M. local sun time. Other characteristics include:

- *Frequency:* 5.3 GHz
- *Wavelength:* C-band (6 cm)
- *Incidence angle:* 23° at midrange (20° at near range; 26° at far range)
- *Polarization:* VV
- *Altitude:* 785 km
- *Spatial resolution:* 30 m
- *Swath width:* 100 km

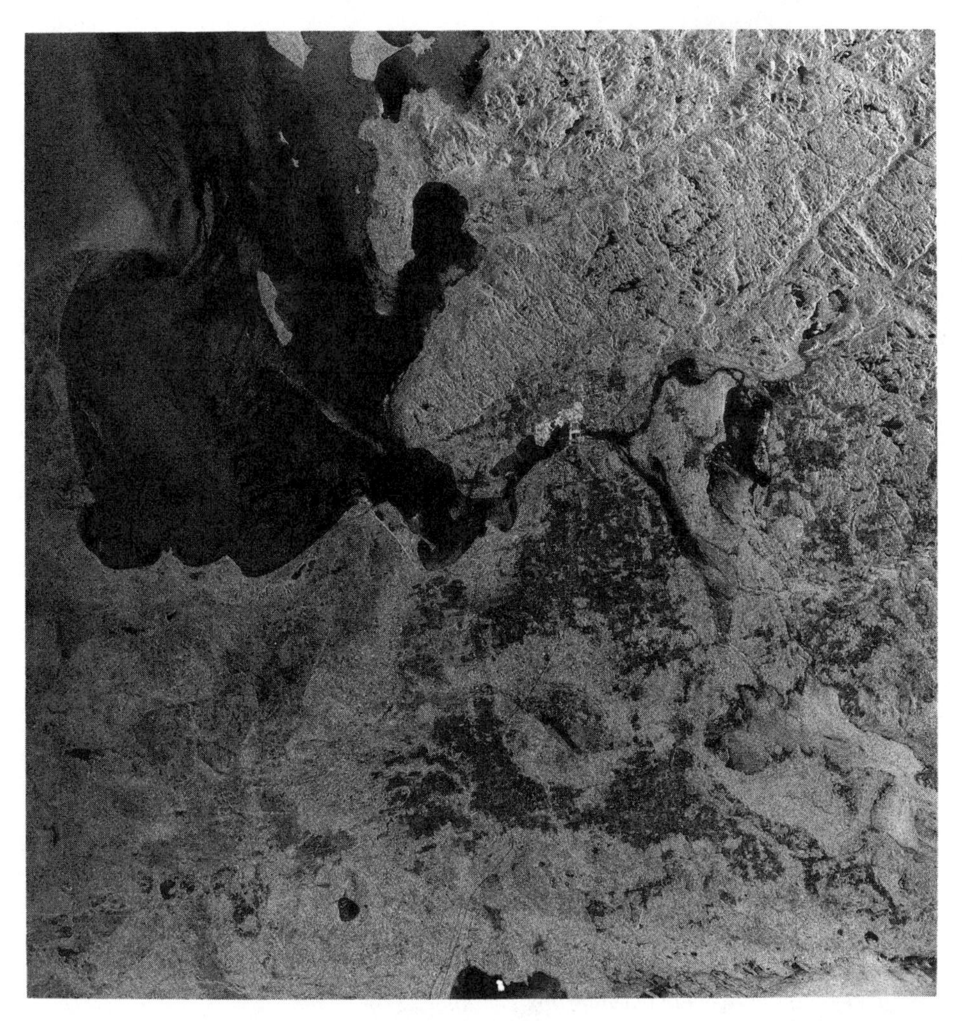

**FIGURE 7.24.** ERS-1 SAR image, Sault St. Marie, Michigan–Ontario, showing forested areas as bright tones, rough texture. Within water bodies, dark areas are calm surfaces; lighter tones are rough surfaces caused by winds and currents. Copyright 1991 by ESA. Received and processed by Canada Centre for Remote Sensing; distributed by RADARSAT International.

ERS data are available from RADARSAT International/MDA (*http://www.rsi.ca;* 604-278-3411) and NASA Alaska SAR Facility (907-474-6166; *http://www.asf.alaska.edu/dataset_documents/ers1* and *_ers2_sar_images.html*)

### *RADARSAT Synthetic Aperture Radar*

RADARSAT is a joint project of the Canadian federal and provincial governments, the United States government, and private corporations. The United States, through NASA, provides launch facilities and the services of a receiving station in Alaska. Canada has special interests in use of radar sensors because of its large territory, the poor illumination prevailing during much of the year at high latitudes, unfavorable weather conditions, the need to monitor sea ice in shipping lanes, and many other issues arising from assessing natural resources, especially forests and mineral deposits. Radar sensors, particularly those carried by satellites, provide capabilities tailored to address many of these concerns.

The RADARSAT's C-band radar was launched on 4 November 1995 into a sun-synchronous orbit at 98.6° inclination, with equatorial crossings at 6:00 A.M. and 6:00 P.M. An important feature of the RADARSAT SAR is the flexibility it offers to select among a wide range of trade-offs between area covered and spatial resolution and to use a wide variety of incidence angles (Figure 7.25). This flexibility increases opportunities for practical applications. Critical specifications include:

- *Frequency:* 5.3 GHz
- *Wavelength:* 5.6 cm (C-band)
- *Incidence angle:* varies, 10°–60°
- *Polarization:* HH
- *Altitude:* 793–821 km
- *Spatial resolution:* varies, 100 × 100 m to 9 × 9 m
- *Swath width:* varies, 45–510 km
- *Repeat cycle:* 24 days

Data are available from:

RADARSAT International/MDA
13800 Commerce Parkway
Richmond, BC, V6V 2J3 Canada
Tel: 604-278-3411
Fax: 604-231-2751
E-mail: *info@mdacorporation.com*
Web site: *http://www.mdacorporation.com/index.shtml*

RADARSAT-2 is planned for launch in 2006. RADARSAT-2 will follow orbital characteristics of RADARSAT-1, and will offer improved resolution; the ability to look to both sides of

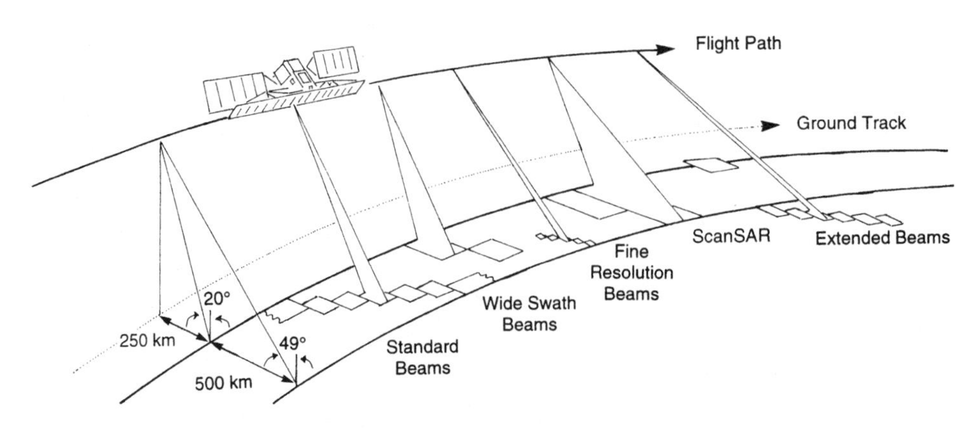


**FIGURE 7.25.** RADARSAT SAR geometry.

the orbital path, thereby improving the revisit interval; improved ability to selected polarization; and improved communication capabilities.

Applications of RADARSAT SAR imagery include geological exploration and mapping, sea ice mapping, rice crop monitoring, flood delineation and mapping, coastal zone mapping, and oil spill/seep detection. Because superstructures of ships form good corner reflectors, SAR data are effective in detecting ships at sea, and RADARSAT imagery has been used in the monitoring of shipping lanes. Figure 7.26 shows a RADARSAT SAR image of Cape Breton Island, Nova Scotia, Canada, on 28 November 1995; this image represents a region extending about 132 km east–west and 156 km north–south. At the time that RADARSAT acquired this image, this region was in darkness and was experiencing high winds and inclement weather, so this image illustrates radar's ability to operate under conditions of poor illumination and unfavorable weather. Sea conditions near Cape Breton Island's Cape North and Aspy Bay are recorded as variations in image tone. Ocean surfaces appearing black or dark are calmer waters sheltered

**FIGURE 7.26.** RADARSAT SAR image, Cape Breton Island, Canada. RADARSAT's first SAR image, acquired 28 November 1995. Image copyright 1995 by Canadian Space Agency. Received by the Canada Center for Remote Sensing; processed and distributed by RADARSAT International.

from winds and currents by the configuration of the coastline or the islands themselves. Brighter surfaces are rougher seas, usually on the windward coastlines, formed as winds and current interact with subsurface topography. The urban area of Sydney is visible as the bright area in the lower center of the image.

### *Japanese Earth Resources Satellite-1*

The Japanese Earth Resources Satellite-1 (JERS-1) is a satellite launched in February 1992 by the National Space Development Agency of Japan, and continued in service till 1998. It carried a C-band SAR, and an optical CCD sensor sensitive in seven spectral regions from 0.52 to 2.40 μm, with resolution of about 18 m. The JERS-1 SAR had the following characteristics:

- *Altitude:* 568 km
- *Orbit:* sun-synchronous
- *Wavelength:* 23 cm (L-band) (1.3 GHz)
- *Incidence angle:* 35°
- *Swath width:* 75 km
- *Spatial resolution:* 18 m
- *Polarization:* HH

The URLs for JERS-1 are

*http://www.eorc.nasda.go.jp/JERS-1/*
*http://southport.jpl.nasa.gov/polar/jers1.html*

### *Envisat*

In 2002, the European Space Agency launched Envisat, designed to monitor a broad suite of environmental variables using 10 different instruments. The sensor of primary interest here is the ASAR—the Advanced Synthetic Aperture Radar—an advanced synthetic aperture imaging radar based upon experience gained from ERS-1 and ERS-2. Envisat completes 14 orbits each day, in a Sun-synchronous orbit. ASAR operates at C-band, with the ability to employ a variety of polarizations, beam widths, and resolutions. Specifics are available at *http://envisat.esa.int/m-s/*.

## 7.12. Radar Interferometry

It is possible to derive accurate topographic information from multiple SAR images of a single region. In a broad sense, SAR interferometry is comparable to the use of stereo photography to determine topography of a region by observation from two different perspectives. However, SAR interferometry is applied not in the optical domain of photogrammetry, but in the realm of radar geometry, to exploit radar's ranging capability. Because SARs can accurately measure differences in slant-range distances to the same feature from separate observation points, the technique can provide very accurate topographic information. Plates 12 and 13 illustrate elevation data that can be acquired using this technique. The symbolization used for Plate 12 is known as *wrapped color*—a common method of depicting elevations derived using interferometric SAR.

If two images are acquired from the same track at different times (e.g., if two antennas are mounted fore and aft in the same aircraft, or if images are acquired at different times within the same orbital track), it is possible to establish a *temporal baseline,* which provides an image pair that can reveal changes that occurred during the interval between the acquisition of the two images. This precise record of change permits recognition of changes in position. In a military context, the objects in motion might be vehicles; in a scientific context, the motion might record ocean currents, ice flow in glaciers, or ice floes in polar oceans. The sensitivity of such analyses depends upon the nature of the temporal baseline, so that very short temporal baselines can record rather rapid motion (such as vehicular motion), whereas longer baselines can detect slower speeds, such as movement of surface ice in glaciers. Paired antennas operated from a single platform are sensitive to velocities of centimeters per second (e.g., suitable for observing moving vehicles or ocean waves). Longer temporal baselines (e.g., separate passes within the same orbital path) are effective in recording slower speeds, perhaps centimeters per day, such as the motion of glacial ice.

A more widespread use of SAR interferometry depends upon the acquisition of pairs of images of the same region acquired by imaging radars following separate tracks. The separation in distance of the two instruments establishes a *spatial baseline,* which permits measurement of topographic relief. The most favorable case arises when two images are acquired on parallel tracks, although analysis is possible with nonparallel tracks provided that the angle of intersection is small.

### *Shuttle Radar Topography Mission*

The most striking application of SAR interferometry was the Shuttle Radar Topography Mission (SRTM) (STS-99 Shuttle Endeavour), launched 11 February 2000, landed 22 February 2000, for a duration of about 11 days, 5½ hours. More information is available on the WWW at *http://www.jpl.nasa.gov/srtm/.*

SRTM was designed to use C-band and X-band interferometric SAR to acquire topographic data over 80% of the Earth's land mass (between 60° N and 56° S). One antenna was mounted, as in previous shuttle SAR missions, in the shuttle's payload bay; another antenna was mounted on a 60-m (200-ft.) mast extended after the shuttle attained orbit (Figure 7.27). The configuration was designed for single-pass interferometry using C-band and X-band SAR imagery.

The interferomic data have been used to produce digital topographic data for very large portions of the world at unprecedented levels of detail and accuracy—for the first time, there will be consistent, high-quality, cartographic coverage for most of the Earth's land areas (Figure 7.28). SRTM data, based upon the C-band imagery, cover approximately 80% of the Earth's landmass.

SRTM data have been used for a broad range of practical applications in geology and geophysics, hydrologic modeling, aircraft navigation, transportation engineering, land-use planning, siting of communications facilities, and military training and logistical planning. Furthermore, they will support other remote sensing applications by facilitating rectification of

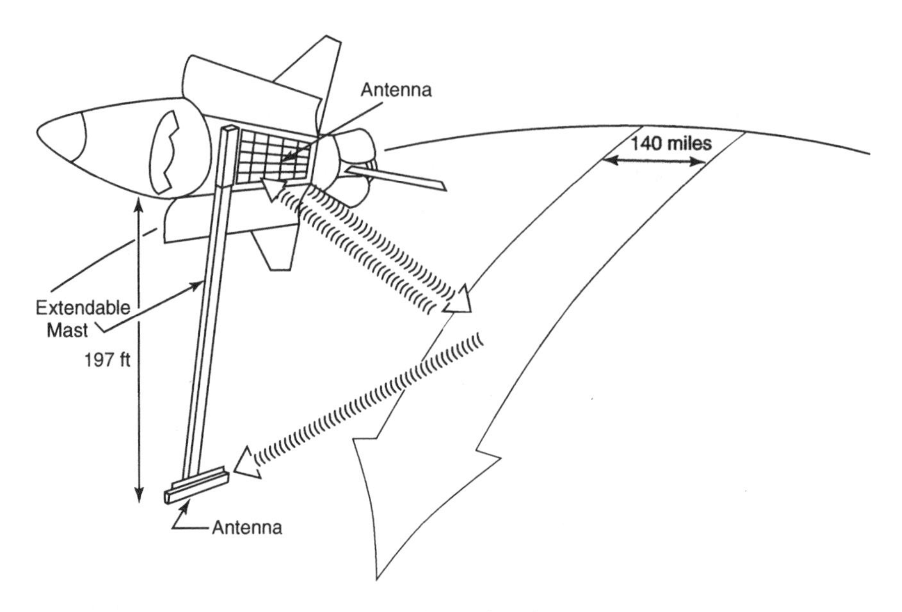


**FIGURE 7.27.** SRTM interferometry.

remotely sensed data and registration of remotely acquired image data. (SRTM data are consistent with U.S. National Map Accuracy Standards, at 30 m × 30 m spatial sampling with less than 16-m absolute vertical height accuracy, less than 10-m relative vertical height accuracy, and less than 20-m absolute horizontal circular accuracy. All accuracies are quoted at the 90% level, consistent with U.S. National Map Accuracy Standards.) SRTM was sponsored by the U.S. National Imagery and Mapping Agency (NIMA), NASA, the German aerospace center

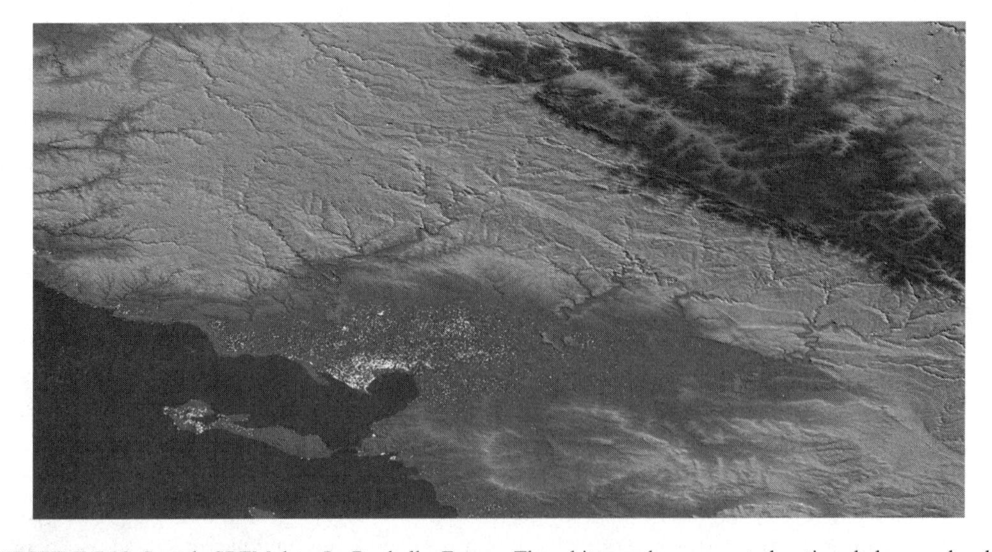

**FIGURE 7.28.** Sample SRTM data, La Rochelle, France. The white specks represent elevations below sea level (portions of this area have been reclaimed, and are protected by dikes and seawalls). These data provide 90 m pixels; within the United States, data are distributed at 30m. From JPL, Pasedena, CA, and the EROS Data Center, Sioux Falls, SD.

(Deutsches Zentrum fur Luft- und Raumfart [DLR]), the Italian space agency (Agenzia Spaziale Italiana [ASI]), and managed by the Jet Propulsion Laboratory, Pasadena, California. Data are distributed by the EROS Data Center, Sioux Falls, South Dakota.

## 7.13. Summary

Radar imagery is especially useful because it complements characteristics of images acquired in other portions of the spectrum. Aerial photography, for example, provides excellent information concerning the distribution and status of the Earth's vegetation cover. The information it conveys is derived from biologic components of plant tissues. However, from aerial photography we learn little direct information about the physical structure of the vegetation. In contrast, although active microwave imagery provides no data about the biologic component of the plant cover, it does provide detailed information concerning the physical structure of plant communities. We would not expect to replace aerial photography or optical satellite data with radar imagery, but we could expect to be able to combine information from microwave imagery to acquire a more complete understanding of the character of the vegetation cover. Thus the value of any sensor must be assessed not only in the context of its specific capabilities, but also in the context of its characteristics relative to other sensors.

## Review Questions

1. List advantages for the use of radar images, relative to aerial photography and Landsat imagery. Can you identify disadvantages?
2. Imaging radars many not be equally useful in all regions of the Earth. Can you suggest certain geographic regions where they might be most effective? Are there other geographic zones where imaging radars might be less effective?
3. Radar imagery has been combined with data from other imaging systems, such as the Landsat TM, to produce composite images. Because these composites are formed from data from two widely separated portions of the spectrum, together they convey much more information than either image can alone. Perhaps you can suggest (from information already given in Chapters 3 and 6) some of the problems encountered in forming and interpreting such composites.
4. Why might radar images be more useful in many less developed nations than in industrialized nations? Can you think of situations in which radar images might be especially useful in the industrialized regions of the world?
5. A given object or feature will not necessarily have the same appearance on all radar images. List some of the factors that will determine the texture and tone of an object as it is represented on a radar image.
6. Seasat was, of course, designed for observation of Earth's oceans. Why are the steep depression angles of the Seasat SAR inappropriate for many land areas? Can you think of advantages for use of steep depression angles in some regions?
7. What problems would you expect to encounter if you attempted to prepare a mosaic from several radar images?

8. Why are synthetic aperture radars required for radar observation of the Earth by satellite?
9. Why is the shuttle imaging radar so important in developing a more complete understanding of interpretation of radar imagery?
10. Compare and contrast satellite sensors that use active microwave energy and those that depend upon energy in the visible and near infrared imagery. Consider scale, resolution, orbit, timing, wavelengths used, satellite altitude, and area represented on a given image.

## References

Born, G. H., J. A. Dunne, and D. B. Lane. 1979. Seasat Mission Overview. *Science,* Vol. 204, pp. 1405–1406.

Brown, W. M., and L. J. Porcello. 1969. An Introduction to Synthetic Aperture Radar. *IEEE Spectrum,* Vol. 6, No. 9, pp. 52–62.

Elachi, C.1982. Radar Images from Space. *Scientific American,* Vol. 247, pp. 54–61

Elachi, C., et al. 1982a. Shuttle Imaging Radar Experiment. *Science,* Vol. 218, pp. 996–1003.

Elachi, C., et al. 1982b. Subsurface Valleys and Geoarcheology of the eastern Sahara. *Science,* Vol. 218, pp. 1004–1007.

Evans, D. L., J. J. Plant, and E. R. Stofan. 1997. Overview of the Spaceborne Imaging Radar-C/Xband Synthetic Aperture Radar (SIR-C/X-SAR) Missions. *Remote Sensing of Environment,* Vol. 59, pp. 135–140.

Ford, J. P., J. B. Cimino, and C. Elachi. 1983. *Space Shuttle Columbia Views the World with Imaging Radar: The SIR-A Experiment* (JPL Publication 82-95). Pasadena, CA: Jet Propulsion Laboratory, 179 pp.

Gens, R. 1999. Quality Assessment of Interferometrically Derived Digital Elevation Models. *International Journal of Applied Earth Observation and Geoinformation,* Vol. 1, pp. 102–108.

Gens, R., and J. L. Vangenderen. 1996. SAR Interferometry: Issues, Techniques, Applications. *International Journal of Remote Sensing,* Vol. 17, pp. 1803–1835.

Henderson, F. M., and A. J. Lewis (eds.). 1998. *Principles and Applications of Imaging Radars (Manual of Remote Sensing,* Vol. 2, R. A. Ryerson, editor-in-chief, American Society for Photogrammetry and Remote Sensing). New York: Wiley, 866 pp.

Jensen, H., L. C. Graham, L. J. Porcello, and E. M. Leith. 1977. Side-Looking Airborne Radar. *Scientific American,* Vol. 237, pp. 84–95.

Jet Propulsion Laboratory. 1982. *The SIR-B Science Plan* (JPL Publication 82-78). Pasadena, CA: Jet Propulsion Laboratory, 90 pp.

Jordan, R. L., B. L. Honeycutt, and M. Werner. 1991. The SIR-C/X-SAR Synthetic Aperture Radar System. *Proceedings of IEEE,* Vol. 79, pp. 827–838.

Kimura, H., and Y. Yamaguchi. 2000. Detection of Landslide Areas Using Satellite Radar Technology. *Photogrammetric Engineering and Remote Sensing,* Vol. 66, pp. 337–344.

Page, R. M. 1962. *The Origin of Radar.* New York: Doubleday, 169 pp.

Raney, R. K., A. P. Luscombe, E. J. Langham, and S. Ahmed. 1991. RADASAT. *Proceedings of IEEE,* Vol. 79, pp. 839–849.

Sabins, F. F. 1983. Geologic Interpretation of Space Shuttle Radar Images of Indonesia. *AAPG Bulletin,* Vol. 67, pp. 2076–2099.

Simpson, R. B. 1966. Radar, Geographic Tool. *Annals, Association of American Geographers,* Vol. 56, pp. 80–96.

Settle, M., and J. V. Taranick. 1982. Use of the Space Shuttle for Remote Sensing Research: Recent Results and Future Prospects. *Science,* Vol. 218, pp. 993–995.

CHAPTER EIGHT

# Lidar

## 8.1. Introduction

Lidar—the name is an acronym for "light detection and ranging"—imagery can be considered analogous to radar imagery, in the sense that both families of sensors are designed to transmit energy in a narrow range of frequencies, then receive the backscattered energy to form an image of the earth's surface. Both families are active sensors: they provide their own sources of energy, which mean they are independent of solar illumination. More importantly, they can compare the characteristics of the transmitted and the returned energy—the timing of pulses, the wavelengths, and the angles—so they can assess not only the brightness of the backscatter, but also its angular position, changes in frequency, and the timing of reflected pulses. Knowledge of these characteristics means that lidar data, much like data acquired by active microwave sensors, can be analyzed to extract information describing the structure of terrain and vegetation features not conveyed by conventional optical sensors.

Because lidars are based upon an application of *lasers,* they use a form of *coherent* light—light that is composed of a very narrow band of wavelengths—very "pure" with respect to color. Whereas ordinary light, even if it is dominated by a specific color, is composed of many wavelengths, with a diverse assemblage of waveforms, a laser produces light that is in phase ("coherent") and composed of a narrow range of wavelengths ("monochromatic") (Figure 8.1). Such light can be transmitted over large distances as a narrow beam that will diverge only slightly, in contrast with most forms of light in our everyday experience.

The laser—the name is an acronym for "light amplification by stimulated emission of radiation"—is an instrument that applies a strong electrical current to a "lasable" material, usually gasses or crystals, such as carbon dioxide, helium–neon, argon, rubies, and many other less familiar materials. Such *lasable* materials have atoms, molecules, or ions that emit light as they return to a normal, ground, state after excitement by a stimulus such as electricity or light. The emitted light forms the coherent beam described above. Each separate material provides a specific laser with its distinctive characteristics with respect to wavelength.

The laser provides an intense beam that does not diverge as its travels from the transmitter, a property that can favor applications involving heating, cutting (including surgery), etching, or illumination. Pointers, printers, CD players, scanners, bar code readers, and many other everyday consumer items are based upon laser technology. Although imaging lasers, of course, do not use intense beams, they do exploit the focused, coherent nature of the beam to produce very

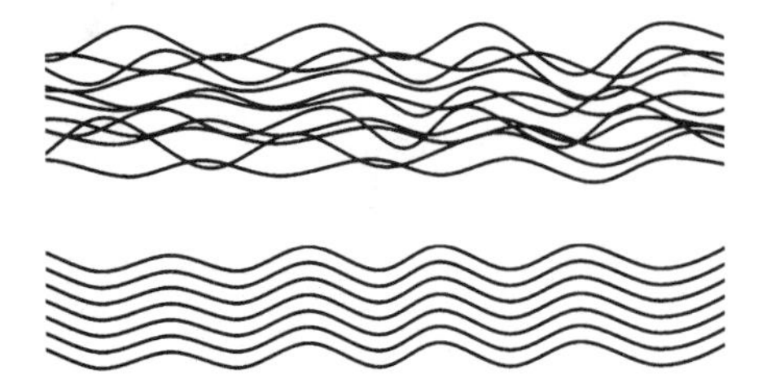

**FIGURE 8.1.** Normal (top) and coherent (bottom) light.

detailed imagery. A laser uses mirrored surfaces to accumulate many pulses to increase the intensity of the light before it leaves the laser (Figure 8.2).

## 8.2. Profiling Lasers

Lasers were invented in the later 1950s. Initially, they were used for scientific inquiry and industrial applications. The first environmental applications of lidars were used principally for *atmospheric profiling*—static lasers can be mounted to point upward into the atmosphere to assess *atmospheric aerosols,* solid particles suspended in the atmosphere. These particles direct a portion of the laser beam back to the ground, where it is measured to indicate their abundance. Because lasers can measure the time delay of the backscatter, they can assess the clarity of the atmosphere over a depth of several kilometers, providing data concerning the altitudes of the layers they detect.

The first airborne lasers were designed as *profiling lasers,* lasers aimed directly beneath the aircraft in the nadir to illuminate a single region in the nadir position. (When used primarily to acquire topographic data, such instruments are known as *airborne laser altimeters.*) The forward motion of the aircraft carries the illuminated region forward to view a single track directly beneath the aircraft. Echoes from repetitive lidar pulses provide an elevation profile of the narrow region immediately beneath the aircraft (Figure 8.3). Although lidar profilers do not pro-

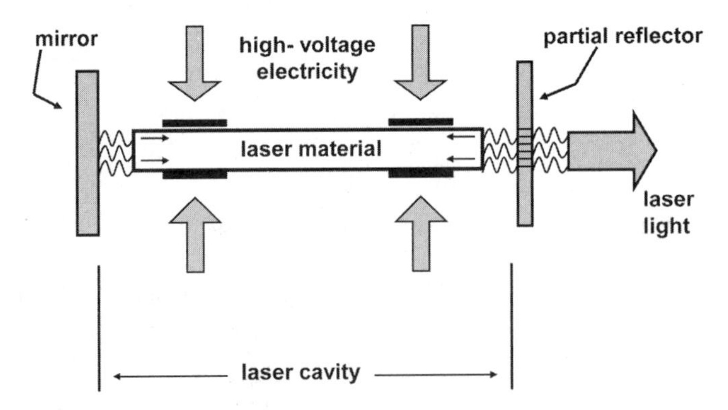


**FIGURE 8.2.** Schematic diagram of a simple laser. Energy, such as electricity, is applied to a substance, such as lasable gasses (e.g., nitrogen, helium–neon) or materials (e.g., ruby crystals). When the materials return to their normal state, they emit coherent light, which is intensified before release by multiple reflections between the mirrored surfaces. Intensified light can then pass through the semitransparent mirror to form the beam of coherent light that is emitted by the instrument.

vide the image formats that we are now expect, they provide a very high density of observations, and are used as investigative tools for researchers studying topography, vegetation structure, hydrography, and atmospheric studies, to list only a few of many applications.

## 8.3. Imaging Lidars

It is only relatively recently that lidars could be considered remote sensing instruments that collect images of the Earth's surface. By the late 1980s, several technologies matured and converged to create the context for the development of the precision scanning lidar systems that we now know. *Inertial measurement units* (IMUs) enabled precise control and recording of the orientation of aircraft (roll, pitch, and yaw). *Global positioning systems* (GPS) could provide accurate records of the geographic location of an aircraft as it acquired data. And development of highly accurate clocks permitted the precise timing of lidar pulses required to create high-performance lidar scanning systems.

A lidar scanner can transmit between 2,000 and 33,000 pulses each second, depending upon the specific design and application. A scanning mirror directs the pulses back and forth across the image swath beneath the aircraft. The width of the swath is determined by the instrument's design and the operating conditions of the aircraft. Imaging lidars usually use wavelengths in the visible (e.g., 0.532 μm, green, for penetration of water bodies) or near infrared (e.g., 1.64 μm, for sensitivity to vegetation, ability to detect open water, and freedom from atmospheric scattering) regions of the spectrum.

Several alternative designs for imaging lidars instruments are in use. Figure 8.4 presents a schematic representation of a typical lidar system: (1) the system's laser (coordinated by the

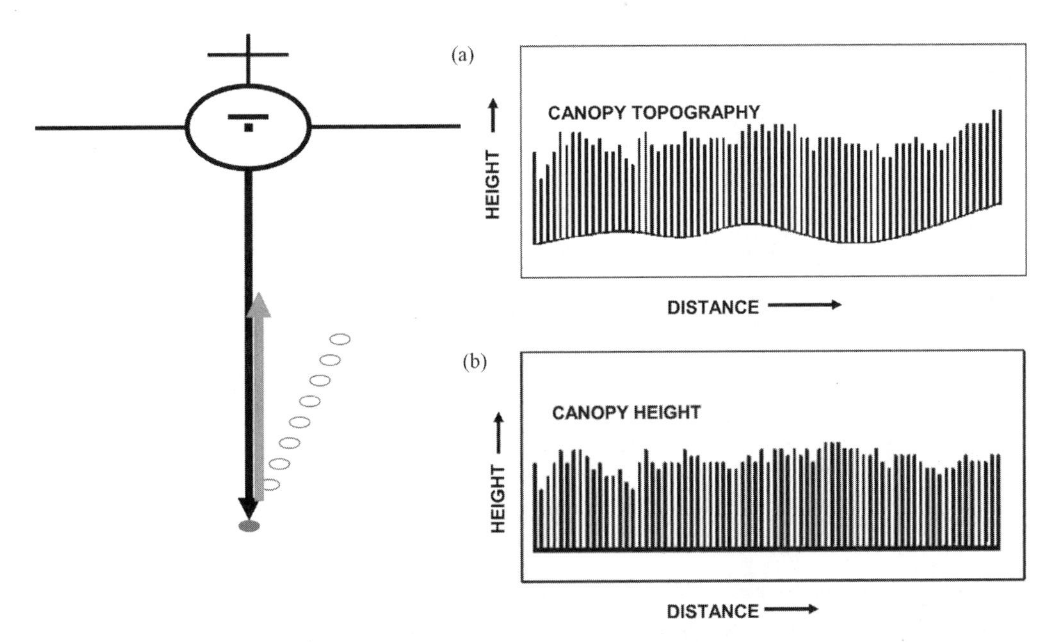


**FIGURE 8.3.** Schematic representation of an airborne laser profiler. (a) Acquisition of laser profiles; (b) Sample data gathered by a laser profiler, illustrating extraction of canopy height from the raw profile data.

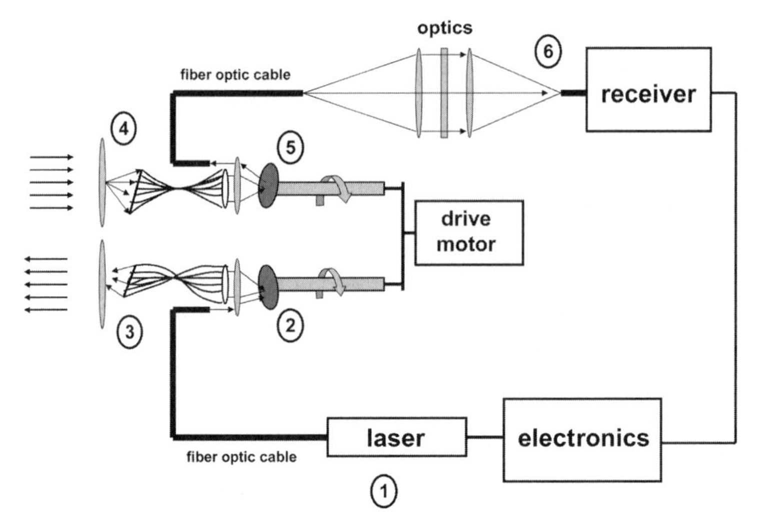


**FIGURE 8.4.** Schematic diagram of a lidar scanner. (1) The system's laser (coordinated by the electronic component) generates a beam of coherent light, transmitted by a fiber optic cable to (2) a rotating mirror, offset to provide a scanning motion. The laser light is directed to a bundle of fiber optic cables that are twisted to provide a linear beam, and then directed through a system of lenses toward the ground. The energy received back from the terrain is received by another system of lenses and processed to form an image.

electronic component) generates a beam of coherent light, transmitted by a fiber optic cable to (2) a rotating mirror, offset to provide a scanning motion. The laser light is directed to a bundle of fiber optic cables that can be twisted to transmit the light as a linear beam. The oscillating motion of the mirror scans the laser beam side to side along the cross-track axis of the image, recording many thousands of returns each second. Because a lidar scanner is well integrated with GPS, IMU, and timing systems, these pulses can be associated with specific points on the Earth's surface.

As the reflected portion of the laser beam reaches the lidar aperture, it is received by another system of lenses and directed through fiber optic cables to another scanning lens (5), then directed through an optical system to filter the light before it is directed (6) to a receiving system to accept and direct the signal to the electronics component. The electronics coordinate timing of the pulses and permit matching of the signal with data from the inertial navigation system and GPS. Together these components permit the system to accurately place each returned signal in its correct geographic position. The system operates at such high speed that a high density of pulses are received from each square meter on the terrain. The timing capability of the lidar permits accurate assessment of distance and elevation, which permits formation of an image with detailed and accurate representation of elevations in the scene.

## 8.4. Lidar Imagery

Lidar imagery is acquired in parallel strips that match to form a continuous image of a region (Figure 8.5). In Figure 8.5 the upper image represents the raw lidar data, with each pixel representing the elevation of a specific point on the ground. Light tones represent higher elevations,

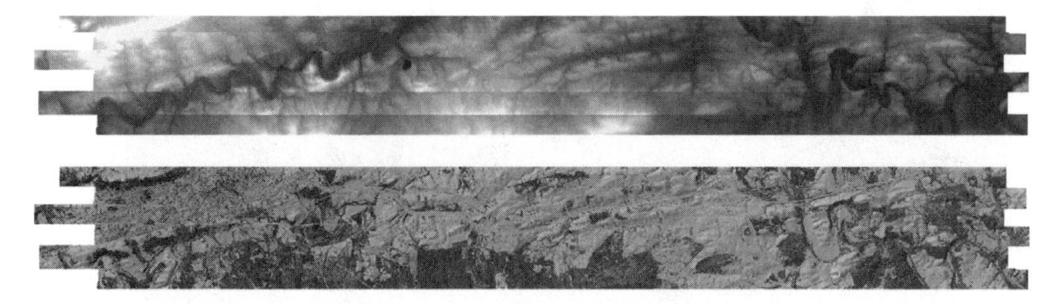

**FIGURE 8.5.** Examples of lidar flight lines. Upper image: raw lidar data, with brightness indicating increasing relative elevation. Lower image: lidar data processed with a hillshading algorithm, revealing terrain texture in Wytheville, VA. From Virginia Department of Transportation. Copyright 2003, Commonwealth of Virginia. Reproduced by permission.

darker tones represent lower elevations. The lower image is formed from the same data, but represented using a hillshading technique that assumes each pixel is illuminated from the upper left corner of the image. This effect creates image texture reminiscent of an aerial photograph, which is usually easier for casual interpretation.

Figures 8.6 and 8.7 show enlargements of portions of the same image, selected to represent some of the distinctive qualities of lidar imagery. Figure 8.6 depicts an interstate interchange, with structures, forests, pastureland, and cropland represented with precision and detail. Figure 8.7 depicts a nearby region. A deep quarry is visible at the center. Open land and forest are again visible. The parallel strips visible near the upper right of this image depict mature corn fields—an indication of the detail recorded by these images.

## 8.5. Types of Imaging Lidars

Lidar instruments differ in their design. Often they are designated by the size of the area illuminated at any given instant and the form of processing used to generate data. The area the sensor illuminates at a given instant is known as the system's "footprint." Sometimes imaging lidars are designated as *small footprint* ("discrete return lidars") or *large footprint* ("waveform lidars"). Small-footprint systems might acquire data from areas as small as 0.5–2 ft. in diameter. Large-footprint systems might view areas as large as 5 m or more.

Thus one family of lidars uses *pulsed lasers* that generate very carefully timed bursts of light (*discrete return lidars*). These lidars generate range information by measuring the time delay (travel time) for emitted and received pulses. Because light travels at a known constant velocity, the time for a pulse to return to the sensor translates directly to a distance, or range, to the ground beneath the aircraft. Data is processed to generate a three-dimensional point cloud that reflects the varied times for pulses to return to the instrument. These times can then be translated into differences in elevations of the surfaces that produced the reflection, as explained below. Small-footprint lidar data are collected from a relatively wide image swath.

Another family of imaging lidars uses *continuous wave lasers* that generate a continuously modulated beam of light. The instrument transmits this radiation to the ground, and receives as

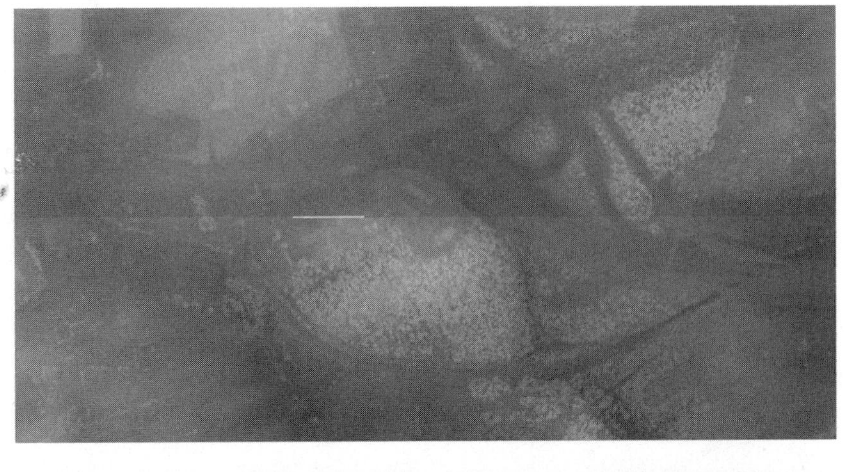

**FIGURE 8.6.** Section of a lidar image of Wytheville, VA, enlarged to depict detail, depicting a highway interchange and forested region. Upper image: raw lidar data, with brightness indicating increasing relative elevation. Lower image: lidar data processed with a hillshading algorithm, revealing terrain texture. From Virginia Department of Transportation. Copyright 2003, Commonwealth of Virginia. Reproduced by Permission.

many as five echoes for each pulse, sometimes with the ability to discern changes in signal amplitude, phase, and intensity. These lidars illuminate rather large areas, so they can be designated as *large-footprint lidars.* Although large-footprint lidars offer advantages for accuracy and convenience of processing, this technology is now still under development, so it is not generally available for operational applications.

Resolutions of specific lidar systems can differ widely. The detail available depends upon system design and operation, including flying height and the beam divergence.

## 8.6. Processing Lidar Image Data

Organizations that collect lidar data usually use their own in-house processing systems specifically designed to manipulate the very large data volumes associated with each lidar mission, to

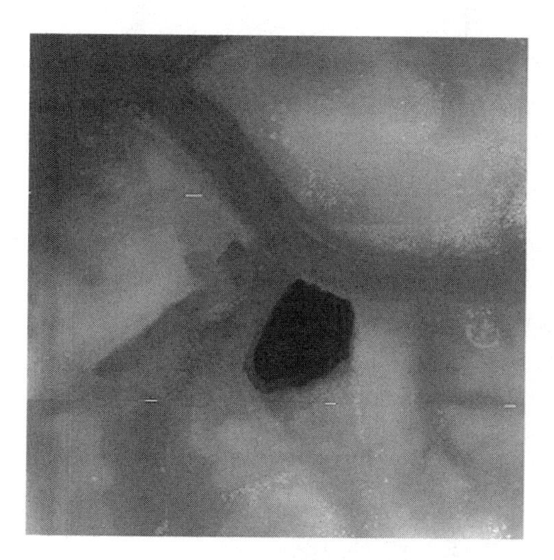

**FIGURE 8.7.** Section of a lidar image of Wytheville, VA, enlarged to depict detail of a limestone quarry. Left: raw lidar data, with brightness indicating increasing relative elevation. Right: lidar data processed with a hill-shading algorithm, revealing terrain texture. From Virginia Department of Transportation. Copyright 2003, Commonwealth of Virginia. Reproduced by Permission.

perform the special operations tailored for each specific instrument, and to meet the requirements for each project. Filtering, detection of anomalies, and verification require reference to existing maps, digital elevation models (explained below), satellite imagery, and aerial photography. Often lidar missions are completed using ancillary photography gathered coincident with the lidar data.

As outlined below, a typical lidar project requires production of (1) a surface elevation model (SEM) representing the first surface intercepted by the lidar pulse, (2) a bare-earth digital elevation model (DEM) representing the terrain surface after removal of vegetation or structures, and (3) a canopy layer, representing the height of the canopy above the terrain surface. Some lidar data may be further processed to identify, isolate, and extract features such as structures and vegetation cover, or to be fused with other forms of remotely sensed data, such as CIR aerial photography. Such products are usually provided in formats suitable for standard applications software, so the results can be further processed by end users.

Because the instrument can accurately record the angle of individual pulses, the attitude of the aircraft (using inertial navigational systems), and the position of the aircraft (using GPS), a lidar system can create an image-like array representing variation of elevation within the areas under observation (i.e., each pulse can be associated with a scan angle and data describing the position and orientation of the aircraft) (Figure 8.8). Depending upon the design of a specific lidar system, the positions of individual returns (known as *postings*) are irregularly spaced in their original unprocessed form. These irregular positions are later interpolated to form a regular grid that provides a more coherent representation of the terrain. Plate 14 shows this effect rather well: the upper portion of the illustration shows the original postings, whereas the lower half shows the systematic grid formed by the interpolation process.

Thus each return from the ground surface can be precisely positioned in xyz space to provide an array that records both position and elevation. For small-footprint lidars, horizontal accuracy might be in the range of 20 to 30 cm and vertical accuracy in the range between 15 and 20 cm.

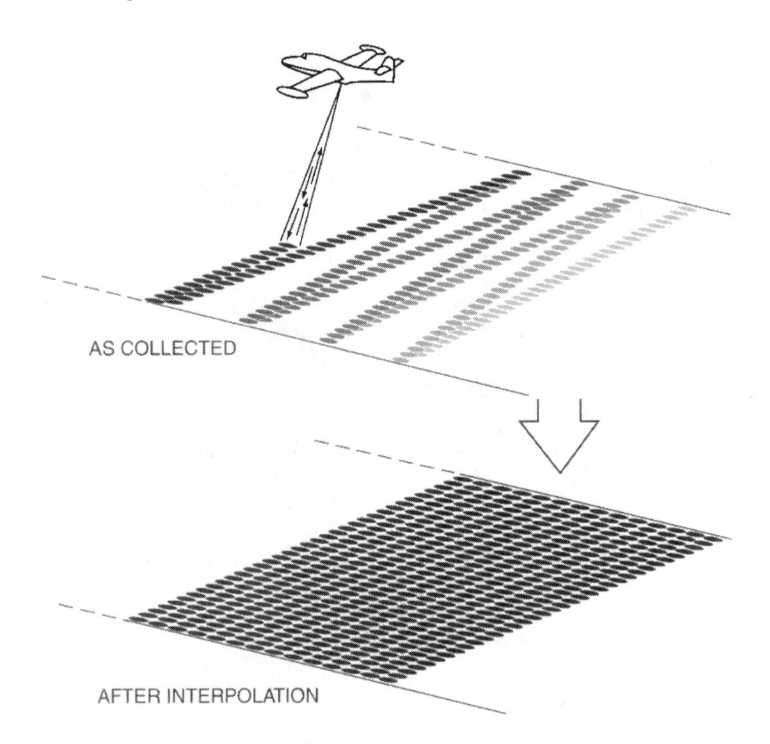


**FIGURE 8.8.** Acquisition of lidar data. Lidar systems acquire data by scanning in the pattern suggested by the top diagram; details vary according to specific systems. The pattern of returns is then interpolated to generate the regular array that forms the lidar image. Examples of actual scan pattern and interpolated data are depicted in Plate 14.

Therefore, the array, or image, forms a detailed DEM. With the addition of ground control (points of known location that can be accurately located within the imaged region, often found using GPS), lidar can provide data comparable in detail and positional accuracy to those acquired by photogrammetric analysis of aerial photographs. A lidar can record different kinds of returns from the terrain. Some returns, known as *primary returns,* originate from the first objects a lidar pulse encounters—often the upper surface of a vegetation canopy (Figure 8.9). In addition, portions of a pulse pass through gaps in the canopy into the interior structure of leaves and branches, to lower vegetation layers, and to the ground surface itself. This energy creates echoes known as *secondary,* or *partial, returns.* Therefore, for complex surfaces, such as forests with multiple canopies, some portions of a pulse might be reflected from upper and middle portions of the canopy and other portions from the ground surface at the base (Figure 8.10).

The total collection of lidar returns for a region can be examined to separate those returns that originated above a specified level from those that originated below that level. This kind of approach can then be used as a kind of filtering process to separate ground returns from non-ground returns, and, with additional analysis, to separate the terrain surface from overlying vegetation cover and structures (Figure 8.11 and Plate 15). Lidars are unique in being the only sensors that can reliably differentiate between multiple imaged layers.

Lidar data may not accurately represent shorelines, stream channels, and ridges. Contours derived from lidar data may not form hydrographically coherent surfaces comparable to those represented on the usual contour maps. For terrain analysis, it is common for the analyst to

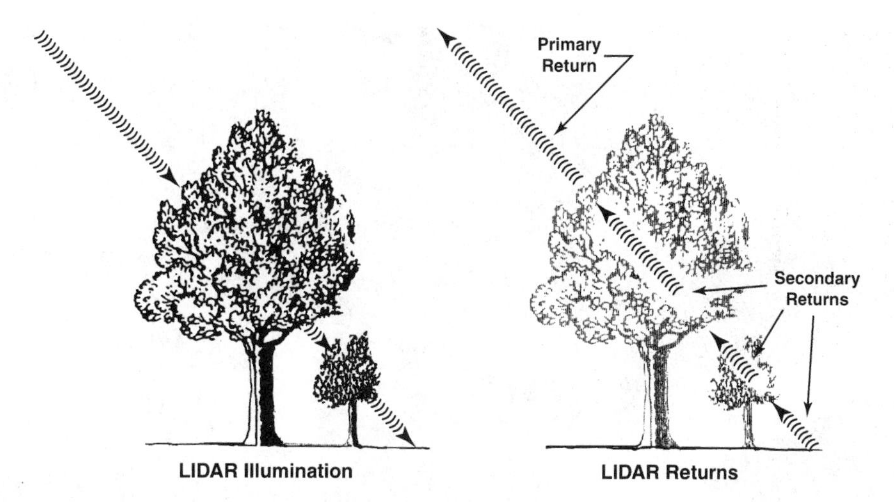


**FIGURE 8.9.** Primary and secondary lidar returns from two forested regions. This illustration represents lidar return from two separate forested areas, shown in profile. The dots near the top of the diagram represent the returns that are received first (primary returns), and the dots at the lower and central portions of the diagram represent the returns received later (secondary returns). Note the contrast between the dome-shaped canopy formed by the crowns of the deciduous forest (left) and the peaked crowns of the coniferous canopy (right). The coniferous forest has only sparse undergrowth, while the deciduous forest is characterized by abundant undergrowth. From Sorin Popepscu and Peter Sforza.

insert breaklines, usually by manual inspection of the data in digital format, to separate the data array into discrete units that can be treated individually. *Breaklines* are interconnected points that define abrupt changes in terrain, such as edges of roads, drainage ditches, and ridgelines. Other boundaries outline *exclusion areas,* such as water bodies and dense forest, that are to be excluded from contouring. The analyst can digitize geomorphic features such as drainageways, road edges, and ditches as "hard breaklines." More subtle variations in topography can be mapped as "soft breaklines." These can be inserted manually, then subsequently are considered while the data are used to generate a digital terrain model.

Often data are processed and organized in units that correspond to flight lines (e.g., Figure

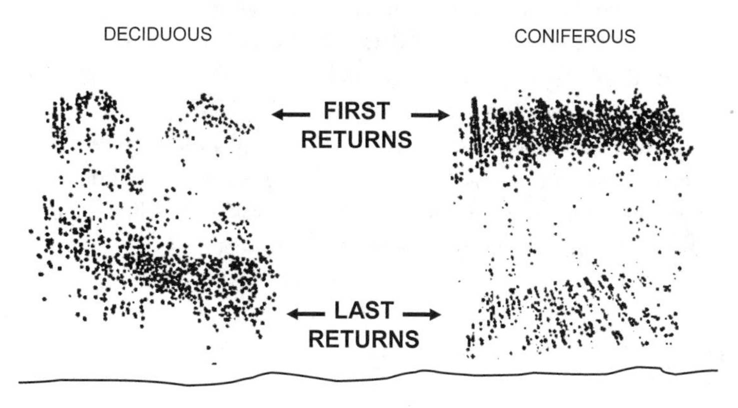


**FIGURE 8.10.** Schematic diagram of primary and secondary lidar returns. From Peter Sforza and Sorin Popescu.

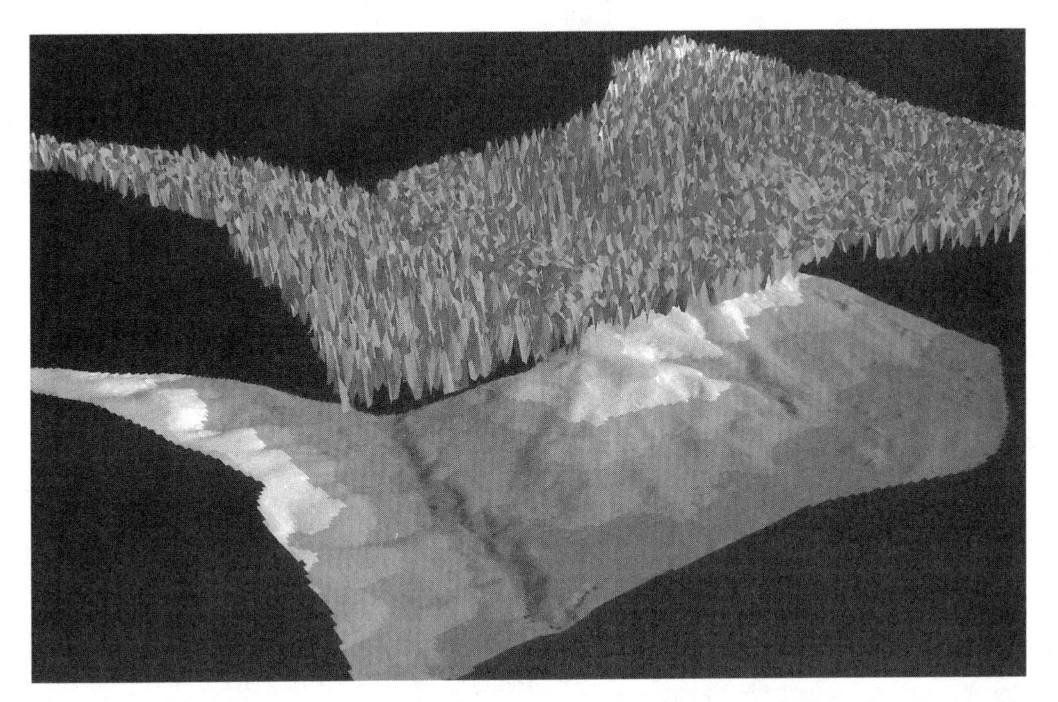

**FIGURE 8.11.** Bare-earth lidar surface with its first reflective surface shown above it. The data is from a multiple-return lidar system and shows the wealth of data that can be extracted from lidar data. These data were collected at an altitude of 12,000 ft., with a nominal post spacing of 15 ft. and a vertical accuracy of about 14 in. RMSE. From EarthData.

8.5). Such units are arbitrary, so some prefer to prepare units that correspond to USGS topographic quadrangles, to facilitate manipulation and storage, especially if the lidar data are to be used in coordination with other data. In contrast, analysts who use the lidar data for modeling may prefer to organize the data in geographic units corresponding to drainage basins, or to place the edges between units at ridgelines, where the seams may be less likely to influence the analysis.

Lidar imagery is increasingly finding a role as a source of data for DEMs. The lidar bare-earth surface isolates topographic information from the effects of dense vegetation, which can be significant in other sources of elevation data. The lidar data can be interpolated to form a DEM, which often displays high accuracy and detail. Photogrammetric methods often encounter difficulties in representing topography covered by dense vegetation and the complex terrain near streams and drainageways, which in many applications may be the very regions where high accuracy is most valued. Although lidar data may be a superior source of elevation data in many situations, many applications require removal of buildings and other structures from the data, and careful consideration of interpolation procedures.

## 8.7. Summary

Lidar data permit reliable, if perhaps imperfect, separation of vegetation from terrain—a capability unique among competing remote sensing instruments. Lidar provides a highly accurate and detailed representation of terrain. Its status as an active sensor permits convenience and flexibility in flight planning due to its insensitivity to variations in weather and solar illumination—both of which are important constraints upon aerial photography.

Lidar data provide detailed direct measurement of surface elevation, with detail and accuracy usually associated only with photogrammetric surveys. Many applications have focused on urban regions, which have a growing need for detailed information concerning building densities, urban structures, and building footprints. Lidar data are used for highway planning, pipeline routing planning, route planning, and for designing wireless communication systems in urban regions (Figure 8.12 and Plate 16). Lidar data have also been used to study forest structure, as the detailed and accurate information describing canopy configuration and structure permits accurate mapping of timber volume. Lidar clearly will find a broad range of environmental applications, in which its detailed representations of terrain will open new avenues of inquiry not practical with coarser data. However applications of lidar still face challenges, especially in defining practical processing protocols that can find wide acceptance across a range of application fields.

## Review Questions

1. Review some of the strengths of lidar data relative to other forms of remotely sensed data discussed thus far.
2. Many observers believe that the increasing availability of lidar will displace the current role of aerial photography for many applications. What are some of the reasons that might lead people to believe that lidar could replace many of the remote sensing tasks now fulfilled by aerial photography?

**FIGURE 8.12.** Yankee Stadium, New York City, as represented by lidar data. This lidar image was collected in the summer of 2000 as part of a citywide project to acquire data to identify line-of-sight obstructions for the telecommunications industry. The lidar data were collected from an altitude of 2,500 m, with a vertical accuracy of 1 m RMSE and a nominal post spacing of 4 m. From EarthData.

3. Can you identify reasons that aerial photography might yet retain a role, even in competition with strengths of lidar data?
4. If aerial photography is largely replaced by lidar, do you believe that there will still be a role for teaching aerial photography as a topic in a university remote sensing course? Explain.
5. Identify some reasons why lidar might be effectively used in combination with other data. What might be some of the difficulties encountered in bringing lidar data and (for example) fine-resolution optical satellite imagery together?
6. Assume for the moment that lidar data becomes much cheaper, easier to use, and in general more widely available to remote sensing practitioners. What kinds of new remote sensing analyses might become possible, or what existing analyses might become more widespread?
7. The text discusses how lidar imagery is based upon the convergence of several technologies. Review your notes to list these technologies. Think about the technologic, scientific, social, and economic contexts that foster the merging of these separate capabilities. How do you think we can prepare now to encourage future convergences of other technologies (now unknown) that might lead to advances in remote sensing instruments?
8. Lidar imagery may not be equally useful in all regions of the Earth. Can you suggest certain geographic regions or environments in which lidar data might not be effective?
9. Discuss some of the considerations that might be significant in deciding what season to acquire lidar data of your region.
10. Identify some of the special considerations that might be significant in planning acquisition of lidar data in urban regions.

## References

Blair, J. B., D. L. Rabine, and M. A. Hofton. 1999. The Laser Vegetation Imaging Sensor: A Medium-Altitude, Digitization-Only, Airborne Laser Altimeter for Mapping Vegetation and Topography. *ISPRS Journal of Photogrammetry and Remote Sensing,* Vol. 54, pp. 115–122.

DeLoach, S. R., and J. Leonard. 2000. Making Photogrammetric History. *Professional Surveyer,* Vol. 20, No. 4, pp. 6–11.

Flood, M. 2001. Laser Altimetry: From Science to Commercial LiDAR Mapping. *Photogrammetric Engineering and Remote Sensing,* Vol. 67, pp. 1209–1217.

Flood, M. 2002. Product Definitions and Guidelines for Use in Specifying LiDAR Deliverables. *Photogrammetric Engineering and Remote Sensing,* Vol. 67, pp. 1209–1217.

Fowler, R. A. 2000, March. The Lowdown on Lidar. *Earth Observation Magazine,* pp. 27–30.

Hill, J. M., L. A. Graham, and R. J. Henry. 2000. Wide-Area Topographic Mapping Using Airborne Light Detection and Ranging (LIDAR) Technology. *Photogrammetric Engineering and Remote Sensing,* Vol. 66, pp. 908–914, 927, 960.

Lefsky, M. A., D. Harding, W. B. Cohen, G. Parker, and H. H. Shugart. 1999. Surface Lidar Remote Sensing of Basal Area and Biomass in Deciduous Forests of Eastern Maryland, USA. *Remote Sensing of Environment,* Vol. 67, pp. 83–98.

Nelson, R., W. Krabill, and J. Tonelli. 1988. Estimating Forest Biomass and Volume Using Airborne Laser Data. *Remote Sensing of Environment,* Vol. 24, pp. 247–267.

Nelson, R., R. Swift, and W. Krabill. 1988. Using Airborne Lasers to Estimate Forest Canopy and stand Characteristics. *Journal of Forestry,* Vol. 86. pp. 31–38

Nilsson, M. 1996. Estimation of Tree Heights and Stand Volume Using an Airborne Lidar System. *Remote Sensing of Environment,* Vol. 56. pp.1–7.

Popescu, S. C. 2002. *Estimating Plot-Level Forest Biophysical Parameters Using Small-Foorprint Airborne Lidar Measurements,* PhD Dissertation, Virginia Tech, Blacksburg, VA, 144 pp.

Romano, M. E. 2004. Innovation in Lidar Processing. *Photogrammtric Engineering and Remote Sensing,* Vol. 70, pp. 1201–1206.

Sapeta, K. 2000. Have You Seen the Light?: LIDAR Technology Is Creating Believers. *GEOWorld,* Vol. 13, No. 10, pp. 32–36.

Wehr, A., and U. Lohr (eds.). 1999a. Airborne Laser Scanning (Theme Issue). *ISPRS Journal of Photogrammetry and Remote Sensing,* Vol. 54, Nos. 2–3.

Wher, A., and U. Lohr. 1999b. Airborne Laser Scanning: An Introduction and Overview. *ISPRS Journal of Photogrammetry and Remote Sensing,* Vol. 54, pp. 68–92.

Zhang, K., and D. Whitman. 2005. Comparison of Three Algorithms for Filtering Airborne Lidar Data. *Photogrammetric Engineering and Remote Sensing,* Vol. 71, pp. 313–324.

CHAPTER NINE

# Thermal Radiation

## 9.1. Introduction

The literal meaning of *infrared* is "below the red," indicating that its frequencies are lower than those in the red portion of the visible spectrum. With respect to wavelength, however, the infrared spectrum is actually "beyond the red," having wavelengths longer than those of red radiation. The infrared is often defined as having wavelengths from about 0.76 μm to about 1,000 μm (1 mm). There are great differences between properties of radiation within this range, as noted below.

The infrared spectrum was discovered in 1800 by Sir William Herschel (1738–1822), a British astronomer who was searching for the relationship between heat sources and visible radiation. Later, in 1847, two Frenchmen, A. H. L. Fizeau (1819–1896) and J. B. L. Foucault (1819–1868), demonstrated that infrared radiation has optical properties similar to those of visible light with respect to reflection, refraction, and interference patterns.

The infrared portion of the spectrum extends beyond the visible region to wavelengths of about 1 mm (Figure 9.1). The shorter wavelengths of the infrared spectrum, near the visible, behave in a manner analogous to visible radiation. This region forms the *reflective infrared spectrum* (also known as the "near infrared"), extending from about 0.7 μm to 3.0 μm. Many of the same kinds of films, filters, lenses, and cameras that we use in the visible portion of the spectrum can also be used, with minor variations, for imaging in the near infrared. The very longest infrared wavelengths are in some respects similar to the shorter wavelengths of the microwave region. This chapter discusses use of far infrared radiation from about 8 to 12 μm for remote sensing of landscapes, and the mid-infrared region from about 3.5 to 4.5 μm.

Remote sensing in the mid- and far infrared is based upon a family of imaging devices that differ greatly from the cameras and films used in the visible and near infrared. The mid- and far infrareds' interaction with the atmosphere is also quite different from that of shorter wavelengths. The far infrared regions are free from the scattering that is so important in the ultraviolet and visible regions, but absorption by atmospheric gases restricts uses of the mid- and far infrared spectrum to specific atmospheric windows. Also, the kinds of information acquired by sensing the far infrared differ from those acquired in the visible and near infrared. Variations in emitted energy in the far infrared provide information concerning surface temperature and thermal properties of soils, rocks, vegetation, and man-made structures. Inferences based upon thermal properties lead to inferences about the identities of surface materials.

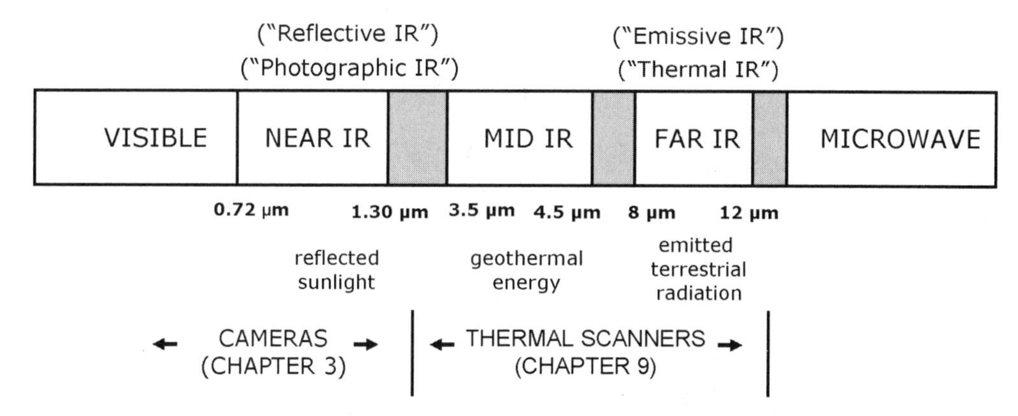


**FIGURE 9.1.** Infrared spectrum. This schematic view of the infrared spectrum identifies principal designations within the infrared region, and approximate definitions, with some illustrative examples. Regions are not always contiguous because some portions of the infrared spectrum are unavailable for remote sensing due to atmospheric effects.

## 9.2. Thermal Detectors

Before the 1940s, the absence of suitable instruments limited use of thermal infrared radiation for aerial reconnaissance. Aerial mapping of thermal energy depends upon use of a sensor that is sufficiently sensitive to thermal radiation that variations in apparent temperature can be detected by an aircraft moving at considerable speed high above the ground. Early instruments for thermographic measurements examined differences in electrical resistance caused by changes in temperature. But such instruments could function only when in close proximity to the objects of interest. Although such instruments were useful in an industrial or laboratory setting, they are not sufficiently sensitive for use in the context of remote sensing. They respond slowly to changes in temperature, and cannot be used at the distances required for remote sensing applications.

During the late 1800s and early 1900s, the development of *photon detectors* (sometimes, called *thermal photon detectors*) provided a practical technology for use of the thermal portion of the spectrum in remote sensing. Such detectors are capable of responding directly to incident photons by reacting to changes in electrical resistance, providing a sensitivity and speed of response suitable for use in reconnaissance instruments. By the 1940s, a family of photon detectors had been developed to provide the basis for electro-optical instruments used in several portions of the thermal infrared spectrum.

*Detectors* are devices formed from substances known to respond to energy over a defined wavelength interval, generating a weak electrical signal with a strength related to the radiances of the features in the field of view of the sensor. (Often, the sensitivity of such materials increases to practical levels when the substances are cooled to very low temperatures to increase sensitivity and reduce noise.) The electrical current is amplified, then used to generate a digital signal that can be used to form a pictorial image, roughly similar in overall form to an aerial photograph (Figure 9.2).

Detectors have been designed with sensitivities for many of the spectral intervals of interest in remote sensing, including regions of the visible, near infrared, and ultraviolet spectra. Detec-

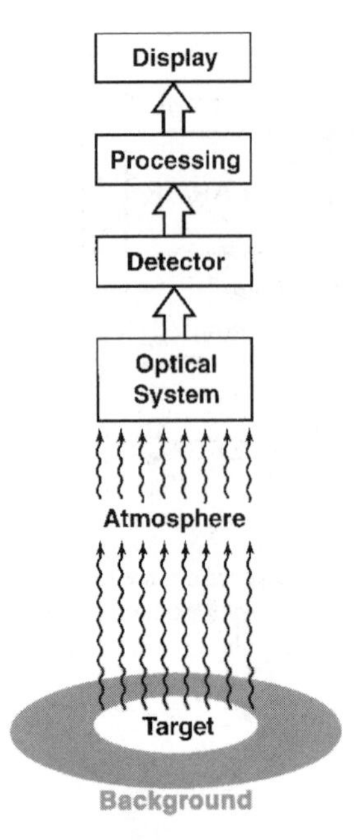


**FIGURE 9.2.** Use of thermal detectors.

tors sensitive in the thermal portion of the spectrum are formed from rather exotic materials, such as *indium antimonide* (InSb), and mercury-doped germanium (Ge:Hg). InSb has a peak sensitivity near 5 µm, in the mid-infrared spectrum, and Ge:Hg has a peak sensitivity near 10 µm, in the far infrared spectrum (Figure 9.3). Mercury cadmium telluride (MCT) is sensitive over the range 8–14 µm. To maintain maximum sensitivity, such detectors must be cooled to very low temperatures (–196°C or –243°C) using liquid nitrogen or liquid helium.

The sensitivity of the detector is a significant variable in the design and operation of the system. Low sensitivity means that only large differences in brightness are recorded ("coarse radiometric resolution") and most of the finer detail in the scene is lost. High sensitivity means that finer differences in scene brightness are recorded ("fine radiometric resolution"). The *signal-to-noise ratio* (SNR or S/N ratio) expresses this concept (Chapter 4). The "signal" in this context refers to differences in image brightness caused by actual variations in scene brightness.

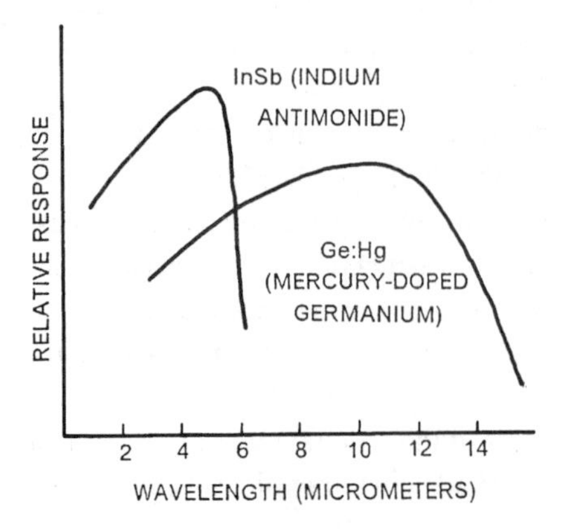


**FIGURE 9.3.** Sensitivity of some common thermal detectors.

"Noise" designates variations unrelated to scene brightness. Such variations may be the result of unpredictable aspects in the performance of the system. (There may also be random elements contributed by the landscape and the atmosphere, but here "noise" refers specifically to that contributed by the sensor.) If noise is large relative to the signal, the image does not provide a reliable representation of the feature of interest. Clearly, high noise levels will prevent imaging of subtle features. Even if noise levels are low, there must be minimum contrast between a feature and its background (i.e., a minimum magnitude for the signal) for the feature to be imaged. Also, note that increasing fineness of spatial resolution decreases the energy incident upon the detector, with the effect of decreasing the strength of the signal. For many detectors, noise levels may remain constant even though the level of incident radiation decreases; if so, the increase in spatial resolution may be accompanied by decreases in radiometric resolution, as suggested in Chapter 4.

## 9.3. Thermal Radiometry

A *radiometer* is a sensor that measures the intensity of radiation received within a specified wavelength interval and within a specific field of view. Figure 9.4 offers a schematic view of a radiometer. A lens or mirror gathers radiation from the ground, then focuses it upon a detector positioned in the focal plane. A field stop may restrict the field of view, and filters may be used to restrict the wavelength interval that reaches the detector. A characteristic feature of radiometers is that radiation received from the ground is compared to a reference source of known radiometric qualities.

A device known as a *chopper* is capable of interrupting the radiation that reaches the detector. The chopper consists of a slotted disk or similar device rotated by an electrical motor so that as the disk rotates it causes the detector to alternately view the target, then view the reference source of radiation. Because the chopper rotates very fast, the signal from the detector consists of a stream of data that alternately measures the radiance of the reference source, then measures

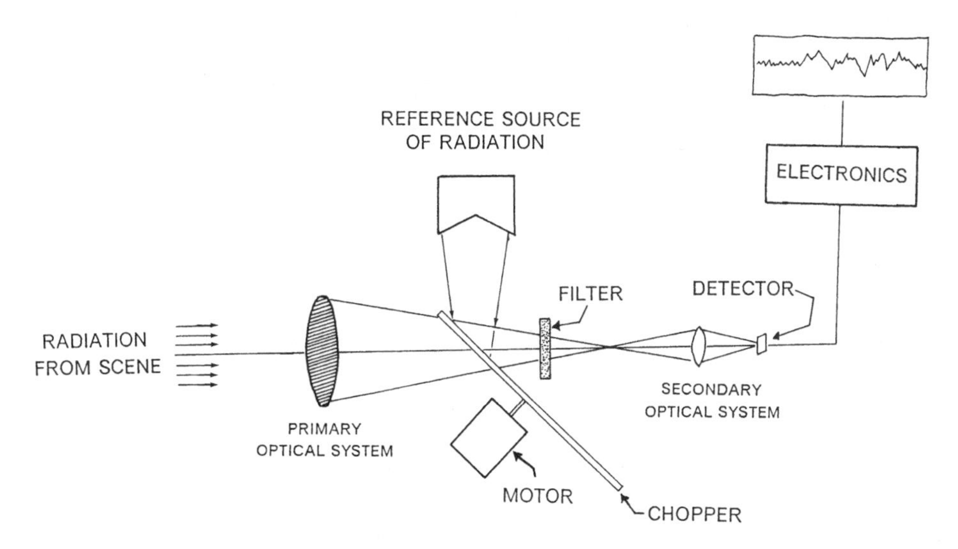


**FIGURE 9.4.** Schematic diagram of a radiometer.

radiation from the ground. The amplitude of this signal can be used to determine the radiance difference between the reference and the target. Because the reference source has known radiance, the radiance of the target can then be estimated.

Although there are many variations on this design, this description identifies the most important components of radiometers. Related instruments include *photometers,* which operate at shorter wavelengths and often lack the internal reference source, and *spectrometers,* which examine radiance over a range of wavelengths. Radiometers can be designed to operate at different wavelength intervals, including portions of the infrared and ultraviolet spectra. By carefully tailoring the sensitivity of radiometers, scientists have been able to design instruments that are very useful in studying atmospheric gases and cloud temperatures. Radiometers used for earth resource study are often configured to view only a single trace along the flight path; the resulting output signal consists of a single stream of data that varies in response to differences in radiances of features along the flight line (Figure 9.5). A scanning radiometer can gather data from a corridor beneath the aircraft; output from such a system resembles outputs from some of the scanning sensors discussed in earlier chapters.

Spatial resolution of a radiometer is determined by an *instantaneous field of view* (IFOV), which is in turn controlled by the sensor's optical system, the detector, and flying altitude. Radiometers often have relatively coarse spatial resolution—for example, satellite-borne radiometers may have spatial resolution of 60 m to 100 km or more—in part because of the desirability of maintaining high radiometric resolution. To assure that the sensor receives enough energy to make reliable measurements of radiance, the IFOV is defined to be rather large; a smaller IFOV would mean that less energy would reach the detector, that the signal would be much too small with respect to system noise, and that the measure of radiance would be much less reliable.

The IFOV can be informally defined as the area viewed by the sensor if the motion of the

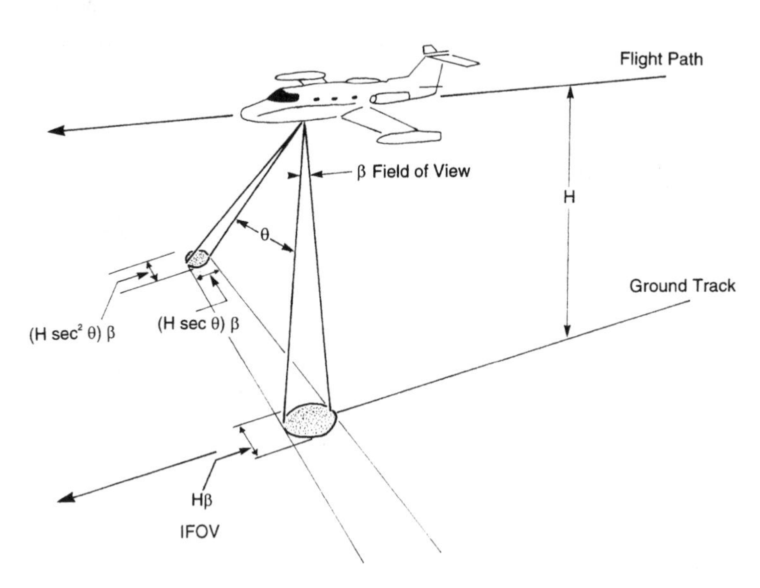


**FIGURE 9.5.** Instantaneous field of view (IFOV). At nadir, the diameter of the IFOV is given by flying altitude ($H$) and the instrument's field of view ($\beta$). As the instrument scans to the side, it observes at angle $\theta$; in the off-nadir position, the IFOV becomes larger and unevenly shaped.

instrument were to be suspended so that it records radiation from only a single patch of ground. The IFOV can be more formally expressed as the angular field of view ($\beta$) of the optical system (Figure 9.5). The projection of this field of view onto the ground surface defines the circular area that contributes radiance to the sensor. Usually for a particular sensor it is expressed in radians (r); to determine the IFOV for a particular image, it is necessary to know the flying altitude ($H$) and to calculate the size of the circular area viewed by the detector.

From elementary trigonometry it can be seen that the diameter of this area ($D$) is given as

$$D = H\beta \qquad \text{(Eq. 9.1)}$$

as illustrated by Figure 9.6. Thus, for example, if the angular field of view is 1.0 milliradians (mr) (1 mr = 0.001 r) and the flying altitude ($H$) is 400 m above the terrain, then:

$$D = H\beta$$

$$D = 400 \times 1.0 \times 0.001$$

$$D = 0.40 \text{ m}$$

Because a thermal scanner views a landscape over a range of angles as it scans from side to side, the IFOV varies in size depending upon the angle of observation ($\theta$). Near the nadir (ground track of the aircraft), the IFOV is relatively small; near the edge of the image, the IFOV

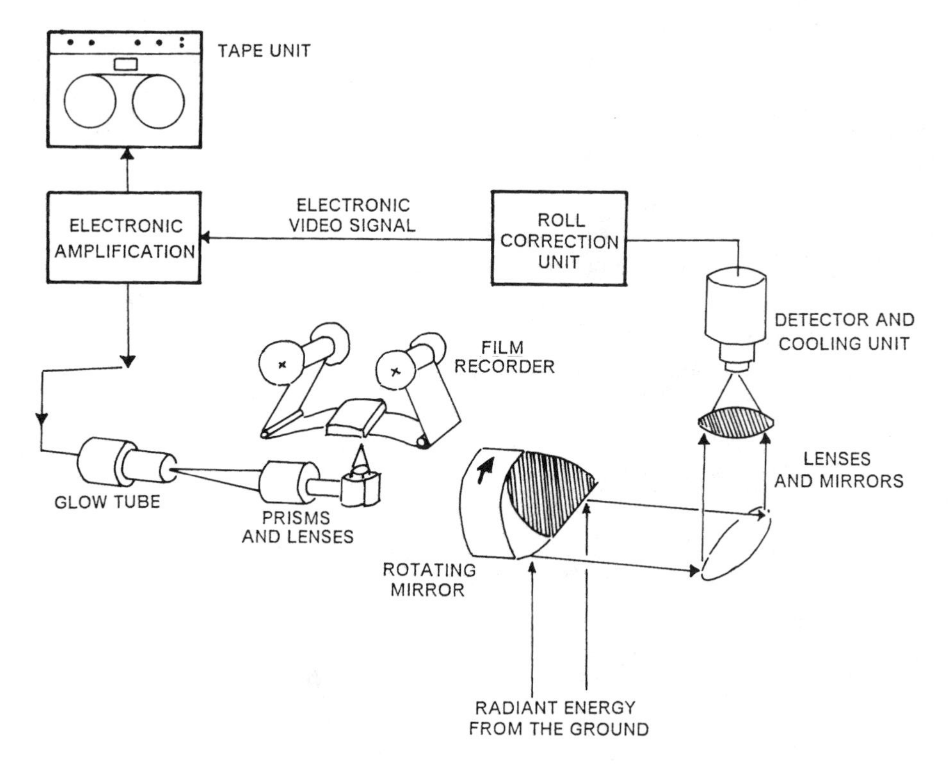


**FIGURE 9.6.** Thermal scanner.

is large. This effect is beneficial in one sense because it compensates for the effects of the increased distance from the sensor to the landscape, thereby providing consistent radiometric sensitivity across the image.

Other effects are more troublesome. Equation 9.1 defines the IFOV at nadir as $H\beta$. Most thermal scanners scan side to side at a constant angular velocity, which means that in a given interval of time they scan a larger cross-track distance near the sides of the image than at the nadir. At angle $\theta$, the IFOV measures $H \sec \theta\beta$ in the direction of flight and $H \sec^2 \theta\beta$ along the scan axis. Thus, near the nadir, the IFOV is small and symmetrical; near the edge of the image, it is larger and elongated in the direction of flight. The variation in the shape of the IFOV creates geometric errors in the representations of features—this problem is discussed in subsequent sections. The variation in size means that radiance from the scene is averaged over a larger area and can be influenced by the presence of small features of contrasting temperature. Although small features can influence data for IFOVs of any size, the impact is more severe when the IFOV is large.

## 9.4. Microwave Radiometers

Microwave emissions from the Earth convey some of the same information carried by thermal (far infrared) radiation. Even though their wavelengths are much longer than those of thermal radiation, microwave emissions are related to temperature and emissivity in much the same manner as is thermal radiation. *Microwave radiometers* are very sensitive instruments tailored to receive and record radiation in the range from about 0.1 mm to 3 cm. Whereas the imaging radars discussed in Chapter 7 are active sensors that illuminate the terrain with their own energy, microwave radiometers are passive sensors that receive microwave radiation naturally emitted by the environment. The strength and wavelength of such radiation is largely a function of the temperature and emissivity of the target. Thus, although microwave radiometers, like radars, use the microwave region of the spectrum, they are functionally most closely related to the thermal sensors discussed in this chapter.

In the present context, we are concerned with microwave emissions from the Earth, which indirectly provide information pertaining to vegetation cover, soil moisture status, and surface materials. Other kinds of studies, peripheral to the field of remote sensing, derive information from microwave emissions from the Earth's atmosphere or from extraterrestrial objects. In fact, the field of microwave radiometry originated with radio astronomy, and some of its most dramatic achievements have been in the reconnaissance of extraterrestrial objects.

A *microwave radiometer* consists of a sensitive receiving instrument typically in the form of a horn- or dish-shaped antenna that observes a path directly beneath the aircraft or satellite; the signal gathered by the antenna is electronically filtered and amplified, and then displayed as a stream of digital data, or, in the instance of scanning radiometers, as an image (Figure 9.7). As with thermal radiometers, microwave radiometers have a reference signal from an object of known temperature. The received signal is compared with the reference signal as a means of deriving the radiance of the target.

Examination of data from a microwave radiometer can be very complex due to many factors that contribute to a given observation. The component of primary interest is usually energy radiated by the features within the IFOV; of course, variations within the IFOV are lost, as the sensor can detect only the average radiance within this area. The atmosphere also radiates energy, so it contributes radiance, depending upon moisture content and temperatures. In addition, solar radiation in the microwave radiation can be reflected from the surface to the antenna.

## 9.5. Thermal Scanners

The most widely used imaging sensors for thermal remote sensing are known as *thermal scanners.* Thermal scanners sense radiances of features beneath the aircraft flight path and produce digital and/or pictorial images of that terrain. There are several designs for thermal scanners. *Object-plane scanners* view the landscape by means of a moving mirror that oscillates at right angles to the flight path of the aircraft, generating a series of parallel (or perhaps overlapping) scan lines that together image a corridor directly beneath the aircraft. *Image-plane scanners* use a wider field of view to collect a more comprehensive image of the landscape; this image is then moved, by means of a moving mirror, relative to the detector. In either instance, the instrument is designed as a series of lenses and mirrors configured to acquire energy from the ground and to focus it on the detector.

An *infrared scanning system* consists of a scanning unit, with a gyroscopic roll connection unit, infrared detectors (connected to a liquid nitrogen cooling unit), and an amplification and control unit (Figure 9.6). A magnetic tape unit records data for later display as a video image; some systems may provide film recording of imagery as it is acquired. Together these units might weigh about 91 kg (200 lbs.), and are usually mounted in an aircraft specially modified to permit the scanning unit to view the ground through an opening in the fuselage.

Infrared energy is collected by a scanning mirror that scans side to side across the flight path, in a manner similar to that described earlier for the Landsat MSS. The typical field of view might be as wide as 77°, so an aircraft flying at an altitude of 300 m (1,000 ft.) could record a strip as wide as 477 m (1,564 ft.). The forward motion of the aircraft generates the along-track dimension of the imagery. The mirror might make as many as 80 scans per second, with each scan representing a strip of ground about 46 cm (18 in.) in width (assuming the 300-m altitude mentioned above).

Energy collected by the scanning mirror is focused first on a parabolic mirror and then on a flat, stationary mirror that focuses energy on the infrared detector unit. The infrared detector unit consists of one of the detectors mentioned above, confined in a vacuum container, and cooled by liquid nitrogen (to reduce electronic noise and enhance the sensitivity of the detector). On some units, the detector can be easily changed by removing a small unit, then replacing it with another. The detector generates an electrical signal that varies in strength in proportion to the radiation received by the mirror and focused on the detector. The signal from the detector is very weak, however, so it must be amplified before it is recorded by the magnetic tape unit that is connected to the scanner. A *roll correction unit,* consisting in part of a gyroscope, senses side-to-side motion of the aircraft and sends a signal that permits the electronic control unit to correct the signal to reduce geometric errors caused by aircraft instability.

After the aircraft has landed, the magnetic tape from the sensor is removed from the aircraft, then taken to a specially equipped laboratory. There data from the tape can be displayed on a cathode ray tube; brightnesses on the screen are proportional to infrared energy received by the scanning mirror. Photographic representations of such images are prepared to form the scenes used for interpretation, or often the digital data may be analyzed directly, as described in Chapters 11 through 14.

## 9.6. Thermal Properties of Objects

All objects at temperatures above absolute zero emit thermal radiation, although the intensity and peak wavelength of such radiation varies with the temperature of the object, as specified by the radiation laws outlined in Chapter 2. For remote sensing in the visible and near infrared, we examine contrasts in the abilities of objects to reflect direct solar radiation to the sensor. For remote sensing in the far infrared spectrum, we sense differences in the abilities of objects and landscape features to absorb shortwave visible and near infrared radiation, then to emit this energy as longer wavelengths in the far infrared region.

Thus, except for geothermal energy, man-made thermal sources, and range and forest fires, the immediate source of emitted thermal infrared radiation is shortwave solar energy. Direct solar radiation (with a peak at about 0.5 μm in the visible spectrum) is received and absorbed by the landscape (Chapter 2). The amount and spectral distribution of energy emitted by landscape features depend upon the thermal properties of these features, as discussed below. The contrasts in thermal brightness, observed as varied gray tones on the image, are used as the basis for identification of features.

A *blackbody* is a theoretical object that acts as a perfect absorber and emitter of radiation; it absorbs and reemits all energy that it receives. Although the blackbody is a theoretical concept, it is useful in describing and modeling the thermal behavior of actual objects. Moreover, it is possible to approximate the behavior of blackbodies in laboratory experiments.

As explained in Chapter 2, as the temperature of a blackbody increases, the wavelength of peak emission decreases in accordance with Wien's displacement law (Eq. 2.5). The Stefan–Boltzmann law (Eq. 2.4) describes mathematically the increase in total radiation emitted (over a range of wavelengths) as the temperature of a blackbody increases.

*Emissivity* ($\varepsilon$) is a ratio between emittance of an object in relation to emittance of a blackbody at the same temperature:

$$\varepsilon_\lambda = \frac{\text{Radiant emittance of an object}}{\text{Radiant emittance of a blackbody at the same temperature}} \qquad \text{(Eq. 9.2)}$$

(See also Eq. 2.3.) The subscript ($\lambda$) sometimes used with $\varepsilon$ signifies that $\varepsilon$ has been measured for specific wavelengths. Emissivity therefore varies from 0 to 1, with 1 signifying a substance with a thermal behavior identical to that of a blackbody. Table 9.1 lists emissivities for some common materials. Note that many of the substances commonly present in the landscape (e.g., soil, water) have emissivities rather close to 1. Note, however, that emissivity can vary with temperature, wavelength, and angle of observation.

### *Graybodies*

An object that has an emissivity of less than 1.0 but has constant emissivity over all wavelengths is known as a *gray body* (Figure 9.7). A *selective radiator* is an object with an emissivity that varies with respect to wavelength. If two objects in the same setting are at the same temperature, but have different emissivities, the one having the higher emissivity will radiate more strongly. Because the sensor detects radiant energy (the apparent temperature) rather than the

TABLE 9.1. Emissivities of Some Common Materials

| Material | Temperature (°C) | Emissivity[a] |
|---|---|---|
| Polished copper | 50–100 | 0.02 |
| Polished brass | 200 | 0.03 |
| Polished silver | 100 | 0.03 |
| Steel alloy | 500 | 0.35 |
| Graphite | 0–3,600 | 0.7–0.8 |
| Lubricating oil (thick film on nickel base) | 20 | 0.82 |
| Snow | –10 | 0.85 |
| Sand | 20 | 0.90 |
| Wood (planed oak) | 20 | 0.90 |
| Concrete | 20 | 0.92 |
| Dry soil | 20 | 0.92 |
| Brick (red common) | 20 | 0.93 |
| Glass (polished plate) | 20 | 0.94 |
| Wet soil (saturated) | 20 | 0.95 |
| Distilled water | 20 | 0.96 |
| Ice | –10 | 0.96 |
| Carbon lamp black | 20–400 | 0.96 |
| Lacquer (matte black) | 100 | 0.97 |

*Note.* Data from Hudson (1969) and Weast (1986).
[a]Measured at normal incidence over a range of wavelengths.

kinetic ("true") temperature, precise interpretation of an image requires knowledge of emissivities of features shown on the image.

### *Heat*

*Heat* is the internal energy of a substance arising from the motion of its component atoms and molecules. *Temperature* measures the relative warmth or coolness of a substance. It is the kinetic temperature or average thermal energy of molecules within a substance. *Kinetic temperature,* sometimes known as the *true temperature,* is measured using the usual temperature scales, most notably the Fahrenheit, Celsius (centigrade), and Kelvin (absolute) scales. *Radiant* (or *apparent*) *temperature* measures the emitted energy of an object. Photons from the radiant energy are detected by the thermal scanner.

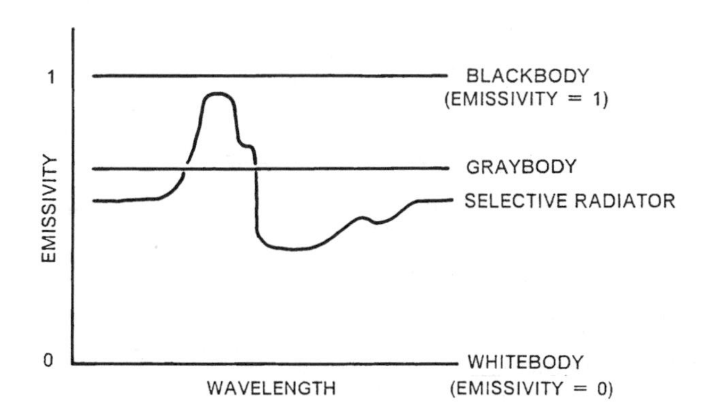


**FIGURE 9.7.** Blackbody, graybody, whitebody.

*Heat capacity* is the ratio of the change in heat energy per unit mass to the corresponding change in temperature (at constant pressure). For example, we can measure the heat capacity of pure water to be 1 calorie (cal) per gram (g), meaning that 1 cal is required for each gram to raise its temperature by 1°C. The specific heat of a substance is the ratio of its heat capacity to that of a reference substance. Because the reference substance typically is pure water, specific heat is often numerically equal to its heat capacity. Because a calorie is defined as the amount of heat required to raise by 1°C the temperature of 1 g of pure water, use of water as the reference means that heat capacity and specific heat will be numerically equivalent. In this context, specific heat can be defined as the amount of heat (measured in calories) required to raise the temperature of 1 g of a substance 1°C.

*Thermal conductivity* is a measure of the rate that a substance transfers heat. Conductivity is measured as calories per centimeter per second per degree Celsius, so it measures calories required to transfer a change in temperature over specified intervals of length and time.

Some of these variables can be integrated into a single measure, called *thermal inertia* ($P$), defined as

$$P = \sqrt{KC\rho} \quad \text{(Eq. 9.3)}$$

where $K$ is the thermal conductivity ($\text{cal} \cdot \text{cm}^{-1} \cdot \text{sec}^{-1} \cdot {}^\circ\text{C}^{-1}$); $C$ is the heat capacity ($\text{cal} \cdot \text{gm}^{-1} \cdot {}^\circ\text{C}^{-1}$), and $\rho$ is the density ($\text{gm} \cdot \text{cm}^{-3}$). *Thermal inertia, P,* is then measured in $\text{cal cm}^{-2} \cdot {}^\circ\text{C}^{-1}\ \text{sec}^{-1/2}$. *Thermal inertia* measures the tendency of a substance to resist changes in temperature, or, more precisely, the rate of heat transfer at the contact between two substances. Table 9.2 gives thermal inertia values for a number of common materials. A substance with low thermal inertia is one with low density, low conductivity, and low specific heat, such that it resists changes in temperature. Such substances—for example, wood, glass, and cork—are insensitive to changes in temperature. In contrast, substances with high values of $P$ will heat and cool

**TABLE 9.2. Thermal Properties of Some Common Substances**

| | $K$ | $\rho$ | $C$ | $P$ | $P^{-1}$ |
|---|---|---|---|---|---|
| Basalt | 0.0050 | 2.8 | 0.20 | 0.053 | 19 |
| Clay soil (moist) | 0.0030 | 1.7 | 0.35 | 0.042 | 24 |
| Dolomite | 0.012 | 2.6 | 0.18 | 0.075 | 13 |
| Granite | 0.0065 | 2.6 | 0.16 | 0.052 | 19 |
| Limestone | 0.0048 | 2.5 | 0.17 | 0.045 | 22 |
| Sandy soil | 0.0014 | 1.8 | 0.24 | 0.024 | 41 |
| Shale | 0.0030 | 2.3 | 0.17 | 0.034 | 29 |
| Slate | 0.0050 | 2.8 | 0.17 | 0.049 | 21 |
| Aluminum | 0.538 | 2.69 | 0.215 | 0.544 | 1.81 |
| Copper | 0.941 | 8.93 | 0.092 | 0.879 | 1.14 |
| Pure iron | 0.18 | 7.86 | 0.107 | 0.389 | 2.57 |
| Lead | 0.083 | 11.34 | 0.031 | 0.171 | 5.86 |
| Silver | 1.00 | 10.42 | 0.056 | 0.764 | 1.31 |
| Carbon steel | 0.150 | 7.86 | 0.110 | 0.360 | 2.78 |
| Glass | 0.0021 | 2.6 | 0.16 | 0.029 | 34 |
| Wood | 0.0005 | 0.5 | 0.327 | 0.009 | — |

*Note.* Selected from values given by Janza (1975). Thermal conductivity, $K$, measured in $\text{cal} \cdot \text{cm}^{-1} \cdot \text{sec}^{-1} \cdot {}^\circ\text{C}^{-1}$; density, $\rho$, measured in $\text{g} \cdot \text{cm}^{-3}$; thermal inertia, $P$, measured in $\text{cal} \cdot \text{cm}^{-2} \cdot {}^\circ\text{C}^{-1} \cdot \text{sec}^{-1/2}$; thermal parameter symbolized by $P^{-1}$.

quickly. Such substances—for example, silver, copper, and lead—have high densities, high conductivity, and high specific heat.

These measures, and the terminology, are established by long-standing convention, even though they cause misunderstanding. Many people find it confusing to find that a low value for thermal inertia indicates high resistance to temperature change, since they interpret "low inertia" to mean the opposite. Therefore, the thermal parameter, $P^{-1}$, is perhaps easier to interpret, as high values indicate high resistance to change and low values indicate low resistance to thermal change.

## 9.7. Geometry of Thermal Images

Thermal scanners, like all remote sensing systems, generate geometric errors as they gather data. These errors mean that representations of positions and shapes of features depicted on thermal imagery do not match to their correct planimetric forms. Therefore, images cannot be directly used as the basis for accurate measurements.

Thermal imagery also exhibits relief displacement analogous to that encountered in aerial photography (Figure 9.8). Thermal imagery, however, does not have the single central perspective of an aerial photograph, but rather a separate nadir for each scan line. Thus the focal point for relief displacement is the nadir for each scan line, or in effect the trace of the flight path on the ground. Thus relief displacement is projected from a line that follows the center of the long axis of the image. At the center of the image, the sensor views objects from directly overhead and planimetric positions are correct. However, as distance from the centerline increases, the

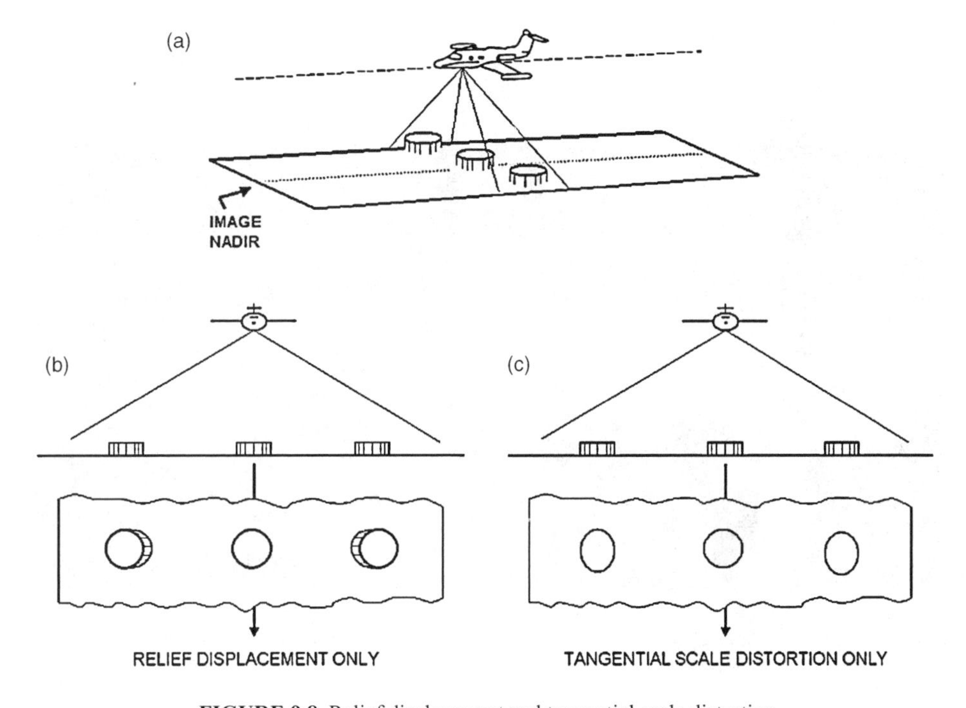


**FIGURE 9.8.** Relief displacement and tangential scale distortion.

sensor tends to view the sides as well as the tops of features, and relief displacement increases. These effects are visible in Figure 9.9; the tanker and the tanks appear to lean outward from a line that passes through the center of the image. The effect increases toward the edges of the image.

Figure 9.9 also illustrates other geometric qualities of thermal line scan imagery. Although the scanning mirror rotates at a constant speed, the projection of the IFOV onto the ground surface does not move (relative to the ground) at equal speed because of the varied distance from the aircraft to the ground. At nadir, the sensor is closer to the ground than it is at the edge of the image; in a given interval of time, the sensor scans a shorter distance at nadir than it does at the edge of the image. Therefore, the scanner produces a geometric error that tends to compress features along an axis oriented perpendicular to the flight line and parallel to the scan lines. In Figure 9.9 this effect, known as *tangential scale distortion,* is visible in the shapes of the cylindrical storage tanks. The images of those tanks nearest the flight line are more circular, whereas the shapes of those tanks furthest from the flight line (nearest the edge of the image) are compressed along an axis perpendicular to the flight line. Sometimes the worst effects can be removed by corrections applied as the film image is generated, although it is often necessary to avoid use of the extreme edges of the image.

## 9.8. The Thermal Image and Its Interpretation

The image generated by a thermal scanner is a strip of black-and-white film depicting thermal contrasts in the landscape as variations in gray tones (Figure 9.10). Usually brighter tones (whites and light grays) represent warmer features; darker tones (dark grays and blacks) represent cooler features. In some applications the black-and-white image may be subjected to level slicing or other enhancements that assign distinctive hues to specific gray tones as an aid for manual interpretation. Often it is easier for the eye to separate subtle shades of color than the

**FIGURE 9.9.** Thermal image of an oil tanker and petroleum storage facilities near the Delaware River, 19 December 1979. This image, acquired at 11:43 P.M., shows discharge of warm water into the Delaware River and thermal patterns related to operation of a large petrochemical facility. Thermal image from Daedulus Enterprises, Inc., Ann Arbor, MI.

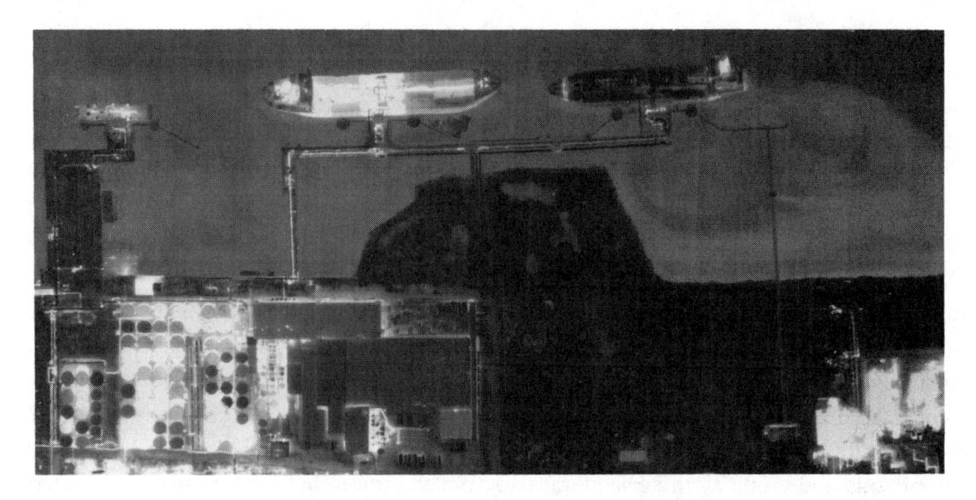

**FIGURE 9.10.** This thermal image shows another section of the same facility imaged in Figure 9.9. Thermal image from Daedulus Enterprises, Inc., Ann Arbor, MI.

variations in gray on the original image. Such enhancements are simply manipulations of the basic infrared image; they do not represent differences in means of acquisition or in the quality of the basic information available for interpretation.

For any thermal infrared image, the interpreter must always determine (1) if the image at hand is a positive or a negative image, and (2) the time of day that the image was acquired. Sometimes it may not be possible to determine the correct time of day from information within the image itself; misinterpretation can alter the meaning of gray tones on the image and render the resulting interpretation useless.

As the sensor views objects near the edge of the image, the distance from the sensor to the ground increases. This relationship means that the IFOV is larger nearer the edges of the image than it is near the flight line.

Thermal scanners are generally uncalibrated, so they show relative radiances rather than absolute measurements of radiances. However, some thermal scanners do include reference sources that are viewed by the scanning system at the beginning and end of each scan. The reference sources can be set at specific temperatures that are related to those expected to be encountered in the scene. Thus each scan line includes values of known temperature that permit the analyst to estimate temperatures of objects within the image.

In addition, errors caused by the atmosphere and by the system itself prevent precise interpretation of thermal imagery. Typical system errors might include recording noise, variations in reference temperatures, and detector errors. Full correction for atmospheric conditions requires information not usually available in detail, so often it is necessary to use approximations, or value-based samples acquired at a few selected times and places, then extrapolated to estimate values elsewhere. Also, the *atmospheric path traveled* by radiation reaching the sensor varies with the angle of observation, which changes as the instrument scans the ground surface. These variations in angle lead to errors in observed values in the image.

Even when accurate measures of radiances are available, it is difficult to derive data for kinetic temperatures from the apparent temperature information within the image. Derivation of kinetic temperatures requires knowledge of emissivities of the materials. In some instances, such knowledge may be available, as the survey may be focused on a known area that must be repeatedly imaged to monitor changes over time (e.g., as moisture conditions change). But many other surveys examine areas not previously studied in detail, and information regarding surface materials and their emissivities may not be known.

*Emissivity* is a measure of the effectiveness of an object in translating temperature into emitted radiation (and in converting absorbed radiation into a change in observed temperature). Because objects differ with respect to emissivity, observed differences in emitted infrared energy do not translate directly into corresponding differences in temperature. As a result, it is necessary to apply knowledge of surface temperature, or of emissivity variations, to accurately study surface temperature patterns from thermal imagery. Because knowledge of these characteristics assumes a very detailed prior knowledge of the landscape, such interpretations should be considered as appropriate for examination of a distribution known already in some detail rather than for reconnaissance of an unknown pattern. (E.g., one might already know the patterns of soils and crops at an agricultural experiment station, but wish to use the imagery to monitor temperature patterns.) Often estimated values for emissivity are used or assumed values are applied to areas of unknown emissivity.

Also, it should be recognized that the sensor records radiances of the surfaces of objects. Because radiances may be determined at the surface of an object by a layer perhaps as thin as 50 m, a sensor may record conditions that are not characteristic of the subsurface mass, which is probably the object of the study. For example, evaporation from a water body or a moist soil surface may cool the thin layer of moisture at the contact point with the atmosphere. Because the sensor detects radiation emitted at this surface layer, the observed temperature may differ considerably from that of the remaining mass of the soil or water body.

Leckie (1982) estimates that calibration error and other instrument errors are generally rather small, although they may be important in some instances. Errors in estimating emissivity and in attempts to correct for atmospheric effects are likely to be the most important sources of error in quantitative studies of thermal imagery.

In most instances, a thermal image must be interpreted to yield qualitative rather than quantitative information. Although some applications do require interpretations of quantitative information, there are many others for which qualitative interpretation is completely satisfactory. An interpreter who is well informed about the landscape represented on the image, the imaging system, the thermal behavior of various materials, and the timing of the flight is prepared to derive considerable information from an image, even though it may not be possible to derive precise temperatures from the image.

The *thermal landscape* is a composite of the familiar elements of surface material, topography, vegetation cover, and moisture. Various rocks, soils, and other surface materials respond differently to solar heating. Thus in some instances the differences in thermal properties tabulated in Table 9.1 can be observed in thermal imagery. However, the thermal behavior of surface materials is also influenced by other factors. For example, slopes that face the sun will tend to receive more solar radiation than slopes that are shadowed by topography. Such differences are, of course, combined with those arising from different surface materials. Also, the presence and nature of vegetation alters the thermal behavior of the landscape. Vegetation tends to heat rather rapidly, but can also shade areas, creating patterns of warm and cool.

Water tends to retain heat, to cool slowly at night, and to warm slowly during daytime. In contrast, many soils and rocks (if dry) tend to release heat rapidly at night and to absorb heat quickly during the daytime. Even small or modest amounts of moisture can greatly alter the thermal properties of soil and rock. Therefore, thermal sensors can be very effective in monitoring the presence and movement of moisture in the environment. In any given image, the influences of surface materials, topography, vegetation, and moisture can combine to cause very complex image patterns. However, often it is possible to isolate the effect of some of these variables and therefore to derive useful information concerning, for example, movement of moisture or the patterns of differing surface materials.

Timing of acquisition of thermal imagery is very important. The optimum times vary according to the purpose and subject of the study, so it is not possible to specify universally applicable rules. Since the greatest thermal contrast tends to occur during the daylight hours, sometimes thermal images are acquired in the early afternoon to capture the differences in thermal properties of landscape features. However in the 3–6 μm range the sensor may record reflected as well as emitted thermal radiation, so daytime missions in this region may not be optimum for thermal information. Also during daytime the sensor may record thermal patterns caused by topographic or cloud shadowing; although shadows may sometimes be useful in interpretation, they are more likely to complicate analysis of a thermal image, so it is usually best to avoid acquiring heavily shadowed images. In a daytime image, water bodies typically appear as cool relative to land, and bare soil, meadow, and wooded areas appear as warm features.

Some of the problems arising from daytime images are avoided by missions planned just before dawn. Shadows are absent, and sunlight, of course, cannot cause reflection (at shorter wavelengths) or shadows. However, thermal contrast is lower, so it may be more difficult to distinguish between broad classes of surfaces based upon differences in thermal behavior. On such an image, water bodies would appear as warm relative to land. Forested areas may also appear to be warm. Open meadows and dry, bare soil are likely to appear as cool features.

The thermal images of petroleum storage facilities (Figures 9.12 and 9.13) show thermal contrasts that are especially interesting. A prominent feature on Figure 9.12 is the bright thermal plume discharged by the tributary to the Delaware River. The image clearly shows the sharp contrast in temperature as the warm water flows into the main channel, then disperses and cools as it is carried downstream. Note the contrast between the full and partially full tanks, and the warm temperatures of the pipelines that connect the tanker with the storage tanks. Many of the same features are also visible in Figure 9.10, which also shows a partially loaded tanker with clear delineation of the separate storage tanks in the ship.

Thermal imagery has obvious significance for studies of heat loss, thermal efficiency, and effectiveness of insulation in residential and commercial structures. Plate 17 shows thermal images of two residential structures observed at night during the winter season. Windows and walls are the principal avenues for escape of heat, and the chimney shows as an especially bright feature. Note that the walkways, paved roads, and parked automobiles show as cool features.

In Figure 9.11, two thermal images depict a portion of the Cornell University campus in Ithaca, New York, acquired in January (left) and again the following November (right). Campus buildings are clearly visible, as are losses of heat through vents in the roofs of buildings and at manholes where steampipes for the campus heating system join or change direction. The left-hand image shows a substantial leak in a steam pipe as it passes over the bridge in the right center of the image. On the right, a later image of the same region shows clearly the effects of repair of the defective section

Figure 9.12 shows Painted Rock Dam, Arizona, as depicted both by an aerial photograph

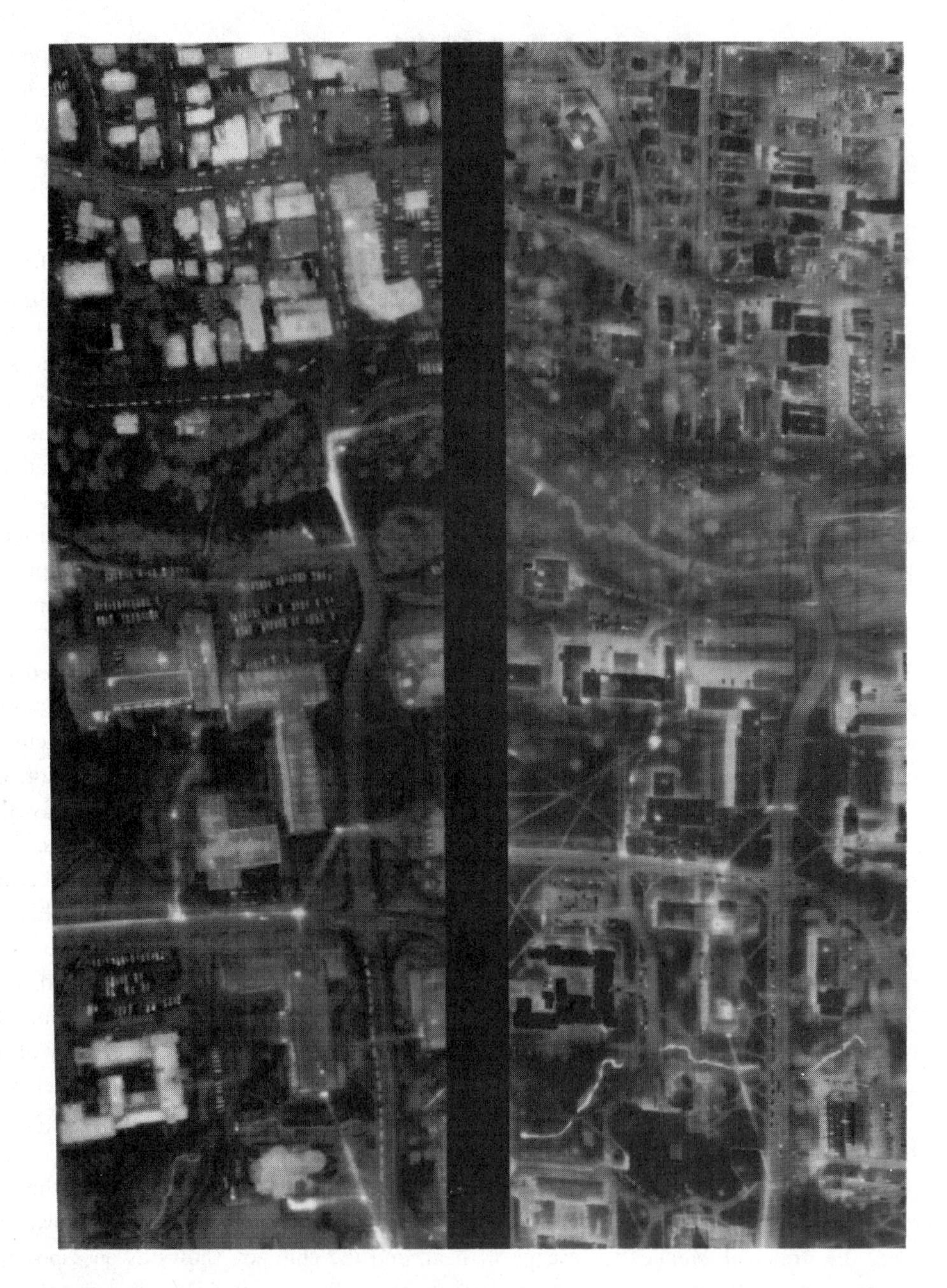

**FIGURE 9.11.** Two thermal images of a portion of the Cornell University campus, in Ithaca, NY. Thermal images from Daedalus Enterprises, Inc., Ann Arbor, MI.

(top) and a thermal infrared image (bottom). The aerial photograph was taken at about 10:30 A.M.; the thermal image was acquired at about 7:00 A.M. the same day. The prominent linear feature is a large earthen dam, with the spillway visible at the lower left. On the thermal image, the open water upstream from the dam appears as a uniformly white (warm) region, whereas land areas are dark (cool)—a typical situation for early morning hours, before solar radiation has

**FIGURE 9.12.** Painted Rock Dam, AZ, 28 January 1979. Aerial photograph (top) and thermal image (below). Thermal image from Daedulus Enterprises, Inc., Ann Arbor, MI.

warmed the Earth. On the downstream side of the dam, the white (warm) regions reveal areas of open water or saturated soil. The open water in the spillway is, of course, expected, but the other white areas indicate places where there may be seepage and potentially weak points in the dam structure.

Figure 9.13 shows thermal images of a power plant acquired at four different stages of the tidal cycle. The discharge of warm water is visible as the bright plume in the upper left of each image. At the top, at low tide (5:59 A.M.), the warm water is carried downstream toward the ocean. The second image, acquired at flood tide (8:00 A.M.), shows the deflection of the plume by rising water. In the third image, taken at high tide (10:59 A.M.), the plume extends down-

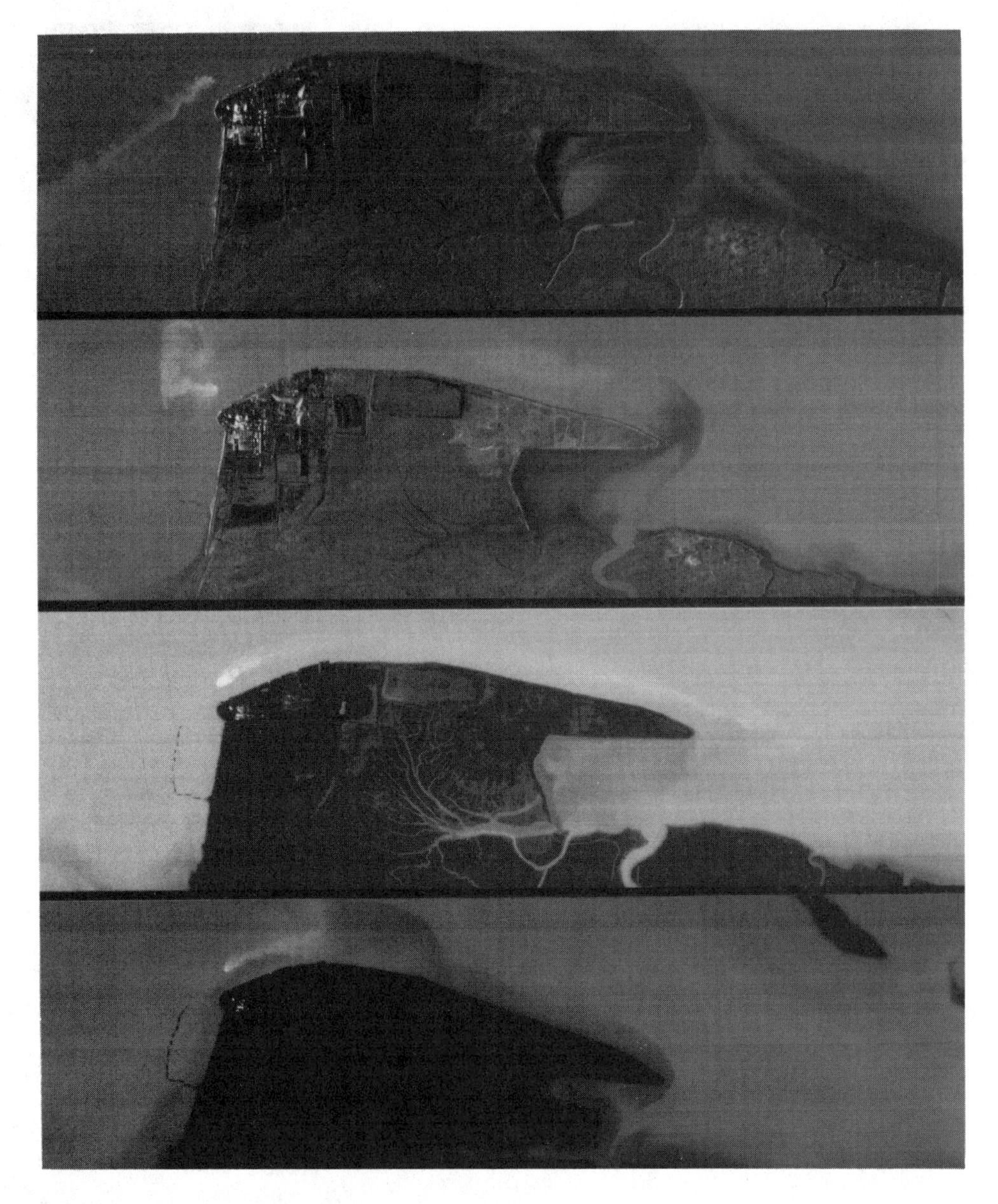

**FIGURE 9.13.** Thermal images of a power plant acquired at different states of the tidal cycle. Thermal images from Daedalus Enterprises, Inc., Ann Arbor, MI.

stream for a considerable distance. Finally, in the bottom image, acquired at ebb tide (2:20 P.M.), the shape of the plume reflects the reversal of tidal flow once again.

If imagery or data for two separate times are available, it may be possible to employ knowledge of thermal inertia as a means of studying the pattern of different materials at the earth's surface. Figure 9.14 illustrates the principles involved. Two images are acquired at times that permit observation of extremes of temperature, perhaps near noontime and again just before dawn. These two sets of data permit estimation of the ranges of temperature variation for each region on the image. Because these variations are determined by the thermal inertias of the sub-

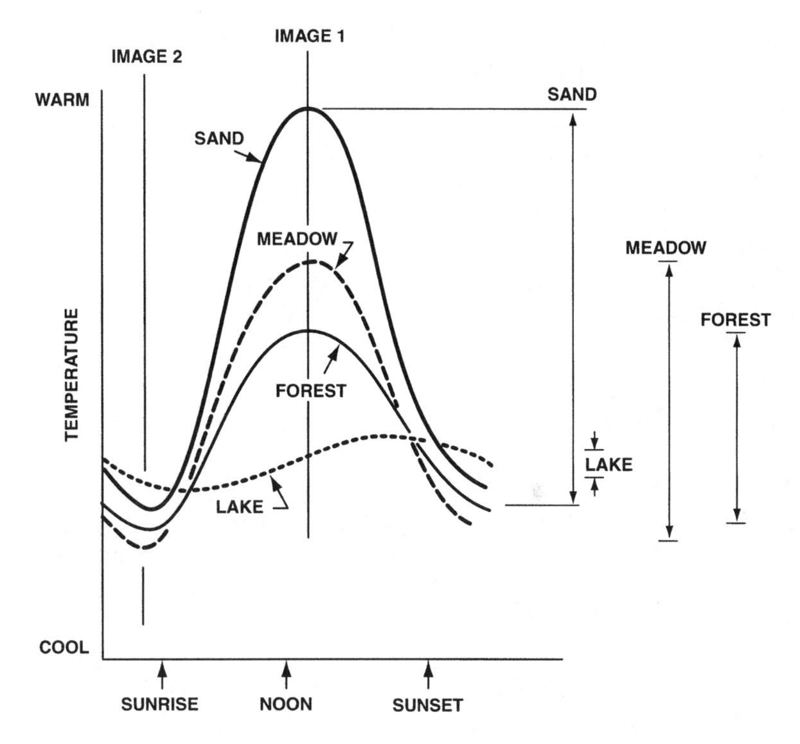


**FIGURE 9.14.** Schematic illustration of diurnal temperature variation of several broad classes of land cover.

stances, they permit interpretation of features represented by the images. Leckie (1982) notes that misregistration can be a source of error in comparisons of day and night images, although such errors are thought to be small relative to other errors.

In Figure 9.14 images are acquired at noon, when temperatures are high, and again just before dawn, when temperatures are lowest. By observing the differences in temperature extremes, it is possible to estimate thermal inertia. Thus the lake, composed of a material that resists changes in temperature, shows rather small changes in temperature, whereas the surface of open sand displays little resistance to thermal changes and exhibits a wider range of temperature. The double-headed arrows on the right of the diagram represent differing daily temperature ranges in this example.

Plate 18 shows another example of diurnal temperature changes. These two images show the landscape near Erfurt, Germany, as observed by a thermal scanner in the region between 8.5 and 12 μm. The thermal data have been geometrically and radiometrically processed, then superimposed over digital elevation data, with reds and yellows assigned to represent warmer temperatures and blues and greens to represent cooler temperatures. The city of Erfurt is positioned at the edge of a limestone plateau, which is visible as the irregular topography in the foreground of the image. The valley of the River Gera is visible in the lower left, extending across the image to the upper center of the image.

The top image represents this landscape as observed just after sunset. The urbanized area and much of the forested topography south of the city show as warm reds and yellows. The bottom image shows the same area observed just before sunrise, when temperatures are at their coolest. As the open land of the rural landscape and areas at the periphery of the urbanized areas have cooled considerably, the forested region on the plateau south of the city is now the warmest surface, due to the thermal effects of the forest canopy.

Figure 9.15 represents a pair of thermal images that illustrate the same principle applied to seasonal temperature variation. At the left is an ASTER image (discussed in Chapter 21) of the thermal variations within a landscape near Richmond, Virginia, acquired in March, when the landscape is cool relative to the water bodies in this region. In contrast, the image of the same region in October shows a reversal of the relative brightnesses of the terrain: the land surface is now warm relative to the water bodies. This example illustrates seasonal, rather than daily, differences in thermal inertia.

## 9.9. Heat Capacity Mapping Mission

The Heat Capacity Mapping Mission (HCMM) (in service from April 1978 to September 1980) was a satellite system specifically designed to evaluate the concept that orbital observations of temperature differences at the Earth's surface at different points in the daily heating/cooling cycle might provide a basis for estimation of thermal inertia and other thermal properties of surface materials. The satellite was in a sun-synchronous orbit, at an altitude of 620 km, bringing it over the equator at 2:00 P.M. local sun time. At 40°N latitude the satellite passed overhead at 1:30 P.M., then again about 12 hours later at 2:30 A.M. These times provided observations at two points on the diurnal heating/cooling cycle (Figure 9.16). (It would have been desirable to use a time later in the morning just before sunrise, in order to provide higher thermal contrast, but orbital constraints dictated the time differences between passes). The repeat cycle varied with latitude—at midlatitude locations the cycle was 5 days. The 12-hour repeat coverage was available for some locations at 16-day intervals, but other locations received only 36-hour coverage.

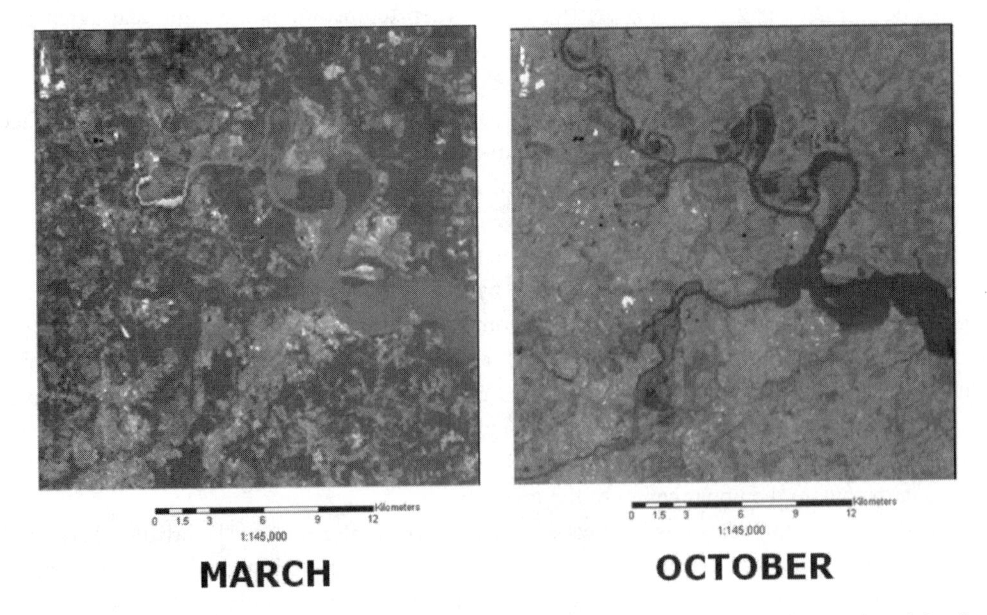


**FIGURE 9.15.** ASTER thermal images illustrating seasonal temperature differences. The March and October images illustrate differences in the responses of land and water to seasonal temperature variations.

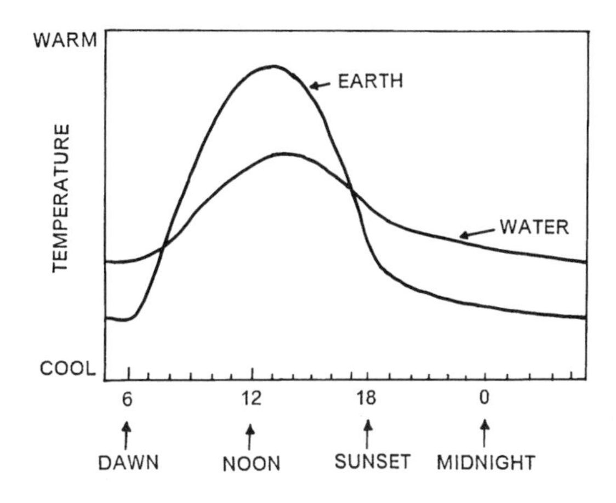


**FIGURE 9.16.** Diurnal temperature cycle.

The HCMM radiometer used two channels. One, in the reflective portion of the spectrum (0.5–1.1 m), had a spatial resolution of about 500 m × 500 m. A second channel was available in the thermal infrared region, from 10.5 to 12.5 m, and had a spatial resolution of 600 m × 600 m. The image swath was about 716-km wide.

Plate 19 shows an HCMM image of a portion of the eastern coast of North America. The image represents a strip of land about 688 km (430 mi.) in width extending from Lake Ontario in the north (top) south toward Cape Hatteras, North Carolina (bottom left). The image uses colors to portray variations in observed temperature so that the sequence purple, blue, green, brown, yellow, orange, red, gray, and white represents increasing temperature. Black areas represent cloud cover; the coldest areas are the larger water bodies and clouds. Major metropolitan centers including New York, Philadelphia, and Washington, D.C., are visible as bright white areas near the coastline. Variations in radiances within land areas are mainly caused by differences in land cover, which in some areas (especially southeastern Pennsylvania) closely follow topography and geologic structure.

Estimation of thermal inertia requires that two HCMM images be registered to observe apparent temperatures at different points on the diurnal heating/cooling cycle. Because the day and night images must be acquired on different passes with different inclinations (Figure 9.17), the registration of the two images is often much more difficult than might be the case with similar images. The rather coarse resolution means that normally distinct landmarks are not visible or are more difficult to recognize. Watson et al. (1982) have discussed some of the problems encountered in registering HCMM images.

NASA's Goddard Space Flight Center received and processed HCMM data, calibrating the thermal data, and performing geometric registration, when possible, for the two contrasting thermal images. Ideally, these two images permit reconstruction of a thermal inertia map of the imaged area. In fact, atmospheric conditions varied during the intervals between passes, and the measured temperatures were influenced by variations in cloud cover, wind, evaporation at the ground surface, and atmospheric water vapor. As a result, the measured differences are *apparent thermal inertia* (ATI) rather than a true measure of thermal inertia. Due to the number of approximations employed and scaling of the values, the values for ATI are best considered to be measures of relative thermal inertia.

As an experimental prototype, HCMM was in service for a relatively short interval and acquired data for rather restricted regions of the Earth. Some of the concepts developed and

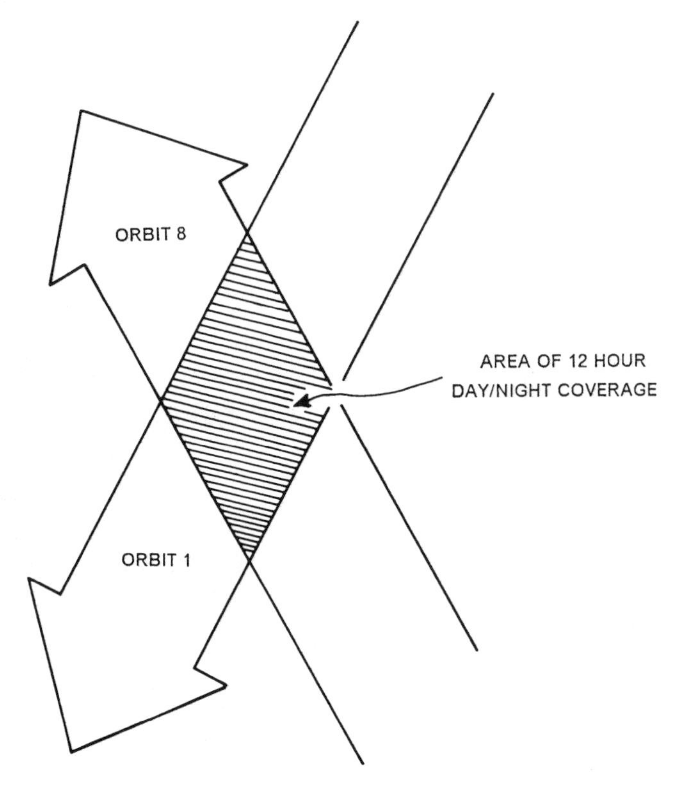


**FIGURE 9.17.** Overlap of two HCCM passes at 40° latitude.

tested by HCMM have been employed in the design of other programs, most notably the NASA Earth Observing System (Chapter 21).

Data are available in both film and digital formats; HCMM scenes show areas about 700 km × 700 km (1,127 mi. × 1,127 mi.) in size. The HCMM archive, maintained by the National Space Science Data Center (World Data Center-A, Code 601, NASA Goddard Space Flight Center, Greenbelt, MD 20771) provides coverage for many midlatitude regions (including North America, Australia, Europe, and northern Africa).

## 9.10. Landsat Multispectral Scanner and Thematic Mapper Thermal Data

Chapter 6 mentioned, but did not discuss in any detail, the thermal data collected by the TM and by the Landsat 3 MSS. The Landsat 3 MSS, unlike those of earlier Landsats, included a channel (MSS band 5, formerly designated as band 8) sensitive to thermal data in the region 10.4–12.6 μm. Due to decreased sensitivity of the thermal detectors, spatial resolution of this band was significantly coarser than that of the visible and near infrared channels (237 m × 237 m vs. 79 m × 79 m). Each of the MSS band-5 pixels therefore corresponds to about nine pixels from one of the other four bands. Landsat 3 band 5 was designed to detect apparent temperature differences of about 15°C. Although initial images were reported to meet this standard, progressive degradation of image quality was observed as the thermal sensors deteriorated over time. In addition, the thermal band was subject to other technical problems. Because of these problems, the MSS thermal imaging system was turned off about 1 year after Landsat 3 was launched.

As a result, relatively few images were acquired using the MSS thermal band and relatively little analysis has been attempted. Lougeay (1982) examined a Landsat 3 thermal image of an alpine region in Alaska with the prospect that the extensive glaciation in this region could contribute to sufficient thermal contrast to permit interpretation of terrain features. The coarse spatial resolution, solar heating of south-facing slopes, and extensive shadowing due to rugged topography and low Sun angle contributed to difficulties in interpretation. Although Lougeay was able to recognize major landscape features, his analysis is more of a suggestion of potential applications of the MSS thermal data than an illustration of their usefulness.

Price (1981) also studied the Landsat MSS thermal data. He concluded that the thermal band provided new information not conveyed by the other MSS channels. Rather broad land-cover classes (including several classes of open water, urban and suburban land, vegetated regions, and barren areas) could readily be identified on the scene that he examined. The time of the Landsat overpass was too early to record maximum thermal contrast (which occurs in early afternoon), and analyses of temperatures are greatly complicated by topographic and energy balance effects at the ground surface and the influence of the intervening atmosphere. Despite the possibility of adding new information to the analysis, Price advised against routine use of the thermal channel in the usual analyses of Landsat MSS data. In general, the experience of other analysts has confirmed this conclusion and these data have not found widespread use.

The Landsat TM includes a thermal band, usually designated as TM band 6, sensitive in the region 10.4–12.5 μm (Figure 9.18). It has lower radiometric sensitivity and coarser spatial resolution (about 120 m) relative to other TM bands. This image reveals the thermal impact of human activities in urban regions, which present a distinct thermal signature (also visible in Plate 19).

## 9.11. Summary

Thermal imagery is a valuable asset for remote sensing because it conveys information not easily derived from other forms of imagery. The thermal behavior of different soils, rocks, and construction materials can permit derivation of information not present in other images. The thermal properties of water contrast with those of many other landscape materials, so that thermal images can be very sensitive to the presence of moisture in the environment. And the presence of moisture is itself often a clue to the differences between different classes of soil and rock.

Of course, use of data from the far infrared region can present its own problems. Like all images, thermal imagery has geometric errors. Moreover, the analyst cannot derive detailed quantitative interpretations of temperatures unless detailed knowledge of emissivity is at hand. Timing of image acquisition can be critical. Atmospheric effects can pose serious problems, especially from satellite altitudes. Because the thermal landscape differs so greatly from the visible landscape, it may often be necessary to use aerial photography to locate familiar landmarks while interpreting thermal images. Existing archives of thermal imagery are not comparable in scope to those for aerial photography or satellite data (such as those of Landsat or SPOT), so it

**FIGURE 9.18.** TM band 6, showing a thermal image of New Orleans, LA, 16 September 1982. This image shows the same area, and was acquired at the same time, as Figures 6.13 and 6.14. Resolution of TM band 6 is coarser than the other bands, so detail is not as sharp. Here dark image tone represents relatively cool areas and bright image tone represents relatively warm areas. The horizontal banding near the center of the image is caused by a defect in the operation of the TM discussed in Chapter 11. From USGS.

may be difficult to acquire suitable thermal data unless it is feasible to purchase custom-flown imagery.

## Review Questions

1. Explain why choice of time of day is so important in planning acquisition of thermal imagery.
2. Would you expect season of the year to be important in acquiring thermal imagery of, for example, a region in Pennsylvania? Explain.
3. In your new job as an analyst for an institution that studies environmental problems in coastal areas, it is necessary for you to prepare a plan to acquire thermal imagery of a tidal marsh. List the important factors you must consider as you plan the mission.

4. Many beginning students are surprised to find that geothermal heat and man-made heat are of such modest significance in the Earth's energy balance and in determining information presented on thermal imagery. Explain, then, what it is that thermal imagery does depict, and why thermal imagery is so useful in so many disciplines.

5. Can you understand why thermal infrared imagery is considered so useful to so many scientists even though it does not usually provide measurements of actual temperatures? Can you identify situations in which it might be important to be able to determine actual temperatures from imagery?

6. Fagerlund et al. (1982) report that it is possible to judge from thermal imagery if storage tanks for gasoline and oil are empty, full, or partially full. Examine Figure 9.12 and find examples of each. How can you confirm from evidence on the image itself that full tanks are warm (bright) and empty tanks are cool (dark)?

7. Examine Figure 9.19. Is the tanker empty, partially full, or full? Examine Figure 9.10; are these tankers empty, partially full, or full? From your inspection of the imagery can you determine something about the construction of tankers and the procedures used to empty or fill tankers?

8. What information would be necessary to plan an aircraft mission to acquire thermal imagery to study heat loss from residential areas in the northeastern United States?

9. In what ways can thermal imagery be important in agricultural research?

10. Outline ways in which thermal imagery would be especially useful in studies of the urban landscape.

## References


Colcord, J. E. 1981. Thermal Imagery Energy Surveys. *Photogrammetric Engineering and Remote Sensing,* Vol. 47, pp. 237–240.

Fagerlund, E., B. Kleman, L. Sellin, and H. Svenson. 1970. Physical Studies of Nature by Thermal Mapping. *Earth Science Reviews,* Vol. 6, pp. 169–180.

Gillespie, A. R., and A. B. Kahle. 1978. Construction and Interpretation of a Digital Thermal Inertia Image. *Photogrammetric Engineering and Remote Sensing,* Vol. 43, pp. 983–1000.

Goddard Space Flight Center. 1978. *Data Users Handbook, Heat Capacity Mapping Mission (HCMM), for Applications Explorer Mission-A (AEM).* Greenbelt, MD: NASA.

Goward, S. N. 1981, February. Longwave Infrared Observation of Urban Landscapes. In *Technical Papers, American Society of Photogrammetry 47th Annual Meeting.* Washington, DC. Bethesda, MD: American Society for Photogrammetry and Remote Sensing.

Hatfield, J. L., J. P. Millard, and R. C. Goettelman. 1982. Variability of Surface Temperature in Agricultural Fields of Central California. *Photogrammetric Engineering and Remote Sensing,* Vol. 48, pp. 1319–1325.

Hudson, R. D. 1969. *Infrared System Engineering.* New York: Wiley, 642 pp.

Janza, F. J. 1975. Interaction Mechanisms. Chapter 4 in *Manual of Remote Sensing* (R. G. Reeves, ed.). Falls Church, VA: American Society of Photogrammetry, pp. 75–179.

Leckie, D. G. 1982. An Error Analysis of Thermal Infrared Line-Scan Data for Quantitative Studies. *Photogrammetric Engineering and Remote Sensing,* Vol. 48, pp. 945–954.

Lougeay, R. 1982. Landsat Thermal Imaging of Alpine Regions. *Photogrammetric Engineering and Remote Sensing,* Vol. 48, pp. 269–273.

Lowe, D. S. (ed.). 1975. Imaging and Nonimaging Sensors. Chapter 8 in *Manual of Remote Sensing* (R. G. Reeves, ed.). Falls Church, VA: American Society of Photogrammetry, pp. 367–398.

Moore, R. K. (ed.). 1975. Microwave Remote Sensors. Chapter 9 in *Manual of Remote Sensing* (R. G. Reeves, ed.). Falls Church, VA: American Society of Photogrammetry, pp. 399–538.

Pratt, D. A., and C. D. Ellyett. 1979. The Thermal Inertia Approach to Mapping of Soil Moisture and Geology. *Remote Sensing of Environment,* Vol. 8, pp. 151–168.

Price, J. C. 1978. Thermal Inertia Mapping: A New View of the *Earth. Journal of Geophysical Research,* Vol. 82, pp. 2582–2590.

Price, J. C. 1981. The Contribution of Thermal Data in Landsat Multispectral Classification. *Photogrammetric Engineering and Remote Sensing,* Vol. 47, pp. 229–236.

Sabins, F. 1969. Thermal Infrared Imaging and Its Application to Structural Mapping, Southern California. *Geological Society of American Geological Society of America Bulletin,* Vol. 80, pp. 397–404.

Schott, J. R., and W. J. Volchok. 1985. Thematic Mapper Infrared Calibration. *Photogrammetric Engineering and Remote Sensing,* Vol. 51, pp. 1351–1358.

Short, N. M., and L. M. Stuart. 1982. *The Heat Capacity Mapping Mission (HCMM) Anthology* (NASA Special Publication 465). Washington, DC: U.S. Government Printing Office, 264.

Toll, D. L. 1985. Landsat-4 Thematic Mapper Scene Characteristics of a Suburban and Rural Area. *Photogrammetric Engineering and Remote Sensing,* Vol. 51, pp. 1471–1482.

Watson, K. 1975. Geologic Applications of Thermal Infrared Images. *Proceedings, Institute of Electrical and Electronic Engineers,* Vol. 63, pp. 128–138.

Watson, K., S. Hummer-Miller, and D. L. Sawatzky. 1982. Registration of Heat Capacity Mapping Mission Day and Night Images. *Photogrammetric Engineering and Remote Sensing,* Vol. 48, pp. 263–268.

Weast, R. C. (ed.). 1986. *CRC Handbook of Chemistry and Physics.* Boca Raton, FL: CRC Press.

CHAPTER TEN

# Image Resolution

## 10.1. Introduction and Definitions

In very broad terms, *resolution* refers to the ability of a remote sensing system to record and display fine spatial, spectral, and radiometric detail. A working knowledge of resolution is essential for understanding both practical and conceptual aspects of remote sensing. Our understanding or lack of understanding of resolution may be the limiting factor in our efforts to use remotely sensed data, especially at coarse spatial resolutions.

For scientists with an interest in instrument design and performance, measurement of resolution is of primary significance in determining the optimum design and configuration of individual elements (e.g., specific lenses, detectors, or photographic emulsions) of a remote sensing system. Here our interest focuses upon understanding image resolution in terms of the entire remote sensing system, regardless of our interests in specific elements of the landscape. Whether our focus concerns soil patterns, geology, water quality, land use, or vegetation distributions, knowledge of image resolution is a prerequisite for understanding the information recorded on the images we examine.

The purpose of this chapter is to discuss image resolution as a separate concept in recognition of its significance throughout the field of remote sensing. It attempts to outline generally applicable concepts, without ignoring special and unique factors that apply in specific instances.

Estes and Simonett (1975) define resolution as "the ability of an imaging system . . . to record fine detail in a distinguishable manner" (p. 879). This definition includes several key concepts. The emphasis upon the imaging *system* is significant because in most practical situations it makes little sense to focus attention upon the resolving power of a single element of the system (e.g., the film) if another element (e.g., the camera lens) limits the resolution of the final image. "Fine detail" is, of course, a relative concept, as is the specification that detail be recorded in a "distinguishable" manner. These aspects of the definition emphasize that resolution can be clearly defined only by operational definitions applicable under specified conditions.

For the present, it is sufficient to note that there is a practical limit to the level of detail that can be acquired from a given aerial or satellite image. This limit we define informally as the "resolution" of the remote sensing system, although it must be recognized that image detail also depends upon the character of the scene that has been imaged, atmospheric conditions, illumination, and the experience and ability of the image interpreter.

Most individuals think of resolution as *spatial resolution*, the fineness of the spatial detail visible in an image. "Fine detail" in this sense means that small objects can be identified on an image. But other forms of resolution are equally important. *Radiometric resolution* can be defined as the ability of an imaging system to record many levels of brightness. *Coarse* radiometric resolution would record a scene using only a few brightness levels or a few bits (i.e., at very high contrast), whereas *fine* radiometric resolution would record the same scene using many levels of brightness. *Spectral resolution* denotes the ability of a sensor to define fine wavelength intervals. Hyperspectral sensors (Chapter 15) generate images composed of 200 or more narrowly defined spectral regions—these data represent an extreme of spectral resolution relative to TM or Landsat MSS images, which convey spectral information in only a few rather broad spectral regions.

Finally, *temporal resolution* is an important consideration in many applications. Remote sensing has the ability to record sequences of images, thereby representing changes in landscape patterns over time. The ability of a remote sensing system to record such a sequence at relatively close intervals generates a data set with *fine* temporal resolution. In contrast, systems that can record images of a given region only at infrequent intervals produce data at *coarse* temporal resolution. In some applications, such as flood or disaster mapping, temporal resolution is a critical characteristic that might override other desirable qualities. Clearly, those applications that attempt to monitor dynamic phenomena, such as news events, range fires, land-use changes, traffic flows, or weather-related events, will have an interest in temporal resolution.

In many situations, there are clear trade-offs between different forms of resolution. For example, in traditional photographic emulsions, increases in spatial resolving power are based upon decreased size of film grain, which produces accompanying decreases in radiometric resolution (i.e., the decreased sizes of grains in the emulsion portray a lower range of brightnesses). In other systems there are similar trade-offs. Increasing spatial detail requires, in scanning systems, a smaller IFOV (i.e., energy reaching the sensor has been reflected from a smaller ground area). If all other variables have been held constant, this must translate into decreased energy reaching the sensor; lower levels of energy mean that the sensor may record less "signal" and more "noise," thereby reducing the usefulness of the data. This effect can be compensated for by broadening the spectral window to pass more energy (i.e., decreasing spectral resolution) or by dividing the energy into fewer brightness levels (i.e., decreasing radiometric resolution). Of course, overall improvements can be achieved by improved instrumentation or by altering operating conditions (e.g., flying at a lower altitude). The general situation, however, seems to demand costs in one form of resolution for benefits achieved in another.

## 10.2. Target Variables

Observed spatial resolution in a specific image depends greatly upon the character of the scene that has been imaged. In complex natural landscapes, identification of the essential variables influencing detail observed in the image may be difficult, although many of the key factors can be enumerated. Contrast is clearly one of the most important influences upon spatial and radiometric resolution. *Contrast* can be defined as the difference in brightness between an object and its background. If other factors are held constant, high contrast favors recording of fine spatial detail; low contrast produces coarser detail. A black automobile imaged against a black asphalt background will be more difficult to observe than a white vehicle observed under the same conditions.

The significance of contrast as an influence on spatial resolution illustrates the interrelationships between the various forms of resolution and emphasizes the reality that no single element of system resolution can be considered in isolation from the others. It is equally important to distinguish between contrast in the original scene and that recorded on the image of that scene; the two may be related, but not necessarily in a direct fashion (see Sections 3.4 and 4.2). Also, it should be noted that contrast in the original scene is a dynamic quality that, for a given landscape, varies greatly from season to season (with changes in vegetation, snow cover, etc.), and within a single day (as angle and intensity of illumination change).

The *shape* of an object or feature is significant. *Aspect ratio* refers to the length of a feature in relation to its width. Usually long thin features, such as highways, railways, and rivers, tend to be visible on aerial imagery, even in circumstances when their widths are much less than the nominal spatial resolution of the imagery. *Regularity of shape* favors recording of fine detail. Features with regular shapes, such as cropped agricultural fields, tend to be recorded in fine detail, whereas complex shapes will be imaged in coarser detail.

The *number* of objects in a pattern also influences the level of detail recorded by a sensor. For example, the pattern formed by the number and regular arrangement of tree crowns in an orchard favors the imaging of the entire pattern in fine detail. Under similar circumstances, the crown of a single isolated tree might not be visible on the imagery.

*Extent and uniformity of background* contributes to resolution of fine detail in many distributions. For example, a single automobile in a large, uniform parking area or a single tree positioned in a large cropped field will be imaged in detail not achieved under other conditions.

## 10.3. System Variables

Remember that the resolution of individual sensors depends in part upon the design of that sensor and in part upon its operation at a given time. For example, resolution of an aerial photograph (Chapter 3) is determined by the quality of the camera lens, the choice of film, flying altitude, scale, and the design of the aerial camera. For scanning systems such as the Landsat MSS/TM (Chapter 6) or thermal scanners (Chapter 9), the IFOV determines many of the qualities of image resolution. The IFOV depends, of course, upon the optical system (the angular field of view) and the operating altitude. Speed of the scanning motion and movement of the vehicle that carries the sensor will also have their effects upon image quality. For active microwave sensors (Chapter 7), image resolution is determined by beamwidth (antenna gain), angle of observation, wavelength, and other factors discussed previously.

## 10.4. Operating Conditions

For all remote sensing systems, the operating conditions, including flying altitude and ground speed, are important elements influencing the level of detail in the imagery. Atmospheric conditions can be included as important variables, especially for satellite and high-altitude imagery.

## 10.5. Measurement of Resolution

### *Ground Resolved Distance*

Perhaps the simplest measure of spatial resolution is *ground resolved distance* (GRD), defined simply as the dimensions of the smallest objects recorded on an image. One might speak of the resolution of an aerial photograph as being "2 m," meaning that objects of that size and larger could be detected and interpreted from the image in question. Smaller objects presumably would not be resolved, and therefore would not be interpretable.

Such measures of resolution may have utility as a rather rough suggestion of usable detail, but must be recognized as having only a very subjective meaning. The objects and features that compose the landscape vary greatly in size, shape, contrast with background, and pattern. Usually we have no means of relating a given estimate of GRD to a specific problem of interest. For example, the spatial resolution of U.S. Department of Agriculture (USDA) 1:20,000 black-and-white aerial photography is often said to be "about 1 m," yet typically one can easily detect on these photographs the painted white lines in parking lots and highways; these lines may be as narrow as 6–9 in. Does this mean that the resolution of this photography should be assessed as 6 in. rather than 1 m? Only if we are interested in the interpretation of long thin features that exhibit high contrast with their background could we accept such an estimate as useful. Similarly, the estimate of 1 m may be inappropriate for many applications.

### *Line Pairs per Millimeter*

*Line pairs per millimeter* (LPM) is a means of standardizing the characteristics of targets used to assess image resolution. Essentially, it is a means of quantifying, under controlled conditions, the estimate of GRD by using a standard target, positioned on the ground, that is imaged by the remote sensing system under specified operating conditions.

Although many targets have been used, the resolution target designed by the U.S. Air Force (USAF) has been a standard for a variety of studies (Figure 10.1). This target consists of parallel black lines positioned against a white background. The width of spaces between lines is

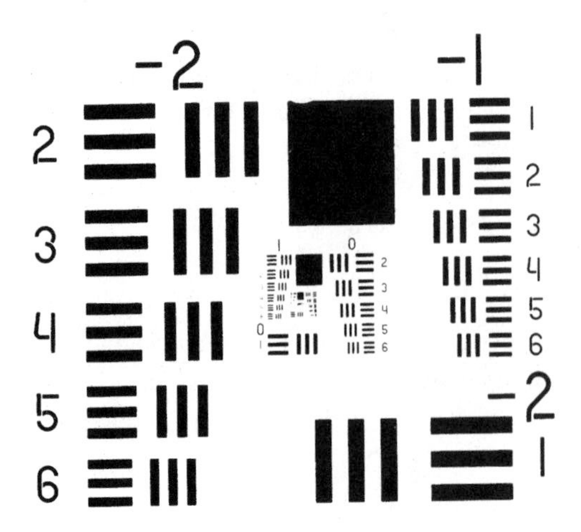


**FIGURE 10.1.** Bar target used in resolution studies.

equal to that of the lines themselves; their length is five times their width. As a result, a block of three lines and the two white spaces that separate them form a square. This square pattern is reproduced at varied sizes to form an array consisting of bars of differing widths and spacings. Sizes are controlled to produce changes in spacing of the bars (spatial frequency) of 12%. Repetition of the pattern at differing scales assures that the image of the pattern will include at least one pattern so small that individual lines and their spaces will not be fully resolved.

If images of two objects are visually separated, they are said to be "spatially resolved." Images of the USAF resolution target are examined by an interpreter to find that smallest set of lines in which the individual lines are all completely separated along their entire length. The analyst measures the width of the image representation of one "line pair" (i.e., the width of the image of one line and its adjacent white space) (Figure 10.2). This measurement provides the basis for the calculation of the number of line pairs per millimeter (or any other length we may choose; "line pairs per millimeter" [LPM] is standard for many applications). For example, in Figure 10.2 the width of a line and its adjacent gap is measured to be 0.04 mm. From 1 line pair/0.04 mm we find a resolution of 25 LPM.

For aerial photography, this measure of resolution can be translated into GRD by the relationship

$$\text{GRD} = \frac{H}{(f)(R)} \qquad \text{(Eq. 10.1)}$$

where GRD is ground resolved distance, in meters; $H$ is the flying altitude above the terrain, in meters; $f$ is the focal length, in millimeters; and $R$ is the system resolution, in line pairs per millimeter.

Such measures have little predictable relationship to the actual size of landscape features that might be interpreted in practical situations because seldom will the features of interest have the same regularity of size, shape, and arrangement and the high contrast of the resolution target used to derive the measures. They are, of course, valuable as comparative measures for assessing the performance of separate systems under the same operating conditions or of a single system under different conditions.

Although the USAF target has been widely used, other resolution targets have been developed. For example, a colored target has been used to assess the spectral fidelity of color films (Brooke, 1974), and bar targets have been constructed with contrast ratios somewhat closer to conditions observed during actual applications. The USGS target is a large array painted on the roof of the USGS National Center, in Reston, Virginia, as a means of assessing aerial imagery under operational conditions from high altitudes. The array is formed from large bar targets,

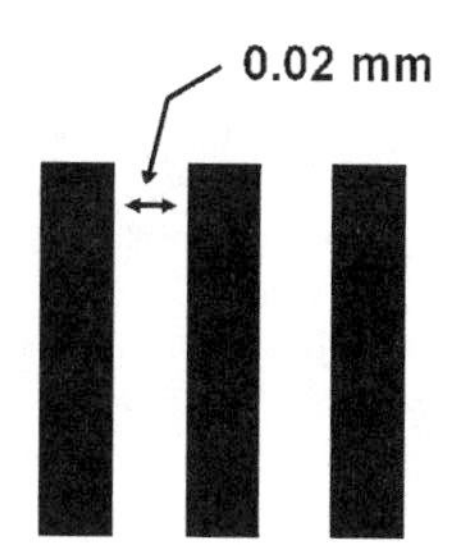


**FIGURE 10.2.** Use of the bar target to find LPM.

about 100 ft. in length, of known contrast, and a star target about 140 ft. in diameter designed for assessment of the resolution of nonphotographic sensors.

### *Modulation Transfer Function*

The *modulation transfer function* (MTF) records system response to a target array with elements of varying spatial frequency (i.e., unlike the bar targets described above, targets used to find MTFs are spaced at varied intervals). Often the target array is formed from bars of equal length spaced against a white background at intervals that produce a sinusoidal variation in image density along the axis of the target.

*Modulation* refers to changes in the widths and spacings of the target. Transfer denotes the ability of the imaging system to record these changes on the image—that is, to "transfer" these changes from the target to the image. Because the target is explicitly designed with spatial frequencies too fine to be recorded on the image, some frequencies (the high frequencies at the closest spacings) cannot be imaged. The "function" then shows the degree to which the image records specified frequencies (Figure 10.3).

Although the MTF is probably the "best" measure of the ability of an imaging system as a whole or of a single component of that system to record spatial detail, the complexity of the method prevents its routine use in many situations. The MTF can be estimated using simpler and more readily available targets, including the USAF target described above (Welch, 1971).

## 10.6. Mixed Pixels

As spatial resolution interacts with the fabric of the landscape, a special problem is created in digital imagery by those pixels that are not completely occupied by a single homogeneous category. The subdivision of a scene into discrete pixels acts to average brightnesses over the entire pixel area. If a uniform or relatively uniform land area occupies the pixel, then similar brightnesses are averaged, and the resulting digital value forms a reasonable representation of the brightnesses within the pixel. That is, the average value does not differ greatly from the values

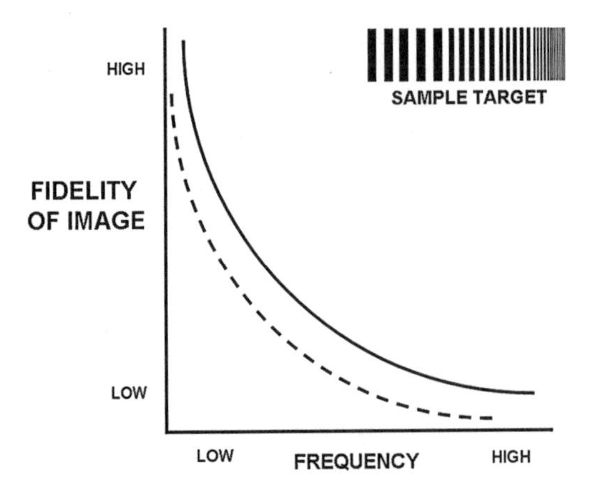


**FIGURE 10.3.** Modulation transfer function. The inset illustrates an example of a target used in estimating the modulation transfer function. The solid line illustrates a hypothecal modulation transfer for an image that records the pattern with high fidelity; the dashed line represents an image that cannot portray the high-frequency variation at the right side of the target.

that contribute to the average. However, when a pixel area is composed of two or more areas that differ greatly with respect to brightness, then the average is composed of several very different values, and the single digital value that represents the pixel may not accurately represent any of the categories present (Figure 10.4).

An important consequence of the occurrence of mixed pixels is that pure spectral responses of specific features are mixed together with the pure responses of other features. The mixed response sometimes known as a *composite signature* does not match the *pure signatures* that we wish to use to map the landscape. Note, however, that sometimes composite signatures can be useful because they permit us to map features that are too complex to resolve individually.

Nonetheless, mixed pixels are also a source of error and confusion. In some instances, the digital values from mixed pixels may not resemble any of the several categories in the scene; in other instances, the value formed by a mixed pixel may resemble those from other categories in the scene but not actually present within the pixel, an especially misleading kind of error.

Mixed pixels occur often at the edges of large parcels, or along long linear features, such as rivers or highways, where contrasting brightnesses are immediately adjacent to one another (Figure 10.5). The edge, or border, pixels then promote opportunities for errors in digital classification. Scattered occurrences of small parcels (such as farm ponds observed at the resolution of the Landsat MSS) may produce special problems because they may be represented *only* by mixed pixels, and the image analyst may not be aware of the presence of the small areas of high contrast because they occur at subpixel sizes. An especially difficult situation can be created by landscapes composed of many parcels that are small relative to the spatial resolution of the sensor. A mosaic of such parcels will create an array of digital values, all formed by mixed pixels (Figure 10.6).

It is interesting to examine the relationships between the numbers of mixed pixels in a given scene and the spatial resolution of the sensor. Studies have documented the increase in numbers of mixed pixels that occurs as spatial resolution decreases. Because the numbers, sizes, and shapes of landscape parcels vary greatly with season and geographic setting, there can be no generally applicable conclusions regarding this problem. Yet examination of a few simple examples may help us understand the general character of the problem.

Consider the same contrived scene that is examined at several different spatial resolutions

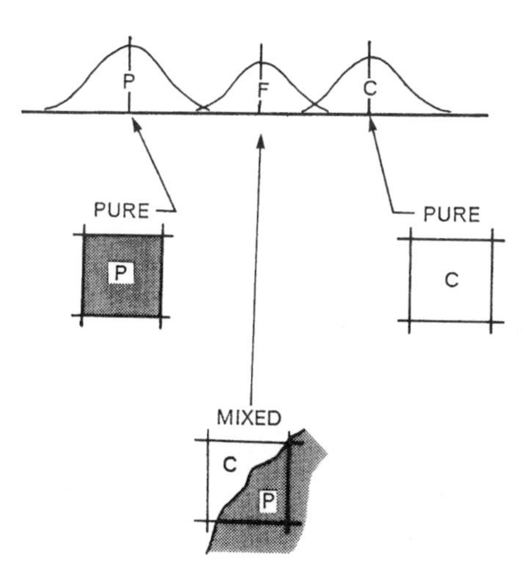


**FIGURE 10.4.** False resemblance of mixed pixels to a third category.

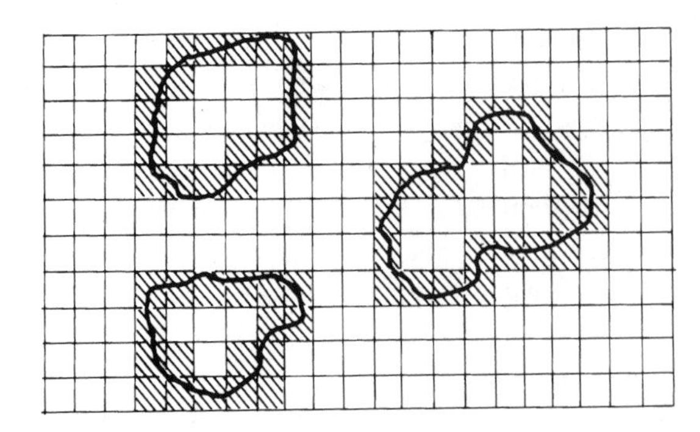

**FIGURE 10.5.** Edge pixels.

(Figure 10.7). This scene consists of two contrasting categories with two parcels of one superimposed against the more extensive background of the other. This image is then examined at four levels of spatial resolution; for each level of detail, pixels are categorized as "background," "interior," or "border." (Background and interior pixels consist only of a single category; border pixels are those composed of two categories.) A tabulation of proportions of the total in each category reveals a consistent pattern (Table 10.1). As resolution becomes coarser, the number of mixed pixels increases (naturally) at the expense of the number of pure background and pure interior pixels. In this example, interior pixels experience the larger loss, but this result is the consequence of the specific circumstances of this example, and is unlikely to reveal any generally applicable conclusions.

If other factors could be held constant, it would seem that fine spatial resolution would offer many practical advantages, including capture of fine detail. Note, however, the substantial increases in the total numbers of pixels required to achieve this advantage; note too that increases in the numbers of pixels produce compensating disadvantages, including increased costs. Also, this example does not consider another important effect often encountered as spatial resolution is increased: the finer detail may resolve features not recorded at coarser detail, thereby increasing, rather than decreasing, the proportions of mixed pixels. This effect may explain some of the results observed by Sadowski and Sarno (1976), who found that classification accuracy decreased as spatial resolution became finer.

Marsh et al. (1980) have reviewed strategies for resolving the percentages of components that compose the ground areas with mixed pixels. The measured digital value for each pixel is

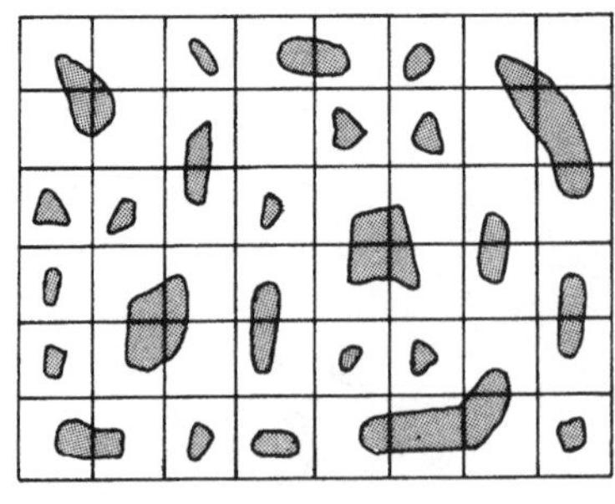


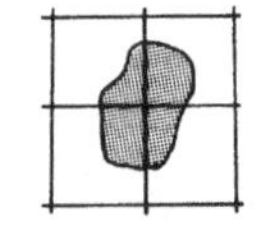


**FIGURE 10.6.** Mixed pixels generated by an image of a landscape composed of small pixels.

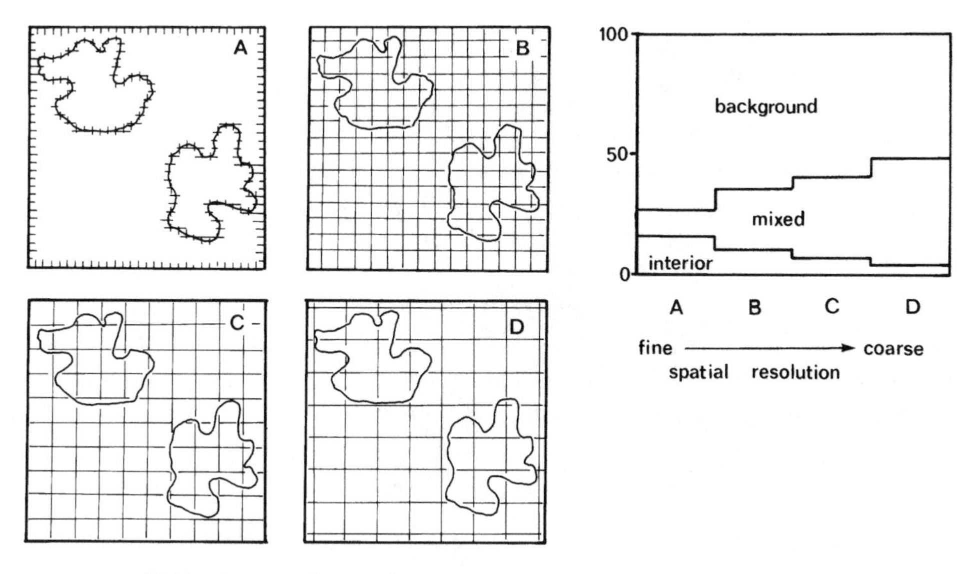


**FIGURE 10.7.** Influence of spatial resolution on proportions of mixed pixels.

**TABLE 10.1. Summary of Data Derived from Figure 10.7**

| Spatial resolution | | Total | Mixed | Interior | Background |
|---|---|---|---|---|---|
| Fine | A | 900 | 109 | 143 | 648 |
| | B | 225 | 59 | 25 | 141 |
| | C | 100 | 34 | 6 | 60 |
| Coarse | D | 49 | 23 | 1 | 25 |

determined by the brightnesses of distinct categories within that pixel area projected on the ground, as integrated by the sensor over the area of the pixel. For example, the projection of a pixel on the Earth's surface may encompass areas of open water (*W*) and forest (*F*). Assume that we know that (1) the digital value for such a pixel is "mixed," not "pure"; (2) the mean digital value for water in all bands is *i* ($W_i$); (3) the mean digital value for forest in all bands is ($F_i$); and (4) the observed value of the mixed pixel in all spectral bands is ($M_i$). We wish then to find the areal percentages *PW* and *PF* that contribute to the observed value $M_i$.

Marsh et al. (1980) outline several strategies for estimating *PW* and *PF* under these conditions. The simplest, if not the most accurate, is the weighted average method:

$$PW = (M_i - F_i)/(W_i - F_i) \quad \text{(Eq. 10.2)}$$

An example can be shown using the following data:

| | Band | | | |
|---|---|---|---|---|
| | 1 | 2 | 3 | 4 |
| Means for the mixed pixel ($M_i$): | 16 | 12 | 16 | 18 |
| Means for forest ($F_i$): | 23 | 16 | 32 | 35 |
| Means for water ($W_i$): | 9 | 8 | 0 | 1 |

Using Equation 10.2, the areal proportion of the mixed pixel composed of the water category can be estimated as follows:

Band 1: $P_W = (16 - 23)/(9 - 23) = -7/-14 = 0.50$
Band 2: $P_W = (12 - 16)/(8 - 16) = -4/-8 = 0.50$
Band 3: $P_W = (16 - 32)/(0 - 32) = -16/-32 = 0.50$
Band 4: $P_W = (18 - 35)/(1 - 35) = -17/-34 = 0.50$

Thus the mixed pixel is apparently composed of about 50% water and 50% forest. Note that in practice we may not know which pixels are mixed, and may not know the categories that might contribute to the mixture. Note also that this procedure may yield different estimates for each band. Other procedures, too lengthy for concise description here, may give more suitable results in some instances (Marsh et al., 1980).

## 10.7. Spatial and Radiometric Resolution: Simple Examples

Some of these effects can be illustrated by contrived examples prepared by manipulation of a black-and-white image of a simple agricultural scene (Figures 10.8 and 10.9). In each case, the upper left image shows the original image, followed by successive images contrived to show effects of coarser spatial resolution or coarser radiometric resolution.

Figure 10.8 shows a sequence in which the radiometric resolution is held constant, but the spatial resolution has been degraded by averaging brightnesses over increasingly large blocks of adjacent pixels, thereby decreasing the spatial resolution by factors of 5, 10, 25, and so on. At the image size shown here, the effects of lower resolutions are not visually obvious until rather substantial decreases have been made. As pixels are averaged over blocks of 25 pixels, the loss in detail is visible, and at levels of 50 and 100 pixels the image has lost the detail that permits

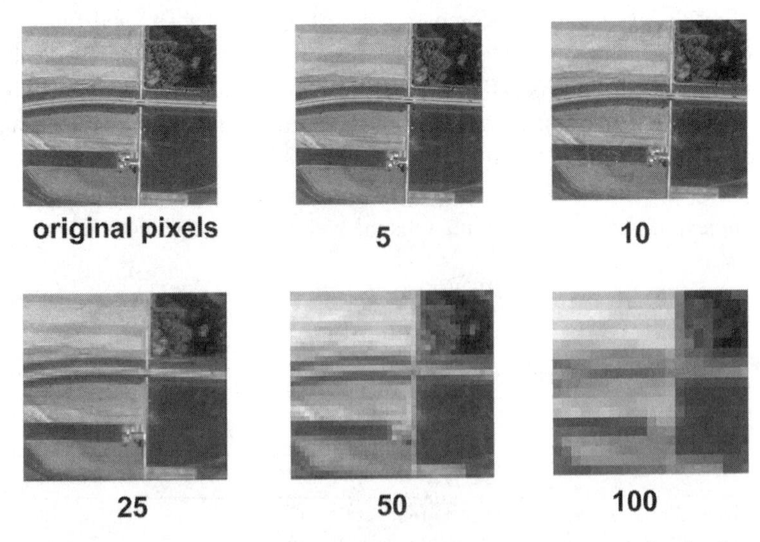


**FIGURE 10.8.** Spatial resolution. A contrived example in which radiometric resolution has been degraded by averaging brightnesses over increasingly larger blocks of adjacent pixels.

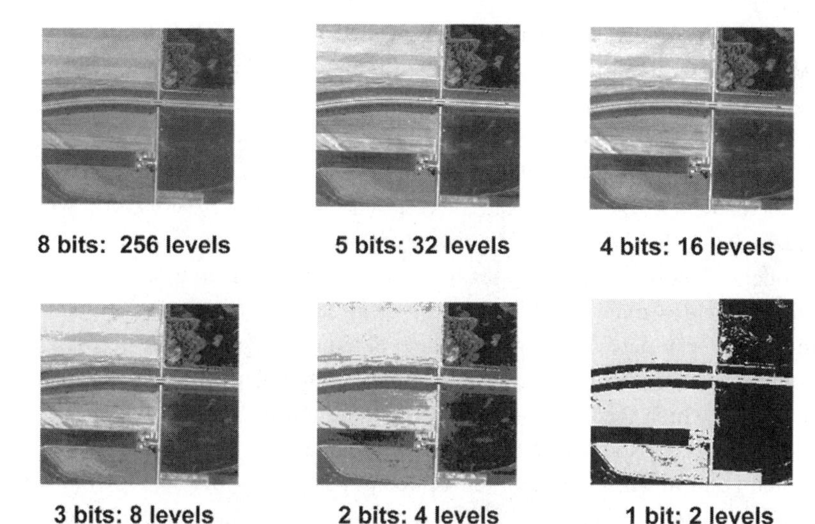


**FIGURE 10.9.** Radiometric resolution. A contrived example in which spatial resolution has been held constant while radiometric resolution has been reduced from 8 bits (256 brightness levels) to 1 bit (2 brightness levels).

recognition and interpretation of the meaning of the scene. Note also the effects of mixed pixels at high-contrast edges, where the sharp changes in brightness blur into broad transitional zones at the coarser levels of detail. At the coarse level of detail of Figure 10.8's final image, only the vaguest suggestions of the original pattern are visible.

For Figure 10.9, the spatial resolution has been held constant while the radiometric resolution has been reduced from 8 bits (256 brightness levels) to 1 bit (2 brightness levels). As is the case for the previous example, changes are visually evident only after the image has been subjected to major reductions in detail. At higher radiometric resolutions, the scene has a broad range of brightness levels, allowing the eye to see subtle features in the landscape. At lower resolutions, the brightnesses have been represented at only a few levels (ultimately, only black and white), depicting only the coarsest outline of the scene.

Keep these examples in mind as you examine images from varied sensors. Any sensor is designed to record specific levels of spatial and radiometric resolution—these qualities determine its effectiveness in portraying features on the landscape. Broader levels of resolution may be adequate for rather coarse-textured landscapes. For example, even very low levels of resolution may be effective for certain well-defined tasks, such as separating open water from land, detecting cloud patterns, and the like. Note also that finer resolution permits more subtle distinctions, but also records detail that may not be relevant to the task at hand, and may tend to complicate analysis.

## 10.8. Interactions with the Landscape

Although most discussions of image resolution tend to focus upon sensor characteristics, understanding the significance of image resolution in the application sciences requires assessment of the effect of specific resolutions upon images of specific landscapes or classes of landscapes. For example, relatively low resolution may be sufficient for recording the essential features of landscapes with rather coarse fabrics (e.g., the broad-scale patterns of the agricultural fields of the North American Great Plains), but inadequate for imaging complex landscapes composed of many small parcels with low contrast.

Podwysocki's studies (1976a, 1976b) of field sizes in the major grain-producing regions of the world is an excellent example of the systematic investigation of this topic. His research can be placed in the context of the widespread interest in accurate forecasts of world wheat production in the years that followed large international wheat purchases by the Soviet Union in 1972. Computer models of biophysical processes of crop growth and maturation could provide accurate estimates of yields (given suitable climatological data), but estimates of total production also require accurate estimates of planted acreage. Satellite imagery would seem to provide the capability to derive the required estimates of area plowed and planted. Podwysocki attempted to define the extent to which the spatial resolution of the Landsat MSS would be capable of providing the detail necessary to provide the required estimates.

He examined Landsat MSS scenes of the United States, China, the Soviet Union, Argentina, and other wheat-producing regions, sampling fields for measurements of length, width, and area. His data are summarized by frequency distributions of field sizes for samples of each of the world's major wheat-producing regions. (He used his samples to find the Gaussian distributions for each of his samples, so he was able to extrapolate the frequency distributions to estimate frequencies at sizes smaller than the resolution of the MSS data.) Cumulative frequency distributions for his normalized data reveal the percentages of each sample that equal or exceed specific areas (Figure 10.10). For example, the curve for India reveals that 99% (or more) of this sample were at least 1 ha in size, and that all were smaller than about 100 ha (247 acres), and that we can expect the Indian wheat fields to be smaller than those in Kansas. These data, and others presented in his study, provide the basis for evaluating the effectiveness of a given resolution in monitoring features of specified sizes. This example is especially instructive because it emphasizes not only the differences in average field size in the different regions

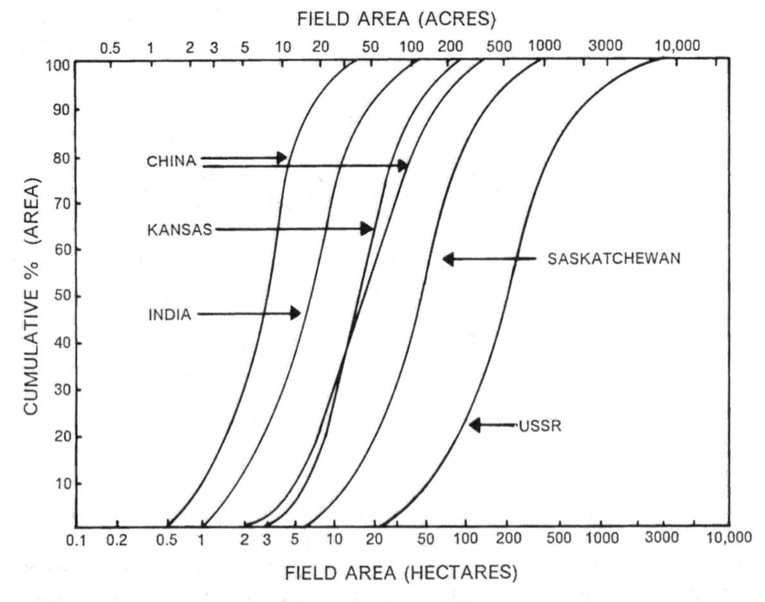


**FIGURE 10.10.** Field size distributions for selected wheat-producing regions. From Podwysocki, 1976a.

**PLATE 1.** Color and color infrared aerial photographs, Torch Lake, Michigan. Top: natural color aerial photograph. Bottom: color infrared aerial photograph of approximately the same region. Photographs courtesy of U.S. Environmental Protection Agency.

**PLATE 2.** Digital color infrared image, Oahu, Hawaii. This digital color infrared image conveys information comparable to the color infrared photographs discussed in Section 3.6. This image was acquired using a proprietary digital camera that provides 1-m detail, and positional accuracy that meets U.S. National Digital Orthophoto specifications. Image courtesy of Emerge, Inc. 

**PLATE 3.** High-altitude aerial photograph, Corpus Christi, Texas, January 31, 1995. Image provided by U.S. Geological Survey.

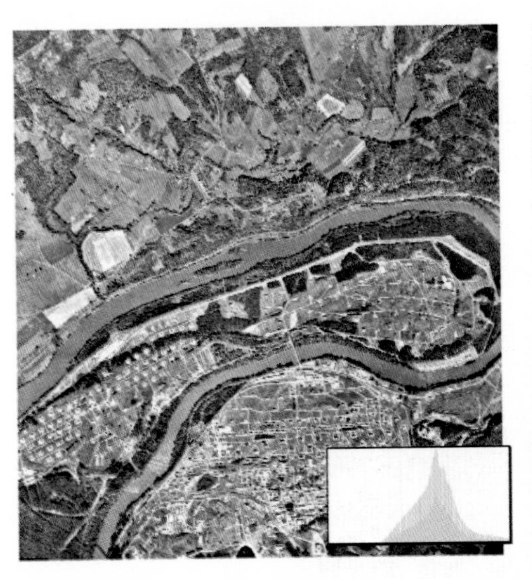

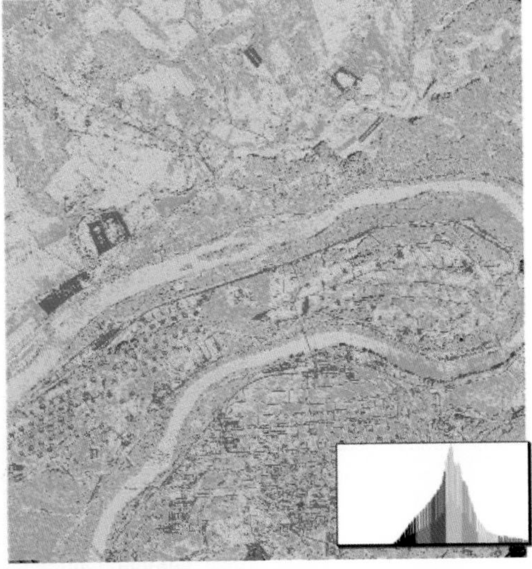

**PLATE 4.** Density slicing. Density slicing is a form of image enhancement that assigns colors to brightness ranges within a black-and-white image. Colors do not add new information, but can assist in enhancing information present within the original image. Left: the original black-and-white image, with a histogram representing the frequencies of brightness values. Right: the same image after application of the density slice, with a histogram colored to represent approximate assignment of colors. The analyst can adjust the levels represented by colors to emphasize those features of greatest interest.

**PLATE 5.** Thematic mapper color composite (bands 1, 3, and 4). This image forms a color composite providing spectral information approximately equivalent to a color infrared aerial photograph (compare with Plates 1 and 2). This image shows the same region depicted by Figures 6.9, 6.10, 6.13, and 6.14: New Orleans, Louisiana, September 16, 1982. Scene ID: 40062-15591. Image provided by U.S. Geological Survey.

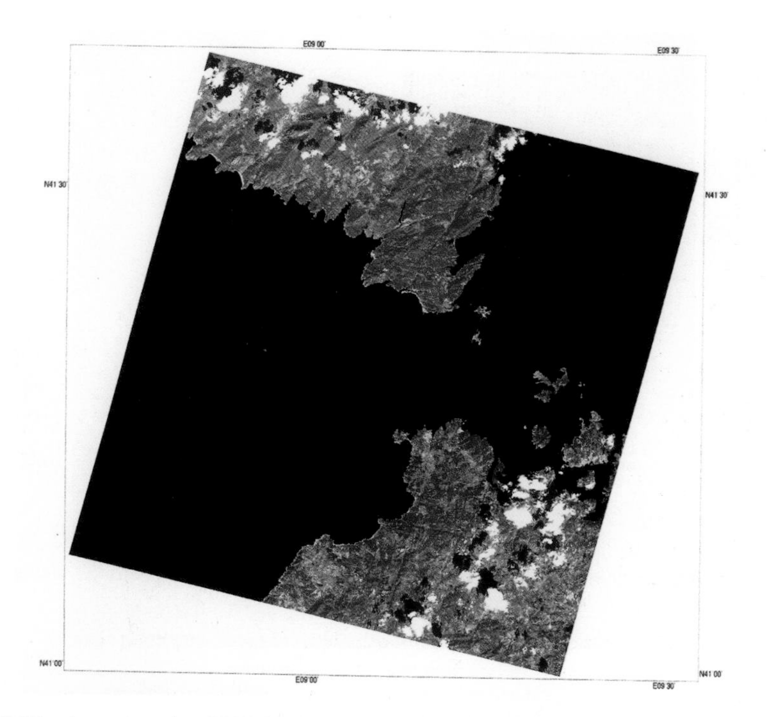


**PLATE 6.** SPOT color composite. SPOT 3 HRV image of the Strait of Bonifacio, separating Corsica (on the north) from Sardinia (on the south), September 27, 1993. © CNES/SPOT Image. Reproduced by permission.

**PLATE 7.** SeaWifS image, February 26, 2000. This image shows a broad-scale view of the western coast of Africa. The tan-colored swirl is dust carried in winds blowing to the west from eastern Africa. Dust is lifted to altitudes as high as 15,000 ft.; some of it will be blown to the west, causing air quality alerts in Caribbean cities and contributing to sedimentation that threatens the health of coral reefs. 

**PLATE 8.** IKONOS multispectral imagery, Montego Bay, Jamaica. Example of high-resolution optical imagery, nominal resolution about 1 m, that conveys spatial detail comparable to some aerial photographs (compare with Plate 3). 

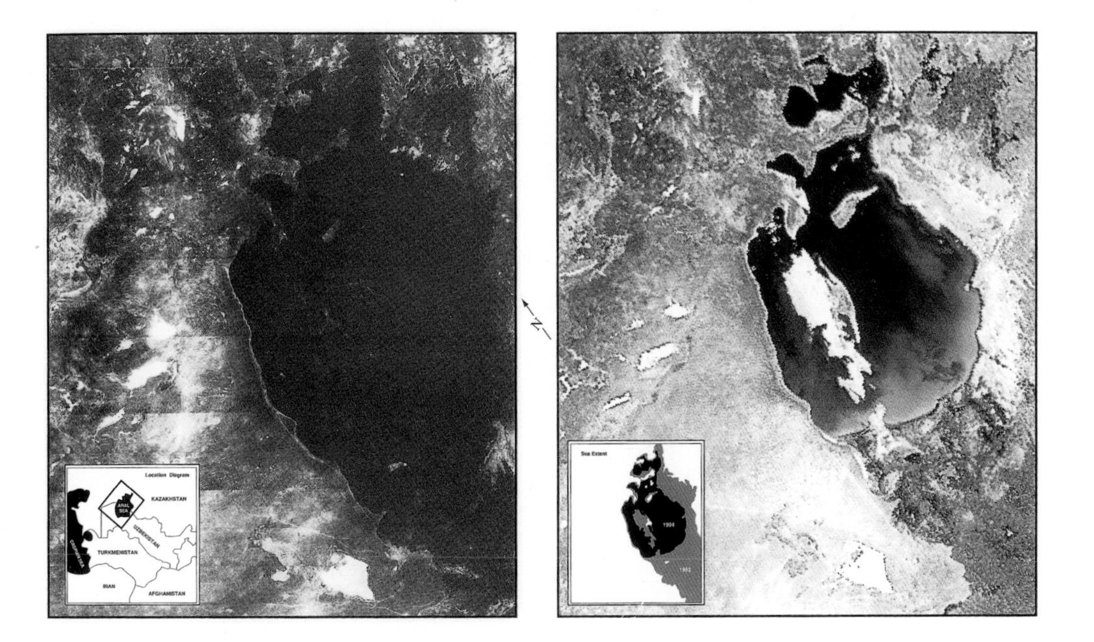


**PLATE 9.** Aral Sea shrinkage, 1962–1994. The left image is a mosaic of CORONA images from 1962, revealing the extent of the Aral Sea. The right image, acquired by AVHRR in 1994, has been prepared to the same scale and orientation as the CORONA image. The comparison reveals the dramatic decrease in limits of the Aral Sea during this interval, due mainly to large-scale diversion for irrigation of waters feeding the lake. Images prepared by U.S. Central Intelligence Agency and U.S. Geological Survey.

**PLATE 10.** Multifrequency SAR image, Barstow, California. This spaceborne SAR image depicts the Mojave Desert near Barstow, California—an area 46 mi. × 30 mi. (75 km × 48 km) using several frequencies and polarizations. It was acquired by spaceborne imaging radar-C/X-Band Synthetic Aperture Radar (SIR-C/X-SAR) onboard the space shuttle Endeavour, April 1994. Colors represent frequencies and polarizations as follows: red is C-band, HV; green is L-band, HH; and blue is the ratio of C-band to L-band, both HH. The city of Barstow is the pinkish area in the lower left center, on the banks of the Mojave River, visible as the diagonal light blue sinuous line. The orange circular and rectangular patches are irrigated agricultural fields. Radar data have been used to evaluate effects of irrigation on soil stability and land degradation. Sparsely vegetated areas of sand and fine gravel appear blue, while rocky hills and rougher gravel deposits appear mostly in shades of orange and brown. Image courtesy of NASA and the Jet Propulsion Laboratory, Pasadena, California (P-4747308).

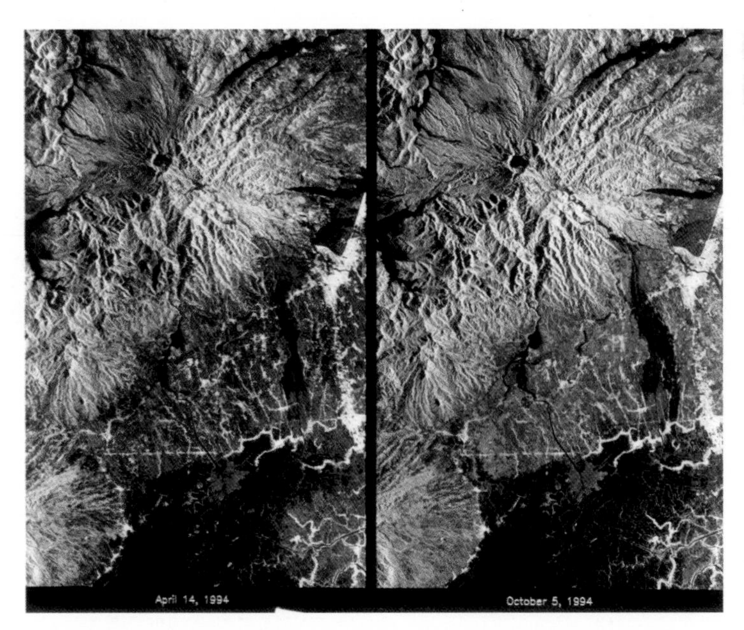


**PLATE 11.** Shuttle imaging radar images (SIR-C/X-SAR) of the region near Mount Pinatubo, Philippines. The left image was acquired on April 14, 1994, the right image on October 5, 1994. Colors are assigned according to the responses of different surfaces to different wavelengths and polarizations, and discussed in Chapter 7, pp. 229–230. Red regions reveal ash deposits from the 1991 eruption; dark strips in the central right region of each image are mudflows caused by movement of the ash during heavy rains. Significant changes have occurred in the interval between the two images. Images courtesy of NASA and the Jet Propulsion Laboratory, Pasadena, California (P44729).

**PLATE 12.** Radar interferometry, Honolulu, Hawaii, February 2000. The Shuttle Radar Topography Mission (SRTM) uses two radar antennae to observe the same topography from separate positions. Processing of the two sets of microwave returns generates topographic data, shown here using the "wrapped color" bands to signify elevation classes. SRTM data have been processed to provide elevation data for a very large proportion of the Earth's surface. Image provided by the Jet Propulsion Laboratory (PIA02718).

**PLATE 13.** Radar interferometry, Missouri River flood plain. This is an image of an area along the Missouri River that experienced severe flooding in the summer of 1993. It represents an area of about 5 km × 10 km (3.1 mi. × 6.2 mi.) centered south of the town of Glasgow in central Missouri.

The Missouri River is seen as the dark curving band on the left. The predominantly blue area is the river's floodplain, completely inundated during the 1993 flood. Low areas are shown in purple; intermediate elevations in blue, green, and yellow; and the highest areas as orange. The total elevation range is 85 m (279 ft.). The yellow and orange area on the right shows higher topography typical of neighboring upland regions. Dark streaks in the floodplain are agricultural areas damaged by levee failures. The region enclosed by the C-shaped bend in the river in the upper part of the image is Lisbon Bottoms; a burst of water from a failed levee on the north side of Lisbon Bottoms scoured a deep channel, which shows as a purple band. As flood waters receded, deposits of sand and silt were left behind, now seen as dark, smooth streaks. Yellow areas in the blue, near the river, are clumps of trees on slightly higher ground in the floodplain. Microwave energy is sensitive to the rough, irregular form of the tree tops, so they are represented as higher (yellow) than the fields.

The image was acquired by the NASA/JPL Topographic Synthetic Aperture Radar system (TOPSAR) in August 1994. The elevations are obtained by radar interferometry, in which the radar signals are transmitted by one antenna, and echoes are received by two antennae aboard the aircraft. The image represents radar backscatter at C-band, as vertically transmitted and received polarization. Radar and interferometry processing was performed at the Jet Propulsion Laboratory; image generation was performed at Washington University, St. Louis (P-46311).

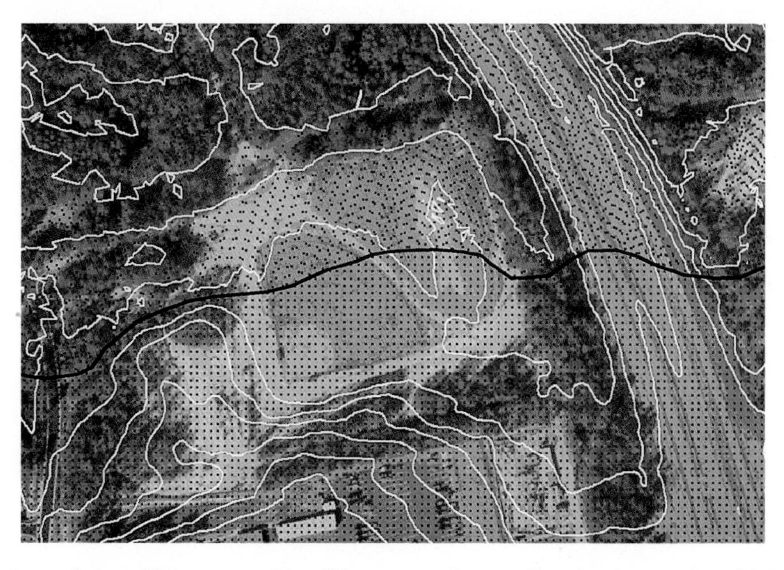

**PLATE 14.** Interpolation of lidar returns. Most lidar systems do not collect data in a regular grid. In the top half of this image, the irregular spacing of the lidar ground points can be seen, as well as the jerky, irregular nature of contours created directly from the lidar. In the bottom half, a regular grid has been interpolated from the lidar data, generalizing and smoothing the surface. These data were collected from an altitude of 12,000 ft., at a nominal post spacing of 15 ft., and a vertical accuracy of about 14 in. RMSE. The contours are at a 5-ft. interval. Image courtesy of EarthData.

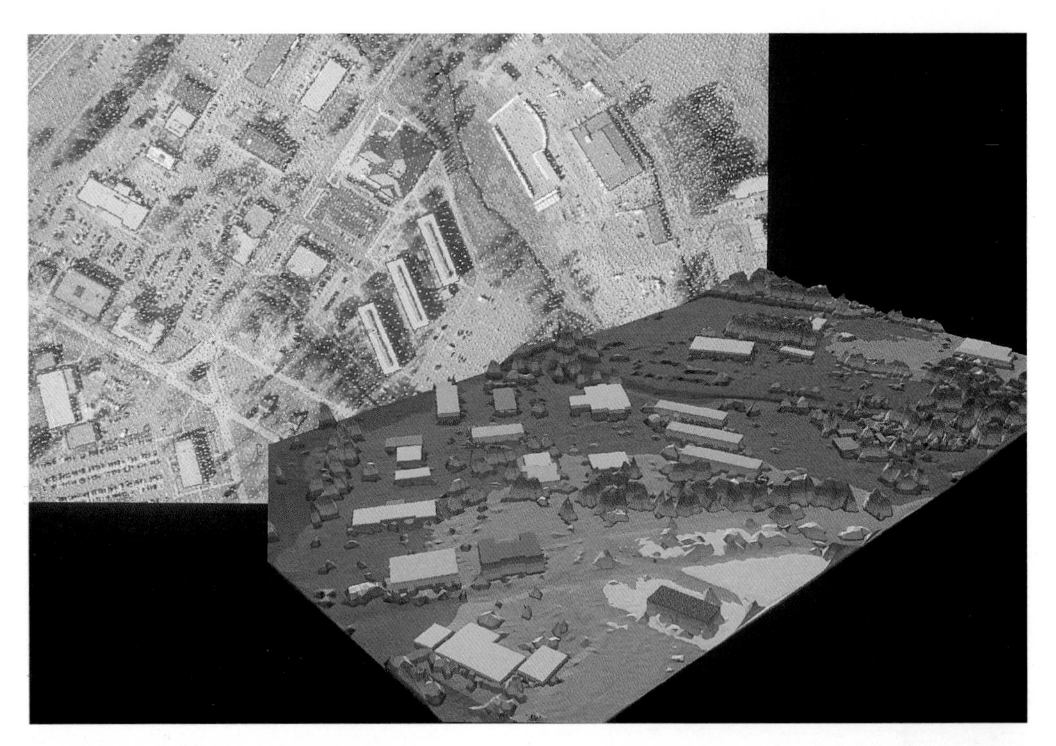

**PLATE 15.** Lidar data used to model the built environment. Data from a multiple-return lidar system can be used to develop advanced surface models. The first reflective surface is used to determine the height of the buildings, while the last returns are used to construct a ground elevation model. Footprints of buildings from the existing GIS are incorporated to define feature edges and enhance the final 3D surface. These data were collected from an altitude of 12,000 ft., at a nominal post spacing of 15 ft., and a vertical accuracy of about 14 in. RMSE. Images courtesy of EarthData.

**PLATE 16.** Glideslope surface estimated from lidar data. This image shows a rapidly growing use for lidar data. Unlike traditional photogrammetric techniques, lidar captures the first reflective surface. This allows for quick and easy assessment of a variety of obstructions that cannot be collected as easily from stereo photography. In this example, the imaginary floor of a runway approach corridor is depicted in 3D as a flat blue surface. First return lidar data are rendered in their true location, and the penetrations of the trees through the "glideslope" can be identified. GIS can be used to map these obstructions for removal. These data were collected from an altitude of 12,000 ft., at a nominal post spacing of 15 ft., and a vertical accuracy of about 14 in. RMSE. Image courtesy of EarthData.

**PLATE 17.** Thermal images of residential structures showing the thermal properties of their separate elements. (March 1979, 11:00 P.M.; air temperature = 2°C.) Top: the original thermal image (white = hot; dark = cool). Bottom: a color-coded version of the same data shown at the top, with each change in color representing a change in temperature of about 1.5°C, over a range from 2°C to 11°C. Colors represent increasing temperatures in the sequence black, magenta, blue, cyan, green, yellow, red, and white. Thermal images provided by Daedalus Enterprises, Inc., Ann Arbor, Michigan.

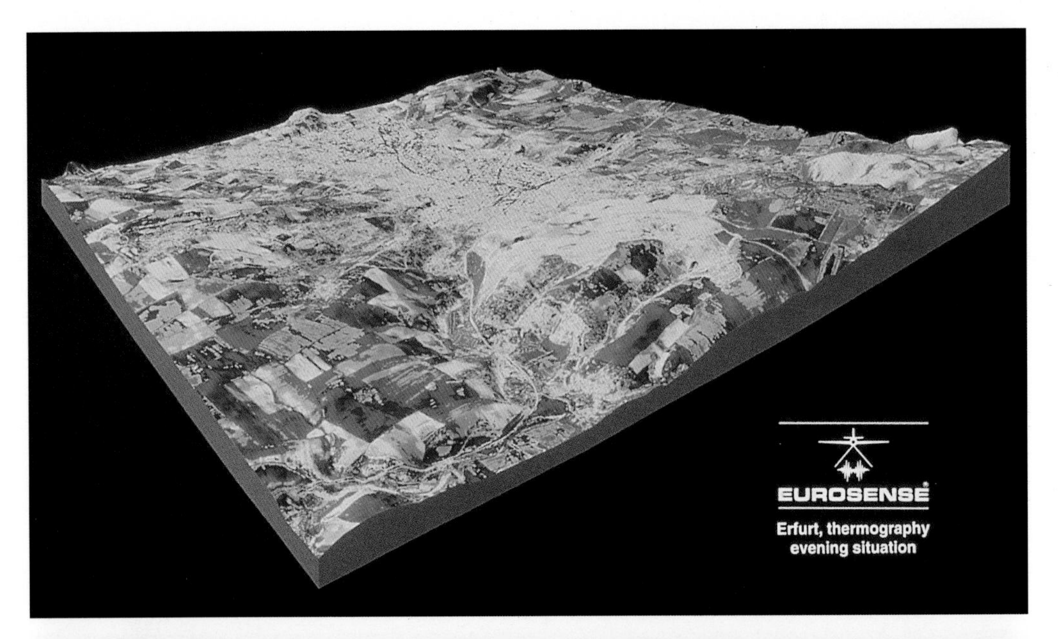


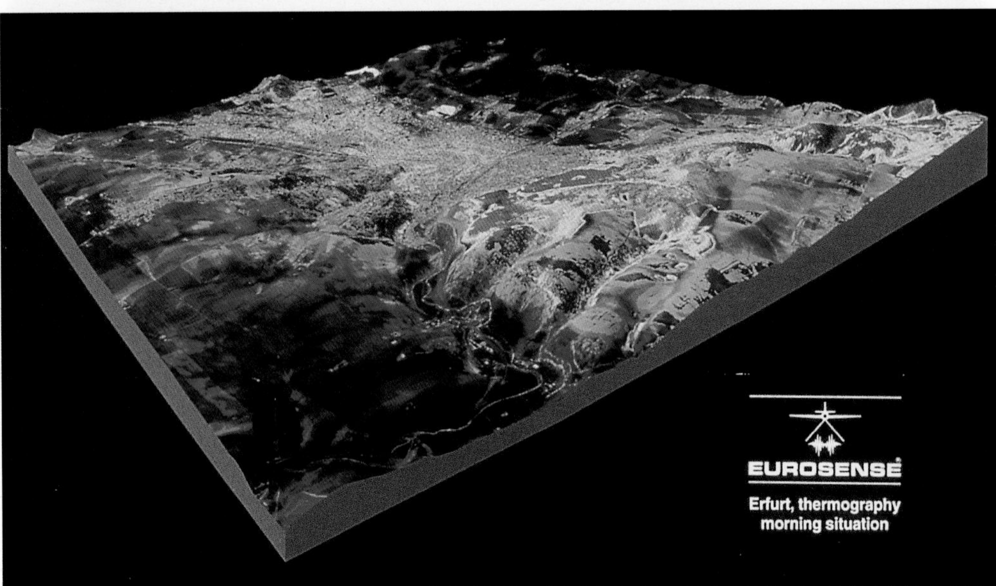


**PLATE 18.** Landscape near Erfurt, Germany, as observed by a thermal scanner in the region between 8.5 and 12 m. The thermal data have been geometrically and radiometrically processed, then superimposed over digital elevation data, with reds and yellows assigned to represent warmer temperatures and blues and greens to represent cooler temperatures. See Chapter 9 for extended discussion. Images courtesy of EUROSENSE-BELFOTOP N.W. and Daedalus Enterprises Inc., Ann Arbor Michigan.

**PLATE 19.** Applications Explorer Mission 1 (HCMM) image of northeastern United States and southeastern Canada, May 11, 1978. This image depicts broad-scale temperature patterns of the eastern seaboard of the United States, from the southern tip of Long Island at the upper right to the North Carolina coast at the lower left. Colors represent different temperatures, with the coolest temperatures (black, dark purple, and green) representing the cold surfaces of high-altitude clouds at the upper left (northeastern Ohio) and the waters near the North Carolina, Virginia, Maryland, Delaware, and New Jersey coastlines. The sinuous green line at the lower right is a flow of warmer water related to the Gulf Stream The urban areas of the Northeast are recognizable as the irregular white spots—these are the urban heat islands of New York, Philadelphia, Baltimore, Washington, and Norfolk, formed by the concentrated and intense human activities and changes to the landscape in larger urban/suburban regions. Yellows and browns represent temperature variations related to rural land use and topography. Compare with similar images at different scales represented in Plates 26 and 28. Image courtesy of NASA.

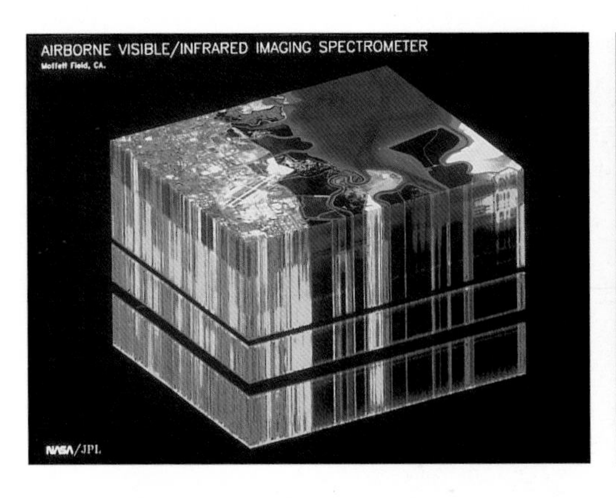


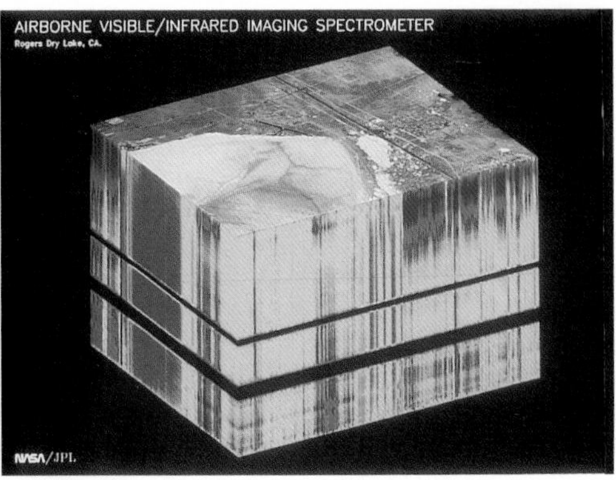


**PLATE 20.** Image cubes. Two examples of AVIRIS images displayed in image cube format. The top portion shows the image as recorded at the shortest wavelength, in conventional format with a perspective view. Sides of the cubes represent edges of each image as recorded in 220 channels, with the longest wavelength at the bottom (see Figure 15.4). Changes in colors along the sides of the cube represent changes in the spectral properties of surfaces as wavelength changes; the horizontal black stripes represent wavelengths of strong atmospheric absorption. Left: Moffet Field, California. Right: Rogers Dry Lake, California. Images courtesy of NASA and the Jet Propulsion Laboratory.

**PLATE 21.** Example of a ratio image. The left image shows a portion of a SPOT HRV image of a landscape in Rondonia, Brazil. The long axis of the rectangle marked on the image represents 2 km on the ground. Red represents forested regions; blue the dry grasses of open pasture; and lighter reds and oranges brushy regrowth in open land. The right image shows the same image as represented using the normalized difference vegetation index, as discussed in the text. White represents pixels dominated by healthy vegetation. Black represents very sparsely vegetated pixels, or, in this region, pixels of grassy regions with seasonally dry vegetation. Gray regions are pixels with varying amounts of brushy regrowth. © CNES/SPOT Image. Reproduced by permission.

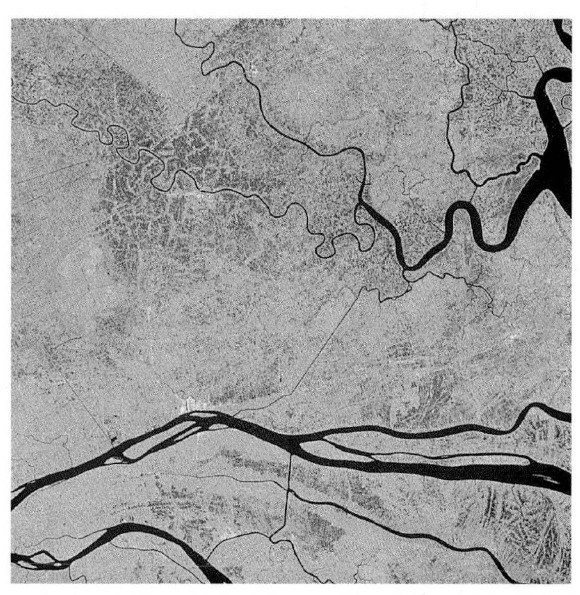

**PLATE 22.** Rice crop-monitoring, Mekong Delta, Vietnam. The Mekong Delta is the largest and most productive rice-growing region in Vietnam. Therefore, observing and monitoring change in this region is vital. This task can be accomplished by comparing RADARSAT scenes acquired over the same area through time. As the appearances of rice fields evolve, from tilled, through smoothed, seeded, flooded, maturing, senescent, to harvested, areas of change can be observed and located by using color: red: May 12, 1999 (Wide 2, Descending); green: May 22, 1999 (Standard 7, Descending); blue: June 15, 1999 (Wide 3, Descending). RADARSAT data © Canadian Space Agency/Agence Spatiale Canadienne 1999. Received by the Canada Centre for Remote Sensing. Processed and distributed by RADARSAT International.

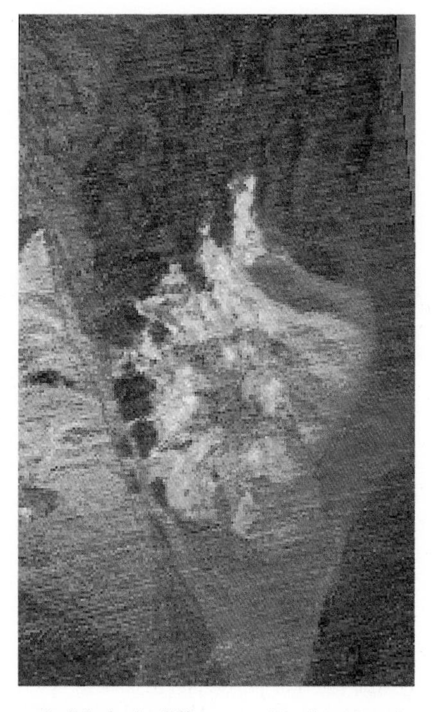

**PLATE 23.** Band ratios used to study lithologic differences, Cuprite, Nevada, mining district. Variations in color represent variations in ratios between spectral bands, as discussed in the text. Image courtesy of GeoSpectra Corporation. From Vincent et al. (1984, p. 226).

**PLATE 24.** Chesapeake Bay, as photographed June 8, 1991 from the space shuttle. Land areas appear in silhouette due to underexposure to record hydrographic features. As a result, the land–water interface is very sharply delineated. Note, for example, the coastal marshes discussed in Chapter 19 (section 19.10, Figures 19.19–19.22). The Chesapeake Bay occupies the left-hand portion of the image; Norfolk, Virginia, and associated naval facilities occupy the lower left corner, but are not visible due to the underexposure. The Chesapeake Bay Bridge–Tunnel is visible as the dark line at the mouth of the bay. Just to the right of the Bridge–Tunnel, a curved front separates the waters of the bay from those of the Atlantic. The coastal waters of the Virginia coast are visually distinct from those of the bay, due to currents, surface oils, temperature, and wind and wave patterns. Near this front, wakes of several ships are visible. Photo courtesy of NASA (STS-40-614-047).

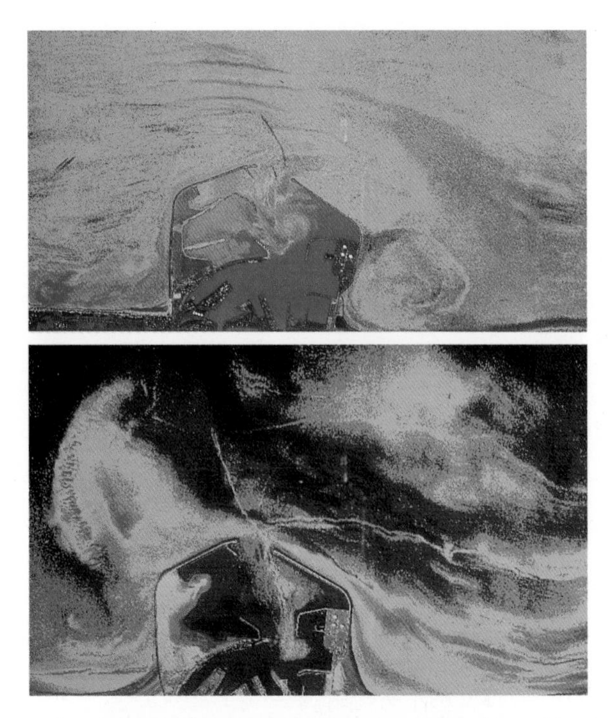

**PLATE 25.** Belgian port of Zeebrugge. The port and nearby waters were observed by the Daedalus digital multispectral Scanner using 12 channels in the visible, near infrared, and thermal spectra. Image data have been processed to provide correct geometry and radiometry, and, with the use of on-site observations, analyzed to reveal sediment content within the upper 1 m of the water column. Reds and yellows indicate high sediment content; blues and greens represent clearer water. The top image shows conditions at low tide; the bottom image conditions at high tide. Images courtesy of EUROSENSE-BELFOTOP N.W. and Daedalus Enterprises Inc., Ann Arbor, Michigan.

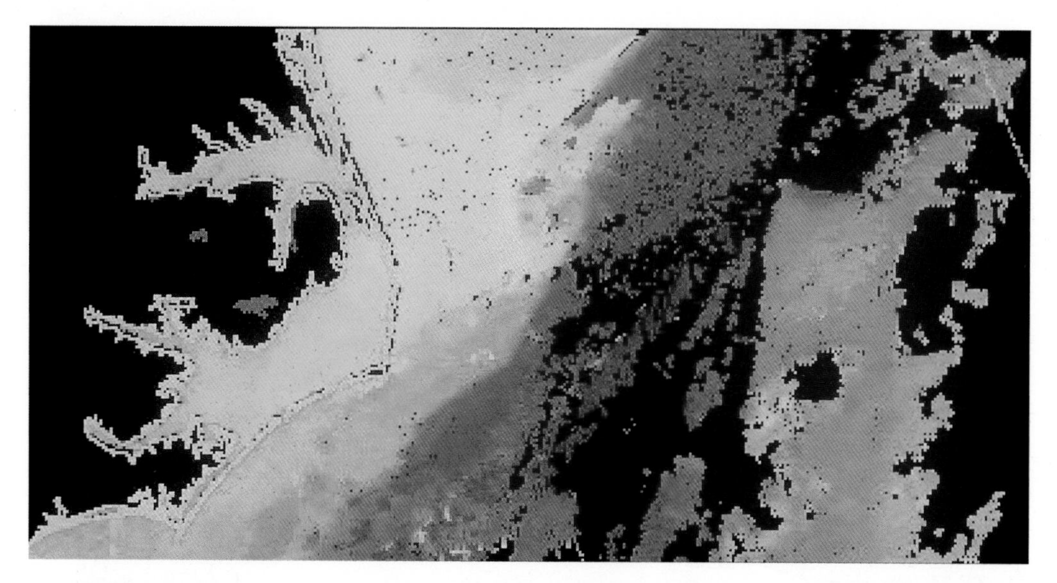

**PLATE 26.** AVHRR sea-surface temperature, Duck, North Carolina. Black areas at the left represent the land areas of coastal North Carolina. The irregularly shaped black regions at the right are clouds. The colors represent varying sea-surface temperatures. See Plates 19 and 28 for examples at other scales. NOAA image.

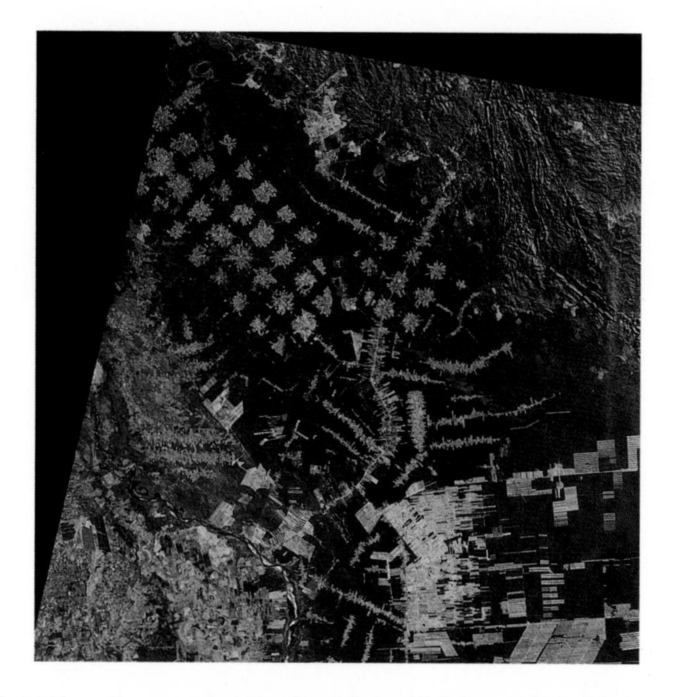

**PLATE 27.** Landsat TM quarter scene depicting Santa Rosa del Palmar, Bolivia (northwest of Santa Bruise), July 1992, TM bands 3, 4, and 5. Here four distinct land-use patterns are visible. The landscape in the southeastern portion of the scene is dominated by broad-scale mechanized agriculture practiced by a Mennonite community. The upper right (northeastern) region of the image is a mountainous area occupied by a diminishing population of Indians, who practice a form of slash-and-burn agriculture, visible here as the dispersed patches of light green. In the images's northwest and central regions, the national government has encouraged broad-scale clearing of forest for agriculture following the construction of highways into the forest. In the upper left, these clearings appear as light green spots aligned northwest to southeast; at the center of each patch is a central facility providing colonists with fertilizers, pesticides, and staples. In the image's southwestern corner, the complex field pattern reflects an established agricultural landscape occupied by Japanese immigrants. GEOPIC image courtesy of Earth Satellite Corporation, Rockville, Maryland. Used by permission.

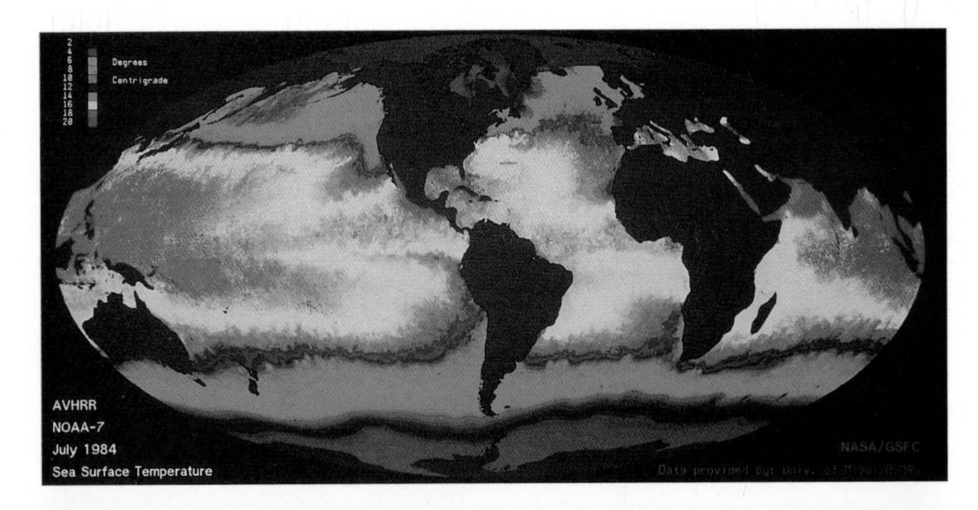


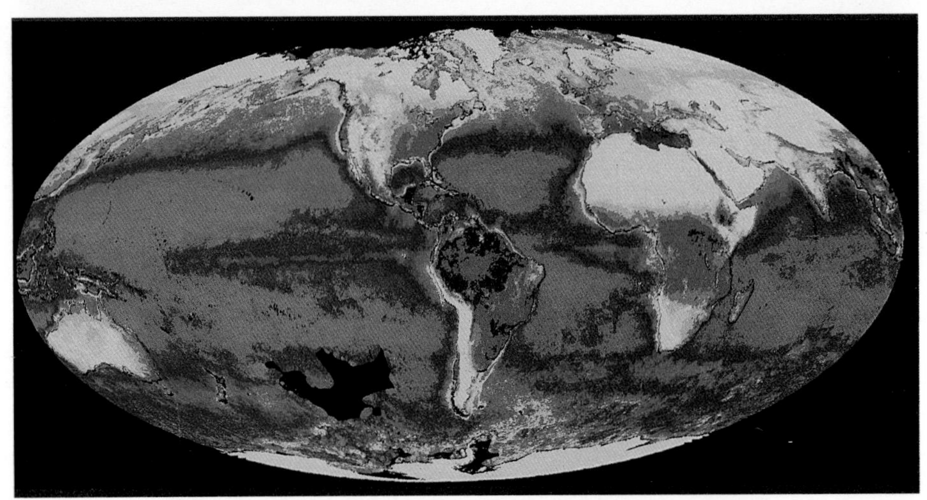

**PLATE 28.** Images depicting global remote sensing data. Top: AVHRR data representing sea surface temperature compiled from data collected during July 1984. The warmest surfaces are shown as red; in decreasing sequence, oranges, yellows, greens, and finally blues represent the coolest temperatures. Bottom: composite data from two separate NASA satellites combined to represent the Earth's biosphere. In the oceans, colors represent phytoplankton concentrations (highest values in browns and yellows, lowest values in the purples and dark blues). On land, greens represent dense vegetation, yellows depict sparsely vegetated regions. Images prepared by NASA Goddard Space Flight Center.

(shown in Figure 10.10 by the point where each curve crosses the 50% line), but also the differences in variation of field size between the varied regions (shown in Figure 10.10 by the slopes of the curves).

In a different analysis of relationships between sensor resolution and landscape detail, Simonett and Coiner (1971) examined 106 sites in the United States, each selected to represent a major land-use region. Their study was conducted prior to the launch of Landsat 1 with the objective of assessing the effectiveness of MSS spatial resolution in recording differences between major land-use regions in the United States. Considered as a whole, their sites represent a broad range of physical and cultural patterns in the 48 coterminous states.

For each site they simulated the effects of imaging with low-resolution imagery by superimposing grids over aerial photographs, with grid dimensions corresponding to ground distances of 800, 400, 200, and 100 ft. Samples were randomly selected within each site. Each sample consisted of the *number* of land-use categories within cells of each size, and thereby formed a measure of landscape diversity, as considered at several spatial resolutions. For example, those landscapes that show only a single land-use category at the 800-ft. resolution have a very coarse fabric, and would be effectively imaged at the low resolution of satellite sensors. Those landscapes that have many categories within the 100-ft. grid are so complex that very fine spatial resolution would be required to record the pattern of landscape variation. Their analysis grouped sites according to their behavior at various resolutions. They reported that natural landscapes appeared to be more susceptible than man-made landscapes to analysis at the relatively coarse resolutions of the Landsat MSS.

Welch and Pannell (1982) examined Landsat MSS (bands 2 and 4) and Landsat 3 RBV images (in both pictorial and digital formats) to evaluate their suitability as sources of landscape information at levels of detail consistent with a map scale of 1:250,000. Images of three study areas in China provided a variety of urban and agricultural landscapes for study, representing a range of spatial detail and a number of geographical settings. Their analysis of modulation transfer functions reveals that the RBV imagery represents an improvement in spatial resolution of about 1.7 over the MSS imagery, and that Landsat 4 TM provided an improvement of about 1.4 over the RBV (for target:background contrasts of about 1.6:1).

Features appearing on each image were evaluated with corresponding representations on 1:250,000 maps in respect to size, shape, and contrast. A numerical rating system provided scores for each image based upon the numbers of features represented and the quality of the representations on each form of imagery. MSS images portrayed about 40–50% of the features shown on the usual 1:250,000 topographic maps. MSS band 2 was the most effective for identification of airfields; band 4 performed very well for identification and delineation of water bodies. Overall, the MSS images achieved scores of about 40–50%. RBV images attained higher overall scores (50–80%), providing considerable improvement in representation of high-contrast targets, but little improvement in imaging of detail in fine-textured urban landscapes. The authors concluded that spatial resolutions of MSS and RBV images were inadequate for compilation of usual map detail at 1:250,000.

## 10.9. Summary

This chapter highlights the significance of image resolution as a concept that extends across many aspects of remote sensing. Although the special and unique elements of any image must always be recognized and understood, many of the general aspects of image resolution can assist us in understanding how to interpret remotely sensed images.

Although there has long been an intense interest in measuring image resolution, especially in photographic systems, it is clear that much of our more profound understanding has been developed through work with satellite scanning systems such as the Landsat MSS. Such data were of much coarser spatial resolution than any studied previously. As more and more attention was focused upon their analysis and interpretation (Chapter 12), it was necessary to develop a better understanding of image resolution and its significance for specific tasks. Now much finer resolution data are available, but we can continue to develop and apply our knowledge of image resolution to maximize our ability to understand and interpret these images.

## Review Questions

1. Most individuals are quick to appreciate the advantages of fine resolution. However, there may well be *disadvantages* to fine-resolution data, relative to data of coarser spatial, spectral, and radiometric detail. Suggest what some of these effects might be.

2. Imagine that the spatial resolution of the digital remote sensing system is increased from about 80 m to 40 m. List some of the consequences, assuming that image coverage remains the same. What would be some of the consequences of *decreasing* detail from 80 m to 160 m?

3. You examine an image of the U.S. Air Force resolution target and determine that the image distance between the bars in the smallest pair of lines is 0.01 mm. Find the LPM for this image. Find the LPM for an image in which you measure the distance to be 0.04 mm. Which image has finer resolution?

4. For each object or feature listed below, discuss the characteristics that will be significant in our ability to resolve the object on a remotely sensed image. Categorize each as "easy" or "difficult" to resolve clearly. Explain.

   a. A white car parked alone in an asphalt parking lot.
   b. A single tree in a pasture.
   c. An orchard.
   d. A black cat in a snow-covered field.
   e. Painted white lines on a crosswalk across an asphalt highway.
   f. Painted white lines on a crosswalk across a concrete highway.
   g. A pond.
   h. A stream.

5. Write a short essay describing how spatial resolution, spectral resolution, and radiometric resolution are interrelated. Is it possible to increase one kind of resolution without influencing the others?

6. Review Chapters 1–9 to identify the major features that influence spatial resolution of images collected by the several kinds of sensors described. Prepare a table to list these factors in summary form.

7. Explain why some objects might be resolved clearly in one part of the spectrum yet resolved poorly in another portion of the spectrum.

8. Although the U.S. Air Force resolution target is very useful for evaluating some aspects of remotely sensed images, it is not necessarily a good indication of the ability of a remote sensing system to record patterns that are significant for environmental studies. List some of the reasons this might be true.
9. Describe ideal conditions for achieving maximum spatial resolution.

## References

Badhwar, G. B. 1984a. Automatic Corn–Soybean Classification Using Landsat MSS Data: II. Early Season Crop Proportion Estimation. *Remote Sensing of Environment,* Vol. 14, pp. 31–37.

Badhawr, G. D. 1984b. Use of Landsat-Derived Profile Features for Spring Small-Grains Classification. *International Journal of Remote Sensing,* Vol. 5, pp. 783–797.

Badhwar, G. D., J. G. Carnes, and W. W. Austen. 1982. Use of Landsat-Derived Temporal Profiles for Corn–Soybean Feature Extraction and Classification. *Remote Sensing of Environment,* Vol. 12, pp. 57–79.

Badhwar, G. D., C. E. Garganti, and F. V. Redondo. 1987. Landsat Classification of Argentina Summer Crops. *Remote Sensing of Environment,* Vol. 21, pp. 111–117.

Brooke, R. K. 1974. *Spectral/Spatial Resolution Targets for Aerial Imagery* (Technical Report ETL-TR-74-3). Ft. Belvior, VA: U.S. Army Engineer Topographic Laboratories, 20 pp.

Chhikara, R. S. 1984. Effect of Mixed (Boundary) Pixels on Crop Proportion Estimation. *Remote Sensing of Environment,* Vol. 14, pp. 207–218.

Crapper, P. F. 1980. Errors Incurred in Estimating an Area of Uniform Land Cover Using Landsat. *Photogrammetric Engineering and Remote Sensing,* Vol. 46, pp. 1295–1301.

Estes, J. E., and D. S. Simonett. 1975. Fundamentals of Image Interpretation. Chapter 14 in *Manual of Remote Sensing* (R. G. Reeves, ed.). Bethesda, MD: American Society for Photogrammetry and Remote Sensing, pp. 869–1076.

Ferguson, M. C., G. D. Badhwar, R. S. Chhikara, and D. E. Pitts. 1986. Field Size Distributions for Selected Agricultural Crops in the United States and Canada. *Remote Sensing of Environment,* Vol. 19, pp. 25–45.

Hall, F. G., and G. D. Badhwar. 1987. Signature Extendable Technology: Global Space-Based Crop Recognition. *IEEE Transactions on Geoscience and Remote Sensing,* Vol. GE-25, pp. 93–103.

Hallum, C. R., and C. R. Perry. 1984. Estimating Optimal Sampling Unit Sizes for Satellite Surveys. *Remote Sensing of Environment,* Vol. 14, pp. 183–196.

Hyde, R. F., and N. J. Vesper. 1983. Some Effects of Resolution Cell Size on Image Quality. *Landsat Data Users Notes,* Issue 29, pp. 9–12.

Latty, R. S., and R. M. Hoffer. 1981. Computer-Based Classification Accuracy Due to the Spatial Resolution Using Per-Point versus Per-Field Classification Techniques. In *Proceedings, 7th International Symposium on Machine Processing of Remotely Sensed Data.* West Lafayette, IN: Laboratory for Applications of Remote Sensing, pp. 384–393.

MacDonald, D. E. 1958. Resolution as a Measure of Interpretability. *Photogrammetric Engineering,* Vol. 24, No. 1, pp. 58–62.

Markham, B. L., and J. R. G. Townshend. 1981. Land Cover Classification Accuracy as a Function of Sensor Spatial Resolution. In *Proceedings, 15th International Symposium on Remote Sensing of Environment.* Ann Arbor: University of Michigan Press, pp. 1075–1090.

Marsh, S. E., P. Switzer, and R. J. P. Lyon. 1980. Resolving the Percentage of Component Terrains within Single Resolution Elements. *Photogrammetric Engineering and Remote Sensing,* Vol. 46, pp. 1079–1086.

Pitts, D. E., and G. Badhwar. 1980. Field Size, Length, and Width Distributions Based on LACIE Ground Truth Data. *Remote Sensing of Environment,* Vol. 10, pp. 201–213.

Podwysocki, M. H. 1976a. *An Estimate of Field Size Distribution for Selected Sites in the Major Grain Producing Countries* (Publication No. X-923-76-93). Greenbelt, MD: Goddard Space Flight Center, 34 pp.

Podwysocki, M. H. 1976b. *Analysis of Field Size Distributions: LACIE Test Sites 5029, 5033, 5039, Anwhei Province, People's Republic of China* (Publication No. X-923-76-145). Greenbelt, MD: Goddard Space Flight Center, 8 pp.

Potdar, M. B. 1993. Sorghum Yield Modeling Based on Crop Growth Parameters Determined from Visible and Near-IR Channel NOAA AVHRR Data. *International Journal of Remote Sensing,* Vol. 14, pp. 895–905.

Sadowski, F., and J. Sarno. 1976. *Forest Classification Accuracy as Influenced by Multispectral Scanner Spatial Resolution* (Report for Contract NAS9-14123: NASA). Houston, TX: Lyndon Baines Johnson Space Center.

Salmonowicz, P. H. 1982. USGS Aerial Resolution Targets. *Photogrammetric Engineering and Remote Sensing,* Vol. 48, pp. 1469–1473.

Simonett, D. S., and J. C. Coiner. 1971. Susceptibility of Environments to Low Resolution Imaging for Land Use Mapping. In *Proceedings, 7th International Symposium on Remote Sensing of Environment.* Ann Arbor: University of Michigan Press, pp. 373–394.

Tucker, C. J. 1980. Radiometric Resolution for Monitoring Vegetation: How Many Bits Are Needed? *International Journal of Remote Sensing,* Vol. 1, pp. 241–254.

Wehde, M. E. 1979. Spatial Quantification of Maps or Images: Cell Size or Pixel Size Implication. In *Joint Proceedings, American Society of Photogrammetry and American Congress of Surveying and Mapping.* Bethesda, MD: American Society for Photogrammetry and Remote Sensing, pp. 45–65.

Welch, R. 1971. Modulation Transfer Functions. *Photogrammetric Engineering,* Vol. 47, pp. 247–259.

Welch, R., and C. W. Pannell. 1982. Comparative Resolution of Landsat 3 MSS and RBV Images of China. *Photogrammetric Record,* Vol. 10, pp. 575–586.

PART THREE

# ANALYSIS

CHAPTER ELEVEN

# Preprocessing

## 11.1. Introduction

In the context of digital analysis of remotely sensed data, preprocessing refers to those operations that are preliminary to the main analysis. Typical preprocessing operations could include (1) radiometric preprocessing to adjust digital values for the effect of a hazy atmosphere, and/or (2) geometric preprocessing to bring an image into registration with a map or another image. Once corrections have been made, the data can then be subjected to the primary analyses described in subsequent chapters. Thus preprocessing forms a preparatory phase that, in principle, improves image quality as the basis for later analyses that will extract information from the image.

It should be emphasized that although certain preprocessing procedures are frequently used, there can be no definitive list of "standard" preprocessing steps because each project requires individual attention and some preprocessing decisions may be a matter of personal preference. Furthermore, quality of image data vary greatly, so some data may not require the preprocessing that would be necessary in other instances. Also, preprocessing changes data. We may assume that such changes are beneficial, but the analyst should remember that preprocessing can create artifacts that are not immediately obvious. As a result, the analyst should tailor preprocessing to the data at hand and the needs of specific projects, using only those preprocessing operations essential to obtain a specific result.

Preprocessing includes a wide range of operations, from the very simple to extremes of abstractness and complexity. Most can be categorized into one of three groups: (1) feature extraction, (2) radiometric corrections, and (3) geometric corrections. Although there are far too many preprocessing methods to discuss in detail here, we will illustrate some of the principles important for each group.

## 11.2. Feature Extraction

In the context of image processing, the term *feature extraction* (or *feature selection*) has specialized meaning. "Features" are not geographical features, visible on an image, but are rather "statistical" characteristics of image data: individual bands or combinations of band values that carry information concerning systematic variation within the scene. Thus feature extraction could also be known as "information extraction," isolation of components within multispectral data that are most useful in portraying the essential elements of an image. In theory, discarded data contain noise and errors present in original data. Thus feature extraction may increase accuracy. In addition, feature extraction reduces the number of spectral channels, or bands, that must be analyzed, thereby reducing computational demands. After feature selection is complete, the analyst works with fewer but more potent channels. The reduced data set may convey almost as much information as does the complete data set. Feature selection may increase speed and reduce costs of analysis.

Multispectral data, by their nature, consist of several channels of data. Although some images may have as few as 3, 4, or 7 channels (Chapter 6), other image data may have many more, possibly 200 or more channels (Chapter 17). With so much data, processing of even modest-sized images requires considerable time. In this context, feature selection assumes considerable practical significance, as image analysts wish to reduce the amount of data while retaining effectiveness and/or accuracy.

Our examples here are based on TM data, which provide enough channels to illustrate the concept, but are compact enough to be reasonably concise (Table 11.1). A variance–covariance matrix shows interrelationships between pairs of bands; some pairs show rather strong correlations—for example, bands 1 and 3 and 2 and 3 both show correlations above 0.9. High correlation between pairs of bands means that the values in the two channels are closely related. Thus, as values in channel 2 rise or fall, so do those in channel 3; one channel tends to duplicate information in the other. Feature selection attempts to identify, then remove, such duplication so that the data set can include maximum information using the minimum number of channels.

For example, for data represented by Table 11.1, bands 3, 5, and 6 might include almost as much information as the entire set of seven channels because band 3 is closely related to bands 1 and 2, band 5 is closely related to bands 4 and 7, and band 6 carries information largely unrelated to any others. Therefore, the discarded channels (1, 2, 4, and 7) each resemble one of the channels that have been retained. So a simple approach to feature selection discards unneeded bands, thereby reducing the number of channels. Although this kind of selection can be used as a kind of rudimentary feature extraction, typically feature selection is a more complex process based upon statistical interrelationships between channels.

A more powerful approach to feature selection applies a method of data analysis called *principal components analysis* (PCA). This presentation offers only a superficial description of PCA, as more complete explanation requires the level of detail provided by Davis (2002) and

**TABLE 11.1. Correlation Matrix for Seven Bands of a TM Scene**

| | Correlation matrix | | | | | | |
|---|---|---|---|---|---|---|---|
| | 1 | 2 | 3 | 4 | 5 | 6 | 7 |
| 1. | 1.00 | | | | | | |
| 2. | 0.92 | 1.00 | | | | | |
| 3. | 0.90 | 0.94 | 1.00 | | | | |
| 4. | 0.39 | 0.59 | 0.48 | 1.00 | | | |
| 5. | 0.49 | 0.66 | 0.67 | 0.82 | 1.00 | | |
| 6. | 0.03 | 0.08 | 0.02 | 0.18 | 0.12 | 1.00 | |
| 7. | 0.67 | 0.76 | 0.82 | 0.60 | 0.90 | 0.02 | 1.00 |

others. In essence, PCA identifies the optimum linear combinations of the original channels that can account for variation of pixel values within an image. Linear combinations are of the form

$$A = C_1 X_1 + C_2 X_2 + C_3 X + C_4 X_4 \quad \text{(Eq. 11.1)}$$

where $X_1$, $X_2$, $X_3$, and $X_4$ are pixel values in four spectral channels, and $C_1$, $C_2$, $C_3$, and $C_4$ are coefficients applied individually to the values in the respective channels. A represents a transformed value for the pixel. Assume, as an example, that $C_1 = 0.35$, $C_2 = -0.08$, $C_3 = 0.36$, and $C_4 = 0.86$. For a pixel with $X_1 = 28$, $X_2 = 29$, $X_3 = 21$, $X_4 = 54$, the transformation assumes a value of 61.48. Optimum values for coefficients are calculated by a procedure that ensures that the values they produce account for maximum variation within the entire data set. Thus this set of coefficients provides the maximum information that can be conveyed by any single channel formed by a linear combination of the original channels. If we make an image from all the values formed by applying this procedure to an entire image, we generate a single band of data that provides an optimum depiction of the information present within the four channels of the original scene.

The effectiveness of this procedure depends, of course, upon calculation of the optimum coefficients. Here our description must be, by intention, abbreviated because calculation of the coefficients is accomplished by methods described by upper-level statistics texts or discussions such as those of Davis (2002) and Gould (1967). For the present, the important point is that PCA permits identification of a set of coefficients that concentrates maximum information in a single band.

The same procedure also yields a second set of coefficients that will yield a second set of values (we could represent this as the *B* set, or *B* image) that will be a less effective conveyor of information, but will still represent variation of pixels within the image. In all, the procedure will yield seven sets of coefficients (one set for each band in the original image), and therefore will produce seven sets of values, or bands (here denoted as *A, B, C, D, E, F,* and *G*), each in sequence conveying less information than the preceding band. Thus in Table 11.2 transformed channels I and II (each formed from linear combinations of the seven original channels) together account for about 93% of the total variation in the data, whereas channels III–VII together account for only about 7% of the total variance. The analyst may be willing to discard the variables that convey 7% of the variance as a means of reducing the number of channels. The analyst still retains 93% of the original information in a much more concise form. Thus feature selection reduces the size of the data set by eliminating replication of information.

The effect is easily seen in Figure 11.1, which shows transformed data for a subset of a Landsat TM scene. Images PC I and PC II are the most potent; PC III, PC IV, PC VI, and PC VII show the decline in information content, such that the images for the higher PC (such as PCs V and VI in Figure 11.1) record artifacts of system noise, atmospheric scatter, topographic shadowing, and other undesirable contributions to image brightness. If such components are excluded from subsequent analysis, it is likely that accuracy can be retained (relative to the entire set of seven channels) while also reducing time and cost devoted to the analysis. A color presentation of the first three components, assigning each to one of the additive primaries (red, green, and blue) is usually effective in presenting a concise, potent portrayal of the information conveyed by a multichannel multispectral image. Note, however, that because each band is a linear combination of the original channels, the analyst must be prepared to interpret the mean-

**TABLE 11.2. Results of Principal Components Analysis of Data in Table 11.1**

| | Component | | | | | | |
|---|---|---|---|---|---|---|---|
| | I | II | III | IV | V | VI | VII |
| | | | | Eigenvectors | | | |
| % var.: | 82.5% | 10.2% | 5.3% | 1.3% | 0.4% | 0.3% | 0.1% |
| EV: | 848.44 | 104.72 | 54.72 | 13.55 | 4.05 | 2.78 | 0.77 |
| | 0.14 | 0.35 | 0.60 | 0.07 | –0.14 | –0.66 | –0.20 |
| | 0.11 | 0.16 | 0.32 | 0.03 | –0.07 | –0.15 | –0.90 |
| | 0.37 | 0.35 | 0.39 | –0.04 | –0.22 | 0.71 | –0.36 |
| | 0.56 | –0.71 | 0.37 | –0.09 | –0.18 | 0.03 | –0.64 |
| | 0.74 | 0.21 | –0.50 | 0.06 | –0.39 | –0.10 | 0.03 |
| | 0.01 | –0.05 | 0.02 | 0.99 | 0.12 | 0.08 | –0.04 |
| | 0.29 | 0.42 | –0.08 | –0.09 | 0.85 | 0.02 | –0.02 |
| | | | | Loadings | | | |
| Band 1 | 0.562 | 0.519 | 0.629 | 0.037 | –0.040 | –0.160 | –0.245 |
| Band 2 | 0.729 | 0.369 | 0.529 | 0.027 | –0.307 | –0.576 | –0.177 |
| Band 3 | 0.707 | 0.528 | 0.419 | –0.022 | –0.659 | –0.179 | –0.046 |
| Band 4 | 0.903 | –0.401 | 0.150 | –0.017 | 0.020 | 0.003 | –0.003 |
| Band 5 | 0.980 | 0.098 | –0.166 | 0.011 | –0.035 | –0.008 | –0.001 |
| Band 6 | 0.144 | –0.150 | 0.039 | 0.969 | 0.063 | 0.038 | –0.010 |
| Band 7 | 0.873 | 0.448 | –0.062 | –0.033 | 0.180 | 0.004 | –0.002 |

ing of the new channels. In some instance this task is relatively straightforward; in other instances; it can be very difficult to unravel the meaning of a PCA image. Further, the student should remember that the PCA transformation applies only to the specific image at hand, and that each new image requires a new recalculation of the PCA—for some applications, this constraint limits the effectiveness of the technique.

## 11.3. Subsets

Because of the very large sizes of many remotely sensed images, analysts typically work with those segments of full images that specifically pertain to the task at hand. Therefore, to minimize computer storage, and the analyst's time and effort, one of the first tasks in each project is to prepare subsets, portions of larger images selected to show only the region of interest.

Although selecting subsets would not appear to be one of remote sensing's most challenging tasks, it can be more difficult than one might first suppose. Often subsets must be "registered" (matched) to other data, or to other projects, so it is necessary to find distinctive landmarks in both sets of data to assure that coverages coincide spatially. Second, since time and computational effort devoted to matching images to maps or other images (as described below) increase with large images, it is often convenient to prepare subsets before registration. Yet if the subset is too small, then it may be difficult to identify sufficient landmarks for efficient registration. Therefore, it may be useful to prepare a preliminary subset, large enough to conduct the image registration effectively, before selecting the final, smaller, subset for analytical use (Figure 11.2).

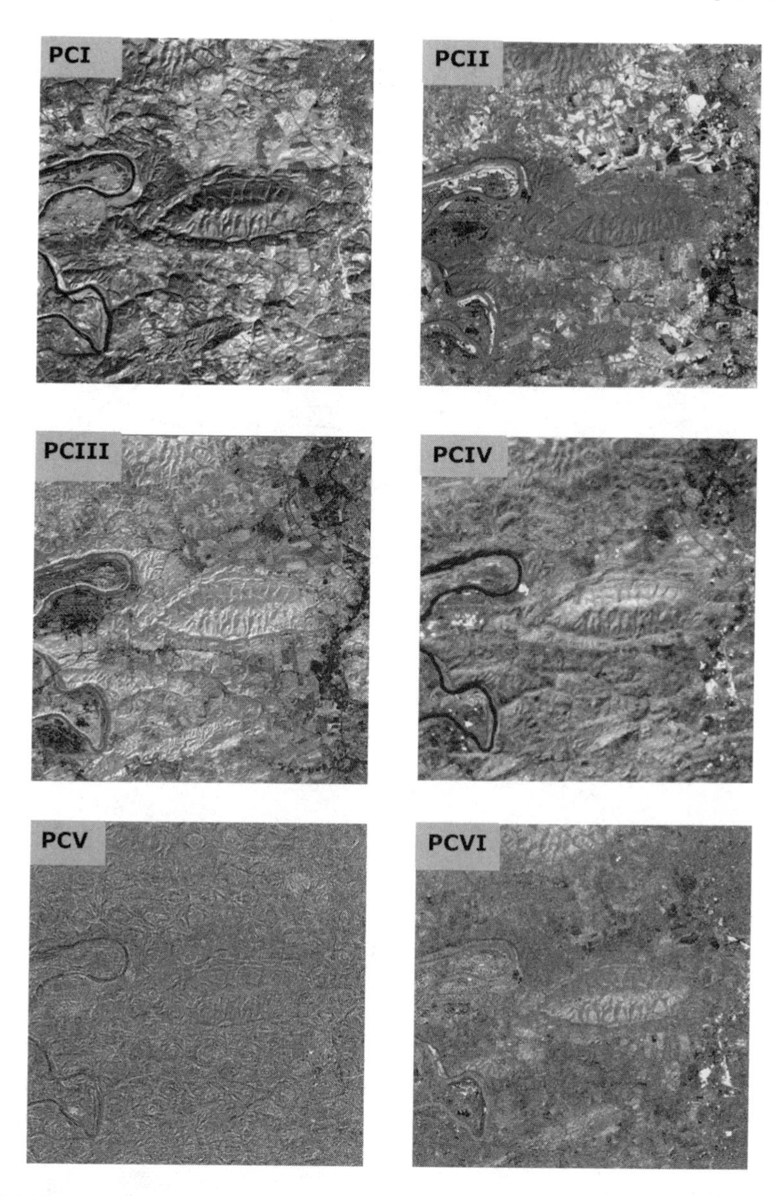


**FIGURE 11.1.** Feature selection by principal components analysis. These images depict six of the seven principal components for the image described by Tables 11.1 and 11.2. The first principal component image (PC I), formed from a linear combination of data from all seven original bands, accounts for over 80% of the total variation of the image data. PC II and PC III present about 10% and 5% of the total variation, respectively. The higher components (e.g., PC V and PC VI) account for very low proportions of the total variation, and convey mainly noise and error, as is clear by the image patterns they show.

The same kinds of considerations apply in other steps of an analysis. Subsets should be large enough to provide context required for the specific analysis at hand. For example, it may be important to prepare subsets large enough to provide sufficient numbers of training fields for image classification (Chapter 12), or a sufficient set of sites for accuracy assessment (Chapter 14).

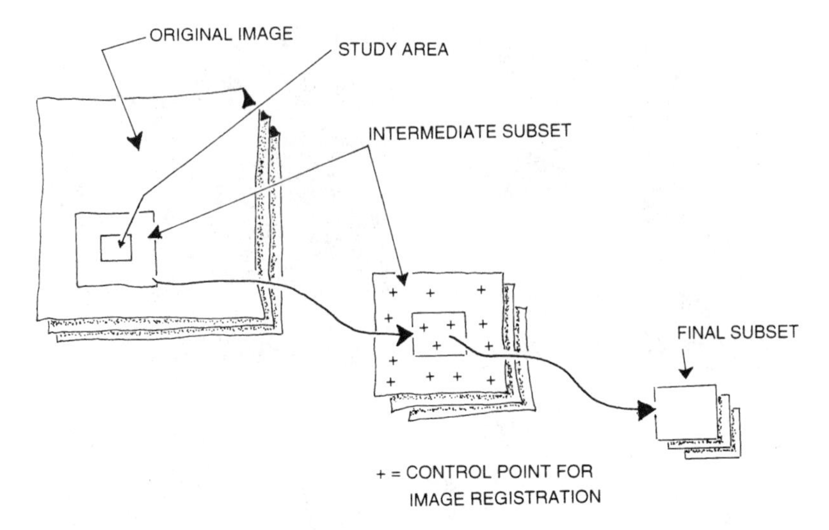


**FIGURE 11.2.** Subsets. Sometimes a subset of a particular area is too small to encompass sufficient points to allow the subset to be accurately matched to an accurate map (discussed in Section 11.6). Selection of an intermediate temporary subset permits accurate registration using an adequate number of control points. After the temporary subset has been matched to the map, the study area can be selected more precisely without concern for the distribution of control points.

## 11.4. Radiometric Preprocessing

Many preprocessing operations fall into the category of *image restoration* (Estes et al., 1983), the effort to remove the undesirable influence of atmospheric interference, system noise, and sensor motion. By applying knowledge of the nature of these effects, it is possible to estimate their magnitude, then to remove or minimize their influence upon the data used in later steps of the analysis. After removing these effects, the data are said to be "restored" to their (hypothetical) correct condition, although we can, of course, never know what the correct values might be, and must always remember that attempts to correct data may themselves introduce errors. So the analyst must decide if the errors removed are likely to be greater than those that might be introduced. Typically, image restoration includes efforts to correct for both radiometric and geometric errors.

Radiometric preprocessing influences the brightness values of an image to correct for sensor malfunctions or to adjust the values to compensate for atmospheric degradation. Any sensor that observes the earth's surface using visible or near visible radiation will record a mixture of two kinds of brightnesses. One brightness is due to the reflectance from the Earth's surface—the brightnesses that are of interest for remote sensing. But the sensor also observes the brightness of the atmosphere itself—the effects of scattering (Chapter 2). Thus an observed digital brightness value (e.g., "56") might be in part the result of surface reflectance (e.g., "45") and in part the result of atmospheric scattering (e.g., "11"). Of course we cannot immediately distinguish the two brightnesses, so one objective of atmospheric correction is to identify and separate these two components so the main analysis can focus upon examination of correct surface brightness (the "45" in this example). Ideally, atmospheric correction should find a separate correction for each pixel in the scene; in practice, we may apply the same correction to an entire band, or apply a single factor to a local region within the image.

Preprocessing operations to correct for atmospheric degradation fall into three rather broad categories. First are those procedures based upon efforts to model the physical behavior of the radiation as it passes through the atmosphere. Application of such models permits observed reflectances to be adjusted to approximate true values that might be observed under a clear atmosphere, thereby improving image quality and accuracies of analyses. Physical models (i.e., models that attempt to model the physical process of scattering at the level of individual particles and molecules) have important advantages with respect to rigor, accuracy, and applicability to a wide variety of circumstances. But they also have significant disadvantages. Often they are very complex, usually requiring detailed data and intricate computer programs. An important limitation is the requirement for detailed meteorological information pertaining to atmospheric humidity and the concentrations of atmospheric particles. Such data may be difficult to obtain in the necessary detail, and may apply only to a few points within a scene. Also, atmospheric conditions vary with altitude and over space. Although meteorological satellites, as well as a growing number of remote sensing systems, collect atmospheric data that can contribute to atmospheric corrections of imagery, procedures for everyday applications of such methods are not now at hand.

A second approach to atmospheric correction of remotely sensed imagery is based upon examination of reflectances from objects of known or assumed brightness recorded by multispectral imagery. From basic principles of atmospheric scattering, we know that scattering is related to wavelength, sizes of atmospheric particles, and their abundance. If a known target is observed using a set of multispectral measurements, the relationships between values in the separate bands can help assess atmospheric effects.

Ideally, the target consists of a natural or man-made feature that can be observed with airborne or ground-based instruments at the time of image acquisition, so the analyst could learn from measurements independent of the image the true brightness of the object when the image was acquired. However, in practice we seldom have such measurements, and therefore we must look for features of known brightness that commonly, or fortuitously, appear within an image. In its simplest form, this strategy can be implemented by identifying a very dark object or feature within the scene. Such an object might be a large water body or possibly shadows cast by clouds or by large topographic features. In the infrared portion of the spectrum both water bodies and shadows should have brightness at or very near zero because clear water absorbs strongly in the near infrared spectrum and because very little infrared energy is scattered to the sensor from shadowed pixels. Analysts who examine such areas, or the histograms of the digital values for a scene, can observe that the lowest values (for dark areas, such as clear water bodies) are not zero, but some larger value. Typically, this value will differ from one band to the next, so, for example, for Landsat band 1 the value might be 12, for band 2 the value might be 7, for band 3 the value might be 2, and for band 4 the value might be 2 (Figure 11.3). These values, assumed to represent the value contributed by atmospheric scattering for each band, are then subtracted from all digital values for that scene and that band. Thus the lowest value in each band is set to zero, the dark black color assumed to be the correct tone for a dark object in the absence of atmospheric scattering. This procedure is one of the simplest, most direct methods for adjusting digital values for atmospheric degradation (Chavez, 1975), known sometimes as the *histogram minimum method* (HMM), or the *dark object subtraction* (DOS) *technique.*

HMM has the advantages of simplicity, directness, and almost universal applicability, as it exploits information present within the image itself. Yet it must be considered as an approxima-

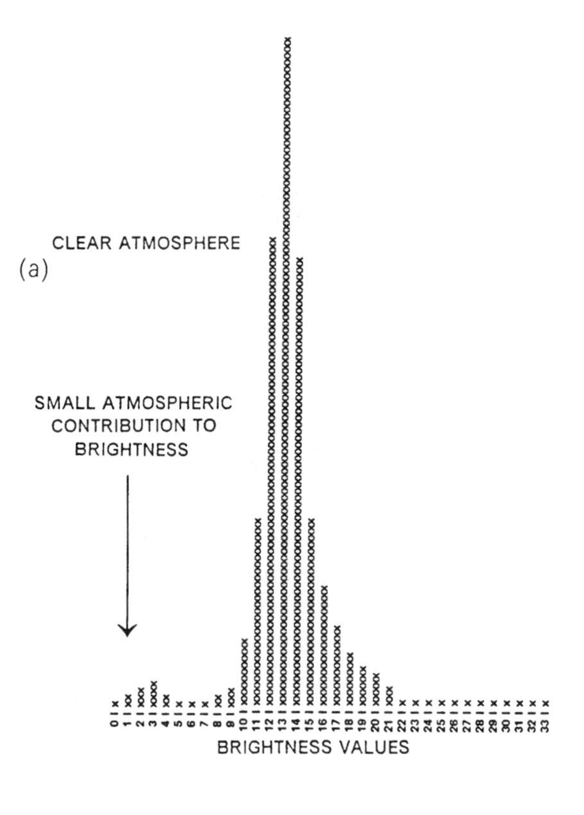


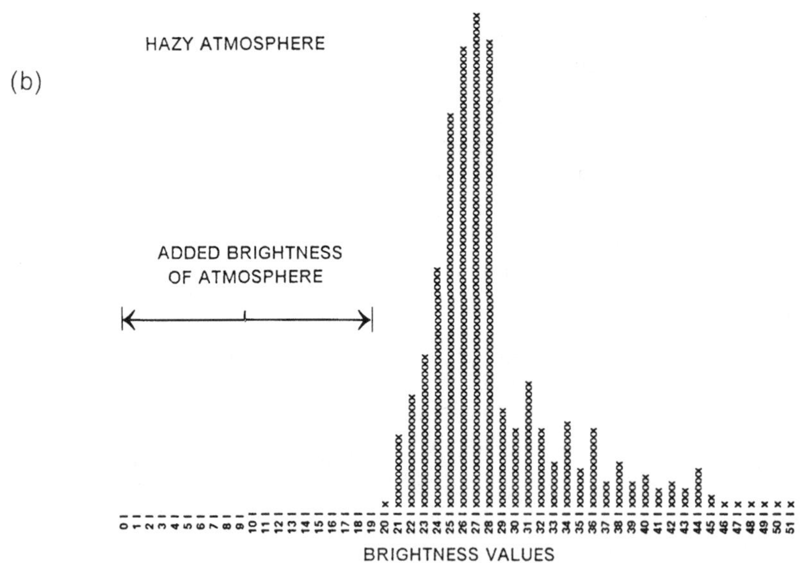


**FIGURE 11.3.** Histogram minimum method for correction of atmospheric effects. The lowest brightness value in a given band is taken to indicate the added brightness of the atmosphere to that band, and is then subtracted from all pixels in that band. (a) Histogram for an image acquired under clear atmospheric conditions; the darkest pixel is near zero brightness. (b) Histogram for an image acquired under hazy atmospheric conditions; the darkest pixels are relatively bright, due to the added brightness of the atmosphere.

tion; atmospheric effects change not only the position of the histogram on the axis, but also its shape (i.e., not all brightnesses are affected equally). (Chapter 2 explained that the atmosphere can cause dark pixels to become brighter and bright pixels to become darker, so application of a single correction to all pixels will provide only a rough adjustment for atmospheric effects.) In addition, in arid regions observed at high sun angles, shadows, clouds, and open water may be so rare or of such small areal extent that the method cannot be applied.

A more sophisticated approach retains the idea of examining brightness of objects within each scene, but attempts to exploit knowledge of interrelationships between separate spectral bands. Chavez (1975) devised a procedure that paired values from each band with values from a near infrared spectral channel. The Y intercept of the regression line is then taken as the correction value for the specific band in question. Whereas the HMM procedure is applied to entire scenes, or to very large areas, the regression technique can be applied to local areas (of possibly only 100–500 pixels each), assuring that the adjustment is tailored to conditions important within specific regions. An extension of the regression technique is to examine the *variance–covariance matrix,* the set of variances and covariances between all band pairs on the data. (This is the *covariance matrix method* [CMM] described by Switzer et al., 1981.) Both procedures assume that within a specified image region, variations in image brightness are due to topographic irregularities and reflectivity is constant (i.e., land-cover reflectivity in several bands is uniform for the region in question). Therefore, variations in brightness are caused by small-scale topographic shadowing, and the dark regions reveal the contributions of scattering to each band. Although these assumptions may not always be strictly met, the procedure, if applied with care and with knowledge of the local geographic setting, seems to be robust and often satisfactory. Campbell and Liu (1994) modified this strategy by using an automatic grouping procedure (as described in Chapter 12) to ensure that regions are defined in such a way that their reflectances are in fact uniform.

How can the analyst decide if atmospheric corrections are necessary? This may be a difficult decision, as effects of atmospheric degradation are not always immediately obvious from casual inspection. The analyst should always examine summary statistics for each scene, inspecting means, variances, and frequency histograms for suggestions of poor image quality and the absence of dark values, especially if the image is known to show large water bodies (Figure 11.4).

Of course, inspection of the image may reveal evidence suggesting a requirement for correction. Loss of resolution and low contrast may indicate poor atmospheric conditions. Sometimes the image date may itself suggest the nature of atmospheric quality. In the central United States, summer dates often imply high humidity, haze, and poor visibility, whereas autumn, winter, and spring dates are often characterized by clearer atmospheric conditions. Thus the image date may provoke further investigation by the analyst to determine if corrections are necessary. Finally, the analyst should examine summary statistics for the scene and especially the frequency histograms for each band.

### *Destriping*

Landsat MSS data sometimes exhibit a kind of radiometric error known as *sixth-line striping,* caused by small differences in the sensitivities of detectors within the sensor. Within a given

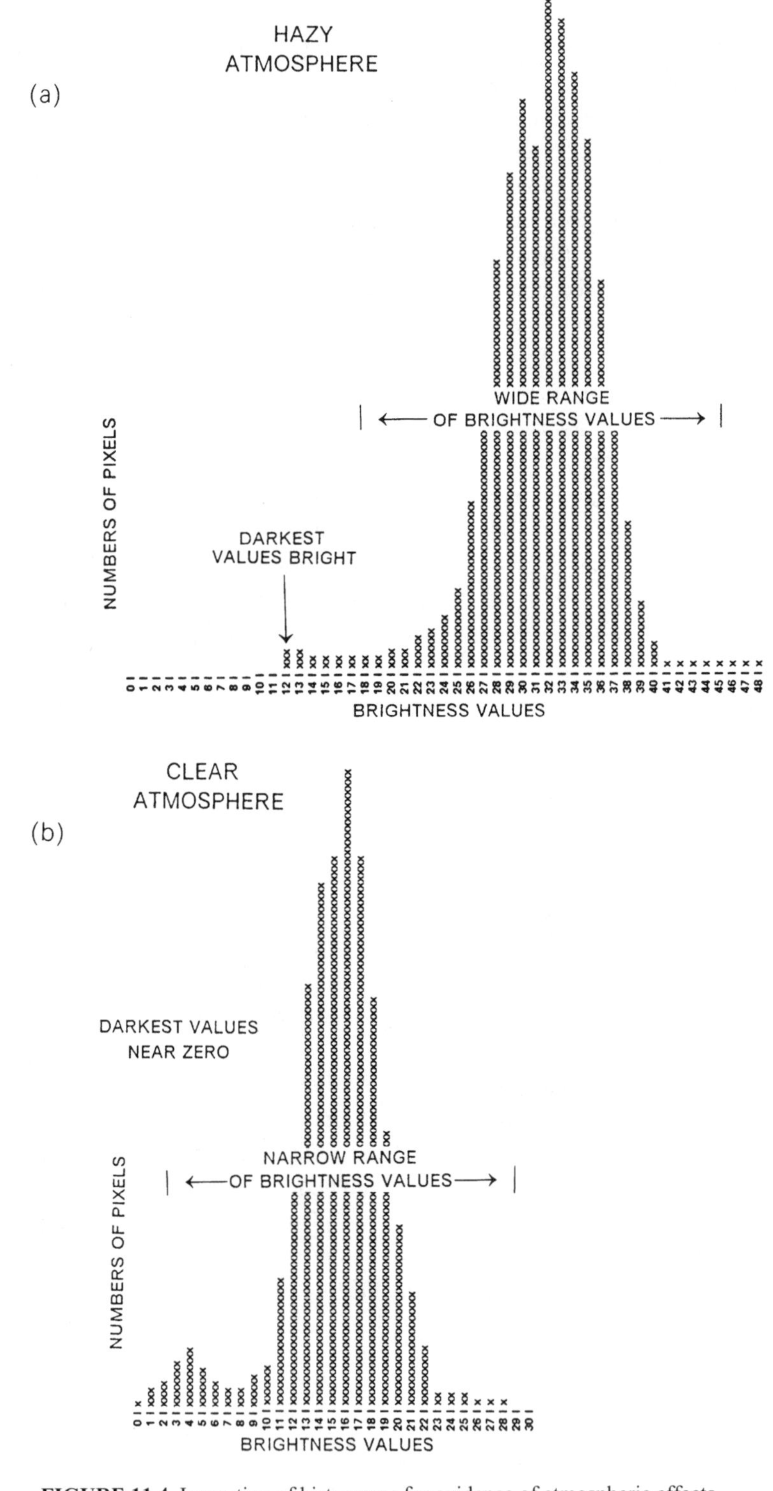


**FIGURE 11.4.** Inspection of histograms for evidence of atmospheric effects.

band, such differences appear on images as a horizontal banding, or "striping," because individual scan lines exhibit unusually brighter or darker brightness values that contrast noticeably with the background brightnesses of the "normal" detectors (Figure 11.5). Because MSS detectors are positioned in arrays of six (Chapter 6), an anomalous detector response appears as linear banding at intervals of six lines. Striping may appear on only one or two bands of a multispectral image or may be severe for only a portion of a band. Other forms of digital imagery sometimes exhibit similar effects, all caused by difficulties in maintaining consistent calibration of detectors within a sensor. Campbell and Liu (1995) found that striping and related defects in digital data imagery seemed to have minimal impact on the character of the data—much less than might be suggested by the appearance of the imagery.

Although sixth-line striping is often clearly visible as an obvious banding (Figure 11.5), it may also be present in a more subtle form that may escape casual visual inspection. *Destriping* refers to the application of algorithms to adjust incorrect brightness values to values thought to be near the correct values. Some image-processing software provides special algorithms to detect striping. Such procedures search through an image line by line to look for systematic differences in average brightnesses of lines spaced at intervals of six; examination of the results permits the analyst to have objective evidence of the existence or absence of sixth-line striping (Table 11.3). If striping is present, the analyst must make a decision. If no correction is applied, the analysis must proceed with brightness values that are known to be incorrect. Conversely, efforts to correct for such serious errors may yield rather rudimentary approximations of the (unknown) true values. Often striping may be so severe that it is obvious the bad lines must be adjusted.

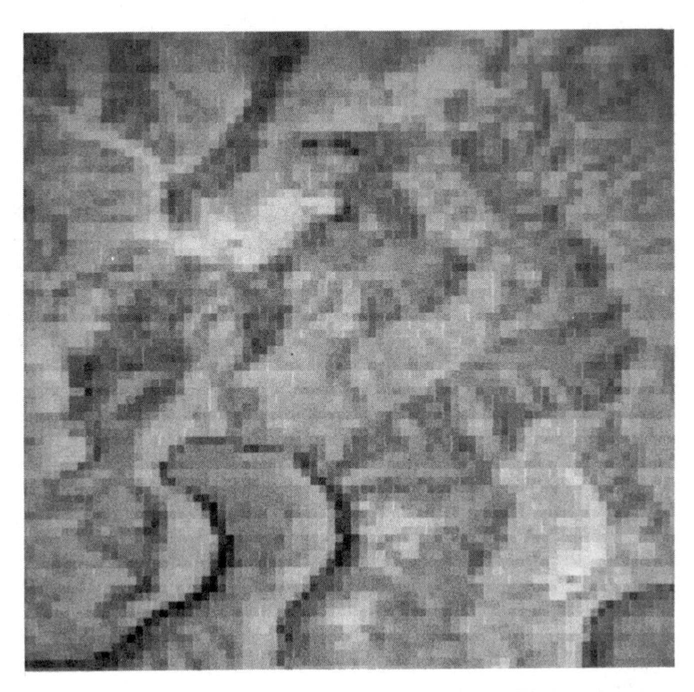

**FIGURE 11.5.** Sixth-line striping in Landsat MSS data.

**TABLE 11.3. Results of a Sixth-Line Striping Analysis**

| | Mean | Standard deviation |
|---|---|---|
| Entire image: | 19.98 | 7.06 |
| Detector 1: | 19.07 | 6.53 |
| Detector 2: | 19.50 | 6.97 |
| Detector 3: | 19.00 | 6.78 |
| Detector 4: | 22.03 | 7.60 |
| Detector 5: | 21.09 | 7.34 |
| Detector 6: | 19.17 | 6.51 |

A variety of destriping algorithms have been devised. All identify the values generated by the defective detectors by searching for lines that are noticeably brighter or darker than the lines in the remainder of the scene. These lines are presumably the bad lines caused by the defective detectors (especially if they occur at intervals of six lines). Then the destriping procedure estimates corrected values for the bad lines. There are many different estimation procedures; most belong to one of two groups. One approach is to replace bad pixels with values based upon the average of adjacent pixels not influenced by striping; this approach is founded upon the notion that the missing value is probably quite similar to the pixels that are nearby (Figure 11.6a). A second strategy is to replace bad pixels with new values based upon the mean and standard deviation of the band in question, or upon statistics developed for each detector (Figure 11.6b). This second approach is based upon the assumption that the overall statistics for the missing data must, because there are so many pixels in the scene, resemble those from the good detectors.

The algorithm described by Rohde et al. (1978) combines elements of both strategies. Their procedure attempts to bring all values in a band to a normalized mean and variance, based on overall statistics for the entire band. Because brightnesses of individual regions within the scene may vary considerably from these overall values, a second algorithm can be applied to perform a local averaging to remove the remaining influences of striping. Some destriping algorithms depend entirely on local averaging; because they tend to degrade image resolution, and introduce statistical dependencies between adjacent brightness values, it is probably best to be cautious in their application if alternatives are available.

Table 11.3 shows the results for an analysis of striping in a Landsat MSS scene, using the histogram normalization approach. The tabulation provides means and standard deviations of lines organized such that all lines collected by detector 1 are grouped together, all those collected by detector 2 are grouped together, and so on. All the lines collected by the first detector are found by starting at the first line, then skipping to line 7, then to line 13, and so on. All the lines collected by the second detector are found by starting at line 2, then skipping to line 8, then to line 14, and so on. For the image reported in Table 11.3, it is clear that detectors 4 and 5 are producing results that differ from the others—the means and standard deviations of the pixels collected by these detectors are higher than those of the pixels collected by the other detectors. The differences seem rather small—only two or three brightness values—yet the effect is usually quite obvious to the eye (Figure 11.5), probably because the brightness differences are organized in distinct linear patterns that attract the attention of the observer's visual system.

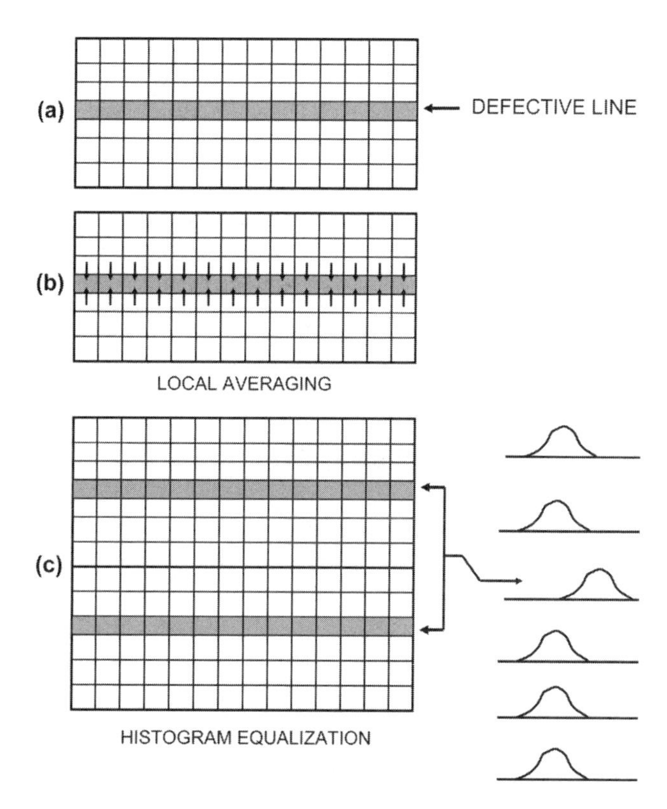


**FIGURE 11.6.** Two strategies for destriping. (*a*) Defective line identified with an array of pixels. (*b*) Local averaging—pixels in the defective line are replaced with an average of values of the the neighboring pixels in adjacent lines. (*c*) Histogram normalization—data from all lines are accumulated at intervals of six lines (for the Landsat MSS); the histogram for defective detectors displays an average different from the others. A correction shifts the values for the defective lines to match the postions for the other lines within the image.

## 11.5. Image Matching

*Image matching* is the process of superimposing two images of the same area, then moving them to find the position at which they best match. Visually, two transparencies have a registration point where the image detail matches. Digitally, there are many separate procedures; some are very complex. One of the simplest and most widely used strategies is to digitally overlay the two images, then calculate a correlation for the area where the two images overlap (or for some smaller zone). Matched positions are systematically shifted, pixel by pixel, until all possible matches have been attempted. At each position a new correlation value is calculated, then saved. The optimum position, presumably the correct registration, is the position that yields the highest correlations. This kind of image-matching procedure is important in many automated processes, including matching of digital stereo images.

A much more common problem is the registration of two images, meaning that two images are brought to registration, but one is usually altered using one of the procedures described below. (In image matching, the two images are not changed—we simply find the position at which the two match.) Typically, we wish to register a Landsat image to a planimetrically correct map, or to register two images from different dates, or to register Landsat and Seasat images of the same area. In each instance, we have the same problem: we must rearrange pixels to match to the new map, or we must estimate pixel values at points that match the new distribution.

The most difficult but most rigorous approach to image registration is to apply knowledge of sensor geometry and motion to derive accurate coordinates for each pixel. To apply this approach, which we will designate as the *analytical* approach, the analyst must know the satellite altitude, its trajectory, the shape of the Earth's surface, its motion relative to the satellite, and the motion of the sensor scanner. Although for satellites such as Landsat these factors are known with some precision, analytical image correction can correct only some of the geometric errors in images.

## 11.6. Geometric Correction by Resampling

A second approach to image registration treats the problem in a completely different manner. No effort is made to apply our knowledge of system geometry; instead, the images are treated simply as an array of values that must be manipulated to create another array with the desired geometry.

This can be seen essentially as an interpolation problem similar to those routinely considered in cartography and other disciplines. In Figure 11.7 the input image is represented as an array of open dots, each representing the center of a pixel in the uncorrected image. Superimposed over this image is a second array, symbolized by the solid dots, which shows the centers of pixels in the image transformed to have the desired geometric properties (the "output" image).

The locations of the output pixels are derived from locational information provided by *ground control points* (GCPs), locations on the input image that can be located with precision on the ground and on planimetrically correct maps. (If two images are to be registered, GCPs must be easily recognized on both images.) The locations of these points establish the geometry of the output image and its relationship to the input image. Thus this first step establishes the framework of pixel positions for the output image using the GCPs.

The next step is to decide how to best estimate the values of pixels in the corrected image, based upon information in the uncorrected image. The simplest strategy from a computational perspective is to assign each "corrected" pixel the value from the nearest "uncorrected" pixel. This is the *nearest-neighbor* approach to resampling (Figure 11.8). It has the advantages of sim-

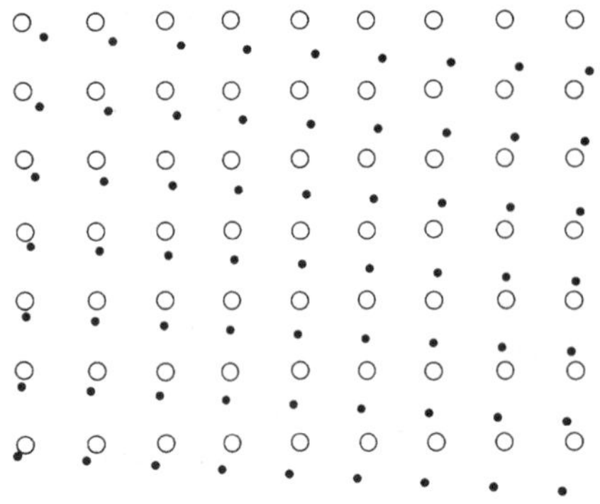

**FIGURE 11.7.** Resampling. Open circles (○) represent the reference grid of known values in the input image. Black dots (●) represent the regular grid of points to be estimated to form the output image. Each resampling method employs a different strategy to estimate values at the output grid, given known values for the input grid.

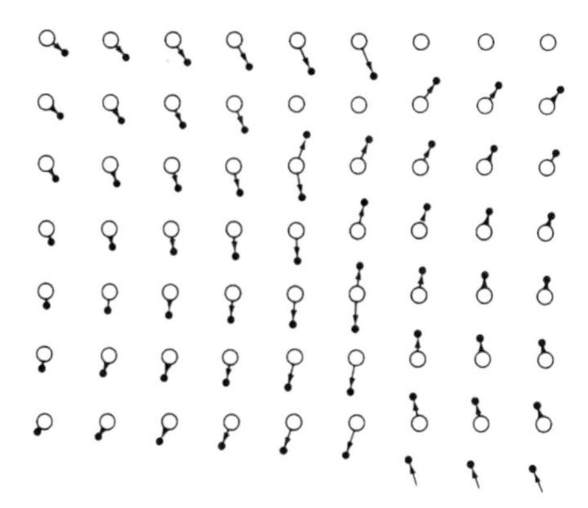

**FIGURE 11.8.** Nearest neighbor resampling. Each estimated value (•) receives its value from the nearest point on the reference grid (○).

plicity and the ability to preserve original values in the unaltered scene. On the other hand, it may creates noticeable positional errors, which may be severe in linear features where the realignment of pixels is obvious. In Kovalick's (1983) study the nearest-neighbor method was computationally the most efficient of the three methods studied.

A second, more complex, approach to resampling is *bilinear interpolation* (Figure 11.9). Bilinear interpolation calculates a value for each output pixel based upon a weighted average of the four nearest input pixels. In this context, "weighted" means that nearer pixel values are given greater influence in calculating output values than are more distant pixels. Because each output value is based upon several input values, the output image will not have the unnaturally blocky appearance of some nearest-neighbor images. The image therefore has a more "natural" look. Yet there are important changes. First, because bilinear interpolation creates new pixel values, the brightness values in the input image are lost. The analyst may find that the range of brightness values in the output image differs from those in the input image. Such changes to digital brightness values may be significant in later processing steps. Second, because the resampling is conducted by averaging over areas (i.e., blocks of pixels), it decreases spatial resolution by a kind of "smearing" caused by averaging small features with adjacent background pixels.

Finally, the most sophisticated, most complex, and (possibly) most widely used resampling method is *cubic convolution* (Figure 11.10). Cubic convolution uses a weighted average of values within a neighborhood that extends about two pixels in each direction, usually encompass-

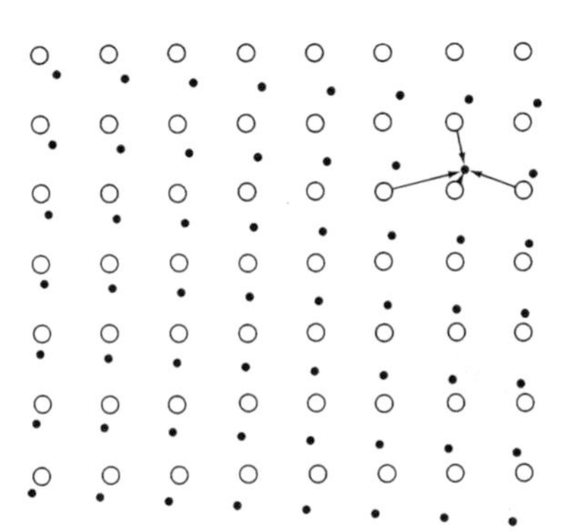

**FIGURE 11.9.** Bilinear interpolation. Each estimated value (•) in the output image is formed by calculating a weighted average of the known values of the four nearest neighbors in the input image (○). Each estimated value is weighted according to its distance from the known values in the input image.

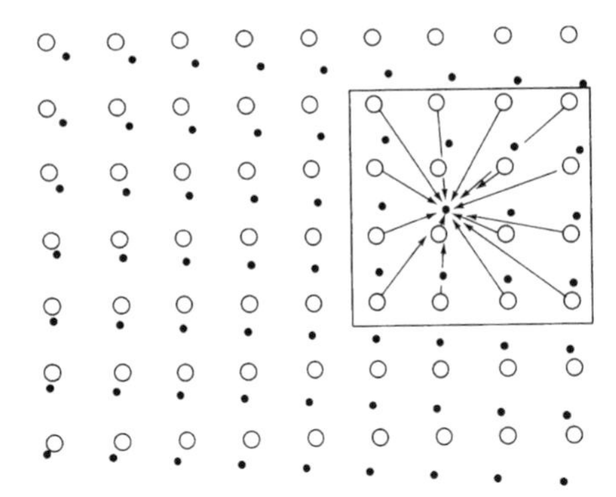

**FIGURE 11.10.** Cubic convolution. Each estimated value in the output matrix (•) is found by assessing values within a neighborhood of 16 pixels in the input image (○).

ing 16 adjacent pixels. Typically, the images produced by cubic convolution resampling are much more attractive than those of other procedures, but the data are altered more than those of nearest-neighbor or bilinear interpolation, the computations are more intensive, and the minimum number of GCPs is larger. Images altered in this manner by resampling should be examined carefully before they are used for subsequent analyses.

### *Identification of Ground Control Points*

A practical problem in applying image registration procedures is the selection of control points (Figure 11.11). GCPs are features that can be located with precision and accuracy on accurate maps yet are also easily located on digital images. Ideally, GCPs could be as small as a single pixel, if one could be easily identified against its background. In practice, most GCPs are likely to be spectrally distinct areas as small as a few pixels. Examples might include intersections of major highways, distinctive water bodies, edges of land-cover parcels, stream junctions, and similar features (Figure 11.12). Although identification of such points may seem to be an easy task, in fact difficulties that might emerge during this step can form a serious roadblock to the entire analytical process, as another procedure may depend on completion of an accurate registration.

Typically, it is relatively easy to find a rather small or modest-sized set of control points. However, in some scenes, the analyst finds it increasingly difficult to expand this set, as one has less and less confidence in each new point added to the set of GCPs. Thus there may be a rather small set of "good" GCPs, points that the analyst can locate with confidence and precision both on the image and on an accurate map of the region.

The locations may also be a problem. In principle, GCPs should be dispersed throughout the

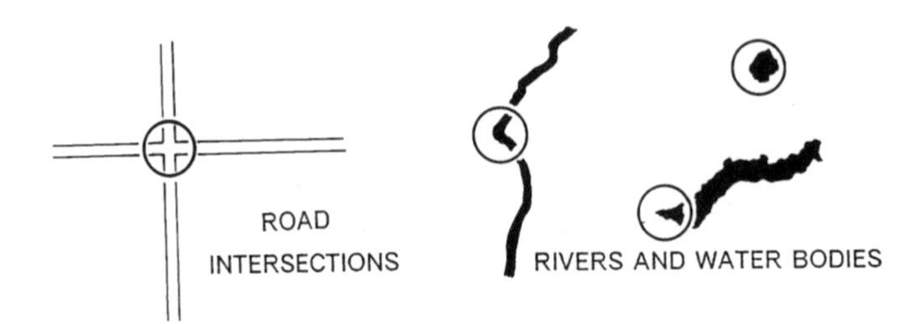


**FIGURE 11.11.** Selection of distinctive ground control points (GCPs).

**FIGURE 11.12.** Examples of ground control points. GCPs must be identifiable both on the image and on a planimetrically correct reference map.

image, with good coverage near edges. Obviously, there is little to be gained from a large number of GCPs if they are all concentrated in a few regions of the image. Analysts who attempt to expand areal coverage to ensure good dispersion are forced to consider points in which it is difficult to locate GCPs with confidence. Therefore the desire to select "good" GCPs and to achieve good dispersion may work against each other such that the analyst finds it difficult to select a judicious balance. Analysts should anticipate difficulties in selecting GCPs as they prepare subsets early in the analytical process. If subsets are too small, or if they do not encompass important landmarks, the analysts may later find that the subset region of the image does not permit selection of a sufficient number of high-quality GCPs.

Bernstein et al. (1983) present information that shows how registration error decreases as the number of GCPs is increased. Obviously, it is better to have more rather than fewer GCPs. But as explained above, the quality of GCP accuracy may decrease as their number increases because the analyst usually picks the best points first. They recommend that 16 GCPs may be a reasonable number if each can be located with an accuracy of one-third of a pixel. This number may not be sufficient if the GCPs are poorly distributed or if the nature of the landscape prevents accurate placement.

Many image-processing programs permit the analyst to anticipate the accuracy of the registration by reporting errors observed at each GCP if a specific registration has been applied. The standard measure of the location error is the *root mean square* (rms) error, which is the standard deviation of the difference between actual positions of GCPs and their calculated positions (i.e., after registration). These differences are known as the *residuals*. Usually rms is reported in units of image pixels for both north–south and east–west directions (Table 11.4). Note that this practice reports locational errors among the GCPs, which may not always reflect the character of errors encountered at other pixels. Nonetheless, it is helpful to be able to select the most useful GCPs. If analysts wish to assess the overall accuracy of the registration, some of the GCPs should be withheld from the registration procedure, then used to assess its success.

**TABLE 11.4. Sample Tabulation of Data for GCPs**

| Point no. | Image *X* pixel | *X* pixel residual | Image *Y* pixel | *Y* pixel residual |
|---|---|---|---|---|
| 1 | 1269.75 | –0.2471E+00 | 1247.59 | 0.1359E+02 |
| 2 | 867.91 | –0.6093E+01 | 1303.90 | 0.8904E+01 |
| 3 | 467.79 | –0.1121E+02 | 1360.51 | 0.5514E+01 |
| 4 | 150.52 | 0.6752E+02 | 1413.42 | –0.8580E+01 |
| 5 | 82.20 | –0.3796E+01 | 163.19 | 0.6189E+01 |
| 6 | 260.89 | 0.2890E+01 | 134.23 | 0.5234E+01 |
| 7 | 680.59 | 0.3595E+01 | 70.16 | 0.9162E+01 |
| 8 | 919.18 | 0.1518E+02 | 33.74 | 0.1074E+02 |
| 9 | 1191.71 | 0.6705E+01 | 689.27 | 0.1127E+02 |
| 10 | 1031.18 | 0.4180E+01 | 553.89 | 0.1189E+02 |
| 11 | 622.44 | –0.6564E+01 | 1029.43 | 0.8427E+01 |
| 12 | 376.04 | –0.5964E+01 | 737.76 | 0.6761E+01 |
| 13 | 162.56 | –0.7443E+01 | 725.63 | 0.8627E+01 |
| 14 | 284.05 | –0.1495E+02 | 1503.73 | 0.1573E+02 |
| 15 | 119.67 | –0.8329E+01 | 461.59 | 0.4594E+01 |
| 16 | 529.78 | –0.2243E+00 | 419.11 | 0.5112E+01 |
| 17 | 210.42 | –0.1558E+02 | 1040.89 | –0.1107E+01 |
| 18 | 781.85 | –0.2915E+02 | 714.94 | –0.1521E+03 |
| 19 | 1051.54 | –0.4590E+00 | 1148.97 | 0.1697E+02 |
| 20 | 1105.95 | 0.9946E+01 | 117.04 | 0.1304E+02 |

*X* rms error = 18.26133
*Y* rms error = 35.33221

Total rms error = 39.77237

| Point no. | Error | Error contribution by point |
|---|---|---|
| 1 | 13.5913 | 0.3417 |
| 2 | 10.7890 | 0.2713 |
| 3 | 12.4971 | 0.3142 |
| 4 | 68.0670 | 1.7114 |
| 5 | 7.2608 | 0.1826 |
| 6 | 5.9790 | 0.1503 |
| 7 | 9.8416 | 0.2474 |
| 8 | 18.5911 | 0.4674 |
| 9 | 13.1155 | 0.3298 |
| 10 | 12.6024 | 0.3169 |
| 11 | 10.6815 | 0.2686 |
| 12 | 9.0161 | 0.2267 |
| 13 | 11.3944 | 0.2865 |
| 14 | 21.6990 | 0.5456 |
| 15 | 9.5121 | 0.2392 |
| 16 | 5.1174 | 0.1287 |
| 17 | 15.6177 | 0.3927 |
| 18 | 154.8258 | 3.8928 |
| 19 | 16.9715 | 0.4267 |
| 20 | 16.3982 | 0.4123 |

## 11.7. Map Projections for Representing Satellite Images and Ground Tracks

Mapping of any distribution on the Earth's surface requires use of a map projection whenever the area considered exceeds a few square miles. This requirement arises from the fundamental problem of projecting locations on the Earth's spherical surface onto the flat surface of the map. A map projection is a system of transformations that enables locations on the spherical Earth to be represented systematically on a flat map. Because of the inherent difference between the two surfaces, there is always some sacrifice in accurate representation of area, shape, scale, or direction on maps relative to globes. However, such errors may be confined to a few of these characteristics or may be very small in certain portions of a map.

Continuous mapping of a region viewed by Landsat or other Earth observation satellites presents serious problems for those using conventional map projections. Both the satellite and the Earth's surface are moving as imagery is acquired, so at best the usual maps are very inconvenient. Therefore, users of early Landsat data encountered problems in representing coverage of Landsat scenes on conventional maps. The ground tracks of Sun-synchronous satellites (Chapter 6) trace curved lines on the usual map projections, greatly complicating the representation of satellite paths and coverage. Snyder (1981) addressed this problem by devising projections specially designed to represent ground tracks of Sun-synchronous satellites as straight lines. For global tracking, his map is based upon a cylindrical projection; for larger scale maps of continents or areas of similar size, he uses a map based upon a conic projection. To show the ground track as a straight line, it is, of course, necessary to sacrifice other map qualities, so these maps cannot show accurate shape and area everywhere.

Furthermore, representation of Landsat data as a rectangular array of pixels does not place them in their correct positions relative to the Earth's surface. Convenient representation of correct positions of pixels requires projects specifically designed to capture the complex geometry of the Landsat image.

Alden P. Colvocoresses is credited with defining the space oblique Mercator (SOM) projection tailored for use with Landsat data, although others contributed to its development and mathematical definition (Colvocoresses, 1974). Research to develop the SOM was conducted by the USGS specifically for the purpose of defining a map projection to provide constant scale for the ground track of an Earth observation satellite for the entire coverage cycle. Although the SOM was designed for use with Landsat, it is equally applicable to other land observation satellites, if tailored to their specific orbital characteristics. For Landsat, the projection must depict coverage from 81° N to 81° S, through a coverage cycle of 251 orbits. The projection shows only a narrow strip parallel to the ground track of the satellite—essentially that area subject to view by the MSS.

The SOM is based upon the map projection devised by Gerhard Kramer (1512–1594), a Dutch scholar and cartographer who used the Latinized version of his name, Gerhardus Mercator. His name has been given to the map projection he devised for an atlas he published in 1569. The Mercator projection was the first projection to attain widespread use, primarily because of its utility for marine navigation.

The Mercator projection can be envisioned as a transformation of the network of lines of latitude and longitude (known as the *graticule*) onto a flat surface such that the meridians of longitude form equally spaced vertical lines and the parallels of latitude form horizontal lines intersecting the meridians at right angles. The creation of this projection is envisioned as wrapping a transparent globe in a cylinder tangent at the equator with a light inside the globe projecting the graticule onto the cylinder, which is then opened up to form a flat, map-like, surface.

On a globe, lines of latitude are parallel to one another, but lines of longitude converge near the poles. The Mercator projection differs significantly from a globe because it shows the converging meridians as parallel lines. But, simply described, the essence of Mercator's projection is that this error is (in part) compensated for by his method of spacing lines of latitude. Although the globe shows lines of latitude as equally spaced, Mercator's map increases this spacing so that as distance from the equator increases, and as divergence of the meridians from their true spacing increases, so does spacing between lines of latitude. Thus the interval between lines of latitude increases dramatically toward the poles, to compensate for converging meridians near the poles. This compensation preserves correct shape in cartographic representation of the Earth's features. The poles cannot be shown on the classic (i.e., centered on the equator) Mercator as parallels of latitude would have to be spaced at an infinite distance from one another at the poles.

If a Mercator projection is centered on the equator, the portion nearest the equator (where the hypothetical cylinder is tangent to the globe) is correct in representation of distance, shape, area, and direction; all are shown accurately or with only minor errors (Figure 11.13a). As distance from the equator increases, errors become increasingly severe. Area, in particular, is shown with large errors at high latitudes. However, shape is shown accurately throughout the map. Furthermore, this projection has an important feature that has been especially valuable throughout its history. A line of constant compass bearing (known as a *loxodrome*) is shown on a Mercator projection as a straight line—a feature that has made the projection especially valuable for navigational use, both now and in the past.

For our present concerns, the Mercator projection is significant in its modified forms, the *transverse Mercator* and the *oblique Mercator* projections. The transverse Mercator can be visualized as a cylinder tangent not at the equator, but at a meridian of longitude (Figure 11.13b). When the cylinder is unfolded, only the most accurate strip centered at the meridian of tangency is selected for use—the outer sections of the map, where the map is inaccurate, are discarded. If the process is repeated successively for different meridians, each time retaining only the accurate center strip, very accurate maps of large areas of the Earth can be constructed. This process forms the basis for the *universal transverse Mercator* (UTM) geographic reference system (described in Chapter 16), which is based upon such a projection.

For the oblique Mercator projection, the line of tangency is shifted from a meridian to a great circle on the Earth oriented at some angle (other than 90°) to the meridians (Figure 11.13c). On oblique Mercator projections the graticule appears as a set of curved lines. For the SOM (Figure 11.14), the line of tangency is defined not as a circle on the Earth's surface, but as a curve that follows the ground track of the satellite. For Landsat and other satellites in Sun-synchronous orbits, this line is a nearly sinusoidal curve.

The scan lines, representing the area depicted by MSS imagery, intersect the ground track at 86° degrees at the equator, but at 90° where the track is closest to the poles. This different is caused by variations in skew (0–4°) caused by interaction of mirror scan motion with movement of the Earth's surface as it rotates beneath the moving satellite. Like other Mercator projections, the SOM has errors in representation of distance and area, but errors are very small near the ground track.

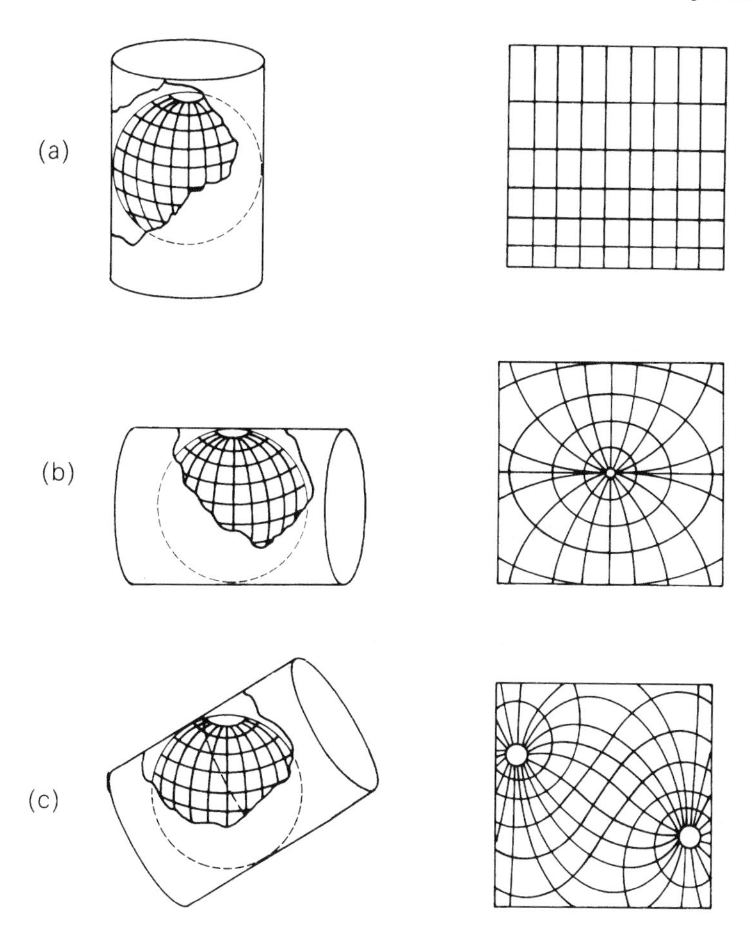


**FIGURE 11.13.** Mercator projection. (*a*) Projection centered on equator; (*b*) Transverse Mercator projection; (*c*) Oblique Mercator. From Snyder (1982).

## 11.8. Effects of Preprocessing

There is indirect evidence that preprocessing operations sometimes have unwanted effects upon digital values, and therefore upon the accuracies of classifications based upon such data. It seems clear from a basic understanding of the preprocessing operations that the data are in fact altered from their original values, and that such operations alter the mean values for each band as well as the variances and covariances. Presumably other qualities, such as correlations between bands, are also influenced. Yet few studies have systematically examined actual effects of preprocessing operations upon final results of analyses.

Kovalick (1983) examined a Landsat MSS scene of a forested wetland in North Carolina to study the effects of resampling and destriping upon accuracies of image classification. He found that resampling tended to reduce class means (of the training data he examined) and to increase variances relative to the original data.

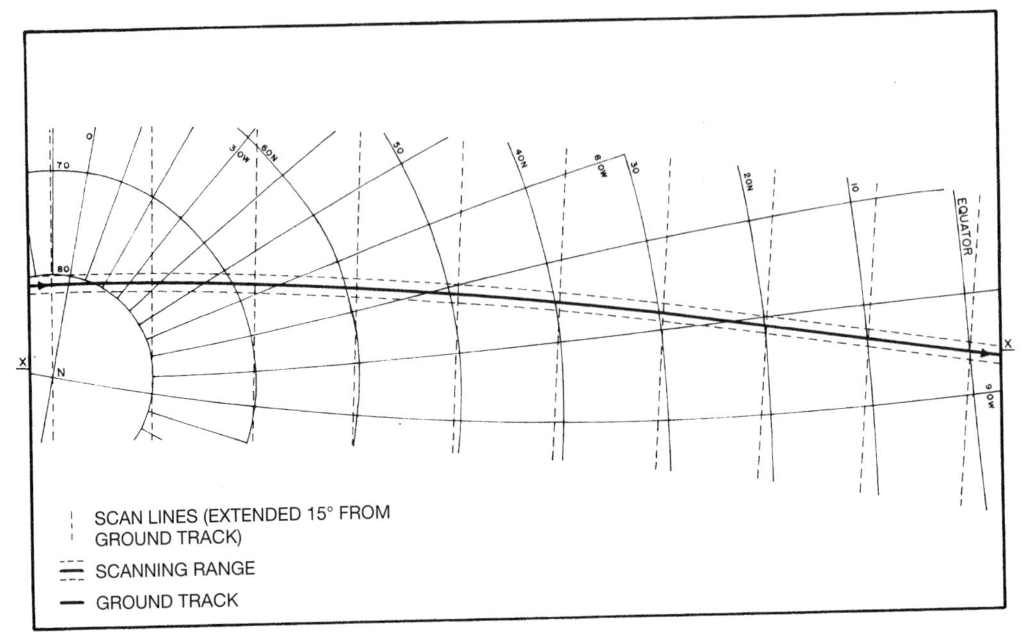


**FIGURE 11.14.** Space oblique Mercator. From Snyder (1982).

## 11.9. Data Fusion

*Data fusion* refers to processes that bring images of varied resolution into a single image that incorporates, for example, the high spatial resolution of a panchromatic image with the multispectral content of a multiband image at coarser resolution. Because of the broad range of characteristics of imagery collected by varied sensors described in previous chapters, and the routine availability of digital data, there has been increasing incentive to bring data from varied systems together into a single image. Such products are often valuable because they can integrate several independent sources of information into a single image. (Because image fusion prepares data for visual interpretation, discussed in Chapter 5, rather than for digital analysis, it differs from the other preprocessing techniques discussed in this chapter. However, placement of this topic in this chapter permits development of concepts in a more progressive sequence than would otherwise be possible.)

Although data fusion can be applied to a varied selection of different forms of remotely sensed data (e.g., merging multispectral data with radar imagery or multispectral images with digital elevation data), the classic example of image fusion involves the merging of a multispectral image of relatively coarse spatial resolution with another image of the same region acquired at finer spatial resolution. Examples might include fusing multispectral SPOT data (20-m spatial resolution) with the corresponding panchromatic SPOT scene (at 10-m spatial resolution) of the same region, or multispectral TM data with corresponding high-altitude aerial photography.

The technique assumes that the images are compatible in the sense that they were acquired on or about the same date and that the images register to each other. The analyst must rectify the images to be sure that they both share the same geometry, and register them to be sure that they match spatially. In this context, the basic task of image fusion is to substitute the spatial detail of the fine-resolution image detail to replace one of the multispectral bands, then apply a technique to restore the lost multispectral content of the discarded band from the coarse-resolution multispectral image. Specifics of these procedures are beyond the scope of this discussion. Chavez et al. (1991), Wald et al. (1997), and Carter (1998) are among the authors who have described these procedures in some detail. Briefly stated, the techniques fall into three classes.

*Spectral domain* procedures project the multispectral bands into spectral data space, then find the new (transformed) band most closely correlated with the panchromatic image. That spectral content can then be assigned to the high-resolution panchromatic image. The *intensity–hue–saturation* (IHS) *technique* (Carper et al., 1990) is one of the spectral domain procedures. Intensity–hue–saturation refers to the three dimensions of multispectral data that we know in the everyday context as "color." *Intensity* is equivalent to brightness, hue refers to dominant wavelength (what we regard as "color"), and *saturation* specifies purity, the degree to which a specific color is dominated by a single wavelength. For the IHS fusion technique, the three bands of the lower resolution image are transformed from normal red–green–blue (RGB) space into IHS space. The high-resolution image is then stretched to approximate the mean and variance of the intensity component. Next, the stretched high-resolution image is substituted for the intensity component of the original image, and then the image is projected back into RGB data space. The logic of this strategy lies in the substitution of the intensity component by the high-resolution image, which is characterized by an equivalent range of intensity, but superior spatial detail.

The *principal components transformation* (PCT) conducts a principal components analysis of the raw low-resolution image. The high-resolution image is then stretched to approximate the mean and variance of the first principal component. Next, the stretched high-resolution image is substituted for the first principal component of the low-resolution image, and then the image is reconstructed back into its usual appearance using the substituted high-resolution image as the first principal component. The logic of this PCT approach rests on the fact that the first principal component of a multispectral image often conveys most of the brightness information in the original image.

*Spatial domain procedures* extract the high-frequency variation of a fine-resolution image and then insert it into the multispectral framework of a corresponding coarse-resolution image. The *high-pass filter* (HPF) *technique* is an example. A high-pass filter is applied to the fine-resolution image to isolate and extract the high-frequency component of the image. It is this high-frequency information that conveys the scene's fine spatial detail. This high-frequency element is then introduced into the low-resolution image (with compensation to preserve the original brightness of the scene) to synthesize the fine detail of the high-resolution image. The ARSIS technique proposed by Ranchin and Wald (2000) applies the wavelet transform as a method of simulating the high-resolution detail within the coarse-resolution image.

*Algebraic procedures* operate on images at the level of the individual pixel to proportion spectral information among the three bands of the multispectral image, so that the replacement (high-resolution) image used as a substitute for one of the bands can be assigned correct spectral brightness. The *Brovey transform* finds proportional brightnesses conveyed by the band to be replaced and assigns it to the substitute. Its objective of preserving the spectral integrity of the original multispectral image is attained if the panchromatic image has a spectral range equivalent to the combined range of the three multispectral bands. Because this assumption is often not correct for many of the usual multispectral images, this procedure does not always preserve the spectral content of the original image.

The *multiplicative model* (MLT) multiplies each multispectral pixel by the corresponding pixel in the high-resolution image. To scale the brightnesses back to an approximation of their original ranges, the square root of the combined brightness are calculated. The result is a combined brightness that requires some sort of weighting to restore some approximation of the relative brightnesses of the individual bands. Although these weights are selected arbitrarily, many have found the procedure to produce satisfactory results.

Composite images created by image fusion are intended for visual interpretation. Comparative evaluations of the different approaches have examined different combinations of fused data using different criteria, so it is difficult to derive definitive reports. Chavez et al. (1991) found that the HPF technique was superior to the IHS and PCT methods for fused TM and high-altitude photographic data. Carter (1988) found that the IHS and HPF methods produced some of the best results. Wald et al. (1997) proposed a framework for evaluating the quality of fused images, and emphasized that the character of the scene recorded by the imagery is an important consideration. Because fused images are composites derived from arbitrary manipulations of data, the combined images are suited for visual interpretation but not for further digital classification or analysis.

## 11.10. Summary

It is important to recognize that many of the preprocessing operations used today have been introduced into the field of remote sensing from the related fields of pattern recognition and image processing. In these other disciplines, the emphasis is usually placed upon detection or recognition of objects as portrayed on digital images. In this context, the digital values have much different significance than they do in remote sensing. Often, the analysis requires no more than recognition of contrasts between different objects or the study of objects against their backgrounds, detection of edges, and reconstruction of shapes from the configuration of edges and lines. Digital values can be manipulated freely to change image geometry or to enhance images without concern that their fundamental information content will be altered.

However, in remote sensing we usually are concerned with much more subtle variations in digital values, and we are concerned when preprocessing operations alter the digital values. Such changes may alter spectral signatures, contrasts between categories, or variances and covariances of spectral bands.

## Review Questions

1. How can an analyst know if preprocessing is advisable? Suggest how you might make this determination.
2. How can an analyst determine if specific preprocessing procedures have been effective?
3. Can you identify situations in which application of preprocessing might be inappropriate? Explain.
4. Discuss the merits of preprocessing techniques that improve the visual appearance of an image but do not alter its basic statistical properties. Are visual qualities important in the context of image analysis?

5. Examine images and maps of your region to identify prospective GCPs. Next, evaluate the pattern of your prospective GCPs. Is the pattern even, or is it necessary to select questionable points to attain an even distribution?

6. Are optimum decisions regarding preprocessing likely to vary according to the subject of the investigation? For example, would optimum preprocessing decisions for a land-cover analysis differ from those for a hydrologic or geologic analysis?

7. Assume for the moment that sixth-line striping in MSS data has a purely visual impact, with no effect on the underlying statistical qualities of the image. In your judgment, should preprocessing procedures be applied? Why or why not?

8. Can you identify analogies for preprocessing in other contexts?

9. Suppose an enterprise offers to sell images with preprocessing already completed. Would such a product be attractive to you? Why or why not?

## References


Berk, A., L. S. Bernstein, and D. C. Robertson. 1989. *MODTRAN: A Moderate Resolution Model for LOWTRAN* 7. Hanscom Air Force Base, MA: U.S. Air Force Geophysics Laboratory, 38 pp.

Bernstein, R., et al. 1983. Image Geometry and Rectification. Chapter 21 in *Manual of Remote Sensing* (R. N. Colwell, ed.). Falls Church, VA: American Society of Photogrammetry, pp. 873–922.

Brach, E. J., A. R. Mack, and V. R. Rao. 1979. Normalization of Radiance Data for Studying Crop Spectra over Time with a Mobile Field Spectro-Radiometer. *Canadian Journal of Remote Sensing,* Vol. 5, pp. 33–42.

Campbell, J. B. 1993. Evaluation of the Dark-Object Subtraction Method of Adjusting Digital Remote Sensing Data for Atmospheric Effects. In *Digital Image Processing and Visual Communications Technologies in the Earth and Atmospheric Sciences II.* (M. J. Carlotto, ed.). *SPIE Proceedings,* Vol. 1819, pp. 176–188.

Campbell, J. B., R. M. Haralick, and S. Wang. 1984. Interpretation of Topographic Relief from Digital Multispectral Imagery. In *Remote Sensing* (P. N. Slater, ed.). *SPIE Proceedings,* Vol. 475, pp. 98–116.

Campbell, J. B., and L. Ran. 1993. CHROM: A C Program to Evaluate the Application of the Dark Object Subtraction Technique to Digital Remote Sensing Data. *Computers and Geosciences,* Vol. 19, pp. 1475–1499.

Campbell, J. B., and X. Liu. 1994. Application of Dark Object Subtraction to Multispectral Data. In *Proceedings, International Symposium on Spectral Sensing Research (ISSSR '94).* Alexandria, VA: U.S. Army Corps of Engineers Topographic Engineering Center, pp. 375–386.

Campbell, J. B., and X. Liu. 1995. Chromaticity Analysis in Support of Multispectral Remote Sensing. In *Proceedings, ACSM/ASPRS Annual Convention and Exposition.* Bethesda, MD: American Society for Photogrammetry and Remote Sensing, pp. 724–932.

Chavez, P. S. 1975. Atmospheric, Solar, and M. T. F. Corrections for ERTS Digital Imagery. In *Proceedings, American Society of Photogrammetry,* Bethesda, MD: American Society for Photogrammetry and Remote Sensing, pp. 69–69a.

Chavez, P. S., G. L. Berlin, and W. B. Mitchell. 1977. Computer Enhancement Techniques of Landsat MSS Digital Images for Land Use/Land Cover Assessment. In *Proceedings of the Sixth Annual Remote Sensing of Earth Resources Conference,* pp. 259–276.

Colvocoresses, A. P. 1974. Space Oblique Mercator: A New Map Projection of the Earth. *Photogrammetric Engineering and Remote Sensing,* Vol. 40, pp. 921–926.

Davis, J. C. 2002. *Statistics and Data Analysis in Geology*. New York: Wiley, 638 pp.

Deetz, C. H., and O. S. Adams. 1945. *Elements of Map Projection with Applications to Map and Chart Construction* (U.S. Coast and Geodetic Survey Special Publication 68). Washington, DC: U.S. Department of Commerce, 226 pp.

Edwards, K., and P. A. Davis. 1994. The Use of Intensity–Hue–Saturation Transformation for Producing Color Shaded Relief Images. *Photogrammetric Engineering and Remote Sensing,* Vol. 56, pp. 1369–1374.

Estes, J. E., E. J. Hajic, L. R. Tinney, et al. 1983. Fundamentals of Image Analysis: Analysis of Visible and Thermal Infrared Data. Chapter 24 in *Manual of Remote Sensing* (R. N. Colwell, ed.). Falls Church, VA: American Society of Photogrammetry, pp. 987–1124.

Franklin, S. E., and P. T. Giles. 1995. Radiometric Processing of Aerial Imagery and Satellite Remote-Sensing Imagery. *Computers and Geosciences,* Vol. 21, pp. 413–423.

Gould, P. 1967. On the Geographical Interpretation of Eigenvalues. *Transactions, Institute of British Geographers,* Vol. 42, pp. 53–86.

Holben, B., E. Vermote, Y. J. Kaufman, D. Tarè, and V. Kalb. 1992. Aerosol Retrieval over Land from AVHRR Data. *IEEE Transactions on Geoscience and Remote Sensing,* Vol. 30, pp. 212–222.

Kovalick, W. M. 1983. *The Effect of Selected Preprocessing Procedures upon the Accuracy of a Landsat-Derived Classification of a Forested Wetland.* MS thesis, Virginia Polytechnic Institute and State University, Blacksburg, 109 pp.

Lam, N. S.-N. 1983. Spatial Interpolation Methods: A Review. *American Cartographer,* Vol. 10, pp. 129–149.

Lambeck, P. F., and J. F. Potter. 1979. Compensation for Atmospheric Effects in LANDSAT Data. In *The LACIE Symposium: Proceedings of Technical Sessions,* Houston, TX: NASA-JSC, Vol. 2, pp. 723–738.

Mather, P. M. 1999. *Computer Processing of Remotely Sensed Images: An Introduction.* Chichester, UK: Wiley, 292 pp.

Mausel, P. W., W. J. Kramber, and J. Lee. 1990. Optimum Band Selection for Supervised Classification of Multispectral Data. *Photogrammetric Engineering and Remote Sensing,* Vol. 56, pp. 55–60.

Pitts, D. E., W. E. McAllum, and A. E. Dillinger. 1974. The Effect of Atmospheric Water Vapor on Automatic Classification of ERTS Data. In *Proceedings of the Ninth International Symposium on Remote Sensing of Environment.* Ann Arbor: Institute of Science and Technology, University of Michigan, pp. 483–497.

Potter, J. F. 1984. The Channel Correlation Method for Estimating Aerosol Levels from Multispectral Scanner Data. *Photogrammetric Engineering and Remote Sensing,* Vol. 50, pp. 43–52.

Rohde, W. G., J. K. Lo, and R. A. Pohl. 1978. EROS Data Center Landsat Digital Enhancement Techniques and Imagery Availability, 1977. *Canadian Journal of Remote Sensing,* Vol. 4, pp. 63–76.

Snyder, J. P. 1978. The Space Oblique Mercator Projection. *Photogrammetric Engineering and Remote Sensing,* Vol. 44, pp. 585–596.

Snyder, J. P. 1981. Map Projections for Satellite Tracking. *Photogrammetric Engineering and Remote Sensing,* Vol. 47, pp. 205–213.

Snyder, J. P. 1982. *Map Projections Used by the U.S. Geological Survey.* Geological Survey Bulletin 1532. Washington, DC: U.S. Geological Survey, 313 pp.

Switzer, P., W. S. Kowalick, and R. J. P. Lyon. 1981. Estimation of Atmospheric Path-Radiance by the Covariance Matrix Method. *Photogrammetric Engineering and Remote Sensing,* Vol. 47, pp. 1469–1476.

Taranick, J. V. 1978. *Principles of Computer Processing of Landsat Data for Geological Applications.* (USGS Open File Report 78-117). Reston, VA: U.S. Geological Survey, 50 pp.

Vishnubhatla, S. S. 1977. *Radiometric Correction of Landsat I and Landsat II MSS Data.* (Technical Note 77-1). Ottawa: Canada Centre for Remote Sensing, 9 pp.

Westin, T. 1990. Precision Rectification of SPOT Imagery. *Photogrammetric Engineering and Remote Sensing,* Vol. 56, pp. 247–253.

## *Data Fusion*

Carper, W. J., T. M. Lillesand, and R. W. Kiefer. 1990. The Use of Intensity Hue Saturation Transformations for Merging SPOT Panchromatic and Multispectral Image Data. *Photogrammetric Engineering and Remote Sensing,* Vol. 56, pp. 459–467.

Carter, D. B. 1998. *Analysis of Multiresolution Data Fusion Techniques.* MS thesis, Virginia Polytechnic Institute and State University, Blacksburg 61 pp.

Chavez, P. S. Jr., S. C. Sides, and J. A. Anderson. 1991. Comparison of Three Different Methods to Merge Multiresolution and Multispectral Data: Landsat TM and SPOT Panchromatic. *Photogrammetric Engineering and Remote Sensing,* Vol. 57, pp. 265–303.

Pohl, C., and J. L. van Genderen. 1988. Multisensor Image Fusion in Remote Sensing: Concepts, Methods, and Applications. *International Journal of Remote Sensing,* Vol. 19, pp. 823–854.

Ranchin, T., and L. Wald. 2000. Fusion of High Spatial and Spectral Resolution Images: The ARSIS Concept and Its Implementation. *Photogrammetric Engineering and Remote Sensing,* Vol. 66, pp. 49–61.

Schowengerdt, R. A. 1980. Reconstruction of Multispatial, Multispectral Image Data Using Spatial Frequency Contents. *Photogrammetric Engineering and Remote Sensing,* Vol. 46, pp. 1325–1334.

Wald, L., T. Ranchin, and M. Mangolini, 1997. Fusion of Satellite Images of Different Spatial Resolutions: Assessing the Quality of Resulting Images. *Photogrammetric Engineering and Remote Sensing,* Vol. 63, pp. 691–699.

CHAPTER TWELVE

# Image Classification

## 12.1. Introduction

*Digital image classification* is the process of assigning pixels to classes. Usually each pixel is treated as an individual unit composed of values in several spectral bands. By comparing pixels to one another, and to pixels of known identity, it is possible to assemble groups of similar pixels into classes that are associated with the informational categories of interest to users of remotely sensed data. These classes form regions on a map or an image, so that after classification the digital image is presented as a mosaic of uniform parcels, each identified by a color or symbol (Figure 12.1). These classes are, in theory, homogeneous: pixels within classes are spectrally more similar to one another than they are to pixels in other classes. In practice, of course, each class will display some diversity, as each scene will exhibit some variability within classes.

Image classification is an important part of the fields of remote sensing, image analysis, and pattern recognition. In some instances, the classification itself may be the object of the analysis. For example, classification of land use from remotely sensed data (Chapter 20) produces a map-like image as the final product of the analysis. In other instances, the classification may be only an intermediate step in a more elaborate analysis in which the classified data form one of several data layers in a GIS. For example, in a study of water quality (Chapter 19), an initial step may be to use image classification to distinguish wetlands and open water within a scene. Later steps may then focus upon more detailed study of these areas to identify influences and to map variations in water quality. Image classification therefore forms an important tool for examination of digital images—sometimes to produce a final product, other times as one of several analytical procedures applied to derive information from an image.

The term *classifier* refers loosely to a computer program that implements a specific procedure for image classification. Over the years scientists have devised many classification strategies. The analyst must select a classification method that will best accomplish a specific task. At present it is not possible to state that a given classifier is "best" for all situations because the characteristics of each image and the circumstances for each study vary so greatly. Therefore, it is essential that each analyst understand the alternative strategies for image classification so that he or she may be prepared to select the most appropriate classifier for the task at hand.

The simplest form of digital image classification is to consider each pixel individually, and to assign it to a class based upon its several values measured in separate spectral bands (Figure

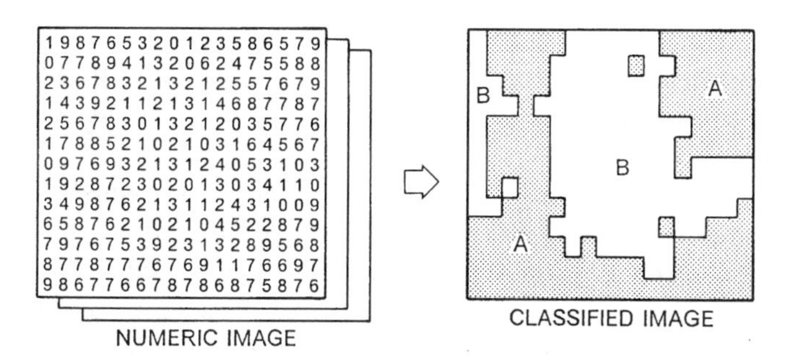


**FIGURE 12.1.** Numeric image and classified image. The classified image (right) is defined by examining the numeric image (left, and then grouping together those pixels that have similar spectral values. Here class "A" is defined by bright values (6, 7, 8, and 9). Class "B" is formed from dark pixels (values of 0, 1, 2, and 3). Usually there are many more classes, and at least three or four spectral bands.

12.2). Sometimes such classifiers are referred to as *spectral* or *point* classifiers because they consider each pixel as a "point" observation (i.e., as values isolated from their neighbors). Although point classifiers offer the benefits of simplicity and economy, they are not capable of exploiting the information contained in relationships between each pixel and those that neighbor it. Human interpreters, for example, could derive little information using the point-by-point approach because humans derive less information from the brightnesses of individual pixels than they do from the context and the patterns of brightnesses of groups of pixels, and from the sizes, shapes, and arrangements of parcels of adjacent pixels. These are the same qualities that we discussed in the context of manual image interpretation (Chapter 5).

As an alternative, more complex classification processes consider groups of pixels within their spatial setting within the image as a means of using the textural information so important for the human interpreter. These are *spatial* or *neighborhood* classifiers, which examine small areas within the image using both spectral and textural information to classify the im-

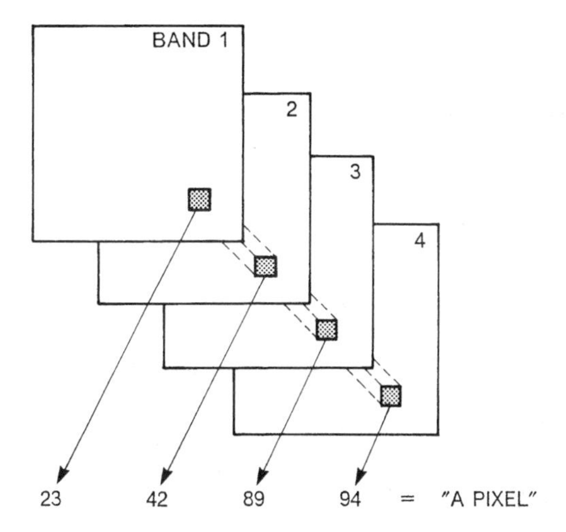


**FIGURE 12.2.** Point classifiers operate upon each pixel as a single set of spectral values, considered in isolation from its neighbors.

age (Figure 12.3). Spatial classifiers are typically more difficult to program and much more expensive to use than point classifiers. In some situations spatial classifiers have demonstrated improved accuracy, but few have found their way into routine use for remote sensing image classification.

Another kind of distinction in image classification separates supervised classification from unsupervised classification. *Supervised classification* procedures require considerable interaction with the analyst, who must guide the classification by identifying areas on the image that are known to belong to each category. *Unsupervised classification,* on the other hand, which proceeds with only minimal interaction with the analyst, in a search for natural groups of pixels present within the image. The distinction between supervised and unsupervised classification is useful, especially for students who are first learning about image classification. But the two strategies are not as clearly distinct as these definitions suggest, for some methods do not fit neatly into either category. These *hybrid classifiers* share characteristics of both supervised and unsupervised methods.

## 12.2. Informational Classes and Spectral Classes

*Informational classes* are the categories of interest to the users of the data. Informational classes are, for example, the different kinds of geological units, different kinds of forest, or the different kinds of land use that convey information to planners, managers, administrators, and scientists who use the information derived from remotely sensed data. These classes are the information that we wish to derive from the data—they are the object of our analysis. Unfortunately, these classes are not directly recorded on remotely sensed images; we can derive them only indirectly, using the evidence contained in brightnesses recorded by each image. For example, the image cannot directly show geological units, but rather only the differences in topography, vegetation, soil color, shadow, and other factors that lead the analyst to conclude that certain geological conditions exist in specific areas.

*Spectral classes* are groups of pixels that are uniform with respect to the brightnesses in their

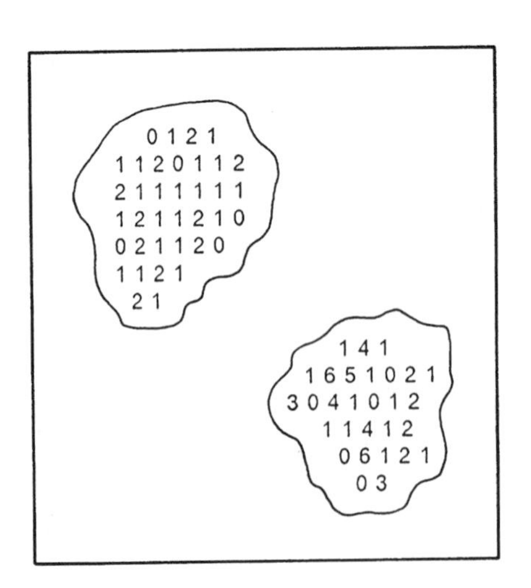


**FIGURE 12.3.** Image texture, the basis for neighborhood classifiers. A neighborhood classifier considers values within a region of the image in defining class membership. Here two regions within the image differ with respect to average brightness and also with respect to "texture," the uniformity of pixels within a small neighborhood.

several spectral channels. The analyst can observe spectral classes within remotely sensed data; if it is possible to define links between the spectral classes on the image and the informational classes that are of primary interest, then the image forms a valuable source of information. Thus remote sensing classification proceeds by matching spectral categories to informational categories. If the match can be made with confidence, then the information is likely to be reliable. If spectral and informational categories do not correspond, then the image is unlikely to be a useful source for that particular form of information. Seldom can we expect to find exact one-to-one matches between informational and spectral classes. Any informational class includes spectral variations arising from natural variations within the class. For example, a region of the informational class "forest" is still "forest," even though it may display variations in age, species composition, density, and vigor, which all lead to differences in the spectral appearance of a single informational class. Furthermore, other factors, such as variations in illumination and shadowing, may produce additional variations even within otherwise spectrally uniform classes.

Thus informational classes are typically composed of numerous *spectral subclasses,* spectrally distinct groups of pixels that together may be assembled to form an informational class (Figure 12.4). In digital classification, we must often treat spectral subclasses as distinct units during classification, but then display several spectral classes under a single symbol for the final image or map to be used by planners or administrators (who are, after all, interested only in the informational categories, not in the intermediate steps required to generate them).

In subsequent chapters we will be interested in several properties of spectral classes. For each band, each class is characterized by a mean, or average, value that of course represents the typical brightness of each class. In nature, all classes exhibit some variability around their mean values: some pixels are darker than the average, others a bit brighter. These departures from the mean are measured as the *variance,* or sometimes by the *standard deviation* (the square root of the variance).

Often we wish to assess the distinctiveness of separate spectral classes, perhaps to determine if they really are separate classes or if they should be combined to form a single, larger class. A

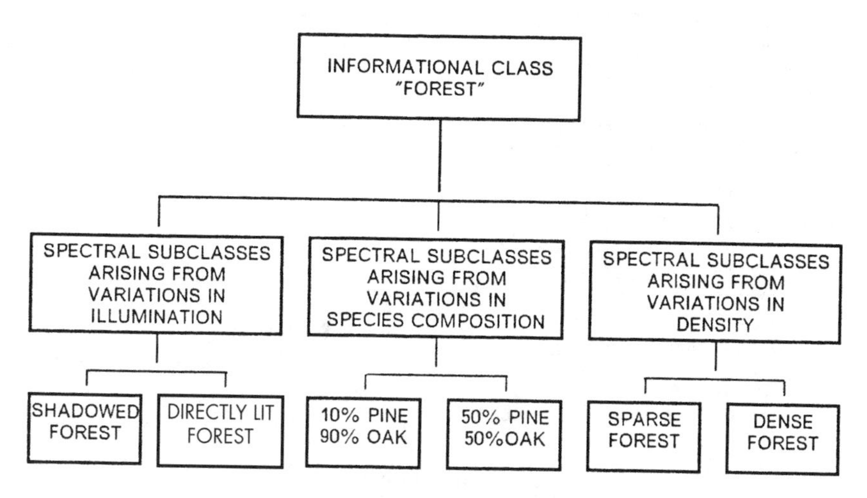


**FIGURE 12.4.** Spectral subclasses.

crude measure of the distinctiveness of two classes is simply the difference in the mean values; presumably classes that are very different should have big differences in average brightness, while classes that are similar to one another should have small differences in average brightness. This measure is a bit too simple because it does not take into account the differences in variability between classes.

Another simple measure of distinctiveness is the *normalized difference* (ND), found by dividing the difference in class means by the sum of their standard deviations. For two classes A and B, with means $a$ and $b$, and standard deviations $s_a$ and $s_b$, the ND is defined as

$$\text{ND} = \frac{|\bar{x}_a - \bar{x}_b|}{s_a + s_b} \quad \text{(Eq. 12.1)}$$

As an example, Table 12.1 summarizes properties of several classes, then uses these properties to show the normalized difference between these classes in several spectral channels. Note that some pairs of categories are very distinct relative to others. Also note that some spectral channels are much more effective than others in separating categories from one another.

## 12.3. Unsupervised Classification

*Unsupervised classification* can be defined as the identification of natural groups, or structures, within multispectral data. The notion of the existence of natural, inherent groupings of spectral values within a scene may not be intuitively obvious, but it can be demonstrated that remotely sensed images are usually composed of spectral classes that are reasonably uniform internally in respect to brightnesses in several spectral channels. Unsupervised classification is the definition, identification, labeling, and mapping of these natural classes.

**TABLE 12.1. Normalized Difference**

| | MSS band | | | | MSS band | | | |
|---|---|---|---|---|---|---|---|---|
| | 1 | 2 | 3 | 4 | 1 | 2 | 3 | 4 |
| | Water | | | | Forest | | | |
| $\bar{x}$ | 37.5 | 31.9 | 22.8 | 6.3 | 26.9 | 16.6 | 55.7 | 32.5 |
| $s$ | 0.67 | 2.77 | 2.44 | 0.82 | 1.21 | 1.49 | 3.97 | 3.12 |
| | Crop | | | | Pasture | | | |
| $\bar{x}$ | 37.7 | 38.0 | 52.3 | 27.3 | 28.6 | 22.0 | 53.4 | 32.9 |
| $s$ | 3.56 | 5.08 | 4.13 | 4.42 | 1.51 | 5.09 | 13.16 | 3.80 |

$$\text{Water} - \text{forest (band 6)}: \frac{55.7 - 22.8}{2.44 + 3.97} = \frac{32.9}{6.41} = 5.13$$

$$\text{Crop} - \text{pasture (band 6)}: \frac{52.3 - 53.4}{4.13 + 13.16} = \frac{1.1}{17.29} = 0.06$$

### *Advantages*

The advantages of unsupervised classification (relative to supervised classification) can be enumerated as follows:

- *No extensive prior knowledge of the region is required.* Or, more accurately, the nature of knowledge required for unsupervised classification differs from that required for supervised classification. To conduct supervised classification, detailed knowledge of the area to be examined is required to select representative examples of each class to be mapped. To conduct unsupervised classification, no detailed prior knowledge is required, but knowledge of the region is required to interpret the meaning of the results produced by the classification process.
- *Opportunity for human error is minimized.* To conduct unsupervised classification, the operator may perhaps specify only the number of categories desired (or possibly, minimum and maximum limits on the number of categories), and sometimes constraints governing the distinctness and uniformity of groups. Many of the detailed decisions required for supervised classification are not required for unsupervised classification, so the analyst is presented with less opportunity for error. If the analyst has inaccurate preconceptions regarding the region, they will have little opportunity to influence the classification. The classes defined by unsupervised classification are often much more uniform with respect to spectral composition than those generated by supervised classification.
- *Unique classes are recognized as distinct units.* Such classes, perhaps of very small areal extent, may remain unrecognized in the process of supervised classification and could inadvertently be incorporated into other classes, generating error and imprecision throughout the entire classification.

### ***Disadvantages and Limitations***

The disadvantages and limitations of unsupervised classification arise primarily from reliance upon "natural" groupings and difficulties in matching these groups to the informational categories that are of interest to the analyst.

- *Unsupervised classification identifies spectrally homogeneous classes within the data that do not necessarily correspond to the informational categories that are of interest to the analyst.* As a result, the analyst is faced with the problem of matching spectral classes generated by the classification to the informational classes that are required by the ultimate user of the information. Seldom is there a simple one-to-one correspondence between the two sets of classes.
- *The analyst has limited control over the menu of classes and their specific identities.* If it is necessary to generate a specific menu of informational classes (e.g., to match to other classifications for other dates or adjacent regions), the use of unsupervised classification may be unsatisfactory.
- *Spectral properties of specific informational classes will change over time* (on a seasonal basis, as well as over the years). As a result, relationships between informational classes and spectral classes are not constant and relationships defined for one image cannot be extended to others.

### *Distance Measures*

Some of the basic elements of unsupervised classification can be illustrated using the data presented in Table 12.2. These values can be plotted on simple diagrams constructed using brightnesses of two spectral bands as orthogonal axes (Figures 12.5). These two-dimensional plots illustrate principles that can be extended to include additional variables (e.g., to analyze Landsat ETM+ bands 1 through 5 simultaneously). The additional variables create three- or four-dimensional plots of points in multidimensional data space, which are difficult to depict in a diagram, but can be envisioned as groups of pixels represented by swarms of points with depth in several dimensions (Figure 12.6).

The general form for such diagrams is illustrated by Figure 12.7. A diagonal line of points extends upward from a point near the origin. This pattern occurs because a pixel that is dark in one band will often tend to be dark in another band, and as brightness increases in one spectral region, it tends to increase in others. (This relationship often holds for spectral measurements in the visible and near infrared; it is not necessarily observed in data from widely separated spectral regions.) In Figure 12.5 specific groupings, or clusters, are evident; these clusters may correspond to informational categories of interest to the analyst. Unsupervised classification is the process of defining such clusters in multidimensional data space and (if possible) matching them to informational categories.

When we consider large numbers of pixels, the clusters are not usually as distinct because pixels of intermediate values tend to fill in the gaps between groups (Figure 12.7). We must therefore apply a variety of methods to assist us in the identification of those groups that may be

**TABLE 12.2. Landsat MSS Digital Values for February**

| | MSS band | | | | | MSS band | | | |
|---|---|---|---|---|---|---|---|---|---|
| | 1 | 2 | 3 | 4 | | 1 | 2 | 3 | 4 |
| 1. | 19 | 15 | 22 | 11 | 21. | 24 | 24 | 25 | 11 |
| 2. | 21 | 15 | 22 | 12 | 22. | 25 | 25 | 38 | 20 |
| 3. | 19 | 13 | 25 | 14 | 23. | 20 | 29 | 19 | 3 |
| 4. | 28 | 27 | 41 | 21 | 24. | 28 | 29 | 18 | 2 |
| 5 | 27 | 25 | 32 | 19 | 25. | 25 | 26 | 42 | 21 |
| 6. | 21 | 15 | 25 | 13 | 26. | 24 | 23 | 41 | 22 |
| 7. | 21 | 17 | 23 | 12 | 27. | 21 | 18 | 12 | 12 |
| 8. | 19 | 16 | 24 | 12 | 28. | 25 | 21 | 31 | 15 |
| 9. | 19 | 12 | 25 | 14 | 29. | 22 | 22 | 31 | 15 |
| 10. | 28 | 29 | 17 | 3 | 30. | 26 | 24 | 43 | 21 |
| 11. | 28 | 26 | 41 | 21 | 31. | 19 | 16 | 24 | 12 |
| 12. | 19 | 16 | 24 | 12 | 32. | 30 | 31 | 18 | 3 |
| 13. | 29 | 32 | 17 | 3 | 33. | 28 | 27 | 44 | 24 |
| 14. | 19 | 16 | 22 | 12 | 34. | 22 | 22 | 28 | 15 |
| 15. | 19 | 16 | 24 | 12 | 35. | 30 | 31 | 18 | 2 |
| 16. | 19 | 16 | 25 | 13 | 36. | 19 | 16 | 22 | 12 |
| 17. | 24 | 21 | 35 | 19 | 37. | 30 | 31 | 18 | 2 |
| 18. | 22 | 18 | 31 | 14 | 38. | 27 | 23 | 34 | 20 |
| 19. | 21 | 18 | 25 | 13 | 39. | 21 | 16 | 22 | 12 |
| 20. | 21 | 16 | 27 | 13 | 40. | 23 | 22 | 26 | 16 |

*Note.* These are raw digital values for a forested area in central Virginia as acquired by the Landsat 1 MSS in February 1974. These values represent the same area as those in Table 12.3, although individual pixels do not correspond.

**TABLE 12.3. Landsat MSS Digital Values for May**

| | MSS band | | | | | MSS band | | | |
|---|---|---|---|---|---|---|---|---|---|
| | 1 | 2 | 3 | 4 | | 1 | 2 | 3 | 4 |
| 1. | 34 | 28 | 22 | 6 | 21. | 26 | 16 | 52 | 29 |
| 2. | 26 | 16 | 52 | 29 | 22. | 30 | 18 | 57 | 35 |
| 3. | 36 | 35 | 24 | 6 | 23. | 30 | 18 | 62 | 28 |
| 4. | 39 | 41 | 48 | 23 | 24. | 35 | 30 | 18 | 6 |
| 5. | 26 | 15 | 52 | 31 | 25. | 36 | 33 | 24 | 7 |
| 6. | 36 | 28 | 22 | 6 | 26. | 27 | 16 | 57 | 32 |
| 7. | 28 | 18 | 59 | 35 | 27. | 26 | 15 | 57 | 34 |
| 8. | 28 | 21 | 57 | 34 | 28. | 26 | 15 | 50 | 29 |
| 9. | 26 | 16 | 55 | 30 | 29. | 26 | 33 | 24 | 27 |
| 10. | 32 | 30 | 52 | 25 | 30. | 36 | 36 | 27 | 8 |
| 11. | 40 | 45 | 59 | 26 | 31. | 40 | 43 | 51 | 27 |
| 12. | 33 | 30 | 48 | 24 | 32. | 30 | 18 | 62 | 38 |
| 13. | 28 | 21 | 57 | 34 | 33. | 28 | 18 | 62 | 38 |
| 14. | 28 | 21 | 59 | 35 | 34. | 36 | 33 | 22 | 6 |
| 15. | 36 | 38 | 48 | 22 | 35. | 35 | 36 | 56 | 33 |
| 16. | 36 | 31 | 23 | 5 | 36. | 42 | 42 | 53 | 26 |
| 17. | 26 | 19 | 57 | 33 | 37. | 26 | 16 | 50 | 30 |
| 18. | 36 | 34 | 25 | 7 | 38. | 42 | 38 | 58 | 33 |
| 19. | 36 | 31 | 21 | 6 | 39. | 30 | 22 | 59 | 37 |
| 20. | 27 | 19 | 55 | 30 | 40. | 27 | 16 | 56 | 34 |

*Note.* These are raw digital values for a forested area in central Virginia as acquired by the Landsat 1 MSS in May 1974. These values represent the same area as those in Table 12.2, although individual pixels do not correspond.

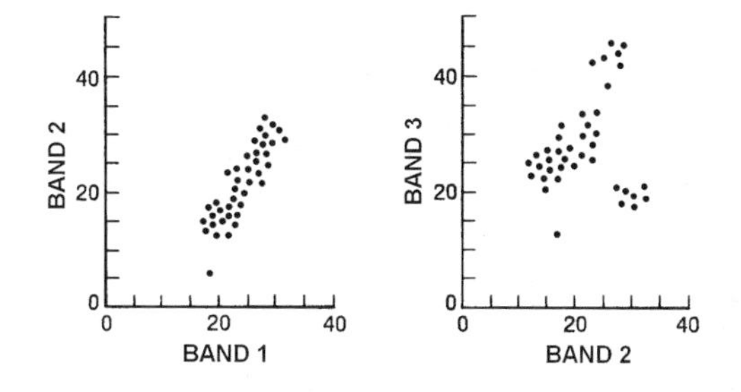


**FIGURE 12.5.** Two-dimensional scatter diagrams illustrating the grouping of pixels within multispectral data space.

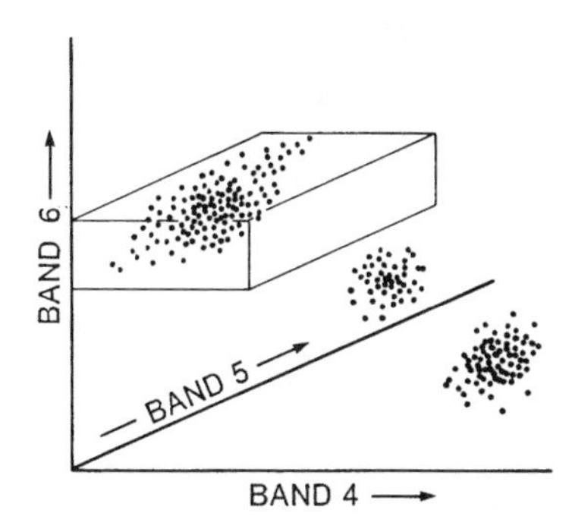


**FIGURE 12.6.** Sketch illustrating multidimensional scatter diagram. Here three bands of data are shown.

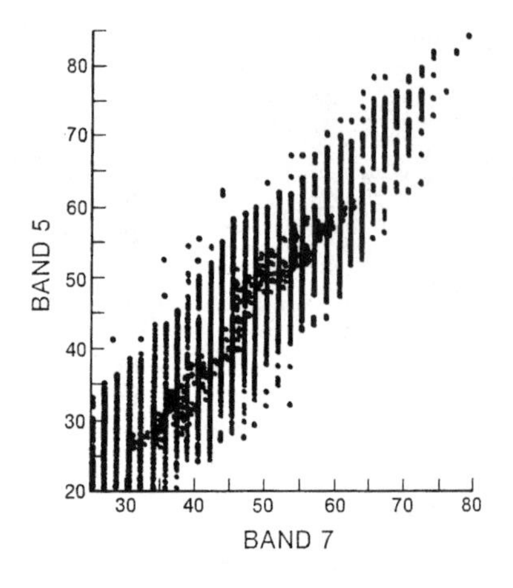


**FIGURE 12.7.** Scatter diagram. Data from two Landsat MSS bands illustrate the general form of the relationship between spectral measurements in neighboring regions of the spectrum. The diagram shows several hundred points; when so many values are shown, the distinct clusters visible in Figures 12.5 and 12.6 are often not visible. The groups may still be present, but may only be able to be detected with the aid of classification algorithms that can simultaneously consider values in several spectral bands. From Todd et al. (1980, p. 511). Copyright 1980 by the American Society for Photogrammetry and Remote Sensing. Reproduced by permission.

present within the data but may not be obvious to visual inspection. Over the years image scientists and statisticians have developed a wide variety of procedures for identifying such clusters—procedures that vary greatly in complexity and effectiveness. The general model for unsupervised classification can, however, be illustrated using one of the simplest classification strategies, which can serve as the foundation for understanding more complex approaches.

Figure 12.8 shows two pixels, each with measurements in several spectral channels, plotted in multidimensional data space in the same manner as those illustrated in Figure 12.5. For ease

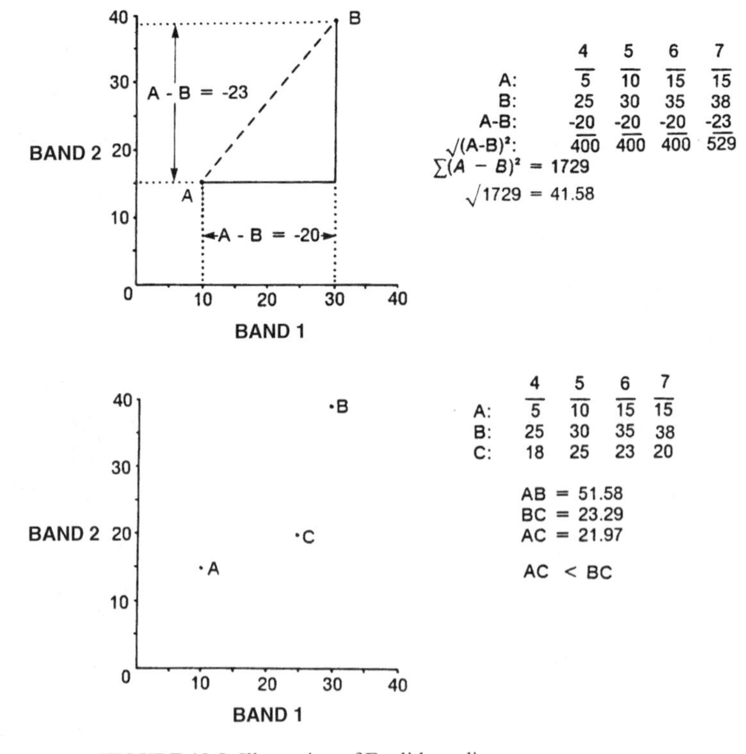


**FIGURE 12.8.** Illustration of Euclidean distance measure.

of illustration, only two bands are shown here, although the principles illustrated extend to as many bands as may be available.

Unsupervised classification of an entire image must consider many thousands of pixels. But the classification process is always based upon the answer to the same question: "Do the two pixels belong to the same group?" For this example, the question is, "Should pixel C be grouped with pixel A or with pixel B?" (see the lower part of Figure 12.8). This question can be answered by finding the distance between pairs of pixels. If the distance between A and C is greater than that between B and C, then B and C are said to belong to the same group and A may be defined as a member of a separate class.

There are thousands of pixels in a remotely sensed image; if they are considered individually as prospective members of groups, the distances to other pixels can always be used to define group membership. How can such distances be calculated? A number of methods for finding distances in multidimensional data space are available. One of the simplest is *Euclidean distance:*

$$D_{ab} = \left[ \sum_{i=1}^{n} (a_i - b_i)^2 \right]^{1/2} \quad \text{(Eq. 12.2)}$$

where $i$ is one of $n$ spectral bands, $a$ and $b$ are pixels, and $D_{ab}$ is the distance between the two pixels. The distance calculation is based upon the Pythagorean theorem (Figure 12.9):

$$c = \sqrt{a^2 + b^2} \quad \text{(Eq. 12.3)}$$

In this instance we are interested in distance $c$; $a$, $b$, and $c$ are measured in units of the two spectral channels.

$$c = D_{ab} \quad \text{(Eq. 12.4)}$$

To find $D_{ab}$, we need to find distances $a$ and $b$. Distance $a$ is found by subtracting values of $A$ and $B$ in MSS channel 7 ($a = 38 - 15 = 23$). Distance $b$ is found by finding the difference between $A$ and $B$ with respect to channel 6 ($b = 30 - 10 = 20$).

$$D_{ab} = c = \sqrt{20^2 + 23^2}$$

$$D_{ab} = \sqrt{400 + 529} = \sqrt{929}$$

$$D_{ab} = 30.47$$

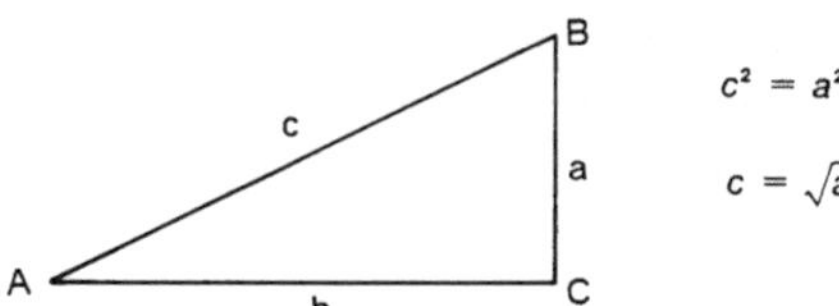


$c^2 = a^2 + b^2$

$c = \sqrt{a^2 + b^2}$

**FIGURE 12.9.** Definition of symbols used for explanation of Euclidean distance.

This measure can be applied to as many dimensions (spectral channels) as might be available, by addition of distances. For example:

| | Landsat MSS band | | | |
|---|---|---|---|---|
| | 1 | 2 | 3 | 4 |
| Pixel A | 34 | 28 | 22 | 6 |
| Pixel B | 26 | 16 | 52 | 29 |
| Difference | 8 | 12 | –30 | –23 |
| $(\text{Difference})^2$ | 64 | 144 | 900 | 529 |

Total of $(\text{differences})^2 = 1{,}637$

$\sqrt{\text{total}} = 40.5$

The lower part of Figure 12.9 shows another worked example.

Thus the Euclidean distance between A and B is equal to 40.45 distance units. This value in itself has little significance, but in relation to other distances it forms a means of defining similarities between pixels. For example, if we find that distance *ab* = 40.45 and that distance *ac* = 86.34, then we know that pixel A is closer (i.e., more nearly similar) to B than it is to C, and that we should form a group from A and B rather than from A and C.

Unsupervised classification proceeds by making thousands of distance calculations as a means of determining similarities for the many pixels and groups within an image. Usually the analyst does not actually know any of these many distances that must be calculated for unsupervised classification, as the computer presents only the final classified image without the intermediate steps necessary to derive the classification. Nonetheless, distance measures are the heart of unsupervised classification.

But not all distance measures are based upon Euclidean distance. Another simple measure of determining distance is the $L_1$ *distance,* the sum of the absolute differences between values in individual bands (Swain and Davis, 1978). For the example given above, the $L_1$ distance is 73 ($73 = 8 + 12 + 30 + 23$). Other distance measures have been defined for unsupervised classification; many are rather complex methods of scaling distances to promote effective groupings of pixels.

Unsupervised classification often proceeds in an interactive fashion to search for an optimal allocation of pixels to categories, given the constraints specified by the analyst. A computer program for unsupervised classification includes an algorithm for calculation of distances as described above (sometimes the analyst might be able to select between several alternative distance measures) and a procedure for finding, testing, and then revising classes according to limits defined by the analyst. The analyst may be required to specify limits upon the number of clusters to be generated, to constrain the diversity of values within classes, or to require that classes exhibit a specified minimum degree of distinctness with respect to neighboring groups. Specific classification procedures may define distances differently. For example, it is possible to calculate distances to the centroid of each group, or to the closest member of each group, or perhaps to the most densely occupied region of a cluster. Such alternatives represent refinements of the basic strategy outlined here, and each may offer advantages in certain situations. Also, there are many variations in the details of how each classification program may operate.

Because so many distance measures must be calculated for classification of a remotely sensed image, most programs use variations of these basic procedures that accomplish the same objectives with improved computational efficiency.

### *Sequence for Unsupervised Classification*

A typical sequence might begin with the analyst specifying minimum and maximum numbers of categories to be generated by the classification algorithm. These values might be based upon the analyst's knowledge of the scene or upon the user's requirements that the final classification display a certain number of classes. The classification starts with a set of arbitrarily selected pixels as cluster centers; often these are selected at random to ensure that the analyst cannot influence the classification and that the selected pixels are representative of values found throughout the scene. The classification algorithm then finds distances (as described above) between pixels and forms initial estimates of cluster centers as permitted by the constraints specified by the analyst. The class can be represented by a single point, known as the "class centroid," which can be thought of as the center of the cluster of pixels for a given class, even though many classification procedures do not always define it as the exact center of the group. At this point, classes consist only of the arbitrarily selected pixels chosen as initial estimates of class centroids. In the next step, all the remaining pixels in the scene are assigned to the nearest class centroid. The entire scene has now been classified, but this classification forms only an estimate of the final result, as the classes formed by this initial attempt are unlikely to be the optimal set of classes and may not meet the constraints specified by the analyst.

To begin the next step, the algorithm finds new centroids for each class, as the addition of new pixels to the classification means that the initial centroids are no longer accurate. Then the entire scene is classified again, with each pixel assigned to the nearest centroid. And again new centroids are calculated; if the new centroids differ from those found in the preceding step, then the process repeats until there is no significant change detected in locations of class centroids and the classes meet all the constraints required by the operator.

Throughout the process the analyst generally has no interaction with the classification, so it operates as an "objective" classification within the constraints provided by the analyst. Also, the unsupervised approach identifies the "natural" structure of the image in the sense that it finds uniform groupings of pixels that form distinct classes without the influence of preconceptions regarding their identities or distributions. The entire process, however, cannot be considered to be truly "objective," as the analyst has made decisions regarding the data to be examined, the algorithm to be used, the number of classes to be found, and (possibly) the uniformity and distinctness of classes. Each of these decisions influences the character and the accuracy of the final product, so it cannot be regarded as a result isolated from the context in which it was made.

Many different procedures for unsupervised classification are available; despite their diversity, most are based upon the general strategy just described. Although some refinements are possible to improve computational speed and efficiency, this approach is in essence a kind of wearing down of the classification problem by repetitive application assignment and reassignment of pixels to groups. The three key components to any unsupervised classification algorithm are chosing effective methods for (1) measuring distances in data space, (2) identifying class centroids, and (3) testing the distinctness of classes. There are many different strategies for accomplishing each of these tasks; an enumeration of even the most widely used methods is outside the scope of this text, but some are described in the articles listed in the references.

### *AMOEBA*

A useful variation on the basic strategy for unsupervised classification has been described by Bryant (1978). The AMOEBA classification operates in the manner of usual unsupervised classification, with the addition of a *contiguity constraint* that considers the locations of values as spectral classes are formed. The analyst specifies a tolerance limit that governs the diversity permitted as classes are formed. As a class is formed by a group of neighboring pixels, adjacent pixels belonging to other classes are considered as prospective members of the class if it occurs as a small region within a larger, more homogeneous background (Figure 12.10). If the candidate pixel has values that fall within the tolerance limits specified by the analyst, the pixel is accepted as a member of the class despite the fact that it differs from other members. Thus locations as well as spectral properties of pixels form classification criteria for AMOEBA and similar algorithms. Although this approach increases the spectral diversity of classes—normally an undesirable quality—the inclusion of small areas of foreign pixels as members of a more extensive region may satisfy our notion of the proper character of a geographic region. Often, in the manual preparation of maps, the analyst will perform a similar form of generalization as a means of eliminating small inclusions within larger regions. This kind of generalization presents cartographic data in a form suitable for presentation at small scale. The AMOEBA classifier was designed for application to scenes composed primarily of large homogeneous regions, such as the agricultural landscapes of the North American prairies. For such scenes, it seems to work well, although it may not be as effective in more complex landscapes composed of smaller parcels (Story et al., 1984).

### *Assignment of Spectral Categories to Informational Categories*

The classification results described thus far provide *uniform groupings* of pixels—classes that are uniform with respect to the spectral values that compose each pixel. These spectral classes

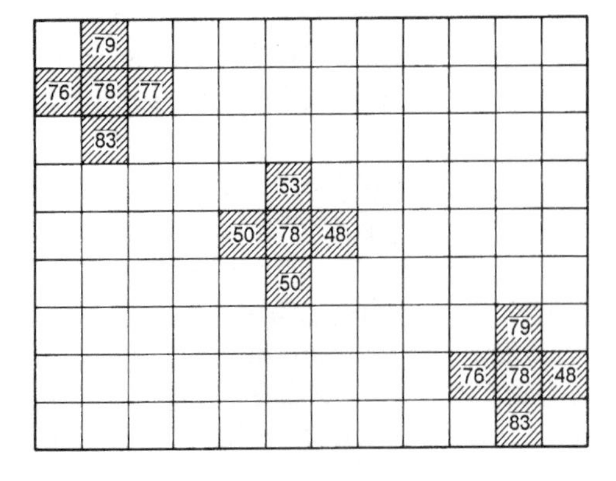


**FIGURE 12.10.** AMOEBA spatial operator. This contrived example shows AMOEBA as it considers classification of three pixels, all with the same digital value of "78." Upper left: The central pixel has a value similar to those of its neighbors; it is classified together with its neighbors. Center: The central pixel differs from its neighbors, but due to its position within a background of darker but homogeneous pixels it would be classified with the darker pixels. Lower left: The central pixel would be classified with the darker pixels at the top, left, and bottom. In all three situations, the classification assignment is based upon values at locations adjacent to the sample pixel.

are significant only to the extent that they can be matched to one or more informational classes that are of interest to the user of the final product. These spectral classes may sometimes correspond directly to informational categories. For example, in Figure 12.11 the pixels that compose the group closest to the origin correspond to "open water." This identification is more or less obvious from the spectral properties of the class: few other classes will exhibit such dark values in both spectral channels. However, seldom can we depend upon a clear identification of spectral categories from spectral values alone. Often it is possible to match spectral and informational categories by examining patterns on the image; many informational categories are recognizable by the positions, sizes, and shapes of individual parcels and their spatial correspondence with areas of known identity.

Often, however, spectral classes do not match directly to informational classes. In some instances, informational classes may occur in complex mixtures and arrangements. Perhaps forested patches are scattered in small areas against a more extensive background of grassland. If these forested areas are small relative to the spatial resolution of the sensor, then the overall spectral response for such an area will differ from either "forest" or "grassland," due to the effect of mixed pixels upon the spectral response. This region may be assigned then to a class separate from either the forest or the grassland classes (Chapter 10). It is the analyst's task to apply his or her knowledge of the region and the sensor to identify the character of the problem and to apply an appropriate remedy.

Or, because of spectral diversity, even nominally uniform informational classes may manifest themselves as a set of spectral classes. For example, a forested area may be recorded as several spectral clusters, due perhaps to variations in density, age, aspect, shadowing, and other factors that alter the spectral properties of a forested region, but do not alter the fact that the region belongs to the informational class "forest." The analyst must therefore examine the output to match spectral categories from the classification with the informational classes of significance to those who will use the results.

Thus a serious practical problem with unsupervised classification is that clear matches

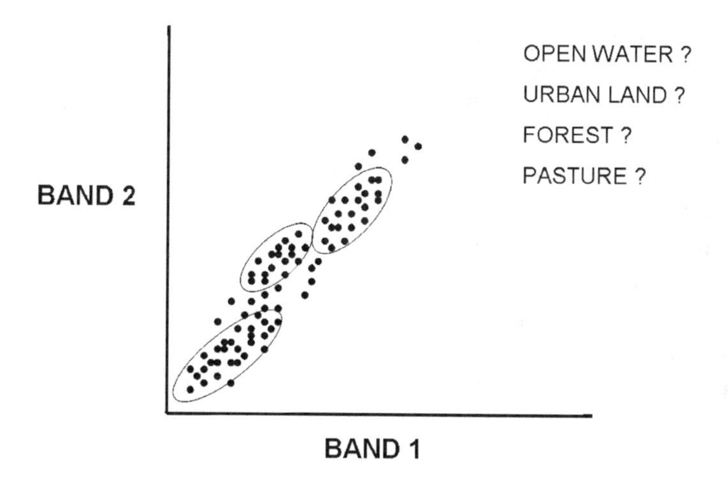


**FIGURE 12.11.** Assignment of spectral categories to image classes. Unsupervised classification defines the clusters defined schematically on the scatter diagram. The analyst must decide which, if any, match to the list of informational categories that form the objective of the analysis.

between spectral and informational classes are not always possible; some informational categories may not have direct spectral counterparts, and vice versa. Furthermore, the analyst does not have control over the nature of the categories generated by unsupervised classification. If a study is conducted to compare results with those from an adjacent region or from a different date, for example, it may be necessary to have the same set of informational categories on both maps or the comparison cannot be made. If unsupervised classification is to be used, it may be difficult to generate the same sets of informational categories on both images.

## 12.4. Supervised Classification

*Supervised classification* can be defined informally as the process of using samples of known identity (i.e., pixels already assigned to informational classes) to classify pixels of unknown identity (i.e., to assign unclassified pixels to one of several informational classes). Samples of known identity are those pixels located within *training areas,* or *training fields.* The analyst defines training areas by identifying regions on the image that can be clearly matched to areas of known identity on the image. Such areas should typify spectral properties of the categories they represent, and, of course, must be homogeneous in respect to the informational category to be classified. That is, training areas should not include unusual regions, nor should they straddle boundaries between categories. Size, shape, and position must favor convenient identification both on the image and on the ground. Pixels located within these areas form the *training samples* used to guide the classification algorithm to assign specific spectral values to appropriate informational classes. Clearly, the selection of these training data is a key step in supervised classification.

### *Advantages*

The advantages of supervised classification, relative to unsupervised classification, can be enumerated as follows: First, the analyst has control of a selected menu of informational categories tailored to a specific purpose and a specific geographic region. This quality may be vitally important if it becomes necessary to generate a classification for the specific purpose of comparison with another classification of the same area at a different date or if the classification must be compatible with those of neighboring regions. Under such circumstances, the unpredictable (i.e., with respect to number, identity, size, and pattern) qualities of categories generated by unsupervised classification may be inconvenient or unsuitable. Second, supervised classification is tied to specific areas of known identity, determined through the process of selecting training areas. Third, the analyst using supervised classification is not faced with the problem of matching spectral categories on the final map with the informational categories of interest (this task has, in effect, been addressed during the process of selecting training data). Fourth, the operator may be able to detect serious errors in classification by examining training data to determine if they have been correctly classified by the classification procedure—inaccurate classification of training data indicates serious problems in the classification or selection of training data, although correct classification of training data does not always indicate correct classification of other data.

### *Disadvantages and Limitations*

The disadvantages of supervised classification are numerous. First, the analyst, in effect, imposes a classification structure upon the data (recall that unsupervised classification searches for "natural" classes). These operator-defined classes may not match the natural classes that exist within the data, and therefore may not be distinct or well defined in multidimensional data space. Second, training data are often defined primarily with reference to informational categories and only secondarily with reference to spectral properties. A training area that is "100% forest" may be accurate with respect to the "forest" designation, but may still be very diverse with respect to density, age, shadowing, and the like, and therefore form a poor training area. Third, training data selected by the analyst may not be representative of conditions encountered throughout the image. This may be true despite the best efforts of the analyst, especially if the area to be classified is large, complex, or inaccessible. Fourth, conscientious selection of training data can be a time-consuming, expensive, and tedious undertaking, even if ample resources are at hand. The analyst may experience problems in matching prospective training areas as defined on maps and aerial photographs to the image to be classified. Fifth, supervised classification may not be able to recognize and represent special or unique categories not represented in the training data, possibly because they are not known to the analyst or because they occupy very small areas on the image.

### *Training Data*

*Training fields* are areas of known identity delineated on the digital image, usually by specifying the corner points of a square or rectangular area using line and column numbers within the coordinate system of the digital image. The analyst must, of course, know the correct class for each area. Usually the analyst begins by assembling and studying maps and aerial photographs of the area to be classified and by investigating selected sites in the field. (Here we assume that the analyst has some field experience with the specific area to be studied, is familiar with the particular problem the study is to address, and has conducted the necessary field observations prior to initiating the actual selection of training data.) Specific training areas are identified for each informational category, following the guidelines outlined below. The objective is to identify a set of pixels that accurately represent the spectral variation present within each informational region (Figure 12.12).

### *Key Characteristics of Training Areas*

#### *Numbers of Pixels*

An important concern is the overall number of pixels selected for each category; as a general guideline, the operator should ensure that several individual training areas for each category provide a total of at least 100 pixels for each category.

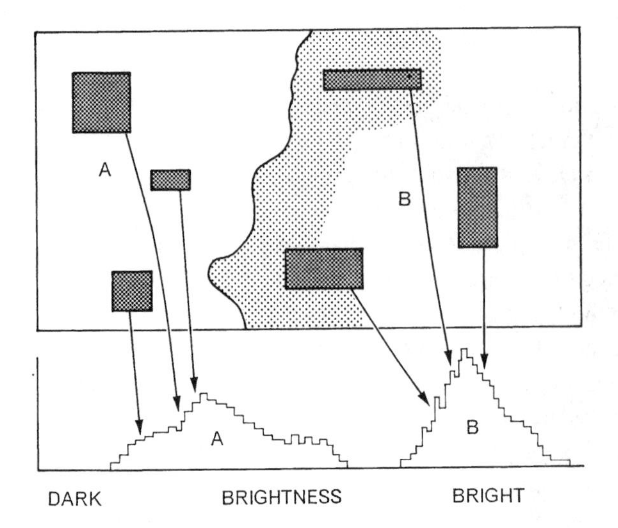


**FIGURE 12.12.** Training fields and training data. Training fields, each composed of many pixels, sample the spectral characteristics of informational categories. Here the shaded figures represent the training fields, each positioned carefully to estimate the spectral properties of each class, as depicted by the histograms. This information provides the basis for classification of the remaining pixels outside the training fields.

### *Size*

Sizes of training areas are important. Each must be large enough to provide accurate estimates of the properties of each informational class. Therefore, they must as a group include enough pixels to form reliable estimates of the spectral characteristics of each class (hence, the minimum figure of 100 suggested above). Individual training fields should not, on the other hand, be too big, as large areas tend to include undesirable variation. (Because the total number of pixels in the training data for each class can be formed from many separate training fields, each individual area can be much smaller than the total number of pixels required for each class.) Joyce (1978) recommends that individual training areas be at least 4 ha (10 acres) in size at the absolute minimum and preferably include about 16 ha (40 acres). Small training fields are difficult to locate accurately on the image. To accumulate an adequate total number of training pixels, the analyst must devote more time to definition and analysis of the additional training fields. Conversely, use of large training fields increases the opportunity for inclusion of spectral inhomogeneities. Joyce (1978) suggests 65 ha (160 acres) as the maximum size for training fields.

Joyce specifically refers to Landsat MSS data. So in terms of numbers of pixels, he is recommending from 10 to about 40 pixels for each training field. For TM or SPOT data, of course, his recommendations would specify different numbers of pixels, as the resolutions of these sensors differ from that of the MSS. Also, since the optimum sizes of training fields vary according to the heterogeneity of each landscape and each class, each analyst should develop his or her own guidelines based on experience acquired in specific circumstances.

### *Shape*

Shapes of training areas are not important, provided that shape does not prohibit accurate delineation and positioning of correct outlines of regions on digital images. Usually it is easiest to define square or rectangular areas; such shapes minimize the number of vertices that must be specified, typically the most bothersome task for the analyst.

*Location*

Location is important, as each informational category should be represented by several training areas positioned throughout the image. Training areas must be positioned in locations that favor accurate and convenient transfer of their outlines from maps and aerial photographs to the digital image. Because as the training data are intended to represent variation within the image, they must not be clustered in favored regions of the image that may not typify conditions encountered throughout the image as a whole. It is desirable for the analyst to use direct field observations in the selection of training data, but the requirement for an even distribution of training fields often conflicts with practical constraints: it may not be practical to visit remote or inaccessible sites that may seem to form good areas for training data. Often aerial observation, or use of good maps and aerial photographs, can provide the basis for accurate delineation of training fields that cannot be inspected in the field. Although such practices are often sound, it is important to avoid development of a cavalier approach to selection of training data that depends completely upon indirect evidence in situations when direct observation is feasible.

*Number*

The optimum number of training areas depends upon the number of categories to be mapped, their diversity, and the resources that can be devoted to delineating training areas. Ideally, each informational category or each spectral subclass should be represented by a number (five to 10 at a minimum) of training areas to ensure that the spectral properties of each category are represented. Because informational classes are often spectrally diverse, it may be necessary to use several sets of training data for each informational category, due to the presence of spectral subclasses. Selection of multiple training areas is also desirable because later in the classification process it may be necessary to discard some training areas if they are discovered to be unsuitable. Experience teaches that it is usually better to define many small training areas than to create only a few large areas.

*Placement*

Placement of training areas may be important. Training areas should be placed within the image in a manner that permits convenient and accurate location with respect to distinctive features, such as water bodies, and boundaries between distinctive features on the image. They should be distributed throughout the image so that they provide a basis for representation of the diversity present within the scene. Boundaries of training fields should be placed well away from the edges of contrasting parcels so that they do not encompass edge pixels.

*Uniformity*

Perhaps the most important property of a good training area is its uniformity, or homogeneity (Figure 12.13) . Data within each training area should exhibit a unimodal frequency distribution for each spectral band to be used (Figure 12.14a). Prospective training areas that exhibit bimodal histograms should be discarded if their boundaries cannot be adjusted to yield more uniformity. Training data provide values that estimate the means, variances, and covariances of spectral data measured in several spectral channels. For each class to be mapped, these esti-

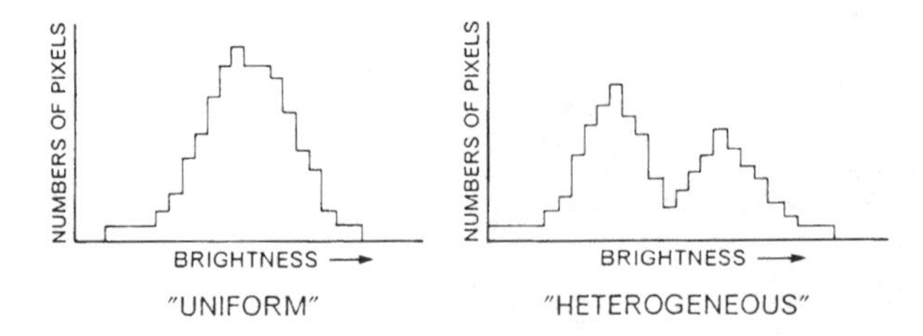


**FIGURE 12.13.** Uniform and heterogeneous training data. On the left, the frequency histogram of the training data has a single peak, indicating a degree of spectral homogeneity. Data from such fields form a suitable basis for image classification. On the right, a second example of training data display a bimodal histogram that reveals that this area represents two, rather than one, spectral classes. These training data are not satisfactory for image classification and must be discarded or redefined.

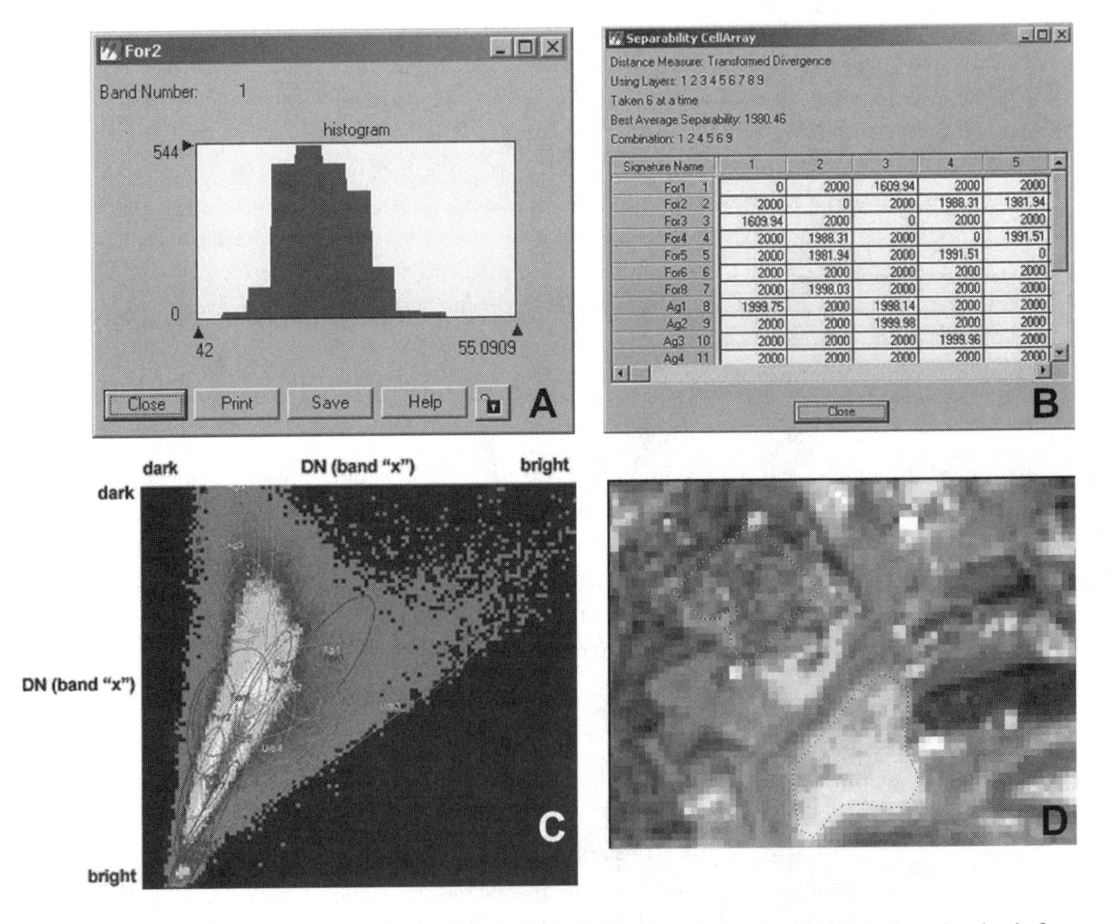


**FIGURE 12.14.** Examples of tools to assess training data. (a) Frequency histogram; (b) ram; (c) a point cloud of pixels in multispectral data space; (d) seed pixels, which allow the analyst to view regions of pixels with values similar to those of a selected pixel, to assist in defining the edges of training fields.

mates approximate the mean values for each band, the variability of each band, and the interrelationships between bands. As an ideal, these values should represent the conditions present within each class within the scene, and thereby form the basis for classification of the vast majority of pixels within each scene that do not belong to training areas. In practice, of course, scenes vary greatly in complexity, and individual analysts differ in their knowledge of a region and in their ability to define training areas that accurately represent the spectral properties of informational classes. Moreover, some informational classes are not spectrally uniform and cannot be conveniently represented by a single set of training data.

### *Significance of Training Data*

Scholz et al. (1979) and Hixson et al. (1980) discovered that selection of training data may be as important as or even more important than choice of classification algorithm in determining classification accuracies of agricultural areas in the central United States. They concluded that differences in the selection of training data were more important influences upon accuracy than were differences among some five different classification procedures.

The results of their studies show little difference in the classification accuracies achieved by the five classification algorithms they were considered, if the same training statistics were used. However, a classification algorithm given two alternative training methods for the same data produced significantly different results. This finding suggests that the choice of training method, at least in some instances, is as important as the choice of classifier. Scholz et al. (1979, p. 4) concluded that the most important aspect of training fields is that all cover types in the scene must be adequately represented by a sufficient number of samples in each spectral subclass.

Campbell (1981) examined the character of the training data as it influences accuracy of the classification. His examples showed that adjacent pixels within training fields tended to have similar values; as a result, the samples that compose each training field may not be independent samples of the properties within a given category. Training samples collected in contiguous blocks may tend to underestimate the variability within each class and to overestimate the distinctness of categories. His examples also show that the degree of similarity varies between land-cover categories, from band to band, and from date to date. If training samples are selected randomly within classes, rather than as blocks of contiguous pixels, effects of high similarity are minimized and classification accuracies improve. Also, his results suggest that it is probably better to use a rather large number of small training fields rather than a few large areas.

### *Idealized Sequence for Selecting Training Data*

Specific circumstances for conducting supervised classification vary greatly, so it is not possible to discuss in detail the procedures to follow in selecting training data, which will be determined in part by the equipment and software available at a given facility. However, it is possible to outline an idealized sequence as a suggestion of the key steps in the selection and evaluation of training data.

1. *Assemble information,* including maps and aerial photographs of the region to be mapped.
2. *Conduct field studies* to acquire firsthand information regarding the area to be studied. The

amount of effort devoted to field studies varies depending upon the analyst's familiarity with the region to be studied. If the analyst is intimately familiar with the region and has access to up-to-date maps and photographs, additional field observations may not be necessary.

3. *Carefully plan collection of field observations* and choose a route designed to observe all regions of the study region. Maps and images should be taken into the field in a form that permits the analyst to annotate them, possibly by using overlays or photocopies. It is important to observe all classes of terrain encountered within the study area, as well as all regions. The analyst should keep good notes, keyed to annotations on the image, and cross-referenced to photographs. If observations cannot be timed to coincide with image acquisition, they should match the season the remotely sensed images were acquired.

4. *Conduct a preliminary examination of the digital scene.* Determine landmarks that may be useful in positioning training fields. Assess image quality. Examine frequency histograms of the data.

5. *Identify prospective training areas,* using guidelines proposed by Joyce (1978) and outlined here. Sizes of prospective areas must be assessed in the light of scale differences between maps or photographs and the digital image. Locations of training areas must be defined with respect to features easily recognizable on the image and on the maps and photographs used as collateral information.

6. *Display the digital image, then locate and delineate training areas on the digital image.* Be sure to place training area boundaries well inside parcel boundaries to avoid including mixed pixels within training areas (Figure 12.14d). At the completion of this step, all training areas should be identified with respect to row and column coordinates within the image.

7. For each spectral class, *evaluate the training data,* using the tools available in the imaging-possessing system of choice (Figure 12.14c). Assess uniformity of the frequency histogram, class separability as revealed by the divergence matrix, the degree to which the training data occupy the available spectral data space, and their visual appearance on the image (Figure 12.14b).

8. *Edit training fields to correct problems identified in Step 7.* Edit boundaries of training fields as needed. If necessary, discard those areas that are not suitable. Return to Step 1 to define new areas to replace those that have been eliminated.

9. Incorporate training data information into a form suitable for use in the classification procedure and proceed with the classification process as described in subsequent sections of this chapter.

### *Specific Methods for Supervised Classification*

A variety of different methods have been devised to implement the basic strategy of supervised classification. All use information derived from the training data as a means of classifying those pixels not assigned to training fields. The following sections outline only a few of the many methods of supervised classification.

#### *Parallelepiped Classification*

Parallelepiped classification, sometimes also known as box decision rule, or level-slice procedures, are based on the ranges of values within the training data to define regions within a multidimensional data space. The spectral values of unclassified pixels are projected into data space; those that fall within the regions defined by the training data are assigned to the appropriate categories.

An example can be formed from data presented in Table 12.4. Here Landsat MSS bands 5 and 7 are selected from a larger data set to provide a concise, easily illustrated example. In practice, four or more bands can be used. The ranges of values with respect to band 5 can be plotted on the horizontal axis in Figure 12.15. The extremes of values in band 7 training data are plotted on the vertical axis, then projected to intersect with the ranges from band 5. The polygons thus defined (Figure 12.15) represent regions in data space that are assigned to categories in the classification. As pixels of unknown identity are considered for classification, those that fall within these regions are assigned to the category associated with each polygon, as derived from the training data. The procedure can be extended to as many bands, or as many categories, as necessary. In addition, the decision boundaries can be defined by the standard deviations of the values within the training areas rather than by their ranges. This kind of strategy is useful because fewer pixels will be placed in an "unclassified" category (a special problem for parallelepiped classification), but it also increases the opportunity for classes to overlap in spectral data space.

Although this procedure for classification has the advantages of accuracy, directness, and simplicity, some of its disadvantages are obvious. Spectral regions for informational categories may intersect. Training data may underestimate actual ranges of classification and leave large areas in data space and on the image unassigned to informational categories. Also, the regions as defined in data space are not uniformly occupied by pixels in each category; those pixels near the edges of class boundaries may belong to other classes. Also, if training data do not encompass the complete range of values encountered in the image (as is frequently the case), large areas of the image remain unclassified, or the basic procedure described here must be modified to assign these pixels to logical classes. This strategy was among the first used in the classification of Landsat data and is still used, although it may not always be the most effective choice for image classification.

**TABLE 12.4. Data for Example Shown in Figure 12.15**

| | Group A | | | | Group B | | | |
|---|---|---|---|---|---|---|---|---|
| | Band | | | | Band | | | |
| | 1 | 2 | 3 | 4 | 1 | 2 | 3 | 4 |
| | 34 | 28 | 22 | 3 | 28 | 18 | 59 | 35 |
| | 36 | 35 | 24 | 6 | 28 | 21 | 57 | 34 |
| | 36 | 28 | 22 | 6 | 28 | 21 | 57 | 30 |
| | 36 | 31 | 23 | 5 | 28 | 14 | 59 | 35 |
| | 36 | 34 | 25 | 7 | 30 | 18 | 62 | 28 |
| | 36 | 31 | 21 | 6 | 30 | 18 | 62 | 38 |
| | 35 | 30 | 18 | 6 | 28 | 16 | 62 | 36 |
| | 36 | 33 | 24 | 2 | 30 | 22 | 59 | 37 |
| | 36 | 36 | 27 | 10 | 27 | 16 | 56 | 34 |
| High | 34 | 28 | 18 | 10 | 27 | 14 | 56 | 28 |
| Low | 36 | 36 | 27 | 3 | 30 | 22 | 62 | 38 |

*Note.* These data have been selected from a larger data set to illustrate parallelepiped classification.

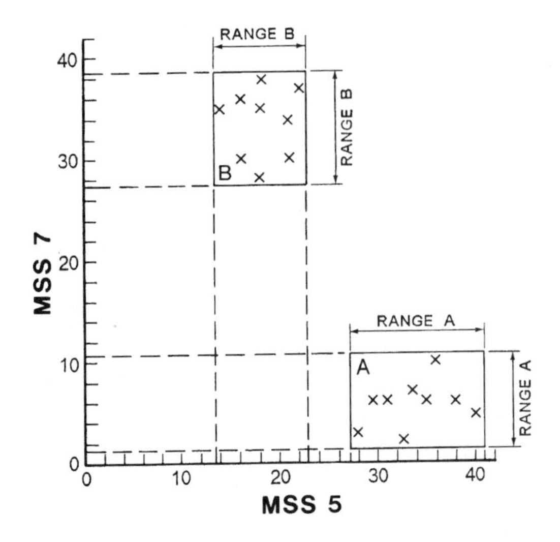


**FIGURE 12.15.** Parallelepiped classification. Ranges of values within training data (Table 12.4) define decision boundaries. Here only two spectral bands are shown, but the method can be extended to several spectral channels. Other pixels, not included in the training data, are assigned to a given category if their positions fall within the polygons defined by the training data.

*Minimum Distance Classification*

Another approach to classification uses the central values of the spectral data that form the training data as a means of assigning pixels to informational categories. The spectral data from training fields can be plotted in multidimensional data space in the same manner illustrated previously for unsupervised classification. Values in several bands determine the positions of each pixel within the clusters that are formed by training data for each category (Figure 12.16). These clusters may appear to be the same as those defined earlier for unsupervised classification. However, in unsupervised classification, these clusters of pixels were defined according to the "natural" structure of the data. Now, for supervised classification, these groups are formed by values of pixels within the training fields defined by the analyst.

Each cluster can be represented by its centroid, often defined as its mean value. As unassigned pixels are considered for assignment to one of the several classes, the multidimensional distance to each cluster centroid is calculated and the pixel is then assigned to the closest cluster. Thus the classification proceeds by always using the "minimum distance" from a given pix-

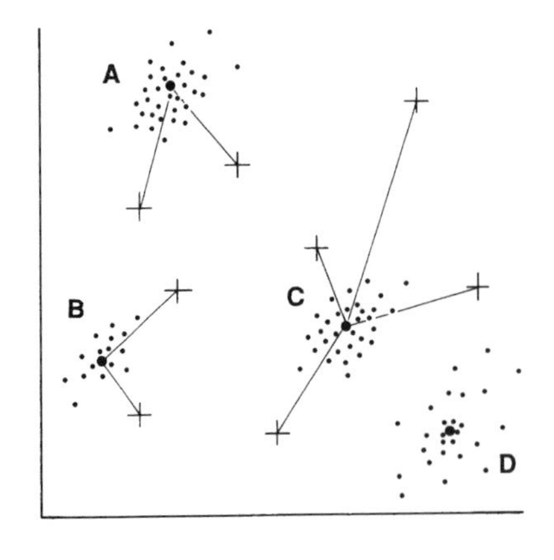


**FIGURE 12.16.** Minimum distance classifier. Here the small dots represent pixels from training fields, and the crosses represent examples of large numbers of unassigned pixels from elsewhere on the image. Each of the pixels is assigned to the closest group, as measured from the centroids (represented by the larger dots) using the distance measures discussed in the text.

el to a cluster centroid defined by the training data as the spectral manifestation of an informational class.

Minimum distance classifiers are direct in concept and in implementation, but are not widely used in remote sensing work. In its simplest form, minimum distance classification is not always accurate; there is no provision for accommodating differences in variability of classes, and some classes may overlap at their edges. It is possible to devise more sophisticated versions of the basic approach just outlined by using different distance measures and different methods of defining cluster centroids.

### *ISODATA*

The ISODATA classifier (Duda and Hart, 1973) is a variation on the minimum distance method; however, it produces results that are often considered to be superior to those derived from the basic minimum distance approach. ISODATA is often seen as a form of supervised classification, although it differs appreciably from the classical model of supervised classification presented at the beginning of this chapter. It is a good example of a technique that shares characteristics of both supervised and unsupervised methods (i.e., "hybrid" classification), and provides evidence that the distinction between the two approaches is not as clear as idealized descriptions imply.

ISODATA starts with the training data selected, as previously described; these data can be envisioned as clusters in multidimensional data space. All unassigned pixels are then assigned to the nearest centroid. Thus far the approach is the same as described previously for the minimum distance classifier. Next, new centroids are found for each group, and the process of allocating pixels to the closest centroids is repeated if the centroid changes position. This process is repeated until there is no change, or only a small change, in class centroids from one iteration to the next. A step-by-step description is as follows:

1. Choose initial estimates of class means. These can be derived from training data; in this respect, ISODATA resembles supervised classification.
2. All other pixels in the scene are assigned to the class with the closest mean, as considered in multidimensional data space.
3. Class means are then recomputed to include effects of those pixels that may have been reassigned in Step 2.
4. If any class mean changes in value from Step 2 to Step 3, then the process returns to Step 2 and repeats the assignment of pixels to the closest centroid. Otherwise, the result at the end of Step 3 represents the final results.

Steps 2, 3, and 4 use the methodology of unsupervised classification, although the requirement for training data identifies the method essentially as supervised classification. It could probably best be considered as a hybrid rather than as a clear example of either approach.

### *Maximum Likelihood Classification*

In nature the classes that we classify exhibit natural variation in their spectral patterns. Further variability is added by the effects of haze, topographic shadowing, system noise, and the effects of mixed pixels. As a result, remote sensing images seldom record spectrally pure classes; more typically, they display a range of brightnesses in each band. The classification strategies considered thus far do not consider variation that may be present within spectral categories, and do not address problems that arise when frequency distributions of spectral values from separate categories overlap. For example, for application of a parallelepiped classifier, the overlap of classes is a serious problem because spectral data space cannot then be neatly divided into discrete units for classification. This kind of situation arises frequently because often our attention is focused upon classifying those pixels that tend to be spectrally similar rather than those that are distinct enough to be easily and accurately classified by other classifiers.

As a result, the situation depicted in Figure 12.17 is common. Assume that we examine a digital image representing a region composed of three-fourths forested land and one-fourth cropland. The two classes "Forest" and "Cropland" are distinct with respect to average brightness, but extreme values (very bright forest pixels or very dark crop pixels) are similar in the region where the two frequency distributions overlap. (For clarity, Figure 12.17 shows data for only a single spectral band, although the principle extends to values observed in several bands and to more than the two classes shown here.) Brightness value "45" falls into the region of overlap, where we cannot make a clear assignment to either "Forest" or to "Cropland." Using the kinds of decision rules mentioned above, we cannot decide which group should receive these pixels unless we place the decision boundary arbitrarily.

In this situation, an effective classification would consider the relative likelihoods of "45 as a member of Forest" and "45 as a member of Cropland." We could then choose the class that would maximize the probability of a correct classification, given the information in the training data. This kind of strategy is known as *maximum likelihood* classification—it uses the training data as a means of estimating means and variances of the classes, which are then used to estimate the probabilities. Maximum likelihood classification considers not only the mean, or average, values in assigning classification, but also the variability of brightness values in each class.

The maximum likelihood decision rule, implemented quantitatively to consider several classes and several spectral channels simultaneously, is a powerful classification technique. It requires intensive calculations, so it has the disadvantage of requiring more computer resources than do most of the simpler techniques mentioned above. Also, it is sensitive to variations in the quality of training data—even more so than most other supervised techniques. Computation of the estimated probabilities is based on the assumption that both training data and the classes themselves display multivariate normal (Gaussian) frequency distributions. (This is one reason

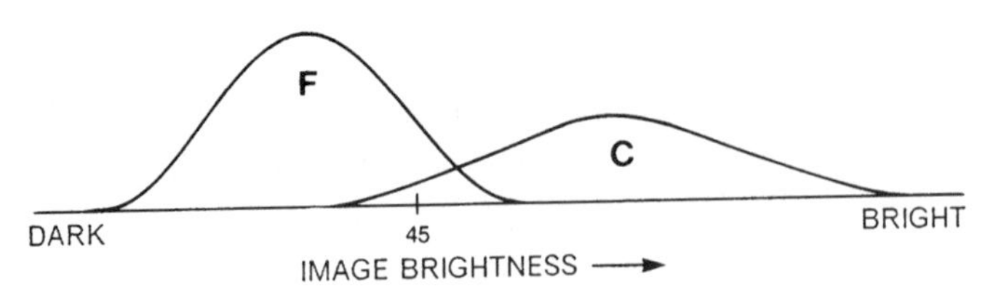


**FIGURE 12.17.** Maximum likelihood classification. These frequency distributions represent pixels from two training fields; the zone of overlap depicts pixels values common to both categories. The relation of the pixels within the region of overlap to the overall frequency distribution for each class defines the basis for assigning pixesl to classes. Here, the relationship between the two histograms indicates that the pixel with the value "45" is more likely to belong to the Forest ("F") class rather than the Crop ("C") class.

that training data should exhibit unimodal distributions, as discussed above.) Data from remotely sensed images often do not strictly adhere to this rule, although the departures are often small enough that the usefulness of the procedure is preserved. Nonetheless, training data that are not carefully selected may introduce error.

### *Bayes's Classification*

The classification problem can be expressed more formally by stating that we wish to estimate the "probability of Forest, given that we have an observed digital value 45," and the "probability of Cropland, given that we have an observed digital value 45." These questions are a form of conditional probabilities, written as "$P(\text{F}|45)$," and "$P(\text{C}|45)$," and read as "The probability of encountering category Forest, given that digital value 45 has been observed at a pixel" and "The probability of encountering category Cropland, given that digital value 45 has been observed at a pixel." That is, they state the probability of one occurrence (finding a given category at a pixel), given that another event has already occurred (the observation of digital value 45 at that same pixel). Whereas estimation of the probabilities of encountering the two categories at random (without a conditional constraint) is straightforward (here $P[\text{F}] = 0.50$, and $P[\text{C}] = 0.50$, as mentioned above), conditional probabilities are based upon two separate events. From our knowledge of the two categories as estimated from our training data, we can estimate $P(45|\text{F})$ ("the probability of encountering digital value 45, given that we have category Forest") and $P(45|\text{C})$ ("the probability of encountering digital value 45, given that we have category Cropland"). For this example, $P(45|\text{F}) = 0.75$, and $P(45|\text{C}) = 0.25$.

However, what we want to know are values for probabilities of "Forest, given that we observe digital value 45" [$P(\text{F}|45)$], and "Cropland, given that we observe digital value 45" [$P(\text{C}|45)$], so that we can compare them to choose the most likely class for the pixel. These probabilities cannot be found directly from the training data. From a purely intuitive examination of the problem, there would seem to be no way to estimate these probabilities.

But, in fact, there is a way to estimate $P(\text{F}|45)$ and $P(\text{C}|45)$ from the information at hand. Thomas Bayes (1702–1761) defined the relationship between the unknowns P(F|45) and $P(\text{C}|45)$, and the known $P(\text{F})$, $P(\text{C})$, $P(45|\text{F})$, and $P(45|\text{C})$. His relationship, now known as *Bayes's theorem,* is expressed as follows for our example:

$$P(\text{F}|45) = \frac{P(\text{F})P(45|\text{F})}{P(\text{F})P(45|\text{F}) + P(\text{C})P(45|\text{C})} \quad \text{(Eq. 12.5)}$$

$$P(\text{C}|45) = \frac{P(\text{C})P(45|\text{C})}{P(\text{C})P(45|\text{C}) + P(\text{F})P(45|\text{F})} \quad \text{(Eq. 12.6)}$$

In a more general form, Bayes's theorem can be written

$$P(b_1|a_1) = \frac{P(b_1)P(a_1|b_1)}{P(b_1)P(a_1|b_1) + P(b_2)P(a_1|b_2) + \ldots} \quad \text{(Eq. 12.7)}$$

where $a_1$ and $a_2$ represent alternative results of the first stage of the experiment, and where $b_1$ and $b_2$ represent alternative results for the second stage.

For our example, Bayes's theorem can be applied as follows:

$$P(\mathrm{F}|45) = \frac{P(\mathrm{F})P(45|\mathrm{F})}{P(\mathrm{F})P(45|\mathrm{F}) + P(\mathrm{C})P(45|\mathrm{C})} \qquad \text{(Eq. 18.8)}$$

$$= \frac{\frac{1}{2} \times \frac{3}{4}}{(\frac{1}{2} \times \frac{3}{4}) + (\frac{1}{2} \times \frac{1}{4})} = \frac{\frac{3}{8}}{\frac{4}{8}} = \frac{3}{4}$$

$$P(\mathrm{C}|45) = \frac{P(\mathrm{C})P(45|\mathrm{C})}{P(\mathrm{C})P(45|\mathrm{C}) + P(\mathrm{F})P(45|\mathrm{F})}$$

$$= \frac{\frac{1}{2} \times \frac{1}{4}}{(\frac{1}{2} \times \frac{1}{4}) + (\frac{1}{2} \times \frac{3}{4})} = \frac{\frac{1}{8}}{\frac{4}{8}} = \frac{1}{4}$$

So we conclude that this pixel is more likely to be "Forest" than "Cropland." Usually data for several spectral channels are considered, and usually we wish to choose from more than two categories, so this example is greatly simplified. We can extend this procedure to as many bands or as many categories as may be necessary, although the expressions become more complex than can be discussed here.

For remote sensing classification, application of Bayes's theorem is especially effective when classes are indistinct or overlap in spectral data space. It can also form a convenient vehicle for incorporating ancillary data (Section 12.5) into the classification, as the added information can be expressed as a conditional probability. In addition, it can provide a means of introducing costs of misclassification into the analysis. (Perhaps an error in misassignment of a pixel to Forest is more serious than a misassignment of a pixel to Cropland.) Furthermore, we can combine Bayes's theorem with other classification procedures, so, for example, most of the pixels can be assigned using a parallelepiped classifier, and then a Baysean classifier can be used for those pixels that are not within the decision boundaries or within a region of overlap. Some studies have shown that such classifiers are very accurate (Story et al., 1984).

Thus Bayes's theorem is an extremely powerful means of using information at hand to estimate probabilities of outcomes related to the occurrence of preceding events. The weak point of the Bayesean approach to classification is the selection of the training data. If the probabilities are accurate, Bayes's strategy must give an effective assignment of observations to classes. Of course, from a purely computational point of view, the procedure will give an answer with any values. But to ensure accurate classification, the training data must have a sound relationship to the categories they represent. For the multidimensional case, with several spectral bands, it is necessary to estimate for each category a mean brightness for each band and a variance–covariance matrix to summarize the variability of each band and its relationships with other bands. From these data we extrapolate to estimate means, variances, and covariances of entire classes. Usually this extrapolation is made on the basis that the data are characterized by multivariate normal frequency distributions. If such assumptions are not justified, the classification results may not be accurate.

If classes and subclasses have been judiciously selected, and if the training data are accurate, Bayes's approach to classification should be as effective as any that can be applied. If the classes are poorly defined, and the training data are not representative of the classes to be mapped, then the results can be no better than those for other classifiers applied under similar circumstances. Use of Bayes's approach to classification is a powerful strategy because it includes information concerning the relative diversities of classes as well as the means and ranges used in the previous classification strategies. The simple example used here is based on only a single channel of data and on a choice between only two classes. The same approach can be extended to consider several bands of data and several sets of categories.

This approach to classification is extremely useful and flexible, and, under certain conditions, provides what is probably the most effective means of classification given the constraints of supervised classification. Note, however, that in most applications, this strategy is limited by the quality of the estimates of the probabilities required for the classification; if these are accurate, the results can provide optimal classification; if they are makeshift values conjured up simply to provide numbers for computation, the results may have serious defects.

### *k-Nearest Neighbors*

*k*-nearest-neighbors classifier (KNN) is simple but efficient application of the distance-based classification approach. Application of KNN is often based upon KNN, although other distance measures are effective. KNN requires training data representing classes of interest within the image. The KNN procedure examines each pixel to be classified (Figure 12.18) in the context of neighboring pixels already labeled, either from the training data, or from the classification process. The algorithm locates the *k* nearest labeled pixels, as positioned in multispectral data space. *k* is set by the analyst, usually to be an odd value. The candidate pixel is then assigned to the class that is represented by the most samples among the *k* neighbors. The effect of distance is usually included by inverse distance weighting, so that nearby neighbors have more influence than more distant neighbors. The decision to assign a given label is usually based upon a majority vote of the neighbors, although other strategies are possible. *k* is preferred to be an odd value, to avoid ties, and classifications are usually cleaner if *k* is set to be relatively small, as larger values may encompass so many neighbors that classes are not clearly defined.

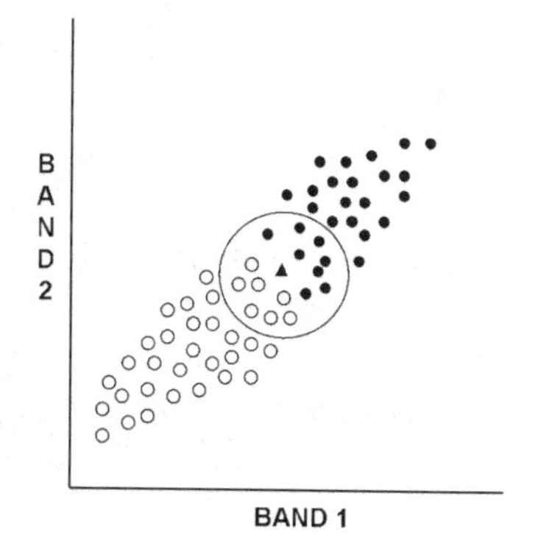


**FIGURE 12.18.** *k*-nearest neighbors classifier. KNN assigns candidate pixels according to a "vote" of the *k* neighboring pixels, with *k* determined by the analyst.

KNN's effectiveness in classifying multi-modal data (data that do not satisfy the homogeneity requirement mentioned previously) provides a practical advantage for many situations. A disadvantage is that all spectral channels assume equal weight in the assignment process, even though some are more potent than others, which can lead to mis-assignments and formation of diffuse clusters (i.e., the procedure weights irrelevant measures equally with those that might be more effective).

### *Idealized Sequence for Conducting Supervised Classification*

The practice of supervised classification is considerably more complex than the practice of unsupervised classification. The analyst must evaluate the situation at each of several steps, then return to an earlier point if it appears that refinements or corrections are necessary to ensure an accurate result.

1. *Prepare the menu of categories to be mapped.* Ideally these categories correspond to those of interest to the final users of the maps and data. But these requirements may not be clearly defined, or the user may require the assistance of the analyst to prepare a suitable plan for image classification.

2. *Select and define training data, as outlined above.* This step may be the most expensive and time-consuming phase of the project. It may be conducted in several stages as the analyst attempts to define a satisfactory set of training data. The analyst must outline the limits of each training site within the image, usually by using a mouse or trackball to define a polygon on a computer screen that displays a portion of the image to be classified.

   This step requires careful comparison of the training fields as marked on maps and photographs with the digital image as displayed on a video screen. It is for this reason that training fields must be easily located with respect to distinctive features visible both on maps and on the digital image. The image-processing system then uses the image coordinates of the edges of the polygons to determine pixel values within the training field to be used in the classification process. After locating all training fields, the analyst can then display frequency distributions for training data as a means of evaluating the uniformity of each training field.

3. *Modify categories and training fields as necessary to define homogeneous training data.* As the analyst examines the training data, he or she may need to delete some categories from the menu, combine others, or to subdivide still others into spectral subclasses. If training data for the modified categories meet the requirements of size and homogeneity, they can be used to enter the next step of the classification. Otherwise, the procedure (at least for certain categories) must start again at Step 1.

4. *Conduct classification.* Each image analysis system will require a different series of commands to conduct a classification, but in essence the analyst must provide the program with access to the training data (often written to a specific computer file) and must identify the image that is to be examined. The results can be displayed on a video display terminal.

5. *Evaluate classification performance.* Finally, the analyst must conduct an evaluation using the procedures discussed in Chapter 14.

This sequence outlines the basic steps required for supervised classification; details may vary from individual to individual and with different image-processing systems, but the basic principles and concerns remain the same: accurate definition of homogenous training fields.

## 12.5. Ancillary Data

*Ancillary,* or *collateral,* data can be defined as data acquired by means other than remote sensing that are used to assist in the classification or analysis of remotely sensed data. For manual interpretation, ancillary data have long been useful in the identification and delineation of features on aerial images. Such uses may be an informal, implicit application of an interpreter's knowledge and experience, or may be more explicit reference to maps, reports, and data. For digital remote sensing, uses of ancillary data are quite different. Ancillary data must be incorporated into the analysis in a structured and formalized manner that connects directly to the analysis of the remotely sensed data. Ancillary data may consist, for example, of topographic, pedologic, or geologic data. Primary requirements are (1) that the data be available in digital form, (2) that the data pertain to the problem at hand, and (3) that the data be compatible with the remotely sensed data.

In fact, a serious limitation to practically applying digital ancillary data is incompatibility with remotely sensed data. *Physical incompatibility* (matching digital formats, etc.) is one such incompatibility, but *logical incompatibility* is a more subtle, but equally important, incompatibility. Seldom are ancillary data collected specifically for use with a specific remote sensing problem; usually they are data collected for another purpose, perhaps as a means of reducing costs in time and money. For example, digital terrain data gathered by the USGS and by the National Geospatial Intelligence Agency (NGA) are frequently used as ancillary data for remote sensing studies. A remote sensing project, however, is not likely to be budgeted to include the resources for digitizing, then editing and correcting, these data, and therefore the data would seldom be compatible with the remotely sensed data with respect to scale, resolution, date, and accuracy. Some differences can be minimized by preprocessing the ancillary data to reduce the effects of different measurement scales, resolutions, and so on, but without that, unresolved incompatibilities, possibly quite subtle, would detract from the effectiveness of the ancillary data.

Choice of ancillary variables is critical. In the mountainous regions of the western United States, elevation data have been very effective as ancillary data for mapping vegetation patterns with digital MSS data, due in part to the fact that there is dramatic variation in local elevation there, and change in vegetation is closely associated with change in elevation, slope, and aspect. In more level settings, elevation differences have subtle influences on vegetation distributions; therefore, elevation data would be less effective as ancillary variables. Although some scientists advocate the use of any and all available ancillary data, in the hope of deriving whatever advantage might be possible, common sense would seem to favor careful selection of variables with conceptual and practical significance to the mapped distributions.

The conceptual basis for the use of ancillary data is that the additional data, collected independently of the remotely sensed data, increase the information available for separating the classes and for performing other kinds of analysis. To continue the example from above, in some regions vegetation patterns are closely related to topographic elevation, slope, and aspect. Therefore the combination of elevation data with remotely sensed data forms a powerful analytical tool because the two kinds of data provide separate, mutually supporting contributions to the subject being interpreted. Some techniques for working with ancillary data are outlined below.

*Stratification* refers to the subdivision of an image into regions that are easy to define using the ancillary data, but might be difficult to define using only the remotely sensed data. For example, topographic data regarding elevation, slope, and aspect would be hard to extract from the spectral information in the remotely sensed image. But ancillary topographic data would provide a means of restricting the classes that would be necessary to do when using the spectral data.

The stratification is essentially an implementation of the layered classification strategy to be discussed in Section 12.7. Some analysts have treated each strata as completely separate, selecting training data separately for each strata, and classifying each strata independent of the others. The several classifications are then merged so as to present the final results as a single image. Hutchinson (1982) emphasized the dangers of ineffective stratification, and warns of the unpredictable, inconsistent effects of classification algorithms across different strata.

Ancillary data have also been used to form an additional channel. For example, digital elevation data could be incorporated with SPOT or Landsat MSS data as an additional band, in the hope that the additional data, independent of the remotely sensed data, would contribute to effective image classification. In general, it appears that this use of ancillary data has not been proven effective in producing substantial improvements in classification accuracy. Strahler (1980) implemented another approach, using ancillary data as a means of modifying prior probabilities in maximum likelihood classification. Digital elevation data, as an additional band of data, were used to refine the probabilities of observing specific classes of vegetation in the Southern California mountains. For example, the probability of observing a specific class, such as "lodgepole pine," varies with topography and elevation. Therefore, a given set of spectral values can have a different meaning, depending on the elevation value in which a given pixel is situated. This approach appears to be effective, at least for how it has been used and studied thus far.

Finally, ancillary data can be used after the usual classifications have been completed, in a process known as *postclassification sorting.* Postclassification sorting examines the confusion matrix derived from a traditional classification. Confused classes are the candidates for postclassification sorting. Ancillary data are then used in an effort to improve discrimination between pairs of classes, thereby improving the overall accuracy by focusing on the weakest aspects of the classification. Hutchinson (1982) encountered several problems with his application of postclassification sorting, but in general favors it for being simple, inexpensive, and convenient to implement.

On the whole, ancillary data, despite their benefits, present numerous practical and conceptual problems for the remote sensing analyst. Sometimes it may be difficult to select and acquire the data that may be most useful for a given problem. The most useful data might not be available in digital form; the analyst might be forced to invest considerable effort in identifying, acquiring, then digitizing the desired information. Therefore, many if not most uses of ancillary data rely on those data that can be acquired "off the shelf," from a library of data already acquired and digitized for another purpose. This presents a danger that the ancillary data may be used as much because of their availability as for their appropriateness for a specific situation.

Because ancillary data are generally acquired for a purpose other than the one at hand, the problem of compatibility can be important. Ideally, ancillary data should possess compatibility with respect to scale, level of detail, accuracy, geographic reference system, and, in instances where time differences are important, date of acquisition. Because analysts are often forced to use off-the-shelf data, incompatibilities form one of the major issues in use of ancillary data. To be useful, ancillary data must be accurately registered to the remotely sensed data. Therefore, the analyst may be required to alter the geometry, or the level of detail, in the ancillary data to bring them into registration with the remotely sensed data. Efforts to alter image geometry or to resample data change the values and therefore are likely to influence the usefulness of the data.

Sometimes ancillary data may already exist in the form of discrete classes—that is, as specific geologic or pedologic classes, for example. Such data are typically represented on maps as discrete parcels, with abrupt changes at the boundaries between classes. Digital remote sensor data are continuous, and therefore differences between the discretely classed and continuous remote sensor data may influence the character of the results, possibly causing sharp changes where only gradual transitions should occur.

To summarize, ancillary data are nonimage information used to aid in classification of spectral data (Hutchinson, 1982). There is a long tradition of both implicit and explicit use of such information in the process of manual interpretation of images, including data from maps, photographs, field observations, reports, and personal experience. For digital analysis, ancillary data often consist of data available in formats consistent with the digital spectral data or in forms that can be conveniently transformed into usable formats. Examples include digital elevation data or digitized soil maps (Anuta, 1976).

Ancillary data can be used in either of two ways. They can be "added to" or "superimposed" over the spectral data to form a single multiband image; the ancillary data are treated simply as additional channels of data. Or the analysis can proceed in two steps using a layered classification strategy (as described below); the spectral data are classified in the first step, then the ancillary data form the basis for reclassification and refinement of the initial results.

### *Classification and Regression Tree Analysis*

*Classification and regression tree analysis* (CART) (Lawrence and Wright, 2001) is a method for incorporating ancillary data into image classification processes. Tools for applying CART are available in many statistical packages, which then can be employed in image-processing packages. In this sense, it is more complicated to implement than many of the other classification techniques employed here. CART requires accurate training data, selected according to usual guidelines for training data delineation, but does not require a priori knowledge of the role of the variables. An advantage to CART, therefore, is that it identifies useful data and separates them from those that do not contribute to successful classification.

CART applies a recursive division of the data to achieve certain specified terminal results, set by the analyst according to the specifics of the particular algorithm employed. In a CART analysis, the dependent variable is the menu of classification classes, and the independent variables are the spectral channels and ancillary variables.

CART recursively examines binary divisions of the explanatory variables to derive an optimal assignment of the classes to the independent variables. Application of CART is sensitive to variations in numbers of pixels: it performs best when numbers of pixels in training data sets are approximately equal.

Often the use of a large number of spectral and ancillary variables leads to very accurate results, but results that are tailored to specific data sets—a condition known as *overfitting*. As a result, the application of CART often includes a pruning step, which reduces the solution to create a more concise, more robust, and generally applicable solution—division of the data into subsets, with results from some subsets cross-validated against others. Pal and Mather (2003) examined consequences of five alternative strategies for attribute selection, and found that differences in accuracy were minimal.

## 12.6. Fuzzy Clustering

*Fuzzy clustering* addresses a problem implicit in much of the preceding material: pixels must be assigned to a single discrete class. Although such classification attempts to maximize correct classifications, the logical framework allows only for direct one-to-one matches between pixels and classes. We know, however, that many processes contribute to prevent clear matches between pixels and classes, as noted in Chapter 10 and in the work of Robinove (1981) and Richards and Kelly (1984). Therefore, the focus upon finding discrete matches between the pixels and informational classes ensures that many pixels will be incorrectly or illogically labeled. Fuzzy logic attempts to address this problem by applying a different classification logic.

Fuzzy logic (Kosko and Isaka, 1993) has applications in many fields, but has special significance for remote sensing. Fuzzy logic permits partial membership, a property that is especially significant in the field of remote sensing, as partial membership translates closely to the problem of mixed pixels (Chapter 10). So whereas traditional classifiers must label pixels as either "Forest" or "Water," for example, a fuzzy classifier is permitted to assign a pixel a membership grade of 0.3 for "Water" and 0.7 for "Forest," in recognition that the pixel may not be properly assigned to a single class. Membership grades typical vary from 0 (nonmembership) to 1.0 (full membership), with intermediate values signifying partial membership in one or more other classes (Table 12.5).

A fuzzy classifier assigns membership to pixels based upon a *membership function* (Figure 12.19). Membership functions for classes are determined either by general relationships or by definitional rules describing the relationships between data and classes. Or, as is more likely in the instance of remote sensing classification, membership functions are derived from experimental (i.e., training) data for each specific scene to be examined. In the instance of remote

**TABLE 12.5. Partial Membership in Fuzzy Classes**

| | Pixel | | | | | | |
|---|---|---|---|---|---|---|---|
| Class | A | B | C | D | E | F | G |
| Water | 0.00 | 0.00 | 0.00 | 0.00 | 0.00 | 0.00 | 0.00 |
| Urban | 0.00 | 0.01 | 0.00 | 0.00 | 0.00 | 0.00 | 0.85 |
| Transportation | 0.00 | 0.35 | 0.00 | 0.00 | 0.99 | 0.79 | 0.14 |
| Forest | 0.07 | 0.00 | 0.78 | 0.98 | 0.00 | 0.00 | 0.00 |
| Pasture | 0.00 | 0.33 | 0.21 | 0.02 | 0.00 | 0.05 | 0.00 |
| Cropland | 0.92 | 0.30 | 0.00 | 0.00 | 0.00 | 0.15 | 0.00 |

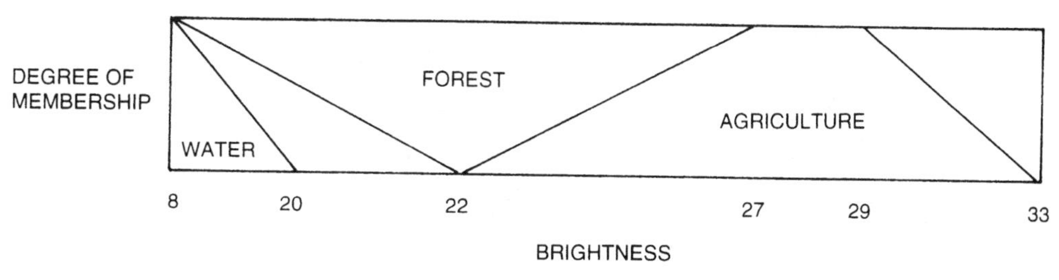


**FIGURE 12.19.** Membership functions for fuzzy clustering. This example illustrates membership functions for the simple instance of three classes considered for a single spectral band, although the method is typically applied to multiple bands. The horizontal axis represents pixels brightness; the vertical axis represents degree of membership, from low near the bottom to high at the top. The class "water" consist of pixels darker than brightness 20, although only pixels darker than 8 are likely to be completely occupied by water. The class "agriculture" can include pixels as dark as 22 and as bright as 33, although pure "agriculture" is found only in the range 27 to 29. A pixel of brightness 28, for example, can only be agriculture, although pixel of brightness of 24 could be partially forested, partially agriculture. (Unlabeled areas on this diagram are not occupied by any of the designated classes.)

sensing data, a membership function describes the relationship between class membership and brightness in several spectral bands.

Figure 12.19 provides contrived examples showing several pixels and their membership grades. (Actual output from a fuzzy classification is likely to form an image that shows varied levels of membership for specific classes.) Membership grades can be hardened (Table 12.6) by setting the highest class membership to 1.0 and all others to 0.0. Hardened classes are equivalent to traditional classifications: each pixel is labeled with a single label and the output is a single image labeled with the identify of the hardened class. Programs designed for remote sensing applications (Bezdek et al., 1984) provide the ability to adjust the degree of fuzziness and thereby adjust the structures of classes and the degree of continuity in the classification pattern.

Fuzzy clustering has been judged to improve results, at least marginally, with respect to traditional classifiers, although the evaluation is difficult because the usual evaluation methods require the discrete logic of traditional classifiers. Thus the improvements noted for hardened classifications are probably conservative as they do not reveal the full power of fuzzy logic.

## 12.7. Artificial Neural Networks

*Artificial neural networks* (ANNs) are computer programs that are designed to simulate human learning processes through establishment and reinforcement of linkages between input data and output data. It is these linkages, or pathways, that form the analogy with the human learning process, in that repeated associations between input and output in the training process reinforce linkages, or pathways, that can then be employed to link input and output in the absence of training data.

ANNs are often represented as composed of three elements. An *input layer* consists of the source data, which in the context of remote sensing are the multispectral observations, perhaps in several bands and from several dates. ANNs are designed to work with large volumes of data,

**TABLE 12.6. "Hardened" Classes for Example Shown in Table 12.5**

| | Pixel | | | | | | |
|---|---|---|---|---|---|---|---|
| Class | A | B | C | D | E | F | G |
| Water | 0.00 | 0.00 | 0.00 | 0.00 | 0.00 | 0.00 | 0.00 |
| Urban | 0.00 | 0.00 | 0.00 | 0.00 | 0.00 | 0.00 | 1.00 |
| Transportation | 0.00 | 1.00 | 0.00 | 0.00 | 1.00 | 1.00 | 0.00 |
| Forest | 0.00 | 0.00 | 1.00 | 1.00 | 0.00 | 0.00 | 0.00 |
| Pasture | 0.00 | 0.00 | 0.00 | 0.00 | 0.00 | 0.00 | 0.00 |
| Cropland | 1.00 | 0.00 | 0.00 | 0.00 | 0.00 | 0.00 | 0.00 |

including many bands and dates of multispectral observations, together with related ancillary data.

The *output layer* consists of the classes required by the analyst. There are few restrictions on the nature of the output layer, although the process will be most reliable when the number of output labels is small, or modest, with respect to the number of input channels. Included are training data in which the association between output labels and input data is clearly established. During the training phase, an ANN establishes an association between input and output data by establishment of weights within one or more *hidden layers* (Figure 12.20). In the context of remote sensing, repeated associations between classes and digital values, as expressed in the training data, strengthen weights within hidden layers that permit the ANN to assign correct labels when given spectral values in the absence of training data.

Further, ANNs can also be trained by *back propagation* (BP). If establishment of the usual training data for conventional image classification can be thought of as "forward propagation," then BP can be thought of as a retrospective examination of the links between input and output

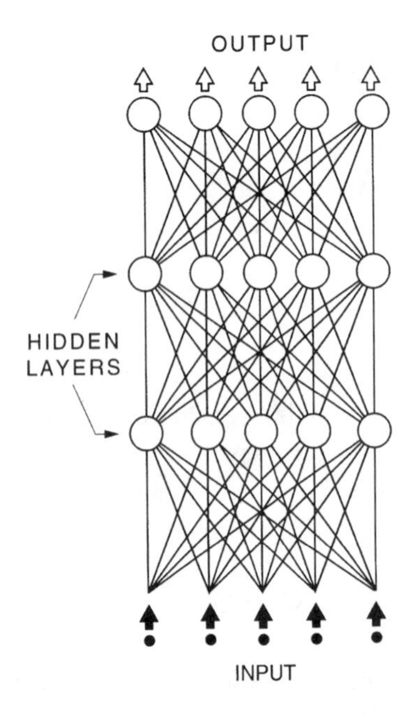


**FIGURE 12.20.** Artificial neural net.

data in which differences between expected and actual results can be used to adjust weights. This process establishes *transfer functions,* quantitative relationships between input and output layers that assign weights to emphasize effective links between input and output layers. (For example, such weights might acknowledge that some band combinations may be very effective in defining certain classes and others effective for other classes.) In BP, hidden layers note errors in matching data to classes and adjust the weights to minimize errors.

ANNs are designed using less severe statistical assumptions than many of the usual classifiers (e.g., maximum likelihood), although in practice successful application requires careful application. ANNs have been found to be accurate in the classification of remotely sensed data, although improvements in accuracies have generally been small or modest. ANNs require considerable effort to train and are intensive in their use of computer time.

## 12.8. Contextual Classification

*Contextual information* is derived from spatial relationships among pixels within a given image. Whereas *texture* usually refers to spatial interrelationships among unclassified pixels within a window of specified size, *context* is determined by positional relationships between pixels, either classified or unclassified, anywhere within the scene (Swain et al., 1981; Gurney and Townshend, 1983).

Although contextual classifiers can operate on either classified or unclassified data, it is convenient to assume that some initial processing has assigned a set of preliminary classes on a pixel-by-pixel basis without using spatial information. The function of the contextual classifier is to operate upon the preliminary classification to reassign pixels as appropriate in the light of contextual information.

Context can be defined in several ways, as illustrated in Figure 12.21a. In each instance the problem is to consider the classification of a pixel or a set of pixels (represented by the shaded pattern) using information concerning the classes of other, related pixels. Several kinds of links define the relationships between the two groups. The simplest link is that of distance. Perhaps the unclassified pixels are agricultural land, which is likely to be "Irrigated Cropland" if positioned within a certain distance of a body of open water; if the distance to water exceeds a certain threshold, the area might be more likely to be assigned to "Rangeland" or "Unirrigated Cropland." The second example in Figure 12.21 illustrates the use of both distance and direction. Contiguity (Figure 12.21c) may be an important classification aid. For example, in urban regions specific land-use types may be found primarily in locations adjacent to a specific category. Finally, specific categories may be characterized by their positions within other categories, as shown in Figure 12.21d.

Contextual classifiers are efforts to simulate some of the higher order interpretation processes used by human interpreters, in which the identity of an image region is derived in part from its location in relation to other regions of specified identity. For example, a human interpreter considers sizes and shapes of parcels in identifying land use, as well as the identities of neighboring parcels. The characteristic spatial arrangement of the central business district, the industrial district, residential had, and agricultural land in an urban region permits the interpreter to identify parcels that might be indistinct if considered only with conventional classifiers.

Contextual classifiers can also operate upon classified data to reclassify erroneously classi-

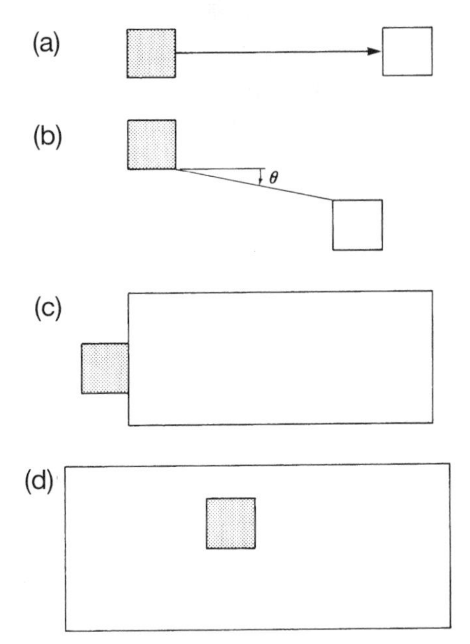


**FIGURE 12.21.** Contextual classification, from Gurney and Townshend (1983). The shaded regions depict pixels to be classified; the open regions represent other pixels considered in the classification decision. (a) Distance is considered; (b) Direction is considered; (c) Contiguity forms a part of the decision process; (d) Inclusion is considered. Copyright 1983 by the American Society of Photogrammetry and Remote Sensing. Reproduced by permission.

fied pixels or to reclassify isolated pixels (perhaps correctly classified) that form regions so small and so isolated that they are of little interest to the user. Such uses may be essentially cosmetic operations, but they could be useful in editing the results for final presentation.

## 12.9. Object-Oriented Classification

Object-oriented classification applies a logic intended to mimic some aspects of the higher order logic employed by human interpreters, who can use the sizes, shapes, and textures of regions, as well as the spectral characteristics used for conventional pixel-based classification.

The first of the two processes is *segmentation* of the image: identification of the edges of homogeneous patches that subdivide the image into interlocking regions. These regions are based on the spectral values of their pixels, as well as analyst-determined constraints. Segmentation is implemented hierarchically—regions are defined at several scales that nest within each other (Figure 12.22). These regions are the "objects" of objected-oriented classification. The analyst has the ability to set criteria that control the measures used to assess homogeneity and distinctness, and the thresholds that apply to a specific classification problem.

These regions form the "objects" indicated by the name "object-oriented"—they, rather than individual pixels, form the entities to be classified. Because the regions are composed of many pixels, they have a multitude of properties, such as standard deviations, maxima, and minima that are possible for regions, but not for individual pixels.

Further, object-oriented classification can consider distinctive topological properties of classes. For example, forest in an urban park and rural forest have similar, if not identical, spectral properties even though they form very different classes. Object-oriented classification can use the forest's spectral properties to recognize it as forest, but can also consider the effect of

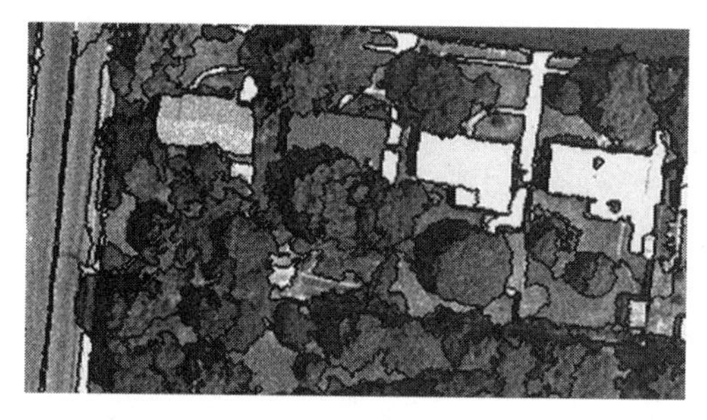


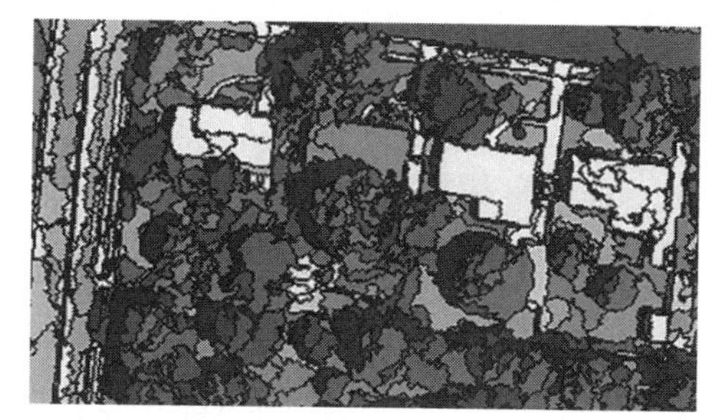


**FIGURE 12.22.** Object-oriented classification. Through the segmentation process, pixels can be organized to form "objects" that correspond to features of interest in the image. These objects can then be treated as entities to be classified in the manner of the classification procedures discussed here. Here, the illustration shows two alternative segmentations, one at a coarse level, which creates large, diverse, objects, and another at fine level of detail, which creates more numerous, homogeneous, objects.

neighboring classes to recognize that regions of urban forest are surrounded by urban land use, whereas rural forest will border rural categories, such as open water, agricultural land, and other forest.

The second process is classification, using conventional classification procedures, usually nearest neighbor or fuzzy classification. Each object or region is characterized by the numerous properties developed as a result of the segmentation process that can then be used in the classification process. The analyst can examine and select those properties that can be useful in classification.

Much of the conceptual and theoretical basis for object-oriented classification was developed in previous decades, but was not widely used because of the difficulties of developing useful user interfaces that enable routine use for practical applications. One of the most successful application programs is eCognition, developed by DEFiNiENS Imaging AG, a Munich-based company that has devoted itself to the development of a practical object-oriented image classification system. eCognition's system provides a systematic approach and a user interface that permits implementation of concepts developed in previous decades. eCognition has been used successfully in defense, natural resources, exploration, and facilities mapping. The latest version of eCognition is known as DEFiNiENS Professional Version 5.0.

Object-oriented classification can be used successfully with any multispectral imagery, but has proven to be especially valuable for fine-resolution imagery, which is particularly difficult to classify accurately using conventional classification techniques.

Implementing object-oriented classification can be-time consuming, as the analyst often must devote considerable effort to trail and error while learning the most efficient approach to classification of a specific image for a specific purpose. However sometimes the segmentation itself is a useful product that can be used in other analytical processes.

## 12.10. Iterative Guided Spectral Class Rejection

*Iterative guided spectral class rejection* (IGSCR) (Wayman et al., 2001) is a classification technique that minimizes user input and the subjectivity characteristic of many other image classification approaches. IGSCR is based upon training data, provided by the analyst, that represent the classes of interests, and application of unsupervised classification procedures. The unsupervised classification groups image pixels together to define uniform spectral classes. IGSCR then attempts to match the spectral classes to classes as defined by the training data.

The IGSCR algorithm evaluates each of the spectral classes with respect to the training data, and accepts or rejects each spectral class based upon the closeness of each spectral class to the training data. IGSCR rejects spectral classes that are candidates for entering the classification if they do not meet set thresholds (often, 90% homogeneity, with the minimum number of samples determined by a binomial probability distribution). Rejected pixels are regrouped into new spectral classes, and considered again during the next iteration. The rejection process assures that all pixels that enter the classification meet criteria set by the analyst. This process assures that all classes in the final classification will meet the analyst's criteria for uniformity. Classification stops when user-defined criteria are satisfied. Those pixels that cannot meet criteria to enter the classification are left unclassified. IGSCR is a hybrid classifier—one that lends itself to accurate classification and ease of replication. The version described by Wayman et al. (2001) was designed to implement a binary (two-class) classification decision, such as forest-nonforest. The authors have since developed a multiclass version.

## 12.11. Summary

This chapter has described a few specific classifiers as a means of introducing the student to the variety of different classification strategies that are available today. Possibly the student may have the opportunity to use some of these procedure, so these descriptions may be the first step in a more detailed learning process. It is more likely, however, that the student who uses this text will never use many of the specific classifiers described here. Nonetheless, those procedures that are available for student use are likely to be based upon the same principles outlined here using specific classifiers as examples. This chapter should not be regarded as a complete catalogue of current image classification methods, but rather as an effort to illustrate some of the primary methods of image classification. Specific details and methods will vary greatly, but if the student has mastered the basic strategies and methods of image classification, he or she will recognize unfamiliar methods as variations on the fundamental approaches described here.

## Review Questions

1. This chapter mentions only a few of the many strategies available for image classification. Why have so many different methods been developed? Why has the remote sensing community not decided simply to select one as the standard?
2. Why might the decision to use or not to use preprocessing be especially significant for image classification?
3. Image classification is not necessarily equally useful for all fields. For a subject of interest to you (geology, forestry, etc.) evaluate the significance of image classification by citing examples of how classification might be used. Also list some subjects for which image classification might be more or less useful.

4. Review Chapters 6 and 11. Speculate on the course of further developments in image classification. Can you suggest relationships between sensor technology and the design of image classification strategies?

5. The table below lists land-cover classes for the pixels given in Tables 12.2 and 12.3. (Note that individual pixels do not correspond for the two dates.) Select a few pixels from Table 12.3; conduct a rudimentary feature selection "by inspection" (i.e., by selecting from the set of four bands). Can you convey most of the ability to distinguish between classes by choosing a subset of the four MSS bands?

| | Feb. | May | | Feb. | May | | Feb. | May | | Feb. | May |
|---|---|---|---|---|---|---|---|---|---|---|---|
| (1) | F | W | (11) | P | C | (21) | F | C | (31) | F | C |
| (2) | F | F | (12) | F | C | (22) | P | P | (32) | W | F |
| (3) | C | W | (13) | W | P | (23) | W | P | (33) | P | F |
| (4) | P | C | (14) | F | P | (24) | W | W | (34) | F | W |
| (5) | P | F | (15) | F | C | (25) | P | W | (35) | W | C |
| (6) | C | W | (16) | F | W | (26) | P | F | (36) | F | C |
| (7) | F | F | (17) | P | F | (27) | F | F | (37) | W | F |
| (8) | F | P | (18) | C | W | (28) | C | F | (38) | P | C |
| (9) | C | F | (19) | F | W | (29) | C | P | (39) | F | P |
| (10) | W | C | (20) | F | F | (30) | P | W | (40) | C | F |

F, forest; C, crop; P, pasture; W, water

6. Using the information given above and in Table 12.3, calculate normalized differences between water and forest, bands 2 and 4. (For this question and those that follow, the student should have access to a calculator.) Identify those classes easiest to separate, and the bands likely to be most useful.

7. For both February and May, calculate normalized differences between forest and pasture, band 4. Give reasons to account for differences between results for the two dates.

8. The results for Questions 5–7 illustrate that normalized difference values vary greatly, depending upon data, bands, and classes considered. Discuss some of the reasons why this is true, both in general and for the specific classes in these examples.

9. Refer to Table 12.1. Calculate the Euclidean distance between the means of the four classes given in the table. Again, explain why there are differences from date to date and from band to band.

## References

### *General*

Anuta, P. E. 1976. Digital Registration of Topographic Data and Satellite MSS Data for Augmented Spectral Analysis. In *Proceedings, 42nd Annual Meeting, American Society of Photogrammetry.* Falls Church, VA.: American Society of Photogrammetry, pp. 180–187.

Bryant, J. 1979. On the Clustering of Multidimensional Pictorial Data. *Pattern Recognition,* Vol. 11, pp. 115–125.

Duda, R. O., and P. E. Hart. 1973. *Pattern Classification and Scene Analysis.* New York: Wiley, 482 pp.

Gurney, C. M., and J. R. Townshend. 1983. The Use of Contextual Information in the Classification of Remotely Sensed Data. *Photogrammetric Engineering and Remote Sensing,* Vol. 49, pp. 55–64.

Hord, R. 1982. *Digital Image Processing of Remotely Sensed Data.* New York: Academic Press, 221 pp.

Hutchinson, C. F. 1982. Techniques for Combining Landsat and Ancillary Data for Digital Classification Improvement. *Photogrammetric Engineering and Remote Sensing,* Vol. 48, pp. 123–130.

Jensen, J. R., and D. L. Toll. 1982. Detecting Residential Land Use Development at the Urban Fringe. *Photogrammetric Engineering and Remote Sensing,* Vol. 48, pp. 629–643.

Kettig, R. L., and D. A. Landgrebe. 1975. Classification of Multispectral Image Data by Extraction and Classification of Homogeneous Objects. In *Proceedings, Symposium on Machine Classification of Remotely Sensed Data.* West Lafayette, IN: Laboratory for Applications in Remote Sensing, pp. 2A-1–2A-11.

Lawrence, R. L., and A. Wright. 2001. Rule-Based Classification Systems Using Classification and Regression Tree Analysis. *Photogrammetric Engineering and Remote Sensing,* Vol. 67, pp. 1137–1142.

Moik, J. G. 1980. *Digital Processing of Remotely Sensed Images* (National Aeronautics and Space Administration Special Publication 431). Washington, DC: U.S. Government Printing Office, 330 pp.

Pal, M., and P. M. Mather. 2003. An Assessment of Decision Tree Methods for Land Cover Classification. *Remote Sensing of Environment,* Vol. 86, pp. 554–565.

Richards, J. A., and D. J. Kelly. 1984. On the Concept of the Spectral Class. *International Journal of Remote Sensing,* Vol. 5, pp. 987–991.

Robinove, C. J. 1981. The Logic of Multispectral Classification and Mapping of Land. *Remote Sensing of Environment,* Vol. 11, pp. 231–244.

Schowengerdt, R. A. 1983. *Techniques for Image Processing and Classification in Remote Sensing.* New York: Academic Press, 272 pp.

Strahler, A. H. 1980. The Use of Prior Probabilities in Maximum Likelihood Classification of Remotely Sensed Data. *Remote Sensing of Environment,* Vol. 10, pp. 135–163.

Swain, P. H. 1984. Advanced Computer Interpretation Techniques for Earth Data Information Systems. In *Proceedings of the Ninth Annual William T. Pecora Remote Sensing Symposium.* Silver Spring, MD: IEEE, pp. 184–189.

### *Training Data*

Campbell, J. B. 1981. Spatial Correlation Effects upon Accuracy of Supervised Classification of Land Cover. *Photogrammetric Engineering and Remote Sensing,* Vol. 47, pp. 355–363.

Joyce, A. T. 1978. *Procedures for Gathering Ground Truth Information for a Supervised Approach to Computer-Implemented Land Cover Classification of Landsat-Acquired Multispectral Scanner Data* (NASA Reference Publication 1015). Houston, TX: National Aeronautics and Space Administration, 43 pp.

Mcaffrey, T. M., and S. F. Frankin. 1993. Automated Training Site Selection for Large-Area Remote-Sensing Image Analysis. *Computers and Geosciences,* Vol. 10, pp. 1413–1428.

### *Image Texture*

Haralick, R. M. 1979. Statistical and Structural Approaches to Texture. *Proceedings of the IEEE,* Vol. 67, pp. 786–804.

Haralick, R. M., et al. 1973. Textural Features for Image Classification. *IEEE Transactions on Systems, Man, and Cybernetics* (SMC-3), pp. 610–622.

Jensen, J. R. 1979. Spectral and Textural Features to Classify Elusive Land Cover at the Urban Fringe. *Professional Geographer,* Vol. 31, pp. 400–409.

### *Comparisons of Classification Techniques*

Chen, K. S., Y. C. Tzeng, C. F. Chen, and W. L. Kao. 1995. Land-Cover Classification of Multispectral Imagery Using a Dynamic Learning Neural Network. *Photogrammetric Engineering and Remote Sensing,* Vol. 61, pp. 403–408.

Gong, P., and P. J. Howarth. 1992. Frequency-Based Contextual Classification and Gray-Level Vector Reduction for Land-Use Identification. *Photogrammetric Engineering and Remote Sensing,* Vol. 58, pp. 423–437.

Hixson, M., D. Scholz, and N. Fuhs. 1980. Evaluation of Several Schemes for Classification of Remotely Sensed Data. *Photogrammetric Engineering and Remote Sensing,* Vol. 46, pp. 1547–1553.

Scholz, D., N. Fuhs, and M. Hixson. 1979. An Evaluation of Several Different Classification Schemes, Their Parameters, and Performance. In *Proceedings, Thirteenth International Symposium on Remote Sensing of the Environment.* Ann Arbor: University of Michigan Press, pp. 1143–1149.

Story, M. H., J. B. Campbell, and G. Best. 1984. An Evaluation of the Accuracies of Five Algorithms for Machine Classification of Remotely Sensed Data. *Proceedings of the Ninth Annual William T. Pecora Remote Sensing Symposium.* Silver Spring, MD: IEEE, pp. 399–405.

### *Fuzzy Clustering*

Bezdek, J. C., R. Ehrilich, and W. Full. 1984. FCM: The Fuzzy c-Means Clustering Algorithm. *Computers and Geosciences,* Vol. 10, pp. 191–203.

Fisher, P. F., and S. Pathirana. 1990. The Evaluation of Fuzzy Membership of Land Cover Classes in the Suburban Zone. *Remote Sensing of Environment,* Vol. 34, pp. 121–132.

Kent, J. T., and K. V. Marida. 1988. Spatial Classification Using Fuzzy Membership Models. *IEEE Transactions on Pattern Analysis and Machine Intelligence,* Vol. 10, pp. 659–671.

Kosko, B., and S. Isaka. 1993. Fuzzy Logic. *Scientific American,* Vol. 271, pp. 76–81.

Wang, F. 1990a. Fuzzy Supervised Classification of Remote Sensing Images. *IEEE Transactions on Geoscience and Remote Sensing,* Vol. 28, pp. 194–201.

Wang, F. 1990b. Improving Remote Sensing Image Analysis through Fuzzy Information Representation. *Photogrammetric Engineering and Remote Sensing,* Vol. 56, pp. 1163–1169.

### *Artificial Neural Networks*

Bischop, H., W. Schnider, and A. J. Pinz. 1992. Multispectral Classification of Landsat Images Using Neural Networks. *IEEE Transactions on Geoscience and Remote Sensing,* Vol. 30, pp. 482–490.

Chen, K. S., Y. C. Tzeng, C. F. Chen, and W. L. Kao. 1995. Land-Cover Classification of Multispectral Imagery Using a Dynamic Learning Neural Network. *Photogrammetric Engineering and Remote Sensing,* Vol. 61, pp. 403–408.

Miller, D. M., E. D. Kaminsky, and S. Rana. 1995. Neural Network Classification of Remote-Sensing Data. *Computers and Geosciences,* Vol. 21, pp. 377–386.

### *Combining Remotely Sensed Data with Ancillary Data*

Davis, F. W., D. A. Quattrochi, M. K. Ridd, N. S.-N. Lam, S. J. Walsh, J. C. Michaelson, J. Franklin, D. A. Stow, C. J. Johannsen, and C. A. Johnson. 1991. Environmental Analysis Using Integrated GIS and Remotely Sensed Data: Some Research Needs and Priorities. *Photogrammetric Engineering and Remote Sensing,* Vol. 57, pp. 689–697.

Ehlers, M., G. Edwards, and Y. Bédard. 1989. Integration of Remote Sensing with Geographic Information Systems: A Necessary Evolution. *Photogrammetric Engineering and Remote Sensing,* Vol. 55, pp. 1619–1627.

Ehlers, M., D. Greenlee, T. Smith, and J. Star. 1992. Integration of Remote Sensing and GIS: Data and Data Access. *Photogrammetric Engineering and Remote Sensing,* Vol. 57, pp. 669–675.

Faust, N. L., W. H. Anderson, and J. H. L. Star. 1992. Geographic Information Systems and Remote Sensing Future Computing Environment. *Photogrammetric Engineering and Remote Sensing,* Vol. 57, pp. 655–668.

Herr, A. M., and L. P. Queen. 1993. Crane Habitat Evaluation Using GIS and Remote Sensing. *Photogrammetric Engineering and Remote Sensing,* Vol. 59, pp. 1531–1538.

Hutchinson, C. F. 1982. Techniques for Combining Landsat and Ancillary Data for Digital Classification Improvement. *Photogrammetric Engineering and Remote Sensing,* Vol. 48, pp. 123–130.

Loveland, T. R., and G. E. Johnson. 1982. The Role of Remote Sensing and Other Spatial Data for Predictive Modeling: The Umatilla, Oregon, Example. *Photogrammetric Engineering and Remote Sensing,* Vol. 49, pp. 1183–1192.

Lunnetta, R. S., R. G. Congalton, L. K. Fenstermaker, J. R. Jensen, K. C. McGwire, and L. R. Tinney. 1991. Remote Sensing and Geographic Information System Data Integration: Error Sources and Research Issues. *Photogrammetric Engineering and Remote Sensing,* Vol. 57, pp. 677–687.

Shelton, R. L., and J. E. Estes. 1979. Integration of Remote Sensing and Geographic Information Systems. *Proceedings of the 13th International Symposium on Remote Sensing of Environment.* Ann Arbor, MI: ERIM, pp. 463–483.

Star, J. L., J. E. Estes, and F. Davis. 1991. Improved Integration of Remote Sensing and Geographic Information Systems: A Background to NCGIA Initiative 12. *Photogrammetric Engineering and Remote Sensing* Vol. 57, pp. 643–645.

Wayman, J. P., R. H. Wynne, and J. A. Scrivani. 1999. Satellite-Assisted Forest Cover Mapping in the Southeastern United States Using Iterative Guided Spectral Class Rejection. In *Proceedings of the Second International Conference-Geospatial Information in Agriculture and Forestry 10–12 January, 2000, Lake Buena Vista, Florida.* Vol, 2, pp. 355–362.

Wayman, J. P., R. H. Wynne, J. A. Scrivani, and G. A. Burns. 2001. Landsat TM-Based Forest Area Estimation Using Iterative Guided Spectral Class Rejection. *Photogrammetric Engineering and Remote Sensing,* Vol. 67, pp. 1155–1166.

CHAPTER THIRTEEN

# Field Data

## 13.1. Introduction

Every application of remote sensing must apply, if only implicitly, field observations in one form or another. Each analyst must define relationships between image data and conditions at corresponding points on the ground. Although use of field data appears, at first glance, to be self-evident, there are numerous practical and conceptual difficulties to be encountered and solved in the course of even the most routine work (Steven, 1987).

*Field data* consist of observations collected at or near ground level in support of remote sensing analysis. Although it is a truism that characteristics of field data must be tailored for the specific study at hand, the inconvenience and expense of collecting accurate field data often lead analysts to apply field data collected for one purpose to other, perhaps very different, purposes. Therefore, a key concern is not only to acquire field data but to acquire data suitable for the specific task at hand. Accurate field data permit the analyst to match points or areas on the imagery to corresponding regions on the ground surface, and thereby to establish with confidence relationships between the image and conditions on the ground. In their simplest form, field data may simply permit an analyst to identify a specific region as *forest.* In more detailed circumstances, field data may identify a region as a specific form of forest, or specify height, volume, leaf area index, photosynthetic activity, or other properties according to the specific purposes of a study.

## 13.2. Kinds of Field Data

Field data typically serve one of three purposes. First, they can be used to verify, to evaluate, or to assess the results of remote sensing investigations (Chapter 14). Second, they can provide reliable data to guide the analytical process, such as creating training fields to support supervised classification (Chapter 12). This distinction introduces the basic difference between the realms of qualitative (nominal labels) and quantitative analyses, discussed later in this chapter. Third, field data can provide information used to model the spectral behavior of specific landscape feature (e.g., plants, soils, or water bodies). For example, analyses of the spectral properties of forest canopies can be based on quantitative models of the behavior of radiation within the canopy—models that require detailed and accurate field observations (Treitz et al., 1992).

Despite its significance, development of sound procedures for field data collection has not been discussed in a manner that permits others to easily grasp the principles that underlie them. Discussions of field data collection procedures typically include some combination of the following components:

- Data to record ground-level or low-level spectral data for specific features and estimates of atmospheric effects (e.g., Lintz et al., 1976; Curtis and Goetz, 1994; Deering, 1989).
- Broad guidelines for collection of data for use with specific kinds of imagery (e.g., Lintz et al., 1976).
- Discipline-specific guidelines for collection and organization of data (e.g., Roy et al., 1991; Congalton and Biging, 1992).
- Guidelines for collection of data for use with specific digital analyses (Joyce, 1978; Foody, 1990).
- Highlighting of special problems arising from methodologies developed in unfamiliar ecological settings (Wikkramatileke, 1959; Campbell and Browder, 1995).
- Evaluation of relative merits of quantitative biophysical data in relation to qualitative designations (Treitz et al., 1992).

The work by Joyce (1978) probably comes closest to defining field data collection as a process to be integrated into a study's analytical plan. He outlined procedures for systematic collection of field data for digital classification of multispectral satellite data. Although his work focused on specific procedures for analysis of Landsat MSS data, he implicitly stressed the need for the data collection plan to be fully compatible with the resolution of the sensor, the spatial scale of the landscape, and the kinds of digital analyses to be employed. Further, his recognition of the need to prepare a specific document devoted to field data collection itself emphasized the significance of the topic.

Field data must include at least three kinds of information. The first consists of attributes or measurements that describe ground conditions at a specific place. Examples might include identification of a specific crop or kind of land use or measurements of soil moisture content. Second, these observations must be linked to location and size (e.g., slope, aspect, and elevation) information so the attributes can be correctly matched to corresponding points in image data. Third, observations must also be described with respect to the time and date. These three elements—attributes, location, time—form the minimum information for useful field data. Complete field data also include other information, such as records of weather, illumination, identities of the persons who collected the data, calibration information for instruments, and other components as required for specific projects.

## 13.3. Nominal Data

*Nominal labels* consist of qualitative designations applied to regions delineated on imagery that convey basic differences with adjacent regions. Simple examples include *forest, urban land, turbid water,* and so on. Despite their apparent rudimentary character, accurate and precise nominal labels convey significant information. Those that are most precise—for example, *evergreen forest, single-family residential housing, maize, winter wheat*—convey more information than broader designations and therefore are more valuable.

Nominal labels originate from several alternative sources. One is an established classification system, such as that proposed by Anderson et al. (1976) (Chapter 20) and comparable systems established in other disciplines. These offer the advantage of acceptance by other workers, comparability with other studies, established definitions, and defined relationships between classes. In other instances, classes and labels have origins in local terminology or in circumstances that are specific to a given study. These may have the advantage of serving well the immediate purposes of the study, but limit comparison with other studies. Collection of data in the field is expensive and inconvenient, so there are many incentives to use a set of field data for as many purposes as possible; this effect means that it is often difficult to anticipate the ultimate application of a specific set of data. Therefore, ad hoc classifications that are unique to a specific application often prove unsatisfactory if the scope of the study expands beyond its initial purpose. Nonetheless, analysts must work to address the immediate issues at hand without attempting to anticipate every potential use for their data—the objective should be to maintain flexibility rather than to cover every possibility.

In the field, nominal data are usually easy to collect at points or for small areas, but difficulties arise as one attempts to apply the labeling system to larger areas. For these reasons, it is usually convenient to annotate maps or aerial photographs in the field as a means of relating isolated point observations to areal units. As the system is applied to broader regions, requiring the work of several observers, it becomes more important to train observers in the consistent application of the system and to evaluate the results to detect inconsistencies and errors at the earliest opportunity. Because physical characteristics of some classes vary seasonally, it is important to ensure that the timing of field observations match those of the imagery to be examined.

## 13.4. Documentation of Nominal Data

The purpose of field data is to permit reconstruction, in as much detail as possible, of ground and atmospheric conditions at the place and time that imagery was acquired. Therefore, field data require careful and complete documentation because they may be required later to answer questions that had not been anticipated at the time they were collected. Nominal data can be recorded as field sketches or as annotated aerial photographs or maps (Figure 13.1). The inherent variability of landscapes within nominal labels means that careful documentation of field data is an important component of the field data collection plan. Field observations should be collected using standardized procedures designed specifically for each project to ensure that uniform data are collected at each site (Figure 13.2). Reliable documentation includes careful notes, sketches, ground photographs, and even videotapes (see Figure 13.8 later in the chapter). Workers must keep a log to identify photographs in relation to field sites and to record dates, times, weather, orientations, and related information. Photographs must be cross-indexed to field sketches and notes. In some instances, small-format aerial photographs might be useful if feasible within the constraints of the project.

## 13.5. Biophysical Data

*Biophysical data* consist of measurements of physical characteristics collected in the field describing, for example, the type, size, form, and spacing of plants that make up the vegetative

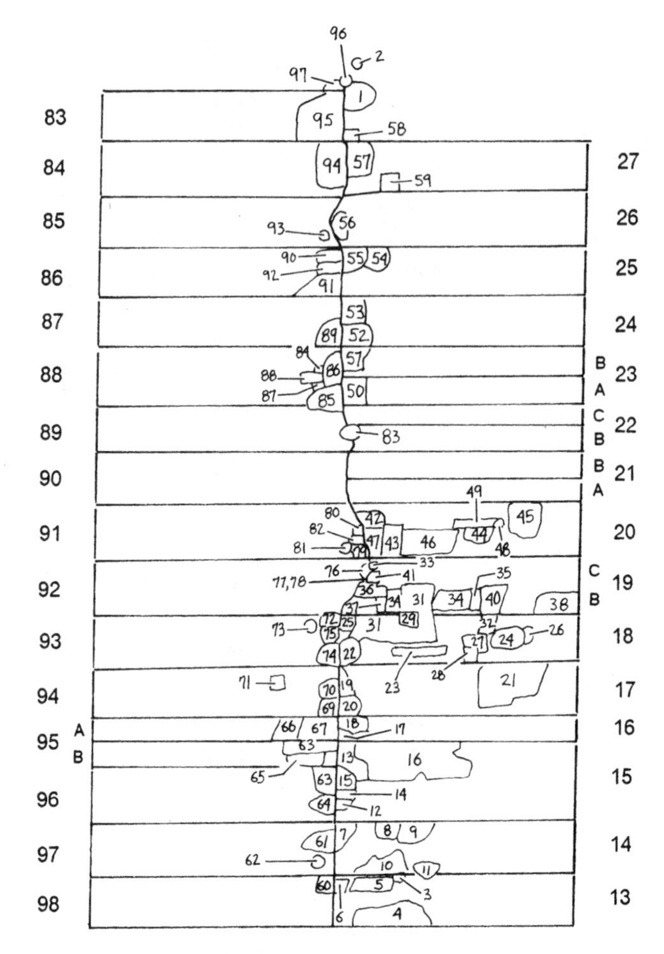


**FIGURE 13.1.** Record of field sites. The rectangular units represent land holdings within the area in Brazil studied by Campbell and Browder (1995). Each colonist occupied a 0.5 km × 2.0 km lot (strip-numbered at the left and right). Within each colonist's lot, the irregular parcels outlined and numbered here show areas visited by field teams who prepared notes and photographs. See Figure 13.2. Copyright 1995 by the International Society for Remote Sensing. Reproduced by permission.

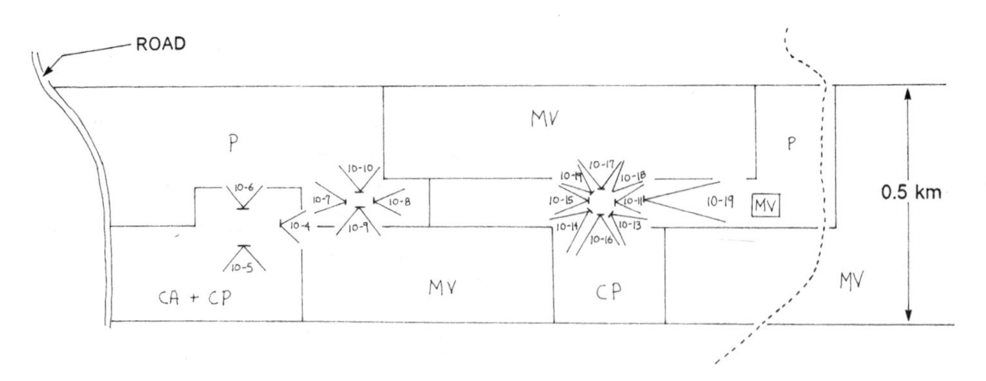


**FIGURE 13.2.** Field sketch illustrating collection of data for nominal labels. The V-shaped symbols represent the fields of view of photographs documenting land cover within cadastral units such as those outlined in Figure 13.1, keyed to notes and maps not shown here. From Campbell and Browder, 1995. Copyright 1995 by the International Society for Remote Sensing. Reproduced by permission.

GROUND DATA INVENTORY FORM

Kansas, Morton County

Data record 7 / 12 / and 7 / 15 / 96

| Field # | Acreage | Land use Crop Code | | | Irrigated | Fertilized | Planting date Month/Day | |
|---|---|---|---|---|---|---|---|---|
| 1 | 160.4 | 7 | 0 | 0 | YES | NO | | WELLSITE SOUTH SIDE |
| 2 | 163.3 | 7 | 0 | 0 | | NO | | |
| 3 | 158.8 | 4 | 0 | 4 | NO | NO | 9-20-95 | HARVESTED |
| 4 | 77.7 | 4 | 0 | 2 | NO | NO | 9-7-95 | GRAZED - STUBBLED PLOWED |
| 5 | 81.9 | 7 | 0 | 0 | YES | NO | | WELLSITE EAST SIDE |
| 6 | 19.6 | 7 | 0 | 0 | | | | |
| 7 | 72.4 | 4 | 0 | 2 | NO | NO | SEPT 95 | HARVESTED |
| 8 | 65.2 | 7 | 0 | 0 | | | | |
| 9 | 153.7 | 4 | 0 | 2 | NO | NO | 9-16-95 | HARVESTED - YIELD 14 BL? - SPRAYED |
| 10 | 156.6 | 7 | 0 | 0 | | | | |
| 11 | 154.6 | 7 | 0 | 0 | | | | WELLSITE IN SW 40 AC |
| 12 | 152.8 | 4 | 0 | 0 | NO | NO | SEPT 95 | STUBBLE BEING PLOWED |
| 13 | 156.8 | 7 | 0 | 0 | NO | NO | SEPT 95 | STUBBLE BEING PLOWED - WELLSITE E SIDE |
| 14 | 73.0 | 4 | 0 | 2 | | | | |
| 15 | 89.4 | 4 | 0 | 0 | NO | YES | 9-15-95 | STUBBLE PLOWED |
| 16 | 162.8 | 4 | 0 | 0 | NO | NO | SEPT 95 | GRAZED JAN-APRIL 96; RESIDUE PLOWED |
| 17 | 120.0 | 5 | 0 | 0 | | | SEPT 95 | DESTROYED BY GREENBUGS; RESIDUE PLOWED |
| 18 | 35.3 | 7 | 0 | 0 | | | | WELLSITE NW CORNER |
| 19 | 163.2 | 6 | 1 | 6 | | | SEPT 95 | HARVESTED - YIELD 16 |
| 20 | 94.8 | 7 | 0 | 0 | YES | YES | 5-28-96 | SPRAYED IN JULY |

**FIGURE 13.3.** Log recording field observations documenting agricultural land use.

cover. Or, as another example, biophysical data might record the texture and mineralogy of the soil surface. The exact character of biophysical data collected for a specific study depends on the purposes of the study, but typical data might include such characteristics as leaf area index (LAI), biomass, soil texture, and soil moisture. Many such measurements vary over time, so they often must be coordinated with image acquisition; in any event, careful records must be made of time, date, location, and weather.

Biophysical data typically apply to points, so often they must be linked to areas by averaging of values from several observations within an area. Further, biophysical data must often be associated with nominal labels, so they do not replace nominal data but rather document the precise meaning of nominal labels. For example, biophysical data often document the biomass or structure of vegetation within a nominal class rather than completely replacing a nominal label. In their assessment of field data recording forest sites, Congalton and Biging (1992) found that purely visual assessment provided accurate records of the species present, but that plot and field measurements were required for accurate assessment of dominant-size classes.

## 13.6. Field Radiometry

*Radiometric data* permit the analyst to relate brightnesses recorded by an aerial sensor to corresponding brightnesses near the ground surface. A *field spectroradiometer* consists of a measuring unit with a handheld probe connected to the measuring unit by a fiber optic cable (Figure

**FIGURE 13.4.** Field radiometry.

13.4). The measuring unit consists of an array of photosensitive detectors, with filters or diffraction gratings to separate radiation into several spectral regions. Radiation received by the probe can therefore be separated into separate spectral regions, and then projected onto detectors similar to those discussed in Chapter 4. Brightnesses can usually be presented either as radiances (absolute physical values) or reflectances (proportions).

Direct measurements can be recorded in physical units, although spectra are often reported as reflectances to facilitate comparisons with other data. Reflectance (relative brightness) can be determined with the aid of a reference target of known reflectance/transmittance over the range of brightnesses examined. Such a target is essentially a pure white surface. Everyday materials, despite their visual appearance, do not provide uniform brightness over a range of nonvisible wavelengths, so reference panels must be prepared from materials with well-known reflectance properties, such as barium sulphate ($BaSO_4$), or proprietary materials, such as Spectralon®, that have been specifically manufactured to present uniform brightness over a wide range of wavelengths. If readings are made from such targets just before and just after readings are acquired in the field, then the field measurements can be linked to measurements of the reference target under the same illumination conditions. Usually, a reported field reading is derived from the average of a set of 10 or more spectroradiometer readings collected under specified conditions to minimize the inevitable variations in illumination and field practice.

Readings from a reflectance standard form the denominator in calculating the *relative reflectance ratio*:

$$\text{Relative reflectance ratio} = \frac{\text{Measured radiation, sample } (\lambda)}{\text{Measured brightness, reference source } (\lambda)} \qquad \text{(Eq. 13.1)}$$

where ($\lambda$) signifies a specific wavelength interval. Because reflectance is an inherent characteristic property of a feature (unlike radiance, which varies according to illumination and weather conditions), it permits comparisons of samples collected under different field conditions.

Some instruments have the ability to match their spectral sensitivities to specific sensors, such as SPOT HRV, MSS, or TM. The results can be displayed as spectral plots or as an array of data. Many units interface to notebook computers, which permit the analyst to record spectra in the field and write data to minidisks, which can then be transferred to office or laboratory computers for analysis or incorporation into databases. Typically, a field spectroradiometer fits within a portable case designed to withstand inclement weather and other rigors of use in the field.

Analysts can select from several designs for probes with differing fields of view, or for immersion in liquids (e.g., to record radiation transmitted by water bodies). Sometimes measuring units are situated on truck-mounted booms that can be used to acquire overhead views that simulate the viewing perspective of airborne or satellite sensors (Figure 13.5).

Field spectroradiometers must be used carefully with specific attention devoted to the field of view in relation to the features to be observed, background surfaces, direction of solar illumination, and the character of diffuse light from nearby objects (Figure 13.6). The field of view for the field unit is determined by the *foreoptic,* a removable attachment that permits the operator to carefully aim the field of view for the instrument. Foreoptics might have fields of view between 1° and 10°; some have pistol grips and sighting scopes. Diffuse light (Chapter 2) is influenced by surrounding features, so measurements should be acquired at sites well clear of features with contrasting radiometric properties. Likewise, the instrument may record indirect radiation reflected from the operator's clothing, so special attention must be devoted to orientation of the probe with respect to surrounding objects.

The most useful measurements are those coordinated with simultaneous acquisition of aircraft or satellite data. Operators must carefully record the circumstances of each set of measurements to include time, date, weather, conditions of illumination, and location. Some instruments permit the analyst to enter such information through the laptop's keyboard as annotations to the radiometric data. As data are collected, the operator should supplement spectral data with photographs or videos to permit clear identification of the region represented by the radiometric observations.

In the field it is often difficult to visualize the relationship between ground features visible at a given location and their representations in multispectral data. Not only is it sometimes difficult to relate specific points in the field to the areas visible in overhead views, but often it is equally difficult to visualize differences between representations in visible and nonvisible

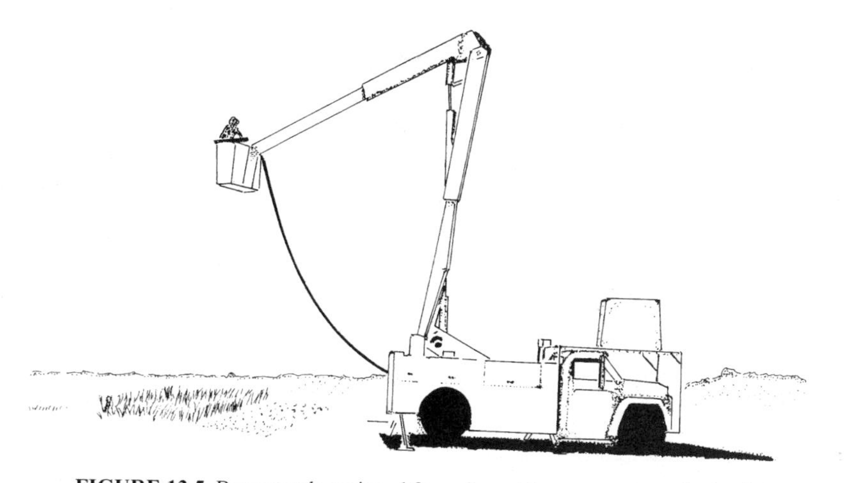

**FIGURE 13.5.** Boom truck equipped for radiometric measurements in the field.

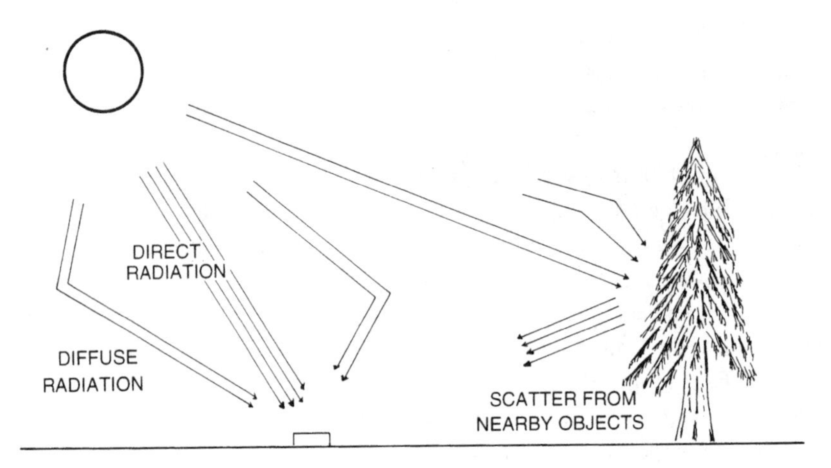


**FIGURE 13.6.** Careful use of the field radiometer with respect to nearby objects. Redrawn from Curtis and Goetz (1994).

regions of the spectrum and at the coarse resolutions of some sensors. Therefore, analysts must devote effort to matching point observations to the spectral classes that will be of interest during image analysis. Robinove (1981) and Richards and Kelley (1984) emphasize the difficulties in defining uniform spectral classes.

## 13.7. Aerial Data Collection

### *Helopolys*

Wayman et al. (1999) describe a procedure for collecting training data that removes many of the errors and uncertainties encountered in the usual methods. Typically training data for image classification and reference data for accuracy assessment are derived using procedures that introduce uncertainty concerning the location of the observations, and /or their characterization.

Wayman et al. used a commercial helicopter and a real-time differential GPS to collect field data for image classification and accuracy assessment. The helicopter was flown at elevations of 500 to 1,000 ft. above the terrain, in a path that outlined specific polygons ("helopolys") to be used as field data, using the GPS receiver to record the co-ordinates that outlined the polygons. The coordinates, believed to provide horizontal accuracy of about 1 m, could then be entered in an image processing system to outline the corresponding area within the multispectral imagery. The path of the helicopter could be displayed in real time on a laptop screen as a map to monitor shapes of polygons in flight, and assure that each polygon was closed before moving to a new site. Depending upon the specific circumstances of each site, the helicopter speed varied from hovering to about 50 mph ground speed. Analysts accompanied the pilot in flight to supervise the flight path, and to monitor the progress of the project.

Their procedure, although expensive, circumvents many of the uncertainties and errors encountered in uses of aerial photography, photointerpretation, and map registration in defining training data.

### *Unmanned Airborne Vehicles*

*Unmanned airborne vehicles* (UAVs) are remotely piloted light aircraft that can carry cameras or other sensors in support of remote sensing applications. Current UAVs extend design concepts originally developed by hobbyists for recreation and by military services for reconnaissance under difficult or dangerous conditions. Although the basic concept of designing small remotely piloted aircraft has been pursued for decades, recent advances in miniaturization, communications, strengths of lightweight materials, and power supplies have recently permitted significant advances in UAV design. This discussion outlines applications of UAVs in support of civil remote sensing operations. There is a much larger body of knowledge and experience that pertains specifically to military UAVs.

UAVs are powered either by small gasoline engines or by electric motors. Smaller UAVs typically have wingspans of perhaps of a meter or so, while wingspans of larger vehicles might reach several meters (Figure 13.7). Navigation is controlled by remote radio signals, usually given by an operator who can directly observe the UAV in flight or use remote television images to view the terrain observed by the UAV.

Recently, a variation of UAVs known as *micro aerial vehicles* (MAVs) have been developed and equipped with GPS receivers and imaging systems (Winkler and Vörsmann, 2004). MAVs have the approximate dimensions of a small bird, with wingspans of about 15 or 16 in. (40 cm), and a weight of about 18 oz. (500 gms). The GPS receivers are used to navigate the MAV remotely, while a miniaturized sensor relays a signal to a ground display. MAVs have been used for aerial reconnaissance, to monitor traffic flows on highways, and to conduct aerial assessments of crop health and vigor. Relative to UAVs, MAVs are easier to launch and control, but are sensitive to weather conditions, especially wind.

Although it may seem incongruous to consider aerial imagery as constituting a form of field data, such imagery can provide detailed data that permits interpretation of other imagery collected at coarser levels of detail. UAVs can carry cameras, digital cameras, or video cameras, as required. Images can be recorded on film or disk for retrieval after the UAV has landed, or they can be transmitted by telemetry to a ground receiver.

Because UAVs can fly close to the surfaces they observe, they can collect ancillary data, such as temperature, carbon dioxide concentrations, humidity, or other information related to the subject under investigation. Furthermore, because they can be deployed in the field at short notice, they have the ability to monitor sites more often than would be possible with other sys-

**FIGURE 13.7.** Unmanned airborne vehicle (UAV).

tems and to respond quickly to unexpected events. Their value resides in the opportunity they provide for investigators to closely coordinate UAV operations with those of broader scale imagery acquired by aircraft or satellite systems.

Despite their potential, there are still obstacles to the use of UAVs for routine remote sensing applications. For example, in rugged terrain or densely vegetated regions, suitable landing surfaces may not be within range of the areas to be observed. Piloting may be difficult, and experienced UAV pilots may not always be at hand. Accidents can lead to loss of expensive equipment, perhaps in inaccessible terrain, or to its destruction. Local weather conditions at or near the ground surface can limit flight operations.

UAVs are unlikely to replace more conventional remote sensing platforms, but they do offer advantages as a means of supplementing conventional field data, especially by providing data that can be coordinated with broad-scale imagery from aircraft and satellite platforms. Civilian applications of UAVs for remote sensing are still under development, principally for applications in agriculture, forestry, and related areas. A classic difficulty encountered in the collection of field data for such applications is assuring that variable data can be collected simultaneously with image data. For example, measurements such as temperature, humidity, or illumination can vary over time and space so rapidly that the values shift as data are collected throughout a large region. Therefore, an investigator who desires to capture a snapshot of temperature conditions, for example, simultaneously with the acquisition of an aircraft or satellite image may find that the temperature values may be changing even as they are being collected. UAVs can assist with this kind of data collection task by facilitating the reliable acquisition of such data rapidly over rather large regions.

### *Aerial Videography*

*Aerial videography* is the use of video images to record aerial imagery, typically with commercial models of video cameras, supported by global positioning receivers and equipment that can record time codes on the imagery as it is collected. Aerial videography is usually acquired using a light aircraft at altitudes of 1,800–2,000 ft. above the terrain. Imagery can be recorded as videotape or in digital form.

Advantages include the ability to acquire aerial imagery quickly and inexpensively. It provides relatively fine spatial resolution and the opportunity to collect imagery at frequent intervals. An especially valuable capability is flexibility in scheduling, the ability to acquire imagery on short notice, without the constraints and long lead times typical for other forms of aerial imagery. Aerial videography is often employed as part of a broader survey using other forms of imagery, to provide additional detail, training data, or ground truth data for other systems (Wulder et al., 2005). Because it usually uses commercial models of common video cameras, equipment costs are inexpensive relative to alternatives.

The principal difficulty in using aerial videography lies in preserving the ability to identify precise locations and to match video images to other maps and imagery of the same region. The use of UAVs and commercial video equipment means that it is difficult to closely control the orientation of the camera, leading to difficulties in registering it to other imagery (Figure 13.8).

There is a range in sophistication, but equipment for aerial videography is basically simple

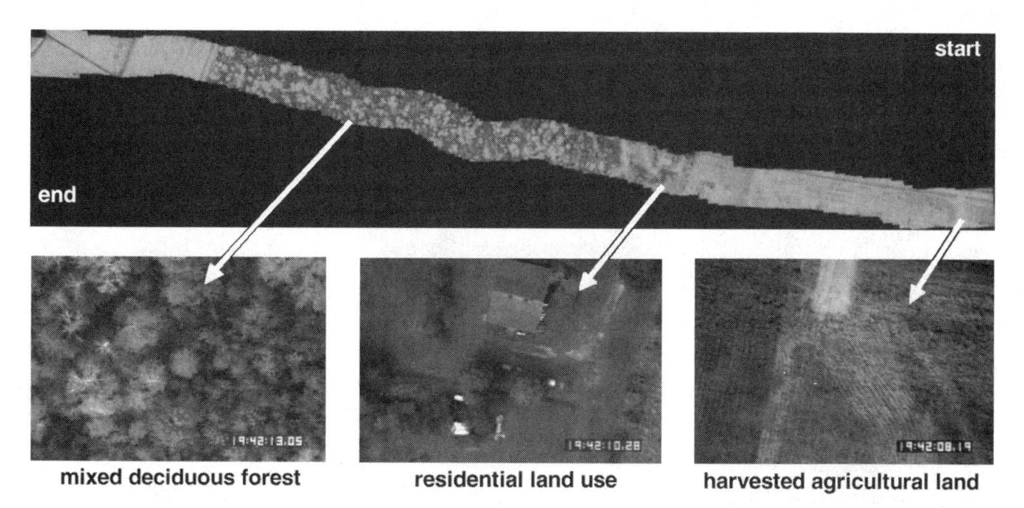


**FIGURE 13.8.** A sample sequence of aerial videography. Top: strip of aerial video. The irregular edges reflect the difficulty of controlling the orientation of the video camera in a light aircraft. Bottom: enlargement of sections illustrating varied landscape features. From Conservation Management Institute.

and inexpensive. Often a system might include two digital video cameras, one used for wide-angle views and the other zoomed to provide fine detail within the wide-angle view. Or alternatively, one camera can be oriented for a vertical view, the other for forward-looking oblique views. Some cameras can provide color infrared images. Depending upon specifics, it may be necessary to use special mounts and other modifications to provide the desired orientation for the cameras.

Because of the difficulty in matching the video imagery to maps and images, special efforts are devoted to recording that permit the analyst to match images to locations. This task can be accomplished using a device that generates a time code that permits the imagery to be matched to GPS signals and a record of the orientation of camera. The real-time GPS receiver and the time code generator send data to a computer and also send the time to the digital video audio track. A gyroscope records the orientation of the aircraft, and writes to the computer. The GPS time code is used to synchronize imagery from the two cameras, and to match imagery with the locational information from the GPS. These data permit extrapolation of GPS positions to the area imaged on the ground, offering the ability to create mosaics.

Aerial videography missions are usually planned to achieve either of two different objectives. If complete coverage of a region is desired, imagery can be collected through parallel transects. This strategy may be used in situations in which it is not possible to acquire higher quality data, or in situations in which high-quality imagery is available, but successive images are required to record patterns that are changing rapidly, such as crops within a growing season, or flooding, fires, and other natural disasters. Or, alternatively, videography can be planned as a series of transects that capture representative regions of interest to the analyst.

For analysis, images and GPS data are synchronized and imported into a GIS, where the video data can be superimposed over maps or images representing the same region, but in coarser detail. The video data then provide a detailed view of selected regions within the lower resolution images.

## 13.8. Locational Information

*Locational data* permit attributes or measurements gathered in the field to be matched to imagery of the same region. Traditionally, many of the principal problems in collection of field data for remote sensing analysis have focused on the difficulty of acquiring locational information that permits field observations to be matched to corresponding points on imagery. Accurate observations have little use unless they can be confidently placed in their correct locations. Because of the difficulty and inaccuracies of deriving absolute locational information (e.g., latitude and longitude, or comparable coordinates) in the field, most field data have been positioned with reference to distinctive landmarks (e.g., water bodies, road intersections, topographic features) visible on imagery. Such methods, of course, can introduce positional errors that negate the effort invested in accurate measurement of other qualities. In well-mapped regions of developed nations, it is often possible to acquire accurate maps to use in positioning of field data, but in unmapped or poorly mapped regions the absence of reliable positional information is a major difficulty for the use of field observations.

In some situations, locational information can be acquired by positioning easily recognized targets within the area to be imaged. If the positions of these targets are known with precision, and if they are easily recognized on the imagery, then it is possible to derive an accurate network of GCPs within an image. For photogrammetric surveys, high-contrast (e.g., black figures against white backgrounds) cloth targets are often employed to mark the positions of known locations. For radar imagery, portable corner reflectors can be positioned in the field to form a network of easily recognizable points within an image.

### *Global Positioning System*

The availability of global positioning system (GPS) technology has permitted convenient, inexpensive, and accurate measurement of absolute location. GPS has greatly enhanced the usefulness of remote sensing data, especially when it is necessary to integrate image data with field data. These instruments now are inexpensive, easy to use, and can be employed in almost any area on the Earth's surface.

A *GPS receiver* consists of a portable receiving unit sensitive to signals transmitted by a network of Earth-orbiting satellites. These satellites are positioned in orbits such that each point on the Earth's surface will be in view of at least four, and perhaps as many as 12, satellites at any given time. A system of 24 satellites is positioned at an altitude of about 13,500 mi. (21,726 km), to circle the Earth at intervals of 12 hrs., spaced to provide complete coverage of the Earth's surface (Figure 13.9).

These satellites continuously broadcast signals at two carrier frequencies within the L-band region of the microwave spectrum (Chapter 7). Although at ground level these signals are very weak, they are designed so that they can be detected even under adverse conditions (e.g., severe weather or interference from other signals). The frequency of each of these carrier signals is modulated in a manner that both (1) identifies the satellite that broadcasts the signal and (2) gives the exact time that the signal was broadcast. A receiver therefore can calculate the time delay for the signal to travel from a specific satellite, and then accurately estimate the distance from the receiver to a specific satellite.

One reason that it is possible to employ such a weak signal is that the time and identification

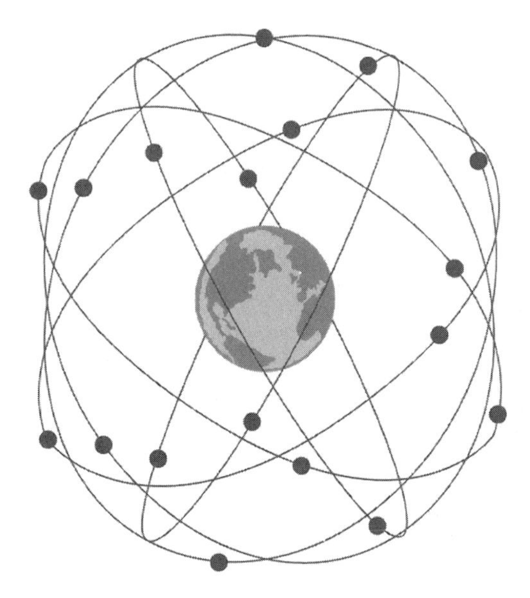

**FIGURE 13.9.** GPS satellites.

information each satellite transmits is very simple, and the receiver can listen for long periods to acquire it accurately. Because a receiver is always within range of multiple satellites, it is possible to combine positional information from three or more satellites to accurately estimate geographic position on the Earth's surface. A network of ground stations periodically recomputes and uploads new positional data to the GPS satellites.

### *GPS Receivers*

A *GPS receiver* consists of an antenna, power supply, electronic clock, and circuitry that can translate signals into positional information. Handheld receivers, intended for recreational or leisure use, typically are about the size of a wireless phone. Survey-grade GPS units, designed for scientific or commercial use, are larger, use large antennae, and include connections to permit use of a keypad and a computer (e.g., laptop or notebook models). Survey-grade units are usually designed to be carried in a small backpack (Figure 13.10). Smaller units achieve portability by conserving power or using less accurate clocks. More sophisticated GPS receivers use very accurate clocks and employ more substantial power supplies. Both handheld and survey-grade units can permit the analyst to enter coding information in the field and to download information to laboratory or office computers for use with GIS or image-processing systems. Other GPS units have been specifically designed as miniaturized circuits (about the size of a credit card) to be used as components in digital cameras or related equipment. Images or data acquired with such instruments can be referenced directly to maps or images.

### *Pseudoranging*

*Pseudoranging* is one of two ways that GPS signals can be used to estimate location. Pseudoranging is based on the use of time differences to estimate distances. GPS satellites are

**FIGURE 13.10.** Field-portable GPS receiver.

equipped with accurate clocks. The system maintains its own time standard, with the capability to update times as signals are received from five ground stations. GPS receivers simultaneously generate codes that match codes generated by GPS satellites. As GPS receivers receive coded signals from satellites, they estimate the temporal displacement required to synchronize the two codes. The amount of displacement forms the basis for an estimate of distances between the receiver and each of the satellites within range. Pseudoranging is less accurate than other methods, but its simplicity, convenience, and cost effectiveness mean that it is widely used for determining positions of fixed points—the kind of information usually required for applications in remote sensing and GIS.

### *Carrier Phase Measurement*

An alternative strategy for using GPS signals to estimate distance is known as *carrier phase measurement,* which is based on a detailed examination of signals broadcast by the satellites. GPS receivers detect either or both of the two L-band signals and add them to a signal generated by the receiver. This process is, in essence, the application of the Doppler principle (Chapter 7), using observed frequency shifts to derive positional information. If a receiver can acquire data from as many as four satellites, it is possible to estimate both vertical and horizontal positions. Survey-grade GPS units often are based upon the use of carrier phase measurement.

### *Precise Positioning Service*

Because of the implications for national security of universal availability of quick, accurate, locational data, some aspects of the satellite signal were encrypted. Exploitation of the system's full accuracy required access to a cryptographic key, provided only to authorized users. The system's most accurate data, known as *precise positioning service* (PPS), is provided to military services and to other governmental agencies with requirements for high levels of positional accuracy.

### *GPS Accuracy*

GPS errors arise from a combination of several sources:

- Orbital position (ephemeris)
- Clock error
- Ionospheric effects (atmospheric effects in the high atmosphere)
- Tropospheric effects (atmospheric effects in the lower atmosphere)
- Noise contributed by the GPS receiver itself
- Multipath error (reflection of GPS signals from nearby surfaces)

In addition, designers of the GPS system deliberately added errors to create *selective availability* (SA), to preserve the advantage of full precision of the GPS system for authorized users, while allowing the civil community access to the degraded signal. SA introduced deliberate errors in both the ephemeris and the clock to significantly increase positional errors. Access to the signal's full accuracy was provided by a cryptographic key that allowed friendly military services and other authorized users to receive the undegraded signal.

In May 2000 the U.S. government announced its decision to turn off SA, allowing the civil GPS community to receive the undegraded GPS signal. This decision leaves ionospheric effects as the largest source of positional error. This policy recognized the growing significance of accurate GPS signals for civil applications and the potential impact if foreign governments were to build competing systems using different technical specifications. The removal of SA opens the possibility that foreign systems would be compatible with the U.S. system, thereby increasing the accuracy of both. Conservative estimates of GPS accuracy suggest that horizontal errors should be within about 22 m 95% of the time, and vertical errors should be about 33 m 95% of the time (Shaw et al., 2000). However, observed accuracies suggest that performance is usually superior to these estimates, perhaps 10 m or less in some instances.

### *Local Differential GPS*

When a GPS receiver can be stationed at a fixed position of known location, it becomes possible to derive estimates of some of the errors that influence GPS accuracy and then to apply these estimates to improve the accuracy of GPS positions at unknown locations. This process is known as *local differential GPS.* Differential GPS requires that a GPS receiver be established at a fixed point for which the geographic location is known with confidence. This location is called the *base station.* Information from the base station can be applied to locational information from GPS receivers to derive more accurate estimates of location. A carefully established differential network can permit estimates of location to within centimeters of the true position.

Careful use of GPS receivers can provide locational information adequate for use in many if not most remote sensing studies, especially if it is possible to employ differential location. Atmospheric effects (Chapter 2) constitute one of the major sources of error in GPS measurements. Atmospheric interference can delay transmission of signals from the satellites, and these delays can be interpreted by the system as a difference in distance. Electrically charged particles in the ionosphere (30–300 mi., 48–483 km, above the Earth's surface) and unusually severe weather in the troposphere (ground level to about 7.5 mi or 12 km) can combine to cause errors of as much as 1–5 m. Some of these errors cancel each other out or can be removed by filtering, while others remain to create errors in the estimated positions.

## 13.9. Using Locational Information

Locational information can serve several purposes. One of the most important is to assist in identification of GCPs (Chapter 11) that allow analysts to resample image data to provide accurate planimetric location and correctly match image detail to maps and other images. Accurate locational information also permits correct positioning of field data with image data so that field data match to the correct regions within the image.

GPS permits analysts to specify coordinates for distinctive landmarks clearly visible both on imagery and on the ground. Examples include highway intersections, water bodies, distinctive terrain features, and the like. A network of such points then provides a locational framework that permits specification of locations of other points that might not be so distinctive. In other instances, field teams may acquire data in remote regions, far from distinctive landmarks. Prior to the availability of GPS, these data could be positioned within imagery only approximately, using relative locations from distant landmarks as the basis for rather rough estimates. GPS provides the capability to specify positions of such observations precisely within the imagery, provided that the image itself can be positioned within a coordinate system. That is, even with the accurate information provided by the GPS, it is still important to devote care to establishing the link between the image data and the ground observations.

## 13.10. Ground Photography

Systematic ground photographs form one of the best means of recording field conditions. Usefulness of photographs depends upon detailed notes that record the location, time, and conditions depicted by the photographs. Stereo photographs acquired using hand-held cameras (Figure 3.33) can be especially effective. Chapter 16 describes how locations of field photographs can be keyed to map locations using GIS software. Figures 13.2 and 13.3 illustrate how careful records can key photographs to specific locations and record the circumstances depicted by the photographs. The star-like symbols in Figure 13.2 show sets of photographs designed to record 360° fields of view from a specific location of significance. If digital photographs are available, specialized software can match the edges to create a 360° panorama (Figure 13.11).

**FIGURE 13.11.** Panoramic ground photograph. This photograph was prepared using specialized software to knit together a sequence of photographs that provide a 360° view. From Peter Sforza.

## 13.11. Geographic Sampling

Although ideally it would be useful to acquire complete information concerning a study area, practical constraints require that analysts sample the patterns they study. Even modest-sized images can show such large areas that complete inventories are not feasible, and large areas of the study area may be inaccessible. As a result, field data must usually be collected for a sample of pixels from the image. Here the word *observation* signifies the selection of a specific cell or pixel; the word *sample* is used here to designate a set of observations that will be used to construct an error matrix.

Any effort to sample a map or any other spatial pattern must consider three separate decisions: (1) determination of the number of observations to be used, (2) choice of the sampling pattern to position observations within an image, and (3) selection for a given sampling pattern of the spacing of observations.

### *Number of Observations*

Clearly, determination of the number of observations to be collected is fundamental. The issue is much more complex than it might seem to be at first consideration, and scientists have employed different strategies to estimate the numbers of observations. However, it is possible to outline the main considerations:

1. The *number of observations* determines the confidence interval of an estimate of the accuracy of a classification. Thus, from elementary statistics, we know that a large sample size decreases the width of the confidence interval of our estimate of a statistic—that is, we can have greater confidence that our estimate is close to the actual value when we employ larger numbers of samples.
2. For most purposes, we must consider not only the observations for the entire image, but those that are allocated to each class, as it is necessary to have some minimum number of observations assigned to each class.

### *Sampling Pattern*

*Sampling pattern* specifies the arrangement of observations. The most commonly used sampling patterns include the *simple random* pattern, the *stratified random* pattern, the *systematic* pattern, and the *systematic stratified unaligned* pattern. A simple random sample (Figure 13.12) positions observations such that the choice of any specific location as the site for an observation

**FIGURE 13.11.** *(cont.)*

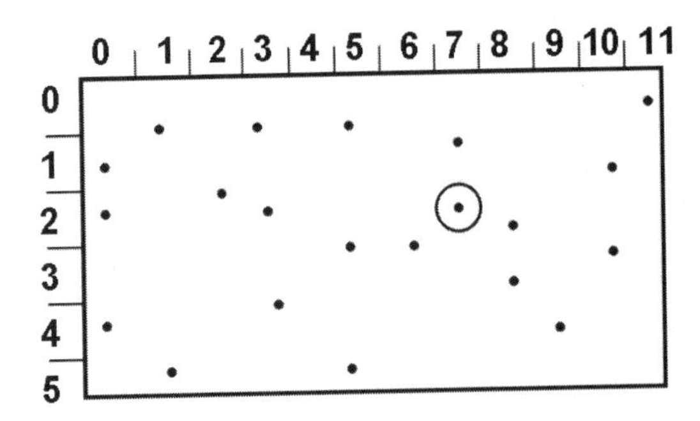


**FIGURE 13.12.** Simple random sampling pattern. Samples (represented by dots) are postioned at random within a study area by selecting pairs of values from a random sequence of values. These pairs then form the coordinates to position the sample within the study area. The circled sample is positioned at location (row 2, column 7).

is independent of the selection of any other location as an observation. Randomness assures that all portions of a region are equally subject to selection for the sample, thereby yielding data that accurately represent the area examined and satisfying one of the fundamental requirements of inferential statistics.

Typically the *simple random* sample of a geographic region is defined by first dividing the region to be studied into a network of cells. Each row and column in the network is numbered, then a random number table is used to select values that, taken two at a time, form coordinate pairs for defining the locations of observations. Because the coordinates are selected at random, the locations they define should be positioned at random.

The random sample is probably the most powerful sampling strategy available because it yields data that can be subjected to analysis using inferential statistics. However, in a geographic context, it has some shortcomings. Often a random sample is not uniform, but instead positions observations in patterns that tend to cluster together. Also, observations are unlikely to be proportionately distributed among the categories on the image, and classes of small areal extent may not be represented at all in the sample. In addition, observations may fall at or near the boundaries of parcels; in the context of accuracy assessment, such observations should probably be avoided, as they may record registration errors rather than classification errors. Thus the pure random sample, despite its advantages in other situations, may not be the best choice for comparing two images. It is, however, important to include an element of randomness in any sampling strategy, as described above.

A *stratified sampling pattern* assigns observations to subregions of the image to ensure that the sampling effort is distributed in a rational manner (Figure 13.13). For example, a stratified

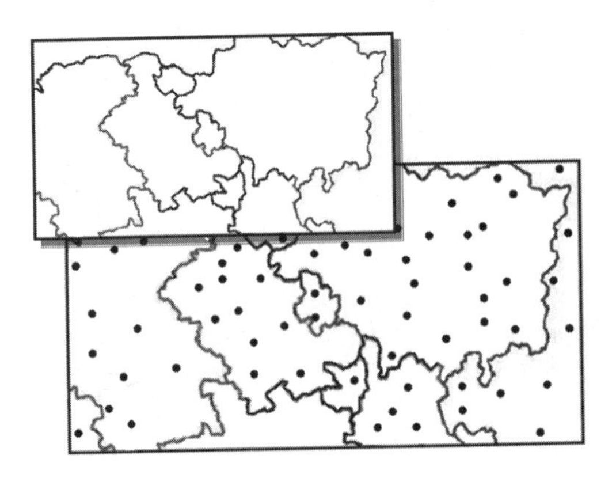

**FIGURE 13.13.** Stratified random sampling pattern. Here a region has been subdivided into subregions (*strata*) shown at the top. Samples (dots) are then allocated to each stratum in proportion to its expected significance to the study, then positioned randomly within each subarea.

sampling effort plan might assign specific numbers of observations to each category on the map to be evaluated. This procedure would ensure that every category would be sampled. If a purely random pattern is used, on the other hand, many of the smaller categories might escape the sampling effort. Often the positions of observations within strata are selected at random—a process that yields the stratified random sampling pattern. Thus stratification allocates observations to categories (the strata, in this instance) in proportion to their size or significance; the random element ensures that observations are located within categories by chance, thereby ensuring a representative sample within each class.

*Systematic sampling* positions observations at equal intervals according to a specific strategy (Figure 13.14). Because selection of the starting point predetermines the positions of all subsequent observations, data derived from systematic samples will not meet the requirements of inferential statistics for randomly selected observations.

Another potential problem in the application of systematic samples is the presence of systematic variation in the underlying pattern that is to be sampled. For example, in the central United States, the Township–Range Survey System introduces a systematic component to the sizes, shapes, and positions of agricultural fields. Therefore, a purely systematic sample of errors in the classification of such landscape could easily place all observations at the centers of fields, thereby missing errors that might occur systematically near the edges of fields and giving an inflated estimate of accuracy. Or systematic samples could oversample the edges and overestimate errors. There may be other, more subtle periodicities in landscapes that could interact with a systematic pattern to introduce other errors. Systematic sampling may nevertheless be useful if it is necessary to ensure that all regions within a study area are represented; a purely random sample, for example, may place observations in clustered patterns and leave some areas only sparsely sampled.

The *stratified systematic nonaligned sampling pattern* combines features of both systematic and stratified samples while simultaneously preserving an element of randomness (Figure 13.15). These features combine to form a sampling strategy that can be especially versatile for sampling geographic distributions.

The entire study area is divided into uniform cells, usually by means of a square grid. The grid cells, of course, introduce a systematic component to the sample, and form the basis for stratification; one observation is placed in each cell. An element of randomness is contributed by the method of placement of observations within each cell. A number is randomly selected as the east–west coordinate for placement of the observation within the first cell; this value forms the east–west coordinate not only within the first cell but for all cells in that column (Figure 13.15). A second number is randomly selected to define the north–south coordinate within the first cell; this value also specifies the north–south placement of observations within all cells in

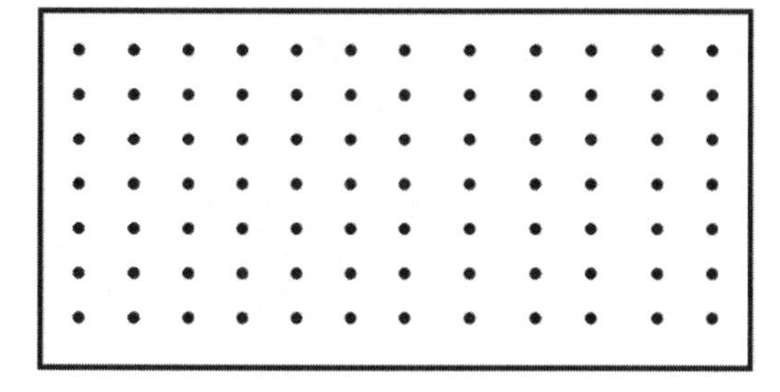

**FIGURE 13.14.** Systematic sampling pattern.

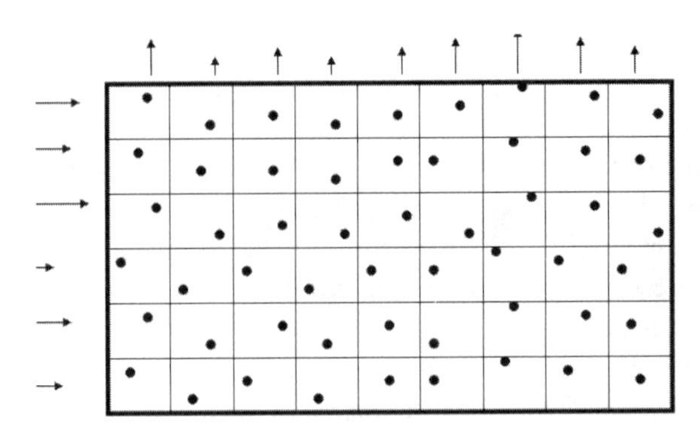

**FIGURE 13.15.** Stratified systematic nonaligned sampling pattern. Each strata (in this application, a cell within the grid) includes a signal sample, represented by the dots. The position of the dot varies according to a value for each row that sets the distance of the sample from the left edge of each cell (represented by the length of each arrow on the left of the grid), and another independent value for each column that specifies the distance above the lower edge of each cell (represented by the length of each arrow at the upper edge of the grid). The values for each row and column are determined randomly.

the first row of cells. Each row and each column is assigned a new random number that is applied to locating observations within each cell.

The resulting placement of observations within the study area has many favorable characteristics. The use of grid cells as the basis for stratification means that observations are distributed evenly within an image; no segment of the area will be unrepresented. The random element destroys the rigid alignment caused by a purely systematic design and introduces an element of chance that increases the probability that the observations will accurately represent the categories within the image. The primary pitfall is that the analyst must specify an appropriate size for the grid, which determines the number of observations and the spacing between them. These factors determine the effectiveness of a given sample in representing a specific pattern. If the analyst knows enough about the region to make a good choice of grid size, the stratified systematic nonaligned sample is likely to be among the most effective.

In some situations, the sampling effort must be confined to certain sites, perhaps because of difficult access to ground observation of remote areas or because financial restrictions confine the analyst's ability to apply the other sampling patterns mentioned above. *Cluster sampling* selects points within a study area, then uses each point as a center to determine locations of additional "satellite" observations placed nearby, so that the overall distribution of observations forms the clustered pattern illustrated in Figure 13.16. Centers can be located randomly or be assigned specifically to certain areas or classes on a reference map. Satellite observations can be located randomly, within constraints of distance. Cluster sampling may be efficient with respect to time and resources devoted to the sampling effort. If the pattern to be sampled is known beforehand, it may provide reasonably accurate results. However, if the distribution to be sampled is not known beforehand, it is probably best not to use a clustered pattern if it is at all possible to employ alternatives. Because observations are constrained to be close to one another, there is a distinct danger that the observations will replicate information rather than provide new information. If so, the analyst receives an inflated confidence in the assessment because the number of observations will suggest that more information is available than is actually the case.

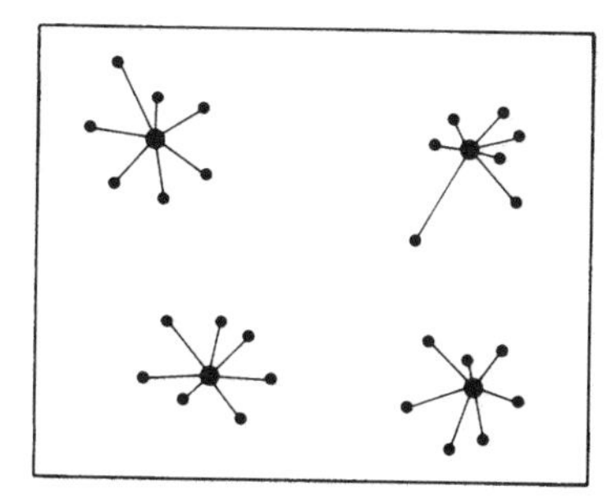

**FIGURE 13.16.** Clustered sampling pattern. Larger dots represent focal points for clusters selected randomly or possibly according to access or availability. Smaller dots represent satellite samples positioned at randomly selected distances and directions from the central point.

## *Spacing of Observations*

All efforts to define geographic sampling patterns must consider the effect of spacing of observations. If observations are spaced too close together, they tend to replicate information already provided by their neighbors and thereby provide the analyst with an unjustified confidence in the reliability of his or her conclusions. Use of 100 observations, if closely spaced, may in fact convey the information of only 50 evenly spaced observations. Closely spaced samples may underestimate variability and provide a poor indication of the number of categories present.

Examination of only a few images should be sufficient to reveal the significance of sample spacing. Some landscapes have a coarse fabric of landscape parcels—for example, the large agricultural fields of the North American plains can be represented by observations spaced at rather large distances. Other landscapes, typically those with complex topography or densely occupied regions, display a finer, more intricate fabric and require more intense sampling patterns.

*Spatial autocorrelation* describes the tendency of measurements at one location to resemble those at nearby locations. Thus a measurement of soil moisture at one place is likely to resemble another measurement made 6 in. distant and less likely to resemble one made 6 ft. away. Spatial autocorrelation provides quantitative estimates of the rate that similarities between paired observations change over distance. Spatial autocorrelation assesses the degree to which samples provide independent information.

Without prior knowledge of the pattern to be studied, it is impossible to decide if a given sampling pattern positions observations too close to one another. However, in some instances we can apply the results of experience acquired in similar situations and avoid using clustered samples unless they are dictated by other considerations. Even a purely random sample will often place observations in clusters, so many investigators prefer the systematic stratified nonaligned pattern, which retains a random component but forces a degree of uniformity in spacing samples. Even when the most effective sampling patterns are used for accuracy assessment, it is usually necessary to augment the sample with observations selected at random from the less significant categories (in terms of area) as a means of ensuring that all categories are represented by an adequate number of independent observations.

## *Sampling Summary*

It is difficult to recommend sampling strategies that might prove best for assessing accuracy because there has been little study of the comparative advantages of sampling procedures in the context of accuracy assessment. Congalton (1984) evaluated five sampling patterns to assess their performance in representing agricultural, range, and forested landscapes (as will be shown in Chapter 14). For the agricultural pattern, he concluded that the random sample performed the best. He found that the stratified random sample was best for the rangeland scene, and that the systematic random sample provided the most accurate information regarding the forest image. In general, he concluded that the systematic and systematic nonaligned sample performed well on the forest scene but poorly on other scenes.

His results highlight the difficulty in making broad recommendations for application of sampling strategies: the optimum sampling pattern is likely to vary according to the nature of the pattern to be sampled. In some applications, we can examine the pattern to be sampled, then select the sampling strategy likely to provide the best results. However, in many instances we are likely to face situations in which it is difficult or impossible to know beforehand the best sampling strategy, and we will be forced to develop experience on a case-by-case basis.

## 13.12. Summary

Full exploitation of field data can be considered one of the unexplored topics in the field of remote sensing. In recent years the availability of GPS, field spectroradiometers, and laptop computers (among other technological advances) has greatly expanded the nature and accuracy of field data. These developments have provided data that are vastly superior to those of previous eras' data with respect to quantity of accuracy by:

- Effective integration of field data within an image-processing system
- Compatibility of varied forms of field data
- Coordination of field acquisition with image data
- Evaluation and assessment
- Questions of efficient and accurate sampling
- Access to remote regions

## Review Questions

1. Typically, one might expect that a thorough field data collection effort might take much longer than the time required to collect image data. During the interval between the two, elements within the scene are likely to change, and certainly weather and illumination will change. How can field data compensate for these changes?
2. Many kinds of field observations necessarily record characteristics of *points*. Yet on remotely sensed data we normally examine *areas*. How can point data be related to areas? How can field observations be planned to accurately record characteristics of areas?
3. Sometimes field data are referred to as "ground truth." Why is this term inappropriate?
4. Prepare a rough estimate of the time and the number of people required to collect field data for a region near your school. Assume they will collect data to support a specific study in your discipline.

5. Prepare an itinerary and schedule for the field crew. Use local road maps or topographical maps to plan a route to observe field sites.
6. Refine your estimates for Questions 4 and 5 to include costs of (a) equipment, (b) transportation, (c) wages, and (d) supplies and materials.
7. Further consider your analyses for Questions 4, 5, and 6, but now include tabulation and analysis of field data. Develop a plan that can permit you to match field data to image data. How can you be sure that the two match accurately? How long will it take to prepare the field observations in a form in which they can be used?
8. Devise a sound sampling plan for field data collection in your discipline and your local area.
9. Some areas are clearly inaccessible, as a practical matter, for field data collection. How should these areas be treated in a field data collection plan?
10. Some features on the ground are highly variable over time and over distance. How can they be accurately observed for field data collection?

## References

### *Field Data: General*

Anderson, J. R., E. E. Hardy, J. T. Roach, and R. E. Witmer. 1976. *A Land Use and Land Cover Classification for Use with Remote Sensor Data* (U.S. Geological Survey Professional Paper 964). Washington, DC: U.S. Government Printing Office, 28 pp.

Campbell, J. B., and J. O. Browder. 1995. Field Data for Remote Sensing Analysis: SPOT Data, Rondonia, Brazil. *International Journal of Remote Sensing,* Vol. 16, pp. 333–350.

Congalton, R. G., and G. S. Biging. 1992. A Pilot Study Evaluating Ground Reference Data for Use in Forest Inventory. *Photogrammetric Engineering and Remote Sensing.* Vol. 58, pp. 1701–1703.

Foody, G. M. 1990. Directed Ground Survey for Improved Maximum Likelihood Classification of Remotely Sensed Data. *International Journal of Remote Sensing,* Vol. 11, pp. 1935–1940.

Joyce, A. T. 1978. *Procedures for Gathering Ground Truth Information for a Supervised Approach to Computer-Implemented Land Cover Classification of Landsat-Acquired Multispectral Scanner Data* (NASA Reference Publication 1015). Houston, TX: NASA 43 pp.

Lintz, J., P. A. Brennan, and P. E. Chapman. 1976. Ground Truth and Mission Operations. Chapter 12 in *Remote Sensing of Environment* (J. Lintz and D. S. Simonett, eds.). Reading, MA: Addison-Wesley, pp. 412–437.

McCoy, R. M. 2005. *Field Methods in Remote Sensing,* New York: Guilford Press.

Peterson, D. L., M. A. Spencer, S. W. Running, and K. B. Teuber. 1987. Relationship of Thematic Mapper Simulator Data to Leaf Area of Temperate Coniferous Forest. *Remote Sensing of Environment,* Vol. 22, pp. 323–341.

Roy, P. S., B. K. Ranganath, P. G. Diwaker, T. P. S. Vophra, S. K. Bhan, I. S. Singh, and V. C. Pandian. 1991. Tropical Forest Type Mapping and Monitoring Using Remote Sensing. *International Journal of Remote Sensing,* Vol. 11, pp. 2205–2225.

Sader, S. A., R. B. Waide, W. T. Lawrence, and A. T. Joyce. 1989. Tropical Forest Biomass and Successional Age Class Relationships to a Vegetation Index Derived from Landsat TM Data. *Remote Sensing of Environment,* Vol. 28, pp. 143–156.

Steven, M. D. 1987. Ground Truth: An Underview. *International Journal of Remote Sensing,* Vol. 8, pp. 1033–1038.

Treitz, P. M., P. J. Howarth, R. C. Suffling, and P. Smith. 1992. Application of Detailed Ground Information to Vegetation Mapping with High Resolution Digital Imagery. *Remote Sensing of Environment,* Vol. 42, pp. 65–82.

Wikkramatileke, R. 1959. Problems of Land-Use Mapping in the Tropics: An Example from Ceylon. Geography, Vol. 44, pp. 79–95.

Winkler, S., and P. Vörsmann, 2004.Bird's-Eye View: GPS & Micro Aerial Vehicles. *GPS World,* Vol. 15, No. 10, pp. 14–22.

### *Field Radiometry*

Curtis, B., and A. Goetz. 1994. Field Spectroscopy: Techniques and Instrumentation. In *Proceedings, International Symposium on Spectral Sensing Research (ISSSR '94).* Alexandria, VA: U.S. Army Corps of Engineers Topographic Engineering Center, pp. 195–203.

Deering, D. W. 1989. Field Measurement of Bidirectional Reflectance. Chapter 2 in *Theory and Applications of Optical Remote Sensing* (G. Asor, ed.). New York: Wiley, pp. 14–65.

Milton, E. J., E. M. Rollin, and D. R. Emery. 1995. Chapter 2 in *Advances in Environmental Remote Sensing* (F. M. Danson and S. E. Plummer, eds.). New York: Wiley, pp. 9–32.

### *GPS*

Berry, J. K. 2000. Capture "Where and When" on Video-Based GIS. *GEOWorld,* Vol. 14, No. 9, p. 26.

Divis, D. 2000. SA: Going the Way of the Dinosaur. *GPS World,* Vol. 11, No. 6, pp. 16–19.

Harrington, A. 2000. How Will Selective Availability Affect GPS/GIS Integration? *GEOWorld,* Vol. 13, No. 8, p. 28.

Hurn, J. 1989. *GPS: A Guide to the Next Utility.* Sunnyvale, CA: Trimble Navigation, 76 pp.

Shaw, M., K. Sandhoo, and D. Turner. 2000. Modernization of the Global Positioning System. *GPS World,* Vol. 11, No. 9, pp. 36–40.

Tralli, D. M. 1993, May–June. A Sense of Where You Are. *Sciences,* pp. 14–19.

### *Spectral Classes*

Richards, J. A., and D. J. Kelley. 1984. On the Concept of the Spectral Class. *International Journal of Remote Sensing,* Vol. 5, pp. 987–991.

Robinove, C. J. 1981. The Logic of Multispectral Classification and Mapping of Land. *Remote Sensing of Environment,* Vol. 11, pp. 231–244.

### *Sampling*

Ayeni, O. O. 1982. Optimum Sampling for Digital Terrain Models: A Trend Towards Automation. *Photogrammetric Engineering and Remote Sensing,* Vol. 40, pp. 1687–1694.

Bellhouse, D. R. 1977. Some Optimal Designs for Sampling in Two Dimensions. *Biometrika,* Vol. 64, pp. 605–611.

Campbell, J. B. 1981. Spatial Correlation Effects upon Accuracy of Supervised Classification of Land Cover. *Photogrammetric Engineering and Remote Sensing,* Vol. 47, pp. 355–363.

Cochran, W. G. 1961. Comparison of Methods for Determining Stratum Boundaries. *Bulletin of the International Statistical Institute,* Vol. 58, pp. 345–358.

Congalton, R. G. 1984. *A Comparison of Five Sampling Schemes Used in Assessing the Accuracy of Land Cover/Land Use Maps Derived from Remotely Sensed Data.* PhD dissertation, Virginia Polytechnic Institute and State University, Blacksburg, 147 pp.

Congalton, R. G. 1988. A Comparison of Sampling Schemes Used in Generating Error Matrices for Assessing the Accuracy Maps Generated from Remotely Sensed Data. *Photogrammetric Engineering and Remote Sensing,* Vol. 54, pp. 593–600.

Franklin, S. E., D. P. Peddle, B. A. Wilson, and C. Blodget. 1991. Pixel Sampling of Remotely Sensed Digital Imagery. *Computers and Geosciences,* Vol. 17, pp. 759–775.

Ginevan, M. E. 1979. Testing Land-Use Map Accuracy: Another Look. *Photogrammetric Engineering and Remote Sensing,* Vol. 45, pp. 1371–1377.

Hay, A. 1979. Sampling Designs to Test Land Use Map Accuracy. *Photogrammetric Engineering and Remote Sensing,* Vol. 45, pp. 529–533.

Rosenfield, G., H., and K. Fitzpatrick-Lins. 1986. A Coefficient of Agreement as a Measure of Thematic Classification Accuracy. *Photogrammetric Engineering and Remote Sensing,* Vol. 52, pp. 223–227.

Rosenfield, G. H., K. Fitzpatrick-Lins, and H. S. Ling. 1982. Sampling for Thematic Map Accuracy Testing. *Photogrammetric Engineering and Remote Sensing,* Vol. 48, pp. 131–137.

Skidmore, A. K., and B. J. Turner. 1992. Map Accuracy Assessment Using Line Intersect Sampling. *Photogrammetric Engineering and Remote Sensing,* Vol. 58, pp. 1453–1457.

Stehman, S. V. 1992. Comparison of Systematic and Random Sampling for Estimating the Accuracy of Maps Generated from Remotely Sensed Data. *Photogrammetric Engineering and Remote Sensing,* Vol. 58, pp. 1343–1350.

Van der Wel, F. J. M., and L. L. F. Janssen. 1994. A Short Note on Pixel Sampling of Remotely Sensed Digital Imagery. *Computers and Geosciences,* Vol. 20, pp. 1263–1264.

Wayman, J. P., R. H. Wynne, and J. A. Scrivani, 1999. Satellite-Assisted Forest Cover Mapping in the Southeastern United States using Iterative Guided Spectral Class Rejection, In: *Proceedings of Second International Conference-Geospatial Information in Agriculture and Forestry,* Lake Buena Vista, Florida, 2:355–362.

Wulder, M., J. White, and S. McDonald. 2005, November. Truth in the Air. *GeoWorld,* pp. 28–31.

CHAPTER FOURTEEN

# Accuracy Assessment

## 14.1. Definition and Significance

Prospective users of maps and data derived from remotely sensed images quite naturally ask about the accuracy of the information they will use. Yet questions concerning accuracy are surprisingly difficult to address in a convincing manner. This chapter describes how the accuracy of a map or image can be evaluated and identifies some of the difficulties that remain to be solved.

### *Accuracy and Precision*

*Accuracy* defines "correctness"; it measures the agreement between a standard assumed to be correct and a classified image of unknown quality. If the image classification corresponds closely with the standard, it is said to be "accurate." There are several methods for measuring the degree of correspondence, all described in later sections.

*Precision* defines "detail" (Figure 14.1). The distinction is important because one may be able to increase accuracy by decreasing precision—that is, by being vague in the classification. For example, as the analyst classifies a stand of trees as *forest, coniferous forest, pine forest, shortleaf pine forest, or mature shortleaf pine forest,* detail increases, and so too does the opportunity for error. (Clearly, it is more difficult to correctly assign detailed classes than to correctly assign general classes.) Evaluation of accuracy seldom explicitly considers precision, yet we must always ask if the precision is appropriate for the purpose at hand. Accuracy of 95% in separating *water* and *forest* is unlikely to be useful if what we really need to know is the distributions of *evergreen* and *deciduous* forests.

In a statistical context, high accuracy means that bias is low (i.e., that estimated values are consistently close to an accepted reference value), and that the variability of estimates is low (Figure 14.1). The usefulness of a map is related not only to its correctness, but also to the precision with which the user can make statements about specific points depicted on the map. A map that offers only general classes (even if correct) enables users to make only vague statements about any given point represented on the map; one that uses detailed classes permits the user to make more precise statements (Webster and Beckett, 1968).

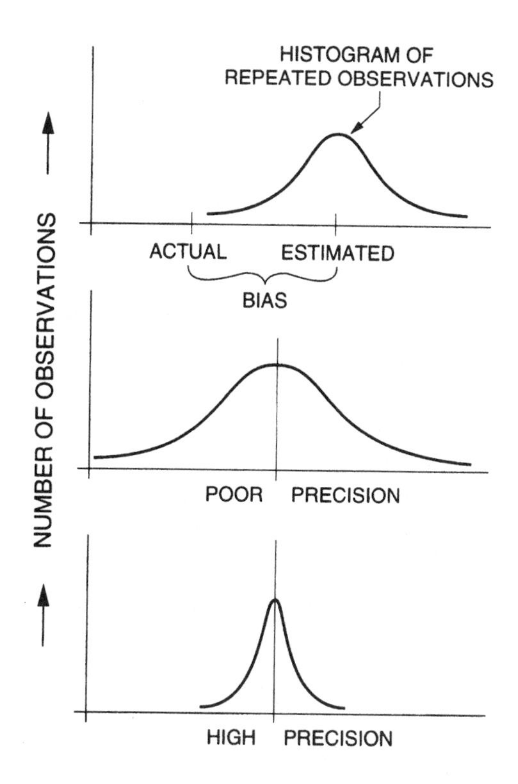


**FIGURE 14.1.** Bias and precision. Accuracy consists of bias and precision. Consistent differences between estimated values and true values create bias (top diagram). The middle and lower diagram illustrate the concept of precision. High variability of estimates leads to poor precision; low variability of estimates creates high precision.

### *Significance*

Accuracy has many practical implications: for example, it affects the legal standing of maps and reports derived from remotely sensed data, the operational usefulness of such data for land management, and their validity as a basis for scientific research. Analyses of accuracies of alternative classification strategies have significance for everyday uses of remotely sensed data. There have, however, been few systematic investigations of relative accuracies of manual and machine classifications, of different individuals, of the same interpreter at different times, of alternative preprocessing and classification algorithms, or of accuracies associated with different images of the same area. As a result, accuracy studies would be valuable research in both practical and theoretical aspects of remote sensing.

Often people assess accuracy from the appearance of a map, from past experience, or from personal knowledge of the areas represented. These can all be misleading, as overall accuracy may be unrelated to the map's cosmetic qualities, and often personal experience may be unavoidably confined to a few unrepresentative sites. Instead, accuracy should be evaluated through a well-defined effort to assess the map in a manner that permits quantitative measure of accuracy and comparisons with alternative images of the same area.

Evaluation of accuracies of information derived from remotely sensed images has long been of interest, but a new concern regarding the accuracies of digital classifications has stimulated research on accuracy assessment. When digital classifications were first offered in the 1970s as replacements for more traditional products, many uses found the methods of machine classification to be abstract and removed from the direct control of the analyst; The concern that their validity could not be accepted without evidence prompted much of the research outlined in this chapter.

Users should not be expected to accept at face value the validity of any map, regardless of its origin or appearance, without supporting evidence. But we shall see in this chapter how difficult it can be to compile the data necessary to support credible statements concerning map accuracy.

## 14.2. Sources of Classification Error

Errors are present in any classification. In manual interpretations, errors are caused by misidentification of parcels, excessive generalization, errors in registration, variations in detail of interpretation, and other factors. Perhaps the simplest causes of error are related to the misassignment of informational categories to spectral categories (Chapter 12). Bare granite in mountainous areas, for example, can be easily confused with the spectral response of concrete in urban areas. However, most errors are probably more complex. Mixed pixels occur as resolution elements fall on the boundaries between landscape parcels; these pixels may well have digital values unlike either of the two categories represented and may be misclassified even by the most robust and accurate classification procedures. Such errors may appear in digital classification products as chains of misclassified pixels that parallel the borders of large, homogeneous parcels (Figure 14.2).

In this manner the character of the landscape contributes to the potential for error through the complex patterns of parcels that form the scene. A very simple landscape composed of large, uniform, distinct categories is likely to be easier to classify accurately than one with small, heterogeneous, indistinct parcels arranged in a complex pattern. Key landscape variables are likely to include:

- Parcel size
- Variation in parcel size
- Parcel identities
- Number of categories
- Arrangement of categories
- Number of parcels per category
- Shapes of parcel
- Radiometric and spectral contrast with surrounding parcels

These variables change from one region to another (Podwysocki, 1976; Simonett and Coiner, 1971) and within a given region from season to season. As a result, errors present in a given

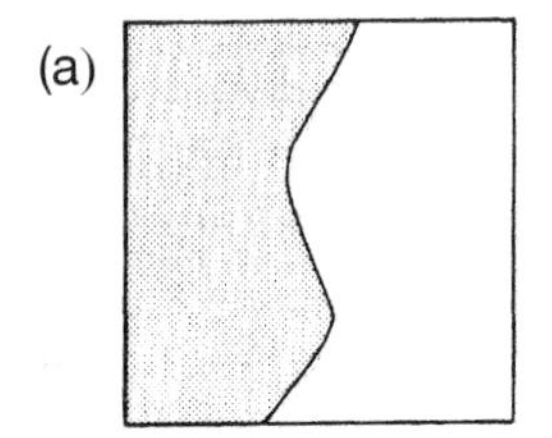


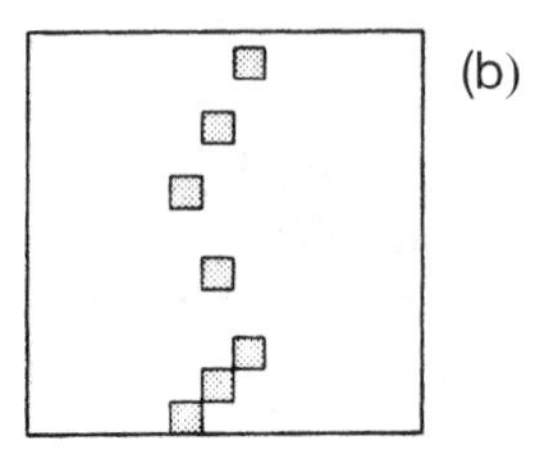


**FIGURE 14.2.** Incorrectly classified border pixels at the edges of parcels. (*a*) Map of edge between two contrasting categories. (*b*) Idealized representation of errors that might occur in a digital classification. The dark cells represent erroneously classified pixels. Usually, of course, we would not be able to know where these errors might occur.

image are not necessarily predictable from previous experience in other regions or on other dates.

## 14.3. Error Characteristics

*Classification error* is the assignment of a pixel belonging to one category (as determined by ground observation) to another category during the classification process. There are few if any systematic studies of geographic characteristics of these errors, but experience and logic suggest that errors are likely to possess at least some of the characteristic listed below:

- Errors are not distributed over the image at random, but display a degree of systematic, ordered occurrence in space. Likewise, errors are not assigned at random to the various categories on the image, but are likely to be preferentially associated with certain classes.
- Often erroneously assigned pixels are not spatially isolated but occur grouped in areas of varied size and shape (Campbell, 1981).
- Errors may have specific spatial relationships to the parcels to which they pertain—for example, they may tend to occur at the edges or in the interiors of parcels.

Figure 14.3 shows three error patterns from Landsat classifications derived by Congalton (1984). Each image shows an area corresponding to a USGA 7.5-minute topographic quadrangle; the distributions show errors in land-cover classifications derived from Landsat images. These three areas were specifically selected to represent contrasting landscapes of predominantly forested, agricultural, and rangeland land use in the rural United States. Because accurate ground observations were available for each region, it was possible to compare the classification based upon the Landsat data with the actual landscapes and to produce a complete inventory of errors for each region on a pixel-by-pixel basis. Dark areas represent incorrectly classified pixels; white areas show correct classification. (Here the scale of the images is so small that we can see only the broad outlines of the pattern without resolving individual pixels.)

We cannot identify the sources of these errors. But we can clearly see the contrasting patterns. None of the images show random patterns: the misclassified pixels tend to form distinctive patterns. In the forest image, errors appear as crescent-shaped strips formed by shadowed terrain. In the agricultural image, center-pivot irrigation systems create ring-shaped patches left fallow or planted with contrasting crops that have been erroneously classified. On the rangeland image, we clearly see the effect of the systematic Township and Range Survey System, probably through the influence of mixed pixels at the edges of rectangular parcels.

In our own image classifications we have no knowledge of the corresponding error patterns because seldom do we have the opportunity to observe the total inventory of classification errors for areas as large as those shown in Congalton's study. Yet these patterns of error do exist, even though we cannot know their abundance or distribution.

## 14.4. Measurement of Map Accuracy

The accuracy assessment task can be defined as one of comparing two maps, one based upon analysis of remotely sensed data (the *map to be evaluated*), and another based upon a different

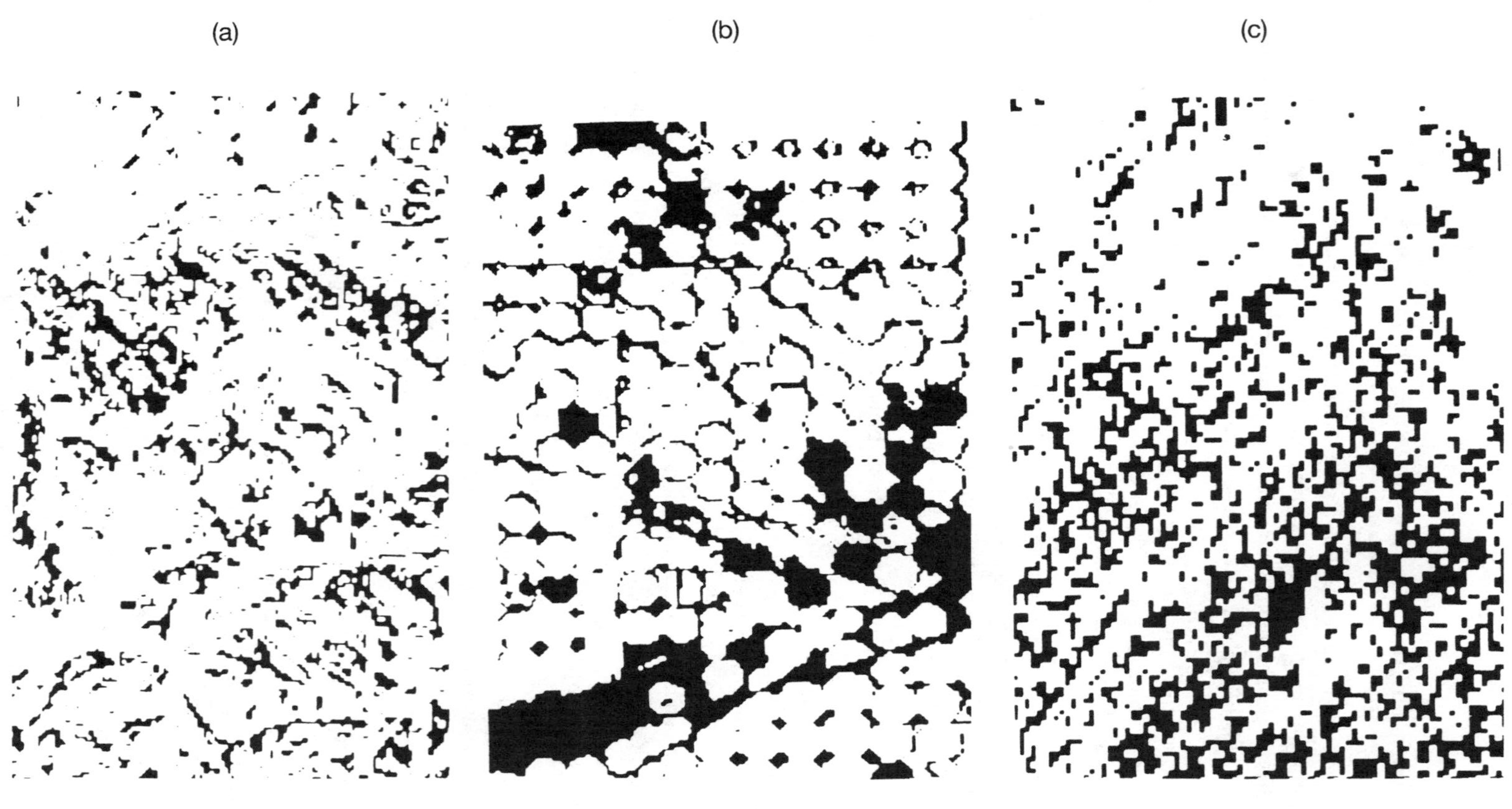


**FIGURE 14.3** .Error patterns. The dark areas are erroneously classified pixels for three areas as given by Congalton (1984): (*a*) Forested region. (*b*) Agricultural scene. (*c*) Rangeland. These images are shown at very small scale; at larger scale, it is possible to recognize the dark areas as individual pixels.

source of information. This second map is designated the *reference map,* assumed to be accurate, that forms the standard for comparison. The reference data are of obvious significance; if they are in error, then the attempt to measure accuracy will be in error. For information that varies seasonally or over time, it is important that the reference image shows information that matches the image to be evaluated with respect to time. Otherwise, the comparison may detect differences between images that are not caused solely by inaccuracies in the classification. For example, some of the differences may not really be errors, but simply changes that have occurred during the interval that elapsed between acquisition of the two images. In other instances, we may examine two images simply to decide if there is a difference, without concluding that one is more accurate than the other. For example, we may compare images of the same area taken by different sensors or classifications made by different interpreters. In such instances, it is not always necessary to assume that one image is more accurate than the other because the objective is simply to determine if and how the two differ.

Usually, however, the reference map is assumed to be the "correct" map. Both maps must be in the form of "parcel" or "mosaic" maps—that is, maps composed of a network of discrete parcels, each designated by a single label from a set of mutually exclusive categories. To assess the accuracy of one map, it is necessary that the two maps register to one another, that they both use the same classification systems, and that they have been mapped at comparable levels of detail. The strategies described here are not appropriate if the two maps differ with respect to detail, number of categories, or meanings of the categories.

The simplest method of evaluation is to compare the two maps with respect to the areas assigned to each category. The result of such a comparison is to report the areal proportions of categories (Figure 14.4). These values report the extent of the agreement between the two maps in respect to total areas in each category, but do not take into account compensating errors in misclassification that cause this kind of accuracy measure to be itself inaccurate. For example, underestimation of "Forest" in one part of the image can compensate for overestimation of "Forest" in another part of the image; serious errors in classification have been made, but are not revealed in the simple report of total areas in each category.

This form of error assessment is sometimes called *non-site-specific accuracy* because it does not consider agreement between the two maps at specific locations, but only the overall figures for the two maps. Figure 14.4 illustrates this point. The two patterns are clearly different, but the difference is not revealed by the overall percentages for each category. Gershmehl and Napton (1982) refer to this kind of error as *inventory error,* as the process considers only the aggregate areas for classes rather than the placement of classes on the map.

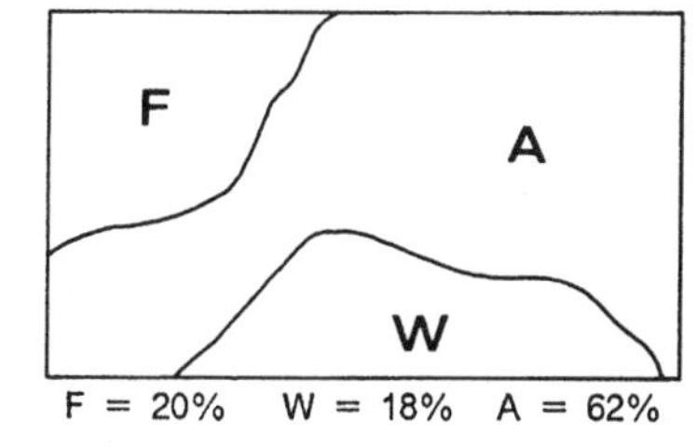


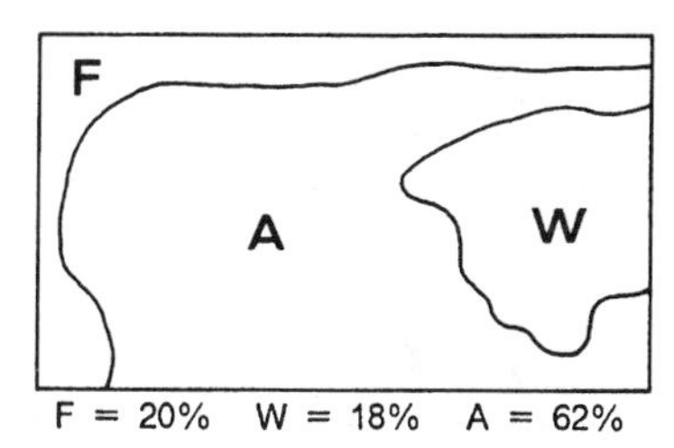


**FIGURE 14.4.** Non-site-specific accuracy. Here two images are compared only on the basis of total areas in each category. Because total areas may be similar even though placement of the boundaries differs greatly, this approach can give misleading results, as shown here.

### *Site-Specific Accuracy*

The second form of accuracy, *site-specific accuracy,* is based upon the detailed assessment of agreement between two maps at specific locations (Figure 14.5). Gershmehl and Napton (1982) refer to this kind of error as *classification error.* In most analyses the units of comparison are simply pixels derived from remotely sensed data, although if necessary a pair of matching maps could be compared using any network of uniform cells. Site-specific accuracy has been measured using several alternative strategies. Sometimes, training data for image classification can be subdivided into two groups, one used to prepare the classification, the other to assess the accuracy of the classification.

Hord and Brooner (1976) distinguish between errors in classification and errors in positioning boundaries. Classification errors are misidentifications of the identities of pixels. Boundary errors are caused by misplacement of boundaries between categories. In automated classifications, boundary errors are caused mainly by geometric errors either in the image or in the reference data, so that the correctly classified pixel may be misregistered to the wrong pixel in the reference image, and thereby counted as an error.

### *The Error Matrix*

The standard form for reporting site-specific error is the *error matrix,* sometimes referred to as the *confusion matrix* because it identifies not only overall errors for each category, but also misclassifications (due to confusion between categories) by category. Compilation of an error matrix is required for any serious study of accuracy. It consists of an $n \times n$ array, where n represents the number of categories (Table 14.1 and Figure 14.6).

The left-hand side (*y* axis) is labeled with the categories on the reference ("correct") classification; the upper edge (*x* axis) is labeled with the same n categories; these refer to those on the map to be evaluated. (Note that the meanings of the two axes can be reversed in some applications, as the convention is not universal.) Here values in the matrix represent numbers of pixels for which the anlyst has been able to compare the evaluated and reference images (in some representations, these numbers might constitute the entire image; other times they might only form a sample of the total). Sometimes the matrix is constructed using percentages rather than absolute values, but here we show absolute values counted directly from the image.

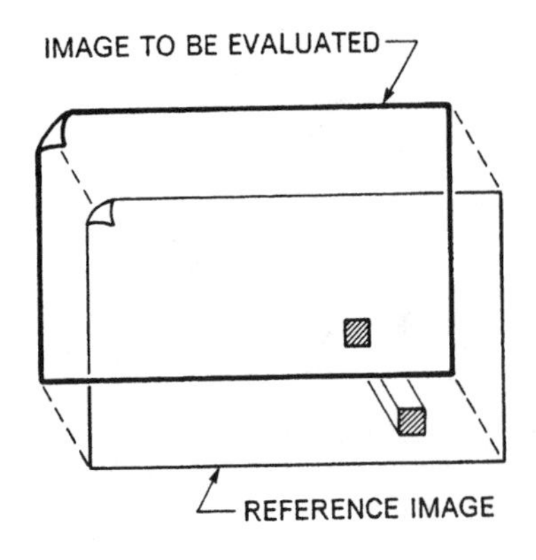


**FIGURE 14.5.** Site-specific accuracy. The two images are compared on a site-by-site ("cell-by-cell," or "pixel-by-pixel") basis to accumulate information concerning the correspondence of the two images. Here only a single pair of cells is shown, although the comparison considers the entire images.

**TABLE 14.1. Example of an Error Matrix**

| | | Image to be evaluated | | | | | | |
|---|---|---|---|---|---|---|---|---|
| | | Urban | Crop | Range | Water | Forest | Barren | Totals |
| Reference image | Urban | 150 | 21 | 0 | 7 | 17 | 30 | 225 |
| | Crop | 0 | 730 | 93 | 14 | 115 | 21 | 973 |
| | Range | 33 | 121 | 320 | 23 | 54 | 43 | 594 |
| | Water | 3 | 18 | 11 | 83 | 8 | 3 | 126 |
| | Forest | 23 | 81 | 12 | 4 | 350 | 13 | 483 |
| | Barren | 39 | 8 | 15 | 3 | 11 | 115 | 191 |
| | Totals | 248 | 979 | 451 | 134 | 555 | 225 | 1,748 |

*Note.* Percentage correct = sum of diagonal entries/total observations = 1,748/2,592 = 67.4%

Inspection of the matrix shows how the classification represents actual areas on the landscape. For example, in Table 14.1, there are 225 pixels of urban land (the far right value in the first row). Of the 225 pixels of urban land, 150 were classified as urban (row 1, column 1); these, of course, are the pixels of urban land correctly classified. Reading succeeding values along the first row, we next find incorrectly classified urban pixels and the categories to which they were assigned: crop, 21; range, 0; water, 7; forest, 17, and barren land, 30. Reading across each row, we then learn how the classification assigned pixels identified by their actual classes as they occur on the landscape. The diagonal from upper left to lower right gives numbers of correctly classified pixels.

Further inspection of the matrix reveals a summary of other information. Column totals on the far right give the total number of pixels in each class as recorded on the reference image. The bottom row of totals shows numbers of pixels assigned to each class as depicted on the image to be evaluated.

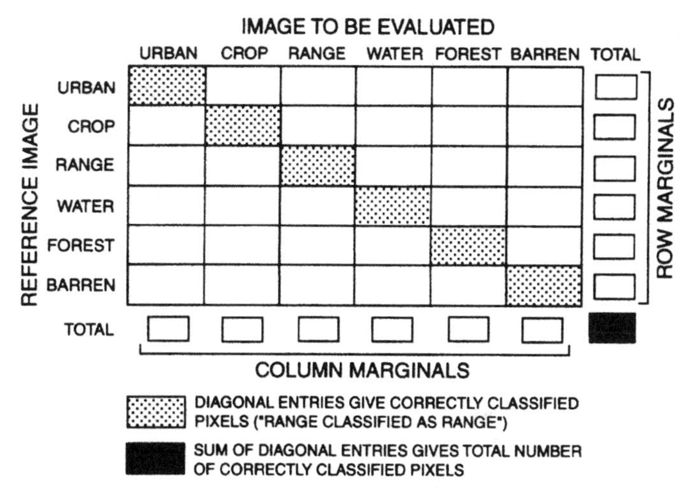


**FIGURE 14.6.** Schematic representation of an error matrix. Here elements of the matrix are represented to correspond to the text and to Table 14.1.

### *Compiling the Error Matrix*

To construct the error matrix, the analyst must compare two images—the reference image and the image to be evaluated—on a point-by-point basis to determine exactly how each site on the reference image is represented in the classification. In this context, the two "images" are not necessarily both remotely sensed images. The comparison could involve remotely sensed images, maps representing the same region, and maps depicting field-verified data. The critical consideration is not the form of the two images, but that they register to one another and use comparable classifications. Errors in registration will appear as errors in classification, so registration problems will create errors in the assessment of accuracy. The analyst must establish a network of uniform cells that form the units of comparison (Figure 14.7); these cells must be small enough to provide enough cells for a statistically valid sample, and also small enough to avoid large numbers of mixed cells (caused by cells that fall on the boundaries between parcels), but large enough to avoid the tedium and expense that accompany use of impractically small units.

The two images are then superimposed, either literally if the compilation is to be conducted manually or figuratively if the compilation is conducted digitally, by computer. For manual compilation, the analyst examines the superimposed images systematically on a cell-by-cell basis, tabulating for each cell the predominant category shown on the reference map and the category of the corresponding cell on the image to be evaluated (Table 14.2). The analyst maintains a count of the numbers in each reference category as they are assigned to classes on the map to be evaluated. A summation of this tabulation then forms the basis for construction of the error matrix.

If both images are in digital form, the unit of compilation is likely to be the pixel as defined by the image sensor, or a unit formed from even multiples of pixels. Matching of the two images, tabulation of the cross-classification, and construction of the error matrix is most difficult when one map, typically the reference map, is in manuscript form and the other, typically the map to be evaluated, is in digital form. It is then necessary to bring them both into the same format, probably by digitizing the manuscript map. Digitization can be time-consuming and expensive, and the analyst then faces questions of optimum registration and resampling (Chap-

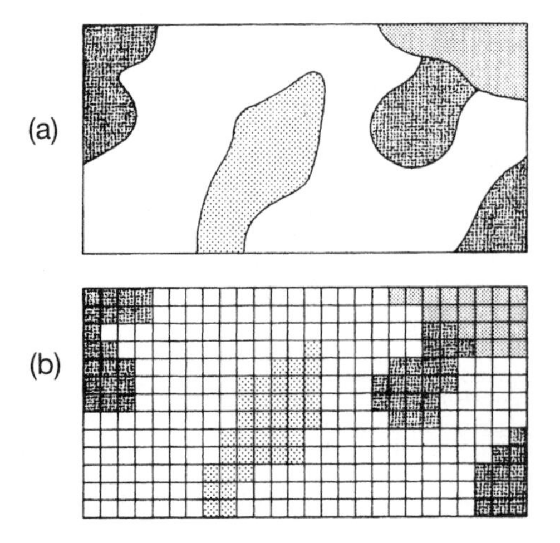


**FIGURE 14.7.** Representation of a category map by uniform cells. The network of uniform cells (*b*) represents the pattern of the original map or image (*a*). These cells then form the units of comparison for site-specific accuracy assessment.

**TABLE 14.2. Tabulation of Data for the Error Matrix**

| Worksheet | | |
|---|---|---|
| Cell | Reference map | Classification |
| 1 | F | F |
| 2 | F | U |
| 3 | W | W |
| 4 | W | W |
| 5 | C | C |
| 6 | C | F |
| 7 | C | C |
| . | . | . |
| . | . | . |
| . | . | . |

| Summation | |
|---|---|
| Forest classified as forest | 350 |
| Forest classified as urban | 23 |
| Forest classified as cropland | 81 |
| Forest classified as water | 4 |
| . . . . | . |
| . . . . | . |
| . . . . | . |
| Cropland classified as cropland | 730 |
| Cropland classified as forest | 115 |
| Cropland classified as water | 14 |
| . . . . | . |
| . . . . | . |
| . . . . | . |

ter 11). Thus, although compilation of the error matrix is straightforward, preparation of maps for comparison may be very difficult.

Table 14.3 illustrates the significance of compatibility of the classification used for the two images. Ideally, the two images use the same classification system. In practice, the two images may differ with respect to classification. Sometimes different classifications may be compatible: categories can be matched to one another in a manner that permits a valid comparison of the two images. If the differences are simply a matter of detail, often the more detailed categories can be collapsed into more general classes, as illustrated in Table 14.3. In other instances, the two images may be fundamentally incompatible; the categories may be based on different classification logics, so the two sets of classes cannot be matched (Table 14.3). In some instances two classifications may use differing definitions but the same names, so it is important to examine each image and the supporting information very closely before attempting a comparison.

### *Errors of Omission and Errors of Commission*

Examination of the error matrix reveals, for each category, *errors of omission* and *errors of commission* (Table 14.4). Errors of omission are, for example, the assignment of errors of forest on the ground to the "Agriculture" category on the map (in other words, an area of "real" forest on the ground has been omitted from the map). Using the same example, an error of commission would be to assign an area of agriculture on the ground to the "Forest" category on the map. The analyst in this instance has actively committed an error by assigning a region of forest

**TABLE 14.3. Examples of Compatible and Incompatible Classifications**

| Reference map | Image to be evaluated |
|---|---|
| | Compatible |
| Water | Water |
| Urban and built-up land | Urban residential<br>Urban commercial<br>Urban industrial |
| Deciduous forest<br>Coniferous forest<br>Mixed forest | Forest |
| Agricultural land | Cropland<br>Pasture |
| | Incompatible |
| Open land | Cropland |
| Dense settlement | Forest |
| Strip development | Cities and towns |
| Water | Lakes and rivers |
| Rough, uneven land | Disturbed land |
| Roads and highways | |

to a wrong category. (This error of commission for one category will, of course, also be tabulated as an error of omission for another category.) The distinction is essential because the interpretation could otherwise achieve 100% accuracy in respect to "Forest" by assigning all pixels to "Forest.“ Tabulations of errors of commission reveal such actions to be meaningless because they are balanced by compensating errors.

By examining relationships between the two kinds of errors, the map user gains insight about the varied reliabilities of classes on the map, and the analyst learns about the performance of the process that generated the maps. Examined from the user's perspective, the matrix reveals *consumer's accuracy* (or *user's accuracy*); examined from the analyst's point of view, the matrix reveals *producer's accuracy.* The difference between the two lies in the base

**TABLE 14.4. Errors of Omission, Errors of Commission**

| Reference image | Image to be evaluated: Urban | Crop | Range | Water | Forest | Barren | Totals | PA% | EO% | EC% |
|---|---|---|---|---|---|---|---|---|---|---|
| Urban | 150 | 21 | 9 | 7 | 17 | 30 | 234 | 64.1 | 35.9 | 39.5 |
| Crop | 0 | 730 | 93 | 14 | 115 | 21 | 973 | 75.0 | 25.0 | 25.4 |
| Range | 33 | 121 | 320 | 23 | 54 | 43 | 594 | 53.9 | 46.1 | 30.4 |
| Water | 3 | 18 | 11 | 83 | 8 | 3 | 126 | 65.9 | 34.1 | 38.1 |
| Forest | 23 | 81 | 12 | 4 | 350 | 13 | 483 | 72.5 | 27.5 | 36.9 |
| Barren | 39 | 8 | 15 | 3 | 11 | 115 | 191 | 60.2 | 39.8 | 48.9 |
| Totals | 248 | 979 | 460 | 134 | 555 | 225 | 1,748 | | | |
| CA (%): | 60.5 | 74.6 | 69.6 | 61.9 | 63.1 | 51.1 | | | | |

*Note.* CA, consumer's accuracy; PA, producer's accuracy; EO, errors of omission; EC, errors of commission.

from which the error is assessed. For producer's accuracy, the base is the area in each class on the final map. Thus, for the example in Table 14.4, producer's accuracy for "Forest" is 350/483, or 73%. For the same class, consumer's accuracy is 350/555, or 63%. Consumer's accuracy is a guide to the reliability of the map as a predictive device—it tells the user of the map that, in this example, of the area labeled "Forest," 63% actually corresponds to forest on the ground. Producer's accuracy informs the analyst who prepared the classification that, of the actual forested landscape, 73% was correctly classified. In both instances the error matrix permits identification of the classes erroneously labeled "Forest" and forested areas mislabeled as other classes.

## 14.5. Interpretation of the Error Matrix

Table 14.1 and Figure 14.6 show an example of an error matrix. Each of the 2,592 pixels in this scene was assigned to one of six land-cover classes. The resulting classification was then compared, pixel by pixel, to a previously existing land-use map of the same area, and the differences were tabulated, category by category, to form the data for Table 14.1. The total number of pixels reported by the matrix (in this instance, 2,592) may constitute the entire image, or may simply be a sample selected from the image as explained below. Also, the land-use classes here simply form examples; the matrix could be based on other kinds of classes (including forest types, geology, etc.) and could be smaller or larger, depending upon the number of classes examined.

The column of sums on the right-hand edge of the matrix gives total numbers of pixels in each category on the reference image; the row of sums at the bottom shows total pixels in each category in the classified scene. These are known, respectively, as row and column *marginals* (Figure 14.6). The sequence of values that extends from the upper left corner to the lower right corner is referred to as the *diagonal* (here we are not interested in the opposite diagonal). These diagonal entries show the number of correctly classified pixels: rangeland classified as rangeland, urban as urban, and so on. Nondiagonal values in each row give errors of omission for row-labeled categories. For example, as we read across the third row (Table 14.1), we see the numbers of pixels of "Rangeland" that have been placed in other categories: urban land, 33; cropland, 121; water, 23; forest, 54, and barren land, 43. These are "errors of omission" because the image analyst has erred by omitting true areas of rangeland from the interpreted image. More specifically, we know that "Rangeland" on the reference map is most likely to be misclassified as "Cropland" on the classified image.

In contrast, nondiagonal values along the columns give errors of commission (for labels at the tops of rows). To continue the example with the "Rangeland" category, errors of commission are caused by active misassignment of other categories to the "Rangeland" class. These errors are found by reading down the third column (Table 14.1): urban land, 0; cropland, 93; water, 11; forest, 12; and barren land, 15. These values reveal that the classification of "Rangeland" erred most often by assigning true rangeland to the "Crop" class. Here errors of omission and commission both reveal that rangeland is most often confused with cropland, and that the classified map image is likely to confuse these two categories.

Table 14.4 summarizes errors of omission and commission for another matrix. For example, there were 234 pixels of "Urban Land" in the land-use map; of these, 150 were correctly classified as "Urban Land," for a percentage of about 64%. The remaining 84 pixels (the sum of the off-diagonal entries from row 1, Table 14.4), forming about 36% of the total urban land, were incorrectly classified, mainly as "Barren Land," but also as other categories. These form the errors of omission for urban land.

Errors of omission (equal to about 42% of the total of urban land in this area) consist mainly of urban land classified as barren land and rangeland, as shown in Table 14.4. For this example, it is clear that the classifier's confusion of urban land and barren land is a major source of error in the classification of urban land use.

For a contrasting relationship between errors of omission and commission, note that rangeland tends to be classified as urban land, whereas very seldom is urban land classified as rangeland. Inspection of the error matrix reveals the kinds of errors generated by the classification process, which may in turn permit improved interpretation of the map's reliability and improved accuracy in future classifications.

### *Percentage Correct*

One of the most widely used measures of accuracy is the *percentage correct,* a report of the overall proportion of correctly classified pixels in the image or in the sample used to construct the matrix. The percentage correct, of course, can be easily found; the number correct is the sum of the diagonal entries. Dividing this value by the total number of pixels examined gives the proportion that have been correctly classified (Table 14.1). This value estimates the unknowable "true" value; the closeness of this value to the true values depends, in part, on the sampling strategy, as explained below. The percentage correct can be reported with a confidence interval (Hord and Brooner, 1976).

Often the percentage correct is used alone, without the error matrix, as a simple measure of accuracy. Reported values vary widely. Anderson et al. (1976) state that accuracies of 85% are required for satisfactory land-use data for resource management. Fitzpatrick-Lins (1978) reports that accuracies of USGS land-cover maps of the central Atlantic coastal region are about 85% (1:24,000), 77% (1:100,000), and 73% (1:250,000). For automated interpretations of land use in the Denver metropolitan area (six level-I categories), Tom et al. (1978) achieved 38% accuracy using only MSS data (including band ratios) and about 78% accuracy using MSS data plus ancillary data.

By itself, the percentage correct may suggest the relative effectiveness of a classification, but in the absence of an opportunity to examine the full error matrix it cannot provide convincing evidence of the classification's accuracy. A full evaluation must consider the categories used in the classification. For example, it would be easy to achieve high values of percentage correct by classifying a scene composed chiefly of open water, a class that is easy to classify correctly. Furthermore, variations in the accuracies of specific classes should be noted, as should the precision of the classes. A classification that used only broadly defined classes could achieve high accuracies, but would not be useful for someone who required more detail. Finally, later sections of this chapter will show that raw values for the percentage correct can be inflated by chance assignment of pixels to classes—yet another reason for careful inspection of the assessment task.

Hay (1979) stated that it is necessary to consider five questions to thoroughly understand the accuracy of a classification:

- What proportion of the classification decision is correct?
- What proportion of assignments to a given category is correct?
- What proportion of a given category is correctly classified?
- Is a given category overestimated or underestimated?
- Are errors randomly distributed?

"Percentage correct" can answer only the first of these questions; the others can be answered only by examination of the full error matrix.

### *Quantitative Assessment of the Error Matrix*

After an initial inspection of the error matrix reveals the overall nature of the errors present, there is often a need for a more objective assessment of the classification. For example, we may ask if the two maps are in agreement—a question that is very difficult to answer because the notion of "agreement" may be difficult to define and implement. The error matrix is an example of a more general class of matrices, known as *contingency tables,* that summarize classifications analogous to those considered here. Some of the procedures that have been developed for analyzing contingency tables can be applied to examination of the error matrix.

Chrisman (1980), Congalton (1983), and Congalton et al. (1983) propose application of techniques described by Bishop et al. (1975) and Cohen (1960) as a means of improving interpretation of the error matrix. A shortcoming of usual interpretations of the error matrix is that even chance assignments of pixels to classes can result in surprisingly good results, as measured by percentage correct. Hord and Brooner (1976) and others have noted that the use of such measures is highly dependent upon the samples, and therefore upon the sampling strategy used to derive the observations used in the analysis.

$\kappa$ (kappa) is a measure of the difference between the observed agreement between two maps (as reported by the diagonal entries in the error matrix) and the agreement that might be attained solely by chance matching of the two maps. Not all agreement can be attributed to the success of the classification. $\kappa$ attempts to provide a measure of agreement that is adjusted for chance agreement. $\kappa$ is estimated by $\hat{\kappa}$ ("$\kappa$ hat"):

$$\hat{\kappa} = \frac{\text{Observed} - \text{expected}}{1 - \text{expected}} \qquad \text{(Eq. 14.1)}$$

This form of the equation, given by Chrisman (1980) and others, is a simplified version of the more complete form given by Bishop et al. (1975). Here "observed" designates the accuracy reported in the error matrix and "expected" designates the correct classification that can be anticipated by chance agreement between the two images.

"Observed" is the overall value for "percentage correct" defined previously: the sum of the diagonal entries divided by the total number of samples. "Expected" is an estimate of the contribution of chance agreement to the observed percentage correct. Expected values are calculated using the row and column totals (the "marginals," as explained above). Products of row and column marginals estimate the numbers of pixels assigned to each cell in the confusion matrix, given that pixels are assigned by chance to each category.

The role of marginals in estimating chance agreement is perhaps clearer if we consider the situation in its spatial context (Figure 14.8). Then the marginals can be seen to represent areas occupied by each of the categories on the reference map and the image, and the chance mapping of any two categories is the product of their proportional extents on the two maps.

In tabular form (Table 14.5), the "expected" correct, based upon chance mapping, is found by summing the diagonals, then dividing by the grand total of products of row and column marginals. This estimate for "expected" correct can be compared to the "percentage correct" or "observed" value defined above. Table 14.5 shows a worked example for estimating $\hat{\kappa}$.

$\hat{\kappa}$ in effect adjusts the "percentage correct" measure by subtracting the estimated contribution of chance agreement. Thus, $\hat{\kappa} = 0.83$ can be interpreted to mean that the classification achieved an accuracy that is 83% better than would be expected from chance assignment of pixels to categories. As the percentage correct approaches 100, and as the contribution of chance agreement approaches 0, the value of $\hat{\kappa}$ approaches +1.0, indicating the perfect effectiveness of the classification (Table 14.6). On the other hand, as the effect of chance agreement increases, and the percentage correct decreases, $\hat{\kappa}$ assumes negative values.

$\hat{\kappa}$ and related coefficients have been discussed by Rosenfield and Fitzpatrick-Lins (1986), Foody (1992), and Ma and Redmond (1995). According to Cohen (1960), any negative value indicates a poor classification, but the possible range of negative values depends upon the specific matrix evaluated. Thus the magnitude of a negative value should not be closely interpreted to indicate the performance of the classification. Values near zero suggest that the contribution of chance is equal to the effect of the classification, and that the classification process yields no better results than would a chance assignment of pixels to classes. $\hat{\kappa}$ is asymptotically normally distributed.

Congalton (1983) and Congalton et al. (1983) described how the large sample variance can be estimated, then used in a Z test to determine if individual $\hat{\kappa}$ scores differ significantly from one another. Students should read the article by Hudson and Ramm (1987) as they consider using these methods. Foody (1992) and Ma and Redmond (1995) have advocated use of the Tau statistic ($\tau$) as an improvement over $\hat{\kappa}$. $\tau$ is analogous to $\kappa$, but is based upon a priori classifica-

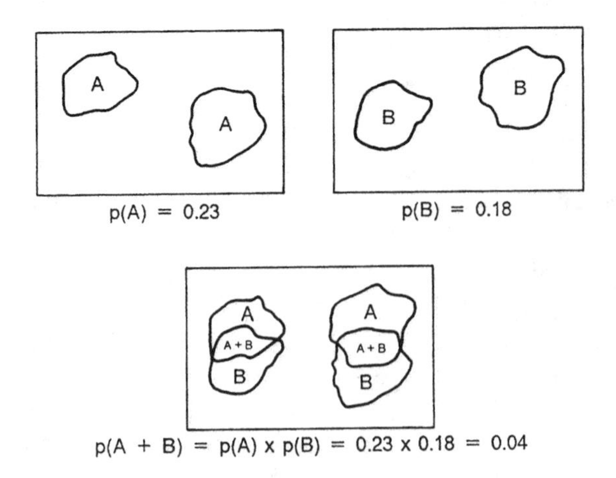


**FIGURE 14.8.** Contrived example illustrating computation of chance agreement of two categories when two images are superimposed. Here $p$(A) and $p$(B) would correspond to row and column marginals; their product estimates the agreement expected by chance matching of the two areas.

**TABLE 14.5. Worked Example Illustrating Calculation of $\hat{\kappa}$**

Part 1. Calculation of "observed correct"

| | Image to be evaluated | | | |
|---|---|---|---|---|
| Reference image | 35 | 14 | 11 | 1 |
| | 4 | 11 | 3 | 0 |
| | 12 | 9 | 38 | 4 |
| | 2 | 5 | 12 | 2 |
| | . | . | . | . |

Grand total = 163
Total correct = 86
"Observed correct" = 86/163 = 0.528
(Percentage correct = 53%)

Part 2. Calculation of "expected correct"

Marginals

| | | | | |
|---|---|---|---|---|
| | | | | 61 |
| | | | | 18 |
| | | | | 63 |
| | | | | 21 |
| 53 | 39 | 64 | 7 | |

Products of row and column marginals

| | | | |
|---|---|---|---|
| 3,233 | 2,379 | 3,904 | 427 |
| 954 | 702 | 1,152 | 126 |
| 3,339 | 2,457 | 4,032 | 441 |
| 1,113 | 819 | 1,344 | 147 |

$$\text{Expected agreement by change} = \frac{\text{Sum of diagonal entries}}{\text{Grand total}} = \frac{8{,}114}{26{,}569} = 0.305$$

Part 3. Calculation of $\hat{\kappa}$ using data from Parts 1 and 2

$$\hat{\kappa} = \frac{\text{Observed} - \text{expected}}{1 - \text{expected}} = \frac{0.528 - 0.305}{1 - 0.305} = \frac{0.223}{0.695} = 0.321$$

tion probabilities rather than the observed occurrences used in calculation of $\hat{\kappa}$. Many image-processing systems provide utilities to calculate error matrices and $\hat{\kappa}$, give a pair of images that meet the requirements outlined previously.

One of the primary purposes of error analyses is to permit comparisons of different interpretations. For a given region, we may have alternative classifications made using images acquired at different dates, classified by different procedures, or produced by different individuals. In such instances, we wish to generate error matrices, then ask which classification might be best for a given purpose. Such an analysis can permit identification of the specific imagery, dates, preprocessing procedures, and other procedures that might be best for a given purpose.

However, comparison of matrices is often complicated by the differing numbers of observations used to compile the matrices. Although the percentage correct and $\hat{\kappa}$ both can assist in evaluation of the matrices, it would be useful to be able to directly compare the matrices class

**TABLE 14.6. Contrived Matrices Illustrating $\hat{\kappa}$**

| 10 | | | | |
|---|---|---|---|---|
| | 10 | | | |
| | | 10 | | |
| | | | 10 | |
| | | | | 10 |

$\hat{\kappa} = +1.00$

| 10 | | | | |
|---|---|---|---|---|
| | 10 | | | |
| | | | 10 | |
| | | 10 | | |
| | | | | 10 |

$\hat{\kappa} = 0.50$

| | | | | 25 |
|---|---|---|---|---|
| | | | | |
| | | | | |
| | | | | |
| 25 | | | | |

$\hat{\kappa} = -1.00$

| | | | | 10 |
|---|---|---|---|---|
| | | | 10 | |
| | | 10 | | |
| | 10 | | | |
| 10 | | | | |

$\hat{\kappa} = 0.00$

| 2 | 2 | 2 | 2 | 2 |
|---|---|---|---|---|
| 2 | 2 | 2 | 2 | 2 |
| 2 | 2 | 2 | 2 | 2 |
| 2 | 2 | 2 | 2 | 2 |
| 2 | 2 | 2 | 2 | 2 |

$\hat{\kappa} = 0.00$

| | | | | 10 |
|---|---|---|---|---|
| | | | 10 | 10 |
| | | | | |
| 10 | | | | |
| 10 | | | | |

$\hat{\kappa} = -0.25$

by class. For example, for a comparison of the accuracies of two image classification algorithms, it would be useful to determine which classifier performed best in classification of agricultural land. From direct inspection of two original matrices, such a comparison is prevented by the differing numbers of samples in the data used to prepare the matrices.

This issue has been addressed by application of procedures that normalize values presented in square matrices. This normalization procedure (called *iterative proportional fitting*) applies an iterative procedure that brings row and column marginals to a common value of +1.0 by incrementally altering values within the matrix. The procedure finds new values for off-diagonal entries that will be greater than 0 in proportion to the magnitude of the errors in each cell of the original matrix, subject to the constraint that row and column marginals each total +1.00.

Error matrices based upon remotely sensed data typically display numerous entries with zeros or low values because some kinds of errors are improbable and occur infrequently. Because normalization can force some zero values in the original matrix to assume nonzero values in the transformed matrix, the practice of normalization has been regarded with skepticism by some. Further, Stehman's (2004) investigation of the use of iterative proportional fitting for error matrices derived from remotely sensed data concludes that normalization of such matrices is not appropriate. His work demonstrates that problems encountered in methodology for comparison of maps from regions of different land-cover distributions have not yet been resolved.

## 14.6. Summary

Accuracy assessment is a complex process with both conceptual and practical difficulties. This chapter cannot address all relevant topics because even the most complete discussion leaves many issues unresolved. Research continues, and despite agreement on many important aspects of accuracy evaluation, other issues are likely to be debated for a long time before they are resolved. For example, there is no clearly superior method of sampling an image for accuracy assessment, and there is disagreement concerning the best way to compare two error matrices. For many problems, while there may be no single "correct" way to conduct the analysis, we may be able to exclude some alternatives and to speak of the relative merits and shortcomings of others.

This chapter should provide the background to assess the accuracies of a classification using procedures that, if not perfect, are at least comparable in quality to those in common use today. Furthermore, the student should now be prepared to read some of the current research on this topic and possibly to contribute to improvements in the study of accuracy assessment. Many of the problems in this field are difficult, but they are not beyond the reach of interested and informed students.

## Review Questions

1. Examine Matrix A. Which category has the greatest areal extent on the image? Which category has the greatest areal extent within the ground area shown on the image?

**Matrix A**

| Actual land use | Interpreted land use | | | | | |
|---|---|---|---|---|---|---|
| | Urban | Agriculture | Range | Forest | Water | Total |
| Urban | 510 | 110 | 85 | 23 | 10 | 738 |
| Agriculture | 54 | 1,155 | 235 | 253 | 35 | 1,732 |
| Range | 15 | 217 | 930 | 173 | 8 | 1,343 |
| Forest | 37 | 173 | 238 | 864 | 27 | 1,339 |
| Water | 5 | 17 | 23 | 11 | 265 | 321 |
| Total | 621 | 1,672 | 1,511 | 1,324 | 345 | 3,724 |

**Matrix B**

| Actual land use | Interpreted land use: Urban | Agriculture | Range | Forest | Water | Total |
|---|---|---|---|---|---|---|
| Urban | 320 | 98 | 230 | 64 | 26 | 738 |
| Agriculture | 36 | 1,451 | 112 | 85 | 48 | 1,732 |
| Range | 98 | 382 | 514 | 296 | 53 | 1,343 |
| Forest | 115 | 208 | 391 | 539 | 86 | 1,339 |
| Water | 28 | 32 | 68 | 23 | 170 | 321 |
| Total | 597 | 2,171 | 1,315 | 1,007 | 383 | 2,994 |

2. Refer again to Matrix A. Which class shows the highest error of commission?
3. Based upon your examination of Matrix A, identify the class that was most often confused with agricultural land.
4. For Matrix A, which class was most accurately classified? Which class has the lowest accuracy?
5. For Matrix A, which class shows the highest number of errors of omission?
6. Compare Matrices A and B, which are error matrices for alternative interpretations of the same area, perhaps derived from different images or compiled by different analysts. Each has 5,473 pixels, but they are allocated to classes in different patterns. Which image is the "most accurate"? Can you see problems in responding with a simple answer to this question?
7. Refer again to Matrices A and B. If you were interested in accurate delineation of agricultural land, which image would you prefer? If accurate classification of forested land is important, which image should be used?
8. What might be the consequence if we had no effective means of assessing the accuracy of a classification of a remotely sensed image?
9. Data for Matrices A and B can be compiled from a complete inventory of each image or by selecting a sample of pixels from both the reference and classified images. List advantages and disadvantages of each method.
10. Make a list of applications of the accuracy assessment technique. That is, think of different circumstances/purposes in which it is valuable to be able to determine which of two alternative classifications is the more accurate.

## References

Anderson, J. R., E. E. Hardy, J. T. Roach, and R. E. Witmer. 1976. *A Land Use and Land Cover Classification for Use with Remote Sensor Data* (U.S. Geological Survey Professional Paper 964). Washington, DC: U.S. Government Printing Office, 28 pp.

Bishop, Y. M. M., S. E. Fienber, and P. W. Holland. 1975. *Discrete Multivariate Analysis: Theory and Practice.* Cambridge, MA: MIT Press, 557 pp.

Campbell, J. B. 1981. Spatial Correlation Effects upon Accuracy of Supervised Classification of Land Cover. *Photogrammetric Engineering and Remote Sensing,* Vol. 47, pp. 355–363.

Card, D.H. 1982. Using Known Map Category Marginal Frequencies to Improve Estimates of Thematic Map Accuracy. *Photogrammetric Engineering and Remote Sensing,* Vol. 48, pp. 431–439.

Chrisman, N. R. 1980. Assessing Landsat Accuracy: A Geographic Application of Misclassification Analysis. In *Second Colloquium on Quantitative and Theoretical Geography.* Cambridge, UK:

Cohen, J. 1960. A Coefficient of Agreement for Nominal Scales. *Educational and Psychological Measurement,* Vol. 20, No. 1, pp. 37–40.

Congalton, R. G. 1988a. A Comparison of Sampling Schemes Used in Generating Error Matrices for Assessing the Accuracy Maps Generated from Remotely Sensed Data. *Photogrammetric Engineering and Remote Sensing,* Vol. 54, pp. 593–600.

Congalton, R. G. 1988b. Using Spatial Autocorrelation Analysis to Explore the Errors in Maps Generated from Remotely Sensed Data. *Photogrammetric Engineering and Remote Sensing,* Vol. 54, pp. 587–592.

Congalton, R. G., and K. Green. 1999. *Assessing the Accuracy of Remotely Sensed Data: Principals and Practices.* New York: Lewis, 137 pp.

Congalton, R. G. *The Use of Discrete Multivariate Techniques for Assessment of Landsat Classification Accuracy.* MS thesis, Virginia Polytechnic Institute and State University, Blacksburg, 147 pp.

Congalton, R. G., and R. A. Mead. 1983. A Quantitative Method to Test for Consistency and Correctness in Photointerpretation. *Photogrammetric Engineering and Remote Sensing,* Vol. 49, pp. 69–74.

Congalton, R. G., R. G. Oderwald, and R. A. Mead. 1983. Assessing Landsat Classification Accuracy Using Discrete Multivariate Analysis Statistical Techniques. *Photogrammetric Engineering and Remote Sensing,* Vol. 49, pp. 1671–1687.

Congalton, R. G. 1984. *A Comparison of Five Sampling Schemes used in Assessing the Accuracy of Land Cover/Land Use Maps Derived from Remotely Sensed Data.* Ph.D. Dissertation. Virginia Polytechnic Institute and State University. Blacksburg, 147 pp.

Fitzpatrick-Lins, K. 1978. Accuracy and Consistency Comparisons of Land Use and Land Cover Maps Made from High-Altitude Photographs and Landsat Multispectral Imagery. *Journal of Research, U.S. Geological Survey,* Vol. 6, pp. 23–40.

Foody, G. M. 1992. On the Compensation for Chance Agreement in Image Classification Accuracy Assessment. *Photogrammetric Engineering and Remote Sensing,* Vol. 58, pp. 1459–1460.

Foody, G. M. 2002. Status of Land Cover Classification Accuracy Assessment. *Remote Sensing of Environment,* Vol. 80, pp. 185–201.

Foody, G. M. 2004. Thematic Map Comparison: Evaluating the Statistical Significance of Differences in Classification Accuracy. *Photogrammetric Engineering and Remote Sensing,* Vol. 70, pp. 627–633.

Gershmehl, P. J., and D. E. Napton. 1982. Interpretation of Resource Data: Problems of Scale and Transferability. In *Practical Applications of Computers in Government: Papers from the Annual Conference of the Urban and Regional Systems Association,* pp. 471–482. Peak Ridge, IL. Urban and Regional Systems Association.

Ginevan, M. E. 1979. Testing Land-Use Map Accuracy: Another Look. *Photogrammetric Engineering and Remote Sensing,* Vol. 45, pp. 1371–1377.

Hay, A. 1979. Sampling Designs to Test Land Use Map Accuracy. *Photogrammetric Engineering and Remote Sensing,* Vol. 45, pp. 529–533.

Hord, R. M., and W. Brooner. 1976. Land Use Map Accuracy Criteria. *Photogrammetric Engineering and Remote Sensing,* Vol. 46, pp. 671–677.

Hudson, W. D., and C. W. Ramm. 1987. Correct Formulation of the Kappa Coefficient of Agreement. *Photogrammetric Engineering and Remote Sensing,* Vol. 53, pp. 421–422.

Ma, Z., and R. L. Redmond. 1995. Tau Coefficients for Accuracy Assessment of Classification of Remotely Sensed Data. *Photogrammetric Engineering and Remote Sensing,* Vol. 61, pp. 435–439.

Podwysocki, M. H. 1976. *An Estimate of Field Size Distribution for Selected Sites in Major Grain Pro-*

*ducing Countries* (Publication No. X-923-76-93). Greenbelt, MD: Goddard Space Flight Center, 34 pp.

Quirk, B. K., and F. L. Scarpace. 1980. A Method of Assessing Accuracy of a Digital Classification. *Photogrammetric Engineering and Remote Sensing,* Vol . 46, pp. 1427–1431.

Rosenfield, G. H., and K. Fitzpatrick-Lins. 1986. A Coefficient of Agreement as a Measure of Thematic Map Classification Accuracy. *Photogrammetric Engineering and Remote Sensing,* Vol. 52, pp. 223–227.

Rosenfield, G. H., K. Fitzpatrick-Lins, and H. S. Ling. 1982. Sampling for Thematic Map Accuracy Testing. *Photogrammetric Engineering and Remote Sensing,* Vol. 48, pp. 131–137.

Simonett, D. S., and J. C. Coiner 1971. Susceptibility of Environments to Low Resolution Imaging for Land Use Mapping. In *Proceedings of the Seventh International Symposium on Remote Sensing of Environment.* Ann Arbor: University of Michigan Press, pp. 373–394.

Skidmore, A. K., and B. J. Turner. 1992. Map Accuracy Assessment Using Line Intersect Sampling. *Photogrammetric Engineering and Remote Sensing,* Vol. 58, pp. 1453–1457.

Stehman, S. V. 1992. Comparison of Systematic and Random Sampling for Estimating the Accuracy of Maps Generated from Remotely Sensed Data. *Photogrammetric Engineering and Remote Sensing,* Vol. 58, pp. 1343–1350.

Stehman, S. V. 2004. A Critical Evaluation of the Normalized Error Matrix in Map Accuracy Assessment. *Photogrammetric Engineering and Remote Sensing,* Vol. 70, pp. 743–751.

Story, M., and R. G. Congalton. 1986. Accuracy Assessment: A User's Perspective. *Photogrammetric Engineering and Remote Sensing,* Vol. 52, pp. 397–399.

Todd, W. J., D. G. Gehring, and J. F. Haman. 1980. Landsat Wildland Mapping Accuracy. *Photogrammetric Engineering and Remote Sensing,* Vol. 46, pp. 509–520.

Tom, C., L. D. Miller, and J. W. Christenson. 1978. *Spatial Land-Use Inventory, Modeling, and Projection: Denver Metropolitan Area, with Inputs from Existing Maps, Airphotos, and Landsat Imagery* (NASA Technical Memorandum 79710). Greenbelt, MD: Goddard Space Flight Center, 225 pp.

Turk, G. 1979. GT Index: A Measure of the Success of Prediction. *Remote Sensing of Environment,* Vol. 8, pp. 65–75.

Van der Wel, F. J. M., and L. L. F. Janssen. 1994. A Short Note on Pixel Sampling of Remotely Sensed Digital Imagery. *Computers and Geosciences,* Vol. 20, pp. 1263–1264.

Van Genderen, J., and B. Lock. 1977. Testing Map Use Accuracy. *Photogrammetric Engineering and Remote Sensing,* Vol. 43, pp. 1135–1137.

Webster, R., and P. H. T. Beckett. 1968. Quality and Usefulness of Soil Maps. *Nature,* Vol. 219, pp. 680–682.

CHAPTER FIFTEEN

# Hyperspectral Remote Sensing

## 15.1. Introduction

Remote sensing involves examination of features observed in several regions of the electromagnetic spectrum. Conventional remote sensing, as outlined in previous chapters, is based upon use of several rather broadly defined spectral regions. *Hyperspectral remote sensing* is based upon examination of many narrowly defined spectral channels. Sensor systems such as SPOT 1 HRV, Landsat MSS, and Landsat TM provide 3, 4, and 7 spectral channels, respectively. The hyperspectral sensors described below can provide 200 or more channels, each only 10-nm wide. In the context of the discussion of image resolution presented in Chapter 10, hyperspectral sensors implement the concept of "spectral resolution" to the extreme. Although hyperspectral remote sensing applies the same principles and methods discussed previously, it requires such specialized data sets, instruments, field data, and software that it forms a specialized field of inquiry.

## 15.2. Spectroscopy

Hyperspectral data have detail and accuracy that permit investigation of phenomena and concepts that greatly extend the scope of traditional remote sensing. For example, analysts can begin to match observed spectra to those recorded in spectral libraries and to closely examine relationships between brightnesses in several spectral channels to estimate atmospheric effects using data within the image itself. Such capabilities present opportunities for much more precise identification of Earth surface phenomena than is possible with broad-band sensors, for investigation of phenomena such as the blue shift (Chapter 18) and the red shift (Chapter 17), and for correction of data in some bands using other bands that convey information about atmospheric transmission.

These capabilities extend the reach of remote sensing into the field of *spectroscopy,* the science devoted to the detailed examination of very accurate spectral data. Classical spectroscopy has its origins in experiments conducted by Isaac Newton (1642–1727), who used glass prisms to separate visible light into the spectrum of colors. Later, William Wollaston (1766–1828), an English physicist, noted that spectra displayed dark lines when light is projected through a narrow slit. The meaning of these lines was discovered through the work of Joseph Fraunhofer

(1787–1826), a German glassmaker who discovered distinctive lines in spectra of light from the Sun and from stars. Dark lines (*absorption spectra*) are observed as radiation passes through gases at low pressure; bright lines (*emission spectra*) form as heated gases (e.g., in the sun's atmosphere) emit radiation. These lines have their origins in the chemical elements present in the gases, a discovery that has permitted astronomers to investigate differences in the chemical compositions of stars and planets. The Danish physicist Neils Bohr (1885–1962) found that the character of Fraunhofer lines is ultimately determined by the atomic structure of gases.

Instruments used in spectroscopy—*spectroscopes, spectrometers, spectrographs*—are designed to collect radiation with a lens and to divide it into spectral regions (using a prism or diffraction grating) that are then recorded on film or measured electronically. This form of spectroscopy is now a standard method not only in astronomy, but also for laboratory analyses to characterize unidentified materials.

## 15.3. Hyperspectral Remote Sensing

*Hyperspectral remote sensing* uses the practice of spectroscopy to examine images of the Earth's surface. Although hyperspectral remote sensing sometimes applies the techniques of classical spectroscopy to the study of atmospheric gases and pollutants, for example, more commonly it applies these techniques to the making of precise, accurate, detailed spectral measurements of the Earth's surface (*imaging spectrometry*). Such data have accuracy and detail sufficient to begin to match observed spectra to those stored in databases known as *spectral libraries*. Instruments for hyperspectral remote sensing differ from those of conventional spectroscopy in that they gather spectra not only for point targets, but for areas—not for stars or laboratory samples, but for regions of the Earth's surface. Instruments for hyperspectral remote sensing differ from other remote sensing instruments in terms of their extraordinarily fine spectral, spatial, and radiometric resolutions and their careful calibration. Some hyperspectral instruments collect data in 200 or more channels at 10–12 bits. Because of their calibration and ability to collect data having fine detail, such instruments greatly extend the reach of remote sensing not only by extending the range of applications but also by defining new concepts and analytical techniques.

Although the techniques of classical spectroscopy can be used in hyperspectral remote sensing to examine, for example, atmospheric gases, hyperspectral remote sensing typically examines very detailed spectra for images of the Earth's surface, applies corrections for atmospheric effects, and matches them to spectra of known features.

## 15.4. The Airborne Visible/Infrared Imaging Spectrometer

One of the first airborne hyperspectral sensors was designed in the early 1980s by the Jet Propulsion Laboratory (JPL; Pasadena, CA). The *airborne imaging spectrometer* (AIS) greatly extended the scope of remote sensing by virtue of its number of spectral bands; their fine spatial, spectral, and radiometric detail; and the accuracy of its calibration. AIS collected 128 spectral channels, each about 10-nm wide in the interval 1.2–2.4 μm. The term *hyperspectral*

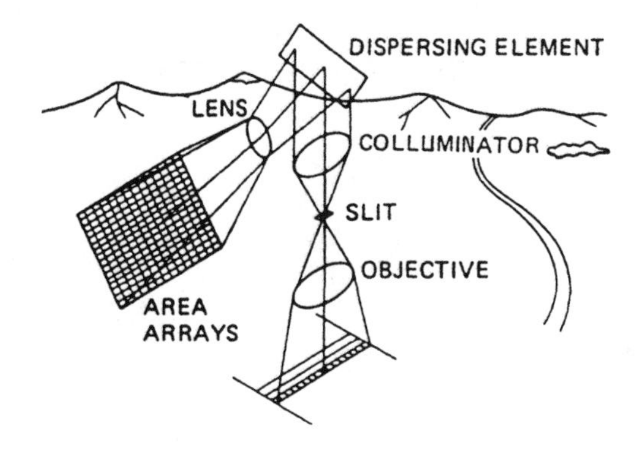


**FIGURE 15.1.** Imaging spectrometer (NASA diagram).

*remote sensing* recognizes the fundamental difference between these data and those of the usual broadband remote sensing instruments. (Sensor systems with even finer spectral resolution, designed primarily to study atmospheric gases, are known as *ultraspectral* sensors.)

Although several hyperspectral instruments are now in operational use, an important pioneer in the field of hyperspectral remote sensing is the airborne visible/infrared imaging spectrometer (AVIRIS; *http://aviris.jpl.nasa.gov/*). AVIRIS was developed by NASA and JPL from the foundations established by AIS. AVIRIS was first tested in 1987, was placed in service in 1989, and has since been modified at intervals to upgrade its reliability and performance. It has now acquired hundreds of images of test sites in North America and Europe.

Hyperspectral sensors necessarily employ designs that differ from those of the usual sensor systems. An objective lens collects radiation reflected or emitted from the scene; a collimating lens then projects the radiation as a beam of parallel rays through a diffraction grating that separates the radiation into discrete spectral bands (Figure 15.1). Energy in each spectral band is detected by linear arrays of silicon and indium antimonide (Chapter 9). Because of the wide spectral range of AVIRIS, detectors are configured in four separate panels (0.4–0.7 μm, 0.7–1.3 μm, 1.3–1.9 μm, and 1.8–2.8 μm), each calibrated independently (Figure 15.2). AVIRIS operates over the spectral range of 400–2500 nm (0.4–2.45 μm), producing 224 spectral channels, each 10-nm wide (Figure 15.3). After processing, data consist of 210 spectral channels. At AVIRIS's usual operating altitude, it records images on a strip 11-km wide, processed to form scenes recording areas about 11 km × 11 km each. Each line of data conveys about 614 pixels. If operated at low altitude, pixels might each represent ground areas about 4 m on a side; if

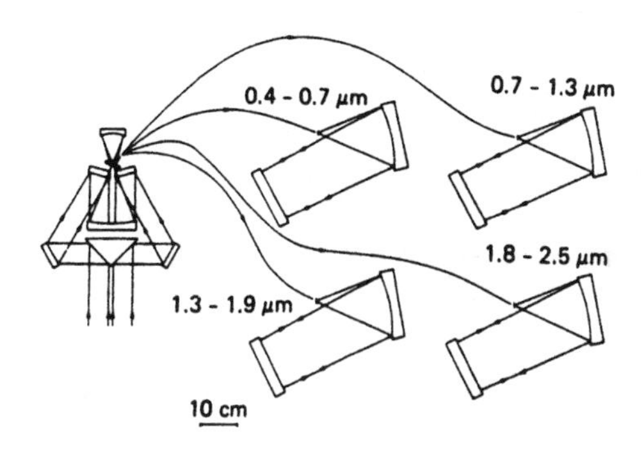


**FIGURE 15.2.** AVIRIS sensor design (NASA diagram).

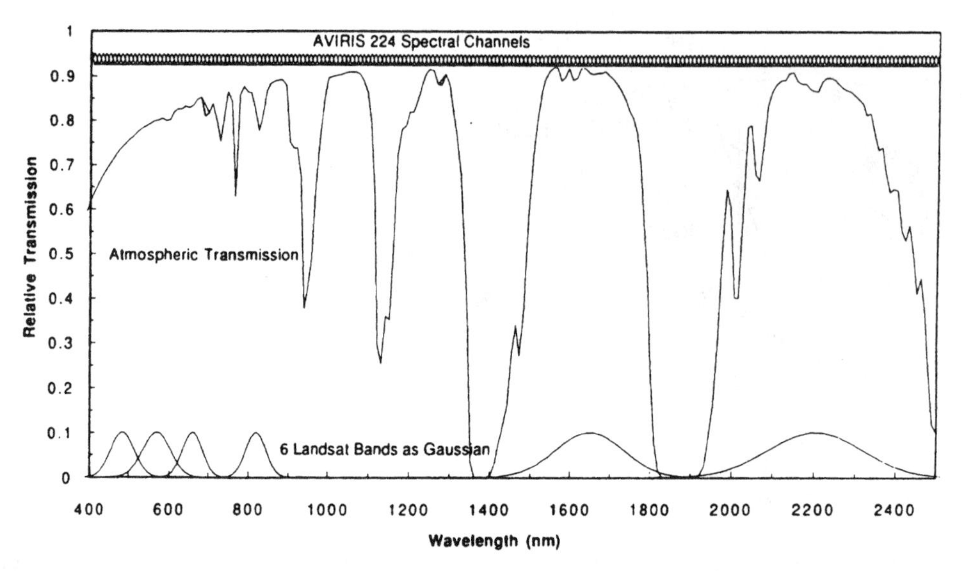


**FIGURE 15.3.** AVIRIS spectral channels compared to Landsat TM spectral channels. The 224 narrow bands at the top represent AVIRIS spectral channels. For comparison, Landsat TM channels are shown at the bottom (Green and Simmonds, 1993).

operated at higher altitudes, as is common practice, pixels each represent ground areas as large as about 20 m on a side.

## 15.5. The Image Cube

The *image cube* refers to the representation of hyperspectral data as a three-dimensional figure, with two dimensions formed by the x and y axes of the usual map or image display and the third (z) formed by the accumulation of spectral data as additional bands are superimposed on each other. In Figure 15.4 and Plate 20, the top of the cube is an image composed of data collected at the shortest wavelength (collected in the ultraviolet), and the bottom of the cube is an image composed of data collected at the longest wavelength (2.5 μm). Intermediate wavelengths are found as horizontal slices through the cube at intermediate positions. Values for a single pixel observed along the edge of the cube form a spectral trace describing the spectra of the surface represented by the pixel.

## 15.6. Spectral Libraries

The development of hyperspectral remote sensing has been accompanied by the accumulation of detailed spectral data acquired in the laboratory and in the field. These data are organized in *spectral libraries,* databases maintained primarily by governmental agencies, but also by other organizations. These libraries assemble spectra that have been acquired at test sites representative of varied terrain and climate zones, observed in the field under natural conditions. Included also are other data describing, for example, construction materials, minerals, vegetation, and fabrics as observed in laboratories under standardized conditions.

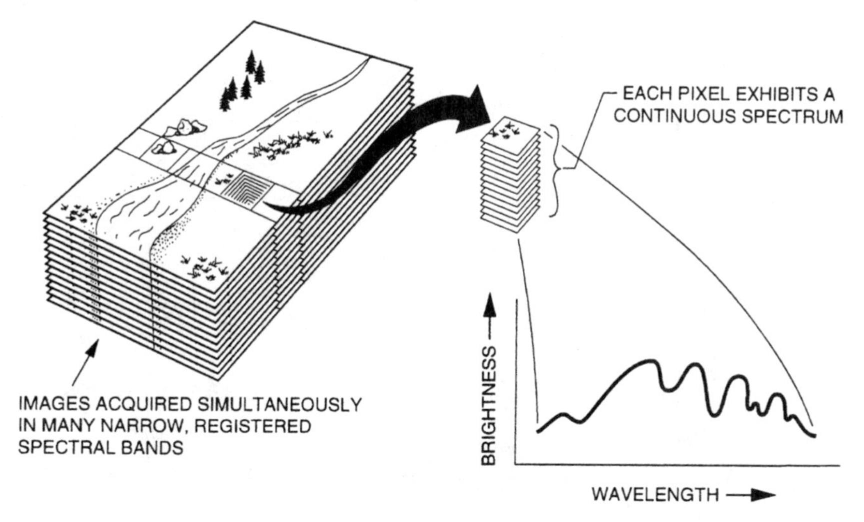


**FIGURE 15.4.** Image cube (based on a NASA diagram).

Such data are publicly available to the remote sensing community and have been incorporated into software designed for use in hyperspectral remote sensing. Maintenance of a spectral library requires specialized effort to bring data into a common format that can be used by a diverse community of users. Spectral data are typically collected by diverse instruments under varied conditions of illumination. These many differences must be resolved to prepare data in a format that permits use by a diverse community.

Because of the fine spectral, spatial, and radiometric detail of hyperspectral analysis, identification and cataloging are special problems for design of spectral libraries. Therefore each spectral record must be linked to detailed information specifying the instruments used, meteorological conditions, nature of the surface, and the circumstances of the measurement. These kinds of ancillary data are required for successful interpretation and analysis of the image data.

## 15.7. Spectral Matching

Figure 15.5, derived from Barr (1994), illustrates a sequence of analysis for hyperspectral data. Analysis begins with acquisition and preprocessing (1) to remove known system errors and assure accurate calibration. Any of several methods can be used to correct for atmospheric effects (2), as outlined in Chapter 11. The lower portion of Figure 15.5 represents the analytical process. Data are organized to permit convenient display and manipulation of specified bands or combinations of bands. Selected bands can be used to prepare color composites for display. *Alarm* (5) means that the analyst can place the cursor on the image to mark a specific pixel or region of pixels, then instruct the computer to highlight (alarm) other regions on the image that are characterized by the same spectral response. Then (6) the analyst can attempt to match the alarmed areas to spectra from a spectral library accessible to the image-processing system. If a match can be made, then the analyst can examine another region and attempt to match it to entries from the spectral library.

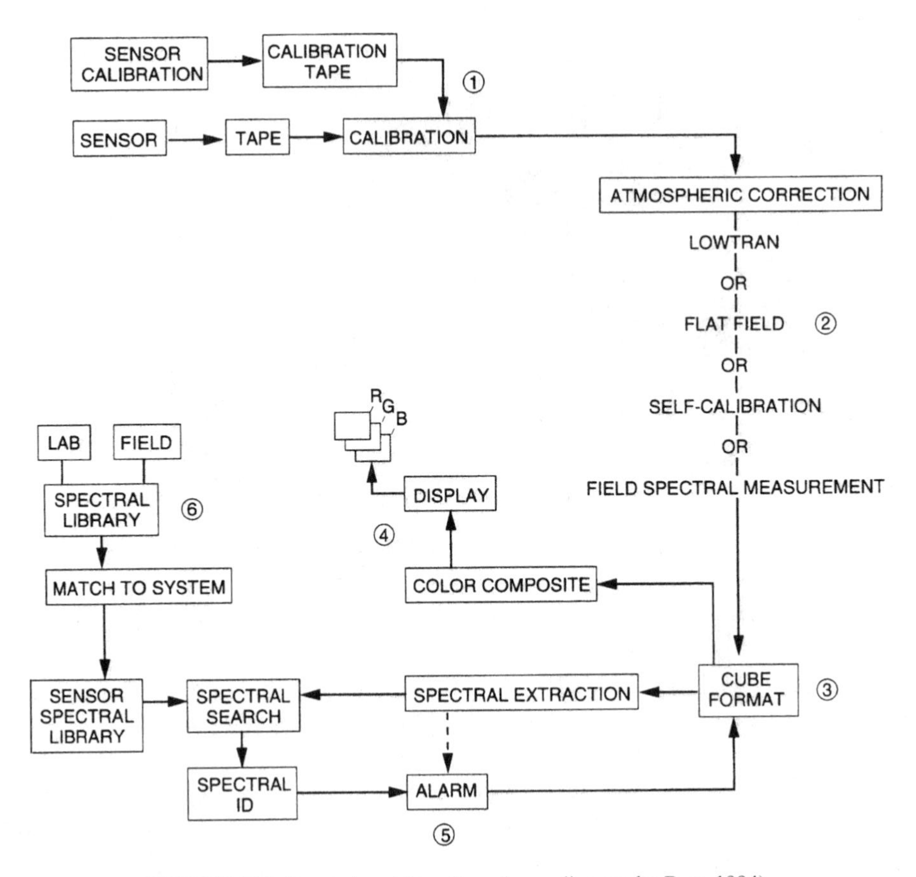


**FIGURE 15.5.** Spectral matching (based on a diagram by Barr, 1994).

## 15.8. Spectral Mixing Analysis

Fine spectral resolution, like fine spatial resolution, does not overcome the enduring obstacles to the practice of remote sensing. Surface materials recorded by the sensor are not always characteristic of subsurface conditions. Atmospheric effects, shadowing, and topographic variations contribute to observed spectra to confuse interpretations. Even when observed at fine spatial and spectral detail, surfaces are often composed of varied materials. Therefore, the sensor observes composite spectra that may not clearly match to the pure spectra of spectral libraries. *Linear mixing* refers to additive combinations of several diverse materials that occur in patterns too fine to be resolved at the resolution of the sensors. This is the effect of mixed pixels outlined in Chapter 10. As long as the radiation from component patches remains separate until it reaches the sensor, it is possible to estimate proportions of component surfaces from the observed pixel brightness using methods such as those illustrated by Equation 10.2. Linear mixing might occur when components of a composite surface are found in a few compact areas (Figure 15.6b). *Nonlinear mixing* occurs when radiation from several surfaces combines before it reaches the sensor. Nonlinear mixing occurs when component surfaces arise in highly dispersed patterns (Figure 15.6c). Nonlinear mixing cannot be described by the techniques addressed here.

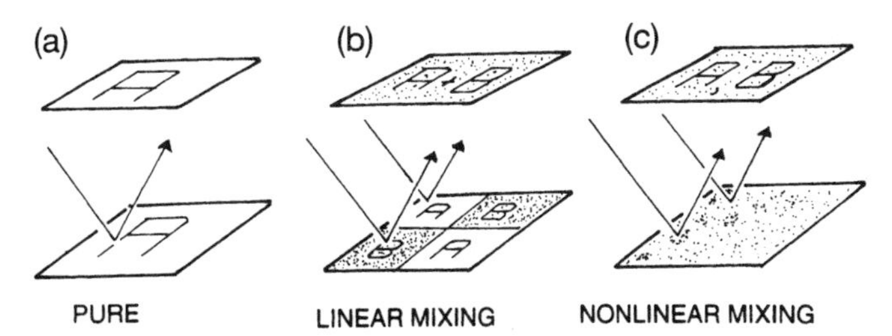


**FIGURE 15.6.** Linear and nonlinear spectral mixing. (*a*) If a pixel represents a uniform ground area at the resolution of the sensor, the pixel represents a pure spectrum. (*b*) If a pixel represents two or more surfaces that occur in patches that are large relative to the sensor's resolution, mixing occurs at the sensor. The pattern of the composite surfaces can never be resolved, but because mixing occurs in a linear manner, proportions of the components can be estimated. (*c*) If the composite occurs at a scale that is fine relative to the resolution of the sensor, mixing occurs before the radiation reaches the sensor, and components of the composite cannot be estimated using the methods described here.

*Spectral mixing analysis* (also known as *spectral unmixing*) is devoted to extracting pure spectra from the complex composites of spectra that by necessity form each image. It assumes that pixels are formed by linear mixing and further assumes that it is possible to identify the components contributing to the mixture. It permits analysts to define key components of a specific scene and forms an essential component in the process of spectral matching, discussed below. Analysts desire to match data from hyperspectral images to corresponding laboratory data to identify surfaces from their spectral data much more precisely than previously was possible. Whereas conventional images analysis (Chapter 12) matches pixels to broad classes of features, hyperspectral image matching attempts to make more precise identifications—for example, to specific mineralogies of soils or rocks.

Therefore, spectral matching requires the application of techniques that enable analysts to separate pure pixels from impure pixels. This problem is well matched to the capabilities of *convex geometry,* which examines multidimensional data envisioned in n dimensions. Individual points (pixels) within this data space can be examined as linear combinations of an unknown number of pure components. Convex geometry can solve such problems provided that the components are linearly weighted, sum to unity, and are positive. We assume also that the data have greater dimensionality (more spectral bands) than the number of pure components.

The illustrations here, for convenience and legibility, show only two dimensions, although the power of the technique is evident only with much higher dimensionality. In Figure 15.7, the three points $A$, $B$, and $C$ represent three spectral observations at the extreme limits of the swarm of data points, represented by the shaded pattern. That is, the shaded pattern represents values of all the pixels within a specific image or subimage, which is simplified by the triangle shape. (Other shapes can be defined as appropriate to approximate the shape of a data swarm, although a key objective is to define the simplest shape [*simplex*] that can reasonably approximate the pattern of the data swarm.)

These three points (for the example in Figure 15.7a) form *endmembers,* defined as the pure pixels that contribute to the varied mixtures of pixels in the interior of the data swarm. Once the simple form is defined, the interior pixels can be defined as linear combinations of the pure endmembers. (The discussion of mixed pixels in Chapter 10 anticipates this concept.) In general, interior points can be interpreted as positive unit-sum combinations of the pure variables represented by endmembers at the vertices. In general, a shape defined by $n + 1$ vertices is the sim-

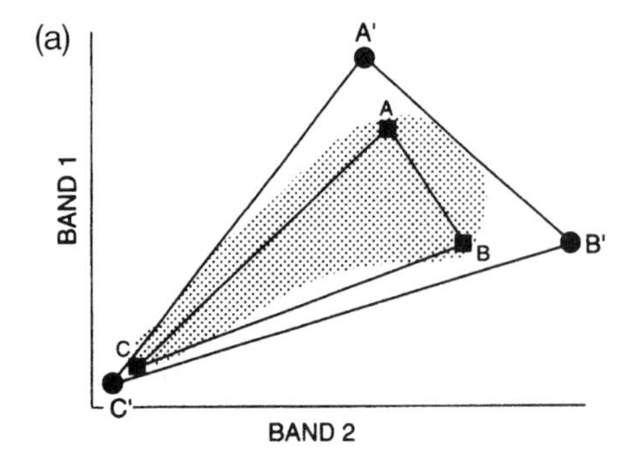


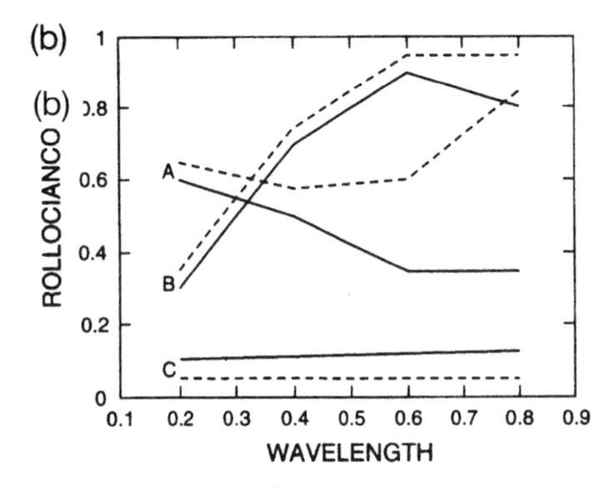


**FIGURE 15.7.** Spectral mixing analysis (Tompkins et al., 1993). (*a*) Simplex; (*b*) endmembers.

plest shape that encompasses interior points (for two dimensions [$n = 2$], the simplex is a triangle [$3 = n + 1$]). The faces of the shape are *facets,* and the exterior surface is a *convex hull.* In Figure 15.7, *A, B,* and *C* are the observed approximations of the idealized spectra *A, B,* and *C* that may not be observed on any specific image.

In the application of convex geometry to hyperspectral data, it is first necessary to define the dimensionality of the data. An image is typically subset to define a region of relative homogeneity and to remove bands with high variance (assumed to include high levels of error). The original hyperspectral data are converted to ground-level reflectances (from radiances) using atmospheric models. The data are condensed by applying a version of principal components analysis (PCA) (Chapter 11). Although hyperspectral data may include many bands (224 for AVIRIS), duplication from one channel to another means that inherent dimensionality is much less (perhaps as few as 3 to 10), depending upon the specifics of each scene.

The analyst then examines the transformed data in a data space to define the smallest simplex that fits the data. This process defines the $n + 1$ facets that permit identification of the $n$ endmembers (Figure 15.7a). These vertices, when projected back into the original spectral domain, estimate the spectra of the endmembers. These spectra are represented by Figure 15.7b. The objective is to match these endmembers to spectra from spectral libraries, then to prepare maps and images that reveal the varied mixtures of surfaces that contribute to the observed spectra in each image.

Typical endmembers in arid regions have included bare soil, water, partially vegetated surfaces, fully vegetated surfaces, and shadows. Endmembers can be investigated in the field to confirm or revise identifications made by computer. Software for hyperspectral analysis often includes provisions for accessing spectral libraries (and for importing additional spectra as acquired in the field or laboratory), and the ability to search for matches with endmembers identified. Although it may not always be possible to uniquely identify matches in spectral libraries, such analyses can narrow the range of alternatives. In some instances mathematical models can assist in defining poorly established endmembers.

## 15.9. Spectral Angle Mapping

*Spectral angle mapping* (SAM) is a classification approach that examines multispectral or hyperspectral data by evaluating the relationships between pixel values projected in multispectral data space. Envision a pixel projected into multispectral data space: its position can be described by a vector with an angle in relation to the measurement axes (Figure 15.8). Its position relative to another pixel (or perhaps a set of reference or training data) can be evaluated by assessing the difference between the angles of the two vectors. Small angles indicate a close similarity, large angles indicate lower similarity.

The effectiveness of this technique is achieved when it is applied to many more pixels and dimensions than can be represented in a single diagram. SAM is most useful when extended to tens or hundreds of dimensions (Kruse et al., 1993). SAMs differ from the usual classification approaches because they compare each pixel in the image with each spectral class, then assign a value between 0 (low resemblance) and 1 (high resemblance) to each pair. Although SAMs are valuable for hyperspectral analyses in many dimensions, their effectiveness is based upon the implicit assumption that each pixel is being compared to a pure spectra—in fact, of course, most images include many mixed pixels (Chapter 10), so the results are influenced by the presence of mixed pixels.

## 15.10. Analyses

Other investigations have attempted to understand relationships between spectral data and specific physical or biological processes. For example, Curran (1994) reviews efforts to use hyper-

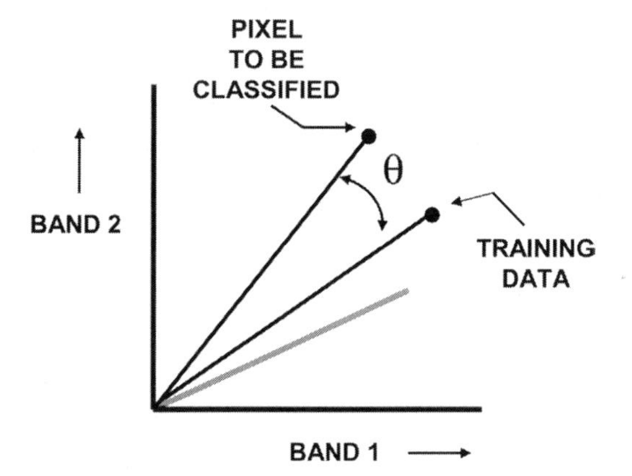


**FIGURE 15. 8.** Spectral angle mapping.

spectral data to monitor botanical variables, such as the chlorophyll, lignin, cellulose, water, and nitrogen content of plant tissues. Others have examined, at high levels of spectral, radiometric, and spatial resolution, observed spectra of laboratory samples of plant tissues influenced by atomic and molecular structures of water and by specific concentrations of organic compounds (such as chlorophyll, lignin and cellulose). These relationships have formed the basis for research devoted to the examination of hyperspectral data of vegetation canopies, to derive estimates of foliar chemistry of plant tissues in situ. Such estimates would support agricultural, forestry, and ecological studies by providing indications of nutrient availability, rates of productivity, and rates of decomposition. Green (1993) and Rivard and Arvidson (1992) report some of the applications of hyperspectral data to lithological and mineralogic analyses

## 15.11. Summary

The vast amounts of data collected by hyperspectral systems, and the problems they present for both collection and analysis, prevent routine use of hyperspectral data in the same way that we might collect TM or SPOT data on a regular basis. More likely, hyperspectral data will provide a means for discovering and refining the knowledge needed to develop improved sensors and analytical techniques that can be applied on a more routine basis. A second important role for hyperspectral data lies in the monitoring of long-term research sites, especially those devoted to study of biophysical processes and other phenomena that change over time. The fine detail of hyperspectral data will provide enhanced data for ecological monitoring and for understanding patterns recorded on lower resolution data collected for the same sites.

Several plans to deploy hyperspectral sensors for satellite observation are underway. One is NASA's EO–1 satellite, launched in November 2000, and designed as a vehicle to test the feasibility of advanced imaging systems. One of EO–1's sensors is the Hyperion, a high-resolution hyperspectral sensor capable of resolving 220 spectral bands from 0.4 to 2.5 μm at a 30-m resolution (Figure 15.9). The instrument can image a 7.5 km × 185 km land area to provide detailed spectral mapping across all 220 channels with high radiometric accuracy. EO–1 has been placed in an orbit that permits both Landsat 7 and EO–1 to image the same ground area at least once a day, thereby enabling both systems to collect images under identical viewing conditions. The paired images are then available for evaluation of the imaging technology.

## Review Questions

1. Discuss the *advantages* of hyperspectral remote sensing in relation to more conventional remote sensing instruments.

2. List some *disadvantages* of hyperspectral remote sensing relative to use of systems such as SPOT or Landsat.

3. Prepare a plan to monitor an agricultural landscape in the Midwest using *both* hyperspectral data *and* SPOT or Landsat data.

4. How would use of hyperspectral data influence collection of field data, compared to similar studies using SPOT or Landsat imagery?

**FIGURE 15.9.** Hyperion image, Mt. Fuji, Japan. Right: A single band from the 220-channel Hyperion image collected in February 2003. Left: A single band from the ALI sensor described in Chapter 6 representing the same region.

5. Discuss how *preprocessing* and *image classification* differ with hyperspectral data, compared to more conventional multispectral data?
6. How would *equipment* needs differ for image processing of hyperspectral data, compared to more conventional multispectral data?
7. The question of choosing between broad-scale coverage at coarse detail and focused coverage at fine detail recurs frequently in many fields of study. How does the availability of hyperspectral data influence this discussion?
8. Can you think of ways that the availability of hyperspectral data will influence the concepts and theories of remote sensing?
9. Hyperspectral data have so much volume that it is unfeasible to accumulate geographic coverage

comparable to that of SPOT or Landsat, for example. What value, then, can hyperspectral data have?

10. Discuss the problems that arise in attempting to design and maintain a spectral library. Consider, for example, the multitude of different materials and surfaces that must be considered, each under conditions of varied illumination.

## References


Barr, S. 1994. Hyperspectral Image Processing. In *Proceedings of the International Symposium on Spectral Sensing Research.* Ft. Belvior, VA: U.S. Army Corps of Engineers, pp. 447–490.

Crippen, R. R., and R. G. Blom. 2001. Unveiling the Lithology of Vegetated Terrains in Remotely Sensed Imagery. *Photogrammetric Engineering and Remote Sensing,* Vol. 67, pp. 935–942.

Curran, P. J. 1989. Remote Sensing of Foliar Chemistry. *Remote Sensing of Environment,* Vol. 29, pp. 271–278.

Curran, P. J. 1994. Imaging Spectrometry. *Progress in Physical Geography,* Vol. 18, pp. 247–266.

Curran, P. J., and J. A. Kupiec. 1995. Imaging Spectrometry: A New Tool for Ecology. Chapter 5 in *Advances in Environmental Remote Sensing* (F. M. Danson and S. E. Plummer, eds.). New York: Wiley, pp. 71–88.

Goa, B. C., K. B. Heidebrecht, and A. F. H. Goetz. 1993. Derivation of Scaled Surface Reflectances from AVIRIS Data. *Remote Sensing of Environment,* Vol. 44, pp. 165–178.

Goetz, A. F. H., G. Vane, J. E. Solomon, and B. N. Rock. 1985. Imaging Spectrometry for Earth Remote Sensing. *Science,* Vol. 228, pp. 1147–1153.

Green, R. O. (ed.). 1993. *Summaries of the Fourth Annual JPL Airborne Geoscience Workshop, October 25–29, 1993.* Pasadena, CA: NASA and the Jet Propulsion Laboratory, 209 pp.

Green, R. O., and J. J. Simmonds. 1993. A Role for AVIRIS in Landsat and Advanced Land Remote Sensing System Program. In *Summaries of the Fourth Annual JPL Airborne Geoscience Workshop, October 25–29, 1993* (R. O. Green, ed.). Pasadena, CA: NASA and the Jet Propulsion Laboratory, pp. 85–88.

Kruse, F. A., A. B. Lefkoff, J. B. Boardman, K. B. Heidebrecht, A. T. Shapiro, P. J. Barloon, and A. F. H. Goetz. 1993. The Spectral Image Processing System (SIPS): Interactive Visualization and Analysis of Imaging Spectrometer Data. *Remote Sensing of Environment,* Vol. 44, pp. 145–163.

Lee, C., and D. A. Landgrebe. 1992. Analyzing High Dimensional Data. In *Proceedings, International Geoscience and Remote Sensing Symposium (IGARS '93).* New York: Institute of Electrical and Electronics Engineers, Vol. 1, pp. 561–563.

Rivard, B., and D. E. Arvidson. 1992. Utility of Imaging Spectrometry for Lithologic Mapping in Greenland. *Photogrammetric Engineering and Remote Sensing,* Vol. 58, pp. 945–949.

Roberts, D. A., M. O. Smith, and J. B. Adams. 1993. Green Vegetation, Nonphotosynthetic Vegetation, and Soils in AVIRIS Data. *Remote Sensing of Environment,* Vol. 44, pp. 255–269.

Tompkins, S., J. F. Mustard, C. M. Peters, and D. W. Forsyth. 1993. Objective Determination of Image End-Members in Spectral Mixture Analysis of AVIRIS Data. In *Summaries of the Fourth Annual JPL Airborne Geoscience Workshop, October 25–29, 1993* (R. O. Green, ed.). Pasadena, CA: NASA and the Jet Propulsion Laboratory, pp. 177–180.

Tsai, F., and W. D. Philpot. 1998. Derivative Analysis of Hyperspectral Data. *Remote Sensing of Environment,* Vol. 66, pp. 41–51.

van der Meer, F. D., and S. M. de Jong (eds.). 2002. *Imaging Spectometry: Basic Principals and Prospective Applications.* Kluwer Academic Publishers.

PART FOUR

# APPLICATIONS

CHAPTER SIXTEEN

# Remote Sensing and Geographic Information Systems

## 16.1. Introduction

*Geographic information systems* (GIS) are specialized computer programs designed to analyze spatially referenced data. A computer provides storage, manipulation, and display of large amounts of geospatial data that have been encoded in digital form. Broadly stated, *geospatial data* describe attributes accurately associated with positions on the Earth's surface, including everyday examples such as highways maps, property records, census surveys, or remotely sensed images. The theory and practice of designing and using GIS forms a body of knowledge known as *geographic information science,* which is relevant not only to GIS, but also to remote sensing, cartography, and spatial statistics. This chapter does not attempt to present a complete view of GIS or geographic information science, but instead outlines some of the most significant contributions of remote sensing to GIS and vice versa.

In essence, a GIS consists of a series of map overlays for a specific geographic region. These overlays may depict raw data (e.g., topographic elevation) or may show thematic information (e.g., soils, land use, or geology), but they must share common geographic qualities (including a common geographic coordinate system) that permit them to be merged so that one can identify and analyze interrelationships between the data. These data can be visualized as a set of superimposed images (Figure 16.1).

A GIS consists of several kinds of information. A detailed *planimetric base* accurately records the positions of primary features on the landscape. This base serves as the locational framework that permits accurate positioning and registration of the varied overlays of thematic data (e.g., forest types, rivers, highways, and power lines) that form the GIS. Some systems may also include a *cadastre,* a record of land parcels, defined by its legal description of ownership. A third element consists of the varied *thematic overlays* that describe physical and cultural features such as streams and rivers, topography, soils, geology, highways, land use, political boundaries, and structures. Many times, remotely sensed imagery can be valuable in contributing thematic information for a GIS.

Sometimes the term *land information system* (LIS) is usde to describe the cadastre and the planimetric base, and the term *geographic information system* is reserved for those systems that focus on resource information (e.g., vegetation, soils, or geology) and/or demographic and economic data. Further, the term *automated mapping/facilities management* (AM/FM) is reserved

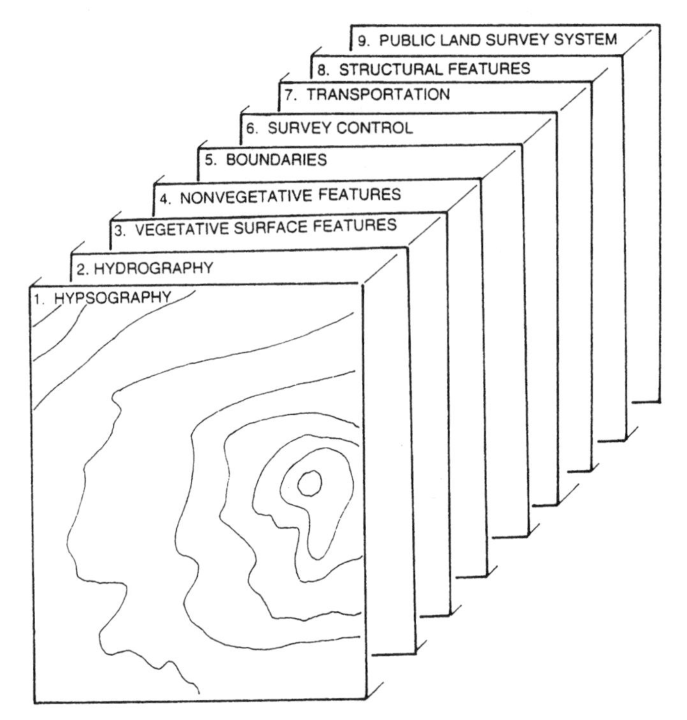


**FIGURE 16.1.** GIS data consist of many compatible data sets for the same geographic region.

for systems devoted to mapping of utilities, transportation networks, and similar infrastructure at large scale, primarily in urban areas. In concept, there is no reason that a single system could not include these varied kinds of information, but as a practical matter the constraints of costs, the varied expertise required to administer and maintain these systems, and the fact that each tends to be used by a specific kind of organization means that in practice they are usually quite separate. Nonetheless, it is still possible for separate systems within an organization to be coordinated and networked, such that information can be shared and passed from one system to another.

Superficially, a GIS may resemble *computer-aided design and display* (CADD) or *computer-aided mapping* (CAD/CAM) systems, which are designed to prepare technical graphics by computer. Quite naturally, CAD/CAMs could be applied not only to prepare engineering diagrams, but also to produce maps and diagrams. However, the defining characteristic of a GIS is not its graphic capability, but rather its ability to conduct analytical operations based on the thematic and geographic characteristics of the data. Therefore, despite the similarity in the appearance of the display used for both GIS and CAD/CAM, there is a fundamental difference in the purposes and capabilities. Although CAD/CAM systems were originally intended specifically for graphic design, in recent years they have included more and more GIS-related capabilities that blur the previously distinct differences between the two kinds of systems.

Many of the basic concepts for GIS originated in the 1960s and 1970s, but recent innovations in computer technology have greatly increased the numbers and kinds of applications of GIS. Especially important are decreases in computer costs and improvements in their reliability and efficiency that have increased availability of computers for businesses and governmental agencies. Such widespread availability of computer equipment has stimulated interest in applications of GIS not only among large organizations, but also among small businesses and governmental units that previously would not have been able to purchase such equipment. In the United States, the near-universal use of 911 emergency telephone systems and the requirement that such systems be based on systematic records of addresses and locations has promoted the introduction of GIS into local governments throughout the nation.

GISs have evolved side by side with remote sensing systems, so there are many close relationships between the two. First, remote sensing systems contribute data to GIS. Remotely sensed data can provide timely information at low cost and in a form that is compatible with the requirements of a GIS. Second, both GIS and digital remote sensing systems use similar equipment and similar computer programs for analysis and display. Therefore, investment of funds and expertise in one field tends to form the foundation for work in the other. Third, the non-remote sensing data from a GIS can be used to assist in the analysis of remotely sensed images; such data form "ancillary" or "collateral" data, analogous to those described in previous chapters. Thus remote sensing and GIS have natural, mutually supporting, relationships to each another. Nonetheless, the links between GIS and remote sensing are incomplete and imperfectly formed, and much more research is necessary to fully exploit the benefits of their interrelationships.

A GIS must include at least three main elements: (1) computer hardware, (2) computer programs, and (3) data. (Some would include the personnel that operate and maintain the GIS as an essential fourth component.) Computer hardware is much like that described in Chapter 4, and can vary in capabilities from the inexpensive personal computers that can easily fit on an office desk to more sophisticated workstations with massive storage capacities and complex peripheral equipment. A computer's capabilities for storage and analysis, of course, are closely related to its size and computing capabilities, and its ability to input, display, and output data are determined by computer hardware. For purposes of discussion in this chapter we assume that the GIS is supported by a substantial workstation, and by peripheral devices for digitizing, plotting, and color display of data.

## 16.2. GIS Software

A GIS requires specialized programs tailored for the manipulation of geographic data. Other kinds of databases may have very large volumes of data, but do not need to retain locational information for data. Therefore, GIS software must satisfy the special needs of the analyst who needs to reference data by geographic location. Furthermore, the GIS must provide the analyst with the capability to solve the special problems that arise whenever maps or images are examined: changing coordinate systems, matching images, bringing different images into registration, and so on. A GIS must be supported by the ability to perform certain operations related to the geographic character of the data. For example, it must be capable of identifying data by location or by specified areas in order to retrieve and to display data in a map-like image. Thus the GIS permits the analyst to display data in a map-like format so that geographic patterns and interrelationships are visible.

Furthermore, software for a GIS must be able to perform operations that relate values at one location to those at neighboring locations. For example, to compile slope information from elevation data, it is necessary to examine not only specific elevation values but also those at neighboring locations in order to calculate the magnitude and direction of the topographic gradient.

Data for GIS is not based upon a single set of information. Rather, GIS data must encompass several data sets that together depict several kinds of information (themes) for the same geographic area (Figure 16.1). Thus a GIS may include data for topographic elevation, streams and rivers, land use, political and administrative boundaries, power lines, and other variables. This combined data set is useful only if the themes register to one another exactly. Therefore, the several kinds of data must share a common coordinate system. Because separate variables are likely to be derived from quite different sources, it is common for different variables to have different scales, different reference systems, and different cartographic projections. Thus a GIS must have special programs to bring input data into registration by changing the scale and geometric qualities of data.

Aside from problems with data acquisition, the capabilities of a GIS are determined largely by the ability it gives the analyst to ask questions concerning the geographic patterns recorded by the data. Although some queries are relatively simple ("How many acres of category A are present?") and can be addressed by relatively simple conventional programs, many other kinds of queries are special for GIS. Therefore, GIS require special computer programs to perform the tasks essential for GIS (Table 16.1). *Image display* permits the analyst to present a specified data set as an image on a display screen, then to manipulate colors, scales, orientation, and other qualities as desired. An *overlay capability* permits the analyst to superimpose two or more data sets for display or analysis. *Visual overlay* refers to the ability to superimpose two overlays on the screen so that the two patterns can be seen together in a single image (Figure 16.2a). *Logical overlay* and *arithmetic overlay* mean that the analyst can define new variables or categories based upon the matching of different overlays at each point on the map (Figure 16.2b).

The ability to overlay different images depends, of course, upon the ability to manipulate the geometric qualities of data sets. *Projection conversion* provides the ability to change from one map projection or geographic reference system to another. For example, geographic position in one overlay may be specified using state plane coordinates, whereas others may be based upon the universal transverse Mercator (UTM) grid system. Unless the GIS has the capability to translate data from one reference system to another, it is not possible to match the separate overlays that form the GIS, and the GIS cannot perform one of its most basic functions: superimposing and comparing different data sets. Thus the GIS must have programs to perform *image registration* and *resampling;* these capabilities permit the analyst to bring two images into registration. These capabilities are among the most important of all those required for a GIS and among the most difficult to perform accurately.

Other operations permit the analyst to examine individual overlays or combinations of individual overlays using logical rules. For example, the analyst may wish to delineate all areas above a certain elevation or to define all areas within a certain soil category that have a certain slope or elevation. Thus the analyst must be able to use the elevation matrix to define a slope overlay; then he or she must be able to find the appropriate soil category within the soil overlay and match it to the elevation and slope overlays to find the slopes and elevations within soil classes. Such operations require a variety of general-purpose programs that permit the analyst to perform operations such as selecting regions from an image, registering two images to identify regions where specific regions match, or calculating slopes from elevations. GIS software should also permit the analyst to define "tailor-made" geographic regions—for ex-

TABLE 16.1. Operations for GIS

| | |
|---|---|
| Data input | Spatial interpolation |
| Display | Raster-to-vector conversion |
| Subset | Data output |
| Overlay | Data storage and retrieval |
| Projection conversion | Buffering |
| Registration/image matching | Network operations |
| Resampling | Data manipulation |
| Logical operations (Boolean) | Data reporting |
| Arithmetic operations | Statistical generation |
| Vector-to-raster conversion | Models |

ample, to delineate all areas within a specified distance of a particular class of stream or highway.

There are a wide range of software packages for GIS analysis, each with its own advantages and disadvantages. A much longer list than can be included here would still omit many useful systems, but it is appropriate to mention a few that have been in relatively wide use for some time.

* **ArcInfo and ArcView** (Environmental Systems Research Institute, Inc., 380 New York Street, Redlands, CA 92373-8100; 909-973-2853; *http://www.esri.com*)
* **Intergraph Corporation** (P.O. Box 240000, 170 Graphics Drive, Madison, AL 35758; 256-730-2000; *http://www.intergraph.com/contact/*)
* **IDRISI** (The Clark Labs, 950 Main Street, Worcester, MA 01610-1477; 508-793-7526; *http://www.clarklabs.org/*)
* **MapInfo** (MapInfo, One Global View, Troy, NY 12180; 518-285-6000; *http://www.mapinfo.com/*)
* **GRASS (Geographic Resources Analysis and Support System)** (Center for Applied Geographic and Spatial Research, Baylor University, P.O. Box 97351, Waco, TX 76798-7351; 254-710-6814; *http://www.baylor.edu/grass/*)

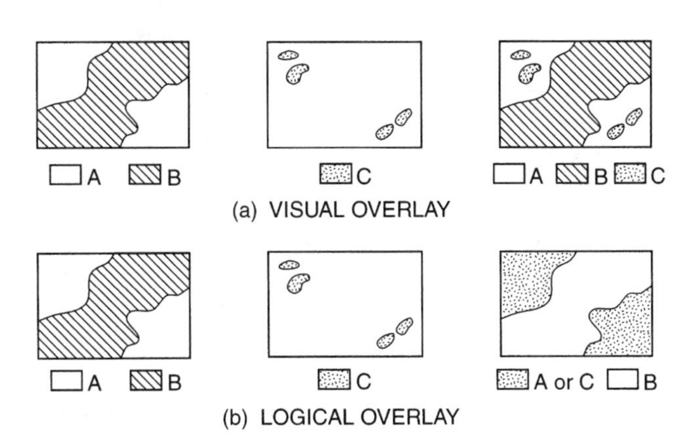


**FIGURE 16.2.** (*a*) Visual overlay; (*b*) logical overlay.

## 16.3. Basic Data Structures

Two alternative GIS data structures offer contrasting advantages and disadvantages. *Raster* (or *cellular*) data structures consist of cell-like units, analogous to the pixels of a TM scene (Chapter 6). The region of interest is subdivided into a network of such cells of uniform size and shape; each unit is then encoded with a single category or value (attribute) (Figure 16.3). Raster structures offer ease of data storage and manipulation, and therefore permit use of relatively simple computer programs. Also, raster structures lend themselves to use with remotely sensed data because digital remote sensing data are collected and presented in raster formats. Disadvantages are primarily related to losses in accuracy and detail due to the coding of each cell with a single category even though several may be present (Figure 16.3). (This is the same kind of error caused by mixed pixels, discussed in Chapter 10). Digital remote sensing data are processed and stored in raster structures.

The alternative format is the *vector* (or *polygon*) format, which records the boundaries, or outlines, of parcels on the source document by listing the coordinates of the boundaries or the coordinates of the vertices of polygons (Figure 16.4). Vector format tends to provide more efficient use of computer storage than does the raster format, finer detail, and more accurate representation of shapes and sizes. Its main disadvantages are (in some instances) the higher costs of encoding data and the greater complexity required for the computer programs that must manipulate data. Today most GIS are designed to use vector format, although most accommodate raster data, or may use raster structures for selected analyses. Raster-to-vector conversations and vector-to-raster conversions permit mixing of the two kinds of data, although unnecessary conversions can create errors.

### *Cellular Data Structures*

Each cell in a GIS theme must be encoded with a single value or a single category that represents a condition at that location (Figure 16.5). For continuously varying information, such as

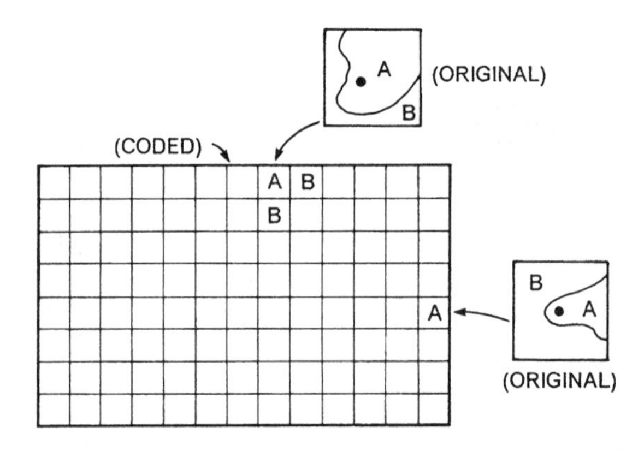


**FIGURE 16.3.** Errors in encoding cell data. Each cell can record only a single identifier, so cells cannot record the fine detail of the original pattern. Here cells illustrate coding by selection of the single class positioned at the center of each cell. The two examples show how cells assigned the same label can correspond to different ground conditions.

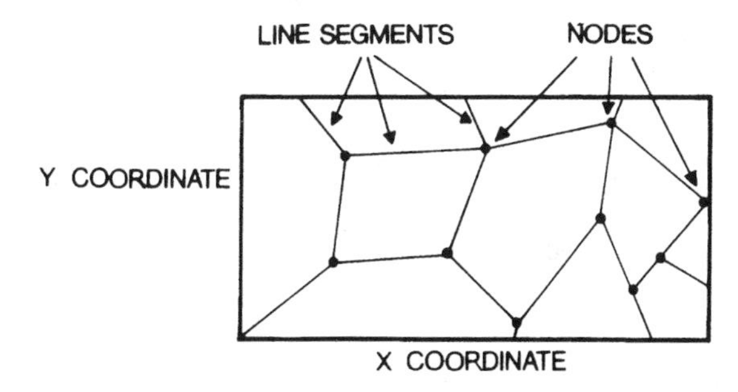


**FIGURE 16.4.** Polygon ("vector") data structure.

topographic elevation, the single value could be the elevation at the center of the cell, or possibly the average value of elevation within the cell. However, once a choice is made, the same procedure is applied uniformly to all cells within the GIS.

A cell can also encode nominal data, such as a land-cover class or a geologic or pedologic category. Because a given cell may include several categories, it is necessary to define a rule to consistently select which category represents the cell in the database. The *predominant category rule* selects the single category that occupies the largest part of the cell (Figure 16.5a). It is easy to understand the major problems inherent to the raster-based GIS. Variation within cells cannot be represented. Sometimes errors are insignificant, but often there are systematic errors due to the fact that some categories occur only as small parcels or as long, narrow shapes that do not occupy large areas within cells, and therefore are systematically excluded from the GIS. Raster-based GISs therefore tend to exaggerate the importance of categories that occur in large, regular parcels, and to underrepresent small, linear, or widely dispersed units. Although using a smaller cell size can reduce these problems, they become insignificant only when a cell size is

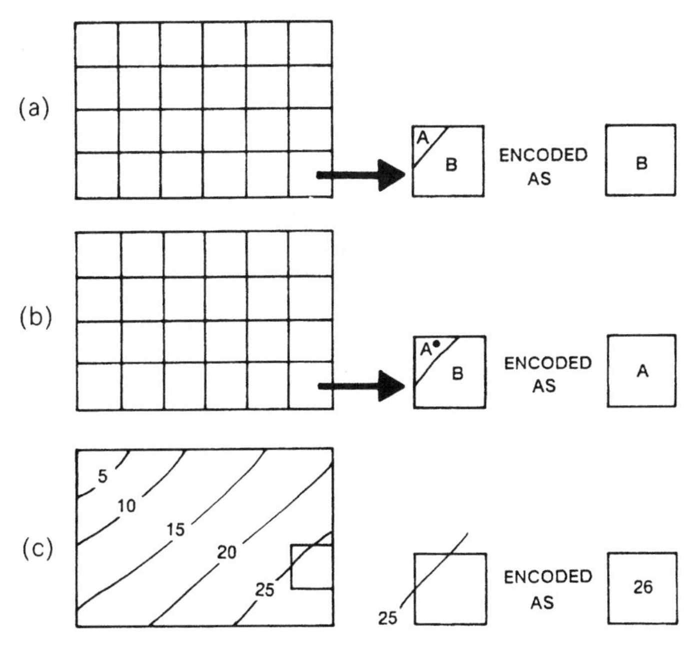


**FIGURE 16.5.** Encoding raster data.

very small in relation to the pattern to be studied. For this reason, costs of using such a small cell size are often prohibitive.

An alternative procedure selects the category by means of positioning a dot in the center of each cell; the category that falls beneath the dot is entered as the category for the cell (Figure 16.5b). A third procedure uses a dot randomly positioned within each cell (Figure 16.5c). (Of course, for a given data set, only a single method will be used.) Relative to the predominant category rule, these procedures improve the accuracy of the database as a whole, but the error at each cell can be quite large.

Gersmehl and Napton (1982) studied these two kinds of errors. *Inventory error* is error for the database as a whole in reporting total areas occupied by specific categories. Inventory error is important for users of the GIS who are interested in a *regional overview,* that is, who wish to learn about the character of specific regions rather than small sites within the region. Use of the dot randomly positioned within each cell reduces inventory error; for large areas, inventory errors may be quite small.

In contrast, *classification error* is the error in reporting the contents of each cell. Classification error is important for uses that require detailed information concerning specific sites. Gersmehl and Napton (1982) found that classification error is usually larger than inventory error (Figure 16.6).

Clearly, classification error increases with use of the random dot method, even though the randomized selection may reduce inventory error. Both kinds of errors can be reduced if cell size is small relative to the pattern to be studied, as illustrated in Figure 16.5 and discussed in Chapter 10. However, even modest decreases in cell size greatly increase the numbers of cells; as a result, costs and practicality soon limit efforts to reduce cell size. Therefore, selection of cell size must be made with great care because once a GIS is committed to a specific cell size, the costs and inconvenience of changes preclude conversion to smaller cell size.

It should be noted that it is possible to interpolate data from one cell size to another (Figure 16.7). Data encoded at a 1-km grid size can be represented, for example, at a 4-km grid, as long as the analyst consistently applies a rule for deciding how to encode data for the larger cell (Figure 16.7a).

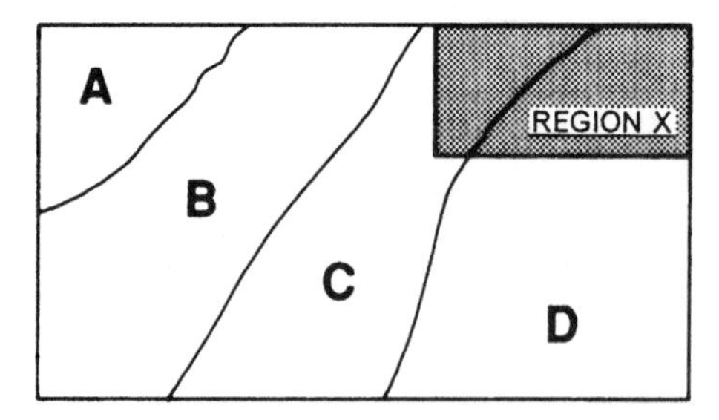


INVENTORY:
"REPORT TOTAL AREA OF CATEGORY A"
(LOCATION IS NOT IMPORTANT)

CLASSIFICATION:
"REPORT CATEGORIES PRESENT WITHIN REGION X"
(LOCATION IS IMPORTANT)

**FIGURE 16.6.** Inventory error and classification error. From Gersmehl and Napton (1982).

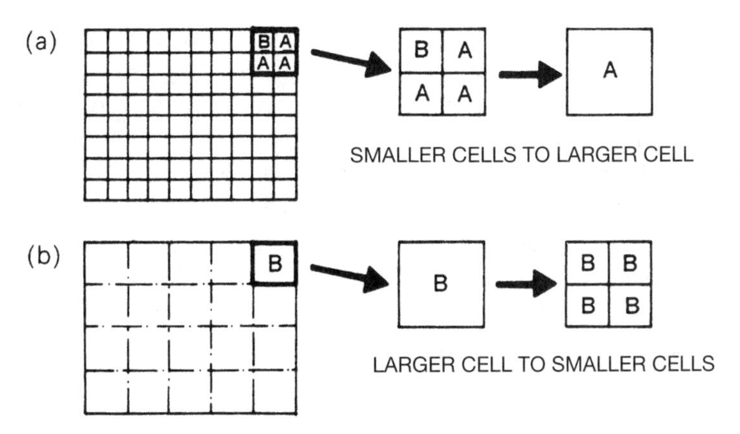


**FIGURE 16.7.** Interpolation from one cell size to another. (*a*) Smaller cell to larger cell. (*b*) Larger cell to smaller cell.

However, interpolation from larger to smaller cells is a different matter. It is operationally possible to create 1-km cells for data encoded at 4-km cells, but the change cannot improve the level of detail in the data—it simply creates a false impression of finer detail. (There can be no method for assigning values for the smaller cells except by replicating information of the larger cells. The visual detail has been improved, but the logical detail remains the same [Figure 16.7b].) Data for raster data systems are entered from paper maps, aerial images, or similar documents depicting the pattern to be encoded. Often a template showing the grid system is superimposed and registered to the map. The spacing of the grid must be appropriate for the scale of the map. For example, to encode 1-km cells from a map at 1:24,000, the template should show cells 4.17 cm in size, whereas to encode from a map at 1:15,840 the template must have cells of 6.31 cm. The analyst then can read the value or category for each cell in sequence (using the specific method adopted for use), recording the information on a special form, or possibly reading them to another operator who enters them directly using a computer keyboard.

The student who is especially interested in GIS may wish to review Chapter 10, with emphasis upon the material pertaining to spatial resolution. Remotely sensed data and data for GIS can differ in many important respects, so not all concepts from the discussion of image resolution apply directly, but the student will benefit from a careful comparison and contrast of the two subjects.

### *National Spatial Data Infrastructure*

The *National Spatial Data Infrastructure* (NSDI) (*Coordinating Geographic Data Acquisition and Access: The National Spatial Data Infrastructure*), established by an executive order signed in April 1994, is intended to form a framework for efficient exchange of geographic data between different organizations and between different computing systems. It establishes standards for geographic data and a plan for establishing a National Digital Geospatial Data Framework. The executive order applies only to the activities of agencies of the federal government.

However, the existence of the federal effort will encourage state governments to implement similar programs, and lead to participation by private industry.

The *Spatial Data Transfer Standard* (SDTS) was developed as part of a broader effort to establish guidelines for information processing within the federal government. SDTS is intended to minimize problems encountered in transferring data between agencies by establishing common formats and standards. SDTS was approved by the U.S. Department of Commerce in 1992 as Federal Information Processing Standard 173. It consists of a family of standards that apply to different forms of spatial data, including Topological Vector Profiles and USGS Digital Line Graphs. Remote sensing data, as well as other raster data, are included within the Raster Profile.

## 16.4. Relationships between Remotely Sensed Data and GIS

It is easy to see some of the advantages for using remotely sensed data in a GIS. Satellite systems such as Landsat or SPOT acquire data for very large areas in a short time period, thereby providing essentially uniform coverage with respect to date and level of detail. Such data are already in digital form, and are provided in more-or-less standard formats, as outlined in Chapter 6. Furthermore, these data are available for almost all of the Earth's land areas and are inexpensive relative to alternative sources. Although satellite data are not planimetrically correct, preprocessing can often bring data to acceptable levels of geometric accuracy with only modest effort. Images formed by analysis or interpretation of the raw data register to the original data. Thus remotely sensed data have the potential to address some of the difficult problems encountered in the formation of GIS.

Nonetheless, there are still numerous problems and difficulties. Data are not always available for desired dates or seasons. If a large area is to be examined, there may be problems in mosaicking data for separate dates. Accuracies of classification and analytical methods required to process the data prior to entry in the GIS may not be consistently reliable. Furthermore, the levels of detail in the satellite data may not match to those from other sources.

There are several avenues for incorporating remotely sensed data into the GIS. The most satisfactory procedure depends upon the specific requirements of a particular project and the kinds of equipment and financial resources that may be available.

1. Manual interpretation of aerial images or satellite images, often viewed on a computer screen in digital form, produces a map or a set of maps that depicts boundaries between a set of categories (e.g., soil or land-use classes). Onscreen digitizing can then provide the digital files suitable for entry into the GIS.
2. Digital remote sensing data are analyzed or classified using automated methods to produce conventional (paper) maps and images, which are then digitized for entry into the GIS.
3. Digital remote sensing data are analyzed or classified using automated methods, then are retained in digital format for entry into the GIS, using reformatting or geometric corrections as required.
4. Digital remote sensing data are entered directly in their raw form as data for the GIS.

Superficially, remote sensing and GIS appear to form separate dimensions of a single activity. They share common data, use many similar or interrelated analytical tools, use similar technology and software, and reflect a similar spatial perspective. Yet the two fields depends upon very different bodies of knowledge and have objectives that differ. Remote sensing is ultimately a means of recording or extracting data from the landscape. Practitioners of remote sensing must master a knowledge of the subject matter at hand, remote sensing instruments, atmosphere, and analytical techniques. More-or-less obvious organization of the interrelationships between GIS and remote sensing can be categorized as (1) contributions of GIS to the practice of remote sensing and (2) contributions of remote sensing to the practice of GIS. Although this distinction oversimplifies the complex relationship between the two fields, it can form the basis for our discussion here.

## 16.5. Contributions of GIS to Remote Sensing

Because a GIS provides a means of organizing diverse spatial data within an accurate planimetric framework, and shares some of the same techniques and principals, GIS is well positioned to support the practice of remote sensing. At least three contributions of GIS to remote sensing deserve discussion here.

### *Mission Planning*

Planning for acquisition of remote sensing data requires use of an accurate planimetric base for the region in question, to permits analysts to position image coverage with accuracy and precision. Mission planning requires representation of landmarks such as transportation, drainage, and settlements to guide placement of image coverage and local topographic elevation. Further, this process can be more effective if the base also includes thematic information, to present information related to the objectives of the project. For example, remote sensing missions to acquire images to support forestry or agricultural analysis can be planned more effectively if the base presents land use information, to permit the analysis to assure that image coverage is planned to encompass the relevant areas. GIS serves this function well because it offers accurate planimetric representations of the landscape, topography, landmarks, and thematic information. Further, the digital format is consistent with the digital mission planning software that can be used for some applications.

### *Ancillary Data*

As outlined in Chapter 12, remote sensing often uses ancillary data to support the analysis of imagery. GIS can provide a framework for integrating ancillary data with remotely sensed data. Because GIS and remote sensing data can share compatible data structures and formats, it is possible to bring them together for analysis. In recent years, remote sensing image-processing systems and GIS software systems have increased their compatibility with respect to ability to share data formats, import and export data ot companion systems, and to perform analyses common to both GIS and remote sensing.

GIS forms an important tool for preclassification segmentation when ancillary data are employed in the image classification process. For example, elevation data, often used as ancillary data for remote sensing analysis, can be conveniently manipulated by GIS as preparation prior to importation into the remote sensing realm.

### *Collection, Organization, and Visualization of Reference Data*

Reference and field data are often difficult to organize and access effectively, due to varied formats (such as photographs, tabulations, written records, and maps) (Chapter 13). Often a GIS can provide an effective framework for organizing and presenting such data. Figure 16.8 shows a portion of an ArcMap application designed to organize and present field data for remote sensing applications. This application uses digital USGS topographic quadrangles as a geographic base to show icons representing the locations of ground photographs, oblique aerial photographs, and the center points of vertical aerial photography. Each offers a way to organize data pertaining to specific portions of a study area for use in selecting training data for image classification (Chapter 12) or for selecting reference data for accuracy assessment (Chapter 13), for example. The analyst can use such applications to refer to varied forms of reference data that pertain to a specific region of the study area. Lunetta et al. (2001) describe a more rigorous application of a similar system for organizing field observations.

### *OakMapper webGIS*

Kelly et al. (2004) used GIS as a way to collect and organize diverse information regarding the occurrence of sudden oak death syndrome (SOD), a serious disease that has influenced oak populations locally in some regions of California, and is now spreading to other regions, with the potential to reach oak populations over wide regions of the United States. Therefore, scientists are monitoring the occurrence of the disease, to permit identification of outbreaks at the earliest possible opportunity, and to guide the deployment of surveillance and treatment.

The OakMapper strategy to be described in Chapter 17 allows local governmental employees and citizens to use GPS receivers to accurately locate trees infected by SODS. These data, recorded within a GIS, provide a systematic and accurate record of outbreaks that enables experts to understand the spread of the disease to plan remote sensing flights to record and monitor the dynamics of SOD, and to guide the planning of ground sampling efforts (Figure 16.9). The patterns depicted by the GIS display permit scientists to understand the dynamics of the diffusion of the disease in relation to the distribution of potential hosts. Password-protected web sites provide common resources for researchers as they develop information. Remotely sensed data include 1-m multispectral imagery, which provides detail sufficient for mapping of the damage to upper-story crowns.

## 16.6. Contributions of Remote Sensing to GIS

Often the vast majority of GIS data are derived from statistical sources, such as census data, marketing information, and environmental measurements. Such data, if employed by themselves, may lack the supporting information that provides the context that enables users to

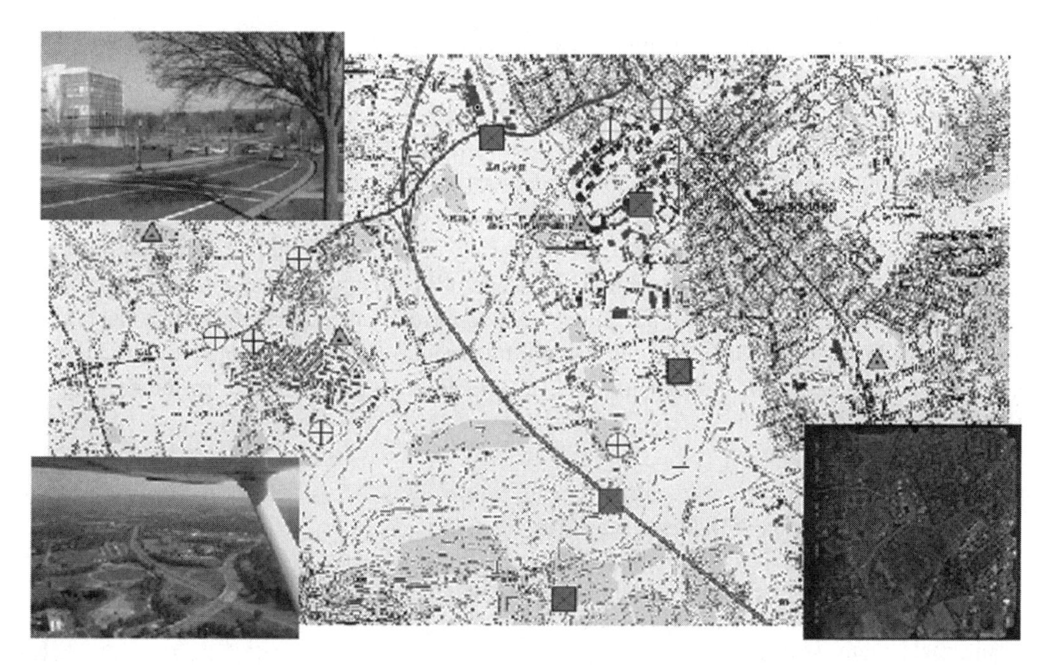

**FIGURE 16.8.** Example of a GIS to organize and display field data for use in remote sensing analysis. The icons on the map permit the analyst to display ground photographs (upper left), vertical aerial photographs (lower right), and oblique photography (lower left) for the area in question.

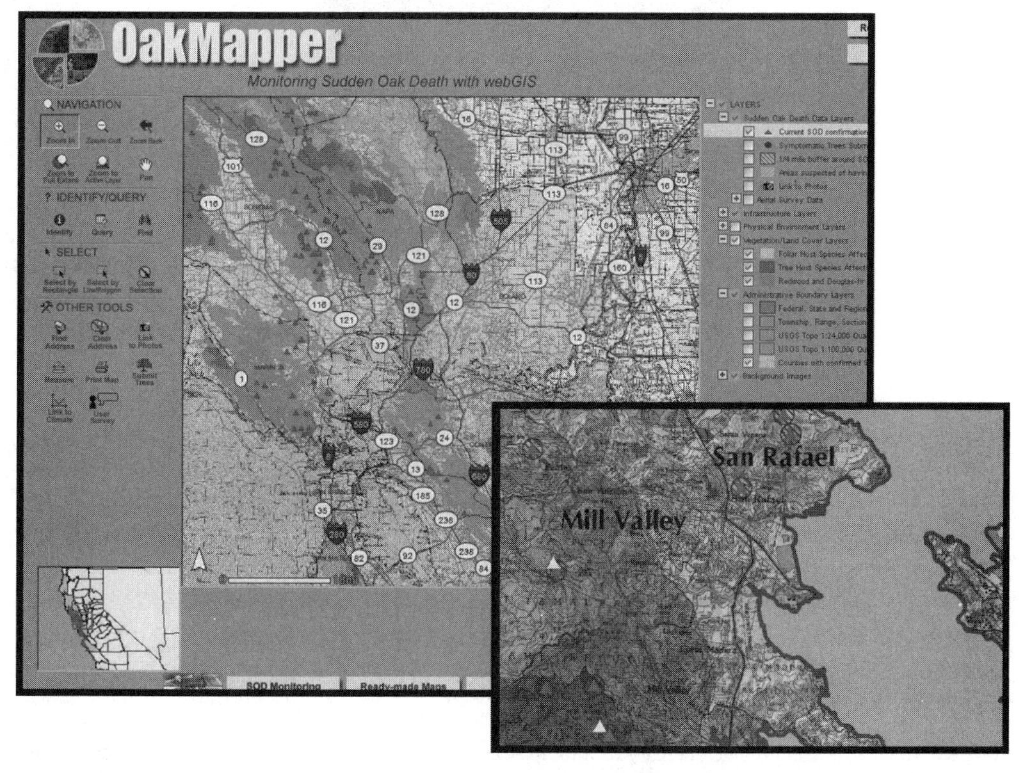


**FIGURE 16.9.** A portion of the OakMapper web page illustrating the use of GIS to provide a system for entering and displaying data collected from, and used to plan, remote sensing flights. From OakMapper, University of California, Berkeley.

assess the significance and interrelationships between variables. Remotely sensed data convey this context, so are valuable supplements for the usual GIS data layers.

### *Remote Sensing Imagery Provides Thematic Layers for GIS*

Data for GIS layers are typically derived from a variety of sources, including demographic censuses, digital elevation data, transportation maps, and utility maps. Remote sensing can provide thematic data not available from other sources. Some forms of thematic data, such as land use, specifics of the road network, and the basic infrastructure of urban regions, are effectively accuired from remote sensing imagery, then integrated with information form other sources (Figure 16.10).

### *Remote Sensing Imagery Provides a Backdrop for GIS*

The utility of a GIS is often enhanced by using a remotely sensed image as a backdrop that provides a locatational reference, with recognizable landmarks, for other data that otherwise would be abstract (Figure 16.11).

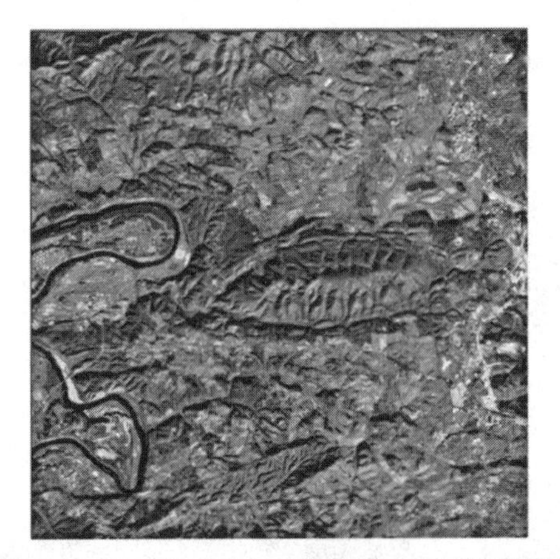

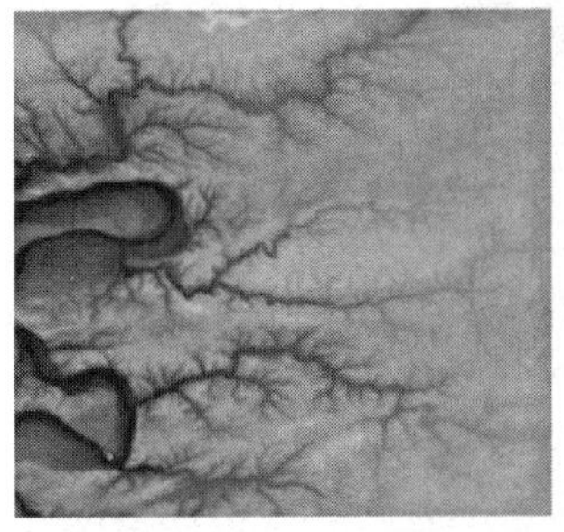

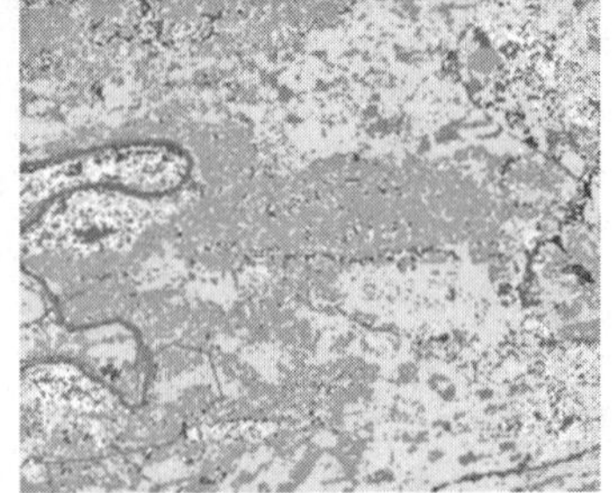

**FIGURE 16.10.** GIS can provide ancillary data to assist in image classification. Top: Remote sensing image. Bottom (left to right): digital data that might be applied as acillary data—digital elevation data, census data, land cover information—for the same area depicted in the image.

***Remote Sensing Imagery Provides a Means of Updating GIS***

The effectiveness of most GIS applications depends upon the timeliness and currency of the data they portray. Therefore updating GIS layers to post changes is an important task to maintain the currency of a GIS without incurring the expense of a complete resurvey of the entire layer (Figure 16.12). Remote sensing is often an important source of recent information through its ability to provide a comprehensive view of a landscape, to offer a spatial perspective, and to identify both those areas that have experienced changes and those that remain unchanged (Figure 16.11).

## 16.7. National Center for Geographic Information and Analysis

In 1988 the National Science Foundation established the National Center for Geographic Information and Analysis (NCGIA). NCGIA is a consortium of three research institutions: the University at Buffalo, the University of California at Santa Barbara, and the University of Maine at Orono. The center was established to lead the effort to conduct fundamental research on the analysis of geographic data using GIS and to develop and expand applications of GIS. NCGIA identifies research topics of significance for continued development and application of GIS technology, and assembles teams of scientists from varied institutions and specializations to pursue these subjects, which have included topics such as the design of very large spatial databases, visualization of spatial data, formalizing cartographic knowledge, and investigating relationships between GIS and remote sensing. In addition to funding research programs, the NCGIA is active in designing university curricula, organizing workshops and conferences, publishing technical papers, and developing computer software. The NCGIA can be accessed on

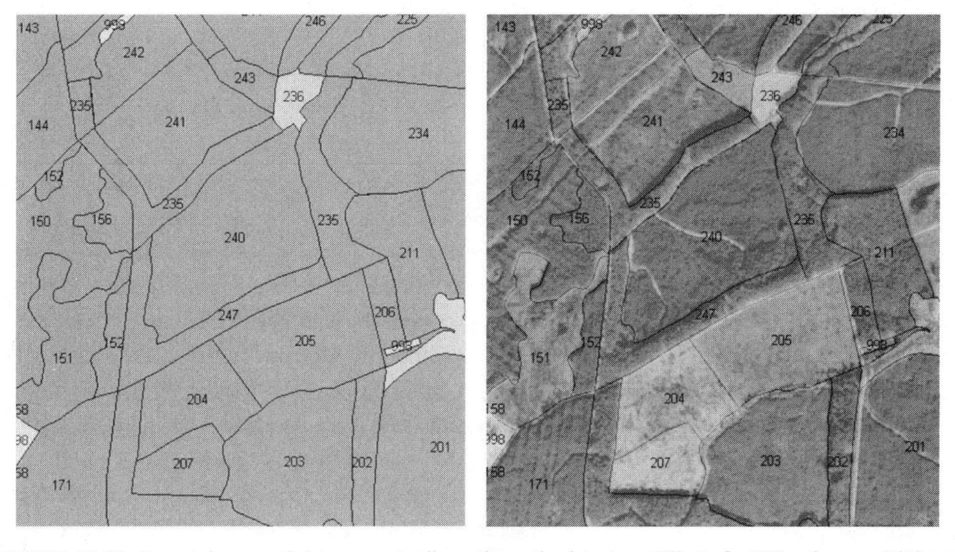


**FIGURE 16.11.** Remotely sensed data can contribute thematic data to a GIS. Left: GIS polygons. Right: the same polygons displayed against an image background provide locational detail with landmarks, transportation, and hydrographical features.

**FIGURE 16.12.** Remotely sensed data can update information for a GIS. Left: a region near the Roanoke Virginia airport as photographed in 1980. Right: the same region in 1987. In the interval between the two photographs, numerous changes have occurred, most notably the airport runway has been extended and the highway has been routed through a tunnel beneath the extended runway. Credit: Virginia Department of Transportation. Copyright Commonwealth of Virginia.

the WWW at *http://www.grad.buffalo.edu/general/research/ncgia.htm* and *http://www.ncgia.ucsb.edu/*. Other related organizations devoted to GIS include the University Consortium for Geographic Information Science: *http://www.ucgis.org,* and the Federal Geographic Data Committee: *http://www.fgdc.gov/*.

## 16.8. Mobile GIS

Recent decreases of the sizes and increases in the computational power of laptop computers and personal data assistants (PDAs) have enabled analysts to develop field-portable GIS (or mobile GIS). Such units can be easily transported in motor vehicles or even carried on foot in the field. Use of GIS in the field permits analysts to enter new data as it is directly observed or to verify or update information already in the system. In some instances, the GIS runs on a laptop as a selfcontained unit; in other cases, it is linked by wireless communications to a GIS maintained in a central location. (Use of PDAs for this purpose requires wireless communications to connect to a central GIS; the PDA can download a small portion of the GIS at a specific time.)

The increasing portability of GPS receivers has enabled laptops and PDAs to link precise, real-time, locational information to a GIS. Such systems are especially valuable when a GIS includes registered digital images or satellite imagery, which permits analysts to display GIS layers against a background that shows recognizable structures or landmarks. Workers can then match their location on the ground with representations of the same location in the GIS as they enter new information.

Mobile GISs is especially valuable for utility systems or other organizations that require workers to have immediate on-site access to spatial data in the field. For example, mobile GIS permits workers to ascertain the position of buried utility lines or the boundaries of lots without use of outdated, often incompatible, paper maps and diagrams. Limitations include the difficulty of viewing computer displays in bright sunlight and difficulties in registering GIS layers with sufficient precision to maintain the integrity of the system at levels of high magnification.

## 16.9. Web-Based GIS

Often a GIS is administered as part of a large organization or network of organizations. Examples include the many administrative departments of a municipal government, including transportation, water and sewer, fire and rescue, law enforcement, environmental quality, highways, and so on. In private industry, organizations might have comparable divisions of responsibility, especially for larger businesses.

Traditional models for organizational use of GIS depend upon each department to update data for the data layers for which they have responsibility, then to forward its data to a central office with responsibility for maintaining the combined data from numerous sources. Revisions and updates are then released to users at appropriate intervals, usually some weeks or months after each department has met a deadline for providing revised data.

Web-based GIS are organized such that each department can post its own data to a common georeferenced database accessible to other departments, other authorized users, or possibly the general public. Although there might be administrative reasons for requiring departments to contribute their revisions according to a certain schedule, there would not be a compelling technical reason. Therefore, each department could, if appropriate, publish its revisions as they are noted and posted. Because users access the GIS through the WWW, they can use the latest versions of overlays as they are updated. This kind of system facilitates distribution of the most current information to those who use a GIS. A disadvantage to use of such a dynamic GIS design is that it does not provide an authoritative reference that is stable over time. Users who examine such a system at different times may see different information—a situation that can itself lead to errors and inefficiencies.

## 16.10. Summary

The concept of the GIS, although implemented in many different forms, is ideally one that permits full utilization of the goals of remote sensing. Remote sensing achieves its maximum usefulness when images from one date or formed from one part of the spectrum can be integrated with those acquired at a different date or from a different portion of the spectrum. Although this kind of integration can be accomplished by many means, the GIS provides maximum flexibility in both display and analysis. The concept of the GIS has also been important in remote sensing because it has formed a focal point for scientists who work in subfields of remote sensing and diverse disciplines. By its nature, a GIS requires expertise from many fields of knowledge, so

**TABLE 16.2. Elements of Data Quality**

| | |
|---|---|
| Age | Thematic accuracy |
| Areal coverage | Thematic detail |
| Scale | Accessibility |
| Detail | Costs |
| Format | Origins |
| Cartographic projection | Continuity with past and future data |
| Positional accuracy | Compatibility with other thematic data |

GIS research has brought scientists from diverse backgrounds together to work on common issues.

GIS research has emphasized the need for further development of capabilities of significance both to GIS and to remote sensing. Image matching, image registration, and related topics are important both for remote sensing and for GIS, and both fields can benefit from improvements. Certainly, future developments in remote sensing will address topics such as data compatibility, improvements in image geometry, and other subjects that will improve the usefulness of remotely sensed data for GIS.

## Review Questions

1. At a superficial level of consideration, GIS and remote sensing may seem to be quite similar fields of knowledge. Yet they are distinct in many important ways. List some of these important differences.

2. Despite the differences you identified in your response to Question 1, GIS remote sensing *do* share many common dimensions. List a few of these common elements. In evaluating the two lists, do you believe that a person experienced in the field of GIS is qualified as a practitioner of remote sensing? Alternatively, do you believe a person experienced only is remote sensing is prepared to serve as a GIS analyst?

3. Assume that you have available three dates of fine-resolution multispectral imagery to support development of a GIS of your region. List some of the GIS layers that might be derived from these data.

4. Image registration, first discussed in Chapter 11, is a critical dimension of bringing remotely sensed data into a GIS. Performs a concise, critical, evaluation of the key steps required to assure that the remote sensed data are properly integrated into the GIS. Identify ways in which the registration could be inadequate, effects upon the quality of the GIS.

5. Assume you have a statewide GIS that includes digital evaluation data and census data—including, of course, population totals. Assume also that you have the ability to accurately superimpose the boundaries of major drainage basins and political boundaries. Discuss some of the ways such information could be used to address issues and problems important at the statewide level of government.

6. A detailed GIS is available for the region you are to analysis using remote sensing data. List some of the ways that the GIS could support your remote sensing analysis.
7. Foresters were among the first to systematically integrate GIS and remote sensing operations, and to use them in their business operations. Can you identify some of the reasons foresters may have been so effective in taking these pioneering steps?
8. The functionality of a GIS depends upon maintenance of high levels of quality and compatibility between data layers within the GIS. Discuss ways in which use of remotely sensed data might contribute to establishment of accuracy and compatibility, and ways in which it might contribute to difficulties with accuracy and compatibility.
9. List ways in which GIS can support accuracy assessment (discussed in Chapter 14).
10. The fields of GIS and remote sensing, although related to one another, have developed as separate fields of inquiry. Can you identify reasons that they have largely developed separately? Can you think of benefits if they were to be more closely integrated? What might be barriers to closer integration?

## References

### *General*

Adams, V. W. 1975. *Earth Science Data in Urban and Regional Information Systems: A Review* (Geological Survey Circular 712). Washington, DC: U.S. Geological Survey, 29 pp.

Broome, F. R., and D. B. Meixler. 1990. The TIGER Data Base Structure. *Cartography and Geographic Information Systems,* Vol. 17, pp. 39–47.

Burrough, P. A. 1998. *Principles of Geographic Information Systems for Land Resources Assessment.* New York: Oxford University Press, 193 pp.

Clarke, K. C. 1985. A Comparative Analysis of Polygon to Raster Interpolation Methods. *Photogrammetric Engineering and Remote Sensing,* Vol. 51, pp. 575–582.

Estes, J. E. 1982. Remote Sensing and Geographic Information Systems Coming of Age in the Eighties. In *Proceedings Pecora VII Symposium* (B. F. Richason, ed.). Falls Church, VA: American Society of Photogrammetry, pp. 23–40.

Homer, C., C. Huang, L. Yang, B. Wylie, and M. Coan. 2004. Development of a 2001. National Land-Cover Database for the United States. *Photogrammetric Engineering and Remote Sensing,* Vol. 70, pp. 829–840.

Kelly, M., K. Tuxen, and F. Kearns. 2004. Geospatial Informatics for Management of a New Forest Disease. *Photogrammetric Engineering and Remote Sensing,* Vol. 70, pp. 1001–1004.

Laurini, R., and D. Thompson. 1992. *Fundamentals of Spatial Information Systems.* New York: Academic Press, 680 pp.

Lunetta, R. S., J. Llames, J. Knight, R. G. Congalton, and T. H. Mace. 2001. An Assessment of Reference Data Variability Using a Virtual Field Reference Data Base. *Photogrammetric Engineering and Remote Sensing,* Vol. 63, pp. 707–715.

Maguire, D. J., M. F. Goodchild, and D. W. Rhind (eds.). 1991. *Geographical Information Systems.* New York. Longman, 2 Vols: 640 pp. (Vol. 1), 447 pp. (Vol. 2).

Maizel, M. S., and R. J. Gray. 1985. *A Survey of Geographic Information Systems for Natural Resource Decision Making*. Washington, DC: American Farmland Trust, 182 pp.

Marble, D. F., and D. J. Peuquet. 1983. Geographic Information Systems and Remote Sensing. Chapter 22 in *Manual of Remote Sensing* (R. N. Colwell, ed.). Falls Church, VA: American Society of Photogrammetry, pp. 923–958.

Marx, R. W. 1990. The TIGER System: Yesterday, Today, and Tomorrow. *Cartography and Geographic Information Systems,* Vol. 17, pp. 89–97.

Tomlin, C. D. 1990. *Geographic Information Systems and Cartographic Modeling.* Englewood Cliffs, NJ: Prentice-Hall, 249 pp.

### *Errors and Data Quality*

Gersmehl, P. J., and D. E. Napton. 1982. Interpretation of Resource Data: Problems of Scale and Transferability. In *Papers of the Conference of the Urban and Regional Information Systems Association.* Atlanta, GA: Urban and Regional Information Systems Association, pp. 471–482.

Mead, D. A. 1982. Assessing Quality in Geographic Information Systems. Chapter 5 in *Remote Sensing for Resource Management* (C. J. Johannsen and J. L. Sanders, eds.). Ankeny, IA: Soil Conservation Society of America, pp. 51–59.

Walsh, S. J., D. R. Lightfoort, and D. R. Butler. 1987. Recognition and Assessment of Error in Geographic Information Systems. *Photogrammetric Engineering and Remote Sensing,* Vol. 53, pp. 1423–1430.

Whede, M. 1982. Grid Cell Size in Relation to Errors in Maps and Inventories Produced by Computerized Map Processing. *Photogrammetric Engineering and Remote Sensing,* Vol. 48, pp. 1289–1298.

CHAPTER SEVENTEEN

# Plant Sciences

## 17.1. Introduction

The Earth's vegetative cover is often the first surface encountered by the energy we use for remote sensing. So, for much of the Earth's land area, remote sensing imagery records chiefly the character of the vegetation at the surface. Therefore, our ability to interpret the Earth's vegetation canopy forms the key to knowledge of other distributions, such as geologic and pedologic patterns that are not directly visible but do manifest themselves indirectly through variations in the character and distribution of the vegetation cover.

In other situations, we have a direct interest in the vegetation itself. Remote sensing can be useful for monitoring areas planted to specific crops, for detecting plant diseases and insect infestations, and for contributing to accurate crop production forecasts. In addition, remote sensing has been used to map forests, including assessments of timber volume, insect infestation, and site quality.

Further, remote sensing provides the only practical means of mapping and monitoring changes in major ecological regions that, although not directly used for production of food or fiber, have great long-term significance for mankind. For example, the tropical forests that cover significant areas of the Earth's surface have never been mapped or studied except in local regions that are unlikely to be representative of the unstudied regions. Yet these regions are of critical importance to mankind due to their role in maintaining the Earth's climate (Rouse et al., 1979) and as a source of genetic diversity. Humans are rapidly destroying large areas of tropical forests: it is only by means of remote sensing that we are ever likely to understand the nature and locations of these changes. Similar issues exist with respect to other ecological zones; remote sensing provides a means to observe such regions at global scales and to better understand the interrelationships between the many factors that influence such patterns.

### *Vegetation Classification and Mapping*

Vegetation classification can proceed along any of several alternative avenues. The most fundamental is simply to separate vegetated from nonvegetated regions or forested from open lands. Such distinctions, although ostensibly very simple, can have great significance in some contexts, especially when data are aggregated over large areas or are observed over long intervals of time. Thus national or state governments, for example, may have an interest in knowing how much of their territories are covered by forest or may want to make changes in forested land from one 10-year period to the next, even though there may be no data available regarding the different kinds of forest.

However, it is usually important to acquire information at finer levels of detail. Although individual plants can be identified on aerial photographs, seldom if ever is it practical to use the individual plant as the unit for mapping vegetation. Instead, it is more useful to define mapping units that represent groups of plants. A *plant community* is an aggregation of plants with mutual interrelationships among each other and with the environment. Thus an *oak–hickory forest* is a useful designation because we know that communities are not formed by random collections of plants, but by consistent associations of the same groups of plants, those that tend to prefer the same environmental conditions and to create the environments that permit certain other plants to exist nearby. A community consists of many *stands,* specific, individual occurrences of a given community (Figure 17.1).

Plants within communities do not occur in equal proportions. Certain species tend to dominate; these species are often used to name communities (e.g., *hickory* forest), although others may be present. Dominant species may also dominate physically, forming the largest plants in a sequence of layers, or *strata,* that are present in virtually all communities. *Stratification* is the tendency of communities to be organized vertically, with some species forming an upper canopy, others a middle stratum, with shrubs, mosses, lichens, and so on forming other layers ever nearer to the ground. Even ostensibly simple vegetation communities in grasslands or arctic tundra, for example, can be shown to consist of distinct strata. Specific plants within each community tend

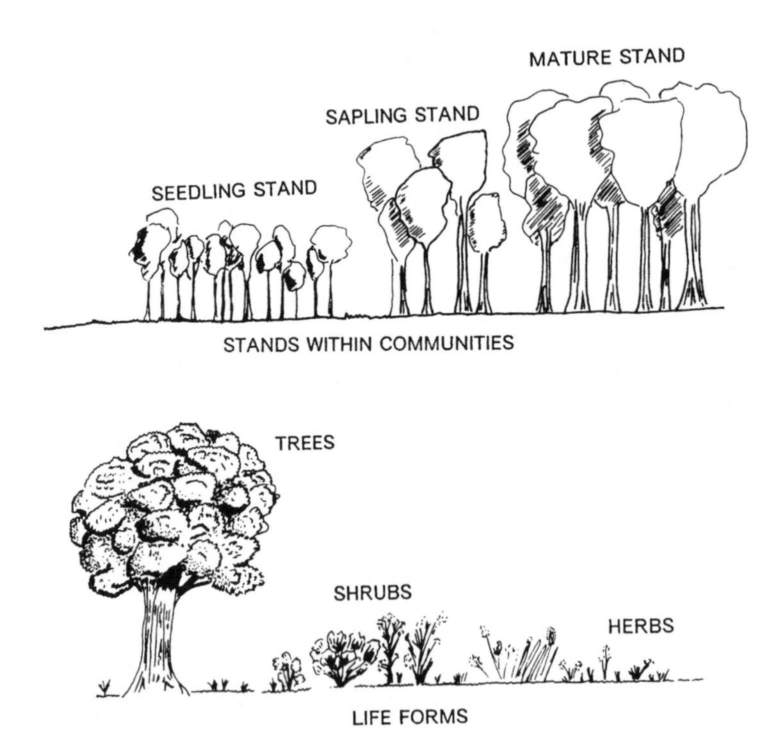


**FIGURE 17.1.** Vegetation communities, stands, vertical stratification, and different life forms.

**TABLE 17.1. Floristic Classification**

| Level | Example[a] |
|---|---|
| Class | Angiospermae (broad/general) |
| Order | Sapindales |
| Family | Aceraccae |
| Genus | *Acer* |
| Species | *saccharum* (narrow/specific) |

[a]Example = sugar maple (*Acer saccharum*).

to favor distinctive positions within each strata because they have evolved to grow best under the conditions of light, temperature, wind, and humidity that prevail in their strata.

*Floristics* refers to the botanical classification of plants, usually based upon the character of their reproductive organs, using the system founded by Carolus Linnaeus (1707–1778). Linnaeus, a Swedish botanist, created the basis for the binomial system of designating plants by Latin or Latinized names that specify a hierarchical nomenclature, of which the *genus* and *species* (Table 17.1) are most frequently used. The *Linnaean system* provides a distinctive name that places each plant in relationship with others in the taxonomy. Floristic classification reveals the genetic character and evolutionary origin of individual plants.

In contrast, the *life form,* or *physiognomy,* form of classification describes the physical form of plants (Table 17.2). For example, common physiognomies might include "tree," "shrub," "herbaceous vegetation," and so on. Physiognomy is important because it reveals the *ecological role* of the plant, the nature of its relationship with other plants and the environment. Floristics and physiognomy often have little direct relationship with one another. Plants that are quite close floristically may have little similarity in their growth form, and plants that are quite similar in their ecological roles may be very different floristically. For example, the rose family (Rosaceae) includes a wide variety of trees, shrubs, and herbaceous plants that occupy diverse environments and ecological settings. Conversely, a single environment can be home to hundreds of different species, for example, the alpine meadows of New England are composed of

**TABLE 17.2. Classification by Physiognomy and Structure**

| | |
|---|---|
| Woody plants | Broadleaf evergreen |
| | Broadleaf deciduous |
| | Needleleaf evergreen |
| | Leaves absent |
| | Mixed |
| | Semideciduous |
| Herbaceous plants | Graminoids |
| | Forbs |
| | Lichens and mosses |
| Special life forms | Climbers |
| | Stem succulents |
| | Tuft plants |
| | Bamboos |
| | Epiphytes |

*Note.* Data from Küchler (1967). Küchler's complete classification specifies plant height, leaf characteristics, and plant coverage.

some 250 species, including such diverse families as Primulaceae (primrose), Labiatae (mint), Araliaceae (ginseng), and Umbelliferae (parsley). Thus a purely floristic description of plants (e.g., "the rose family") seldom conveys the kind of information we would like to know about plants, their relationships with one another, and their relationships with the environment.

Although it is often possible to identify specific plants and to assign taxonomic designations from large-scale imagery, vegetation studies founded on remotely sensed images typically employ the structure and physiognomy of vegetation for classification purposes. That is, usually we wish to separate forest from grassland, for example, or to distinguish between various classes of forest. Although it is important to identify the dominant species for each class, our focus is usually upon separation of vegetation communities based upon their overall form and structure rather than upon floristics alone.

No single approach to vegetation classification can be said to be universally superior to others. At given levels of detail, and for specific purposes, each of the approaches mentioned above serves important functions. Floristic classification is useful when scale is large and mapping is possible in fine detail that permits identification of specific plants. For example, analyses for forest management often require a large scale, both for measurement of timber volume and for identification of individual trees. Physiognomy or structure is important whenever image scale is smaller, detail is coarser, and the analyst focuses more clearly upon the relationships of plants to the environment. Ecological classification may be used at several scales for analyses that require consideration of broader aspects of planning for resource policy, wildlife management, or inventory of biological resources.

Another approach to classification of vegetation, the *ecological* format classification, considers vegetation as the most easily observed component of an environmental complex including vegetation, soil, climate, and topography. This approach classifies regions as ecological zones, usually in a hierarchical system comparable to that shown in Table 17.3. Bailey (1998b, p. 145) defines his units as *ecoregions:* "major ecosystem[s], resulting from large-scale predictable patterns of solar irradiation and moisture, which in turn affect the kinds of local ecosystems and animals and plants found there." At the very broadest scales, ecological classification is based on long-term climate and very broad-scale vegetation patterns, traditionally derived from information other than remotely sensed data. However, later sections of this chapter will show how it is now possible to use remotely sensed data to derive these classifications with much more precision and accuracy than was previously possible.

At finer levels of detail, remotely sensed imagery is essential for delineating ecoregions. The

**TABLE 17.3. Bailey's Ecosystem Classification**

| Level | Example |
|---|---|
| Domain | Humid temperate |
| Division | Hot continental |
| Province | Eastern deciduous forest |
| Section | Mixed mesophytic forest |
| District | |
| Land-type association | |
| Land type | |
| Land-type phase | |
| Site | |

*Note.* Data from Bailey (1978).

interpreter considers not only vegetation cover, but also elevation, slope, aspect, and other topographic factors in defining units on the map.

### *Kinds of Aerial Imagery for Vegetation Studies*

Aerial imagery enables the analyst to conduct quick and accurate delineation of major vegetation units, providing at least preliminary identification of their nature and composition. Note, however, that interpretation from aerial images cannot replace ground observations; an accurate interpretation assumes that the analyst has field experience and knowledge of the area to be examined and will be able to evaluate his or her interpretation in the field.

Küchler (1967) recommends use of vertical photographs that provide stereoscopic coverage. Optimum choice of image scale, if the analyst has control over such variables, depends upon the nature of the map and the complexity of the vegetation pattern. Detailed studies have been conducted using photography at scales as large as 1:5,000, but scales from about 1:15,000 to 1:24,000 are probably more typical for general-purpose vegetation studies. Of course, if photographs at several different scales, dates, or seasons are available, multiple coverage can be used to good advantage to study changes. Small-scale images can be used as the basis for delineation of extents of major regions; the greater detail of large-scale images can be used to identify specific plants and plant associations.

Panoramic photographs have been successfully used for a variety of purposes pertaining to vegetation mapping and forest management. Modern panoramic cameras permit acquisition of high-resolution imagery over very wide regions, such that large areas can be surveyed quickly. Such images have extreme variations in scale and perspective, so they cannot be used for measurements, but their wide areal coverage permits a rapid, inexpensive inventory of wide areas, and identification of areas that might require examination using more detailed imagery.

Timing of flights, which may not always be under the control of the analyst, can be a critical factor in some projects. For example, mapping the understory in forested areas can be attempted only in the early spring when shrubs and herbaceous vegetation have already bloomed, but the forest canopy has not fully emerged to obscure the smaller plants from overhead views. Because not all plants bloom at the same time, a succession of carefully timed photographic missions in the spring can sometimes record the sequence in which specific species bloom, thereby permitting mapping of detail that could not be reliably determined by a single image showing all trees in full bloom.

Infrared films and color infrared films have obvious advantages for vegetation studies of all kinds, as emphasized previously. The CIR photography acquired at high altitudes for the United States (usually during leaf-off conditions) is routinely available at modest cost through the NAPP (Section 3.13). Custom-flown CIR photography is likely to be considered expensive; many users will find black-and-white photographs satisfactory for many purposes.

## 17.2. Structure of the Leaf

Many applications of remote sensing to vegetation patterns depend on a knowledge of the spectral properties of individual leaves and plants. These properties are best understood by examining leaf structure at a rather fine level of detail.

The cross section of a typical leaf reveals its essential elements (Figure 17.2). The uppermost layer, the *upper epidermis,* consists of specialized cells that fit closely together without openings or gaps between cells. This upper epidermis is covered by the *cuticle,* a translucent waxy layer that prevents moisture loss from the interior of the leaf. The underside of the leaf is protected by the *lower epidermis,* similar to the upper epidermis except that it includes openings called *stomates* (or *stomata*), which permit movement of air into the interior of the leaf. Each stomate is protected by a pair of *guard cells* that can open and close as necessary to facilitate or prevent movement of air to the interior of the leaf. The primary function of stomata is apparently to allow atmospheric carbon dioxide ($CO_2$) to enter the leaf for photosynthesis. Although the guard cells and the epidermis appear to be small and inefficient, they are in fact very effective in transmitting gases from one side of the epidermis to the other. Their role in permitting $CO_2$ to enter the leaf is obviously essential for the growth of the plant, but they also play a critical role in maintaining the thermal balance of the leaf by permitting movement of moisture to and from the interior of the leaf. The guard cells can close to prevent excessive movement of moisture, and thereby economize moisture use by the plant. Apparently the positions of stomata on the lower side of the leaf favor maximum transmission of light through the upper epidermis and minimize moisture loss when stomata are open.

On the upper side of the leaf, just below the epidermis, is the *palisade tissue* consisting of vertically elongated cells arranged in parallel, at right angles to the epidermis. Palisade cells include *chloroplasts,* cells composed of chlorophyll and other ("accessory") pigments active in photosynthesis, as described below. Below the palisade tissue is the spongy *mesophyll tissue,* which consists of irregularly shaped cells separated by interconnected openings. The mesophyll has a very large surface area; it is the site for the oxygen ($O_2$) and $CO_2$ exchange necessary for photosynthesis and respiration. Although leaf structure is not identical for all plants, this description provides a general outline of the major elements common to most plants, especially those that are likely to be important in agricultural and forestry studies.

In the visible portion of the spectrum, chlorophyll controls much of the spectral response of the living leaf (Figure 17.3). *Chlorophyll* is the green pigment that is chiefly responsible for the green color of living vegetation. Chlorophyll enables the plant to absorb sunlight, thereby making photosynthesis possible; it is located in specialized lens-shaped structures, known as *chloroplasts,* found in the palisade layer. Light that passes through the upper tissues of the leaf is received by chlorophyll molecules in the palisade layer, which is apparently specialized for photosynthesis, as it contains the largest chloroplasts, in greater abundance than other portions of the plant. $CO_2$ (as a component of the natural atmosphere) enters the leaf through stomata on the underside of the leaf, and then diffuses throughout cavities within the leaf. Thus photosyn-

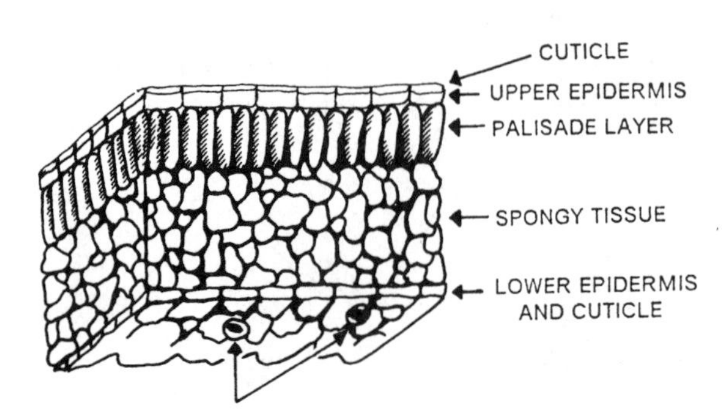


**FIGURE 17.2.** Diagram of a cross section of a typical leaf.

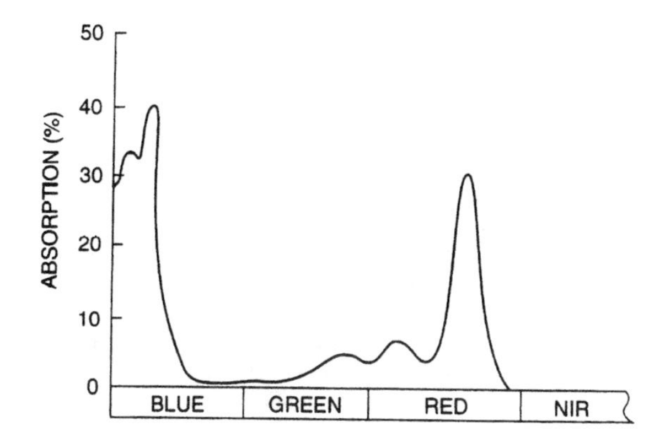


**FIGURE 17.3.** Absorption spectrum of a typical leaf. The leaf absorbs blue and red radiation and reflects green and near infrared radiation.

thesis creates carbohydrates from $CO_2$ and $H_2O$, using the ability of chloroplasts to absorb sunlight as a source of energy. Chloroplasts include a variety of pigments, some known as *accessory pigments,* that can absorb light, then pass its energy to chlorophyll. Chlorophyll occurs in two forms. The most common is *chlorophyll a,* the most important photosynthetic agent in most green plants. A second form, known as *chlorophyll b,* has a slightly different molecular structure; it is found in most green leaves, but also in some algae and bacteria.

## 17.3. Spectral Behavior of the Living Leaf

Chlorophyll does not absorb all sunlight equally. Chlorophyll molecules preferentially absorb blue and red light for use in photosynthesis (Figure 17.4). They may absorb as much as 70–90% of incident light in these regions. Much less of the green light is absorbed and more is reflected, so the human observer, who can see only the visible spectrum, sees the dominant reflection of green light as the color of living vegetation (Figures 17.3 and 17.5).

In the near infrared spectrum, reflection of the leaf is controlled not by plant pigments but by the structure of the spongy mesophyll tissue. The cuticle and epidermis are almost completely transparent to infrared radiation, so very little infrared radiation is reflected from the outer portion of the leaf. Radiation passing through the upper epidermis is strongly scattered by meso-

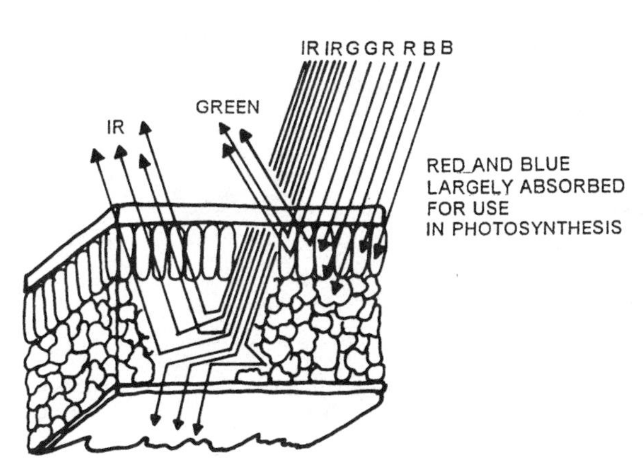


**FIGURE 17.4.** Interaction of leaf structure with visible and near infrared radiation.

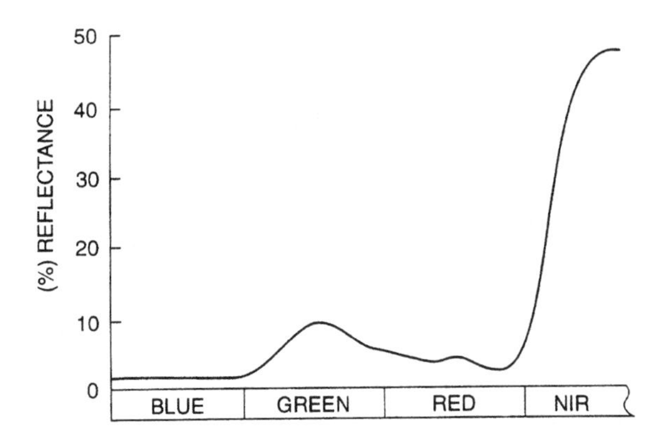


**FIGURE 17.5.** Typical spectral reflectance from a living leaf. The leaf is brightest in the near infrared region, although in the visible region the maximum is in the green region.

phyll tissue and cavities within the leaf. Very little of this infrared energy is absorbed internally—most (up to 60%) is scattered upward (which we call "reflected energy") or downward (which we call "transmitted energy"). Some studies suggest that palisade tissue may also be important in infrared reflectance. Thus the internal structure of the leaf is responsible for the bright infrared reflectance of living vegetation.

At the edge of the visible spectrum, as the absorption of red light by chlorophyll pigments begins to decline, reflectance rises sharply (Figure 17.5). Thus if reflectance is considered not only in the visible but across the visible and the near infrared, peak reflectance of living vegetation is not in the green but in the near infrared. This behavior explains the great utility of the near infrared spectrum for vegetation studies, and facilitates separation of vegetated from non-vegetated surfaces, which are usually much darker in the near infrared (Plate 21). Furthermore, differences in reflectivities of plant species often are more pronounced in the near infrared than they are in the visible, meaning that discrimination of vegetation classes is sometimes made possible by use of the near infrared reflectance (Figure 17.6).

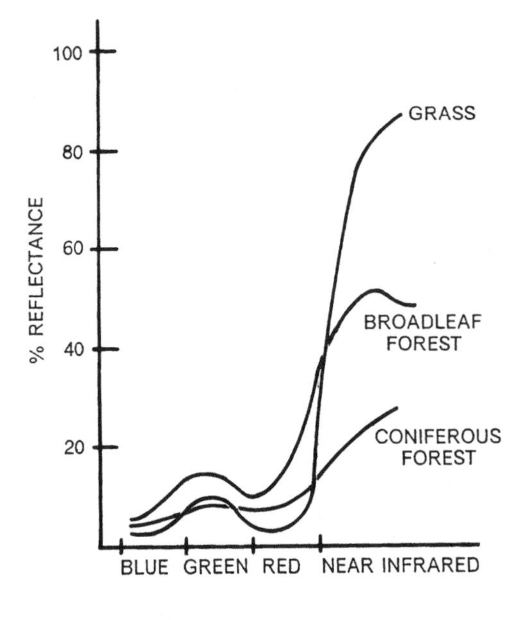


**FIGURE 17.6.** Differences between vegetation classes are often more distinct in the near infrared than in the visible.

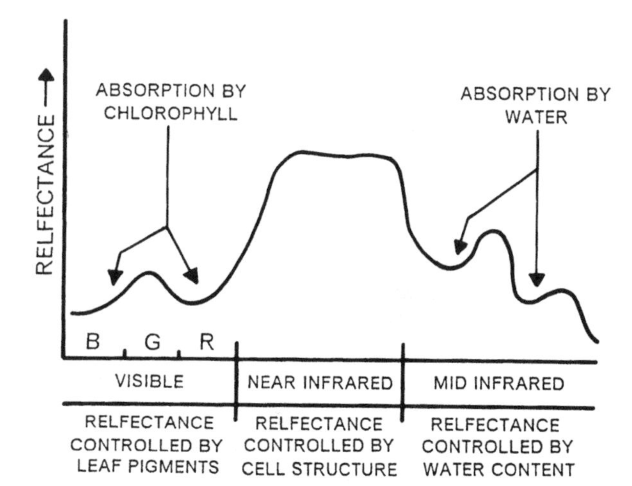


**FIGURE 17.7.** Changes in leaf water content may be pronounced in the mid-infrared region. This diagram illustrates *differences* in reflectances between equivalent water thicknesses of 0.018 cm and 0.014 cm. Changes in reflectance are most evident in the mid-infrared region. Diagram based on simulated data reported by Tucker (1979, p. 10).

As a plant matures or is subjected to stress by disease, insect attack, or moisture shortage, the spectral characteristics of the leaf may change. In general, these changes apparently occur more or less simultaneously in both the visible and the near infrared regions, but changes in near infrared reflectance are often more noticeable. Reflectance in the near infrared region is apparently controlled by the nature of the complex cavities within the leaf and internal reflection of infrared radiation within these cavities. Although some scientists suggest that moisture stress or natural maturity of a leaf causes these cavities to "collapse" as a plant wilts, others maintain that it is more likely that decreases in near infrared reflection are caused by deterioration of cell walls rather than physical changes in the cavities themselves. Thus changes in infrared reflectance can reveal changes in vegetative vigor; infrared images have been valuable in detecting and mapping the presence, distribution, and spread of crop diseases and insect infestations. Furthermore, changes in leaf structure that accompany natural maturing of crops are subject to detection with infrared imagery, so that it is often possible to monitor the ripening of crops as harvest time approaches. CIR film is valuable for observing such changes because of its ability to show spectral changes in both the visible and near infrared regions and to provide clear images that show subtle tonal differences.

In the longer infrared wavelengths (beyond 1.3 μm) leaf water content appears to control the spectral properties of the leaf (Figure 17.7). The term *equivalent water thickness* (EWT) has been proposed to designate the thickness of a film of water that can account for the absorption spectrum of a leaf within the interval 1.4–2.5 μm.

### *Reflection from Canopies*

Knowledge of the spectral behavior of individual leaves is important for understanding the spectral characteristics of vegetation canopies, but cannot itself completely explain reflectance from areas of complete vegetative cover. Vegetation canopies are composed of many separate leaves that may vary in their size, orientation, shape, and coverage of the ground surface. In the field, a vegetation canopy (e.g., in a forest or a cornfield) is composed of many layers of leaves;

the upper leaves form shadows that mask the lower leaves, creating an overall reflectance created by a combination of leaf reflectance and shadow.

Shadowing tends to decrease canopy reflectance below the values normally observed in the laboratory for individual leaves. Knipling (1970) cited the following percentages reported in several previously published studies:

| | Percent reflected | |
|---|---|---|
| | Visible | Near infrared |
| Single leaf | 10% | 50% |
| Canopy | 3–5% | 35% |

Thus the reflectance of a canopy is considerably lower than reflectances measured for individual leaves. But the relative decrease in the near infrared region is much lower than that in the visible. The brighter canopy reflection for the near infrared is attributed to the fact that plant leaves transmit near infrared radiation, perhaps as much as 50–60%. Therefore, infrared radiation is transmitted through the upper layers of the canopy, reflected in part from lower leaves, and retransmitted back through the upper leaves, resulting in bright infrared reflectance (Figure 17.8). Physicists and botanists have been able to develop mathematical models for canopy reflectances by estimating the optical properties of the leaves and the canopy as a whole. Because some of the transmitted infrared radiation may in fact be reflected from the soil surface below the canopy, such models have formed an important part of the research in agricultural remote sensing.

### *The Red Shift*

Collins (1978) reports the results of studies that show changes in the spectral responses of crops as they approach maturity. His research used high-resolution multispectral scanner data of numerous crops at various stages of the growth cycle. Collins's research focused upon examina-

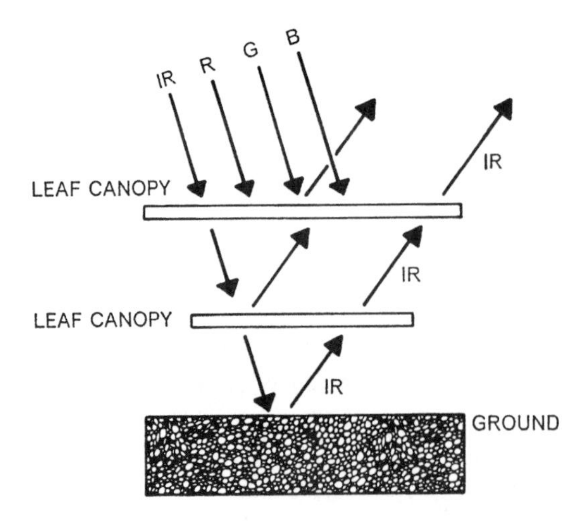


**FIGURE 17.8.** Simplified cross-sectional view of behavior of energy interacting with a vegetation canopy. (See Figures 17.3 and 17.4.) In some portions of the spectrum, energy transmitted through the upper layer is available for reflection from lower layers (or from the soil surface).

tion of the far red region of the spectrum, where chlorophyll absorption decreases and infrared reflection increases (Figure 17.9). In this zone, the spectral response of living vegetation increases sharply as wavelength increases—in the region from just below 0.7 μm to just above 0.7 μm brightness increases by about 10 times (Figure 17.9).

Collins also noticed that as crops approach maturity, the position of the chlorophyll absorption edge shifts toward longer wavelengths, a change he called the *red shift* (Figure 17.8). The red shift is observed not only in crops but also in other plants. The magnitude of the red shift varies with crop type (it is a pronounced and persistent feature in wheat).

Collins observed the red shift along the entire length of the chlorophyll absorption edge, although it was most pronounced near 0.74 μm, in the infrared region, near the shoulder of the absorption edge. He suggests that very narrow bands at about 0.745 μm and 0.780 μm would permit observation of the red shift over time, and thereby provide a means of assessing differences between crops and the onset of maturity of a specific crop.

Causes of the red shift appear to be very complex, and are not understood in detail. Chlorophyll a appears to increase in abundance as the plant matures; increased concentrations change the molecular form in a manner that adds absorption bands to the edge of the chlorophyll a absorption region, thereby producing the red shift. (In Chapter 18 we shall see that certain factors can alter the spectral effect of chlorophyll, thereby shifting the edge of the absorption band toward shorter wavelengths—this is the "blue shift" observed in geobotanical studies.)

## 17.4. Forestry

Identification of individual plants can be accomplished by close examination of crown size and shape (Figure 17.10). At the edges of forested areas, the shadows of trees can form clues to their identification. At smaller scales, individual plants are not recognizable, and the interpreter must examine patterns formed by the aggregate appearance of individual stands, in which individual crowns blend to form a distinctive tone and texture. Identification may be relatively straightforward if stands are pure or are composed of only a few species that occur in consistent propor-

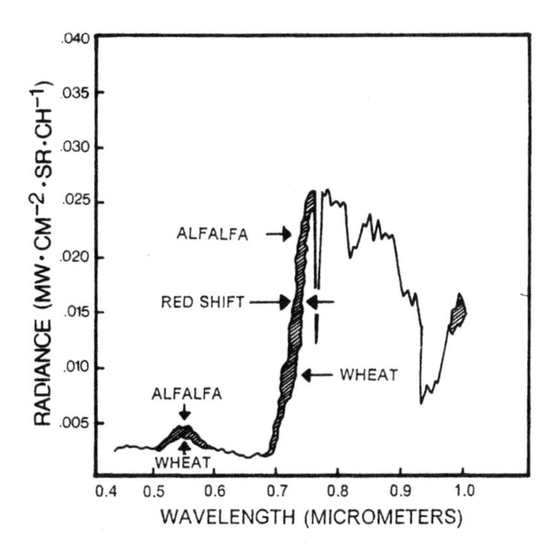


**FIGURE 17.9.** Red shift. The absorption edge of chlorophyll shifts toward longer wavelengths as plants mature. The shaded area represents the magnitude of this shift as the difference between the spectral response of headed wheat and mature alfalfa. Similar but smaller shifts have been observed in other plants. The red shift is important for distinguishing the headed stage of a crop from earlier stages of the same crop and for distinguishing headed grain crops from other green, nongrain, crops. From Collins (1978, p. 47). Copyright 1978 by the American Society of Photogrammetry and Remote Sensing. Reproduced by permission.

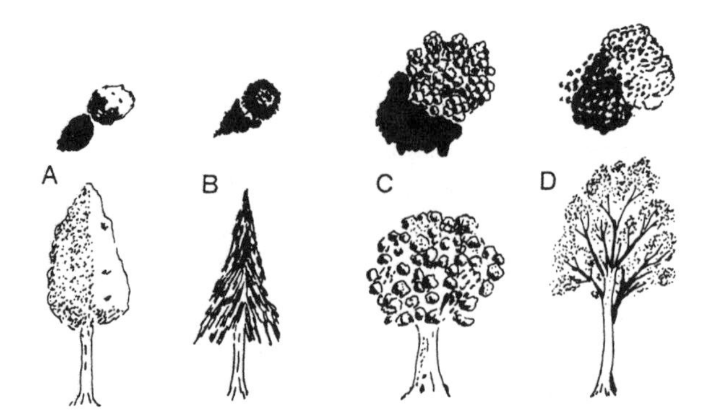


**FIGURE 17.10.** Identification of individual plants by crown size and shape. Crowns and their shadows provide distinctive clues for identification of trees.

tions. If many species are present and their proportions vary, then specific identification may not be possible, and the use of such broad designations as *mixed deciduous forest* may be necessary.

At these smaller scales, an interpreter identifies separate cover types from differences in image tone, image texture, site, and shadow. Identification of cover types can be considered the interpreter's attempt to implement one of the classification strategies mentioned above in the context of a specific geographic region, using a specific image. Cover types may not match exactly to the categories in a classification system, but they form the interpreter's best approximation of these categories for a specific set of imagery. Cover classes can be considered as rather broad vegetation classes, based perhaps upon the predominant species present, their age, and their degree of crown closure. Thus forest-cover-type classes might include *aspen/mixed conifer* and *Douglas fir,* indicating the predominant species without precluding the residence of others. Subclasses or secondary descriptors could indicate sizes of the trees, crown closure, and presence of undergrowth.

Such classifications can be prepared by manual interpretation of aerial photographs if the interpreter has field experience and knowledge of the area to be mapped. Each region has only a finite number of naturally occurring cover types, which usually prefer specific topographic sites, so the interpreter can bring several kinds of knowledge to bear upon the problem of recognition and classification. Image tone, texture, shadow, and other photographic features permit separation of major classes. Rather broad classes can be separated using small-scale photography, but finer distinctions require high-quality, large-scale, stereo photography. For example, the classes given in Table 17.4a might be appropriate for information interpreted from aerial photographs at 1:60,000, whereas the classes in Table 17.4b could be interpreted only from much larger scale photographs, perhaps at 1:15,840 to 1:24,000. At larger scales, stereoscopic coverage is important for identifying and distinguishing between classes.

### *Forest Photogrammetry*

Foresters desire to identify specific *stands* of timber, areas of forest with uniform species composition, age, and density that can be treated as homogeneous units. Stands are the basic unit of forest management, so forest managers wish to monitor their growth time to detect the effects of

**TABLE 17.4. Examples of Cover Types**

*a.* At small scale (1:1,000,000 to 1:50,000), cover types must be defined at coarse levels of detail because it will not be possible to distinguish or to delineate classes that are uniform at the species level:
"evergreen forest"
"deciduous forest"
"chaparral"
"mixed broadleaf forest"
"mangrove forest"

*b.* At large scale (larger than about 1:20,000), cover types may be defined at fine levels of detail. Sometimes, when stands are very uniform, cover types can be mapped at the detail of individual species:
"balsam fir"
"shortleaf pine"
"aspen and white birch"
"oak–hickory forest"

disease, insect damage, and fire or drought. Even when stands have been planted from seedlings by commercial foresters, aerial photography, by virtue of its map-like perspective and its wealth of environmental information, provides accurate and economical information concerning the condition of the stand at a specific time.

If large-scale, high-quality images are available (preferably in stereo), it is possible to apply the principles of photogrammetry to measurement of factors of significance in forestry. Tree height can be measured using lengths of shadows projected onto level, open ground, although there are numerous practical problems that make this method inconvenient for routine use (level, open ground is not often easy to find; it is necessary to determine sun elevations for the latitude, day of the year, and time of day; and often shadows are too short for reliable measurements). As a result, the parallax bar (Chapter 5) is more commonly used for routine work. *Crown diameters* can be measured using transparent rules or graticules with tube magnifiers (Chapter 5).

### *Crown Closure*

One of the most important variables that contributes to estimates of stand density is assessment of *crown closure,* the proportion of the area of a stand that is covered by crowns of trees. Crown closure measures the density of trees in a stand, and indicates the degree of competition between trees as a stand matures. Because crown closure it is also related to stand volume, it assists predictions of economic dimensions of a stand at harvest. Because the size of the crown of a tree is closely related to many of its physiological characteristics (such as its ability to conduct photosynthesis), crown closure, monitored over time, permits foresters to assess the growth of the forest.

Although crown closure can be measured at ground level using any of several procedures, aerial photography is a particularly valuable tool for assessing crown closure. It assesses the percentage of a stand that is covered by the crowns of the dominant and codominant species. (Crowns of understory trees are not included in the estimate.) Crown closure is reported as a percentage (e.g., 60%) or as a decimal (in the form 0.6).

With experience, crown closure can be accurately estimated by eye. Use of stereo is especially useful when understory crowns are visible. The quality of the photography can be important, including scale, shadows, and haze. So also can variables related to the stand, including the presence of irregular topography, the nature of spacing of trees, and the backgrounds of other vegetation. Some examples illustrate varied degrees of crown closure.

*Crown density* scales provide a visual guide for the interpreter to judge the approximate density by visual comparison (Figure 17.11). The interpreter compares the photo with the image on the template to select the closure class nearest to that of the area to be interpreted.

### *Timber Volume Estimation*

Foresters are routinely interested in estimation of timber volume for a specific stand as a means of monitoring its growth over time, assessing management practices, and determining the

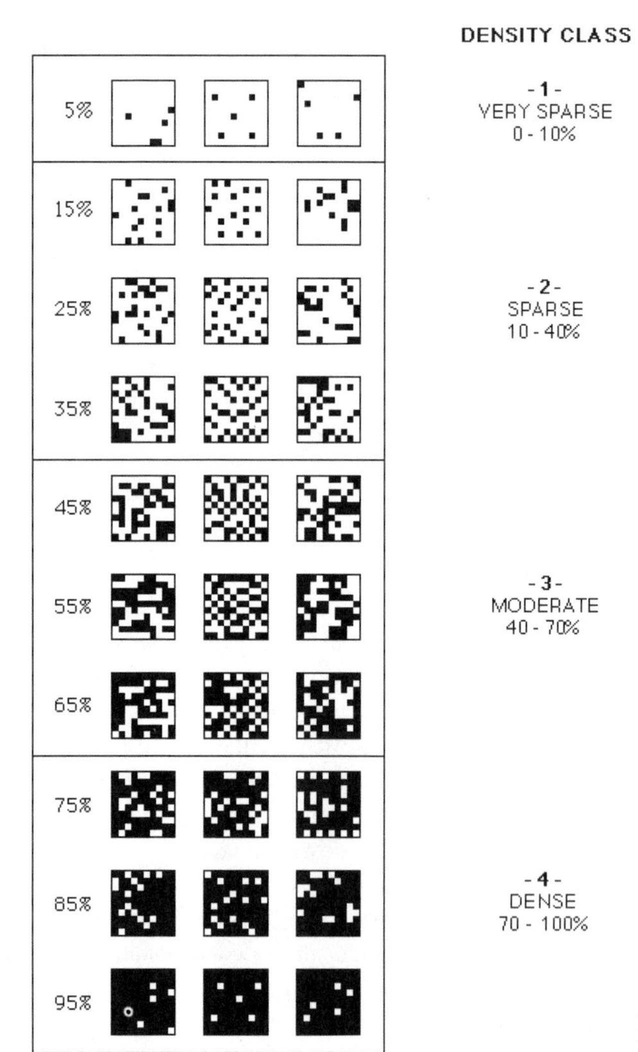


**FIGURE 17.11.** Crown density scale. From USFS Central State Forest Experiment Station.

amount of timber to be obtained at harvest. Volume measurement consists of estimates of board-foot or cubic-foot volumes to be obtained from a specific tree or, more often, from a specific stand. In the field, the forester measures the diameter of the tree at breast height (dbh) and the height of the straight section of the trunk (bole) as the two basic components to volume estimation. Measurements made for each tree are summed to give volume estimates for an entire stand.

There are several approaches to estimation of volume from measurements derived from aerial photographs; there can be no universally applicable relationship between photo measurements and timber volume because as species composition, size, age, soil, and climate vary so greatly from place to place. Typically, the interpreter must apply a *volume table,* a tabular summary of the relationship between timber volume and photo-derived measurements such as crown diameter and tree height (Table 17.5). The interpreter can make measurements directly from the photograph, then use the table to find volumes for each tree or for entire stands. New

**TABLE 17.5. Example of Aerial Stand Volume Table**

| Average stand height (ft.) | Crown cover (%) | | | | |
|---|---|---|---|---|---|
| | 15 | 35 | 55 | 75 | 95 |
| **8- to 12-ft. crown diameter** | | | | | |
| 30 | 15 | 50 | 70 | 95 | 135 |
| 40 | 35 | 70 | 95 | 125 | 175 |
| 50 | 50 | 95 | 125 | 170 | 215 |
| 60 | 90 | 150 | 195 | 235 | 265 |
| 70 | 160 | 230 | 270 | 300 | 330 |
| 80 | 250 | 310 | 345 | 380 | 420 |
| **13- to 17-ft. crown diameter** | | | | | |
| 30 | 20 | 55 | 80 | 105 | 135 |
| 40 | 40 | 80 | 100 | 145 | 195 |
| 50 | 60 | 100 | 135 | 185 | 235 |
| 60 | 100 | 160 | 205 | 245 | 285 |
| 70 | 170 | 240 | 275 | 310 | 350 |
| 80 | 260 | 320 | 360 | 390 | 430 |
| 90 | 360 | 415 | 450 | 485 | 525 |
| 100 | 450 | 515 | 555 | 595 | 635 |
| **18- to 22-ft. crown diameter** | | | | | |
| 30 | 30 | 70 | 90 | 120 | 160 |
| 40 | 50 | 90 | 120 | 160 | 200 |
| 50 | 75 | 120 | 160 | 205 | 245 |
| 60 | 115 | 180 | 225 | 265 | 305 |
| 70 | 190 | 255 | 290 | 330 | 370 |
| 80 | 275 | 330 | 370 | 410 | 450 |
| 90 | 370 | 430 | 470 | 505 | 550 |
| 100 | 470 | 530 | 570 | 610 | 650 |
| 110 | 600 | 660 | 700 | 740 | 780 |

*Note.* Volume (tens of cubic feet) per acre, given average stand height, average crown diameter, and crown cover, for Rocky Mountain conifer species. Derived from field measurements in test plots. Abbreviated from a more extensive table given in Wilson (1960, pp. 480–481). Copyright 1960 by the American Society for Photogrammetry and Remote Sensing. Reproduced by permission.

volume tables must be compiled as stand composition and the environmental setting change from place to place. Many variations on this basic strategy have been developed, but most rely upon the same kinds of estimates from the photograph.

## 17.5. Agriculture

Aerial imagery provides a powerful tool for understanding dynamics of agricultural systems, especially if employed with knowledge of the regional agricultural context.

### *Crop Calendar*

The *crop calendar* is the cycle of crops grown during the year, in harmony with the regional climate, local practices, and economic incentives. Each farmer adopts a specific sequence of preparing the field, planting, and harvesting to minimize labor and risk and to maximize production. Although each farmer decides specifics for a given farm, over broad regions farmers as a group tend to follow a specific sequence of agricultural activities for their region.

As an example, see Table 17.6, in which the crop calendar for the principal crops of western Kansas is broadly sketched. Analysts who examine images of agricultural landscapes must apply knowledge of a region's agricultural calendar to understand the meaning of the patterns they see on the imagery. At any given time, the landscape shows a variety of crops, each at different growth stages, due to varied planting dates, as determined by the local agricultural calendar.

The opportunities for regular coverage provided by Earth observation satellites mean that agricultural areas can be observed over time, and that knowledge of the crop calendar can be applied in a manner that is not always practical if there is imagery for only a single date. For example, the winter wheat crop is characterized by bare fields in September and October, then by a mixture of soil and newly emerged vegetation in mid- to late autumn. During spring and early summer plant cover increases, completely masking the soil by late May or early June. Then, as the crop matures, spectral evidence of senescence records the approach of harvest in

**TABLE 17.6. Crop Calendar for Winter Wheat, Western Kansas**

| Month | Activity |
|---|---|
| September, October | Plowing and planting |
| October, November | Emergence and early growth |
| December, January, February, March | Dormancy |
| April, May | Growth resumes |
| June, July | Heading |
| | Harvest |
| August | |

*Note.* Variations in timing are caused by local differences in weather and climate.

late June or early July. In contrast, the corn crop (Table 17.7) is not planted until spring, does not attain complete coverage until June, and is not mature until August. Thus each crop displays a characteristic behavior over time, which permits tracking of the progress of the crop. Knowledge of the local crop calendar permits selection of a single date that will provide maximum contrast between two crops. In the instance of discrimination between corn and wheat, for example, selection of a date in late spring should show complete vegetative cover in wheat fields, while corn fields show only newly emerged plants against a background of bare soil.

### *Crop Identification*

The *growth stage* of any crop influences its appearance on aerial imagery. Specific crops do not display distinct spectral signatures because their appearance changes throughout the agricultural calendar. Agricultural scientists often define many stages for specific crops—these vary from crop to crop, and not all stages can be determined form aerial photography. However, it is often possible to interpret principal growth stages using large-scale aerial photography. Typically, when the crop is planted, the field is mainly bare tilled soil. Initially, there is little evidence of the presence of plants. *Emergence* indicates that seedlings have sprouted, providing visible evidence of the plants. Although plants at this stage are usually too small to identify as individual crops, their identities may be indicated by planting pattern, the presence or absence of irrigation, and so on. Further growth progressively presents larger areas of leaf, shielding the soil from view, and indicating the increasing maturity of the crop.

Many grains will exhibit *senescence,* the aging of the plant, usually indicated by drying of the leaves, as the crop reaches *maturity. Harvest* occurs, weather permitting, at or near maturity. After harvest, the field is usually partially covered with residue from the harvest process; this residue typically remains on the field until the field is tilled just prior to planting of the next crop.

For many purposes, photointerpreters need only to distinguish between cereals (such as wheat, rye, and oats) with small seeds and coarser grains (such as barley, maize, and sorghum) as they are represented on aerial photography (Figure 17.12). Often, it may not be possible to distinguish between specific crops within each class, so remotely sensed imagery is seldom the principal source of information regarding the specific identity of crops within an area. However, some broad distinctions can be made using aerial imagery, provided the scale is large enough and plants are approaching maturity. Once the crop approaches maturity, coarse grains usually can be recognized by their rougher textures as they appear in the field, due to their larger stalks, coarser leaves, complex structure, and increased microshadowing. Small grains, once mature,

**TABLE 17.7. Crop Calendar for Corn, Central Midwestern United States**

| Month | Activity |
|---|---|
| April, May | Plowing and planting |
| May, June | Emergence and early growth |
| July | Tasseling |
| August, September | Maturity and harvest |

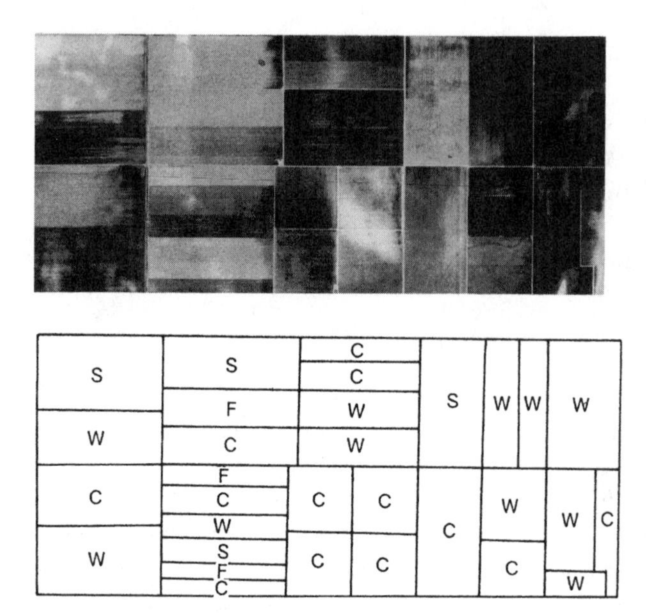


**FIGURE 17.12.** Crop identification. S = sorghum; W = wheat; C = corn; F = summer fallow.

have an even, smooth texture that tends to mask rows and irregularities in terrain. Row crops, such as soybeans, are usually recognizable by the latent linear appearance that remains even after the crops are mature enough to mask the rows themselves.

Although remotely sensed imagery can provide valuable information about the agricultural landscape, it rarely is used as the only or main means of identifying specific crops. Thus, aside from the ability of aerial imagery to provide a broad sketch of the kinds of crops grown in a region, its value is usually to assess acreages planted to croplands or left in fallow, the progress of crops as they mature through the growing season, detection of disease and insect infestation, effects of droughts, and other data relating to crop status.

## *Crop Status*

Throughout the growing season, agronomists monitor the progression of the crop, regionally, through the agricultural calendar, to assess disruptions from the usual timetable. Aerial imagery can be a valuable aid to documenting crop damage for insurance purposes, guiding response to infestations, and assisting commodity markets.

As crops emerge after planting and germination, they present leafy growth above ground. Initially, this growth is usually barely detectable on aerial photography, (on CIR photography, possibly by a light pink tint). At this stage the field, as seen from above is primarily bare soil. As the crop matures, its leaves begin the shade the soil formerly visible between plants, presenting an increasingly complete cover as seen from above. Depending upon the specific crop and its method of cultivation, the linear pattern of rows from initial planting may be visible. Assuming that the crop matures without drought or attack by disease or insects, usually the ground surface will be completely covered as the crop matures. At this stage, on CIR photography, such fields will appear as bright red.

At maturity, plant growth slows or stops, plant tissues begin to lose moisture, and the mesophyll tissue starts to deteriorate. As seen on CIR aerial photography, mature crops eventually attain a lighter color, often a grayish or greenish cast as the NIR reflectance decreases. After harvest, stubble and vegetative debris left on the soil give fields the same greenish cast, although the bare soil usually brightens the overall tone of harvested fields.

Aerial imagery can be a significant capability for observing effects of disease, weather damage, and insect infestation. A severe form of storm damage occurs as lodging. Lodging damages the stems of mature wheat (for example) as they are beaten down by heavy rain, hail, or winds. Mature crops are especially vulnerable to lodging, as crops approach maturity, the plants are larger, and their stems weaker. Even if the lodged plants survive to continue growth, the broken stems prevent harvesting. *Hail damage* for some crops completely shreds the leaves, physically destroying the plant beyond hope of recovery. Yet some plants can survive hail damage, especially if they are young enough, because smaller, more compact leaves and stems are stronger than those of mature plants.

Aerial photography and photointerpretation can be especially important tools for assessing crop status. Although farmers and agricultural analysts usually know the crops planted in a particular region, it is more difficult to know the status of the crops at a given time because of the difficulty of observing variations within such large regions. The spatial perspective and the use of the nonvisible spectrum offered by aerial photography provide the agricultural community with a valuable means for understanding crop status.

## 17.6. Vegetation Indices

*Vegetation indices* (VIs), which are based upon digital brightness values, attempt to measure biomass or vegetative vigor. A VI is formed from combinations of several spectral values that are added, divided, or multiplied in a manner designed to yield a single value that indicates the amount or vigor of vegetation within a pixel. High values for a VI identify pixels covered by substantial proportions of healthy vegetation. The simplest form of VI is a ratio between two digital values from separate spectral bands. Some band ratios have been defined by applying knowledge of the spectral behavior of living vegetation.

*Band ratios* are quotients between measurements of reflectance in separate portions of the spectrum. Ratios are effective in enhancing or revealing latent information when there is an inverse relationship between two spectral responses to the same biophysical phenomenon. If two features have the same spectral behavior, ratios provide little additional information, but if they have quite different spectral responses, the ratio between the two values provides a single value that concisely expresses the contrast between the two reflectances.

For living vegetation, the ratioing strategy can be especially effective because of the inverse relationship between vegetation brightness in the red and infrared regions. That is, absorption of red light (R) by chlorophyll and strong reflection of infrared (IR) radiation by mesophyll tissue assures that the red and near infrared values will be quite different, and that the IR/R ratio of actively growing plants will be high. Nonvegetated surfaces, including open water, man-made features, bare soil, and stressed or dead vegetation, will not display this specific spectral response, and the ratios will decrease in magnitude. Thus the IR/R ratio can provide a measure of photosynthetic activity and biomass within a pixel.

The IR/R ratio is only one of many measures of vegetation vigor and abundance. The green/red (G/R) ratio is based upon the same concepts used for the IR/R ratio, but it is considered less effective. Although ratios can be applied with digital values from any remote sensing system, much of the research on this topic has been conducted using Landsat MSS data. In this context, the IR/R ratio is implemented for Landsat 4 and 5 as MSS 4/MSS 2, although some have preferred to use MSS 3 in place of MSS 4. One of the most widely used VIs is known as the *normalized difference vegetation index* (NDVI):

$$\text{NDVI} = \frac{\text{IR} - \text{R}}{\text{IR} + \text{R}} = \frac{\text{MSS 4} - \text{MSS 2}}{\text{MSS 4} + \text{MSS 2}} \qquad \text{(Eq. 17.1)}$$

In principle, this index conveys the same kind of information as the IR/R and G/R ratios, but it is constrained to vary within limits that provide desirable statistical properties in the resulting values. The studies of Tucker (1979a, 1979b) and Perry and Lautenschlager (1984) suggest that in practice there are few practical differences between the many VIs that have been proposed.

Although such ratios have been shown to be powerful tools for studying vegetation, they must be used with care if the values are to be rigorously (rather than qualitatively) interpreted. Values of ratios and VIs can be influenced by many factors external to the plant leaf, including viewing angle, soil background, and differences in row direction and spacing in the case of agricultural crops. Ratios may be sensitive to atmospheric degradation. Because atmospheric degradation typically influences some bands much more than others, atmospheric effects can greatly alter the value of the ratio from its true value (Figure 17.13). Because atmospheric path length varies with viewing angle, values calculated using off-nadir satellite data (Chapter 6) vary according to position within the image. Clevers and Verhoef (1993) found that the relationship between the leaf area index (discussed in the next section) and the VI they studied was very sensitive to differences in leaf orientation. Although preprocessing can sometimes address such problems, it may still be difficult to compare values of VIs over a period of time because of variation in external factors. Price (1987) and others have noted that efforts to compare ratios or indices over time, or from one sensor to another, should reduce digital values to radiances before calculating ratios, to account for differences in calibration of sensors.

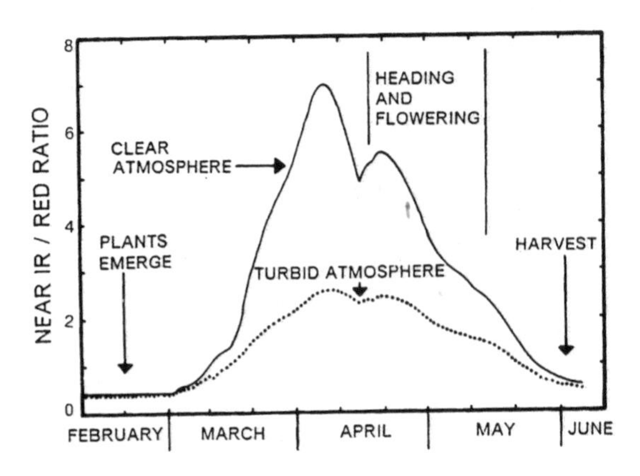


**FIGURE 17.13.** Influence of atmospheric turbidity upon IR/R ratio, as estimated from simulated data. From Jackson et al. (1983, p. 195). Copyright 1983 by Elsevier Science Publishing Co., Inc. Reproduced with permission.

## 17.7. Applications of Vegetation Indices

VIs have been employed in two separate kinds of research. Many of the first studies defining applications of VIs attempted to "validate" their usefulness by establishing that the values of the VIs are closely related to biological properties of plants. Typically, such studies examined test plots during an entire growing season, then compared values of the VIs, measured throughout the growing season, to samples of vegetation taken at the same times. The objective of such studies is ultimately to establish use of VIs as a means of remote monitoring of the growth and productivity of specific crops or of seasonal and yearly fluctuations in productivity.

Often values of the VIs have been compared to in situ measurements of the *leaf area index* (LAI), which is the area of leaf surface per unit area of soil surface. LAI is an important consideration in agronomic studies because it measures leaf area exposed to the atmosphere, and therefore is significant in studies of nutrient transport and photosynthesis. Values of VIs have also been compared to *biomass,* the weight of vegetative tissue. A number of VIs appear to be closely related to LAI (at least for specific crops), but no single VI seems to be equally effective for all plants and all agricultural conditions. Results of such studies have in general confirmed the utility of the quantitative uses of VIs, but details vary with the specific crop considered, atmospheric conditions, and local agricultural practices.

A second category of applications uses VIs as a mapping device, that is, much more as a qualitative rather than a quantitative tool. Such applications use VIs to assist in image classification, to separate vegetated from nonvegetated areas, to distinguish between different types and densities of vegetation, and to monitor seasonal variations in vegetative vigor, abundance, and distribution (Plate 21). Yool et al. (1985) studied MSS imagery of a forested area in southern California and concluded that the ratios they examined were not superior to the original MSS values or to simple linear combinations of these values.

## 17.8. Phenology

*Phenology* is the study of the relationship between vegetative growth and environment; often phenology refers specifically to seasonal changes in vegetative growth and decline. Many phenological changes can be monitored by means of remote sensing because plants change in appearance and structure during their growth cycle. Of special significance are spectral and physiological changes that occur as a plant matures. Each season, plants experience chemical, physical, and biological changes, known as *senescence,* that result in progressive deterioration of leaves, stems, fruit, and flowers. In the midlatitudes, annual plants typically lose most or all of their roots, leaves, and stems each year. Woody perennial plants typically retain some or all of their roots, woody stems, and branches, but shed leaves. Evergreens, including tropical plants, experience much more elaborate phenological cycles; individual leaves may experience senescence separately (i.e., trees do not necessarily shed all leaves simultaneously), and individual trees or branches of trees may shed leaves on cycles quite distinct from others in the same forest (Koriba, 1958).

During the onset of senescence, deterioration of cell walls in the mesophyll tissue produces a distinctive decline in infrared reflectance; an accompanying increase in visible brightness may be the result of decline in the abundance and effectiveness of chlorophyll as an absorber of visible radiation. Changes in chlorophyll produce the red shift mentioned above. Such changes can be observed spectrally, so remote sensing imagery can be an effective means of monitoring seasonal changes in vegetation.

The phenology of a specific plant defines its seasonal pattern of growth, flowering, senescence, and dormancy. Remotely sensed images can expand the scope of study to include overviews of vegetation communities or even of entire biomes. (*Biomes* are broad-scale vegetation regions that correspond roughly to the Earth's major climate regions.) Dethier et al. (1973), for example, used several forms of imagery and data to observe the geographic spread of the emergence of new growth in spring, as it progressed from south to north in North America. This phenomenon has been referred to as the *green wave.* Then in late summer, the *brown wave* sweeps across the continent as plant tissues mature, dry, and are harvested.

Figure 1.1, first discussed at the beginning of Chapter 1, illustrates local phenological differences in the spread of the green wave in the early spring. The image shows a Landsat MSS band 4 (near infrared) radiation, which is sensitive to variations in the density, type, and vigor of living vegetation. The region represented by Figure 1.1, and shown again in somewhat different form in Figure 17.14, has uneven topography covered by a mixture of forested and open land, including large regions of cropland and pasture. When this image was acquired in mid-April, vegetation in the open land was just starting to emerge. Grasses and shrubs in these regions have bright green leaves, but leaves on the larger shrubs and trees have not yet emerged. There-

**FIGURE 17.14.** Landsat MSS band 4 image of southwestern Virginia. From EROS Data Center.

fore, the pattern of white depicts the regions occupied primarily by early blooming grasses and shrubs in lower elevations. Within a week or so after this image was acquired, leaves on trees began to emerge, first at the lower elevations, then later at higher elevations. Thus a second image acquired only 2 weeks or so after the first would appear almost completely white, due to the infrared brightness of the almost complete vegetation cover. If it were possible to observe this region on a daily basis, under cloud-free conditions, we could monitor the movement of the green wave upward from lower to higher elevations, and from south to north, as springtime temperatures prevail over more and more of the region. In reality, of course, we can see only occasional snapshots of the movement of the green wave, as the infrequent passes of Landsat and cloud cover prevent close observation during the short period when we must watch its progress.

In an agricultural context, phenology manifests itself through the local crop calendar, which will vary from region to region in response to interactions between the genetic character of the crop, the local climate, local agricultural technology, and local practices. Plate 22 illustrates this principle in a different geographic setting. RADARSAT images of the Mekong Delta, one of the world's principal rice-producing regions, show variations in the status of rice fields. Rice cultivation depends upon careful control of water and close monitoring of the status of the crop in each field. This can be accomplished by comparing RADARSAT–1 scenes acquired over the same area through time. As the appearance of rice fields evolve, from tilled; through smoothed, seeded, flooded, maturing, and senescent; to harvested, areas of change can be observed and located by using color to represent the scenes at each date:

- Red: 12 May 1999 (Wide 2, Descending)
- Green: 22 May 1999 (Standard 7, Descending)
- Blue: 15 June 1999 (Wide 3, Descending)

Areas in each color represent fields with standing crops on specific dates. Combinations indicate fields with crops that extend across two dates—yellow regions, for example, indicate fields with crops on 12 May and also on 22 May (i.e., the red plus the green). White regions indicate fields with crops on all three dates.

## 17.9. Advanced Very-High-Resolution Radiometer

VIs and knowledge of phenology can also be applied in a much different context, using data acquired by a meteorological satellite first introduced in Chapter 6. The advanced very-high-resolution radiometer (AVHRR) is a multispectral radiometer carried by a series of meteorological satellites operated by NOAA in near-polar, sun-synchronous orbits. They can acquire imagery over a swath width of approximately 2,800 km, providing global coverage on a daily basis. Although AVHRR was designed primarily for meteorological studies, it has been successfully used to monitor vegetation patterns over very broad geographic regions. Areal coverage is extensive enough that entire biomes, or major ecological zones, can be directly observed and monitored in ways that were not previously feasible.

The satellite makes about 14 passes in a 24-hr. period, collecting data for each 2,800-km swath twice daily, at 12-hr. intervals. Resolution at the nadir is about 1.1 km, but an onboard computer can generalize data to 4-km resolution before data are transmitted to ground stations to permit broader geographic coverage for a given volume of data. Data are recorded at 10 bits.

The wide angular view means that areas recorded near the edges of images suffer from severe geometric and angular effects. As a result, AVHRR data selected from the regions near the nadir provide the most accurate information. Although designed initially for much narrower purposes, Tucker et al. (1984) report that 800–900 km of the 2800-km swath are usable. (The unusable data are from the edges of the image, where angular effects of perspective and atmospheric path length present problems.)

Details of spectral coverage vary with the specific mission, but in general AVHRR sensors have been designed to attain meteorological objectives, including discrimination of clouds, snow, ice, land, and open water. Nonetheless, AVHRR has been employed on an ad hoc basis for land resource studies. AVHRR collects five channels of data at a spatial resolution of 1.1 km in the visible, near infrared, and the thermal infrared (Table 17.8). One channel is in the visible, one in the near infrared, and three in the thermal infrared. Channel 1 was originally proposed as 0.55–0.90 m, but after use on the prototype was redefined to 0.58–0.68 m to improve separation of snow-free and snow-covered land areas. For agricultural scientists, this change provided the benefit of offering a channel positioned in the red region that permits calculation of vegetation indices.

The visible and near infrared data can be ratioed to form a greenness index or VI similar to those discussed above:

$$\text{NDVI} = \frac{\text{Channel 2} - \text{Channel 1}}{\text{Channel 2} + \text{Channel 1}} \qquad \text{(Eq. 17.2)}$$

Although the spectral bands differ from those mentioned earlier, the meaning of the ratio is the same: high values reveal pixels dominated by high proportions of green biomass. The resolution is much coarser than that of Landsat MSS and TM data, but areal coverage is much broader and the opportunity for repeat coverage is much greater. The frequent repeat coverage permits gathering of data for areas that may be obscured by clouds on one date but not the next. Therefore, over time, cloud-free coverage of continental or subcontinental areas can be acquired. Such images do not show the detail of MSS or TM data, but they do give a broad geographic perspective that is not portrayed by other images.

### *Compiled AVHRR Data*

Full-resolution data, at about 1.1 km at nadir, can be acquired during any given day from a restricted region of the Earth, usually only when the satellite is within line of sight of a receiv-

**TABLE 17.8. Spectral Channels for AVHRR**

| Channel | Spectral limits | Region |
|---|---|---|
| 1 | 0.58–0.68 μm | Visible |
| 2 | 0.72–1.10 μm | Near infrared |
| 3 | 3.55–3.93 μm | Thermal infrared |
| 4 | 10.30–11.30 μm | Thermal infrared |
| 5 | 11.5–12.5 μm | Thermal infrared |

*Note.* Spectral definitions differ between different AVHRR missions.

ing station. This process produces local area coverage (LAC). LAC archives have been compiled for selected regions of the Earth.

Eidenshink (1992) describes the AVHRR data set for the United States derived from 1990 AVHRR data, since repeated for other years. These composites are designed to record phenological variations throughout a growing season. Full-resolution AVHRR scenes were selected to form a set of 19 biweekly composites from March to December. Compositing over the 2-week period permits selection of a least one cloud-free date for each pixel. A special processing program was designed to select data based upon viewing geometry, solar illumination, sensor calibration, and cloud cover, and then to prepare geometrically registered NDVI composites. The process produces composite images composed of the original AVHRR bands, NDVI data, data describing the compositing process, and a statistical summary.

*Global area coverage* (GAC) is compiled by onboard sampling and processing, transmitted to ground stations on a daily basis. A GAC pixel is formed by averaging the first four pixels in a row, skipping the fifth pixel, averaging the next four pixels, skipping the next, and so on to the end of the line. The algorithm skips the next two lines, then processes the fourth line in the same manner as it did the first line, and so on, until the entire scene is processed. These data provide spatial resolution of about 4 km.

*Global vegetation index* (GVI) data are created from the daily AVHRR data by examining the differences between the red and near infrared channels; the highest values for a 7-day interval are then used to select the dates for which to calculate an NDVI value that represents a pixel for the entire 7-day period. The preliminary selection using the red–near infrared difference is designed both to select cloud-free pixels (if possible) for the 7-day interval and to minimize atmospheric effects. NDVI data are processed in such a manner that they represent about 15-km resolution. These data have been compiled since 1982, although algorithms and formats have varied considerably during this interval (Townshend, 1994).

### *Some AVHRR Examples*

These broad-scale images provide an opportunity to observe major ecological zones and seasonal changes in a manner that was not previously possible. Figures 17.15 and 17.16 show such images for two dates for North America. Darker areas show high values for the VI (a reversal of the usual convention of representing high values as bright tones). The summer scene shows the continent blanketed with a broad cover of vegetation. The lighter tones indicate sparsely vegetated arid regions in the Great Basin and southwestern United States and the tundra zone north of the arctic tree line. The March scene shows (in addition to snow, ice, and cloud cover in northern areas) much lower biomass, except for localized patches in tropical central America, southern California, and the Gulf Coast of the United States, where winter temperature and moisture permit year-round vegetative growth.

Such products are produced on a weekly basis by NOAA. Although they are not used to study specific sites, they provide an excellent overview of ecological conditions and zones, and permit identification of regions worthy of more detailed study with more detailed data.

AVHRR data have been archived to form a global database and have been processed to provide a variety of products depicting seasonal changes in vegetative cover over areas of continental size. Some of these data are described in Chapter 21; Internet access to some of them was noted in Chapter 4.

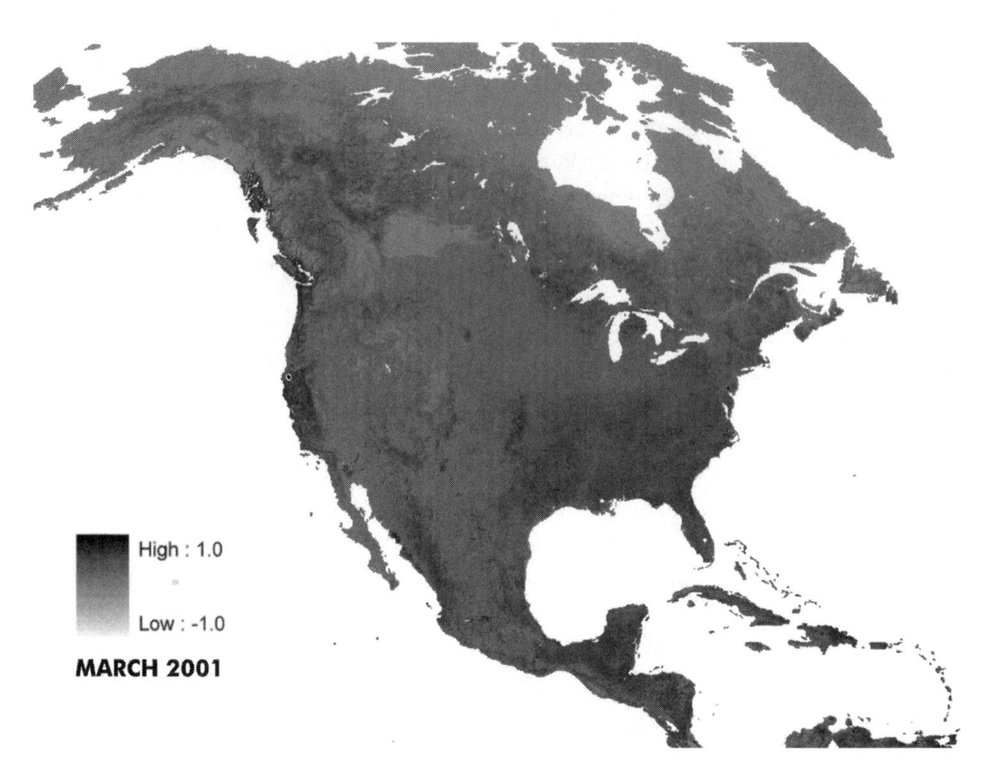


**FIGURE 17.15.** AVHRR image for North America, March 2001. This image shows the values of the NDVI calculated from the red and near infrared bands of the AVHRR (Table 17.8). The key shows that the highest values of the NDVI are shown as dark tones; the lowest values are bare earth, snow, ice, or sparse vegetation. This image shows much of the continent to be covered by dormant vegetation, and the far north by snow and ice. Areas along the Gulf Coast and the central valley of California and the tropical forests of Mexico, Central America, and the Caribbean Islands are covered by green or evergreen vegetation. The western deserts and midcontinental grasslands are shown to be sparsely vegetated. From National Space Science Data Center.

## 17.10. Broad-Scale Phenological Survey

Broad-scale remote sensing systems have been applied to the study of phenological variations over broad regions. NDVI, as observed by satellite sensors, permits tracking the seasonal rise and decline of photosynthetic activity within large regions, over several growing seasons. Reed et al. (1994) were among the first to systematically apply broad- scale remote sensing instruments to derive phenology for very large regions. They accumulated cloud-free AVHRR composites over a 4 year period, calculating NDVI for each pixel. Examination of NDVI over such long periods and such large areas permitted inspection of phenological patterns of varied land-cover classes and their responses to climatic and meteorological events, including drought, floods, and freezes. They used AVHRR 14-day composites, basically collected twice a month during the March–October growing season and once a month during the winter season. Such products are useful for monitoring agricultural regions and assessing climatic impact upon ecosystems—for example, the production of food and cover for wildlife. Agricultural systems can be observed at a broad scale to assess crops' response to meteorological and climatic variations throughout the growing season, forecast yields. Subsequent applications of the same

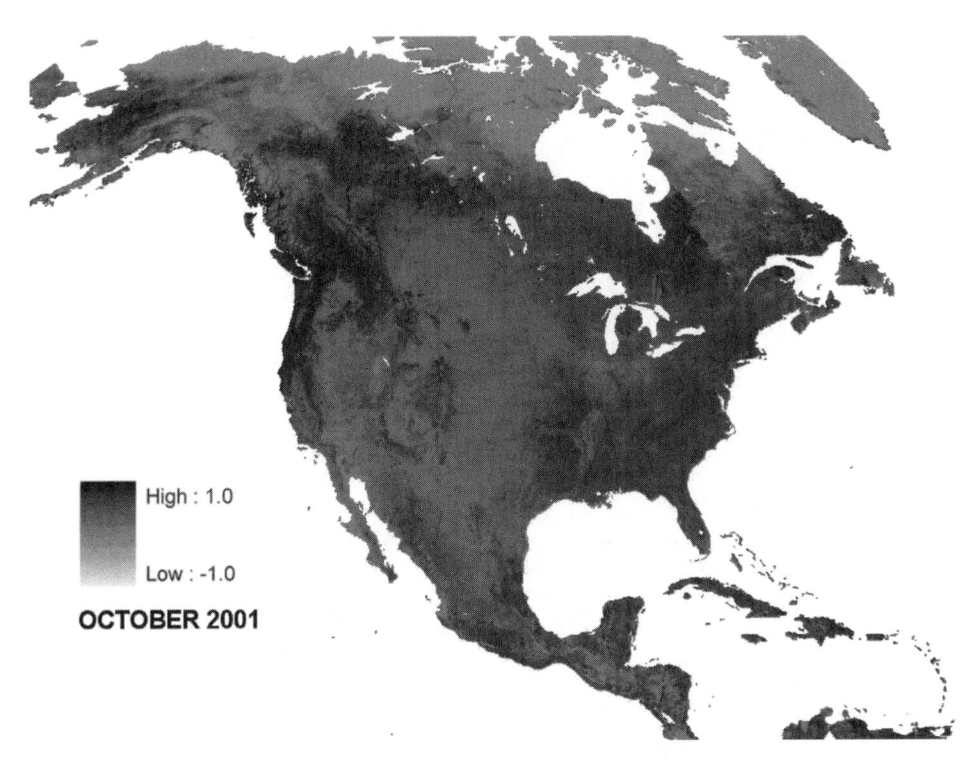


**FIGURE 17.16.** AVHRR image for North America, October 2001. Green vegetation has emerged through most of the humid climates of the continent. At this date, the central grasslands, western deserts, and northern barren land are represented at low NDVI values. From National Space Science Data Center.

approach, some using MODIS data (Chapters 7 and 21) to provide additional spatial detail within the framework of broad coverage. (Stockli and Vidale, 2004; Kanga et al., 2003)

There are some precautions necessary in making detailed interpretations of such data. NDVI is, of course, subject to atmospheric effects, and the coarse resolution of AVHRR data encompasses varied land cover classes (Chapter 10) with varied phenological responses, especially in regions where landscapes are locally complex and diverse. Atmospheric contamination creates noise, so it is necessary to use a moving average to filter out noisy effects of atmospheric noise, snow cover, and the like upon NDVI values (Figure 17.17).

Observed over time, values of the NDVI depict a seasonal pattern of increase and decrease. Values rise as spring approaches, marked by the onset of longer days, warmer temperatures, and spring rains. The seasonal biological response in NDVI is emergence of buds, and then leaves, is known as *green-up.* Eventually green-up peaks. Mature plants invest energy in growing the seeds and fruits necessary to prepare the species for success in the next year's growing season—green-up slows, stops, and then declines. The decline from mid- to late summer, known as *brown-down,* records the decline in vigor as senescence, and harvest time, approaches. The interval between the onset of green-up and the end of brown-down forms one definition of a region's growing season. Thus, these data can provide a means of assessing place-to-place variation in the length of growing seasons. The result produces a typical seasonal pattern as depicted in Figures 17.17 and 17.18.

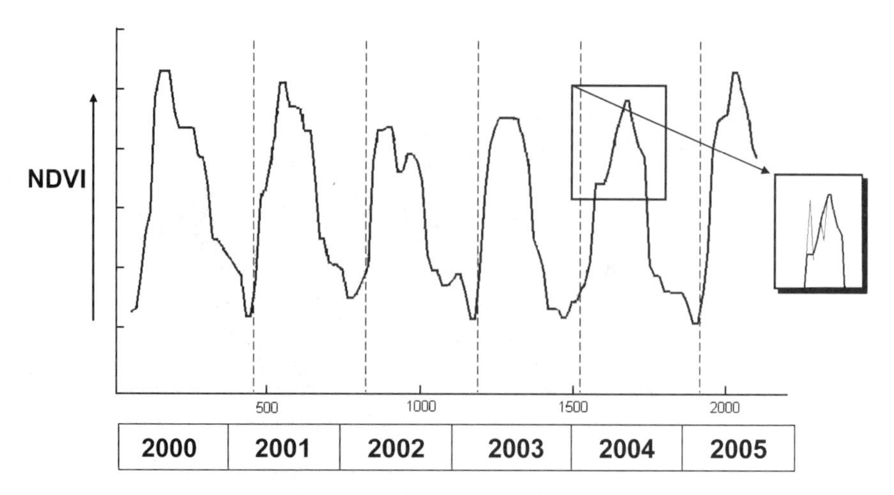


**FIGURE 17.17.** Seasonal phenological variation, derived from NDVI for a single MODIS pixel tracked over the interval 2000–2005. The pattern shows the seasonal variation in NDVI in response to phonological behavior of the plant cover within the pixel. Because of variations in weather and atmosphere, the NDVI values can vary, so the values are statistically filtered and smoothed to improve the continuity of the record. The inset shows a portion of the original data, before filtering, superimposed over the filtered curve. From Aaron Dalton and Suming Jin, VirginiaView.

## 17.11. Separating Soil Reflectance from Vegetation Reflectance

Digital data collected by satellites, especially at the coarse resolutions of the Landsat MSS and AVHRR, mix vegetation reflectance with that of the soil surface. Some mixing of soil and vegetation reflectances occurs because of mixed pixels (Chapter 10). In agricultural scenes, however, reflections from individual plants or individual rows of plants are closely intermingled with the bare soil between plants and between rows of plants, so that reflectances are mixed even at the finest resolutions. Of course, this mixing is especially important after plants have emerged because large proportions of soil are exposed to the sensor. But even after leaves have fully emerged, soil can still contribute to reflectance because of penetration of some wavelengths through the vegetation canopy. Mixing of soil and vegetation reflectances can therefore be a serious barrier to implementation of the concepts presented earlier in this chapter. Hutchinson (1982) found that vegetation information is difficult to extract when vegetation cover is less than 30%. In arid and semiarid regions, vegetative cover may always be this low, but even in agricultural settings plant cover is low when crops are emerging early in the growing season.

If many spectral observations are made of surfaces of bare soil, the values of red and near infrared brightnesses resemble those along axis *ABC* in Figure 17.19. The soil brightnesses tend to fall on a straight line—as a soil becomes brighter in the near infrared, so does it tend to get brighter in the red. Dry soils tend to be bright in both spectral regions and appear at the high end of the line (C); wet soils tend to be dark, and are positioned at the low end (B).

Richardson and Wiegand (1977) defined this relationship, known as the *soil brightness line,* and recognized that the spectral response of living vegetation will always have a consistent relationship to the line. Soils typically have high or modest response in the red and infrared regions, whereas living vegetation must display low values in the red (due to the absorption spectra of

**FIGURE 17.18.** Phenological map of Virginia and neighboring areas compiled from MODIS data. From Peter Sfroza, VirginiaView. Darker pixels represent areas of increasing greeness, here principally near the lower coastal plain and Piedmont, and the Central Valley.

chlorophyll) and high values in the near infrared (due to IR brightness of mesophyll tissue). Thus points representing "pure" vegetation response will be positioned in the upper left of Figure 17.19, where values on the red axis are low and those on the IR axis are high. Furthermore, Richardson and Wiegand defined an index to portray the relative magnitudes of soil background and vegetative cover to a given spectral response. Thus point *X* typifies a "pure" vegetation pixel, with a spectral response determined by vegetation alone, with no spectral contribution from soil. In contrast, point *Y* typifies a response from a partially vegetated pixel—it is brighter in the red and darker in the near infrared than is *X*. Richardson and Wiegand quantified this difference by defining the *perpendicular vegetation index* (PVI) as a measure of the distance of a pixel (in

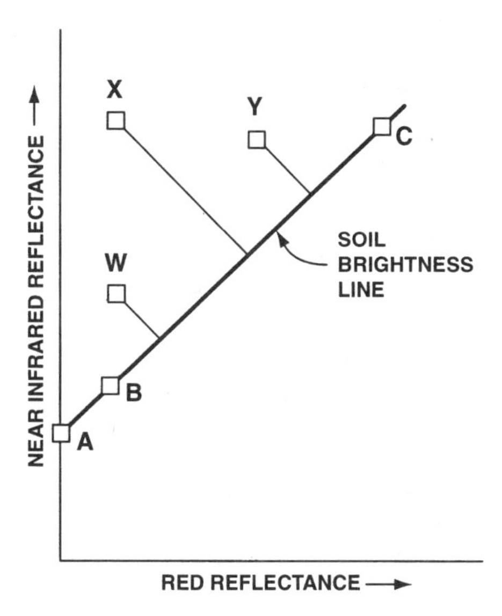


**FIGURE 17.19.** Perpendicular vegetation index. Based on Richardson and Weigand (1977, p. 1547). Copyright 1977 by the American Society for Photogrammetry and Remote Sensing. Reproduced with permission.

spectral data space) from the soil brightness line. The PVI is simply a Euclidean distance measure similar to those discussed in Chapter 12:

$$\mathrm{PVI} = \sqrt{(S_R - V_R)^2_{IR} + (S_R - V_{IR})^2} \qquad \text{(Eq. 17.3)}$$

where $S$ is the soil reflectance, $V$ is the vegetation reflectance, subscript R represents red radiation, and subscript IR represents near infrared radiation.

In practice, the analyst must identify pixels known to be composed of bare soil to identify the local soil brightness line and pixels known to be fully covered by vegetation to identify the local value for full vegetative cover (point $X$ in Figure 17.19). Then, intermediate values of PVI indicate the contributions of soil and vegetation to the spectral response.

Baret et al. (1993) investigated the soil brightness line from both experimental and theoretical perspectives. They concluded that the theoretical basis for the soil brightness line is sound, although the effects of variations in soil moisture and surface roughness are still not well understood. Their experimental data revealed that it is not feasible to define a single, universally applicable soil brightness line, as local variations in soil types lead to spectral variations. However, these variations were found to be minor in the red and infrared regions. They concluded that the use of a single soil brightness line is a reasonable approximation, especially in the context of analysis of course-resolution satellite data.

## 17.12. Tasseled Cap Transformation

The "tasseled cap" transformation (Kauth and Thomas, 1976) is a linear transformation of Landsat MSS data that projects soil and vegetation information into a single plane in multispectral data space—a plane in which the major spectral components of an agricultural scene are displayed in two dimensions. Although defined initially for MSS data, subsequent research (Crist and Cicone, 1984a, 1984b) has extended the concept to the six nonthermal bands of the TM. The transformation can be visualized as a rotation of a solid multidimensional figure (representing all spectral bands) in a manner that permits the analyst to view the major spectral components of an agricultural scene as a two-dimensional figure.

The transformation consists of linear combinations of the original spectral channels to produce a set of four new variables, each describing a specific dimension of the agricultural scene. The following equations present the transformation for the Landsat 2 MSS (Kauth and Thomas, 1976), and are offered here as a concise example.

$$\text{TC 1} = +0.433 \text{ MSS 4} + 0.632 \text{ MSS 5} + 0.586 \text{ MSS 6} + 0.264 \text{ MSS 7} \qquad \text{(Eq. 17.4)}$$

$$\text{TC 2} = -0.290 \text{ MSS 4} - 0.562 \text{ MSS 5} + 0.600 \text{ MSS 6} + 0.491 \text{ MSS 7} \qquad \text{(Eq. 17.5)}$$

$$\text{TC 3} = -0.829 \text{ MSS 4} + 0.522 \text{ MSS 5} - 0.039 \text{ MSS 6} + 0.194 \text{ MSS 7} \qquad \text{(Eq. 17.6)}$$

$$\text{TC 4} = +0.223 \text{ MSS 4} + 0.012 \text{ MSS 5} - 0.543 \text{ MSS 6} + 0.810 \text{ MSS 7} \qquad \text{(Eq. 17.7)}$$

(The equations use the Landsat 2 designations for MSS channels, MSS 4 = MMS 1, MMS 5 = MSS 2, MSS 6 = MSS 3, MSS 7 = MSS 4.) Here the new bands are designated as "TC1" for "tasseled cap band 1," and so on. The coefficients are calculated by means of an iterative procedure (Jackson, 1983). The transformation is based upon the calibration information for each specific sensor, so it requires a dedicated effort to derive the values for a specific sensor. Once the values are available, they can be applied to imagery from that sensor—some are reported in the scientific literature, and some are incorporated into imaging-processing packages.

Although these four new tasseleled cap bands do not match directly to observable spectral bands, they do carry specific information concerning the agricultural scenes. Kauth and Thomas interpret TC1 as *brightness,* a weighted sum of all four bands. TC2 is designated as *greenness,* a band that conveys information concerning abundance and vigor of living vegetation. It can be considered analogous to the PVI mentioned above. TC3 has been designated as *yellowness,* derived from the contrast between the red and green bands; it is an axis oriented at right angles to the other two. Finally, TC4 is referred to as *nonesuch* because it cannot be clearly matched to observable landscape features and is likely to carry system noise and atmospheric information.

The first two bands (TC1, brightness, and TC2, greenness) usually convey almost all the information in an agricultural scene—often 95% or more. Therefore, the essential components of an agricultural landscape are conveyed by a two-dimensional diagram, using TC1 and TC2 (Figure 17.20), which is in many respects similar to that defined by Richardson and Wiegand (Figure 17.19).

Over the interval of an entire growing season, TC1 and TC2 values for a specific field follow a stereotyped trajectory (Figure 17.20). Initially, the spectral response of a field is dominated by soil, as the field is plowed, disked, and planted (point 1 in Figure 17.20). The field has a position near the soil brightness line. As the crop emerges (2) and grows, it simultaneously increases in greenness and decreases in soil brightness as the green canopy covers more and more of the soil surface (2 to 3). Then as senescence, maturity, and harvest occur, the field decreases in greenness and increases in soil brightness to return the field back near its original position on the diagram (3 to 4).

A plot of data for an entire growing season, or for an image that shows many fields at various points in the crop cycle, has a distinctive shape similar to that shown in Figure 17.21; the resemblance of this shape to a tasseled ("Santa Claus-type") cap provides the basis for the name that Kauth and Thomas gave to their technique. The analogy holds as the data are examined in other TC dimensions (Figure 17.21b). Crist and Cicone (1984a) show that the intermediate values are chiefly vegetated pixels in which plants are extending their canopy cover as they mature. Pixels

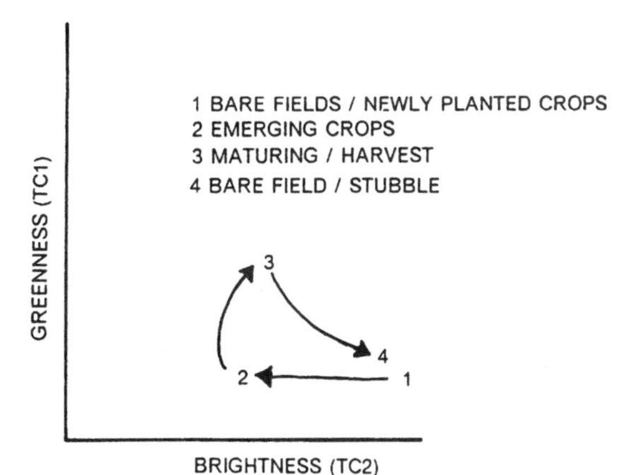


**FIGURE 17.20.** Seasonal variation of a field in data space defined by the greenness and brightness axes. From Crist (1983). Copyright 1983 by Purdue Research Foundation, West Lafayette, IN 47907. Reproduced with permission.

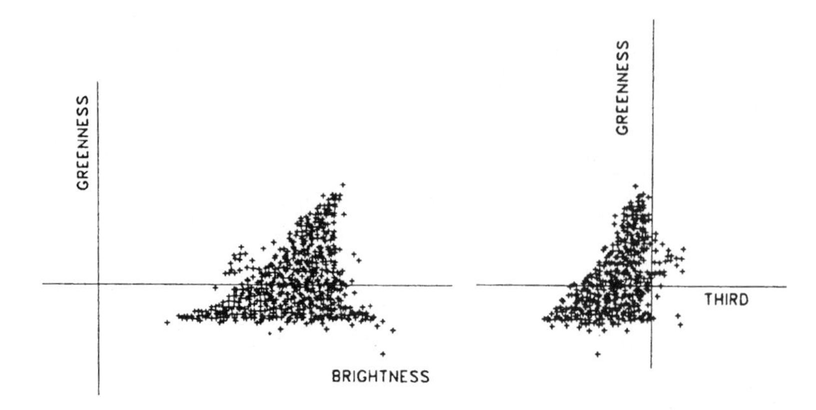


**FIGURE 17.21.** Tasseled cap viewed in multidimensional data space. Spectral values from an agricultural scene have been plotted on TC1 and TC2 axes (left), and TC2 and TC3 axes (right). These data are simulated TM data for agricultural regions in the midwestern United States. From Crist and Cicone (1984b, p. 347). Copyright 1984 by the American Society for Photogrammetry and Remote Sensing. Reproduced with permission.

representing senescent plant cover retain high greenness values until the canopy cover is so low that bare soil is again exposed.

Crist and Cicone (1984b) extended the transformation to the TM. Because the TM employs spectral definitions that differ from the MSS, they report a different set of coefficients. The resulting transformation is analogous (but not identical) to that reported above for the MSS, but differs in detail. For the TM, as for the MSS, TC1 corresponds to brightness, and TC2 to greenness. However, TC3, weighted by the TM's mid-infrared bands, portrays *wetness,* based upon soil moisture and moisture in plant tissues.

Figure 17.22 illustrates the application of the TM tasseled cap model to TM imagery of an irrigated agricultural region in California, showing the original image, TC1 (brightness), TC2 (greenness), and TC3 (wetness). TC4 and TC6, although they do not have assigned interpretations, are shown here to illustrate the declining information content of the higher TC bands.

Unlike the principal components transformation described in Chapter 11, which must be calculated individually for each new image to be examined, and reinterpreted for each individual image, the tasseled cap once calculated for a specific sensor, applies to all images acquired by that sensor. Therefore the meanings of the TC components is consistent across images. As a result, the tasselled cap transformation has been used for understanding not only images of agricultural scenes, but also a variety of ecological settings.

## 17.13. Foliar Chemistry

The development of operational hyperspectral sensors (Chapter 15) has opened possibilities for application of remote sensing to observation of characteristics of vegetated surfaces not feasible using instruments previously at hand. Hyperspectral data have created the capability to investigate *foliar chemistry,* the chemical composition of living leaves. If such a capability could be applied in an operational context, it could assist in the identification of plants at levels of detail not previously feasible and in the monitoring of the growth and health of crops and forests.

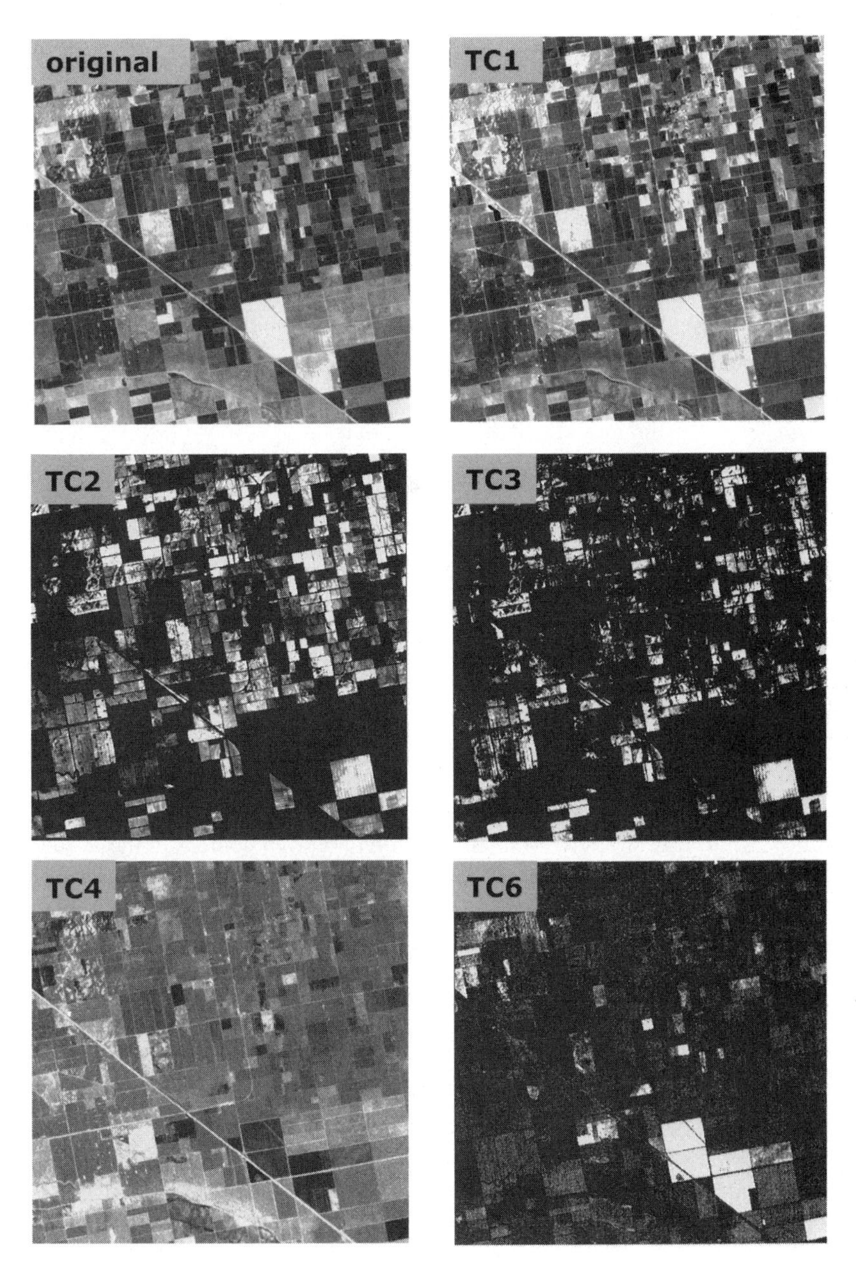


**FIGURE 17.22.** The tasseled cap transformation applied to a TM image of an irrigated agricultural landscape in California. Top left shows visible bands from the orginal image. Top right, tasselled cap band 1, brightness, then in sequence, TC2 (greenness), TC3 (wetness), and TC4 and TC6, which are not named, but show decreasing levels of information relative to TC1, TC2, and TC3.

Although there is a considerable body of research devoted to remote sensing of foliar chemistry, much of this work is specific to particular sensors, individual test sites, or specific processing algorithms, and has been acquired by varied means (laboratory analyses, field observations, and airborne data). Therefore, much work remains to be completed to develop the suggestions offered by existing studies.

The following paragraphs outline some of the efforts investigating hyperspectral sensing of foliar chemistry.

- *Lignin and starch content.* Kokaly and Clark (1999) applied a normalization procedure that permitted detection of lignin and cellulose in dried ground leaves in a laboratory setting. Peterson et al. (1988) found that the spectral region in the range 1,500 nm–1,750 nm in AIS data was linked to levels of lignin and starch content. Martin and Aber (1997) reported positive results for predicting lignin using four AVIRIS bands in the range 1,660 nm –2,280 nm, at test sites in the northeastern United States.
- *Nitrogen.* Card et al. (1988) found promising relationships between spectroradiometer data and nitrogen content of ground dried laboratory samples. Wessman et al. (1989) investigated forest canopy chemistry using AIS imagery; they found strong relationships between foliar nitrogen and reflectance in bands between 1,265 nm and 1,555 nm. Martin and Aber (1997) examined AVIRIS data (20-m resolution) of a mixed broadleaf forest of oak (*Quercus rubra*) maple (*Acer rubrum*), and needle-leaved species; they found a relationship between nitrogen content and brightness in the intervals 750 nm and 2140 nm, and in the intervals 950 and 2290 nm.
- *Chlorophyll concentration.* Kupiec and Curran (1993) examined AVIRIS data to find that brightness at 723 nm was strongly correlated with chlorophyll content in the needles of slash pine (*Pinus elliotti*).

## 17.14. Lidar Data for Forest Inventory and Structure

Chapter 8 introduced the application of laser altimetry to acquisition of data characterizing vegetation structure. An example is the SLICER (Scanning Lidar Imager of Canopies by Echo Recovery) program (Blair et al., 1994; Harding et al., 1994) that applies lidar altimetry to provide detailed information about the vertical structure of vegetation canopies—information that is essential for understanding the function of ecosystems, but very difficult to obtain by the usual field measurements. Details are avaialble at:

*http://ltpwww.gsfc.nasa.gov/eib/projects/airborne_lidar/slicer.html*
*http://www.fsl.orst.edu/~lefsky/slicerpage.html*

The SLICER instrument uses radiation in the near infrared (1.06 μm) to illuminate regions 10–15 m in diameter. Some of the radiation is reflected from the canopy, and some reaches the ground through gaps in the canopy. The complete returned laser signal is examined to reveal the vertical distribution of the backscatter of laser illumination from all canopy elements (foliar and woody) and the ground reflection. The relatively large footprints (5–15 m) are designed to simultaneously recover returns from the top of the canopy and from the ground surface, yet record details of the structure of individual crowns. The time difference between initial and last returns provides the basis for estimation of overall canopy height (Nelson et al., 1988a, 1988b). Figure 17.23 presents the results of the SLICER system flown over a 1k-m section of deciduous forest. The base of each line represents the varying topography at ground level—the light fringe depicts the height of the tallest trees.

Because lidar data directly record data characterizing the physical structure of a forest (heights, crown closure, crown size, etc.), they provide an opportunity to directly assess the three-dimensional structure of vegetation formations in ways that are not possible with other sensors. For example, Means et al. (2000) examined lidar data of Douglas fir (*Pseuotsuga taxi folia*) stands in the western Cascades of Oregon using airborne lidar with a footprint of 0.2 m to 0.9 m. They found that lidar data were effective in estimating basal area, height, volume, and canopy closure for a range of growth states, and they concluded that lidar survey should be an effective inventory tool for forest management.

## 17.15. Precision Agriculture

Traditional agricultural practices treat entire fields as single management units. A farmer applies seed, fertilizer, and pesticides at the same rate for an entire field. This practice prevails despite inherent variation in natural agronomic components, including soil, solar radiation, moisture, and drainage, which are known to vary markedly within fields. Although farmers understand well that such characteristics vary considerably even within short distances, tradi-

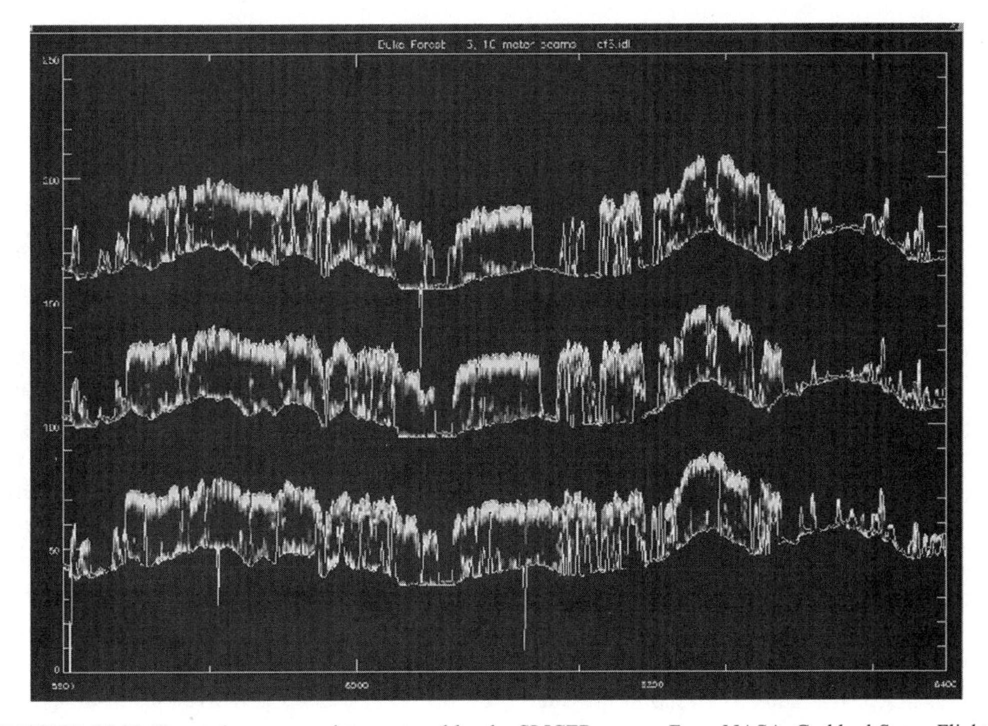


**FIGURE 17.23.** Vegetation canopy data captured by the SLICER sensor. From NASA–Goddard Space Flight Center.

tionally they have lacked both the information they need to understand the details and the means to implement any such understanding.

*Precision agriculture* is based on the premise that farmers can control agricultural management practices (e.g., seeding rate, fertilization, tillage, pesticides, irrigation) at varying rates within fields to optimize production. Thus fertilizer, for example, would be applied at heavier rates where soil conditions require intense fertilization, but at lower rates elsewhere within the same field, thereby economizing on fuel, materials, and labor, and minimizing unintended environmental effects.

Such measures are possible because of the increased precision and detail of information that can be acquired in part from remotely sensed data and its ability to monitor changes throughout the growing season, as moisture conditions, insect infestations, and the like vary within fields, and over time, as the growing season progresses. The practice of precision agriculture is based on use of agricultural equipment with GPS to determine accurate position, multispectral imagery to provide current information on crop conditions, and agricultural machinery designed to employ varying rates of application and even to sense field conditions on the spot.

Remote sensing forms an important component of precision agriculture, first because it provides a map-like representation of agricultural landscapes, and second, because it can provide information pertaining to growth and health of crops throughout the growing season. Remote sensing can reveal variations in surface soil texture (Post et al., 1988), guiding collection of soil samples required to support the precision agriculture effort. Remotely sensed images can detect anomalies, revealing, for example, localized effects of drought or insect infestations. In some instances, data from calibrated instruments can provide direct measurement of physical values, such as brightness in specific portions of the spectrum.

## 17.16. Remote Sensing for Plant Pathology

### *The Disease Triangle*

*Plant pathology* is the study of plant diseases and their occurrence. *Plant pathologists* study the biochemistry and biology of pathogens and the plants they infect to understand the life cycle of the pathogens and the means by which they infect hosts. Plant pathologists counter the effects of pathogens by proposing management techniques, designing pesticides, and recommending countermeasures to minimize effects of the disease, and to prevent, or slow, its spread to other regions.

A few well-known examples can remind us of the significance of plant diseases. During the years 1846–1850, the Irish Potato famine killed at least 1 million Irish citizens, caused immense social disruption in Ireland, and led to massive emigration that itself had a major influence on social systems in the United States and elsewhere. In the United States, Dutch Elm disease and the Chestnut blight have destroyed trees that once constituted North America's most distinctive and beautiful trees. In specific, the decline of the American Chestnut tree changed the ecology of large regions of Appalachian forests. The tree once constituted perhaps one-fourth of the trees of Appalachians forests, and provided food for both humans and wildlife. Both species exist now only as isolated remnants. Often such catastrophes are connected closely to human behavior, especially through the introduction of plants and diseases from other regions; the tendency to depend upon single crops, usually densely planted; and the uses of genetically uniform hybrids that may increase susceptibility to attacks by pathogens.

Infections depend upon the cooccurrence of three interrelated factors known as the *disease triangle*:

1. A pathogen (bacteria, viruses, fungi, mycoplasmas, nematodes)
2. A susceptible host plant (a plant's resistance to disease)
3. Favorable environmental conditions (including temperature, wind, solar illumination, soil)

Pathogens spread through the environment (via wind, water, and soil), and by biologic transmission, including insects, animals, and humans.

Remote sensing offers many ways to supplement and reenforce the efforts of the detailed, ground-based expertise of the plant pathologist. Applications of remote sensing employ the plant pathologist's understanding of the dynamics of an infestation in its spatial context, at the landscape level, and the use of ground-based surveys to understand the behavior of the pathogen at the level of the plant and the cell.

Remote sensing offers the ability to conduct systematic surveys to detect mortality associated with diseases, or declines in vegetation health and document spread from one region to another and the effectiveness of countermeasures. The contributions of remotely sensed imagery can be summarized concisely:

- Detection of infestation symptoms from aerial imagery
- Detection of host plants, identification of crop residue and volunteer growth that can harbor pathogens
- Phenology—understanding the onset of seasonal growth in hosts or vectors
- Assessment of environmental conditions favorable for the pathogen
- Spatial perspective on the occurrence and spread of the disease, and monitoring of the spread of the disease

A few case studies can highlight some of the issues pertaining to the role of remote sensing in assessing and monitoring the occurrence of plant diseases.

### *Corn Blight Watch Experiment, 1971*

One of the earliest applications of remote sensing to a broad-scale infestation occurred during the 1971 Corn Blight Watch Experiment. In previous years, corn breeders had introduced varieties of maize that had been optimized for hybridization, but were vulnerable to specific varieties of the fungus that causes southern corn leaf blight (SCLB). (*Bipolaris maydis*). SCLB exists in nature, but mainly as a minor disease of corn. The fungus produces elongated brown spots on leaves and stalks during the interval mid-June to mid-October. If the disease attacks before the ears have formed, crop losses can be severe. The fungus resides in soil and in crop debris left after harvest, and survives the winter season to infest the crops of the succeeding year. SCLB thrives in warm, moist weather, so it spreads easily in the summer weather of central North America. Its occurrence is very difficult to control.

Once the vulnerable hybrid was widely used, SCLB rose from a minor nuisance to become a major threat to U.S. agriculture. The 1970 corn crop was reduced by 15%, creating havoc in commodities markets and political pressures for agricultural officials. By the early months of 1971, seed carrying the genetic weakness had already been widely distributed to farmers throughout the continent. As the 1971 growing season began, agricultural scientists became increasingly concerned about the vulnerability of the North American corn crop. Because of the widespread use of the vulnerable seed, and the dense occurrence of maize cultivation throughout central North America, the weather favoring spread of SCLB could promote rapid spread of the blight and destruction of the nation's corn production, with ominous implications for the agricultural economy and a major portion of the nation's food production.

The Corn Blight Watch Experiment was based upon uses of remotely sensed data to monitor the occurrence and spread of SCLB during the 1971 growing season. A series of flightlines were designed to assess the spread of the infection throughout the growing season. Biweekly flights collected CIR photography, as well as multispectral scanner data, for the principal corn-growing regions within the United States and Canada (Figure 17.24). Ground data were collected in coordination with the collection of aerial imagery. The sampling design allowed ground data to be extended to broad-scale estimates for the principal corn-producing regions of North America.

Photointerpretation estimates for the Corn Blight Watch Experiment varied in accuracy when compared against the data collected at ground level. Photointerpretation was effective in distinguishing between healthy fields and severely infected fields, but was not effective in assessing slight or mild levels of infection. Early in the growing season, there was a large discrepancy between the estimates of crop damage based upon ground-based observations and those derived from photointerpretation (Figure 17.25). Some difficulties were experienced in assessing the effects of blight when other causes influenced the health of the corn crop.

The Corn Blight Watch Experiment is significant today because it is the only example of a broad-scale attempt to use remote sensing as a tool to contribute to the understating of the spatial behavior of a plant pathogen. As globalization and concerns about bioterrorism raise awareness about the possible impacts of introduced pathogens, this experiment from several decades past may deserve reexamination.

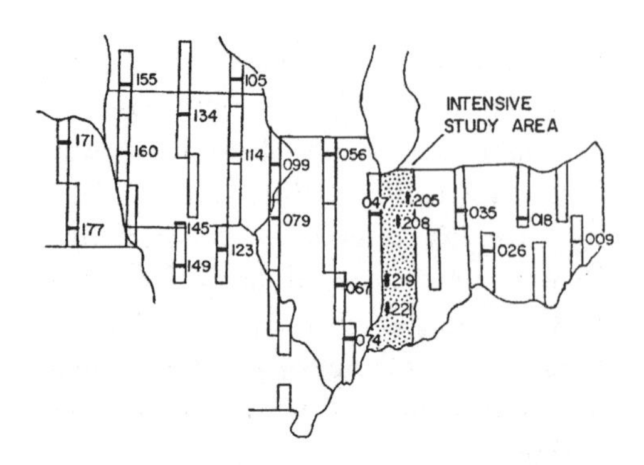


**FIGURE 17.24.** Flight lines for the 1971 Corn Blight Watch Experiment From Sharples (1973). Reproduced by permission of Laboratory for Applications of Remote Sensing, Purdue University.

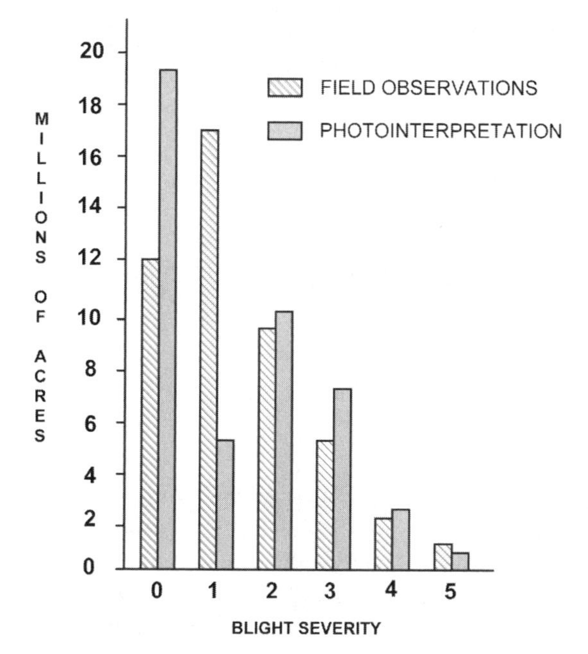


**FIGURE 17.25.** Photointerpretation accuracy for the Corn Blight Watch Experiment. From Johannsen et al. (1972). Reproduced by permission of Laboratory for Applications of Remote Sensing, Purdue University.

### *Sudden Oak Death Syndrome*

A newly defined pathogen (*Phytophthora ramorum,* which causes sudden oak death syndrome, or SODS) has been killing a variety of oak species in coastal California and Oregon since the mid-1990s. Previously, this fungus had been known principally as an infection of European rhododendron. In eastern North America, forest species known or suspected to be hosts include red oak (*Quercus rubra*) and mountain laurel (*Kalmia latifolia*). Studies completed by the USDA Forest Service estimate that over half of the forests in the eastern United States contain some susceptible tree hosts. The greatest impact would be on forests in areas that contain large percentages of red or live oak species, including the Ozark–Ouachita Highlands, the pin oak sand flats in the Lake states, and live oak areas in Florida. Several other species, including some ornamental plants grown in nurseries, are known to be carriers of the infection, although it is not now clear if they are harmed by the disease. SODS is known to spread through shipment of infected nursery plants.

The most obvious evidence of infection on trees is the presence of cankers that ooze sap through the bark on trunks or large limbs of infected trees. Later the infection will cause the trees to shed bark, often girdling trunks, and causing crowns to suddenly turn brown. Trees die within a few months to several years after infection is evident. Infected trees are often subject to attack by insects and other pathogens, which may themselves be the immediate cause of death.

Kelly et al. (2004) report their use of remotely sensed data to assess and monitor the occurrence of SODS. They also established a convenient system to encourage county-level workers to use GPS receivers to record the locations of infected trees, and to report them to a web site. Reporting is coordinated with a training program for field samplers, and the submission of samples to a laboratory for analysis. The web-based system permits workers in local governmental offices, as well as individual citizens, to submit timely reports of the occurrence and spread of SODS. These data are used to guide the planning for remotely sensed data that contribute additional information concerning SODS.

*http://nature.berkeley.edu/comtf/*

Their system forms a model of how remotely sensed data, geospatial data in general, and the Internet can contribute to a framework that permits integration of a landscape-level perspective with plant-level data.

## 17.17. Summary

For a large proportion of the Earth, vegetation cover forms the surface observed by remote sensing instruments. In some instances, we have a direct interest in examining this vegetative blanket, to map the patterns of different forests, rangelands, and agricultural production. In other instances, we must use vegetation as a means of understanding those patterns that may lie hidden beneath the plant cover. In either instance, it is essential that we be able to observe and understand the information conveyed by the vegetated surface.

Given the obvious significance of agriculture and forestry, and the immense difficulties of monitoring use of these resources, the techniques outlined in this chapter are likely to form one of the most significant contributions of remote sensing to the well-being of mankind. The ability to examine vegetation patterns using the VIs described here, combined with the synoptic view and repetitive coverage of satellite sensors, provides an opportunity to survey agricultural patterns in a manner that was not possible even a few years ago. If the information gathered by remote sensing systems can be integrated into the decision-making and transportation/distribution systems so important for agricultural production, there is a prospect for improvements in the effectiveness of food production.

Chapter 18 demonstrates how some of the information introduced here has significance in geological remote sensing. Full pursuit of extraction of geological information from an image requires intimate knowledge of how the spectral behavior of living plants responds to variations in the geologic substratum—a good example of how the fabric of remote sensing is so closely woven that it is often not possible to isolate individual components as separate units.

## Review Questions

1. Summarize differences between classification of vegetation from ground observations and classification from aerial images. Consider such factors as (a) the basis of the classification and (b) the units classified. Identify distinctions for which remote sensing is especially well suited, and those for which it is not likely to be useful.
2. Remote sensing to monitor crop development is much more difficult than it might appear initially. List some of the practical problems you might encounter as you plan an experiment to use satellite data to study the development of the winter wheat crop in western Kansas (or another crop and crop region specified by your instructor).

3. Return to Tables 12.2 and 12.3 and to the category labels tabulated in the review questions to Chapter 12. Use at least two of the vegetation indices discussed in this chapter (as specified by your instructor) to assess the forested, pasture, and crop pixels. Examine differences between classes and between dates. Do some indices seem more effective than others?
4. To apply some of the knowledge presented in this chapter (e.g., the red shift), it is necessary to have data with very fine spectral, radiometric, and spatial detail. From your knowledge of remote sensing, discuss how this requirement presents difficulties for operational applications.
5. List some of the reasons why an understanding of the spectral behavior of an individual plant leaf is not itself sufficient to conduct remote sensing of vegetation patterns.
6. List some reasons why multispectral satellite images might be especially well suited for observation of vegetation patterns. Or, if you prefer, list some reasons why it is not quite so useful for vegetation studies. Briefly compare the MSS, TM, and SPOT with respect to utility for vegetation studies.
7. Write a short description of a design for a multispectral sensor tailored specifically for recording information about living vegetation and vegetation patterns, disregarding all other applications. Suggest optimum timing for a satellite to carry the sensor in a sun-synchronous orbit.
8. Describe some of the ways that image classification (Chapter 12) might be useful in the study of vegetation patterns. Identify also some of the limitations of such methods in the study of vegetation.
9. Study of vegetative patterns by remote sensing is greatly complicated by the effects of soil background. Summarize contributions of soil background to spectral responses of a variety of natural and man-made landscapes, including cultivated crops, dense forest, orchards, and pasture. How might effects differ as season changes?
10. How do changes in Sun angle and Sun azimuth (due to differences in season and time of day) influence the way in which vegetative patterns are recorded on remotely sensed images?
11. List some of the ways in which monthly phenologic information might be useful for study of (a) cropped agricultural landscapes and (b) rangeland or forested land.

## References

### *General*

Atherton, B. C., M. T. Morgan, S. A. Sheraer, T. S. Stombaugh, and A. D. Ward. 1999. Site-Specific Farming: A Perspective on Information Needs, Benefits, and Limitations. *Journal of Soil and Water Conservation,* Second Quarter, pp. 339–355.

Badhwar, G. D. 1980. Crop Emergence Date Determination from Spectral Data. *Photogrammetric Engineering and Remote Sensing,* Vol. 46, pp. 369–377.

Badhwar, G. D., J. G. Carnes, and W. W. Austin. 1982. Use of Landsat-Derived Temporal Profiles for Corn–Soybean Feature Extraction and Classification. *Remote Sensing of Environment,* Vol. 12, pp. 57–79.

Bailey, Robert G. 1976. *Ecoregions of the United States* (Map at 1:7,500,000). Ogden, UT: U.S. Forest Service.

Bailey, R. G. 1978. *Description of Ecoregions of the United States.* Ogden, UT: U.S. Forest Service, 77 pp.

Bailey, R. G. 1996. *Ecosystem Geography.* New York: Springer-Verlag, 204 pp.

Bailey, R. G. 1998a. *Ecoregions Map of North America.* (Misc Publication 1548). Washington, DC: U.S. Forest Service, 10 pp.

Bailey, R. G. 1998b. *Ecoregions: The Ecosystem Geography of the Oceans and the Continents.* New York: Springer-Verlag, 176 pp.

Baret, F., S. Jacquemond, and J. F. Hanocq. 1993. The Soil Line Concept in Remote Sensing. *Remote Sensing Reviews,* Vol. 7, pp. 65–82.

Blair, J. B., D. B. Coyle, J. L. Bufton and D. J. Harding. 1994. Optimization of an Airborne Laser Altimeter for Remote Sensing of Vegetation and Tree Canopies. *Proceedings IGARS 94,* Vol. 2, pp. 939–941.

Collins, W. 1978. Remote Sensing of Crop Type and Maturity. *Photogrammetric Engineering and Remote Sensing,* Vol. 44, pp. 43–55.

Condit, H. R. 1970. The Spectral Reflectance of American Soils. *Photogrammetric Engineering,* Vol. 36, pp. 955–966.

Eidenshink, J. C. 1992. The 1990 Conterminous U.S. AVHRR Data Set. *Photogrammetric Engineering and Remote Sensing,* Vol. 58, pp. 809–813.

Gausman, H. W. 1974. Leaf Reflectance of Near-Infrared. *Photogrammetric Engineering,* Vol. 40, pp. 183–191.

Harding, D. J., J. B. Blair, J. B. Garvin, and W. T. Lawrence. 1994. Laser Altimetry Waveform Measurement of Vegetation Canopy Stucture. *Proceedings IGARS '94,* Vol. 2, pp. 1251–1253.

Howard, J. A. 1970. *Aerial Photo-Ecology.* New York: Elsevier. 325 pp.

Jackson, R. D. 1983. Spectral Indices in N-Space. *Remote Sensing of Environment,* Vol. 13, pp. 409–421.

Jackson, R. D., P. J. Pinter, S. B. Idso, and R. J. Reginato. 1979. Wheat Spectral Reflectance: Interactions between Crop Configuration, Sun Elevation, and Azimuth Angle. *Applied Optics,* Vol. 18, pp. 3730–3732.

Jasinski, M. F., and P. S. Eagleson. 1990. Estimation of Subpixel Vegetation Cover Using Red-Infrared Scattergrams. *IEEE Transactions on Geoscience and Remote Sensing,* Vol. 28, pp. 253–267.

Jensen, J. R. 1978. Digital Land Cover Mapping Using Layered Classification Logic and Physical Composition Attributes. *American Cartographer,* Vol. 5, pp. 121–132.

Jensen, J. R. 1983. Biophysical Remote Sensing. *Annals of the Association of American Geographers,* Vol. 73, pp. 111–132.

Kanemasu, E. T. 1974. Seasonal Canopy Reflectance Patterns of Wheat, Sorghum, and Soybean. *Remote Sensing of Environment,* Vol. 3, pp. 43–57.

Kelly, M., D. Shaari, Q. Guo, and D. Liu. 2004. A Comparison of Standard and Hybrid Classifier Methods for Mapping Hardwood Mortality in Areas Affected by Sudden Oak Death. *Photogrammetric Engineering and Remote Sensing,* Vol. 70, pp. 1229–1239.

Kimes, D. S., W. W. Newcomb, R. F. Nelson, and J. B. Schutt. 1986. Directional Reflectance Distributions of a Hardwood and Pine Forest Canopy. *IEEE Transactions on Geoscience and Remote Sensing,* Vol. GE–24, pp. 281–293.

Kimes, D. S., J. A. Smith, and K. J. Ranson. 1980. Vegetation Reflectance Measurements as a Function of Solar Zenith Angle. *Photogrammetric Engineering and Remote Sensing,* Vol. 46, pp. 1563–1573.

Knipling, E. B. 1970. Physical and Physiological Basis for the Reflectance of Visible and Near Infrared Radiation from Vegetation. *Remote Sensing of Environment,* Vol. 1, pp. 155–159.

Küchler, A. W. 1967. *Vegetation Mapping.* New York: Ronald Press, 472 pp.

McClain, E. P. 1980. Environmental Satellites. Entry in *McGraw-Hill Encyclopedia of Environmental Science.* New York: McGraw-Hill, pp. 15–30.

Means, J. E., S. E. Acker, B. J. Fitt, M. Renslow, L. Emerson, and C. J. Hendrix. 2000. Predicting Forest

Stand Characteristics with Airborne Scanning Lidar. *Photogrammetric Engineering and Remote Sensing,* Vol. 66, pp. 1367–1371.

Meyers, V. I. 1975. Crops and Soils. Chapter 22 in *Manual of Remote Sensing* (R. Reeves, ed.). Falls Church, VA: American Society of Photogrammetry, pp. 1715–1816.

National Research Council. 1997. *Precision Agriculture in the 21st Century.* Washington, DC: National Academy Press, 149 pp.

Phillips, T. L., and Staff. 1972. *1971 Corn Blight Watch Experiment Data Processing, Analysis, and Interpretation* (LARS information Note 012272). West Lafayette, IN: Laboratory for Applications of Remote Sensing, Purdue University, 21 pp. *http://www.lars.purdue.edu/home/references/LTR_012272.pdf*

Post, D. F., C. Mack, P. D. Camp, and A. S. Suliman. 1988. Mapping and Characterization of the Soils on the University of Arizona Maricopa Agricultural Center. *Proceedings, Hydrology and Water Resources in Arizona and the Southwest,* Vol. 18. Tucson: Arizona–Nevada Academy of Science, pp. 49–60.

Ray, P. M. 1963. *The Living Plant.* New York: Holt, Rinehart & Winston, 127 pp.

Richardson, A. J., and C. L. Wiegand. 1977. Distinguishing Vegetation from Soil Background. *Photogrammetric Engineering and Remote Sensing,* Vol. 43, pp. 1541–1552.

Rouse, J. W., R. H. Haas, J. A. Schell, D. W. Deering, J. C. Sagan, O. B. Toon, and J. B. Pollock. 1979. Anthropogenic Changes and the Earth's Climate. *Science,* Vol. 206, pp. 1363–1368.

Tucker, C. J., H. H. Elgin, and J. E. McMurtrey. 1979. *Relationship of Red and Photographic Infrared Spectral Radiances to Alfalfa Biomass, Forage Water Content, Percentage Canopy Cover, and Severity of Drought Stress* (NASA Technical Memorandum 80272). Greenbelt, MD: Goddard Space Flight Center, 13 pp.

Tucker, C. J., J. R. G. Townshend, and T. E. Goff. 1985. African Land-Cover Classification Using Satellite Data. *Science,* Vol. 227, pp. 369–374.

Vanden Heuvel, R. M. 1996. The Promise of Precision Agriculture. *Journal of Soil and Water Conservation,* Vol. 51, pp. 38–40.

Wilson, R. C. 1960. Photo Interpretation in Forestry. Chapter 7 in *Manual of Photographic Interpretation* (R. N. Colwell, ed.). Falls Church, VA: American Society of Photogrammetry, pp. 457–520.

Yool, S. R., D. W. Eckhardt, J. L. Star, T. L. Becking, and J. E. Estes. 1985. Image Processing for Surveying Natural Vegetation: Possible Effects on Classification Accuracy. In Technical Papers, *American Society of Photogrammetry,* Falls Church, VA: American Society of Photogrammetry, pp. 595–603.

### *Phenology*

Dethier, B. E., M. D. Ashley, B. O. Blair, J. M. Caprio, and J. R. Rouse. 1973. *Phenology Satellite Experiment (detection of brown wave and green wave in north-south corridors of the United States).* Progress Report Contract NASS-21781 Ithaca NY: Cornell University.

Harlan, C. 1974. *Monitoring the Vernal Advancement and Retrogradation (Greenwave Effect) of Natural Vegetation* (Type III Final Report). Greenbelt, MD: Goddard Space Flight Center, 371 pp.

Hopkins, A. D. 1938. *Bioclimatics: A Science of Life and Climate Relations* (U.S. Department of Agriculture Miscellaneous Publication 36). Washington, DC: U.S. Department of Agriculture, 188 pp.

Kang, S., S. W. Running, J. Lim, M. Zhao, and C. Park. 2003. A Regional Phenology Model for Detecting Onset of Greenness in Temperate Mixed Forests, Korea: An Application of MODIS Leaf Area Index (LAI). *Remote Sensing of Environment,* Vol. 86, pp. 232.

Koriba, K. 1958. On the Periodicity of Tree-Growth in the Tropics, with Special Reference to the Mode of Branching, the Leaf Fall, and the Formation of the Resting Bud. *Gardens' Bulletin* (Singapore), Series 3, Vol. 17, pp. 11–81.

Loveland, T. R., J. W. Merchant, J. F. Brown, D. O. Ohlen, B. Read, and P. Olsen. 1995. Seasonal Land Cover Regions of the United States. *Annals of the Association of American Geographers,* Vol. 85, pp. 339–355.

Reed, B. C., J. Brown, D. VanderZee, T. Loveland, J. Merchant, and D. Ohlen. 1994a. Measuring Phenological Variability from Satellite Imagery. *Journal of Vegetation Science,* Vol. 3, pp. 703–714.

Reed, B. C., J. Brown, D. VanderZee, T. Loveland, J. Merchant, and D. Ohlen. 1994b. European Plant Phenology and Climate as Seen in a 20-year AVHRR Land Surface Parameter Dataset. *Journal of Vegetation Science,* Vol. 17, pp. 3303–3330.

Reed, B. C., and L. Yang. 1997. Seasonal Vegetation Characteristics of the United States. *GeoCarto International,* Vol. 12, No. 2, pp. 65–71.

Rouse, J. W., R. H. Haas, J. A. Schell, and D. W. Deering. 1973. Monitoring Vegetation Systems in the Great Plains with ERTS. In *Third ERTS Symposium* (NASA Special Publication SP–351I). pp. 309–317.

Schwartz, and B. C. Reed. 1999. Surface Phenology and Satellite Sensor-Derived Onset of Greenness: An Initial Comparison. *International Journal of Remote Sensing,* Vol. 17, pp. 3451–3457.

Stockli, R.,and P. L. Vidale. 2004. European plant phenology and climate as seen in a 20-year AVHRR Land-Surface Parameter Dataset. *International Journal of Remote Sensing.* Vol. 17, pp. 3303–3330.

Weber, K. T. 2001. A Method to Incorporate Phenology into Land Cover Change Analysis. *Journal of Range Management,* Vol. 54, pp. A1–A7.

Wilson, J. S., T. S. Brothers, and E. J. Marcano. 2001. Remote Sensing of Spatial and Temporal Vegetation Dynamics in Hispaniola: A Comparison of Haiti and the Dominican Republic. *GeoCarto International,* Vol. 16, No. 2, pp. 5–16.

Zhang, X., M. A. Friedl, C. B. Schaaf, A. H. Strahler, J. C. F. Hodges, F. Gao, B. C. Reed, and A. Huete. 2003. Monitoring Vegetation Phenology using MODIS. *Remote Sensing of Environment,* Vol. 84. pp. 471–475.

### *Vegetation Indices*

Clevers, J. G. P. W., and W. Verhoef. 1993. LAI Estimation by Means of the WDVI: A Sensitivity Analysis with a Combined PROSPECT-SAIL Model. *Remote Sensing Reviews,* Vol. 7, pp. 43–64.

Cohen, W. B. 1991. Response of Vegetation Indicies to Changes in Three Measures of Leaf Water Stress. *Photogrammetric Engineering and Remote Sensing,* Vol. 57, pp. 195–202.

Curran, P. 1980. Multispectral Remote Sensing of Vegetation Amount.*Progress in Physical Geography,* Vol. 4, pp. 315–341.

Hutchinson, C. 1982. Techniques for Combining Landsat and Ancillary Data for Digital Classification Improvement. *Photogrammetric Engineering and Remote Sensing,* Vol. 48, pp. 123–130.

Jackson, R. D., P. N. Slater, and P. J. Pinter. 1983. Discrimination of Growth and Water Stress in Wheat by Various Vegetation Indices through Clear and Turbid Atmospheres. *Remote Sensing of Environment,* Vol. 13, pp. 181–208.

Morain, S. A. 1978. *A Primer on Image Processing Techniques* (TAC TR #78-009). Albuquerque: Technology Applications Center, University of New Mexico, 54 pp.

Perry, C. R., and L. F. Lautenschlager. 1984. Functional Equivalence of Spectral Vegetation Indices. *Remote Sensing of Environment,* Vol. 14, pp. 169–182.

Price, J. C. 1987. Calibration of Satellite Radiometers and Comparison of Vegetation Indicies. *Remote Sensing of Environment,* Vol. 18, pp. 35–48

Tucker, C. J. 1979a. Red and Photographic Infrared Linear Combinations for Monitoring Vegetation. *Remote Sensing of Environment,* Vol. 8, pp. 127–150.

Tucker, C. J. 1979b. *Remote Sensing of Leaf Water Content in the Near Infrared* (NASA Technical Memorandum 80291). Greenbelt, MD: Goddard Space Flight Center, 17 pp.

### *Hyperspectral Sensing of Vegetation*

Card, D. H., D. L. Peterson, and P. A. Matson. 1988. Prediction of Leaf Chemistry by the Use of Visible and Near Infrared Reflectance Spectroscopy. *Remote Sensing of Environment,* Vol. 26, pp. 123–147.

Curran, P. J. 1989. Remote Sensing of Foliar Chemistry. *Remote Sensing of Environment,* Vol. 30, pp. 217–178.

Kokaly, R. F.; and R. N. Clarke. 1999. Spectroscopic Determination of Leaf Biochemistry Using Band-Depth Analysis of Absorption Features and Stepwise Multiple Linear Regression. *Remote Sensing of Environment,* Vol. 67, pp. 267–287.

Kupiec, J., and P. J. Curran. 1993. AVIRIS Spectra Correlated with the Chlorophyll Concentration of a Forest Canopy. In *Summaries of the Fourth Annual JPL Airborne Geoscience Workshop* (JPL Publication 93-26). Pasadena, CA: Jet Propulsion Laboratory, Vol. 1, pp. 105–108.

Martin, M. E., and J. D. Aber. 1997. High Spectral Resolution Remote Sensing of Forest Canopy Lignin, Nitrogen, and Ecosystem Processes. *Ecological Applications,* Vol. 7, pp. 431–443.

Peterson, D. L., J. D. Aber, P. A. Matson, D. H. Card, N. Swanberg, C. Wessman, and M. Spanner. 1988. Remote Sensing of Forest Canopy and Leaf Biochemistry Contents. *Remote Sensing of Environment,* Vol. 24, pp. 85–108.

Wessman, C. A., J. D. Aber, and D. L. Peterson. 1989. An Evaluation of Imaging Spectrometry for Estimating Forest Canopy Chemistry. *Remote Sensing of Environment,* Vol. 10, pp. 1293–1316.

### *Plant Pathology*

Johannsen, C. J., M. E. Bauer, and Staff. 1972. *Corn Blight Watch Experiment Results* (LARS information Note 012372). West Lafayette, IN: Laboratory for Applications of Remote Sensing, Purdue University, 24 pp. *http://www.lars.purdue.edu/home/references/LTR_012372.PDF*

Kelly, M., D. Shaari, Q. Guo, and D. Liu. 2004. A Comparison of Standard and Hybrid Classifier Methods for Mapping Hardwood Mortality in Areas Affected by Sudden Oak Death. *Photogrammetric Engineering and Remote Sensing,* Vol. 70, pp. 1229–1239.

Kelly, M., K. Tuxin, and F. Kearns. 2004. Geospatial Informatics for Management of a New Forest Disease: Sudden Oak Death. *Photogrammetric Engineering and Remote Sensing,* Vol 70, pp. 1001–1004.

Sharples, J. A. 1973. *The Corn Blight Watch Experiment: Economic Implications for Use of Remote Sensing Data Collecting Data on Major Crops,* (LARS Information Note 110173). West Lafayette, IN: Laboratory for Applications of Remote Sensing, Purdue University, 12 pp. *http://www.lars.purdue.edu/home/references/LTR_110173.PDF*

Ullstrup, A. J. 1972. The Impacts of the Southern Corn Leaf Blight Epidemics of 1970–1971. *Annual Review of Phytopathology,* Vol. 10, pp. 37–50.

### *Lidar and Laser Altimetry*

Lefsky, M. A., D. Harding, W. B. Cohen, G. Parker, and H. H. Shugart. 1999. Surface Lidar Remote Sensing of Basal Area and Biomass in Deciduous Forests of Eastern Maryland, USA. *Remote Sensing of Environment,* Vol. 67, pp. 83–98.

Marshall, G. F. (ed.). 2005. *Handbook of Optical and Laser Scanning.* New York: Marcel Dekker, 792 pp.

Nelson, R., W. Krabill, and J. Tonelli. 1988a. Estimating Forest Biomass and Volume Using Airborne Laser Data. *Remote Sensing of Environment,* Vol. 24, pp. 247–267.

Nelson, R., R. Swift, and W. Krabill. 1988b. Using Airborne Lasers to Estimate Forest Canopy and Stand Characteristics. *Journal of Forestry,* Vol. 86, pp 31–38

Nilsson, M. 1996. Estimation of Tree Heights and Stand. Volume Using an Airborne Lidar System. *Remote Sensing of Environment,* Vol. 56, pp. 1–7.

### *Tasseled Cap*

Crist, E. P. 1983. The Thematic Mapper Tasseled Cap: A Preliminary Formulation. In *Proceedings, Symposium on Machine Processing of Remotely Sensed Data 1983.* West Lafayette, IN: Laboratory for Applications of Remote Sensing, Purdue University, pp. 357–364.

Crist, E. P., and R. C. Cicone. 1984a. A Physically-Based Transformation of Thematic Mapper Data: The TM Tasseled Cap. *IEEE Transactions on Geoscience and Remote Sensing,* Vol. GE-22, pp. 256–263.

Crist, E. P., and R. C. Cicone. 1984b. Application of the Tasseled Cap Concept to Simulated Thematic Mapper Data. *Photogrammetric Engineering and Remote Sensing,* Vol. 50, pp. 343–352.

Kauth, R. J., and G. S. Thomas. 1976. The Tasseled Cap: A Graphic Description of the Spectral–Temporal Development of Agricultural Crops as Seen by Landsat. In LARS: *Proceedings of the Symposium on Machine Processing of Remotely Sensed Data.* West Lafayette, IN: Perdue University Press, pp. 4B-41–4B-51.

CHAPTER EIGHTEEN

# Earth Sciences

## 18.1. Introduction

This chapter addresses applications of remote sensing in the earth sciences, which are loosely defined here to include geology, geomorphology, and soil science. Despite their many differences, these disciplines share a common focus on the earth's shape and structure and on the nature of the soils and sediments at its surface. Applications of remote sensing in the Earth sciences are especially difficult. Often the subjects of investigation are geologic structures, soil horizons, and other features entirely or partially hidden beneath the Earth's surface. Seldom can remotely sensed images directly record such information—instead, we must search for indirect evidence that may be visible at the surface. Thus interpretations of geologic, pedologic, and geomorphic information are often based upon inference rather than direct sensing of the qualities to be studied. Furthermore, interpretations of geoscience information are frequently based upon subtle distinctions in tone, texture, and spectral response. Even direct examination of many geologic and pedologic materials is subject to error and controversy, so it should be no surprise that applications of remote sensing in these fields can be equally difficult.

Finally, we must often depend upon composite signatures formed as individual "pure" signatures of various types of soil, rock, and vegetation are mixed together into a single, composite spectral response that may be unlike any of its components (Chapter 10). Field observations are made by necessity at points such as outcrops, soil profiles, or boreholes. Because remote sensing instruments record radiation from ground *areas* rather than from ground *points,* it may be difficult to relate remotely sensed data to the field observations traditional for the earth sciences. Such problems present both practical and conceptual difficulties for applications of remote sensing in the earth sciences.

Remote sensing can, however, provide opportunities not normally available to geoscientists. The ability to observe reflectance and emittance over a range of wavelengths opens the door to the study of subjects that would not otherwise be possible. Also, the synoptic view of satellite images gives a broad-scale perspective of patterns not discernible within the confines of the large-scale, close-up view of ground observation. A single satellite image can record an integrated signature formed of many kinds of information, including shadow, soil, rock, and vegetation; although these mixtures may present problems, they also portray subtle differences in terrain that cannot be easily derived from other sources. Finally, although remotely sensed images cannot replace field and laboratory studies, they can be valuable supplements to more traditional methods, and sometimes provide information and a perspective not otherwise available.

## 18.2. Photogeology

Geologists study many aspects of the Earth's surface in an effort to understand the Earth's structure, to guide the search for minerals and fuels, and to understand geologic hazards. Remote sensing contributes to several dimensions of the geological sciences by providing information concerning lithology, structure, and vegetation patterns. *Lithology* refers to fundamental physical and chemical properties of rocks, including, for example, the gross distinctions between sedimentary, igneous, and metamorphic rocks. *Structure* defines the kinds of deformation experienced by rocks, including folding, fracturing, and faulting. *Geobotanical studies* focus upon relationships between plant cover at the Earth's surface and the lithology of underlying rocks.

*Photogeology* is the derivation of geological information from interpretation of aerial photographs. Photogeology originated early in the development of air photo interpretation. Many of its basic techniques were developed in the 1920s, then refined and applied into the 1950s and 1960s, when they seemed to approach the limits of their capabilities and were assimilated into newly developing research in geological remote sensing (Ray, 1960). Today photogeology is routinely applied to good effect. However, since aerial photographs now form only one of many forms of aerial images routinely available to the geologist, most research and innovation is now likely to occur in the broader context of geological remote sensing.

The practice of photogeology is based upon direct application of the principles of image interpretation (Chapter 5) to geological problems. Image texture, size, shape, tone, shadow, and so on each have special significance in the realm of the geologist's view of the terrain. Likewise, the principles of photogrammetry (Chapter 3) have special applications in the context of photogeology for calculation of thicknesses of beds and determination of strike and dip from aerial photography. Such measurements permit derivation of structural information from aerial images.

Remotely sensed images can provide valuable information about structure and terrain, especially if stereoscopic views are available. Figures 18.1 and 18.2 show an arid landscape in New Mexico characterized by lava flows lying uncomfortably on gently dipping sedimentary rocks. This landscape is positioned just east of the Rio Grande River, between Albuquerque and Santa Fe. During the Quaternary and late Tertiary, this region experienced extensive volcanism.

At the upper edge of the photograph, upturned sandstone beds are visible as slightly curving arcs, dipping to the left, and continuing toward the bottom of the photograph. In stereo, the vertical exaggeration creates distinctive relief, revealing the resistant character of the sandstone strata that form the ridges. These features are known as hogbacks (i.e., resistant strata upended to form steep-sided ridges with jagged crests). Resistant sandstones form the ridges, with less resistant shales eroded to form intervening valleys. Figure 18.3 presents a ground view of the hogbacks, taken from a position near the upper right-hand corner of Figure 18.2.

These photographs also show evidence of the volcanism that has shaped this landscape. Volcanic dikes cut across established topography as narrow, pronounced, steep-sided ridges. Notice that these dikes cut across the north—south oriented ridges, confirming the recent age of the volcanism relative to the other structures.

The lower edge of Figure 18.2 shows a portion of an extensive basaltic flow that forms the surface of a resistant, steep-sided mesa. Viewed in stereo, the resistant character of the basaltic surface is evident from its sharp relief and precipitous cliffs at its margins. This region is a resistant bedrock surface covered with a thin cover of gravel, sparsely vegetated. As viewed on the

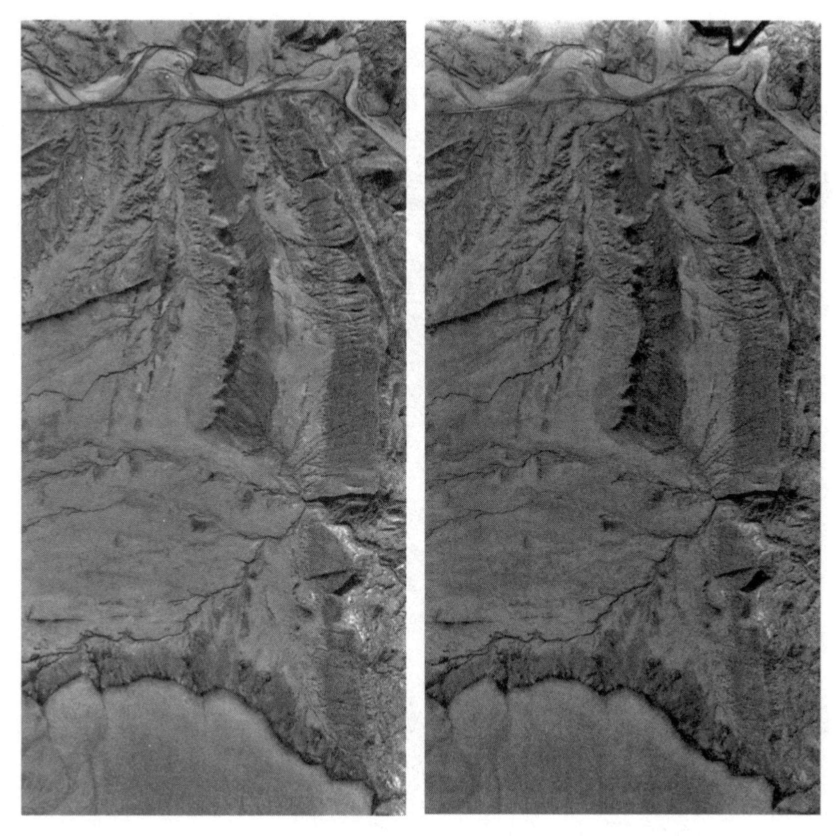

**FIGURE 18.1.** Stereo aerial photographs of the Galisteo Creek region, north-central New Mexico. From USGS and Army Map Service.

aerial photograph, it has a distinctive appearance, even in tone, slightly rough in texture, and with a puffy appearance when viewed in stereo. Note that the basaltic plateau has covered preexisting sedimentary strata, especially where the flow has been superimposed over the rightmost hogback, creating a clear unconformity.

At the base of the cliff, the aerial photograph shows the contrasting textures of a thick sheet of stone, sand, and gravel, shaped by fluvial processes that transport geologic debris across the slope at the base of the cliff toward the stream flowing left to right across the photograph. This stream has maintained its course across the hogbacks as they were formed, carving a distinct V-shaped notch through the right-hand ridge. As the stream passes through the notch, it enters a terrain formed by erosion of shales exposed at the surface. Here surface erosion is much more effective than on the surfaces of the basaltic flows, for example, and the topography is much more disorganized and chaotic relative to the surface visible on the left-hand side of the image.

Although the discussion of this example is necessarily quite brief, it illustrates the power of aerial imagery to provide information and insight to the earth scientist, both to derive new information and to organize and interpret information already at hand. Subsequent sections will explore some other dimensions of photogeology and the value of remotely sensed imagery for the earth sciences.

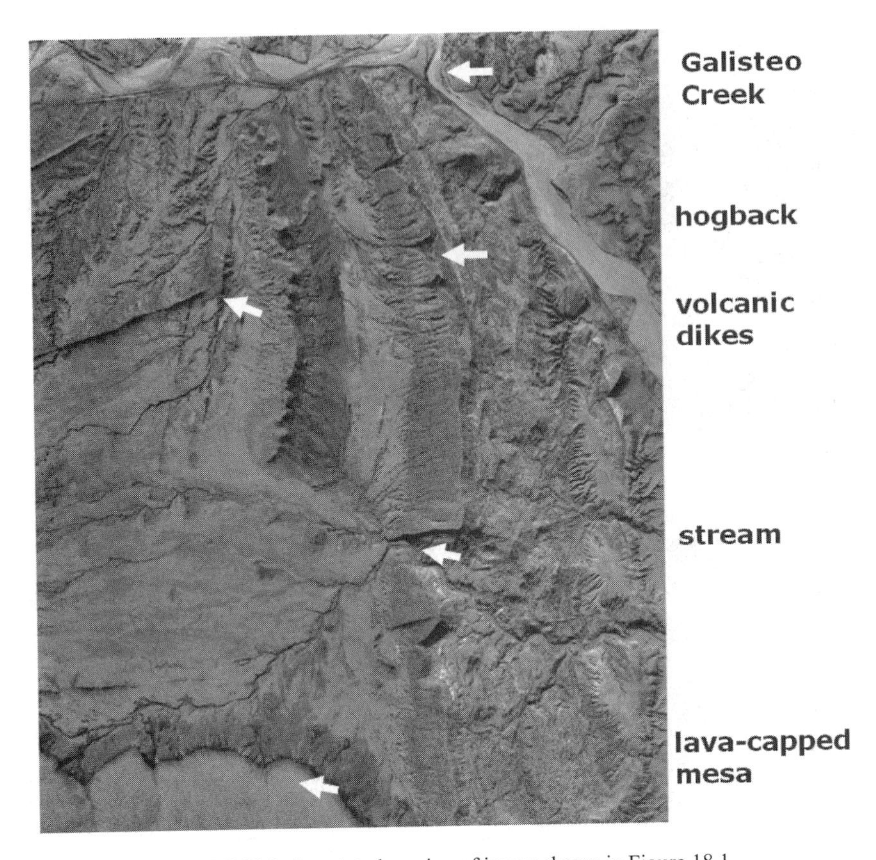


**FIGURE 18.2.** Annotated version of image shown in Figure 18.1.

**FIGURE 18.3.** Ground photograph of the Galisteo Creek region, New Mexico, showing some of the same features represented in Figure 18.1. This photograph was taken from a location near the top right-hand corner of Figure 18.1, looking across the creek towards the hogbacks. Photograph by W.T. Lee, 1910. From USGS Photographic Library, Denver, CO.

## 18.3. Drainage Patterns

Drainage patterns are often the most clearly visible features on remote sensed imagery, and are also among the most informative indicators of surface materials and processes. The character of drainage patterns permits geologists to infer valuable information about the surface materials and the geologic structure that prevail for a specific landscape (Figure 18.4 and 18.5). Drainage patterns as observed in nature may display characteristics of two or more of these patterns, so they may not match precisely to the illustrations used to describe them.

A common drainage pattern is characterized by even branching of tributaries, similar to the patterns of veins on an oak or maple leaf—this is called a *dendritic* drainage pattern. Dendritic drainage patterns suggest uniformly resistant surface materials, gentle regional slopes, and the absence of major faults or structural systems. A *parallel* drainage pattern is also found on uniform materials, but on landscapes that have pronounced regional slopes. Parallel drainage resembles the dendritic pattern, but is distinctive because of its elongated form derived from the increased topographic slope. In fact, there are many transitional drainage pattern forms that fall between these two forms.

If a landscape has been uplifted by tectonic forces, or if the base level has been lowered (e.g., by a decline in sea levels), then streams may have an *entrenched,* or *incised,* drainage pattern, such that their channels are cut well below the surface of the surrounding landscape. The nature of the incision can reveal the character of the bedrock—for example, sharp, deep, well-defined edges suggest the presence of strong, cohesive surface materials. More gently sloping terrain near the channel suggests the presence of less cohesive, weaker, strata.

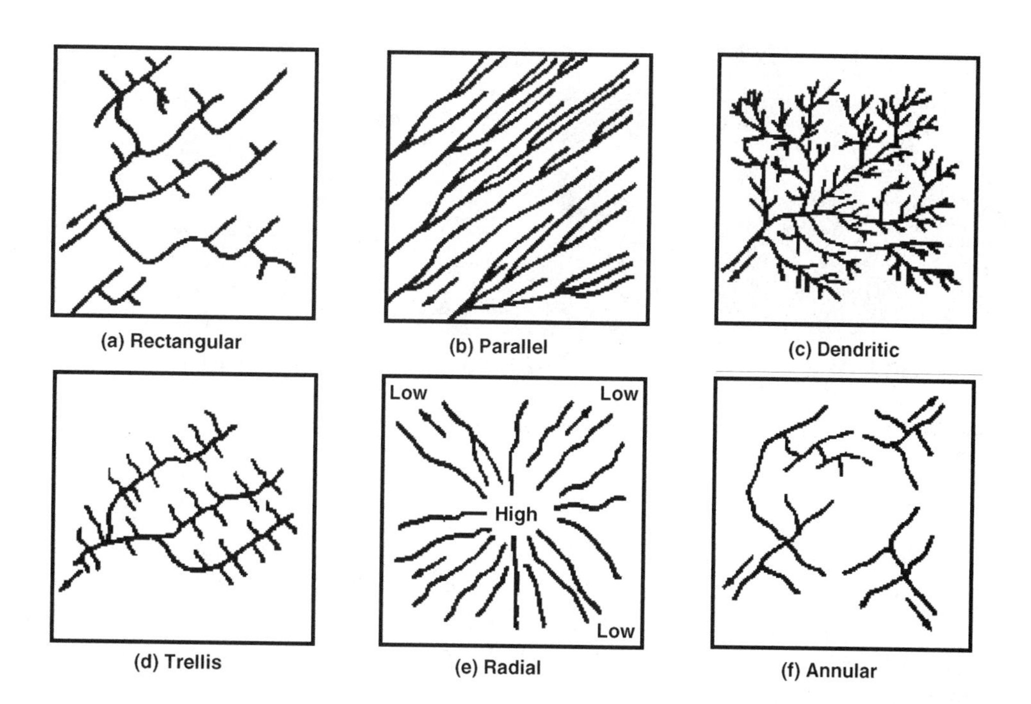


**FIGURE 18.4.** Sketches of varied drainage patterns.

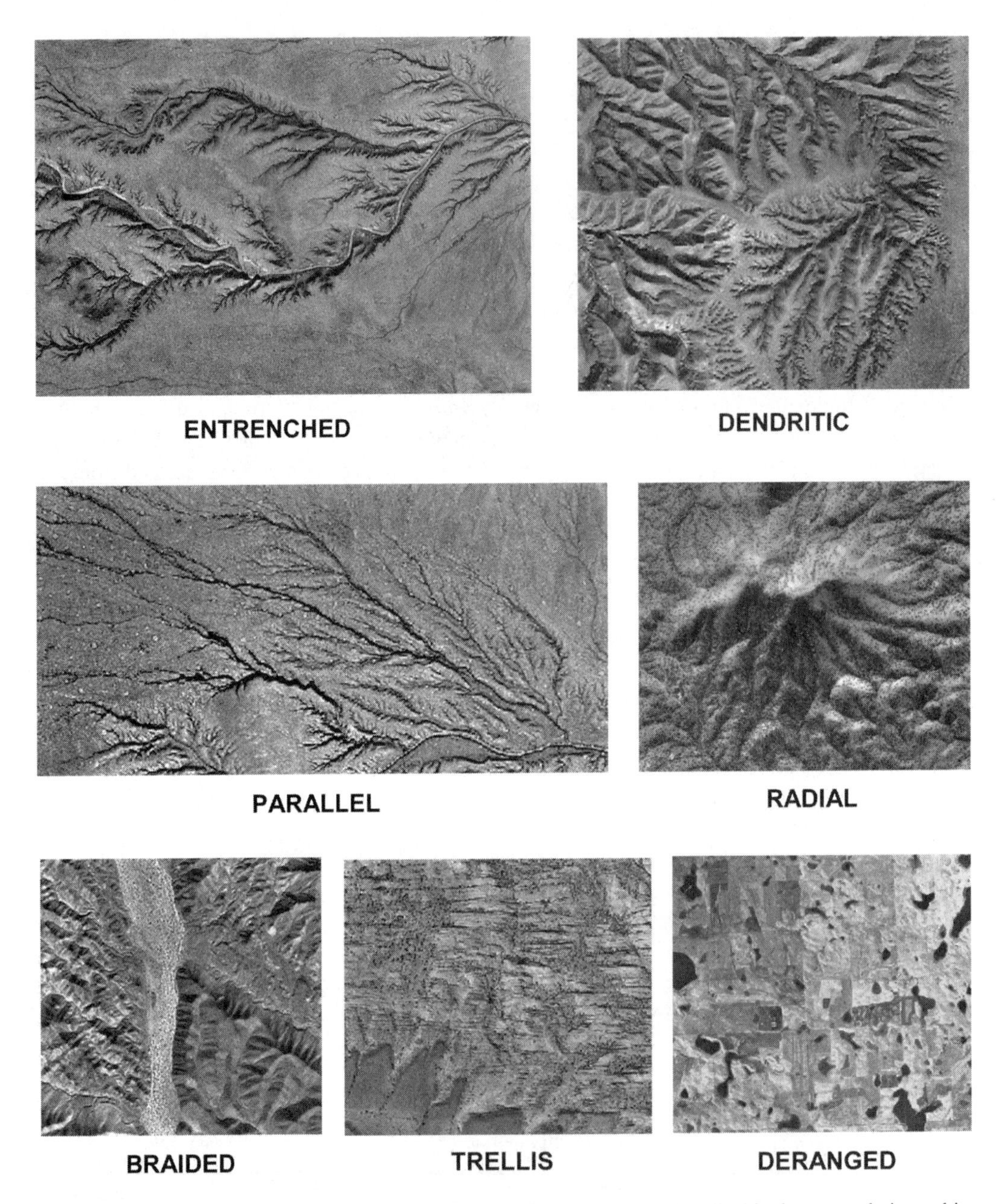


**FIGURE 18.5.** Aerial photographs illustrate some of the drainage patterns described in the text and pictured in Figure 18.4.

If the landscape is characterized by linear structural or lithologic features, drainage often develops to form a *trellis* drainage pattern. Tributaries often follow the strike of the structure, while the main streams cut across the principal structures. Therefore, tributaries join main streams at right angles, to form a system in which the main streams and the tributaries are oriented perpendicular to each other. When a landscape is dominated by a large central peak, the drainage system is organized to drain water away from the central upland, forming a *radial* drainage pattern.

When landscape are disturbed by broad-scale processes such as glaciation, faulting, surface mining, or volcanic deposition, the established drainage system is often buried or disturbed. The resulting landscape is characterized by incoherent drainage systems—lakes and marshes are connected by a confused system of streams. As time passes, an organized stream network will reestablish itself, but in the meantime the drainage system will be confused and disorganized.

Finally, streams with *braided* drainage patterns have several channels that divide and rejoin, separated by ephemeral bars and islands. Braided streams are often found in locales that have variable stream discharge (e.g., those that are in arid regions, or in glacial meltwater) and high sediment loads. Their appearance is usually quite distinctive: a wide, sparsely vegetated strip of open sand or gravel, with a network of ephemeral, anastomosing channels and elongated bars and islands.

Closer inspection of details of the drainage system can reveal specifics of the textures of the surface materials. Steep slopes are characteristic of coarse, loose surface materials. Noticeably flat, rounded slopes may indicate cohesive, plastic surface materials. The absence of surface drainage or a very simple drainage pattern often indicates well-drained pervious soil. Highly integrated drainage patterns often indicate impervious, plastic sediments that lose their strength when wet.

The shape of the cross section of a drainageway is controlled largely by the cohesiveness of the surface materials. Abrupt changes in stream grade, channel direction, or its cross section indicates changes in the underlying bedrock or the surface materials. Generally, short, V-shaped gullies with steep gradients are typical of well-drained, loosely structured sediments, such as coarse sands. U-shaped stream profiles indicate the presence of deep, uniform silt deposits, especially the wind-deposited sediments known as *loess.* Poorly drained, fine textured surface materials usually form shallow drainageways with shallow, rounded, saucer-like profiles.

On black-and-white aerial photography, the tones of surface materials can indicate the character of the drainage. Soft, light tones generally indicate pervious, well-drained soils. Large, flat areas of sand are frequently indicated by uniform light-gray color tones, a flat appearance, and a lack of conformity; this indicates a natural surface drainage. Clays and organic soils frequently appear as dark-gray to black areas. In general, a sharp change in colors or tones signals a change in soil texture.

## 18.4. Lineaments

*Lineaments* is the name given by geologists to lines or edges, of presumed geologic origin, visible on remotely sensed images. Such features have also been referred to as *linears,* or *lineations,* although O'Leary et al. (1975) established *lineaments* as the preferred term. Use of the term *lineament* in a geological context dates to 1904, and apparently has even earlier analogues in other languages. These early uses, prior to the availability of aerial images, applied to specific geologic or geomorphic features, such as topographic features (ridgelines, drainage systems, or coastlines), lithologic contacts, or zones of fracture.

As early as the 1930s photogeologists studied fracture patterns visible on aerial photographs as a means of inferring geologic structure. These photo features apparently corresponded rather closely to faults and fractures defined in the field. More recently, in the context of geological remote sensing, the term has assumed a broader meaning. Any linear feature visible on an aerial image can be referred to as a *lineament.* A problem arises because it is difficult to establish a clear link between the features on images and corresponding features, if any, on the ground.

The strength of this link depends in part upon the nature of the imagery. In the instance of interpretations from aerial photographs and photomosaics, scales are usually large and ample detail is visible. Linear features visible on such images are likely to match to those that can be confirmed in the field. However, more extensive and subtle features cannot be easily detected on aerial photography. Each photograph shows only a small area; if mosaics are formed to show larger regions, variations in illumination and shadowing (Figure 3.30) obscure more subtle lineaments that might extend over many photographs. Furthermore, in the era prior to routine availability of the CIR films now used for high-altitude photography, high-altitude photographs (which *can* show large areas under uniform illumination) were not of good visual quality due to the effects of atmospheric scatter.

The advent of nonphotographic sensors and the availability of the broad-scale view of satellite images changed this situation because it then became possible to view large areas with images of good visual clarity. Early radar images depicted large areas illuminated from a single direction at rather low illumination angles—conditions that tended to increase topographic shadowing in a manner that enhanced the visibility of linear features (Figure 18.6). Locations of these features did not always match previously known faults or fractures, so the term *lineament* was evoked to avoid an explicit statement of a geological origin. Later, similar features were again detected on Landsat MSS imagery. Initially, the detection of lineaments and the interpretation of their meaning generated some controversy—the broad scale and coarse resolution of the early satellite imagery caused lineaments to be subtly expressed, and made them difficult to relate to previously known geological features. During the intervening years,

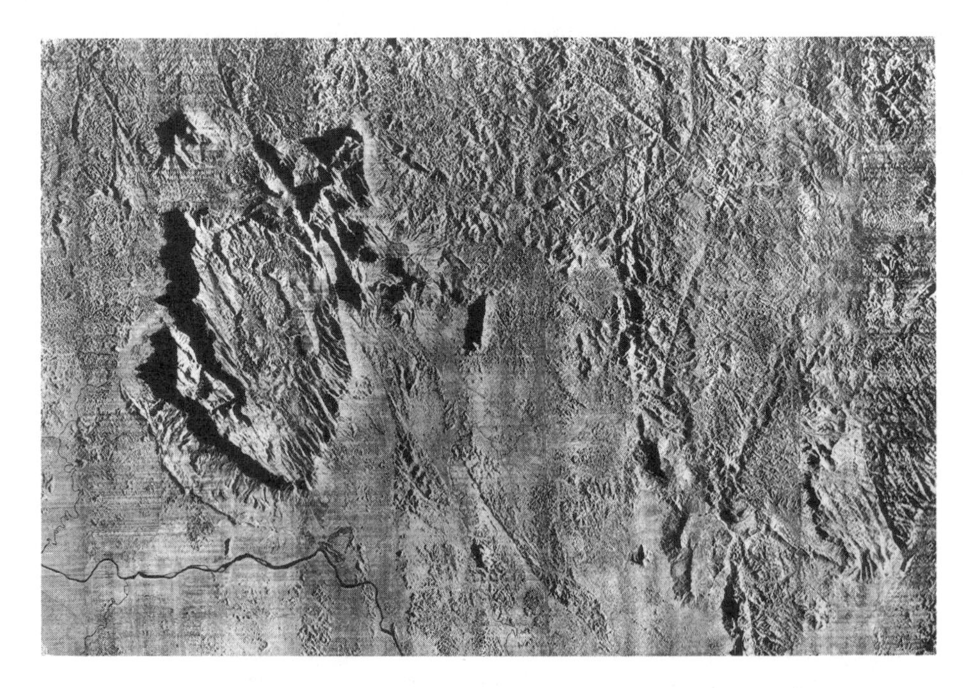

**FIGURE 18.6.** Radar mosaic of southern Venezuela. Here topography and drainage are clearly visible, as are the linear features in the upper right that illustrate the kinds of features important in lineament analysis. From Goodyear Aerospace Corporation.

investigation and debate as confirmed the value of lineaments as guides to understanding geologic phenomena.

Lineaments are, of course, "real" features if we consider them simply to be linear features detected on aerial imagery. The uncertainty arises when we attempt to judge their geological significance. There are sound reasons for assigning a geological meaning to some lineaments, even if they do not always correspond to clearly observable physical features at a specific point. In the simplest instances, a dip–slip fault may leave a subtle topographic feature that is visible as a linear shadow on aerial photography when illuminated obliquely from the elevated side (Figure 18.7). The axis of displacement for a dip–slip fault is along the dip of the fault plane, to create a difference in relief, as depicted in Figure 18.7. In the alternative instance, a strike–slip fault, the axis of the fault displacement follows the strike of the fault (along the trace of the fault on the surface), creating fewer opportunities for surface expression. Of course, a fault that is oriented parallel to the direction of illumination may be indistinct or invisible. Not all faults are expressed topographically, but the fault plane may offer preferred avenues of movement for moisture and for the growth of plant roots (Figure 18.8). Therefore, the trace of the fault may be revealed by vegetation patterns even though it has no obvious topographic expression. Faults of any form can alter drainage in a manner that creates linear drainage segments, which are then visible on aerial images. Such features, when clearly expressed, may have structural origins, and it may be possible to verify their existence through field observations. However, more subtle features may be observable only as broad-scale features, not easily verifiable at a given point. Such lineaments may be genuine structural features, but not be easily confirmable as such.

Linear features that are clearly of structural origin are significant because they indicate zones of fracturing and faulting. It is often assumed that regions of intersection of lineaments of differing orientation are of special significance, as in theory they might identify zones of mineral-

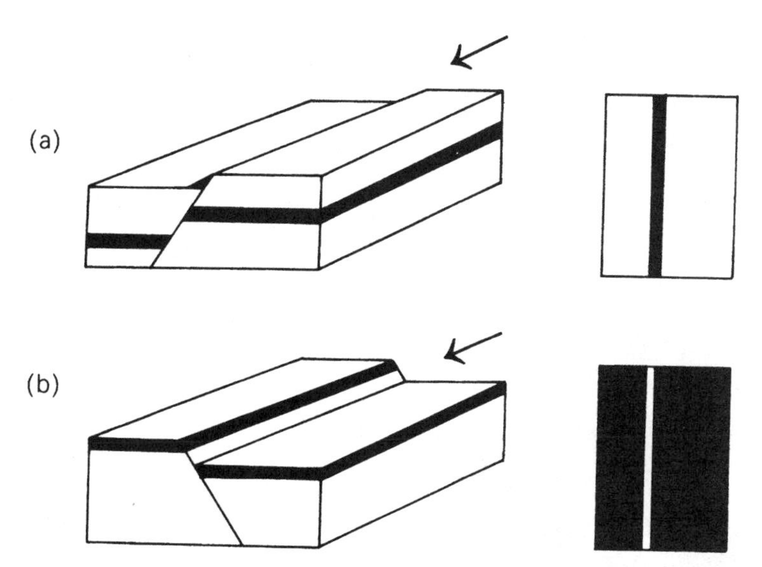


**FIGURE 18.7.** Schematic illustration of a dip–slip fault. (a) Illumination from the elevated side produces a strip-like shadow on the image. (b) Illumination from the downthrust side may produce a bright strip as the Sun is reflected from the linear facet that now faces the solar beam.

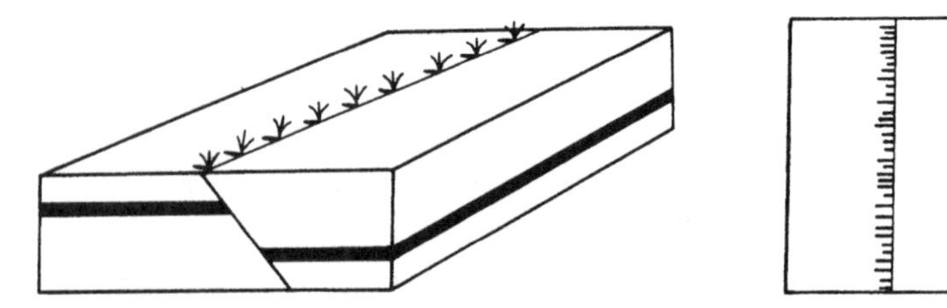

**FIGURE 18.8.** Hypothetical situation in which a fault with little or no surface expression forms a preferred zone of movement for moisture and for the growth of plant roots. Such a zone may create a favorable habitat for certain species or speed the growth of more prevalent plants located near the fault. The linear arrangement of such plants and their shadows may form a linear feature on an image.

ization, stratigraphic traps, regions of abundant groundwater, or zones of structural compression. For this reason, much is often made of the orientation and angles of intersection of lineaments.

Other lineaments may not have structural origins. It is conceivable that some may be artifacts of the imaging system or of preprocessing algorithms. Some linear features may be purely surface features, such as wind-blown sediments, that do not reflect subsurface structure (Figure 18.9). Cultural patterns, including land-cover boundaries, edges of land ownership parcels, and political borders, can all have linear form and can be aligned in a manner that creates the linear features observed on aerial images. Because these features may not be readily distinguished from those of geologic origin, and because of the inconsistency of individual delineations of lineaments from the same image, lineament analysis has been regarded with skepticism by a significant proportion of the geological community (Wise, 1982).

Because the human visual system tends to generalize, manual interpretations may identify continuous linear features that are later found to be formed of separate segments of unrelated origin. Lineaments that are found to correspond to geologic features often are extensions of known fault systems rather than completely new systems. Results of manual interpretations vary greatly from interpreter to interpreter (Podwysocki et al., 1975). As a result, other research has attempted to automate the interpretation of lineaments. Because detection of edges and lines has long been an important task in the fields of image analysis and pattern recognition, the fields of image processing and machine vision were well equipped with techniques to apply to this inquiry. There is a rich literature that has investigated procedures for enhancing images to

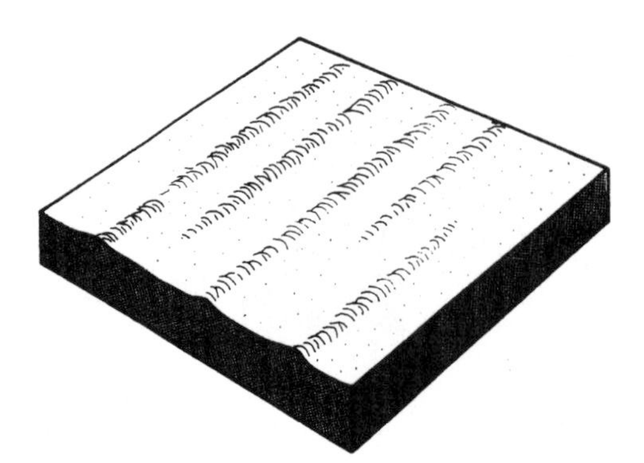

**FIGURE 18.9.** Example of lineaments that do not reflect geologic structure. Wind-blown sand has been aligned in linear ridges that merely indicate the nature of the surface sediments.

highlight linear features and then analyzing patterns to extract linear features that might be present.

### *Fracture Density Studies*

Often images and field data can be analyzed to assess the abundance of lineaments by plotting their occurrence within cells used to partition the image. Analysis of the distribution and orientation of lineaments and fractures, in the context of field studies and other databases, can lead to assessment of groundwater or mineral potential.

Abrams et al. (1983) studied lineaments interpreted from a Landsat MSS color composite of a region in southwestern Arizona. Their analysis focused upon the Helvetia–Rosemont area, known to be rich in porphyry copper. They attempted to determine if their techniques would locate previously discovered mineralized zones. Paleozoic limestones, quartzites, shales, and dolomites cover Precambrian granites and shists in the study area. The Paleozoic deposits themselves were altered by Mesozoic uplift and erosion, then covered by further deposition. The Laramide orogeny during the late Cretaceous produced intrusion of granitic rocks with thrust faulting along a predominant northeast–southwest trend. This tectonic activity, together with later Paleocene intrusions, caused mineralization along some of the faults. Tectonic activity continued into the late Tertiary, producing extensive faulting and a complex pattern of lineaments.

Abrams et al. examined orientations of lineaments they interpreted from the Landsat image, then plotted their results in the form of a rose diagram (Figure 18.10). Each wedge in the circular pattern represents the numbers of lineaments oriented along specific compass azimuths. Because each lineament has two azimuths (e.g., a "north–south" line is oriented as much to the south as it is to the north), the diagram is symmetric, and some investigators prefer to show only half of the diagram.

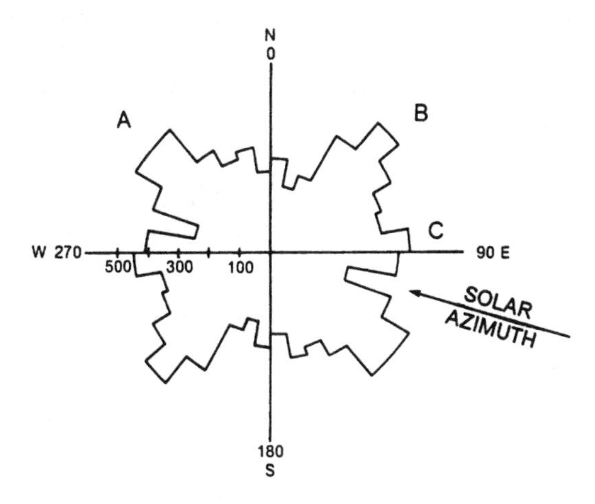


**FIGURE 18.10.** Strike-frequency diagram of lineaments for a region of southeastern Arizona. Features, as discussed in the text, include (A) predominant northwest–southeast orientation; (B) predominant northeast–southwest orientation; and (C) low frequency along the axis that parallels the solar beam at the time of image acquisition. Based on Abrams et al. (1983, p. 593). Reproduced by permission.

Several features are noteworthy. First, the low numbers of lineaments oriented at (approximately) 110 (and 290), roughly east–southeast to west–northwest, correspond to the axis that parallels the orientation of the solar beam at the time that the Landsat data were acquired—the lineaments that might have this orientation will not be easily detected because shadowing is minimized. The northeast–southwest trend (*B* in Figure 18.10) is said to represent the predominant trend of Precambrian faulting that is observed throughout Arizona. Superimposed over this pattern is the northwest–southeast trend (*A*) arising from faulting in the Mesozoic rocks mentioned above. Finally, the east–west trend (*C*) is interpreted as the result of the Laramide faulting known to be associated with the mineralization that produced the copper deposits present in this region. Abrams and his colleagues counted the frequencies of the lineaments in each cell of a 10-km grid superimposed over the image and concluded that the highest frequencies, which they interpreted to form favorable locations for intrusion of magmas and for mineralization, correspond to zones of known mineral deposits.

## 18.5. Geobotany

*Geobotany* is the study of the relationships between plants and the underlying geologic substratum (Ustin et al., 1999). Such relationships include variations in the presence, abundance, vigor, and appearance of plants, indicator specifies, and distinctive phonological behavior; often it can refer more specifically to examination of the relationships between plants and their response to unusual levels of nutrients or toxins released to the soil by weathering of geologic material (Figure 18.11). Regional geobotany focuses upon the study of vegetation patterns as indicators of lithologic variations. Specific plant species may have local significance as *indicator species,* those that tend to avoid or to favor certain lithologic units. In other instances, variations in abundance or vigor may signal the occurrence of certain lithologies.

Geologic processes may concentrate trace elements in specific strata or regions. Geologic weathering may release trace elements in a form that can be absorbed by vegetation. Plants that absorb these trace elements at higher than normal levels may display abnormal spectral signatures, thereby signaling the existence and location of the anomalous concentration of elements.

Geobotany depends upon knowledge of how geologic materials release elements into the nutrient pool, how these elements are absorbed by the soil and concentrated in plant tissues, and how they can alter the spectral signatures of plants. Because specific kinds of plants may thrive under certain unusual conditions of soil fertility, detailed knowledge of the spectral characteristics of such plants can be very valuable. Geobotanical studies are especially valuable in heavily vegetated regions where soil and rock are not exposed to direct view of the sensor, but they may also be useful in sparsely vegetated regions where it may be possible to identify individual plants, or where vegetation patterns may be especially sensitive to subtle environmental variations.

The practice of geobotanical reconnaissance is restricted by several factors. First, geobotanical studies may depend upon observations of subtle distinctions in vegetation vigor and pattern, so successful application may require data of very fine spatial, spectral, and radiometric resolution—resolution much finer than that of Landsat or SPOT data, for example. Thus some of the concepts of geobotanical exploration cannot be routinely applied with the imagery and data most commonly available. Hyperspectral remote sensing (Chapter 15) provides the necessary capabilities for developing geobotanical knowledge. This work, now conducted mainly in a

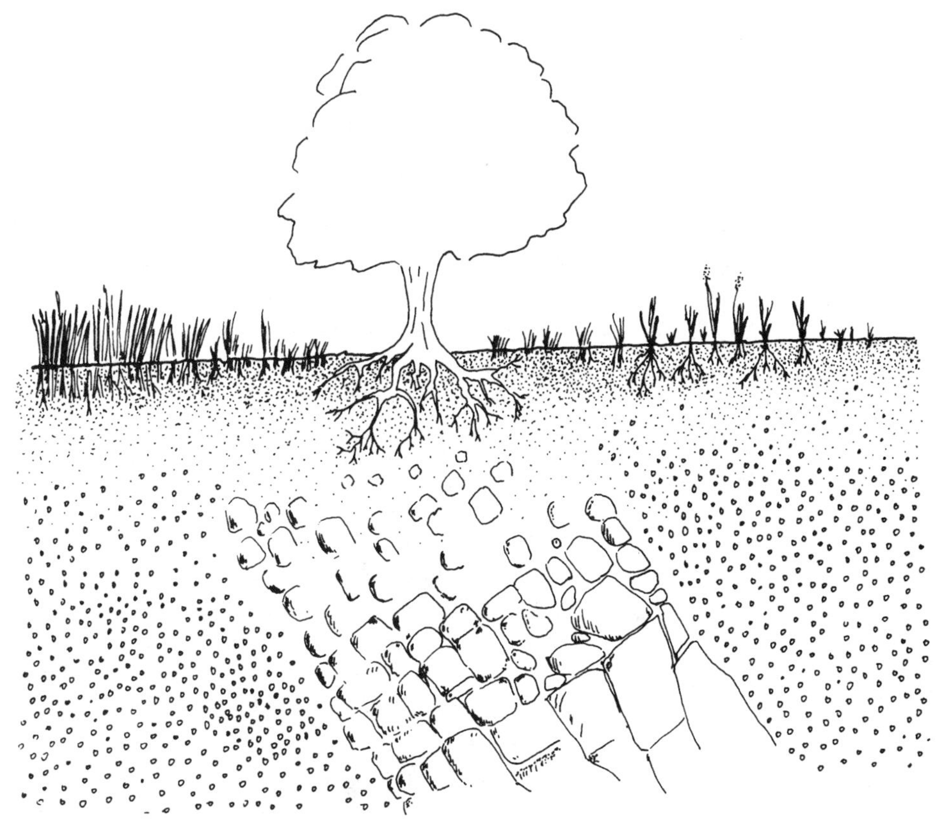

**FIGURE 18.11.** General model for geobotanical studies. Abundance, vigor, and patterns of plants are related to the geologic substratum. If geologists can isolate these geologic factors from others that influence vegetation patterns, botanical information can be used (with other data) to contribute to geological knowledge.

research context, may provide the basis for development of sensors and analytical technques applicable on an operational basis. Second, geology is only one of the many factors that influence plant growth. Site, aspect, and disturbance (both by mankind and by nature) influence vegetation distributions in very complex patterns. Effects of these several causes cannot always be clearly separated.

Timing of image acquisition is critical. Some geobotanical influences may be detectable only at specific seasons or may be observable as, for example, advances in the timing of otherwise normal seasonal changes in vegetation coloring. The issue of timing may be especially important because the remote sensing analyst may not have control over the timing of image acquisition or may have imagery only for a single date, whereas the analyst needs images from several dates to observe the critical changes. Finally, many geobotanical anomalies may have a distinctively local character—specific indicators may have meaning only for restricted regions, so intimate knowledge of regional vegetation and geology may be necessary to fully exploit knowledge of geobotany.

Geobotanical knowledge can be considered first in the context of the individual plant and its response to the geological substratum. Growth of most plants is sensitive not only to the availability of the *major nutrients* (phosphorus [P], potassium [K], and nitrogen [N]), but also to the *micronutrients,* those elements (including barium [B], magnesium [Mg], sulphur [S], and calcium [Ca]) that are required in very small amounts. It is well established that the growth and health of plants depends upon the availability of elements present in very low concentrations, and that sensitivity may be high in some plant species. In contrast, other elements are known to have toxic effects, even at very low concentrations, if present in the soil in soluble form. Heavy metals, including nickel (Ni), copper (Cu), chromium (Cr), and lead (Pb), may be present at sufficient levels to reveal their presence through stunted plant growth, or by the localized absence of specific species that are especially sensitive to such elements. From such evidence, it may be possible to indirectly use the concentrations of these elements to reveal the locations of specific geologic formations or zones of mineralization worthy of further investigation.

Hydrocarbons may significantly influence plant growth if they are present in the root zone. Hydrocarbon gases may migrate from subsurface locations; at the surface, concentrations may be locally sufficient to influence plant growth. Hydrocarbons may be present in greater concentrations at the surface as petroleum seeps or coal seams. The presence of hydrocarbons in the soil may favor plant growth by increasing soil organic matter, but may also alter soil atmosphere and soil biology in a manner that is toxic to some plants.

Specific geobotanical indicators include the following, which of course may have varied geologic interpretation depending upon the local setting. Variations in biomass, either significantly higher or lower than expected, may indicate geobotanical anomalies. Or the coloring of vegetation may be significant. The term *chlorosis* refers to a general yellow discoloration of leaves due to a deficiency in the abundance of chlorophyll. Chlorosis can be caused by many of the geologic factors mentioned above, but it is not uniquely of geologic origin, so its interpretation requires knowledge of the local geologic and biologic setting.

Observation of these indicators may be facilitated by the use of the vegetation indices and ratios mentioned in Chapter 17. Manual interpretation of CIR film often provides especially good information concerning the vigor and distribution of vegetation. Of special significance in the context of geobotany is the *blue shift* in the chlorophyll absorption spectra. The characteristic spectral response for living vegetation was discussed in Chapter 17. In the visible, peak reflectance is in the green region; the absorption spectra of chlorophyll decreases reflectance in both the blue and the red regions. However, in the near infrared, reflectance increases markedly (Chapter 17).

Collins et al. (1981) and others have observed that geochemical stress is most easily observed in the spectral region from about 0.55 to 0.75 μm (which includes portions of the green, red, and infrared regions). Most notably, the position and slope of the line that portrays the increase in reflectance at the edge of the visible region, although constant for healthy green plants, is especially sensitive to geochemical influences. As geochemical stress occurs, the position of this line shifts toward shorter wavelengths (i.e., toward the blue end of the spectrum; hence the term *blue shift*) and its slope becomes steeper (Figure 18.12).

It must be emphasized that, by everyday standards, the change is very subtle (0.007–0.010 μm), and has been observed only in data recorded by high-resolution sensors processed to filter out background reflectance. Such data are not acquired by remote sensing instruments available for routine use, although increased spectral, spatial, and radiometric resolution provided by hyperspectral remote sensing (Chapter 15) opens up opportunities to exploit this kind of knowledge. Nonetheless, Collins et al. (1983) report that the blue shift has been observed in all plant species studied thus far; they conclude that this effect appears to be "universal for most green plants" (p. 739), so it is not confined to a few species.

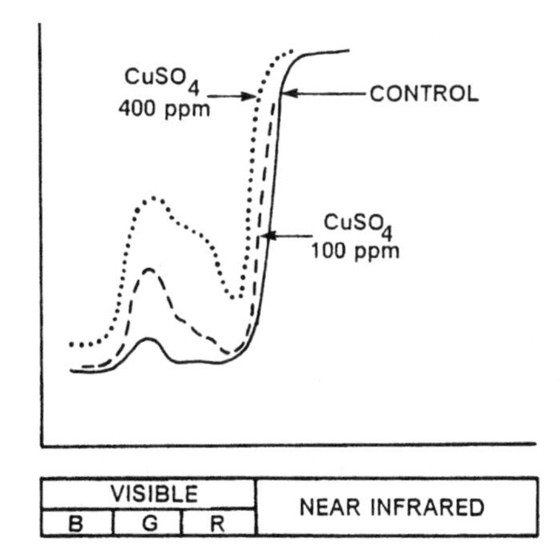


**FIGURE 18.12.** Blue shift in the edge of the chlorophyll absorption band. As heavy mineral concentration in plant tissue increases, the edge of the absorption band shifts toward shorter wavelengths (i.e., toward the blue end of the spectrum). The solid line shows the spectrum for a control plant. The dotted and dashed lines show how the magnitude of the shift increases as concentration of the heavy minerals increases. From Chang and Collins (1983, p. 727). Reproduced by permission.

The cause of the blue shift is not clearly understood at present. Unusual concentrations of heavy metals in the soil are absorbed by plants and apparently tend to be transported toward the actively growing portions of the plant, including the leaves. Concentrations of such metals can cause chlorosis, often observed by more conventional methods. The relationship between heavy metals and the blue shift is clear, but work by Chang and Collins (1983) indicates that the presence of heavy metals in the plant tissues does not alter the chlorophyll itself—the metallic ions apparently do not enter the structure of the chlorophyll. The heavy metals appear to retard the development of chlorophyll, thereby influencing the abundance, but not the quality, of the chlorophyll in the plant tissue.

Chang and Collins (1983) conducted laboratory and greenhouse experiments to confirm the results of spectral measurements in the field and to study the effects of different metals and their concentrations. Their experiments used oxides, sulfates, sulfides, carbonates, and chlorides of copper (Cu), zinc (Zn), iron (Fe), nickel (Ni), manganese (Mn), molybdenum (Mo), and vanadium (V). Selenium (Se), Ni, Cu, and Zn produced stress at concentrations as low as 100 ppm. Fe and Pb appeared to have beneficial effects on plant growth; Mo and V produced little effect at the concentrations studied. In combination, some elements counteract each other, and some vary in effect as concentrations change.

## 18.6. Direct Multispectral Observation of Rocks and Minerals

A second broad strategy for remote sensing of lithology depends upon accurate observation of spectra from areas of soil and rock exposed to observation. Color and, by extension, the spectral characteristics of rocks and minerals can be distinctive identifying features in the direct examination of geologic samples. Geologists use color for identification of samples in the field. In the laboratory they are sensitive instruments to measure spectral properties across a range of wavelengths in the ultraviolet, visible, and infrared regions. Spectral emittance and reflectance of rocks and minerals is an important property that is often closely related to their physical and chemical properties. In the laboratory, spectral characteristics can be observed in sufficient detail that they can sometimes serve as diagnostic tests for the presence of specific minerals. In addition, small radiometers can be taken to the field to permit in situ observation of rocks and minerals in nonvisible portions of the spectrum. As a result, geologists have considerable experience in observing spectral properties of geologic materials and have developed an extensive body of knowledge concerning the spectral properties of rocks and minerals.

The application of this knowledge in the context of remote sensing can be very difficult. In the laboratory, spectra can be observed without the contributions of atmospheric attenuation, vegetation, or shadowing to the observed spectra. In the laboratory setting, scientists can hold secondary properties (such as moisture content) constant from one sample to the next or can compensate for their effects. In remotely sensed data, the effects of such variables often cannot be assessed. The usual remote sensing instruments do not have the fine spectral and radiometric resolutions required to make many of the subtle distinctions this strategy requires. Furthermore, the relatively coarse spatial resolutions of many remote sensing systems means that the analyst must consider composite signatures formed by the interaction of many landscape variables rather than the pure signatures that can be observed in the laboratory. Whereas photogeology distinguishes between units without attempting to make fine lithologic distinctions, much of the research in multispectral remote sensing in geology is devoted to more precise identification of specific minerals thought both to be spectrally distinctive and to be especially valuable in mineral exploration.

Locations of zones of hydrothermal alteration may be revealed by the presence of limonite at the surface. *Limonite* refers broadly to minerals bearing oxides and oxyhydrites of ferric iron, including goethite and hematite; such minerals tend to exhibit broad absorption bands in the near infrared, visible, and ultraviolet regions. Typically their spectra decline below 0.5 µm, producing a decrease in the ultraviolet not normally observed in other minerals (Figure 18.13). A broad, shallow absorption region from 0.85 µm to 0.95 µm is observed in the near infrared. The presence of limonite, either as a primary mineral or secondarily as the product of geologic weathering, may identify the location of a zone of hydrothermal alteration, thereby suggesting the possibility of mineralized zones. However, the identification of limonite is not definitive evidence of hydrothermal activity, as limonite can be present without hydrothermal alteration and hydrothermal activity may occur without the presence of limonite. Thus the observation of these spectra, or any spectra, is not uniquely specific, and the analyst must always consider such evidence in the context of other information and his or her knowledge of the local geologic setting.

Clay minerals often decrease in reflectance beyond 1.6 µm; this behavior is common to enough minerals and occurs over a spectral region that is broad enough that it has been used to locate zones of hydrothermal alteration by means of the surface materials rich in clay minerals. At finer spectral resolution, these clay minerals and carbonates characterized by AlOH and MgOH structures display spectra with a narrow but distinctive absorption band from 2.1 to 2.4 µm (Figure 18.13).

Emittance in the spectral region from 8 to 12 µm (in the mid-infrared) permits the identification of some silicate minerals and the separation of silicate minerals from nonsilicate minerals. Silicate minerals typically exhibit emittance minima in this region, whereas nonsilicate minerals have minima at longer wavelengths.

## 18.7. Photoclinometry

*Photoclinometry,* loosely defined, is the understanding of the relationship between image brightness and the orientation of the surface that generated the brightness. If an irregular surface

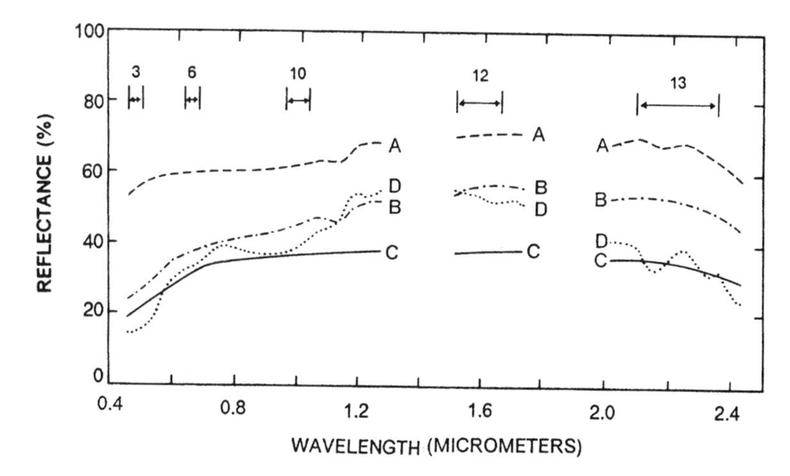


**FIGURE 18.13.** Spectra of some natural geologic surfaces. (A) white Joe Lott Tuff member of Mt. Belknap volcanics; (B) limonitic Joe Lott Tuff member of Mt. Belknap volcanics; (C) tan soil on rhyolite; (D) orange layered laterite. Data were collected in the field near the surface using portable spectroradiometers. Limonitic minerals (B, C, D) show low reflectances at wavelengths below 0.5 µm, relative to nonlimonitic materials (A). From Geotz et al. (1975), as given by Williams (1983, p. 1748).

of uniform reflectance is illuminated at an angle, variations in image brightness carry information concerning the slopes of individual facets on the ground. Therefore, the full image, composed of many such pixels, depicts the shape of the terrain (Eliason et al., 1981).

The Earth's surface does not reflect uniformly, so the brightness caused by irregular terrain is mixed with the bightnesses caused by different surface materials. Remote sensing is usually concerned with the spectral differences that we observe at different places on the Earth's surfaces as clues to the local abundance of minerals and other resources. Therefore, the brightnesses caused by surface orientation are often regarded as complicating factors in our effort to extract and interpret spectral information.

As a result, the field of photoclinometry contributes to remote sensing by providing tools that permit analysts to separate the spectral information from the brightness that arises from the orientation of the surface. A diffuse reflector (Chapter 2) will have a brightness that is predictable from the angle of illumination (Figure 18.14). It is intuitively obvious that as the angle of illumination changes, so also will the brightness of the surface. The relationship between brightness and angle of illumination can be expressed more precisely by using *Lambert's cosine law* (Chapter 2):

$$G(x, y) = I \cos \theta(x, y) \qquad \text{(Eq. 18.1)}$$

where $G$ is the image gray tone value; $(x, y)$ are the image pixel row and column coordinates; $I$ is the intensity of the solar illumination; and $\theta$ is the angle of illumination, measured from the zenith.

In the context of remote sensing, the surface and the source of illumination have a fixed geometric relationship, but the scene itself is composed of many individual facets, each with specific orientation with respect to the solar beam. That is, the solar beam illuminates the Earth at a fixed angle at any given instant, but topographic irregularities cause the image to be formed of

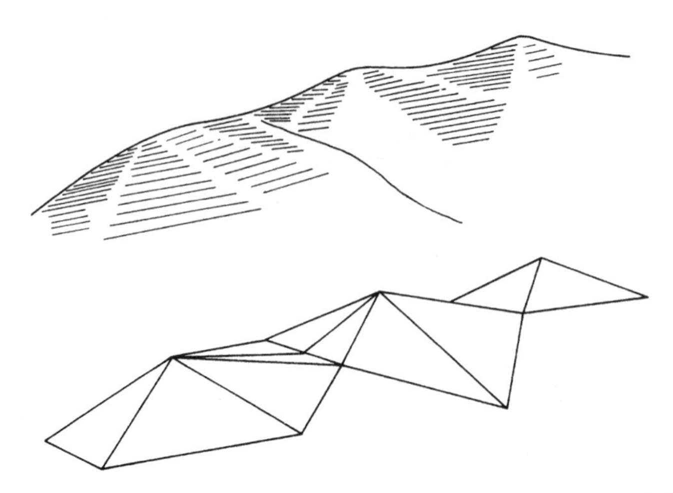

**FIGURE 18.14.** Brightnesses of surfaces depend upon the orientation of the surface in relation to the direction of illumination. If these surfaces all have the same surface material and behave in accordance with Lambert's cosine law, their brightnesses are related to the angles at which they are oriented.

many varied brightnesses. Thus, for a uniform surface of irregular topography, variations in $G$ portray variations in surface slope and orientation. Here, as a simplification, we will refer to such an image as the *topographic image* ($T$).

Although few surfaces on the Earth meet the assumption of uniform reflectance, some extraterrestrial surfaces display only small variations in reflectivity, primarily because of the absence of vegetative cover and differences in color caused by weathering of geologic materials. Photogeologists who have studied the Moon's surface and those of the planets have formalized the methods necessary to derive topographic information from images (Wildey and Pohn, 1964). Photoclinometry encompasses elements of remote sensing, photogrammetry, photometry, and radiometry. Although both the theory and the practice of photoclinometry must still deal with many unsolved issues, it has important applications in remote sensing.

For the present discussion, let us note simply that it is possible to approximate the brightnesses of the terrain, measured in a specific pair of bands, by a ratio between them:

$$\frac{G_{\text{band 1}}}{G_{\text{band 2}}} = c\,\frac{R_1}{R_2} \qquad \text{(Eq. 18.2)}$$

$G$ represents the brightness of the grey tone on an image of spectral band selected from a multiband image, and $c$ represents an arbitrary constant. This ratio image ($R_1/R_2$) has several useful properties.

First, the relationship holds both for shadowed pixels and for directly illuminated pixels. Therefore, a ratio image shows pure reflectance information without effects of topography. This result permits geologists to examine spectral properties of surfaces without the confusing effects of mixed brightnesses of topography and material reflectances. In the raw image, a difference in pixel brightness can be caused by a difference in slope, by shadowing, or by differences in the color of the surface material. In a ratio image, the geologist knows that differences in brightness are caused by differences in reflectance only. This result is illustrated by the example in Figure 18.15. In the upper diagram the brightness of the scene is caused by com-

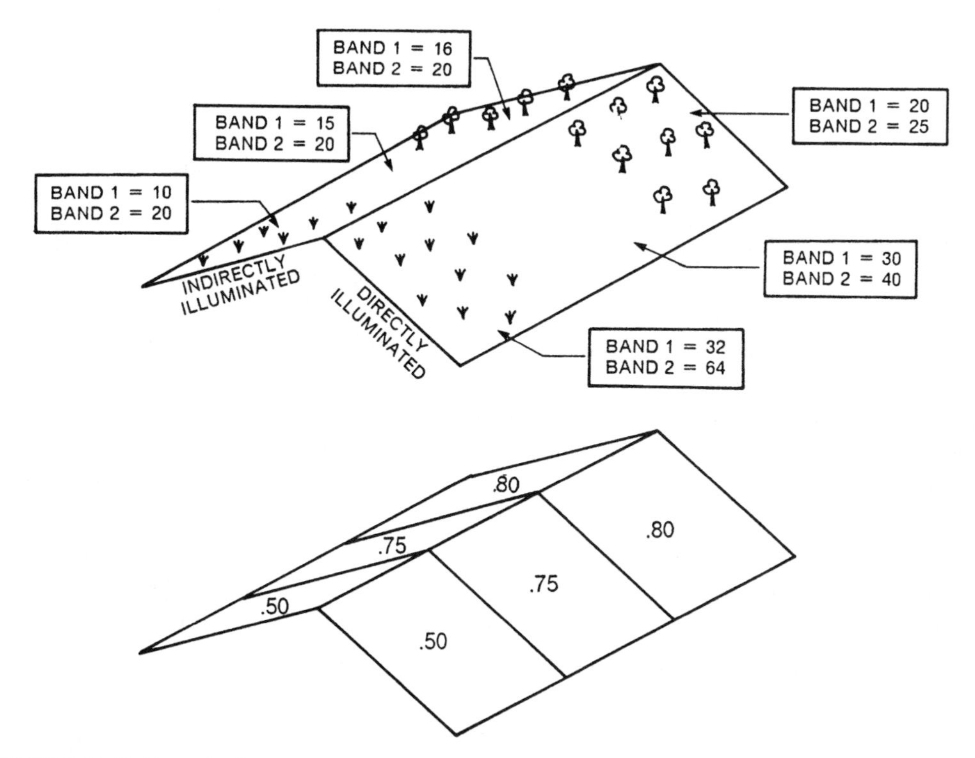


**FIGURE 18.15.** Example illustrating use of band ratios. In this contrived example, a topographic surface illuminated from the right side of the illustration consists of three contrasting surface materials, each partially illuminated by the solar beam, and on the reverse slope illuminated indirectly by diffuse light. Top: The raw digital values convey the mixed information of surface reflectance and shadowing. Because of shadowing, surface materials cannot be clearly separated into their distinctive classes. Bottom: Ratios of the two bands isolate the effects of the different surface materials (removing the topographic effects) to permit recognition of distinct spectral classes. Based on Taranick (1978, p. 28).

bined differences in topography and surface reflectance, and the analyst cannot resolve differences between the two effects. In the lower diagram the ratios clearly separate the separate categories of surface reflectance, even though each is present on two different topographic surfaces.

In the context of photoclinometry, ratio images have a special application. An unsupervised classification of the several ratio images derived from a multispectral image permits identification of regions on the Earth's surface that have uniform spectral behavior. Remember that the ratio images convey only spectral, not topographic, information, so a classification performed on ratio images, not on the original digital values, will produce classes based upon uniform spectral properties, regardless of slope.

## 18.8. Band Ratios

Band ratios have further significance in geologic remote sensing. By removing effects of shadowing, which otherwise are mixed with the spectral information necessary to make lithologic discriminations, they can convey purely spectral information. Furthermore, ratios may minimize differences in brightness between lithologic units (i.e., ratios tend to emphasize color information and to, deemphasize absolute brightness) and may facilitate comparisons of data collected on different dates, which will differ in solar angle.

Goetz et al. (1975) examined rock reflectances in Utah using a portable spectrometer capable of collecting high-resolution spectra for a variety of natural surfaces. They found several ratios to be useful in discriminating between lithologic units. Because their bands are based upon high-resolution data that do not match to familiar subdivisions of the spectrum, their ratios must be defined using their own designations, given in Figure 18.13. For example, a ratio between their bands 6 and 10 (roughly, red/near infrared) provided a good distinction between bare rock and vegetation-covered areas. Their 12/3 ratio (mid-IR/blue) was successful in distinguishing between limonitic and nonlimonitic rocks. The spectrum for limonite decreases sharply in channel 3 (blue) because of the presence of ferric iron, which absorbs strongly in this region, whereas nonlimonitic rocks tend to be brighter in this region. Thus the 12/3 ratio tends to accentuate differences between limonitic and nonlimonitic rocks. Hydrothermally altered rocks containing clay minerals display absorption bands in channel 13, so the 12/13 ratio tended to identify hydrothermally altered regions.

Plate 23 shows a ratio image of the Cuprite region, a mining area in southwestern Nevada near the California border. This image depicts a sparsely vegetated region of known geology and proven mineral resources about 4.5 km × 7.0 km (about 2.8 mi. × 4.3 mi.) in size. The geology here is composed essentially of tufts and basalts within an area otherwise composed chiefly of Cambrian limestones and clastic sediments.

This image was produced for a study proposed by Vincent et al. (1984), in which data from two airborne multispectral scanners were obtained to evaluate geological uses of 12 different spectral regions. Plate 23 is formed from a selection of these data. In this image, the intensity of each of the three primaries is determined by the value of the ratio between two spectral channels. As the value of the ratio varies from pixel to pixel, so does the intensity of the color at each pixel. Because the image is formed from three primaries, and each primary represents data from two bands, the image is derived from a total of six different bands.

Full description of the selection of spectral channels and their interpretation of the image would require an extensive summary of Vincent et al. (1984). However, a concise summary can be presented as follows:

1. The red band in the image is determined by the ratio (0.63–0.69 μm) : (0.52–0.60 μm); greater intensity of red indicates high levels of ferric oxide.
2. The green band is a function of the ratio (1.55–1.75 μm) : (2.08–2.35 μm); the intensity of green indicates areas where there is high clay content in the surface soil.
3. The blue band is controlled by the value of the ratio (8.20–8.54 μm) : (8.60–8.95 μm); the intensity of blue indicates levels of silica at the soil surface.

Thus a yellow color (both red and green) indicates the presence of both ferric oxide and clay, an indication of hydrothermal alteration, and the possible presence of a zone of mineralization. The blue region near the center of the image delineates a region of silification (high silica); it is surrounded by a zone of hydrothermal alternation, marked by yellow regions surrounding the blue center.

This example illustrates the power of using ratios between carefully selected spectral regions as aids for geological investigation. However, it is important to identify aspects of this study that differ from those that might be feasible on a routine basis. Although these data have 30-m resolution, they were collected at aircraft altitudes, so they are much less affected by atmospheric effects than would be the case for comparable satellite data. Also, this region has a sparsely vegetated and even topography, so it favors isolation of geological information that otherwise can be mixed with vegetation, shadowing, and topographic effects, as noted in the discussion of composite signatures in Chapter 15.

The use of ratio images carries certain risks, as mentioned in Chapter 17. In a geologic context, it is important that data be free of atmospheric effects or that such effects be removed by preprocessing. The significance of the atmosphere can be appreciated by reexamining the earlier discussion, where the diffuse light values in the two band values were estimated to have predictable values that would cancel each other out. If severe atmospheric effects are present, they will differ from one band to the other, and the value of the ratio will no longer portray only spectral properties of the ground surface but instead will have values greatly altered by the varied atmospheric contributions to the separate bands.

## 18.9. Soil and Landscape Mapping

Soil scientists study the mantle of soil at the Earth's surface in an effort to understand soil formation and to map patterns of soil variation. *Soil* is a complex mixture of inorganic material weathered from the geologic substratum and mixed with decayed organic matter from tissues of plants and animals. Typically, a "soil" consists of three layers (known as *soil horizons*) of varying thickness and composition (Figure 18.16). Nearest the surface is the A horizon, usually dark in color and rich in decayed organic matter. The A horizon is sometimes known popularly as *topsoil*—the kind of soil one would like to have for a lawn or garden. In this layer plant roots are abundant, as are microorganisms, insects, and other animals. Below the A horizon is the second layer, the B horizon, usually lighter in color and more compact, where plant roots and biologic activity are less abundant. Sometimes the B horizon is known as the *subsoil*—the kind of hard infertile soil one would prefer *not* to have at the surface of a lawn or garden. At even greater depth is the C horizon, the deepest pedologic horizon, which consists of weathered geologic material, decayed or fractured into material that is softer and looser than the unweathered geologic strata below. This is the material that usually forms the raw material for the A and B horizons. Finally, below the C horizon is the R horizon, consisting of unaltered bedrock.

The exact nature of the horizons at a given place is determined by the interaction between local climatic and topographic, geologic, and biologic elements as they act over time. Because varied combinations of climate, vegetation, and topography produce different soils from similar geologic materials, soil science is distinct from, although closely related to, geology and geomorphology.

The landscape is covered by a mosaic of patches of different kinds of soil, each distinct from its neighbors with respect to character and thickness (Figure 18.17). Soil surveyors outline on maps those areas covered by specific kinds of soil as a means of showing the variation of soils on the landscape. Each symbol represents a specific kind of soil, or, when the pattern is very complex, sometimes two or three kinds of soil that occur in an intimate pattern (Figure 18.18). Soil maps portray distributions of pedologic units, which together with other maps and data convey valuable information concerning topography, geology, geomorphology, hydrology, and other landscape elements. In the hands of a knowledgeable reader, they convey a comprehensive picture of the physical landscape. As a result, they can be considered to be among the most

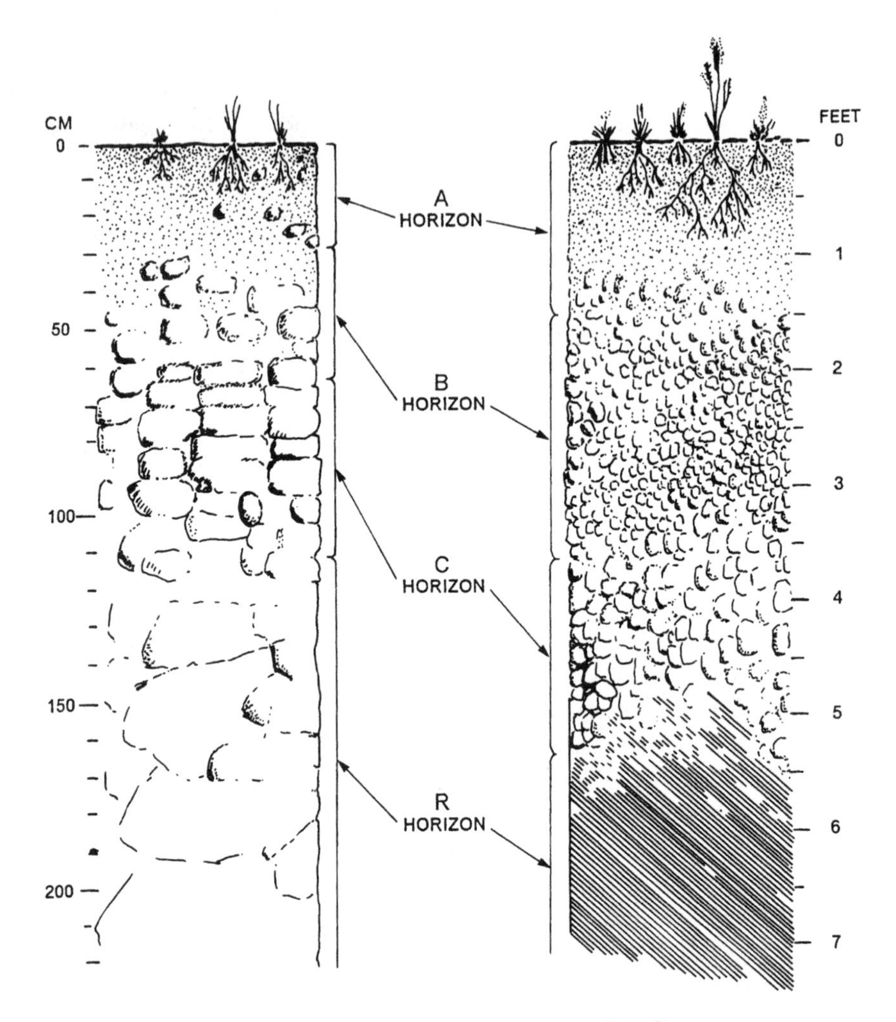


**FIGURE 18.16.** Sketches of two contrasting soil profiles.

practical of all forms of landscape maps, and are used by farmers, planners, and others who must judge the best locations for specific agricultural activities, for community facilities, or for construction of buildings and highways.

Soil mapping is conducted routinely by national soil surveys (or their equivalents) in most countries, and by their regional counterparts at lower administrative levels. Details of mapping technique vary from one organization to another, but the main outlines of the procedure are common to most soil survey organizations. A soil map is formed by subdividing the landscape into a mosaic of discrete parcels. Each parcel is assigned to a class of soil—a *mapping unit*—that is characterized by specific kinds of soil horizons. Each mapping unit is represented on the map by a specific symbol—usually a color or an alphanumeric designation. In theory, mapping units are homogeneous with respect to pedologic properties, although experienced users of soil maps acknowledge the existence of internal variation as well as the presence of foreign inclusions. Mapping units are usually defined with links to a broader system of soil classification defined by the soil survey organization, so that mapping units are consistently defined across,

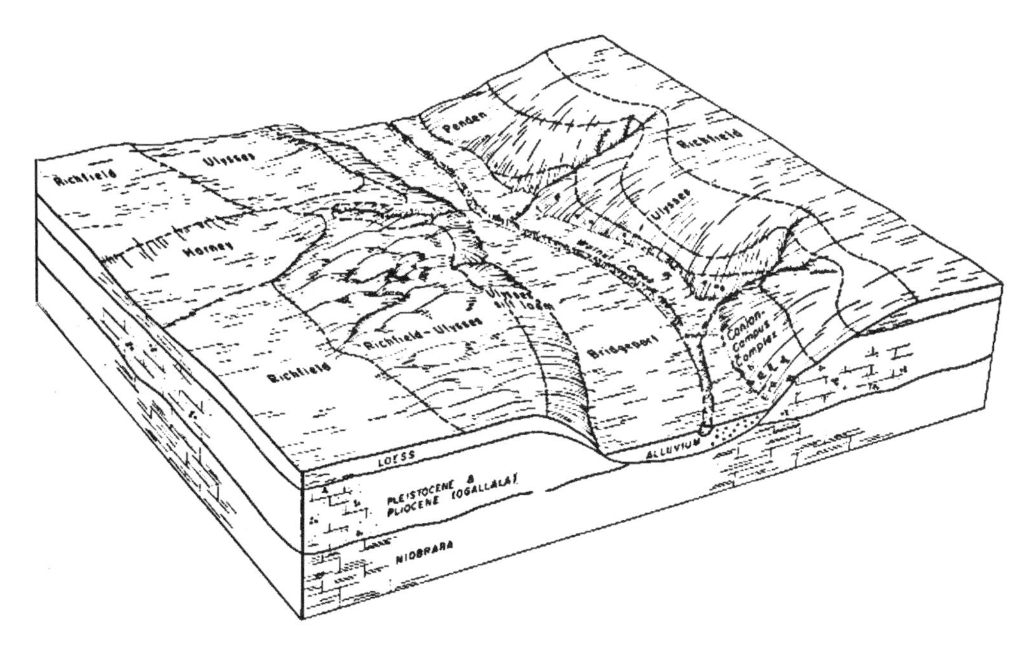


**FIGURE 18.17.** The soil landscape. From USDA Natural Resources Conservation Service.

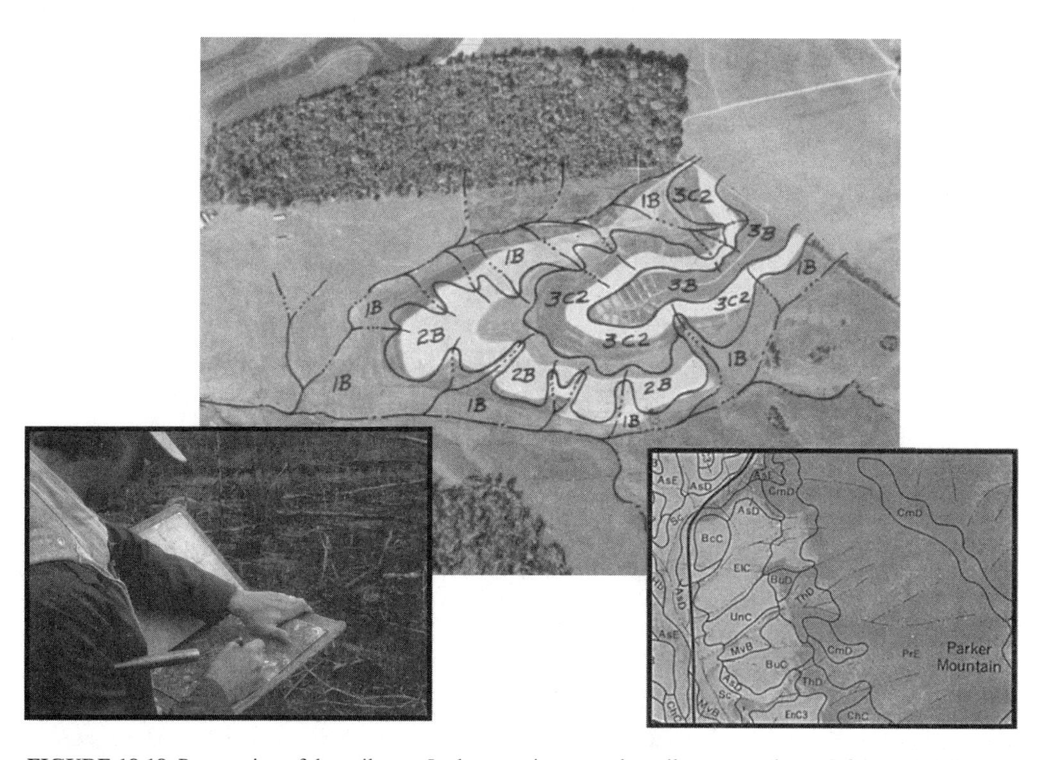


**FIGURE 18.18.** Preparation of the soil map. In the mapping step, the soil surveyor (lower left image) delineates mapping units on aerial photographs (central image); eventually these mapping units are refined and combined to form the units presented on the final soil map (lower right image). From Virginia Cooperative Soil Survey (lower left and central image) and USDA Natural Resources Conservation Service (lower right image).

for example, county and state borders. For each region, a mapping unit can be evaluated with respect to the kind of agriculture that is important locally, so the map can serve as a guide to select the best uses for each soil. The following paragraphs outline an example of the soil survey process, although details may vary from one organization to the next.

The first step in the actual preparation of the draft map is the *mapping* process (Figure 18.18). The term *mapping* in this context means specifically to designate the delineation of landscape units on an aerial photograph or map base, rather than in the broader sense that refers to the entire set of activities necessary to generate the map. In the mapping step, the soil scientist examines aerial photographs, often using a stereoscope, to define the major boundaries between classes of soil. The soil scientist has already learned much about the region's soils through field observations, so he or she has knowledge of the kinds of soils present and their approximate locations. The soil scientist can use this field knowledge in interpretation of the photograph, using breaks in slope, boundaries between vegetation classes, and drainage patterns to define boundaries between soil mapping units. Seldom if ever can the soil classes actually be identified from the photographs—identification must be based upon field observations—but in the mapping step, the scientist can subdivide the landscape into a mosaic of soil parcels, which are then each treated as independent units in the later steps.

Completion of the mapping step requires considerable field experience within the region to be mapped, as the surveyor must acquire knowledge of the numbers and kinds of soils present within a region, their properties and occurrence, and their uses and limitations. As a result, completion of this step requires a much longer time than that actually devoted to marking the aerial photographs if we include the time required to learn about the local landscape. Morphological descriptions of each mapping unit are made in the field, and mapping units are sampled for later analysis.

The second step is *characterization.* Samples are collected from each prospective mapping unit, then subjected to laboratory analysis for physical, chemical, and mineralogical properties.

These measurements form the basis for *classification,* the third step. In the United States, classification is the implementation of the classification criteria specified by Soil Taxonomy, the official classification system of the Soil Conservation Service, U.S. Department of Agriculture. In other countries, classification is conducted by applying the classification criteria established by each national soil survey organization.

*Correlation,* the fourth, and possibly the most difficult, step, matches mapping units within the mapped region to those in adjacent regions and to those in ecologically analogous areas. Whereas the other steps in a soil survey may be conducted largely on a local basis, correlation requires the participation of experienced scientists from adjacent regions, and even those from national or international levels in the organization, to provide the broader experience and perspective often required for successful correlation.

The fifth and final step is *interpretation,* in which each mapping unit is evaluated with respect to prospective agricultural and engineering uses. The interpretation step provides the user of the map with information concerning the likely suitability of each mapping unit to the land uses most common within the region.

Within this rather broad framework there are, of course, many variations in details of technique and in overall philosophy and strategy. The result is a map that shows the pattern of soils within a region and a report that describes the kind of soil that is encountered within each mapping unit. Each mapping unit is evaluated with regard to the kinds of uses that might be possible for the region, so that the map serves as a guide to wise use of soil resources.

## 18.10. Integrated Terrain Units

Definition of *integrated terrain units* is based upon the concept that the varied and complex assemblages of soil, terrain, vegetation, and so on form distinctive spectral responses that can be recognized and mapped. This general approach has long been applied, under many different names, to the interpretation of resource information from aerial photographs and other forms of imagery. The best known, and oldest, of these methods is known as *land system mapping.*

Land system mapping subdivides a region into sets of recurring landscape elements based on comprehensive examination of distributions of soils, vegetation, hydrology, and physiography. The formalization of land system mapping procedures dates from the period just after World War II, although some of the fundamental concepts had been defined earlier. Today a family of similar systems are in use; all are based on similar principles and methods, although the terminology and details vary considerably. One of the oldest, best known, and most formally defined systems is that developed by the Australian CISRO Division of Land Research and Regional Survey (Christian, 1959); variations have been developed and applied throughout the world.

The method is based on a hierarchical subdivision of landscapes. *Land systems* are recurring contiguous associations of landforms, soils, and vegetation, composed of component *land units.* The basic units of the system (which are assigned varied definitions and names in alternative versions of this basic strategy) are areas of uniform lithology with relatively uniform soil and drainage. A characteristic feature of all versions of the land system method is a hierarchical spatial organization, so that subcategories are nested within the broader categories defined at higher levels. Designation of separate levels within the hierarchy differs among alternative versions. For example, Thomas (1969) defines the following sequence: *site* (the smallest units), *facet, unit landform, landform complex, landform system,* and *landform region.* Wendt et al. (1975) define a system using *landtypes* (the smallest regions), *landtype associations, landtype phases, subsections, sections,* and *provinces.* Although many versions imply that both land systems and their component elements recur in widely separated geographic regions (although under analogous ecological circumstances), closer examination of the method (Mitchell and Perrin, 1970) reveals that land systems are most effective when defined to be essentially local units.

Implicit, if not explicit, in most applications is the use of aerial photography as a primary tool for land system mapping, together with direct observation in the field. Aerial photography provides the broad overview and the map-like perspective favoring convenient definition and delineation of land systems, and portrays the complex spatial patterns of topography, vegetation, and drainage in an integrated form that is compatible with the assumptions, methods, and objectives of this approach to terrain analysis. "There would be no point in defining any terrain class if its chances of being recognized from the air photographs and background information were small" (Webster and Beckett, 1970, p. 54).

An assumption of the procedure is that easily identified landscape features (e.g., vegetation and physiography) form surrogates for more subtle soil features not as easily defined from analysis of aerial photographs. Although the procedure has been applied at a range of scales from 1:1,000,000 to 1:25,000, most applications are probably at fairly broad scales, for reconnaissance surveys, or mapping of rather simple landscape divisions under circumstances where the ultimate use will be rather extensive.

The method and its many variations have been criticized for their subjectivity and the variability of the results obtained by different analysts. Compared to most large-scale soil surveys, land system mapping presents a rather rough subdivision of the landscape, as the mapping units display much greater internal variability and are not as carefully defined and correlated as one would expect in more intensive surveys. As a result, the procedure may be best suited for application in reconnaissance mapping, where broad-scale, low-resolution mapping is required as the basis for planning more detailed surveys.

The method of integrated terrain units has also been applied to analysis of digital data. At the coarse resolution of satellite sensors, individual spectral responses from many landscape components are combined to form composite responses. In this context, the usual methods of image classification for mapping of individual classes of soil, vegetation, or geology may be extraordinarily difficult. Therefore, many analysts have applied the integrated terrain unit strategy to digital classification in an effort to define image classes in a more realistic manner. In the ideal, we might prefer to define pure categories that each show only a single thematic class. But, given the complexity of the natural landscape, composite categories are often well suited to representation of the gradations and mixtures that characterize many environments (Robinove, 1981; Green, 1986).

## 18.11. Wetlands Inventory

*Wetlands* are areas of land characterized by saturated or inundated soil. In the United States, wetlands are legally defined by the presence of (1) hydrophytic vegetation (plants that occur predominantly in wetlands environments); (2) a wetlands hydrologic regime, including evidence of inundation or saturation; and (3) diagnostic pedologic features caused by low oxygen levels, most notably gray or black colors, often with distinctive mottling indicative of prolonged saturation. Wetlands are not defined merely by proximity to bodies of open water, but are encountered in a wide variety of ecological and physiographic settings, including uplands.

Understanding the occurrence of wetlands is significant for a broad spectrum of policy issues, including land-use regulation, water quality, flood abatement, agriculture, and carbon sequestration. As a result, mapping their occurrence and extent is an important responsibility of local, state, and federal governments. For some purposes, field studies are necessary to identify and delineate the extent of wetlands.

However, for other purposes, remotely sensed imagery is an important tool for inventory and monitoring of wetlands. At a national level in the United States, the National Wetlands Inventory (NWI), conducted by the U.S. Fish and Wildlife Service, prepares a national inventory based upon photointerpretation of NHAP imagery. Wetlands vary greatly in where they occur and in their other distinctive properties (Lyon, 2001). Also, optical, thermal, and SAR data are each capable of capturing wetlands characteristics that the others can't. Since wetlands do vary so greatly, and since they vary seasonally, remote sensing of wetlands cannot be adequately performed by a single approach. It seems more likely that remotely sensed imagery may be the central resource for a suite of data that might be useful in delineating and monitoring wetlands.

## 18.12. Radar Imagery for Exploration

As a conclusion to this chapter, Figure 18.19 illustrates applications of RADARSAT SAR data to offshore petroleum exploration. Natural seepage of petroleum from the ocean floor causes a thin film of petroleum to float on the ocean surface (a "slick"), that creates a locally smooth

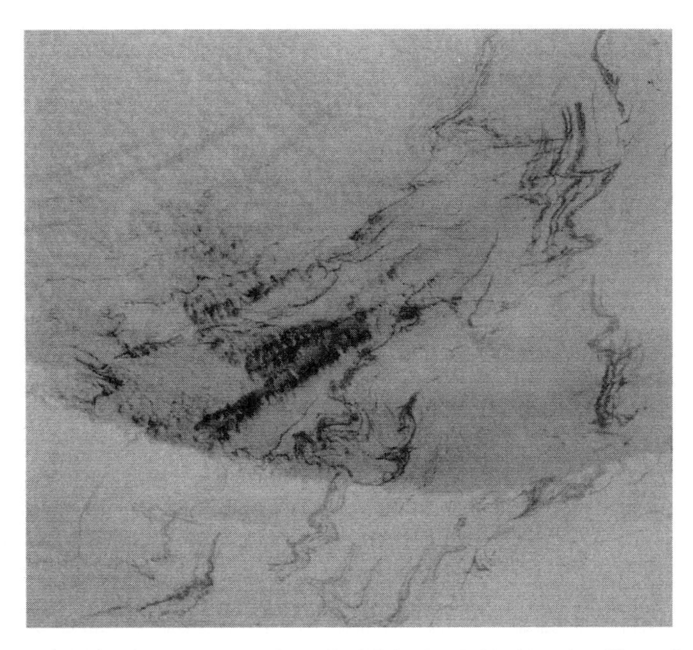

**FIGURE 18.19.** RADARSAT SAR imagery of an oil slick in the Gulf of Mexico, illustrating applications of petroleum exploration. Image copyright by Canadian Space Agency. Received by the Canada Center for Remote Sensing; processed and distributed by RADARSAT International.

water surface, that in turn induces conditions that promote specular reflection over that region of the ocean. Oil slicks occur when molecules of oil reach the sea surface to form a thin layer of petroleum that dampens the ocean-surface capillary waves. SAR technology is sensitive to differences in surface roughness, so it can easily discriminate between the smoother oil slick and the surrounding rough water. Thus oil slicks are regions of little to no backscatter, characterized by distinct areas of darkness on the radar image. Therefore, the dark returns assist in identifying regions where such slicks might originate. Because SAR has the ability to observe such regions day and night, and over a wide a wide range of meteorological conditions, SAR data assist in identifying regions where such slicks form on a regular basis. Critical exploration maps can be produced by using a GIS system to overlay other information such as coastline, bathymetry, gravimetric data, shipping lanes, and existing oil rigs on the RADARSAT image. These maps allow decision makers to reduce magnetic field exploration risk and increase the cost-effectiveness of offshore drilling efforts.

## 18.13. Summary

Aerial photography and remote sensing have long been applied to problems in the earth sciences. Geology, topographic mapping, and related topics formed one of the earliest routine applications of aerial photography, and today remote sensing, used with other techniques, continues to form one of the most important tools for geologic mapping, exploration, and research.

Research for geological remote sensing spans the full range of subjects within the field of remote sensing, including use of additional data in the thermal, near infrared, visible, and microwave regions of the spectrum. More than most other subjects, efforts such as geobotanical research are especially interesting because they bring knowledge of so many different disciplines together to bear on a single problem. In addition, geological investigations have formed important facets of research and applications programs for Landsat, SPOT, RADARSAT, ERS–1, and other systems. Geological studies are said to form a major economic component of the practice of remote sensing. Certainly, they form one of the most important elements in both theoretical and practical advances in remote sensing.

## Review Questions

1. Why is the timing of overpasses of Earth observation satellites such as Landsat or SPOT likely to be of special significance for geologic remote sensing?
2. Explain why the ability to monitor the presence of moisture (both from open water bodies and soil moisture at the ground surface) might be of special significance in geologic, geomorphic, and pedologic studies.
3. In the past there has been discussion of conflicts between earth scientists and plant scientists concerning the design of spectral sensitivity of sensors for the MSS and TM, based upon the notion that the two fields have quite different requirements for spectral information. Based upon information in this chapter, explain why such a distinct separation of information requirements may not be sensible.
4. Why should Landsat imagery have been such an important innovation in studies of geologic structure?
5. Summarize the significance of lineaments in geologic studies, and the significance of remote sensing to the study of lineaments.
6. Compare the relative advantages and disadvantages of photogeology and geologic remote sensing, as outlined in this chapter.
7. Some might consider lineament analysis, application of geobotany, and other elements of geologic remote sensing to be modern, state-of-the-art techniques. If so, why should photogeology, which is certainly not as technologically sophisticated, be so widely practiced today?
8. Full application of the principles and techniques of geologic remote sensing requires much finer spectral, radiometric, and spatial resolution than most operational sensors have at present. What difficulties can you envision in attempts to use such sensors for routine geologic studies?
9. Refer back to the first paragraphs of Section 12.1; compare and contrast the concept of image classification with that of integrated terrain units (Section 18.9). Refer also to Section 5.10. Write two or three paragraphs that summarize the major differences between the two strategies.
10. In what ways might radar imagery (Chapter 7) be especially useful for studies of geology and other earth sciences?
11. Write a short description of a design for a multispectral remote sensing system tailored specifically for recording geologic information. Disregard all other applications. Suggest how you would select the optimum spatial and radiometric resolution and the most useful spectral regions.

## References

### *General*


Abrams, M. J., D. Brown, L. Lepley, and R. Sadowski. 1983. Remote Sensing for Porphyry Copper Deposits in Southern Arizona. *Economic Geology,* Vol. 78, pp. 591–604.

Chang, S., and W. Collins. 1983. Confirmation of the Airborne Biogeophysical Mineral Exploration Technique Using Laboratory Methods. *Economic Geology,* Vol. 78, pp. 723–736.

Clarke, J. I. 1966. Morphometry from Maps. In *Essays in Geomorphology* (G. H. Dury, ed.). New York: American Elsevier, pp. 235–274.

Cole, M. M. 1982. Integrated Use of Remote Sensing Imagery in Mineral Exploration. *Geological Society of America Transactions,* Vol. 85, Part 1, pp. 13–28.

Crippen, R. E., and R. G. Blom. 2001. Unveiling the Lithology of Vegetated Terrains in Remotely Sensed Imagery. *Photogrammetric Engineering and Remote Sensing,* Vol. 67, pp. 935–943.

Goetz, A. F. H. 1984. High Spectral Resolution Remote Sensing of the Land. In *Remote Sensing* (P. N. Slater, ed.). Bellingham, WA: SPIE, pp. 5668.

Goetz, A. F. H., B. N. Rock, and L. C. Rowan. 1983. Remote Sensing for Exploration: An Overview. *Economic Geology,* Vol. 78, pp. 573–590.

Goetz, A. F. H., G. Vane, J. E. Solomon, and B. N. Rock. 1985. Imaging Spectrometry for Earth Remote Sensing. *Science,* Vol. 228, pp. 1147–1153.

Gupta, R. P. 1991. *Remote Sensing Geology.* New York: Springer-Verlag, 356 pp.

King, R. B. 1970. A Parametric Approach to Land System Classification. *Geoderma,* Vol. 4, pp. 37–46.

Lyon, J. G. 1993. *Practical Handbook for Wetland Identification and Delineation.* Lewis.

Lyon, J. G. 2001. *Wetland Landscape Characterization: GIS, Remote Sensing and Image Analysis.* Lewis.

Lyon, J. G. and J. McCarthy. 1995. *Wetland and Environmental Applications of GIS.* Lewis Publishers.

Meyer, B. S., D. B. Anderson, R. H. Bohning, and D. G. Fratianne. 1973. *Introduction to Plant Physiology.* New York: Van Norstrand, 565 pp.

Mitchell, C. W. 1973. *Terrain Evaluation.* London: Longman, 221 pp.

Peel, R. F., L. F. Curtis, and E. C. Barrett. 1977. *Remote Sensing of the Terrestrial Environment.* London: Butterworth, 275 pp.

Rast, M., J. S. Hook, C. D. Elvidge, and R. E. Alley. 1991. An Evaluation of Techniques for the Extraction of Mineral Absorption Features from High Spectral Resolution Remote Sensing Data. *Photogrammetric Engineering and Remote Sensing,* Vol. 57, pp. 1303–1309.

Rowan, L. C., A. F. H. Goetz, and R. P. Ashley. 1977. Discrimination of Hydrothermally Altered and Unaltered Rocks in Visible and Near Infrared Multispectral Images. *Geophysics,* Vol. 42, pp. 522–535.

Schneider, S. R., D. F. McGinnis, and J. A. Pritchard. 1979. Use of Satellite Infrared Data for Geomorphology Studies *Remote Sensing of Environment,* Vol. 8, pp. 313–330.

Settle, M. 1982. Use of Remote Sensing Techniques for Geologic Mapping. In *Proceedings, Harvard Computer Graphics Week.* Cambridge, MA: Graduate School of Design, Harvard University, pp. 1–12.

Siegal, B. S., and A. R. Gillespie (eds.). 1980. Remote Sensing in Geology. New York: Wiley, 702 pp.

Taranick, J. V. 1978. *Characteristics of the Landsat Multispectral Data System.* (USGS Open File Report 78-187).

Watson, K. 1975. Geologic Applications of Thermal Infrared Imagery. *Proceedings of the IEEE,* Vol. 63, pp. 128–137.

Wildey, R. L., and H. A. Pohn. 1964. Detailed Photoelectric Photometry of the Moon. *Astronomical Journal,* Vol. 69, pp. 619–634.

### *Photogeology*

Lueder, D. R. 1959. *Aerial Photographic Interpretation: Principles and Applications.* New York: McGraw-Hill, 462 pp.

Miller, V. C. 1961. *Photogeology.* New York: McGraw-Hill, 248 pp.

Ray, R. G. 1960. *Aerial Photographs in Geologic Interpretation and Mapping* (Geological Survey Professional Paper 373). Washington, DC: U.S. Government Printing Office, 230 pp.

### *Lineaments*

Berlin, G. L., P. S. Chavez, T. E. Grow, and L. A. Soderblom. 1976. Preliminary Geologic Analysis of Southwest Jordon from Computer Enhanced Landsat I Image Data. In *Proceedings, American Society of Photogrammetry.* Bethesda, MD: American Society of Photogrammetry, pp. 543–563.

Casas, A. M., A. L. Cortés, A. Maestro, M. Asunción Soriano, A. Riaguas, and J. Bernal. 2000. LINDENS: A Program for Lineament Length and Density Analysis. *Computers and Geosciences,* Vol. 26, pp. 1011–1022.

Chavez, P. S., G. L. Berlin, and A. V. Costa. 1976. Computer Processing of Landsat MSS Digital Data for Linear Enhancements. In *Proceedings of the Second Annual William T. Percora Memorial Symposium.*

Clark, C. D., and C. Watson. 1994. Spatial Analysis of Lineaments. *Computers and Geosciences,* Vol. 20, pp. 1237–1258.

Ehrich, R. W. 1977. Detection of Global Edges in Textured Images. *IEEE Transactions on Computers,* Vol. C–26, pp. 589–603.

Ehrich, R. W. 1979. Detection of Global Lines and Edges in Heavily Textured Images. In *Proceedings, Second International Symposium on Basement Tectonics.* Newark, DE: pp. 508–513.

Moore, G. K., and F. A. Waltz. 1983. Objective Procedures for Lineament Enhancement and Extraction. *Photogrammetric Engineering and Remote Sensing,* Vol. 49, pp. 641–647.

O'Leary, D. W., J. D. Friedman, and H. A. Pohn. 1976. Lineament, Linear, Lineation: Some Proposed New Standards for Old Terms. *Geological Society of America Bulletin,* Vol. 87, pp. 1463–1469.

Podwysocki, M. H., J. G. Moik, and W. D. Shoup. 1975. Quantification of Geologic Lineaments by Manual and Machine Processing Techniques. In *Proceedings, NASA Earth Resources Survey Symposium.* Houston, TX: NASA, pp. 885–903.

Raghavan, V., K. Wadatsumi, and S. Masummoto. 1994. SMILES: A FORTRAN–77 Program for Sequential Machine Interpreted Lineament Extraction Using Digital Images. *Computers and Geosciences,* Vol. 20, pp. 121–159.

Ustin, S. L., M. D. Smith, S. Jacquemond, M. Verstrate, and Y. Govaerts. 1999. Geobotany: Vegetation Mapping for the Earth Sciences Chapter 3 in *Manual of Remote Sensing* (*Remote Sensing for the Earth Sciences*) 3rd edition,Volume 3. New York: John Wiley and Sons, pp. 198–248.

Vanderbrug, G. J. 1976. Line Detection in Satellite Imagery. *IEEE Transactions on Geoscience Electronics,* Vol. GE–14, pp. 37–44.

Vaughan, R. W. 1983. *A Topographic Approach for Lineament Recognition in Satellite Images.* Unpublished M. S. thesis, Department of Computer Science, Virginia Polytechnic Institute, Blacksburg, 70 pp.

Wheeler, R. L., and D. U. Wise. 1983. Linesmanship and the Practice of Linear Geo-Art: Discussion and Reply. *Geological Society of America Bulletin,* Vol. 94, pp. 1377–1379.

Wise, D. U. 1982. Linesmanship and the Practice of Linear Geo-Art. *Geological Society of America Bulletin,* Vol. 93, pp. 886–888.

Yamaguchi, Y. 1985. Image-Scale and Look-Direction Effects on the Detectability of Lineamants in Radar Images. *Remote Sensing of Environment,* Vol. 17, pp. 117–127.

## *Integrated Terrain Units*

Christian, C. S. 1959. The Eco-Complex and Its Importance for Agricultural Assessment. Chapter 36 in *Biogeography and Ecology in Australia. Monographiae Biologicae,* Vol. 8, pp. 587–605.

Green, G. M. 1986. Use of SIR-A and Landsat MSS Data in Mapping Shrub and Intershrub Vegetation at Koonamore, South Australia. *Photogrammetric Engineering and Remote Sensing,* Vol. 52, pp. 659–670.

Robinove, C. J. 1981. The Logic of Multispectral Classification and Mapping of Land. *Remote Sensing of Environment,* Vol. 11, pp. 231–244.

Wendt, G. E., R. A. Thompson, and K. N. Larson. 1975. *Land Systems Inventory: Boise National Forest, Idaho. A Basic Inventory for Planning and Management.* Ogden, UT: U. S. Forest Service Intermountain Region, 49 pp.

## *Soils and Geomorphology*

Agbu, P. A., and E. Nizeyimana. 1991. Comparisons between Spectral Mapping Units Derived from SPOT Image Texture and Field Mapping Units. *Photogrammetric Engineering and Remote Sensing,* Vol. 57, pp. 397–405.

Bleeker, P., and J. G. Speight. 1978. Soil–Landform Relationships at Two Localities in Papua New Guinea. *Geoderma,* Vol. 21, pp. 183–198.

Goudie, A., et al. 1981. *Geomorphological Techniques.* London: Allen & Unwin, 395 pp.

Post, D. F., E. H. Horvath, W. M. Lucas, S. A. White, M. J. Ehasz, and A. K. Batchily. 1994. Relations between Soil Color and Landsat Reflectance on Semiarid Rangelands. *Soil Science Society of America Journal,* Vol. 58, pp. 1809–1816.

Soil Survey Staff, 1966. *Aerial Photointerpretation in Classifying and Mapping Soils* (Agricultural Handbook 294). Washington DC: U.S. Department of Agriculture, 89 pp.

Thornbury, W. D. 1967. *Regional Geomorphology of the United States.* New York: Wiley, 609 pp.

Townshend, J. R. G. (ed.). 1981. *Terrain Analysis and Remote Sensing.* London: Allen & Unwin, 232 pp.

Webster, R., and P. H. T. Beckett. 1970. Terrain Classification and Evaluation by Air Photography: A Review of Recent Work at Oxford. *Photogrammetria,* Vol. 26, pp. 51–75.

Weismiller, R. A., and S. A. Kaminsky. 1978. An Overview of Remote Sensing as Related to Soil Survey Research. *Journal of Soil and Water Conservation,* Vol. 33, pp. 287–289.

Wright, J. S., T. C. Vogel, A. R. Pearson, and J. A. Messmore. 1981. *Terrain Analysis Procedural Guide for Soil* (ETL-0254). Ft. Belvoir, VA: U. S. Army Corps of Engineers, Geographic Services Laboratory, Engineering Topographic Laboratories, 89 pp.

## *Interferometric SAR*

Clark, C. D., and C. Wilson. 1994. Spatial Analysis of Lineaments. *Computers and Geosciences,* Vol. 20, No. 7–8, pp. 1237–1258.

Eliason, P. T., L. A. Soderblom, and P. S. Chavez. 1981. Extraction of Topographic and Spectral Albedo Information from Multispectral Images. *Photogrammetric Engineering and Remote Sensing,* Vol. 48, pp. 1571–1579.

Franklin, S. E., M. B. Lavigne, E. R. Hunt Jr., B. A. Wilson, D. R. Peddle, G. J. McDermid, and P. T. Giles. 1995. Topographic Dependence of Synthetic Aperture Radar Imagery. *Computers and Geosciences,* Vol. 21, No. 4, pp. 521–532.

Glenn, Nancy F., and J. R. Carr. 2003. The use of geostatistics in relating soil moisture to RADARSAT-1 SAR Data Obtained over the Great Basin, Nevada, USA. *Computers and Geosciences,* Vol. 29, No. 5, pp. 577-586.

Kervyn, F. 2001. Modelling Topography with SAR Interferometry: Illustrations of a Favorable and Less Favorable Environment. *Computers and Geosciences,* Vol. 27, pp. 1039–1050.

Massonnet, D., and K. L. Feigl. 1998. Radar Inteferometry and application to changes in the Earth's Surface. *Reviews of Geophysics,* Vol. 36, no. 4. pp. 441–500.

Raptis, V. S., R. A. Vaughn, and G. G. Wright. 2003. The Effect of Scaling on Land Classification from Satellite Data. *Computers and Geosciences,* Vol. 29, No. 6, pp. 705–714.

Strozzi, T., U. Wegmüller, L. Tosi, G. Bitelli, and V. Spreckels. 2001. Land Subsidence Monitoring with Differential SAR Interferometry. *Photogrammetric Engineering and Remote Sensing,* Vol. 67, pp. 1261–1270.

CHAPTER NINETEEN

# Hydrospheric Sciences

## 19.1. Introduction

Open water covers about 74% of our planet's surface. Oceans account for about 95% of the surface area of open water, but freshwater lakes and rivers (about 0.4%) have a significance for humans that exceeds their small areal extent (Table 19.1). In addition, soil and rock near the Earth's surface hold significant quantities of freshwater (but only about 0.01% by volume of the Earth's total water), as do the ice and snow of the polar regions. Moisture in the form of "permanent" ice (about 5% of the earth's surface) is largely beyond the reach of mankind, although the seasonal accumulation and melt of snowpack in temperate mountains is an important source of moisture for some agricultural regions in otherwise arid zones (such as the Great Basin of North America). Hydrologists and meteorologists monitor water as it occurs in all these forms, as it changes from liquid to vapor, condenses to rain and snow, and moves on and under the earth's surface (Figure 19.1). In addition, studies of sea ice, movement of pollutants, and ocean currents are among the many other important subjects to attract the attention of the many scientists who study hydrology, oceanography, and related subjects.

Most traditional means of monitoring the Earth's water depend largely upon measurements made at specific points or collections of samples from discrete locations. Oceans, lakes, and rivers can be studied by collecting samples at the surface or by using special devices that fill a collection bottle with water from a specified depth. Groundwater can be studied by collecting samples from wells or boreholes. Samples can be subjected to chemical and physical tests to measure levels of pollutants, to detect bacteria and other biological phenomena, and to determine oxygen content, sediment content, and many other qualities of the water.

Such measurements or samples, of course, can only provide information about discrete points within the water body, whereas the analyst usually is interested in examining entire water bodies or regions within water bodies. Although measurements can be collected at several locations to build up a record of place-to-place variation within the water body, such efforts are at best a piecemeal approach to studying the complex and dynamic characteristics of water bodies that are of interest to hydrologists.

Therefore, remote sensing can provide a valuable perspective concerning broad-scale, dynamic patterns that can be difficult to examine in detail using point measurements only. Careful coordination and choice of places to collect surface samples permits establishment of the relationship between the sample data and those collected by the remote sensor. Remotely

**TABLE 19.1. Water on Earth**

| | Percentage | |
|---|---|---|
| | By surface area[a] | By volume[b] |
| Oceans | 94.90 | 97.1 |
| Rivers and lakes | 0.40 | 0.02 |
| Groundwater | — | 0.60 |
| Permanent ice cap | 4.69 | 2.20 |
| Earth's atmosphere | — | 0.001 |

*Note.* Calculated from values given by Nace (1967).
[a]Percentage by area of Earth's total water surface.
[b]Percentage by volume of Earth's water.

sensed data can be especially valuable in studying phenomena over large areas. Satellite sensors provide the opportunity for regular observation of even the most remote regions. Although remotely sensed images seldom replace the usual sources of information concerning water resources, they can provide valuable supplements to field data by revealing broad-scale patterns not recognizable at the surface, recording changes over time, and providing data for inaccessible regions.

## 19.2. Spectral Characteristics of Water Bodies

The spectral qualities of water bodies are determined by the interaction of several factors, including the radiation incident to the water surface, the optical properties of the water, the roughness of the surface, angles of observation and illumination, and, in some instances, reflection of light from the bottom (Figure 19.2). As incident light strikes the water surface, some is

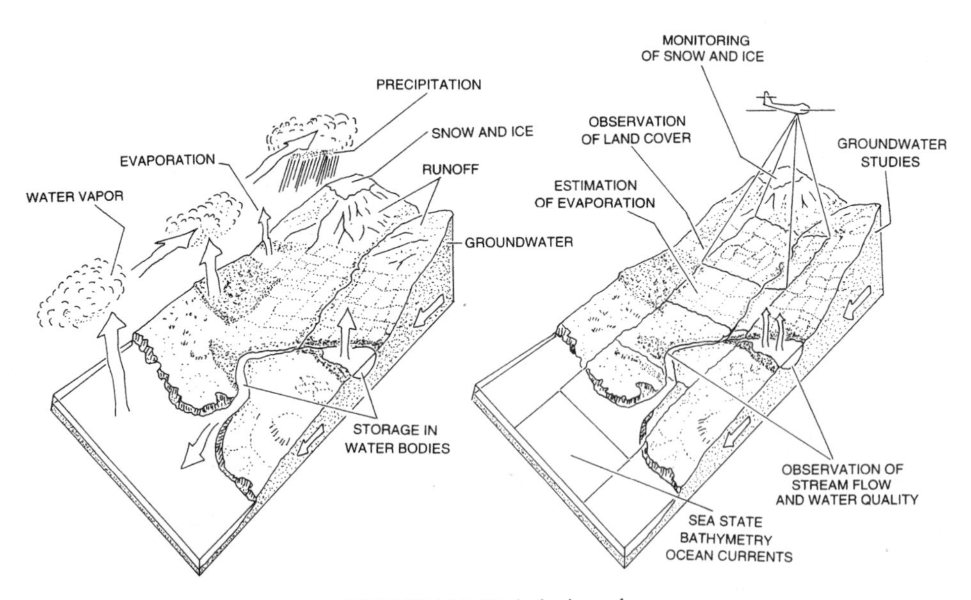


**FIGURE 19.1.** Hydrologic cycle.

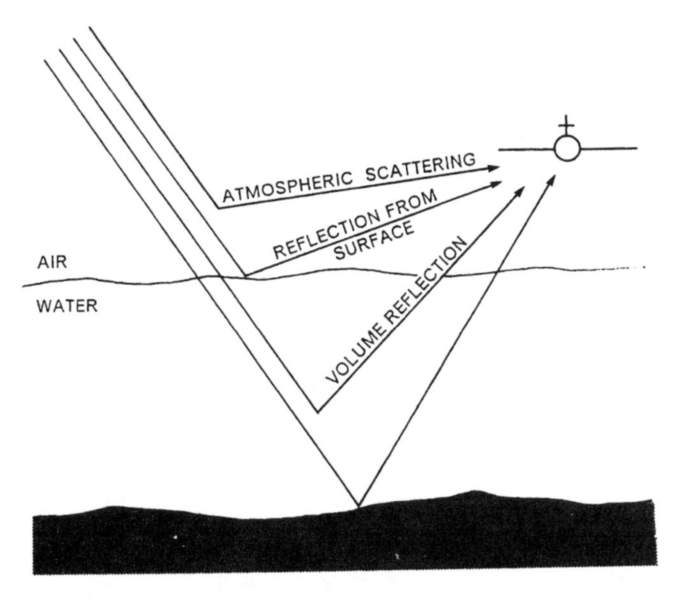


**FIGURE 19.2.** Diagram illustrating major factors influencing spectral characteristics of a water body. Modified from Alföldi (1982, p. 318).

reflected back to the atmosphere; this reflected radiation carries little information about the water itself, although it may convey information about the roughness of the surface, and therefore about wind and waves. Instead, the spectral properties (i.e., "color") of a water body are determined largely by energy that is scattered and reflected within the water body itself, known as volume reflection because it occurs over a range of depths rather than at the surface. Some of this energy is directed back toward the surface, where it again passes through the atmosphere, and then to the sensor (Figure 19.2). This light, sometimes known as underlight, is the primary source of the color of a water body.

The light that enters a water body is influenced by (1) absorption and scattering by pure water itself, and (2) scattering, reflection, and diffraction by particles that may be suspended in the water. For pure water, some of the same principles described previously for atmospheric scattering apply. Scattering by particles that are small relative to wavelength (*Rayleigh scattering*) causes shorter wavelengths to be scattered the most. Thus, for deep water bodies, we expect (in the absence of impurities) water to be blue or blue-green in color (Figure 19.3). Maximum transmittance of light by clear water occurs in the range 0.44 μm–0.54 μm, with peak transmittance at 0.48 μm. Because the color of water is determined by volume scattering, rather than by surface reflection, the spectral properties of water bodies (unlike those of land features) are determined by transmittance rather than by surface characteristics alone. In the blue region, light penetration is not at its optimum, but at slightly longer wavelengths, in the blue-green region, penetration is greater. At these wavelengths, the opportunity for recording features on the bottom of the water body are greatest. At longer wavelengths, in the red region, absorption of sunlight is much greater, and only shallow features can be detected. Finally, in the near infrared region, absorption is so great that only land–water distinctions can be made.

As impurities are added to a water body, its spectral properties change. Sediments are introduced both from natural sources and by mankind's activities. Such sediments consist of fine-

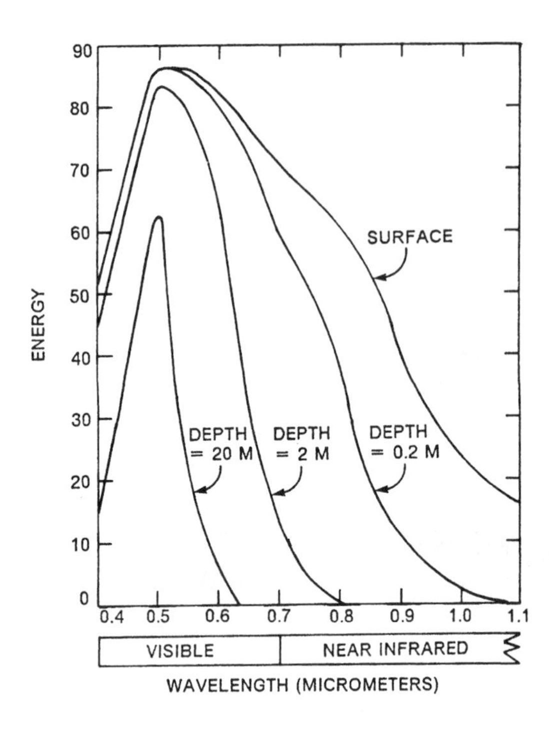


**FIGURE 19.3.** Light penetration within a clear water body. From Moore (1978, p. 458). Reproduced by permission of the Environmental Research Institute of Michigan.

textured silts and clays eroded from stream banks or by water running off disturbed land that are fine enough to be carried in suspension by moving water. As moving water erodes the land surface or the shoreline, it carries small particles as suspended sediment; faster flowing streams can erode and carry more and larger particles than can slower moving streams. As a stream enters a lake or ocean, the decrease in velocity causes coarser materials to be deposited. But even slow-moving rivers and currents can carry large amounts of fine-textured sediments, such as clays and silts, which can be found even in calm water bodies. Sediment-laden water is referred to as *turbid water;* scientists can measure *turbidity* by sampling the water body or by using devices that estimate turbidity by the transparency of the water. One such device is the *Secchi disk,* a white disk of specified diameter that can be lowered on a line from the side of a small boat. Because turbidity decreases the transparency of the water body, the depth at which the disk is no longer visible can be related to sediment content. Another indication of turbidity, *nephometric turbidity units* (NTU), are measured by the intensity of light that passes through a water sample. A special instrument uses a light beam and a sensor to detect differences in light intensity. Water of high turbidity decreases the intensity of the light in a manner that can be related to sediment content.

Thus, as sediment concentration increases, the spectral properties of a water body change. First, its overall brightness in the visible region increases, so the water body ceases to act as a "dark" object. Second, the wavelength of peak reflectance shifts from a maximum in the blue region toward the green. The presence of larger particles means that the wavelength of maximum scattering shifts toward the blue-green and green regions (Figure 19.4). Therefore, as sediment content increases there tends to be a simultaneous increase in brightness and a shift in peak reflectance toward longer wavelengths, and the peak itself becomes broader, so that at high levels of turbidity the color becomes a less precise indicator of sediment content. As sedi-

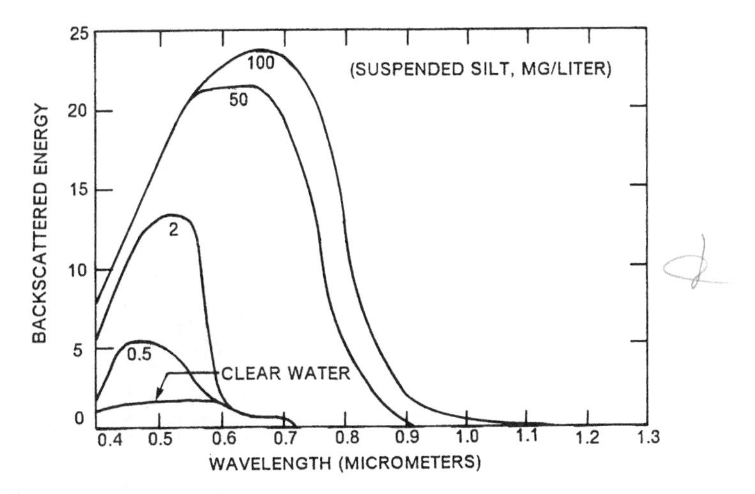


**FIGURE 19.4.** Effects of turbidity on spectral properties of water. From Moore (1978, p. 460). Reproduced by permission of the Environmental Research Institute of Michigan.

ment content approaches very high levels, the color of the water begins to approach that of the sediment itself.

## 19.3. Spectral Changes as Water Depth Increases

Figure 19.3 shows the spectral characteristics of sunlight as it penetrates a clear water body. Near the surface, the overall shape of the curve resembles the spectrum of solar radiation, but the water body increasingly influences the spectral composition of the light as depth increases. At a depth of 20 m, little or no infrared radiation is present because the water body is an effective absorber of these longer wavelengths. At this depth, only blue-green wavelengths remain—these wavelengths are therefore available for scattering back to the surface, from the water itself, and from the bottom of the water body.

The attenuation coefficient ($k$) describes the rate at which light becomes dimmer as depth increases. If $E_0$ is the brightness at the surface, then the brightness at depth $z$ ($E_z$) is given by:

$$E_z = E_0 e^{-kz} \quad \text{(Eq. 19.1)}$$

In hydrologic studies the influence of the atmosphere can be especially important. The atmosphere, of course, alters the spectral properties of incident radiation and also influences the characteristics of the reflected signal. Although these influences are also present in remote sensing of land surfaces, they assume special significance in hydrologic studies, in part because such studies often depend upon subtle spectral differences (easily lost in atmospheric haze) and also perhaps because much of the hydrologic information is carried by the short wavelengths that are most easily scattered by the atmosphere.

Water bodies are typically dark, so the analyst must work with a rather restricted range of brightnesses relative to those available for study of land surfaces (Figure 19.5). As a result, analysts who specialize in remote sensing of water bodies must devote special attention to the

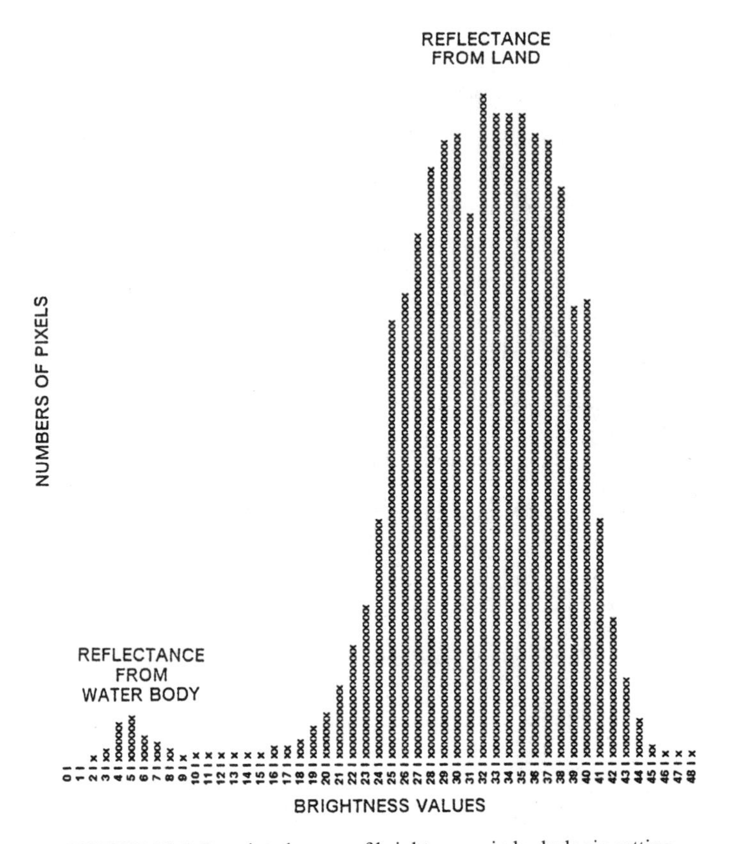


**FIGURE 19.5.** Restricted range of brightnesses in hydrologic setting.

radiometric qualities of the remotely sensed data. Typically, data to be analyzed for hydrologic information are examined carefully to assess their quality, the effects of the atmosphere, and sun angle. Geometric preprocessing is used with caution, to avoid unnecessary alteration of the radiometric qualities of the data. In some instances, analysts calculate average brightness over blocks of contiguous pixels to reduce transient noisy effects of clouds and whitecaps. Sometimes several scenes of the same area, acquired at different times, can be used to isolate permanent features (e.g., shallows and shoals) from temporary features (e.g., waves and atmospheric effects). Also, it is often advantageous to estimate original radiometric brightnesses from digital values as a means of accurately assessing differences in color and brightness.

## 19.4. Location and Extent of Water Bodies

Remote sensing provides a straightforward means to map the extent of water bodies, to inventory area occupied by open water, and to monitor changes in water bodies over time. Although such tasks may seem very basic, there are numerous situations in which simple determination of the land–water boundary can be very important. Study of coastal erosion requires accurate determination of shoreline position and configuration at several dates. Likewise, comparisons of shoreline position before and after flooding permits measurement of areas flooded, as well as determination of locations of flooded areas. Such information can be difficult to acquire by conventional means.

Determination of the land–water body border is usually easiest in the near infrared region, where land, especially if vegetated, is bright, and open water is dark. Usually it is possible to determine a sharp contact between the two (Figure 19.6). With Landsat 1–3 MSS data, the contrast is usually clear on bands 6 and 7. With photography, black-and-white infrared film is usually satisfactory. CIR film is suitable if the water is free of sediment. However, because CIR film is sensitive to green light, turbid water will appear bright, thereby preventing clear discrimination of the land–water boundary.

In the early 1970s Landsat MSS data were used to inventory impounded water throughout the United States. The failure of earth dams at Buffalo Creek, West Virginia, in February 1972 and at Rapid City, South Dakota, in June 1972 produced catastrophic loss of life and property. In both instances, poorly constructed earth dams failed after periods of unusually heavy rainfall, which both increased the volume of impounded water and weakened the earthen dams. In August 1972 the U.S. Congress passed Public Law 92-367 requiring federal inspection of all dams impounding water bodies exceeding specified sizes. Because there existed no comprehensive record of such water bodies, NASA conducted a study to determine the feasibility of using Landsat MSS data as the means of locating water bodies impounded by dams requiring inspection under Public Law 92-367. NASA's method required selection of GCPs (Chapter 11) to register the data to accurate maps. A density slice technique separated water bodies from land areas; a printed output showing water pixels could then be matched to maps to locate impoundments not already delineated on the maps.

Philipson and Hafker (1981) studied application of Landsat MSS data to delineation of flooded areas. They concluded that simple visual interpretation of MSS band 4 images was as accurate as manual interpretation of multiband composite images and as accurate as digital analysis of MSS band 4 and of combinations of both bands 2 and 4. They experienced problems in applying the digital approach because of difficulties in registering the flooded and unflooded scenes. In addition, turbidity associated with flooding increased brightness in the green region

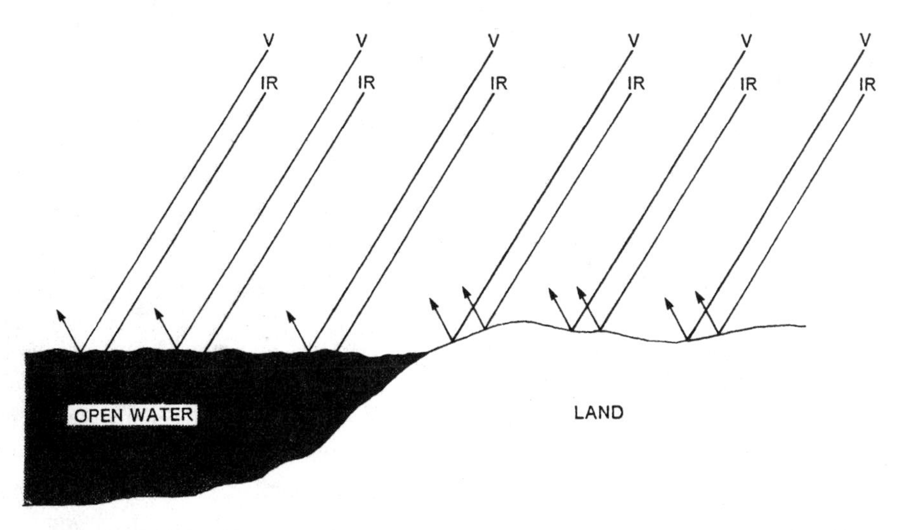


**FIGURE 19.6.** Determining the location and extent of land–water contact.

(band 1) to the extent that there was insufficient contrast between the two dates to unambiguously separate flooded and unflooded areas.

For their digital interpretation, they defined a ratio that compares the brightness values in near infrared (band 4) on the two dates:

$$\frac{\text{(MSS 1, unflooded)} - \text{(MSS 4, flooded or unflooded)}}{\text{(MSS 1, unflooded)}} \quad \text{(Eq. 19.2)}$$

Because open clear water is very dark in band 4, the ratio approaches 1 for flooded areas and 0 for unflooded areas. For example, if dry land has digital values near 35 in both bands, then for flooded areas the ratio is (35 – 0)/35 = 1, and the ratio for unflooded areas is (35 – 35)/35 = 0. Although Philipson and Hafker found that when water turbiditity was high their procedure was not always successful, there are other procedures that can identify turbid water and separate such areas from unflooded land. In later studies, Frazier and Page (2000) found that density slicing of Landsat TM band 5 (mid-infrared) was effective in identifying water bodies on a riverine flood plain.

Some laboratories have specifically dedicated their work to applications of multispectral data to observation of hydrologic events, notably coastal and riverine floods. For example, the widespread flooding of the upper midwestern United States 1993 was examined in detail using remotely sensed imagery (Brakenridge, et al., 1994). Remotely sensed data are the foundation for research laboratories dedicated specifically to monitoring floods using such imagery, in coordination with in situ data when it is available. Because of the routine widespread availability of satellite imagery through the world, such organizations can rapidly report the extent and severity of events in both local and remote locations (Figure 19.7). The capabilities of the varied remote sensing systems discussed here complement each other, providing the detail of fine-scale satellite systems, the synoptic view of broad-scale systems, and the all-weather capabilities of imaging radars. Examples of such research laboratories include the Dartmouth Flood Observatory:

*http://www.dartmouth.edu/~floods/*

and the USGS Rocky Mountain Mapping Center:

*http://rmmcweb.cr.usgs.gov/* and *http://nhss.cr.usgs.gov/*

## 19.5. Roughness of the Water Surface

Figure 19.8 shows spectra for calm and wave-roughened surfaces in the visible and the near infrared. Wave-roughened surfaces are brighter than smoother surfaces. Calm, smooth water surfaces direct only volume-reflected radiation to the sensor, but rough, wavy water surfaces direct a portion of the solar beam directly to the sensor. As a result, wavy surfaces are much brighter, especially in the visible portion of the spectrum. McKim et al. (1984) describe a procedure that uses polarizing filters to separate the surface brightness from that of the water body itself, using high-resolution data.

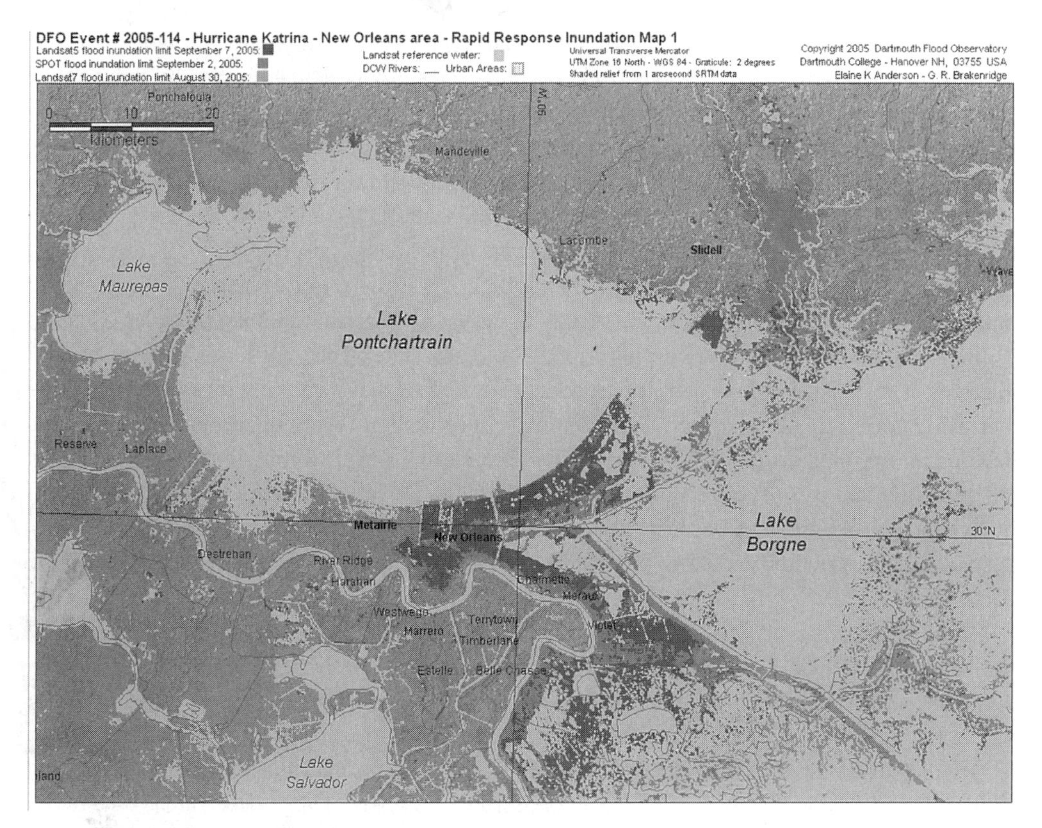


**FIGURE 19.7.** Flood mapping using remotely sensed data. Flooded areas, New Orleans region in the aftermath of Hurricane Katrina, prepared from SPOT and Landsat data. From Brakenridge, Anderson, and Caquard (2005) (original in color). Dark tones indicate severe flooding. USA—Hurricane Katrina—Gulf Coast—New Orleans. DFO 2005-114. Digital media. *www.dartmouth.edu/~floods.* © Dartmouth Flood Observatory. Reproduced by permission.

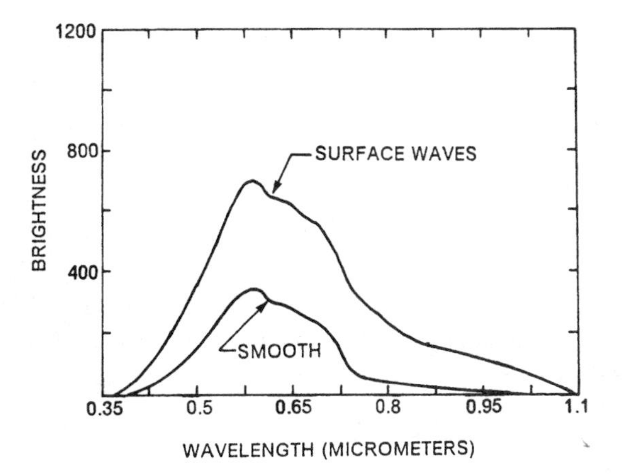


**FIGURE 19.8.** Spectra of calm and wind-roughened water surfaces (from laboratory experiments). Based on McKim et al. (1984, p. 358). Copyright 1984 by the American Society for Photogrammetry and Remote Sensing. Reproduced by permission.

The most intensive studies of ocean roughness have used active microwave sensors (Chapter 7). *Sea state* refers to the roughness of the ocean surface, which is determined by wind speed and direction, as they interact locally with currents and tides. Sea state is an important oceanographic and meteorological quality because if studied over large areas and over time, it permits inference of wind speed and direction—valuable information for research and forecasts. Radars provide data concerning ocean roughness: the backscattering coefficient increases as wave height increases. Radars therefore provide a means of indirectly observing sea state over large areas and a basis for inferring wind speed and direction at the water's surface. If the imaging radar is carried by a satellite, as is the case with Seasat and the shuttle imaging radar (Chapter 7), then the analyst has the opportunity to observe sea state over very large areas at regular intervals. These observations permit oceanographic studies that were impractical or very difficult using conventional data. Furthermore, timely information regarding sea state has obvious benefits for marine navigation, and can, in principle, form the basis for inferences of wind speed and direction that can contribute to meteorological data. Direct observation of sea state is relatively straightforward, but because ocean areas are so very large and conditions so rapidly changeable, the usual observations from ships in transit are much too sparse to provide reliable data.

Rigorous study of sea state by radar started soon after World War II, and continued during the 1950s and 1960s, culminating with the radars carried by Seasat and the space shuttle. Although many experiments have been conducted using a variety of microwave instruments, one of the most important broad-scale sea state experiments was based upon the Seasat A project, which used several instruments to monitor the Earth's oceans. The Seasat SAR and RADARSAT (Chapter 7) have provided an opportunity to observe large ocean areas on a repetitive basis. Calm ocean surfaces, with waves that are small relative to the radar wavelength, act as specular reflectors and appear as dark regions on the image, as energy is reflected away from the antenna. As wind speed increases, the ocean surface becomes rougher and acts more like a diffuse reflector, causing bright regions on the imagery (Figures 19.8 and 19.9). Because radar wavelength and system geometry are known, the received signal can be used to estimate wave height and velocity.

In a test in the Gulf of Alaska, scientists found agreement between data derived from the Seasat SAR and those observed at the ocean surface. Wave direction, height, and wavelength were all estimated from the SAR imagery. Waves were most accurately observed when their direction of travel was within 25° of the look direction. Other studies have shown the SAR data to be useful for studying other kinds of ocean swells and currents.

## 19.6. Bathymetry

*Bathymetry,* information concerning water depth and the configuration of the ocean floor, is one of the most basic forms of hydrographic data. Bathymetry is especially important near coastlines, in harbors, and near shoals and banks, where shallow water can present hazards to navigation, and where changes can occur rapidly as sedimentation, erosion, and scouring of channels alters underwater topography.

Water depth can be measured by instruments carried on vessels, especially by acoustic (sonar) instruments that measure depths directly below the vessel. Ideally, bathymetric maps should be compiled from a more or less uniform network of depth measurements rather than

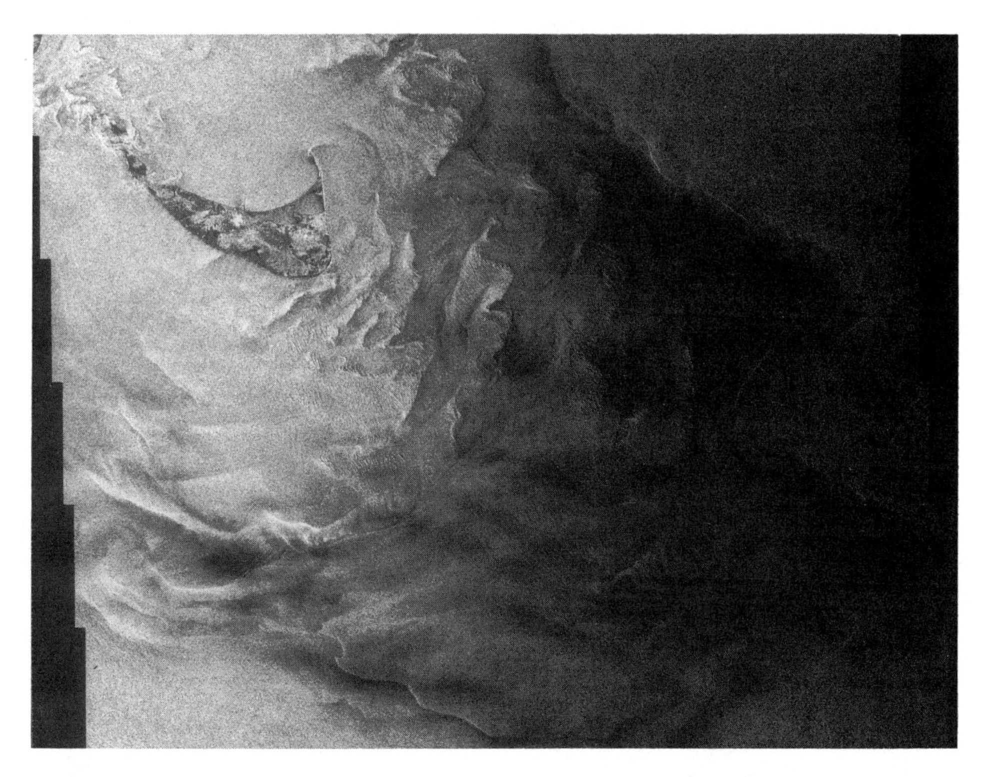

**FIGURE 19.9.** Seasat SAR image illustrating rough ocean surface, Nantucket Shoals, 27 August 1978. From Jet Propulsion Laboratory, Pasadena, CA.

from a limited set of data from traverses. Although modern side-scanning sonar can yield a continuous surface of depth information, aerial imagery is an important means of mapping subsurface topography in shallow water.

Photogrammetry can be applied to bathymetric measurements if high-quality, large-scale photography is available, and if the water is clear. Filters have been used to separate radiation in the spectral region 0.44–0.54 μm, where solar energy is most easily transmitted by clear water. As Figure 19.3 indicates, in principle sunlight can penetrate to depths of about 20 m. Aerial photographs record information only from relatively shallow depths, although depths as deep as 16 m have been mapped using aerial photographs.

Special problems in applications of photogrammetry to bathymetry include estimation of differences in refraction between air and water (Figure 19.10) and the difficulty in acquiring a good set of underwater control points. Furthermore, mapping is difficult in regions far removed from the shoreline because of the difficulty in extending control across zones of deep water. Nonetheless, photogrammetric methods have been successfully used for mapping zones of shallow water.

Water depth can also be estimated by observing refraction of waves as they break in shallow water. As a wave approaches the shore, it begins to break at a depth that is determined by its velocity, wavelength, and period. If velocities and wavelengths can be measured on successive overlapping photographs, it is possible to calculate an estimate of depth. The method works well only when the water body is large, and if the bottom drops slowly with increasing distance

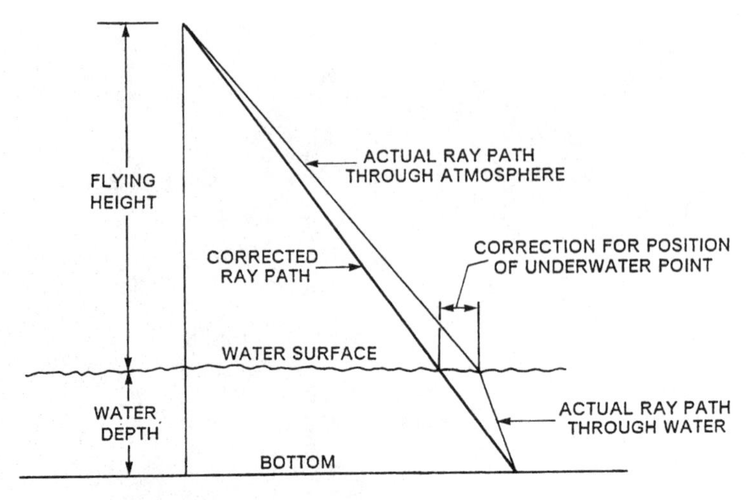


**FIGURE 19.10.** Bathymetry by photogrammetry.

from shore. The *wave refraction method* is not as accurate as other procedures, but it does have the advantage of applicability in turbid water, when other methods cannot be applied.

### *Multispectral Bathymetry*

Depth of penetration of solar energy varies with wavelength. Longer wavelengths are absorbed most strongly; even modest depths—perhaps 2 m or so—of clear water will absorb all infrared radiation, so that longer wavelengths convey no information concerning depth of water or the nature of the bottom. Visible radiation penetrates to greater depths, but is still influenced by wavelength-dependent factors. For clear water, maximum penetration occurs in the blue-green region, which is approximated by Landsat MSS 1 (using the Landsat 4 and 5 band designation), although TM1 is likely to be more effective for bathymetric studies (Chapter 6). Polcyn and Lyzenga (1979) report that for their study region in the Bahamas, under optimum conditions, maximum penetration for MSS 1 is as deep as 25 m, and about 6 m for MSS 2. (For consistency in designation, Landsat 4 and 5 MSS numbering is employed here.)

Intensity of radiation decreases exponentially with depth (i.e., brightness decreases as depth increases). Thus use of MSS 4 (or TM1; both are positioned at or near the spectral region of maximum penetration) should be useful for estimating depth. Dark values suggest that the bottom is deep, beyond the 20-m range represented in Table 19.2. Bright values suggest that the bottom is near the surface (Figure 19.11). Many depth-extraction algorithms use a logarithmic transformation of MSS 4 brightness as a means of incorporating our knowledge that brightness decreases exponentially with depth.

Polcyn and Lyzenga (1979) defined a depth-finding algorithm:

$$L_{sen} = L_{atm} + T_{atm} (L_{surf} + L_{wat}) \quad \text{(Eq. 19.3)}$$

where $L_{atm}$ is the radiance contributed by the atmosphere; $T_{atm}$ is atmospheric transmission; $L_{surf}$

TABLE 19.2. Logarithmic Transformation of Brightnesses

| $x$ | ln ($x$) | Brightness | Depth |
|---|---|---|---|
| 2 | 0.69 | Dim | Deep |
| 6 | 1.79 | | |
| 10 | 2.30 | | |
| 14 | 2.64 | | |
| 18 | 2.89 | | |
| 22 | 3.09 | | |
| 26 | 3.26 | Bright | Shallow |

is radiance from the water surface; and $L_{wat}$ is radiance from beneath the water surface. $L_{wat}$ for a specific band $i$ is usually designated as $L_i$, and has been estimated as:

$$L_i = L_{w(i)} + (L_{b(i)}L_{w(i)})e^{-2k_i z} \quad \text{(Eq. 19.4)}$$

where $i$ is a specific spectral band; $L_{w(i)}$ is the deep-water radiance for band $i$; $L_{b(i)}$ is the bottom reflectance for band $i$; and $z$ is water depth.

$L_w$ is defined as the energy reflected from a water body so deep that incident solar radiation is absorbed, and the energy observed by the sensor is so dim that it cannot be distinguished from sensor and environmental noise. Values for $L_w$ vary depending upon sensor sensitivity, spectral band, and sensor resolution; but for a given sensor, $L_w$ is influenced also by atmospheric clarity, solar angle, and other factors. Therefore, for each specific study, $L_w$ must be estimated, usually by selecting a set of pixels known to correspond to depths well beyond the range of penetration of solar energy.

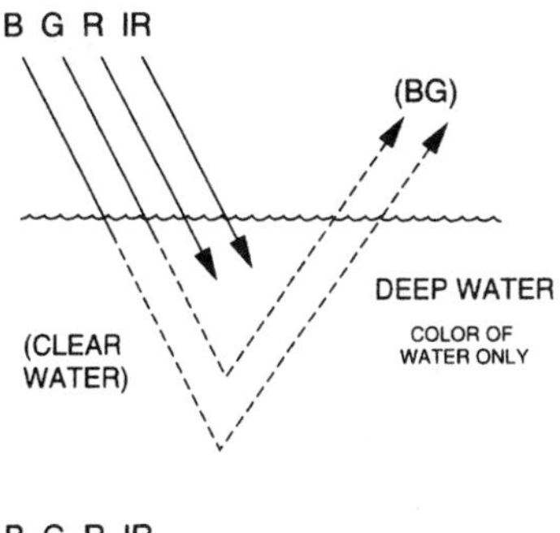


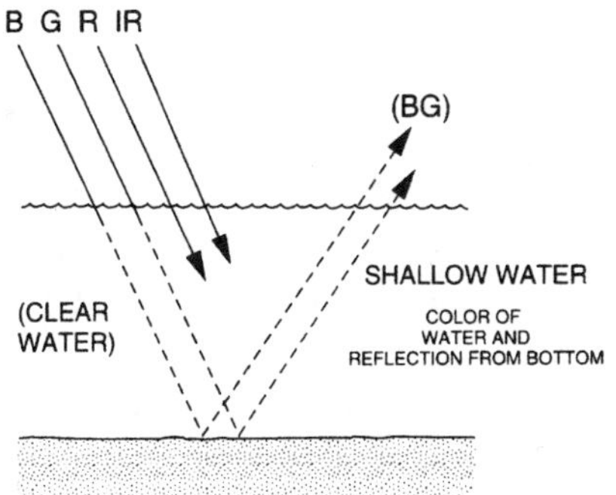


**FIGURE 19.11.** Multispectral bathymetry. Shallow water bodies filter radiation reflected from the bottom; the balance between spectral bands observed at the sensor reveals the approximate depth of the water body. See Figure 19.12.

Jupp et al. (1985) have summarized published values to produce the following estimates for maximum penetration of solar radiation within clear, calm ocean water under a clear sky:

- MSS band 1: 15–20 m (50–60 ft.)
- MSS band 2: 4–5 m (13–17 ft.)
- MSS band 3: 0.5 m (1.5 ft.)
- MSS band 4: 0

As expected from inspection of Figure 19.4, penetration is greatest within the blue-green region of MSS band 1, and least in MSS band 4, where solar radiation is totally absorbed. It is possible to use this information to designate each MSS pixel as a member of a specific depth class.

If we find the numeric difference between the observed brightness in band $I$ ($L_i$) and the deep-water signal for the same band ($L_{w(i)}$), then the quantity ($L_I - L_{w(i)}$) measures the degree to which a given pixel is brighter than the dark pixels of open water.

If the value for band 1 exceeds the band 1 deep-water limit, but all other pixels are at deep-water values for their respective bands, then it is known that water depth is within the 15-m to 20-m range (Figure 19.12). If both band 1 and band 2 are brighter than their respective values for $L_w$, but bands 3 and 4 are at their $L_w$ values, then water depth is within the 4-m to 5-m range suggested above. If a pixel displays values in bands 1, 2, and 3 that all exceed their respective thresholds for $L_{w,}$ then water depth is likely to fall within the 0.5-m range. Finally, if all four bands are at their threshold values, then the pixel corresponds to a depth beyond the 20-m limit. Because band 4 values for water pixels should always be very dim, bright values in this band suggest that water areas may have been delineated incorrectly, marsh vegetation rises above the water surface, or other factors have altered the spectral properties of the water.

This procedure assigns pixels to depth classes but does not provide quantitative estimates of depth, which, of course, would be more useful. Because of the exponential effect of the attenuation coefficient (Equation 19.1), brightnesses of water bodies are not linearly related to depth. That is, we cannot observe a given brightness in band i, then directly match that brightness to a specific depth. Lyzenga (1981) has attempted to solve this problem by devising a relationship that creates a new variable, $x_{i,}$ for each band $i$; $x_i$ is linearly related to water depth and therefore can be used to estimate depth within a given depth zone:

$$x_i = \ln (L_i - L_{w(i)}) \qquad \text{(Eq. 19.5)}$$

Thus $x_i$ is defined as the natural logarithm of the difference between observed brightness in land $i$ ($L_i$) and the deep-water reflectance ($L_{w(i)}$) in the same band. Table 19.2 shows that the logarithmic transformation enhances the significance of dim values from deep water. In Table 19.2, $x$ represents a hypothetical set of digital brightnesses observed by a sensor; ln($x$) does not itself measure depth, but is linearly related to water depth, so that the investigator can define the local relationship between $x_i$ and water depth.

Some methods have relied upon estimates of $k$, which varies with wavelength and depends upon local effects that influence the clarity of the water. Polcyn and Lyzenga (1979) used $k$ = 0.075m$^{-1}$ for MSS1 and $k$ = 0.325 m$^{-1}$ for MSS 2 for their studies of water depths in the Bahamas. They were able to use this algorithm to achieve accuracies of 10% at depths of about 20 m (70 ft). Under ideal conditions, they measured depths as deep as 40 m (130 ft.).

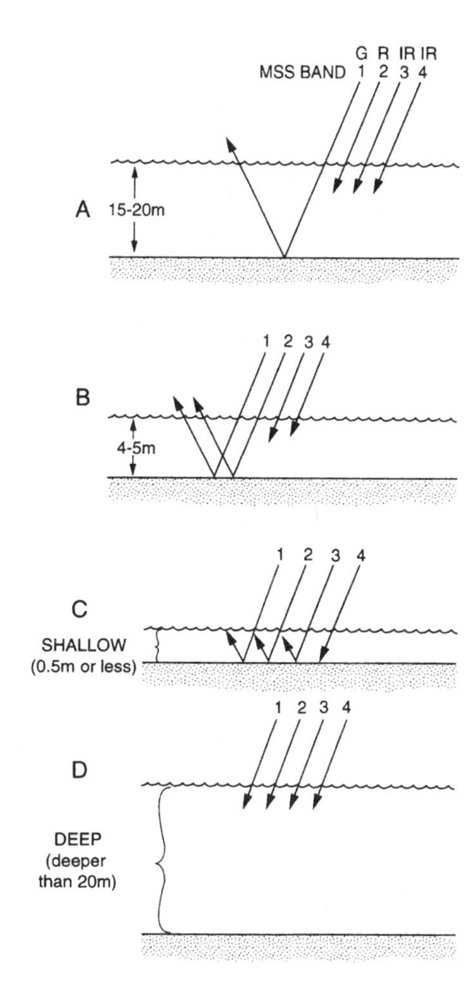


**FIGURE 19.12.** Multispectral bathymetry using Landsat MSS data. The relative brightnesses of the several spectral bands indicate depth classes, as discussed in the text.

These procedures seem to be effective, provided that the water body is clear (otherwise turbidity contributes to brightness) and the bottom reflectance is uniform. If attenuation of the water and bottom reflectivity is known, accuracies as high as 2.5% have been achieved (although typical accuracies are lower). If the bottom reflectance is not uniform, then differences in brightness will be caused not only by differences in depth but also by differences in reflectivity from the bottom materials. Because such differences are commonly present, due to contrasting reflectances of different sediments and vegetation, it is often necessary to apply procedures to adjust for differing reflectances from the bottom of the water body. The effectiveness of this procedure varies with sun angle (due to variations in intensity of illumination), and is most effective when data of high radiometric resolution are available.

In some instances, ratios of two bands may remove differences in bottom reflectivity. Lyzenga (1979) defines an index that separates different reflectivities using different spectral bands. His index, calculated for each pixel, can be used to classify pixels into different bottom reflectivity classes, which can then each be examined, as classes, to estimate water depth.

## 19.7. Landsat Chromaticity Diagram

*Colorimetry,* the precise measurement of color, is of significance in several aspects of remote sensing, but is especially significant in applications of remote sensing to studies of water quality. Even casual observation reveals that water bodies differ significantly in color, yet it can be very difficult to accurately measure these color differences and to relate color to the variables that we wish to study.

Specification of color is not as self-evident as it might seem. Scientists have established several alternative systems for describing color. Most establish a three-dimensional color space in which three primaries (red, green, and blue) form three orthogonal axes. Unique specification of a color requires identification of a location within this color space, giving hue (dominant wavelength), saturation (purity), and value (brightness) of a given color (Chapter 2). These axes define *tristimulus* space, in which any color can be represented as a mixture of the three primaries—that is, as a location within the coordinate system represented in Figure 19.13. Such a system defines a color solid, which in principle permits specification of any color.

For many purposes, it is convenient to define a two-dimensional plane within the color solid by setting:

$$R = \frac{X}{X+Y+Z} \qquad G = \frac{Y}{X+Y+Z} \qquad B = \frac{Z}{X+Y+Z} \qquad \text{(Eq. 19.6)}$$

The values *R, G,* and *B* define locations within a two-dimensional surface that show variations in hue and saturation but do not portray differences in brightness. In effect, *R, G,* and *B* have been standardized by dividing by total brightness in each primary without conveying informa-

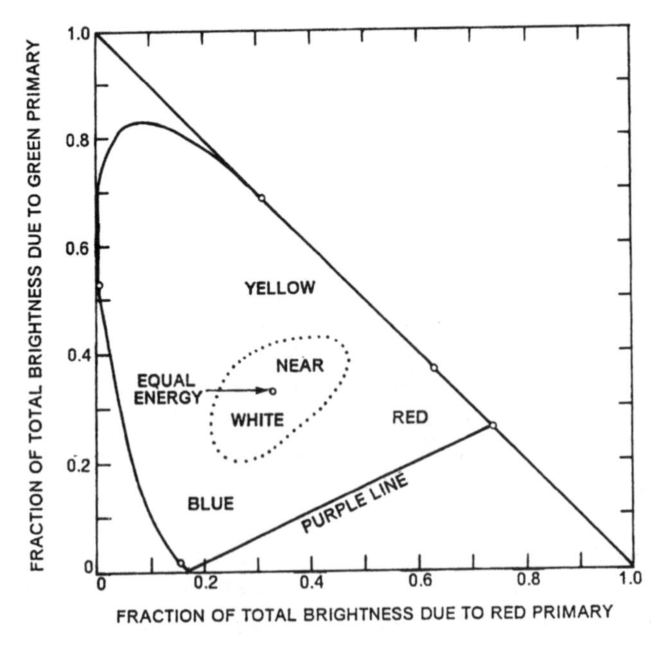


**FIGURE 19.13.** CIE chromaticity diagram.

tion concerning absolute brightness. Thus colors represented by this convention vary only with respect to hue and saturation, because brightness information has been removed by standardization. For example, $R$ specifies "percent red," but does not convey information concerning whether it is a percentage of a very bright color or of a dark color.

In 1931 the Commission Internationale de l'Eclairage (CIE; International Commission for Illumination) met in England to define international standards for specifying color. The CIE system defines a tristimulus space as follows:

$$x = \frac{X}{X+Y+Z} \qquad y = \frac{Y}{X+Y+Z} \qquad z = \frac{Z}{X+Y+Z} \qquad \text{(Eq. 19.7)}$$

in which $x$, $y$, and $z$ correspond approximately to the red, green, and blue primaries, and $X$, $Y$, and $Z$ are the CIE chromaticity coordinates. Also, $x + y + z = 1$.

The three CIE primaries are "artificial" because, as a matter of convenience, they are defined to facilitate mathematical manipulation and do not correspond to physically real colors. Because $x$, $y$, and $z$ sum to 1.0, only two are necessary to uniquely specify a given color, as the third can be determined from the other two. Therefore the CIE chromaticity diagram has only two axes, using the x and y coordinates to locate a specific color in the diagram.

Figure 19.13 shows the CIE chromaticity diagram. The vertical axis defines the $y$ (green) axis, and the horizontal axis defines the $x$ (red) primary, each scaled from 0 to 1.0 (to portray the proportion of total brightness in each band). At coordinates $x = 0.333$ and $y = 0.333$ ($z$ must then equal 0.333), all three primaries have equal brightness, and the color is pure white. Colors to which the human visual system is sensitive are inside the horseshoe-shaped outline in Figure 19.13. The white colors near the center are "without color" as there is no dominant color; purity (*saturation*) increases toward the edges of the horseshoe where colors are "fully saturated." The line at the base of the horseshoe, known as the *purple line,* locates *nonspectral* colors. (Nonspectral colors are those that do not appear in a spectrum of colors because they are formed by mixtures of colors from opposite ends of the spectrum.) Spectral colors are positioned in sequence around the curved edge of the horseshoe.

Because each of the $x$, $y$, and $z$ primaries has been defined by dividing by the total brightness to yield the percentage brightness in each primary, they convey hue information without brightness information. That is, colors have been standardized to remove brightness, leaving only hue and saturation information. Because $x$, $y$, and $z$ sum to 1.0, only two are necessary to specify a specific hue, as the third can be determined from the other two. Therefore the CIE color diagram uses only two axes, usually the $x$ and $y$ primaries, as a means of specifying color. The CIE diagram is useful for precisely specifying color and for transferring color specifications from one context to another. Note, once again, that the CIE diagram conveys differences in color without portraying differences in brightnesses of colors—that is, it shows only hue and saturation.

Alföldi and Munday (1977) have applied the chromaticity concept to analysis of Landsat data. They defined a *Landsat chromaticity diagram,* analogous to the CIE chromaticity diagram, except that it defines a color space using MSS bands 4, 5, and 6 instead of the red, green, and blue primaries used for the CIE diagram. Thus infrared data from band 3 are substituted for the missing primary. Because MSS digital counts do not accurately portray absolute brightnesses, it is necessary to use the appropriate calibration data (Table 19.3 and Chapter 4) to estimate the brightnesses in each band.

For example, digital data from the Landsat 1 MSS require the following transformation:

$$\frac{\text{MSS 1 digital value}}{127} \times 2.48 = x \qquad \text{(Eq. 19.8)}$$

$$\frac{\text{MSS 2 digital value}}{127} \times 2.00 = y \qquad \text{(Eq. 19.9)}$$

$$\frac{\text{MSS 3 digital value}}{127} \times 1.76 = z \qquad \text{(Eq. 19.10)}$$

where 127 represents the maximum digital value possible for the sensor, and the appropriate constants are taken from Table 19.3; $x$, $y$, and $z$ now represent coordinates for the Landsat chromaticity diagram.

The Landsat chromaticity coordinates can be used to assess the turbidity of water bodies imaged by the MSS. Figure 19.14 shows the Landsat chromaticity diagram, which is identical to the CIE chromaticity diagram except for the use of MSS bands 4, 5, and 6 as substitutes for the CIE primaries. "E" signifies the *equal radiance point,* where radiances in MSS bands 4, 5 and 6 are equal. (Note that the labels in Figure 19.13 do not apply to the Landsat chromaticity diagram, as it does not use the CIE primaries.) The curved line extending to the right of the equal radiance point is the experimentally defined locus that defines the positions of water pixels as their radiances are plotted on the diagram. Pixels with positions near the right end of this line represent water bodies with clear water. As chromaticity coordinates shift closer to the other end of the line, near the equal radiance point, the spectral properties of the pixels change, indicating increases in turbidity as the pixel position approaches the left-hand end of the line. This line was defined experimentally by Alföldi and Munday (1977) for their study area, using sediment samples collected at the time the imagery was acquired. (For detailed studies, it is

**TABLE 19.3. Landsat MSS Calibration Data**

| | | Band | | | |
|---|---|---|---|---|---|
| | | 1 | 2 | 3 | 4 |
| Landsat 1 | Min. | 0.00 | 0.00 | 0.00 | 0.00 |
| | Max. | 24.8 | 20.0 | 17.6 | 15.3 |
| Landsat 2[a] | Min. | 1.0 | 0.7 | 0.7 | 0.5 |
| | Max. | 21.0 | 15.6 | 14.0 | 13.8 |
| Landsat 2[b] | Min. | 0.8 | 0.6 | 0.6 | 0.4 |
| | Max. | 26.3 | 17.6 | 15.2 | 13.0 |
| Landsat 3[c] | Min. | 0.4 | 0.3 | 0.3 | 0.1 |
| | Max. | 22.0 | 17.5 | 14.5 | 14.7 |
| Landsat 3[d] | Min. | 0.4 | 0.3 | 0.3 | 0.1 |
| | Max. | 25.9 | 17.9 | 14.9 | 12.8 |

*Note.* Values in MW · $cm^{-2}$ · SR · $\mu m^{-1}$. From Markham and Barker (1986). MSS bands 1, 2, 3, and 4 were formerly designated as 4, 5, 6, and 7, respectively.
[a]For data processed before 16 July 1975.
[b]For data processed after 16 July 1975.
[c]For data processed before 1 June 1978.
[d]For data processed after 1 June 1978.

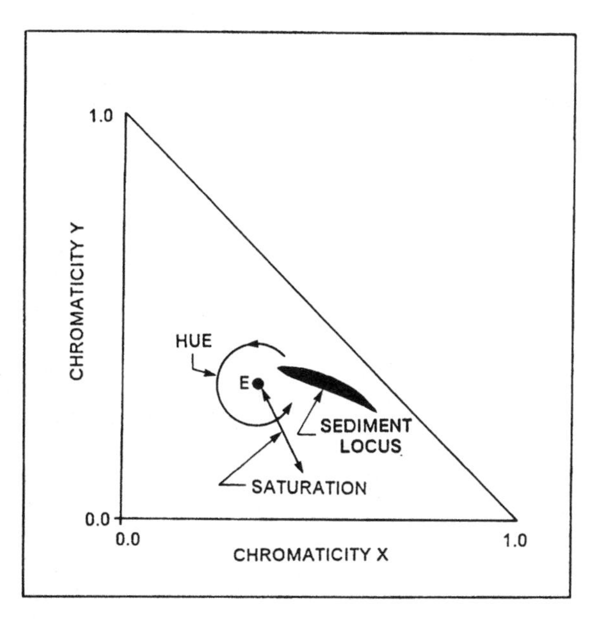


**FIGURE 19.14.** Landsat MSS chromaticity space. The sediment locus is experimentally derived and may vary slightly from region to region or over time within a given region. The radial dimension (outward from E) presents spectral purity information, and radial shifts in the position of the locus are attributed to atmospheric scattering, thin clouds, air pollution, or whitecaps. Diagram based on Alföldi and Munday (1977).

important to define a specific locus for each individual study region, but for illustrative purposes it is sufficient to use their line as a general-purpose locus.)

Thus, in Figure 19.15, pixel A reveals a clear water body, and pixel B reveals a turbid water body. As pixels known to represent water bodies drift away from the line (pixel C is an example), we then suspect that atmospheric haze or thin clouds have altered their radiances. Although our inspection of the chromaticity diagram is essentially a qualitative analysis of pixel positions on the chromaticity diagram, Alföldi and Munday have conducted more quantitative studies, using observed values of suspended sediment to calibrate the technique.

Table 19.4 lists Landsat MSS data for the same water body observed at several dates. The first two pixels are worked through as examples, but the others are left for the student to work as problems and to plot on Figure 19.16. In practice, of course, these calculations are easily accomplished by computer. Also, it is often convenient to use groups of contiguous pixels as a means of reducing noisy variability that might be caused by system noise, whitecaps, and other spectral anomalies.

## 19.8. Drainage Basin Hydrology

A less obvious component of water resources concerns the management of land areas to promote orderly retention and flow of moisture over the land surface, and then within the stream system, so that flooding and erosion are minimized and an even, steady flow of moisture is maximized. The nature and distribution of land use and land cover within drainage basins influences runoff from the land surface, and eventually also influences the flow of water within

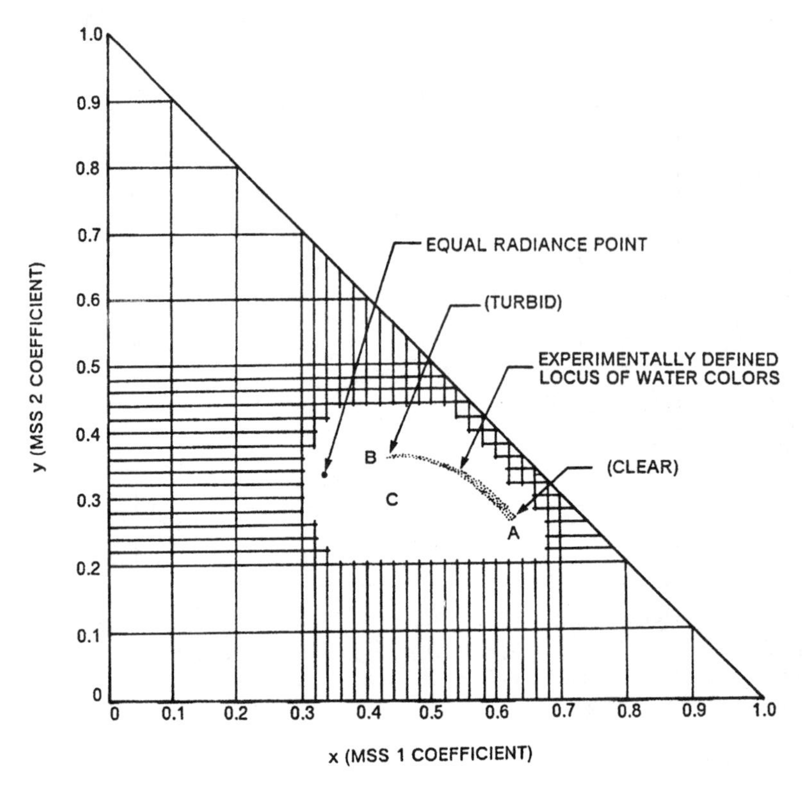


**FIGURE 19.15.** Landsat MSS chromaticity diagram. From Alföldi and Munday (1977, p. 336). Reproduced by permission of the Canadian Aeronautics and Space Institute.

stream channels. Although there is no set relationship between these elements that can be said to form a universally "normal" balance, each climatic and ecological situation tends to be characterized by an established pattern of rainfall, evapotranspiration, and streamflow. Extensive changes to natural land-cover patterns can disrupt this local balance and cause accelerated soil erosion, flooding, and other undesirable consequences.

Hydrologists attempt to study such processes by constructing *hydrologic models:* mathematical simulations that use measurements of local rainfall, together with information concerning patterns of land use, soil, topography, and drainage, to predict the nature of the streamflow that might be caused by storms of a given magnitude. Such models provide a means of assessing effects of proposed changes in land use (e.g., new construction, harvesting of timber) or in channel configuration because they can estimate the effects of such changes before they actually occur. Often such models can be evaluated and refined using data from watersheds that have very accurate records of rainfall and streamflow.

Hydrologic models require large amounts of detailed and accurate data, so many of the costs involved in applying a given model to a specific region arise from the effort required to collect accurate information concerning soil and land-cover patterns—two of the important variables required for such models. Some of the information concerning land use can be derived from remotely sensed data. Manual interpretation of land use and land cover from aerial photographs (Chapter 20) is a standard procedure, and, of course, aerial photographs are a source of much of the information regarding topography and soils. Many studies have evaluated the use of image

TABLE 19.4. Data for Chromaticity Diagram

| Date | MSS digital counts | Radiance | Sum | Chromaticity coordinates |
|---|---|---|---|---|
| May[a] | 36 33 24 07 | 7.02 5.20 3.33 | 15.55 | 0.451 0.334 |
| May[a] | 36 36 27 08 | 7.02 5.67 3.74 | 16.43 | 0.427 0.345 |
| Feb. | 30 29 19 03 | | | |
| Feb. | 29 32 17 03 | | | |
| Feb. | 21 16 22 12 | | | |
| Apr. | 28 21 14 04 | | | |
| Apr. | 28 21 14 05 | | | |
| Apr. | 31 22 15 03 | | | |
| May | 40 44 24 05 | | | |
| May | 40 44 25 06 | | | |
| May | 42 39 22 05 | | | |
| June | 37 28 26 08 | | | |
| June | 37 30 31 11 | | | |
| June | 37 30 29 12 | | | |
| Sept. | 26 25 16 03 | | | |
| Sept. | 28 25 16 03 | | | |
| Sept. | 27 25 16 02 | | | |
| Nov. | 18 09 05 01 | | | |
| Nov. | 16 11 06 02 | | | |
| Nov. | 18 11 07 01 | | | |

[a]As an illustration, positions of these two pixels are plotted on Figure 19.16. The student should use the remaining data as practice for application of the chromaticity technique.

classification of digital remote sensing data (Chapter 12) as a source of land-cover information for such models. Advantages include reduced costs and the ability to quickly prepare new information as changes occur.

Jackson et al. (1977) applied this kind of approach in a study of a small watershed in northern Virginia. Their study, using both manual interpretations of aerial photography and computer classifications of Landsat MSS data, was able to compare results based upon MSS data with those of conventional estimates. Their results indicated that the use of Landsat data was much cheaper than the manual method but tended to produce results of comparable reliability.

## 19.9. Evapotranspiration

*Evapotranspiration* is defined as combined *evaporation,* moisture lost to the atmosphere from soil and open water, and *transpiration,* moisture lost to the atmosphere by plants. Evapotranspiration measures the total moisture lost to the atmosphere from the land surface, and as such it

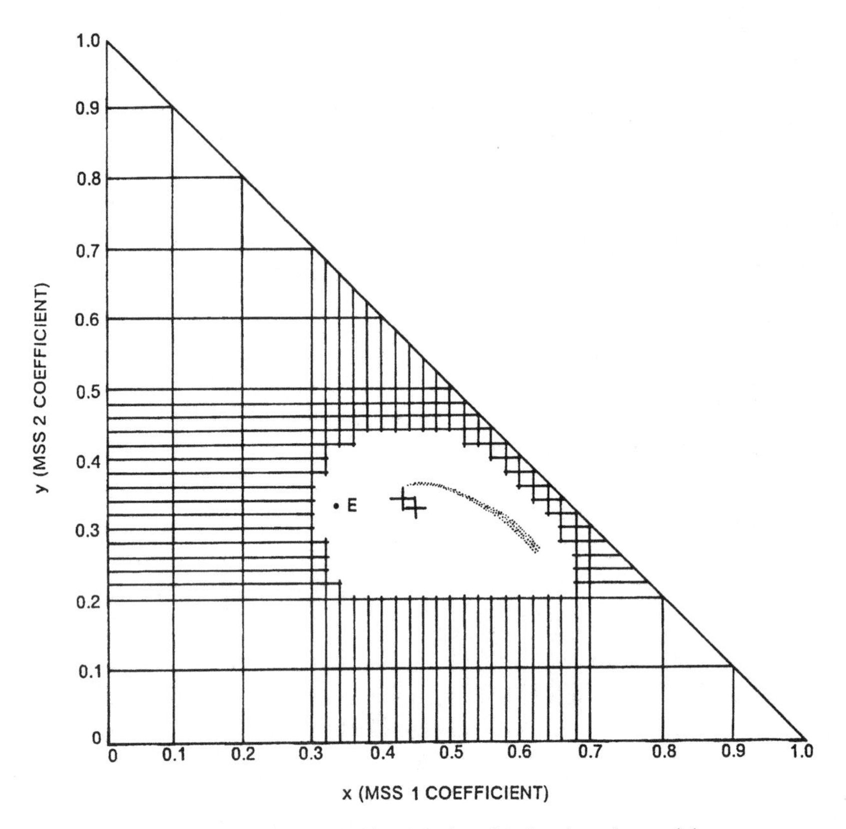


**FIGURE 19.16.** Data from Table 19.4 plotted in Landsat chromaticity space.

(together with other variables) forms an important variable in understanding the local operation of the hydrologic cycle.

However, accurate measurement of evapotranspiration is difficult. Reliable data are usually available only for well-equipped meteorological stations and agricultural research facilities that are regularly attended by a trained staff. Most of the Earth's land area is not near such installations, so reliable measurements of evapotranspiration are not available. Evapotranspiration can, of course, be estimated using measurements of air temperature, wind speed, plant cover, cloudiness, and other variables, but such estimates often lack the detail and accuracy required for good environmental studies, and also require data that are themselves difficult to obtain.

Remotely sensed data can provide the foundation for *spatially referenced estimates* of evapotranspiration—that is, estimates that apply to specific areas of small areal extent (Jackson, 1985). In contrast, estimates based upon meteorological measurements can usually provide only a single estimate for a large region served by a climate station. In effect, spatially referenced estimates provide detailed, independent estimates of evapotranspiration for each pixel.

Direct sensing of the soil surface using remote sensing instruments (especially passive) can provide estimates of soil moisture, or calculation of such estimates can be based upon digital elevation data (described in Chapter 16) that represent topographic elevation as a matrix of values, similar in concept to digital remote sensing images. For the elevation data, each pixel value represents the topographic elevation at the center of each pixel rather than the brightness of the energy reflected from the land at that pixel. From such data a computer program can calculate the topographic slope at each pixel and can also determine the aspect, the compass direction at which the slope faces.

With the elevation data in hand, it is possible to overlay Landsat data and data from meteorological satellites, such as MODIS and AVHRR (Chapters 6, 17, and 21), so that all three forms of data register to one another. A classification of the Landsat data (as described in Chapter 12) can provide a map of the major vegetation classes (forest, pasture, bare ground, etc.) present. The data gathered by meteorological satellites can be used to estimate temperature and total solar radiation received during a 24-hr. interval. These vary depending upon the slope and aspect of each pixel, as derived from the elevation data. From these separate forms of data, it is then possible to estimate evapotranspiration for each pixel. With independent estimates of evapotranspiration for each pixel within a study area, the analyst can then prepare much more detailed hydrologic data.

Jackson (1985) describes the estimation of evapotranspiration from remotely sensed data and observes that some difficulties still remain. Atmospheric degradation of the remotely sensed data can cause errors in the estimates, and the usual procedures cannot consider effects of factors such as wind speed or surface roughness. Also, estimates apply only to a specific time (when the remotely sensed data were acquired), although usually the requirement is to estimate evapotranspiration over intervals of a day, several days, or even longer. Therefore, it is necessary to extend the estimates from a specific time to a longer interval—a process that is likely to introduce error. Nonetheless, it should be emphasized that further development of such procedures is one of the most interesting and practical applications of remote sensing, as they can make major contributions to studies of hydrology, climatology, and agriculture in many of the Earth's remote and inaccessible regions.

## 19.10. Manual Interpretation

The following examples illustrate how techniques of manual image interpretation (Chapter 5), when combined with a knowledge of films and filters (Chapter 3) and of the spectral behavior of water bodies (as outlined earlier in this chapter), can be used to derive information concerning water bodies and coastal features. Figures 19.17 and 19.18 show aerial photography of the same region of a coastline near Pensacola, Florida, as recorded by different film and filter combinations. Figure 19.17 depicts the scene using panchromatic film. Vegetation is depicted in light and medium gray tones, with patches of evergreen forest represented in darker tones. At the coastline, the beach is visible as a bright strip, and the open water is recorded as smooth textures with medium gray tones. The coastline at the left side of the image has a brighter tone where turbidity is higher—in the surf zone, currents carry sediments towards the left edge, and green and blue light can penetrate deep enough to reflect from shallow subsurface features. (Figure 19.3 shows the deep penetration of blue and green light as it is recorded, along with longer wavelengths of the visible region, by panchromatic films.)

Figure 19.18 uses black and white infrared film to portray the same scene (although at a different date). Here, the forested regions have a lighter tone than was visible in the companion image, although it is not as bright as in the usual near infrared image. Open water is shown here as a uniform deep black tone, caused by the strong absorption of near infrared radiation by open water. Figure 19.3 shows that near infrared radiation can penetrate water only to very shallow depths, so the image shows none of the subsurface features and sediment transport vis-

**FIGURE 19.17.** Black-and-white panchromatic photograph of the same area shown in Figure 19.18. Image courtesy of USGS. Note the record of sediment transport and shallow subsurface features at the coastline.

**FIGURE 19.18.** Black-and-white infrared photograph of the same area depicted in Figure 19.17. Note the absence of detail in the open water. Image Courtesy of USGS.

ible in the companion image. As a result, the land–water contact is shown to be especially sharp.

Plate 24 shows the mouth of the Chesapeake Bay as photographed on 8 June 1991 from the space shuttle. Land areas appear in silhouette due to underexposure to record hydrographic features. As a result, the land–water interface is very sharply delineated. The Chesapeake Bay occupies the left-hand portion for the image; Norfolk, Virginia, and associated naval facilities occupy the lower left corner, but are not visible due to the underexposure of land areas. The Chesapeake Bay bridge-tunnel is visible as a dark line at the mouth of the bay. Just to right of the bridge-tunnel, a curved front separates the waters of the bay from those of the Atlantic. Virginia's coastal waters are visually distinct from those of the bay, due to currents, surface oils, temperature, and wind and wave patterns. Near this front, wakes of several ships are visible.

Figures 19.19 through 19.22 represent this same region from a different perspective. Figure 19.19 shows the Eastern Shore of Virginia—a long, low, sandy peninsula formed by marine and aeolian deposition. Figure 19.20 presents a schematic view of the barrier islands and tidal marshes—the most dynamic components of this landscape shown the images in Figures 19.21

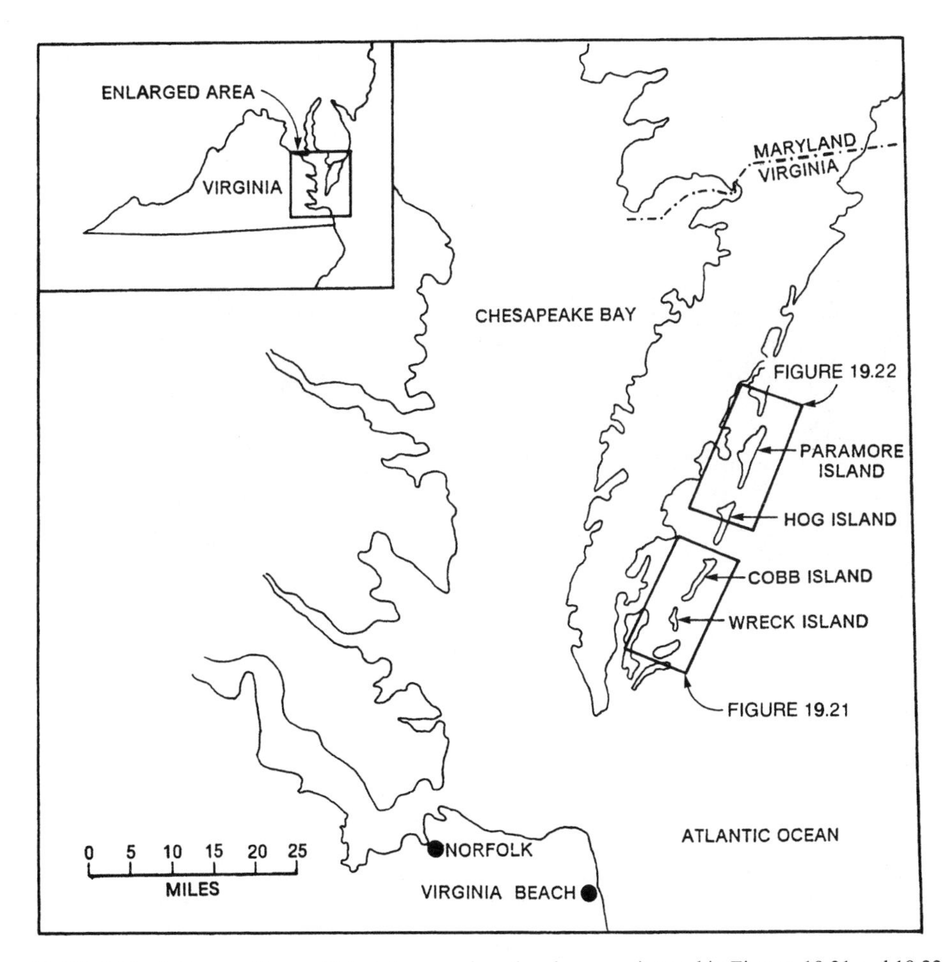


**FIGURE 19.19.** Eastern Shore of Virginia: overview showing areas imaged in Figures 19.21 and 19.22.

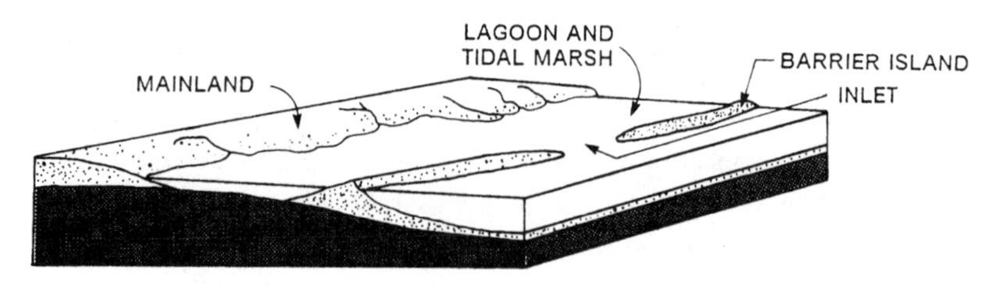


**FIGURE 19.20.** Eastern Shore of Virginia: diagram of barrier islands and tidal lagoons.

and 19.22. Thermal images of the Atlantic coast were acquired just after noon on 23 July 1972. The setting is depicted in a broader context in Plate 24 and Figures 19.21 and 19.22; the coastline is protected by a series of barrier islands formed by sediment transported from the north and periodically reshaped by currents and storms. The islands themselves are long, low strips of sand shaped by the action of both water and wind. At the edge of the ocean, a long beach slopes toward the water; the lowest sections of the beach are influenced each day by the effects of tides and currents. The upper, higher sections are affected only by the highest tides and strongest storms, so these regions may experience major changes only once or twice a year or even less often. Inland from the upper beach ridges is wind-blown sand formed into dunes that are generally above the reach of waves, although very strong storms may alter these regions. The dunes are reshaped by wind, but their general configuration is often stabilized by grasses and small shrubs that cover much of this zone. Inland from the dunes, elevations decrease, and water

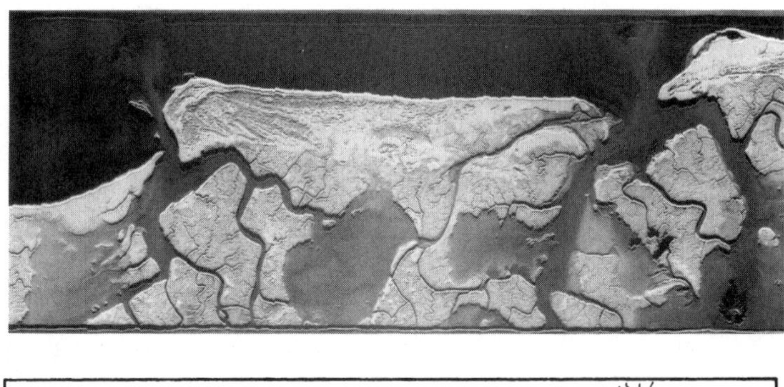

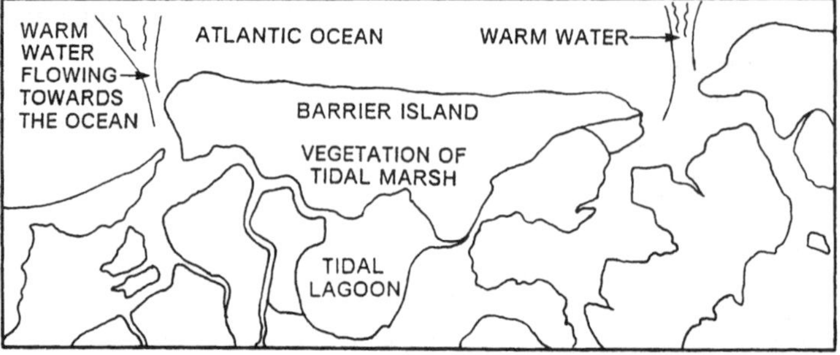


**FIGURE 19.21.** Thermal infrared image of Eastern Shore of Virginia, 23 July 1972, approximately 12:50 P.M.: Cobb Island and Wreck Island. From NASA.

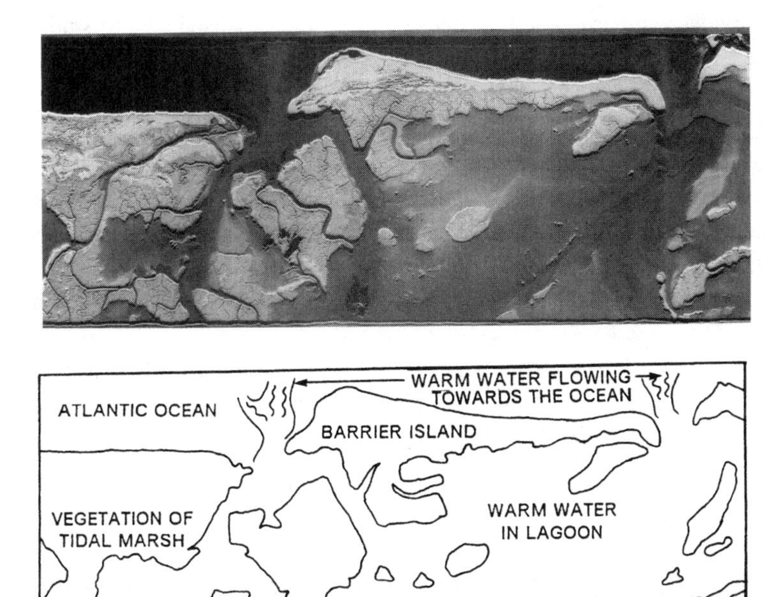


**FIGURE 19.22.** Thermal infrared image of Eastern Shore of Virginia, 23 July 1972, approximately 12:50 P.M.: Paramore Island. From NASA.

again assumes the dominant influence in shaping the ecosystems. Water from the tidal marshes between the islands and the mainland rises and falls with the tides, but is without the strong waves that characterize the seaward side of the islands. These tidal flats are covered with vegetation adapted to the brackish water, experience tidal fluctuations, and have poorly drained soils.

Many of these features are clearly visible on the infrared images of this region (Figures 19.21 and 19.22). The bright white strip along the seaward side of the island contrasts sharply with the cool dark of the open ocean. This bright strip is, of course, the hot surface of the open sandy beach, which has received the direct energy of the solar beam for several hours, and now, at midday, is very hot. Inland from the beach the dunes are visible as a slightly darker region on the image with darker areas caused by shadowing. The topographic structure of the beach ridges are clearly visible in several regions. In the tidal marshes, the image appearance is controlled largely by vegetative cover, which gives a clear delineation of the edges of the open channel not usually visible on other images. The open water within the tidal lagoons has a lighter tone than that of the open ocean, indicating warmer temperatures. Here the shallow water and the restricted circulation have enabled the water in these areas to absorb the solar radiation it has been receiving now for several hours. Nonetheless, this water is still cooler than water flowing from the land surface; it is possible to see the bright (warm) plumes of water from the streams that flow into the tidal marshes from the mainland. Finally, tidal currents are clearly visible as the bright (warm) stream of water passing between the barrier islands to enter the darker (cooler) water of the open ocean. This image captures the tide flowing outwards toward the open ocean (ebb tide).

Plate 25 provides another example of the use of remotely sensed data to study currents and tidal flow. Here the images show a portion of the Belgian port of Zeebrugge as observed using 12 multispectral channels encompassing portions of the visible, near infrared, and thermal regions. Image data have been processed to provide correct geometry and radiometry, and, with the use of on-site observations, analyzed to reveal sediment content within the upper 1 m of the water column. Reds and yellows indicate high sediment content; blues and greens indicate low sediment content. The top image shows conditions at low tide; the bottom image shows conditions at high tide. From such images it is possible to estimate the broad-scale patterns of sediment transport and deposition, and to plan efficient dredging and construction activities.

Applications of active microwave imagery to understanding marine resources are illustrated by Figure 19.23, a RADARSAT image of Manila Harbor, Philippines (30 March 1999). This scene represents aquaculture and fishing activities in the harbor area. The smooth black surface

**FIGURE 19.23.** RADARSAT SAR imagery of Manila Harbor. RADARSAT data © Canadian Space Agency/Agence Spatiale Canadienne 1999. Received by the Canada Centre for Remote Sensing. Processed and distributed by RADARSAT International.

is the relatively calm surface of open coastal waters, with distinctive angular lines formed by structures that hold nets designed to entrap fish in the tidal flow. The urban and periurban landscape is visible as the bright responses typical of abundant corner reflectors present in an urban region. The fine network of thin bright lines is created by the network of dykes that separate open ponds devoted to aquaculture.

## 19.11. Sea Surface Temperature

Remote sensing has become one of the most effective means of assessing *sea surface temperature* (SST), the average temperature of the ocean surface, indicative of temperatures within the upper meter or so of the ocean surface. SST is an important variable for oceanographic studies, including characteristics and dynamics of currents and distributions and movements of marine life. Because SST is a major influence for fish and their sources of food, SST maps are important for marine fishing fleets. Commercial services have been successful in providing timely interpretations of satellite data and images for fishing fleets at sea. Because of the significance of oceans for absorbing, radiating, and transporting thermal energy, SST has profound meteorological significance not only for marine weather and climate, but also for understanding general atmospheric processes. Therefore, satellite sensing of SST has formed an important component of attempts to model global climate.

The radiation detected by sensors is determined by the temperature of the upper 1 mm or so of the ocean surface. Although there is mixing of water in the upper portion of the water column, these values may not accurately represent true temperature of water volume at the upper meter of the ocean. Nonetheless, estimates of SST temperature based on observations from AVHRR and related satellite sensors provide some of the most valuable and widely used sources of SST data.

To reduce or eliminate effects of cloud cover, images with differing patterns of cloud cover or cover acquired at different times (but within a short interval) are used to create composites that show cloud-free views of the ocean surface. To correct for contributions of atmospheric moisture upon SST estimates, two spectral channels in different atmospheric windows are used—the two-channel split-window method (McClain et al., 1985). Wavelengths selected for this purpose are positioned close to the peak of the blackbody radiation expected from the Earth (roughly, 10 μm), and lie within well-defined atmospheric windows, so the radiation will be transmitted to the atmosphere.

Plate 26 shows AVHRR-derived SST for waters offshore from Duck, North Carolina, a region near the edge of the warm, north-flowing Gulf Stream, where its waters mix with coastal currents that bring cooler water from the north along the east coast of North America's coastline. Reds and yellows represent warmer temperatures; blues and black represent cooler temperatures. This imagery is formed from the data collected over a 3-day interval.

At broader scales, satellite estimates of SST detail the world's major ocean currents, including the Gulf Stream, and their many related features, such as cold and warm core eddies. Satellites such as AVHRR provide synoptic views of the ocean and a high frequency of repeat views, allowing the examination of basinwide upper ocean dynamics. A given area may be viewed four or more times in a day, collecting data at an intensity that would be infeasible using other methods.

## 19.12. The Marine Food Chain

The marine food chain depends upon phytoplankton, microscopic plants that live in the ocean. They are consumed by zooplankton, which are then consumed by barnacles, mussels, and the ascending chain of larger organisms. Because phytoplankton require sunlight, nutrients, and other favorable environmental conditions, they are not evenly distributed within the world's oceans, but concentrated where conditions are best suited to their growth. In general, phytoplankton is found within the upper few meters of the ocean, where sunlight can penetrate, in cloud-free regions where sunlight is abundant, and where ocean currents bring cold, nutrient-rich waters near the surface. Because phytoplankton contain chlorophyll, multispectral remote sensing instrument can detect their abundance under favorable circumstances. The Coastal Zone Color Scanner (CZCS) and SeaWifs systems have been applied to place-to-place and temporal variations in phytoplankton. This in turn helps in examining patterns in the occurrence of marine life that depend upon these microorganisms.

Further, remote sensing instruments can be applied to detect the presence of nutrient-rich materials known collectively as seston. *Seston* is a generic term that encompasses varied forms of particulate matter such as plankton, bacteria, fungi, algae, protozoa, organic detritus, and inorganic particles (such as silts and clays) suspended in seawater. Water reflectance, as measured in the visible and near infrared regions, has been shown to be effective in recording provisional measures of sediment and seston. Stumpf (1992) reports uses of AVHRR data to monitor the sediment and seston of Lake Pontchartrain in Louisiana.

## 19.13. Summary

Hydrologic studies in general cover a very broad range of subject matter, from the movement of currents in bodies of open water to the evaporation of moisture from a soil surface. Such studies can be very difficult, to carry out especially if the goal is to examine the changes in hydrologic variables as they occur over time and space, because the usual methods of surface observation involve gathering data at isolated points or at specific times. Two of the great advantages of remote sensing in this context are the synoptic view of the aerial perspective and the opportunity to examine dynamic patterns at frequent intervals.

Yet remote sensing encounters difficulties when applied to hydrologic studies. Many of the standard sensors and analytical techniques have been developed for study of land areas and are not easily applied to the special problems of studying water bodies. Analyses often depend upon detection of rather subtle differences in color, which are easily lost by effects of atmospheric degradation of the remotely sensed energy. Currents and other dynamic features may change rapidly, requiring frequent observation to record the characteristics of significance to the analyst. Some important hydrologic variables, such as ground water, are not usually directly visible, and others, such as evapotranspiration, are not at all visible, but must be estimated through other quantities that may themselves be difficult to observe. These problems, and others, mean that further development of hydrologic remote sensing is likely to be one of the most challenging research areas in remote sensing.

## Review Questions

1. List qualities of water bodies that present difficulties for those who study them using only surface observations collected from a ship. Identify those difficulties that are at least partially alleviated by use of some form of remote sensing.

2. Review Chapter 7 to refresh your memory on the qualities of Seasat, and Chapter 6 to find corresponding information for Landsat. List differences between the two systems (orbit, frequency of repeat observation, spectral bands, etc.), identifying those qualities that are particularly well or especially poorly suited for hydrographic applications.

3. It is probably best to compile bathymetric information using directly observed surface data because of their significance for navigation. Yet there are some special situations in which use of remotely sensed data may be especially advantageous. Can you identify at least two such circumstances?

4. Can you think of important applications of accurate delineation of edges of water bodies (i.e., simple separation of land vs. open water)? Be sure to consider observations over time as well as use of images from a single data point.

5. Contrast SPOT (Chapter 6) and Seasat (Chapter 7) with respect to their usefulness for oceanographic studies. (Chapter 7 discusses only some of Seasat's sensors, but it provides enough information to conduct a partial comparison.)

6. Plot the data listed in Table 19.4 on the chromaticity diagram (Figures 19.14 and 19.15) and assesss both the degree of turbidity (high, moderate, low) and the atmospheric clarity for each entry.

7. Write a short description of a design for a multispectral remote sensing system tailored specifically for collecting hydrologic information and accurate location of the edges of water bodies. Disregard all other applications. Suggest some of the factors that might be considered in choosing the optimum time of day for using such a system.

8. Outline the dilemma faced by scientists who wish to use preprocessing in the application of MSS or TM data for bathymetric information. Describe the reasons why a scientist would very much like to use preprocessing in some instances, as well as the counterbalancing reasons why he or she would prefer to avoid preprocessing.

9. Outline some of the ways that the methods of image classification (Chapter 12) might be useful for hydrologic studies. Outline also problems and difficulties that limit the usefulness of these methods for studies of water bodies.

## References

### *General*

Alföldi, T. T. 1982. Remote Sensing for Water Quality Monitoring. Chapter 27 in *Remote Sensing for Resource Management* (C. J. Johannsen and J. L. Sanders, eds.). Ankeny, IA: Soil Conservation Society of America, pp. 317–328.

Bhargava, D. S., and D. W. Mariam. 1991. Effects of Suspended Particle Sizes and Concentrations on Reflectance Measurements. *Photogrammetric Engineering and Remote Sensing,* Vol. 57, pp. 519–529.

Brakenridge, G. R., E. Anderson, and S. Caquard. 2005. *Flood Inundation Map DFO 2003–282.* Hanover NH: Dartmouth Flood Observatory, Hanover, USA, digital media, *http://www.dartmouth.edu/~floods/2003282.html.*

Brakenridge, G. R., J. C. Knox, F. J. Magilligan, and E. Paylor. 1994. Radar Remote Sensing of the 1993 Mississippi Valley Flood. *EOS: Transactions of the American Geophysical Union.*

Curran, P. J., and E. M. M. Novo. 1988. The Effects of Suspended Particle Size and Concentration on Reflectance Measurements: A Review. *Journal of Coastal Research,* Vol. 4, pp. 351–368.

Fishes, L. T., F. Scarpace, and R. Thomson. 1979. Multidate Landsat Lake Quality Monitoring Program. *Photogrammetric Engineering and Remote Sensing,* Vol. 45, pp. 623–633.

Frazier, P. S., and K. L. Page. 2000. Water Body Detection and Delineation with Landsat TM Data. *Photogrammetric Engineering and Remote Sensing,* Vol. 66, pp. 1461–1465.

Han, L. 1997. Spectral Reflectance with Varying Suspended Sediment Concentrations in Clear and Algae-Laden Waters. *Photogrammetric Engineering and Remote Sensing,* Vol. 63, pp. 701–705

Han, L., and D. C. Rundquest. 1997. Comparison of NIR/RED Ratio and First Derivative of Reflectance in Estimating Algal-Chlorophyll Concentration: A Case Study in a Turbid Reservoir. *Remote Sensing of Environment,* Vol. 62, pp. 253–261.

Han, L., and D. C. Rundquest. 2003. The Spectral Response of Ceratophyllum demersum at varying depths in an experimental tank. *International Journal of Remote Sensing,* Vol. 24, pp. 859–864.

Holden, H., and E. Ledrow. 1999. Hyperspectral Identification of Coral Reef Features. *International Journal of Remote Sensing,* Vol. 20, pp. 2545–2563.

Jackson, R. D. 1985. Evaluating Evapotranspiration at Local and Regional Scales. *Proceedings of the IEEE,* Vol. 73, pp. 1086–1096.

Jackson, T. J., R. M. Ragan, and W. N. Fitch. 1977. Test of Landsat-Based Urban Hydrologic Modeling. *Journal of Water Resources Planning and Management Division, American Society of Civil Engineers,* Vol. 103, No. WRI, pp. 141–158.

Khorram, S. 1980. Water Quality Mapping from Landsat Digital Data. *International Journal of Remote Sensing,* Vol. 2, pp. 143–153.

Klemas, V., R. Sicna, W. Treasure, and M. Otley. 1973. Applicability of ERTS-1 Imagery to the Study of Suspended Sediment and Aquatic Forms. In *Symposium on Significant Results Obtained from Earth Resources Technology Satellite-1.* Greenbelt, MD: Goddard Space Flight Center, pp. 1275–1290.

Liedtke, T. A., A. Roberts, and J. Lutenauer. 1995. Practical Remote Sensing of Suspended Sediment Concentration. *Photogrammetric Engineering and Remote Sensing,* Vol. 61, pp. 167–175.

Lillesand, T. M., E. L. Johnson, R. L. Deuell, O. M. Lindstrom, and D. E. Miesner. 1983. Use of Landsat Data to Predict the Trophic State of Minnesota Lakes. *Photogrammetric Engineering and Remote Sensing,* Vol. 49, pp. 219–229.

McKim, H. L., C. J. Merry, and R. W. Layman. 1984. Water Quality Monitoring Using an Airborne Spectroradiometer. *Photogrammetric Engineering and Remote Sensing,* Vol. 50, pp. 353–360.

Moore, G. K. 1978. Satellite Surveillance of Physical Water Quality Characteristics. In *Proceedings of the Twelfth International Symposium on Remote Sensing of Environment.* Ann Arbor: Environmental Research Institute of Michigan, pp. 445–462.

Nace, R. L. 1967. *Are We Running Out of Water?* (U.S. Geological Survey Circular 586). Reston, VA: U.S. Geological Survey, 7 pp.

Philipson, W. R., and W. R. Hafker. 1981. Manual versus Digital Landsat Analysis for Delineating River Flooding. *Photogrammetric Engineering and Remote Sensing,* Vol. 47, pp. 1351–1356.

Schwab, D. J., G. A. Leshkevich, and G. C. Muhr. 1992. Satellite Measurements of Surface Water Temperatures in the Great Lakes. *Journal of Great Lakes Research,* Vol. 18, pp. 247–258.

Strandberg, C. 1966. Water Quality Analysis. *Photogrammetric Engineering,* Vol. 32, pp. 234–249.

Stumpf, R. P. 1992. Remote Sensing of Water Clarity and Suspended Sediment in Coastal Waters. In Proceedings of the First Thematic Conference on Remote Sensing for Marine and Coastal Environments. *Proceedings of the International Society of Optical Engineering,* Vol. 1930, pp. 1436–1473.

Tolk, B. L., L. Han, and D. C. Rundquist. 2000. The Impact of Bottom Brightness on Spectral Reflectance of Suspended Sediments. *International Journal of Remote Sensing,* Vol. 21, pp. 2259–2268.

Westawy, R. M., S. N. Lane, and D. M. Hicks. 2001. Remote Sensing of Clear-Water, Shallow, Gravel-Bed Rivers Using Digital Photogrammetry. *Photogrammetric Engineering and Remote Sensing,* Vol. 67, pp. 1271–1282.

## *Chromaticity Analysis*

Alföldi, T. T., and J. C. Munday. 1977. Progress toward a Landsat Water Quality Monitoring System. In *Proceedings, Fourth Canadian Symposium on Remote Sensing.* Quebec, Canada: Canadian Remote Sensing Society of the Canadian Aeronautics and Space Institute, pp. 325–340.

Markam, B. L., and J. L. Barker. 1986. Landsat MSS and TM Post-Calibration Dynamic Ranges, Exoatmospheric Reflectance, and At-Satellite Temperatures. *Landsat Technical Notes,* No. 1, pp. 3–8.

Munday, J. C. Jr., and T. T Alföldi. 1975. Chromaticity Changes from Isoluminous Techniques Used to Enhance Multispectral Remote Sensing Data. *Remote Sensing of Environment,* Vol. 4, pp. 221–236.

Munday, J. C. Jr., T. T. Alföldi, and C. L. Amos. 1979. Bay of Funday Verification of a System for Multidate Landsat Measurement of Suspended Sediment. In *Satellite Hydrology* (M. Deutsch, D. R. Wiesner, and A. Rango, eds.). Minneapolis, MN: American Water Resource Association, pp. 622–640.

Nelson, R. 1985. Reducing Landsat MSS Scene Variability. *Photogrammetric Engineering and Remote Sensing,* Vol. 51, pp. 583–593.

Stimson, A. 1974. *Photometry and Radiometry for Engineers.* New York: Wiley, 466 pp.

## *Bathymetric Mapping*

Hallada, W. A. 1984. Mapping Bathymetry with Landsat 4 Thematic Mapper: Preliminary Findings. In *Proceedings, Ninth Canadian Symposium on Remote Sensing.* Quebec, Canada: Canadian Aeronautics and Space Institute, pp. 629–643.

Han, L. 1997. Spectral Reflectance with Varying Suspended Sediment Concentrations in Clear and Algae-Laden Waters. *Photogrammetric Engineering and Remote Sensing,* Vol. 63, pp. 701–705.

Jupp, D. L. B., K. K. Mayo, D. A. Kucker, D. Van, R. Classen, R. A. Kenchinton, and P. R. Guerin. 1985. Remote Sensing for Planning and Managing the Great Barrier Reef of Australia. *Photogrammetria,* Vol. 40, pp. 21–42.

Lyzenga, D. R. 1979. Shallow-Water Reflectance Modeling with Applications to Remote Sensing of the Ocean Floor. In *Proceedings, Thirteenth International Symposium on Remote Sensing of Environment.* Ann Arbor: Environmental Research Institute of Michigan, pp. 583–602.

Lyzenga, D. R., R. A. Shuchman, and R. A. Arnone. 1979. Evaluation of an Algorithm for Mapping Bottom Features under a Variable Depth of Water. In *Proceedings, Thirteenth International Symposium on Remote Sensing of Environment.* Ann Arbor: Environmental Research Institute of Michigan, pp. 1767–1780.

Polcyn, F. C., and D. R. Lyzenga. 1979. Landsat Bathymetric Mapping by Multispectral Processing. In *Proceedings, Thirteenth International Symposium on Remote Sensing of Environment.* Ann Arbor: Environmental Research Institute of Michigan, pp. 1269–1276.

Satzman, B. (ed.). 1985. *Satellite Oceanic Remote Sensing: Advances in Geophysics.* New York: Academic Press, 511 pp.

Tanis, F. J., and W. A. Hallada. 1984. Evaluation of Landsat Thematic Mapper Data for Shallow Water Bathymetry. In *Proceedings, Eighteenth International Symposium on Remote Sensing of Environment.* Ann Arbor: Environmental Research Institute of Michigan, pp. 629–643.

### *Sea Surface Temperature*

McClain, E. P., W. G. Pichel, and C. C Walton. 1985. Comparative Performance of AVHRR-Based Multichannel Sea Surface Temperatures. *Journal of Geophysical Research,* Vol. 90, pp. 11587–11601.

Reynolds, R. W. 1988. A Real-Time Global Sea Surface Temperature Analysis. *Journal of Climate,* Vol. 1, pp. 75–86.

Reynolds, R. W., C. K. Folland, and D. E. Parker. 1989. Biases in Satellite-Derived Sea Surface Temperature Data. *Nature,* Vol. 341, pp. 728–731.

Reynolds, R. W., and D. C. Marsico. 1993. An Improved Real-Time Global Sea Surface Temperature Analysis. *Journal of Climate,* Vol. 6, pp. 114–119.

Reynolds, R. W., and T. M. Smith. 1994. Improved Global Sea Surface Temperature Analyses Using Optimum Interpolation. *Journal of Climate,* Vol. 7, pp. 929–948.

CHAPTER TWENTY

# Land Use and Land Cover

## 20.1. Introduction

*Land use* describes use of the land surface by humans. Normally, use of land is defined in an economic context, so we think of land as it is used for agricultural, residential, commercial, and other purposes. However, strictly speaking, we can seldom really see the use of the land, except upon the very closest inspection, so we consider also the *land cover*—the visible features of the Earth's surface—including in the vegetative cover, natural and as modified by humans, its structures, transportation and communications, and so on (Figure 20.1). As a practical matter, we must consider land use and land cover together, although also recognizing the distinction between the two.

Modern society depends upon accurate land-use data for both scientific and administrative purposes. It forms an essential component of local and regional economic planning, to assure that various activities are positioned on the landscape in a rational manner. For example, accurate knowledge of land-use patterns permits planning to avoid placing residential housing adjacent to heavy industry, or in floodplains. Accurate land-use information can assure that residential neighborhoods are logically placed with respect to commercial centers and access to transportation services. In another context, land use is an important component of climatic and hydrologic modeling, to estimate the runoff of rainfall from varied surfaces into stream systems. And accurate land-use data are important for transportation planning, so traffic engineers can estimate the flow of vehicles form one region to another, and design highways with appropriate capacities.

Land-use patterns also reflect the character of a society's intimate interaction with its physical environment. Although land-use patterns within our own society seem self-evident and natural, other societies often organize themselves in contrasting patterns. This fact becomes obvious when it is possible to observe different economic and social systems occupying the similar environments. Plate 27 shows a Landsat TM quarter scene depicting Santa Rosa del Palmar, Bolivia (northwest of Santa Bruise) acquired in July 1992. Here four distinct land-use patterns are visible, each reflecting its distinctive relationship with the landscape it occupies. In the southeastern portion of the scene, the land-use pattern is dominated by broad-scale mechanized agriculture practiced by a Mennonite community. The upper right (northeastern) region of the image is a mountainous area occupied by a diminishing population of Indians, who practice a form of slash-and-burn agriculture, visible as dispersed patches of light green. Throughout the

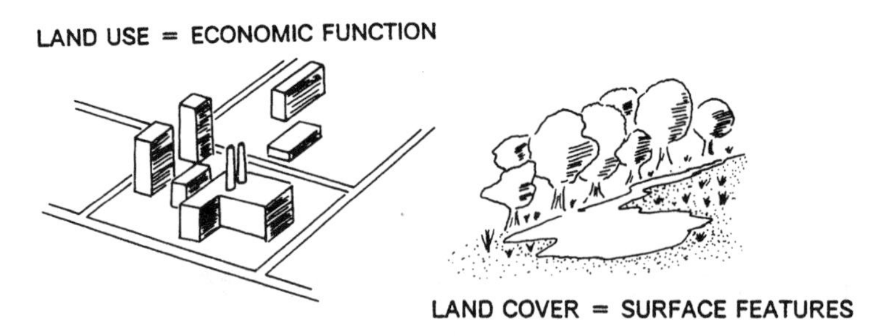


**FIGURE 20.1.** Land use and land cover. Land use refers to the principal economic enterprises that characterize an area of land: agriculture, manufacturing, commerce, and residential. Land cover indicates the physical features that occupy the surface of the Earth, such as water, forest, and urban structures.

northwestern and central portions of the image, the Bolivian government has encouraged broadscale clearing of forest for agriculture. In the upper left region of the image, these clearings take the form of light green spots aligned northwest to southeast; at the center of each patch of cleared land is a central facility providing colonists with fertilizers, pesticides, and staples. In the southwestern corner of the image, the complex field pattern reflects an established landscape occupied by Japanese immigrants. This image clearly depicts, with a single image, the unique patterns that each cultural perspective has created as humans have occupied the Earth's surface.

## 20.2. Aerial Imagery for Land-Use Information

Although some forms of land-use data must be acquired by direct observation by officials who visit each site in person (e.g., for local zoning or tax data), the vast majority of land-use and land-cover data are acquired by interpretation of aerial photography and similar imagery. Aerial photography provides the overhead view, the spatial perspective, and the comprehensive detail that permits accurate, systematic, and effective study of land-use patterns. Early proponents of the use of aerial photography in civil society envisioned that aerial photographs would provide a rich source of information regarding the structure of cities and landscapes. By the late 1930s, the staff of the Tennessee Valley Authority was among the first to pioneer methods for using aerial photography for land-use survey. Later, the USGS developed classification systems specifically tailored for use with aerial photography. These techniques form the foundations of today's methods of applying aerial imagery to land-use survey.

Remotely sensed images lend themselves to accurate land-cover and land-use mapping in part because land-cover information can be visually interpreted more or less directly from evidence visible on aerial images and objects can be seen in the context of their neighboring features.

Land-use maps are routinely prepared at a wide variety of scales, typically ranging from 1:12,500 and larger to 1:250,000 and smaller. At one end of this spectrum (the larger scale maps), remotely sensed imagery may itself contribute relatively little information to the survey; its main role may be to form a highly detailed base for recording data gathered by other means (Figure 20.2). At such large scales the land-use map may actually serve as a kind of reference map, having little cartographic generalization. Such products are often used at the lowest levels

**FIGURE 20.2.** A community planning official uses an aerial photograph to discuss land use policy relating to planning land use near Boston's Logan Airport, 1973. From National Archives and Records Administration, ARC548443.

of local governments, perhaps mainly urban areas, that have requirements for such detailed information and the financial resources to acquire it.

As the scale of the survey becomes broader (i.e., as map scale becomes smaller), the contribution of the image to the informational content of the map becomes greater, although even at the smallest scales there must always be some contribution from collateral information. Differing scales and levels of detail serve different purposes and different users. For the regional planner, the loss of resolution and detail at smaller scales may actually form an advantage in the sense that often analysts sometimes prefer an integration and a simplification of the information that must be examined. For the medium- and small-scale land-use surveys that are most often compiled by use of remotely sensed images, the product is a thematic map that depicts the predominant land cover within relatively homogeneous areas, delineated subject to limitations of scale, resolution, generalization, and other constraints accepted (or at least recognized) by both compilers and users of thematic maps.

## 20.3. Land-Use Classification

Preparation of a land-use/land-cover map requires that the mapped area be subdivided into discrete parcels, each labeled with a distinct, mutually exclusive, nominal label. That is, each parcel must be identified using a single, distinct label in a classification system. "Nominal" simply means that the labels are names, rather than values. So, for example, when we use a symbol such as "22" to designate a category, we do not mean that it has twice the value of "11," but simply that the separate designations convey different qualitative meanings, such as "urban" and "agricultural" land uses. Usually, we also use colors, or colors and symbols, to designate land-use classes on maps.

Perhaps the most widely used classification system for land-use and land-cover data derived from aerial photography is the U.S. Geological Survey's Land Use and Land Cover classification system, developed during the 1970s (see Table 20.1). This classification system has many advantages over previous systems. Whereas previous systems did not consider the unique advantages of aerial imagery as a source of land-use and land-cover data, the USGS system was specifically prepared for use with aerial photography and related imagery. Its categories are tailored for use with information derived from aerial images. It has a hierarchical design that lends itself to use with images of varied scales and resolutions. Level I categories, for example, are designed for broad-scale, coarse-resolution imagery (e.g., broad-scale satellite imagery or high-altitude aerial photography).

Although the USGS system specifies Level I and II categories, more detailed classes at Level III and below must be defined by the analyst to meet the specific requirements of a particular study and a specific region. When an interpreter defines Level III categories, the Level I and II categories should be used as a framework for the more detailed Level III classes.

## 20.4. Visual Interpretation of Land Use and Land Cover

Image interpreters apply the elements of image interpretation to delineate separate land-cover classes, applied in an organized and systematic manner. Land-use interpretation proceeds by marking boundaries between categories as they occur on the imagery. Often interpreters mark transparent overlays that register to the image, or digitize directly as the image is displayed on a computer screen. As each land-use parcel is outlined, its identity is marked with a symbol (usually consisting of one to three numerals) matching to the classification system (Figure 20.3).

A land-use map then is prepared by delineating regions of consistent land use, and assigning them to appropriate classes in land-use classification. An example can show an application of land-use mapping to an aerial photograph, using principles outlined in Chapter 5.

Steps can be seen in Figure 20.4: (1) the original aerial photograph (A); (2) the land-use boundaries, as defined by a photointerpreters to separate land-use parcels at Anderson Level II (B and D); (3) the separate parcels are then assigned to classes and symbolized with colors or symbols (C); and finally (4) a map is displayed with a legend (E). Land-use/land-cover information must be presented to the user in a planimetrically correct form, so the imagery that is used for the interpretation must be either corrected beforehand (e.g., in the case of a DOQ), or the boundaries must later be rectified.

Identification of land-cover parcels is based upon the elements of image interpretation, as presented in Chapter 5. Although sometimes land use may be characterized by specific objects, usually the primary task is one of consistent delineation of uniform parcels that match the classification system. As an example, *cropped agricultural land* (Anderson symbol 21) (as it occurs in midlatitude landscapes) is usually easily recognizable by the systematic division of fields and the smooth, even textures. Tone varies with the crop and the growth stage (Figure 20.5). *Pasture* (Anderson symbol 21) usually has more irregular boundaries, a mottled texture, medium tones, and is often characterized by isolated trees or small groves both at the edges and in the

**TABLE 20.1. USGS Land-Use and Land-Cover Classification**

1 Urban or built-up land
- 11 Residential
- 12 Commercial and services
- 13 Industrial
- 14 Transportation, communications, and utilities
- 15 Industrial and commercial complexes
- 16 Mixed urban or built-up land
- 17 Other urban or built-up land

2 Agricultural land
- 21 Croplands and pasture
- 22 Orchards, groves, vineyards, and nurseries
- 23 Confined feeding operations
- 24 Other agricultural land

3 Rangeland
- 31 Herbaceous rangeland
- 32 Shrub and brush rangeland
- 33 Mixed rangeland

4 Forest land
- 41 Deciduous forest land
- 42 Evergreen forest land
- 43 Mixed forest land

5 Water
- 51 Streams and canals
- 52 Lakes
- 53 Reservoirs
- 54 Bays and estuaries

6 Wetland
- 61 Forested wetland
- 62 Nonforested wetland

7 Barren land
- 71 Dry salt flats
- 72 Beaches
- 73 Sandy areas other than beaches
- 74 Bare exposed rock
- 75 Strip mines, quarries, and gravel pits
- 76 Transitional areas
- 77 Mixed barren land

8 Tundra
- 81 Shrub and brush tundra
- 82 Herbaceous tundra
- 83 Bare ground tundra
- 84 Wet tundra
- 85 Mixed tundra

9 Perennial snow or ice
- 91 Perennial snowfields
- 92 Glaciers

*Note.* From Anderson et al. (1976, p. 8). The single-digit designations signify "Level I" classes for small-scale, broadly defined classes. The two-digit designations signify "Level II" classes, applied to larger-scale delineations with finer detail.

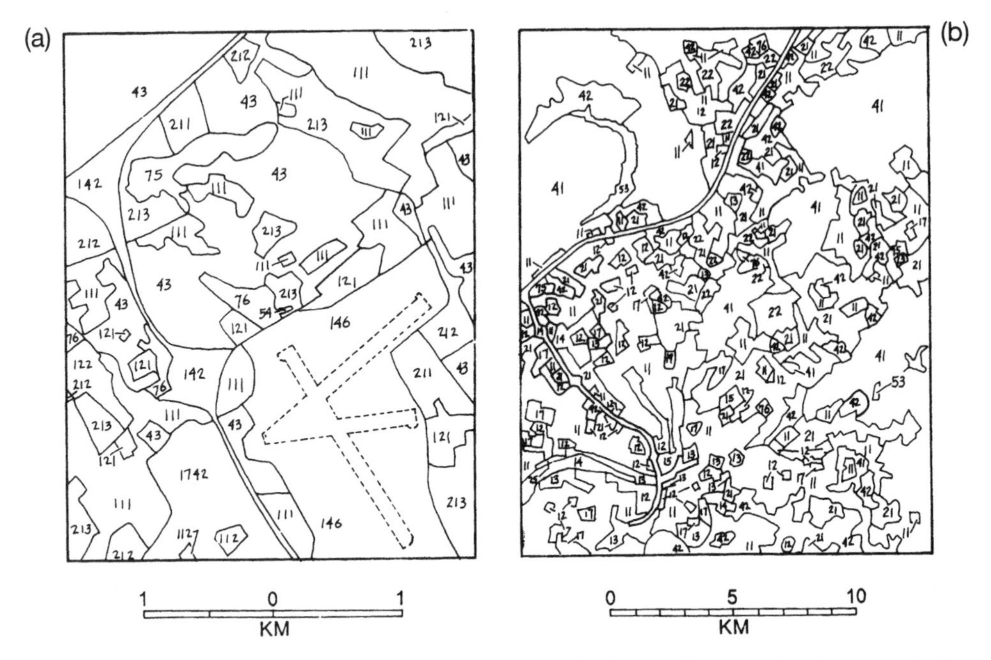


**FIGURE 20.3.** Land-use and land-cover maps. (a) Large scale, fine detail; (b) small scale, coarse detail.

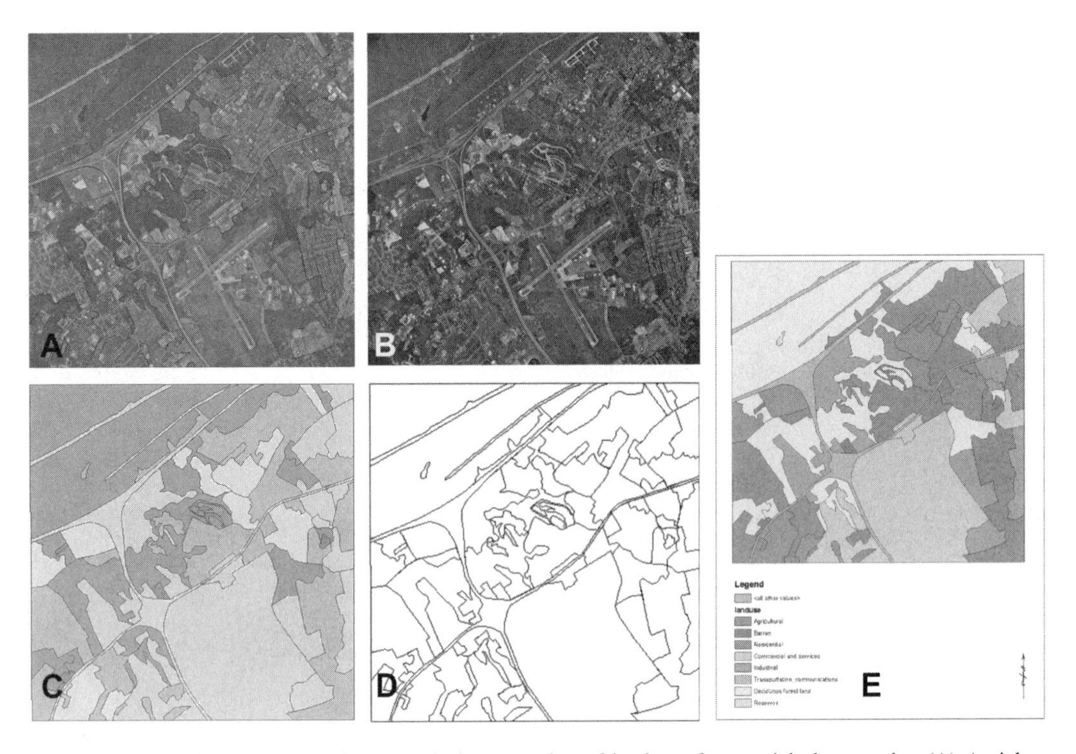


**FIGURE 20.4.** Representation of key steps in interpretation of land use from aerial photography. (A) Aerial photograph only; (B) Delineation of boundaries; (C) Symbolization of principal classes; (D) Boundaries without the backdrop of the areal photograph; (E) Map with symbols and key. From Jessica Dorr.

**FIGURE 20.5.** Visual interpretation of aerial imagery for land-use/land-cover information: cropped agricultural land (Anderson symbol 21). Distinctive characteristics include smooth texture and the contrasting tones that separate varied crops and growth stages. Irrigated cropland, western Kansas, principally wheat, alfalfa, maize, and sorghum. From USDA.

interiors of parcels (Figure 20.6). *Deciduous forest* (Anderson symbol 41) is characterized by rough textures and medium dark tones, and usually occurs in relatively large parcels with irregular edges (Figure 20.7). Roads and clearings are common.

*Land in transition* (Anderson symbol 76) is usually recognizable by the bright tones characteristic of bare soil exposed during construction and irregular outlines. Sometimes, the outlines of roadways and foundations for buildings and partially completed structures are visible (Figure 20.8).

**FIGURE 20.6.** Visual interpretation of aerial imagery for land-use/land-cover information: pasture (Anderson symbol 21). Pastureland is recognizable by its medium gray tones, often without sharp boundaries, and the trees that occur in small groves or as isolated individuals. Fields at top center are cropped agricultural land. Copyright, Virginia Department of Transportation. Used by permission..

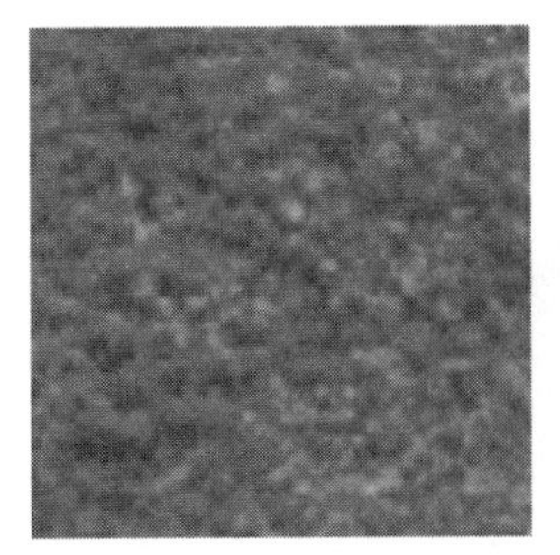
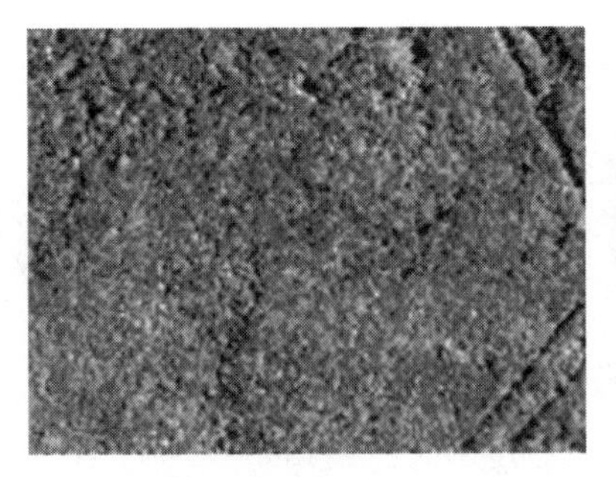

**FIGURE 20.7.** Visual interpretation of aerial imagery for land use/land cover information: deciduous forest (Anderson symbol 41). In the growing season, deciduous forests are usually characterized by a coarse texture, dark tones, and large, uneven parcels. Copyright, Virginia Department of Transportation. Used by permission.

*Transportation* (Anderson symbol 14) is often recognizable by linear patterns that cut across the predominate land use pattern (Figure 20.9) and by distinctive loops of interchanges. *Residential* (Anderson symbol 11) land uses, especially in suburban developments, are often distinctive because of the even placement of structures, the curving street pattern, and the even background of lawns and ornamental trees (Figure 20.10). *Commercial and services* (Anderson symbol 12) is characterized by bright tones (because of the predominance of pavement, parking lots, and rooftops), uneven textures (because of shadowing of buildings), and close proximity to transportation and other high-density land use classes (Figure 20.11).

As the map is prepared, remember that some land-use maps have legal standing, and that some maps may be intended to match data collected previously or by a neighboring jurisdiction. Therefore, the interpreter should not make off-the-cuff adjustments in procedures to solve local interpretation problems.

Interpreters must take care to define delineations that are consistent, clear, and legible. Each parcel is completely enclosed by a boundary, and is labeled with a symbol keyed to the classification system. The interpreter delineates only those features that occupy areas at the scale of the

**FIGURE 20.8.** Visual interpretation of aerial imagery for land-use/land-cover information: land in transition (Anderson symbol 76). Land in transition is often characterized by bright tones, uneven texture, and irregular parcels. Copyright, Virginia Department of Transportation. Used by permission.

**FIGURE 20.9.** Visual interpretation of aerial imagery for land-use/land-cover information: transportation (Anderson symbol 14). Divided highways display the linear forms and the distinctive loops at interchanges shown in this photograph. Copyright, Virginia Department of Transportation. Used by permission.

final map, so usually linear features (e.g., streams and rivers, railways, power lines) or point data are not mapped.

Some parcels will encompass categories other than those named by the parcel label. Such inclusions are permissible, but must be clearly described in an accompanying report, and must be consistent throughout the map. When several interpreters work on the same project, consis-

**FIGURE 20.10.** Visual interpretation of aerial imagery for land-use/land-cover information: residential land (Anderson symbol 11). Suburban residential land, such as that shown here, is often characterized by evenly spaced rooftops and systematic pattern of paved streets. Copyright, Virginia Department of Transportation. Used by permission.

**FIGURE 20.11.** Visual interpretation of aerial imagery for land-use/land-cover information: commercial and services (Anderson symbol 12). In this instance, the identifying characteristics are the large commercial buildings positioned in a linear form, with large parking lots nearby. Copyright, Virginia Department of Transportation. Used by permission.

tency is especially important. Interpreters must coordinate their work with those assigned to neighboring areas, so that detail is uniform throughout the mapped area.

The entire area devoted to a specific use is delineated on the overlay. Thus the delineation of an airfield normally includes not only the runway, but also the hanger, passenger terminals, parking areas, access roads, and in general all features inside the limits of the perimeter fence (i.e., the outline of the parcel encompasses areas occupied by all of these features, even though they are not shown individually on the map). In a similar manner, the delineation of an interstate highway includes not only the two paved roadways, but also the median strip and the right of way.

An aerial image of a quarry can illustrate this practice (Figure 20.12). The quarry itself consists of several features, each with a distinctive appearance on the aerial photograph. Even though it would be possible to individually delineate the transportation and loading structures, the milling and administration buildings, the quarries, and the spoil, these features constitute a single land use, all these features are outlined and labeled as a single parcel. Only at the very largest scales, possibly required for planning and management of the facility itself, would the individual components be delineated as separate features.

The issue of *multiple use* occurs because we assign a single label to each parcel, even though we know that there may in fact be several uses. For example, a forested area may simultaneous-

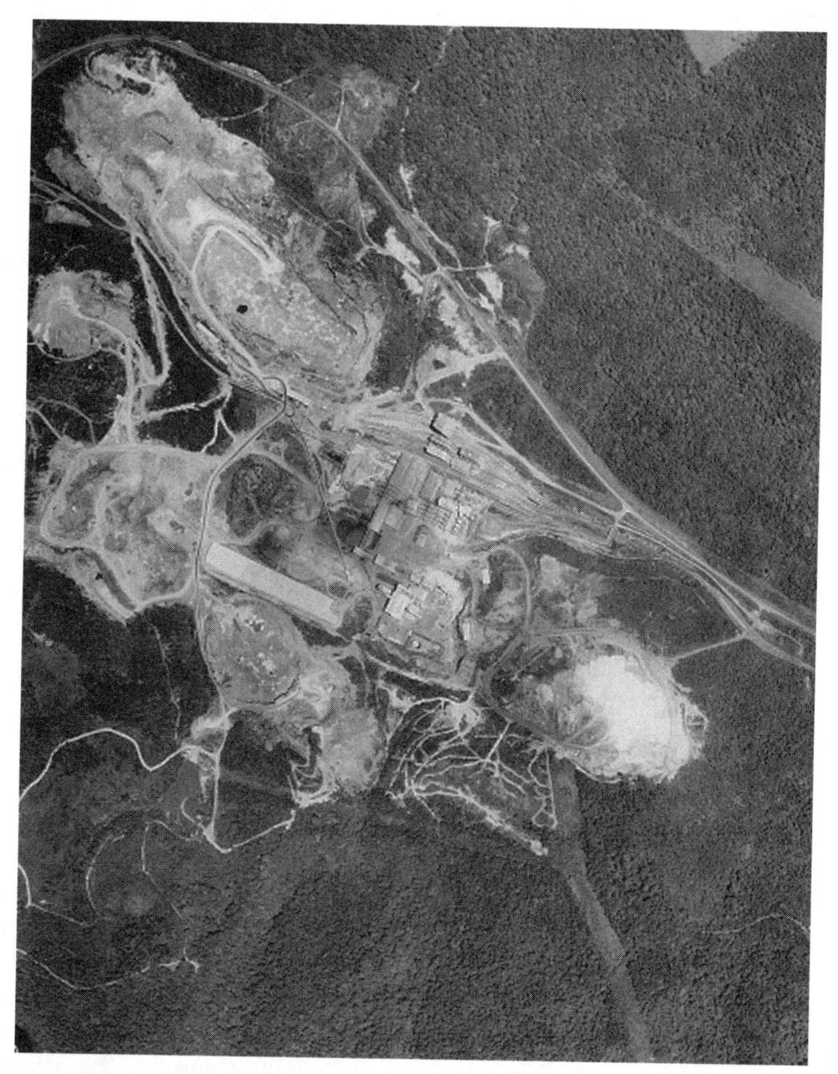

**FIGURE 20.12.** Delineation of the entire area devoted to a given use. This quarry consists of a complex of individual activities that together constitute a single enterprise, so the individual components need not be delineated separately. Copyright, Virginia Department of Transportation. Used by permission.

ly serve as a source of timber, as a recreational area for hunters and hikers, and as a source of runoff that supplies water for an urban region.

The interpreter must select an appropriate minimum size for the smallest parcels to be represented on the final map. Areas smaller than this size will be omitted. Even though they are visible to the interpreter, they are too small to depict legibly.

Usually the label of each category identifies the *predominant category* present within each parcel. Although at small scales there may be unavoidable inclusions of other categories, the interpreter should aspire to define categories that include consistent mixtures of such inclusions.

## 20.5. Land-Use Change by Visual Interpretation

Land-use patterns change over time in response to economic, social, and environmental forces. The practical significance of such changes is obvious. For planners and administrators, they reveal the areas that require the greatest attention if communities are to develop in a harmonious and orderly manner. From a conceptual perspective, study of land-use changes permits identification of long-term trends in time and space and the formation of policy in anticipation of the problems that accompany changes in land use (Estes and Senger, 1972; Anderson, 1977; Jensen and Toll, 1982).

Remote sensing and photointerpretation provide the primary vehicles for compiling *land-use change maps,* maps that represent changes in land use from one date to another. Such maps are important tools for planning land-use policy, for hydrologic management of watersheds, for transportation planning (to understand changes in the population patterns that generate traffic flow), and for environmental studies.

Land-use and land-cover change is very simple in concept: two maps representing the same region, prepared to depict land-use patterns at different dates, are compared, point by point, to summarize differences between the two dates (Figure 20.13). In practice, compiling land-use and land-cover change maps requires mastery of several practical procedures that often reveal previously hidden difficulties.

First, the two maps must use the same classification system, or at least two classification systems that are compatible. *Compatible classifications* mean that the classes can be clearly matched to one another without omission or ambiguity. For example, if one map represents forested land as a single category, "forest," and a second uses three classes, "coniferous," "deciduous," and "mixed coniferous/deciduous," then the two maps are *compatible* in the sense that the three categories of the second map can be combined into a single forested class that is logically compatible with the "forested" class on the first map. As long as the classes are matched at that level of generalization, one can be confident that the comparisons would reveal changes in the extent of forested land.

However, if the second map employed classes such as "densely forested," "partially forested," and "sparsely forested," we could not reasonably match these classes to the "forested" class on the first map because the two sets of classes are not logically compatible. The meanings conveyed by these classes, as represented on the two maps, are not equivalent, so they cannot be compared to determine if the differences represent true change, or simply differences in the way the maps were prepared.

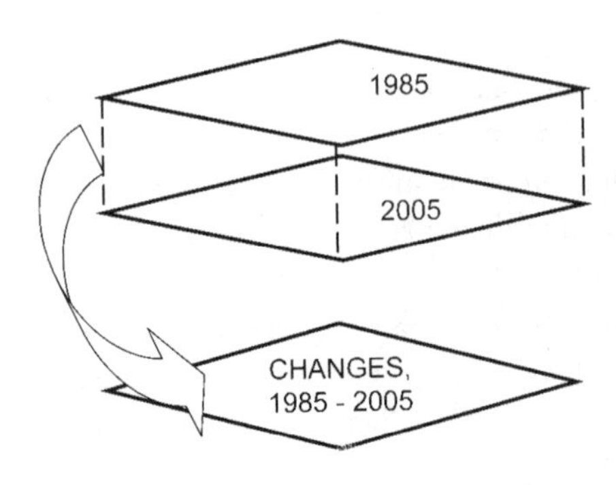


**FIGURE 20.13.** Land-use change.

Second, the two maps must be compatible with respect to scale, geometry, and level of detail. Can we match one point on one map with the corresponding point on the second, and be sure that both points refer to the same place on the Earth's surface? If not, we cannot conduct a change analysis using the two maps because we cannot be sure that differences in the two maps reflect genuine differences in land use, and not simply differences in the projection, scale, or geometric properties of the two maps.

Thus two basic conditions must be satisfied before change data can be compiled using any two maps or images: informational compatibility and geometric compatibility.

Often, land-use maps are prepared to meet the very specific objectives of the sponsoring organization. Therefore, it may be very difficult to reuse that map in a land-use change study many years later because objectives may have changed, or perhaps the level of detail, the classification system, or some other key feature are not compatible. It is for this reason that it is common in land-use change studies for maps to be prepared at the time of the land-use change study itself, using consistent materials, methods, and procedures, even if earlier studies already exist and are available.

## 20.6. Historical Land-Cover Interpretation for Environmental Analysis

A special application of land-use change analysis has been developed to address environmental hazards arising from land-use change in regions where hazardous industrial materials have been abandoned. Here aerial imagery is especially valuable because it offers a spatial perspective, the ability to provide a detailed historical record, and the ability to observe outside the visible spectrum. These capabilities allow analysts to examine problems over time, to detect the existence of specific environmental problems, and to see them in their geographical context. Image analysis can provide a historical record of conditions and actions at specific sites, to assess the nature and severity of current conditions, often when other sources of information are not available.

Over the decades, in industrial regions of the United States, hazardous industrial wastes have been deposited in all sorts of disposal sites, including ponds, lagoons, and landfills that were not designed to safely store dangerous substances. Although many of these areas were once positioned in rural or sparsely populated regions, urban growth has brought many of these sites close to residential neighborhoods. As a result, many such sites are now near populated regions, and, in some instances, have been converted to other uses, including residential housing, school yards, and recreational areas. Deterioration of containers has released these wastes to contaminate nearby soils and enter groundwater systems to migrate beyond their original site. Often, the use of the site for waste disposal was not properly recorded or the nature of the wastes was never documented, so remediation is limited by ignorance of specifics of the situation.

Because of the difficulty of establishing the historical pattern of use at such sites, and the absence or inaccuracy of official records, aerial photography has been one of the most valuable resources for investigating them. Many areas in the United States have been aerially photographed, going back to the 1930s. Thus it may be possible to assemble photographs to form a sequence of snapshots that record the use of the site over time. Such images, when combined with field data, official records, and health surveys, can permit development of an understanding of the sequence of events and assessment of the risks to the environment and to nearby populations. In some instances use of photography permits estimation of the amount of waste present and even specification of the class of materials by identification of the kinds of containers used. Aerial photography permits development of an understanding of the relationship of each site with respect to nearby drainage, population, economic activities, and other waste disposal sites.

Figure 20.14 illustrate the EPA's ability to track the history of waste disposal sites. These photographs show the Elmore Waste Disposal site and the Sunnyside Dump, in Spartanburg County, South Carolina. These sites are outlined on each photograph. The Elmore Waste Disposal site, about one-half acre in area, is situated near a residential neighborhood. The Elmore site was designated as a superfund site, so it has been monitored by the Environmental Protection Agency.

It is now known that a large number of drums containing liquid wastes were placed at this latter site between 1975 and 1977. In 1977, the property owner signed a consent order with the State of South Carolina for cleanup of the site. However, state health and environmental officials judged the actions taken to be inadequate for the situation, and ordered the owner to stop use of the site. Between 1981 and 1984, state and federal EPA officials investigated the site, and found arsenic, chromium, and other heavy metals, as well as a number of *volatile organic compounds* (VOCs), in site soils.

At various times during this interval, there were between 150 and 300 drums present on-site, as well as a 6,000-gallon tank, partially buried, containing contaminated waste. When the owner died in 1983, ownership passed to his heirs, one of whom continued to operate the site and accept waste drums. After many failed efforts to compel this owner to clean the site, in June 1986 state officials completed a state-funded removal of 5,500 tons of contaminated soil and 16,800 pounds of liquid wastes, which were taken to an appropriate hazardous waste facility.

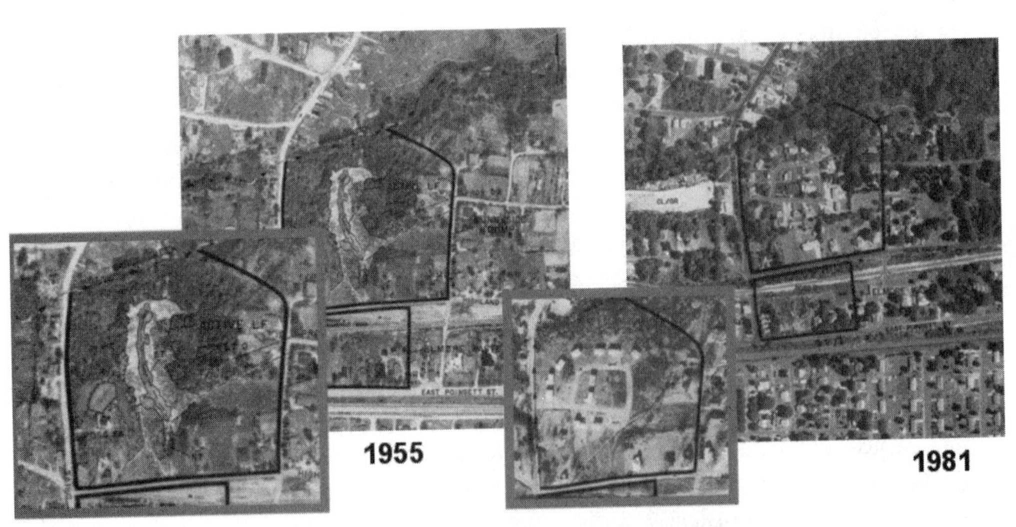


**FIGURE 20.14.** Aerial photographs depicting land-use changes at the Elmore/Sunnyside waste disposal area. These photographs, selected from a longer sequence, show the active landfill (1955, left), newly constructed residential housing at the same site (1965, center), and the mature residential neighborhood (1981, right). Historical aerial photography permits unraveling of the historical sequence that is no longer visible. From Environmental Protection Agency, Environmental Photographic Interpretation Center.

After this removal, groundwater monitoring wells and EPA's remedial investigation during 1991–1992 identified a contaminated groundwater plume extending some 700 feet north from the site. The estimated area underlain by this plume (north of the Elmore property) is 6–10 acres. Although no private water wells are located near this plume, the groundwater discharges to a creek. Additionally, surface soil in a one-quarter acre area at one end of the Elmore property was found to be contaminated by lead and arsenic at levels exceeding health-based residential standards for those metals.

Aerial photography provides documentation of these events, and an overview that places the disposal site, the residential areas, and the drainage system in their spatial context.

## 20.7. Other Land-Use Classification Systems

A land-use classification system is not simply a list of categories—it must organize its classes according to an underlying logic that defines the relationships between classes. The organizing logic must impose a consistent organization upon the classes. Classes must be mutually exclusive, and subclasses must nest within each other to form the hierarchal structure that we encountered in the Anderson system, for example. Further, the distinctions between the classes must be consistent from class to class. For example, if some classes are defined on the basis of economic function, then the entire classification structure must reflect that logic.

### *General-Purpose Land-Use Classification*

The Anderson (USGS) land-use classification system (Table 20.1) is designed specifically for use with aerial photography and related imagery. It is a good example of a *general-purpose land-use classification system,* intended to provide a comprehensive classification of land use. General-purpose classifications are among the most widely used, and are probably the classifications most likely to be encountered by photointerpreters. General-purpose land-use classification attempts to provide a classification that serves many purposes, although it is not tailored for any specific application.

There were several predecessors to the Anderson system, with respect to scope and purpose.

- L. Dudley Stamp, a British economist, organized a national survey of the land use of Britain during the interval 1931–1938 (Table 20.2). He gathered large-scale land-use information for Britain using volunteers, who recorded land use near their residences by annotating large-scale topographic maps, which were then mailed to a central office, where information was edited and posted to maps at 1:63,360. Although completion of this ambitious project was delayed by World War II, the information was still valuable (despite the delay) when it was used in planning Britain's economic recovery after the war.
- In the United States, during the same interval, photointerpreters at the *Tennessee Valley Authority* (TVA) were defining procedures to interpret land use from aerial photography. TVA photointerpreters devised a system (not now in use) that classified each parcel according to both its economic use and its physical properties.
- During the mid-1960s scientists at Cornell University designed *New York's Land Use and Natural Resources* (LUNR) Survey, a database to record land use and related information

**TABLE 20.2. Land Utilization Survey of Britain**

- Forest and woodland
  - High forest (specified as coniferous, deciduous, or mixed)
  - Coppice
  - Scrub
  - Forest (cut and not replanted)
- Meadowland and permanent grass
- Arable or tilled land
- Fallow land
- Heathlands, moorland, commons, and rough hill pasture
- Gardens
- Land agriculturally unproductive, buildings, yards, and mines
- Ponds

*Note.* From Stamp (1951).

for the entire state of New York. Much of the information was interpreted from aerial photography (1:24,000 panchromatic aerial photography). Photointerpreters coded information according to predominant land use within 1-km cells, using 100 classes defined specifically by the LUNR project (Table 20.3). This project was one for the first computerized land data inventories, forming an early precedent for the Geographic Information Systems now in use today.

### *Special-Purpose Land-Use Classification*

Another approach to land-use classification, special-purpose classifications, are also significant. Special-purpose land-use classifications are designed to address a specific classification issue, with no attempt to provide comprehensive scope.

A good example of a special-purpose land-use classification system, and one that is of practical significance to photointerpreters, is the wetlands classification developed by Cowardin (1979). This classification is devoted specifically to classification of wetlands (Table 20.4). Its logic therefore reflects the concerns of scientists who wish to portray the distribution of wetlands in detail, and to indicate specific characteristics of wetlands that are of significance for the hydrologist and the ecologist. For example, in the Anderson system, wetlands at Levels I and II are subdivided simply as the distinction between forested and nonforested wetlands, whereas in Cowardin's system, wetlands are distinguished first by hydrologic criteria ("Marine," "Riverine," etc., and "Tidal," "Subtidal," etc.), and then by geomorphic and ecologic criteria ("Reef," "Streambed," etc.). Special-purpose systems such as Cowardin's can be seen as specialized alternatives to the general-purpose strategy, or, in some instances, might serve as a Level III or Level IV within a hierarchical system such as the Anderson system.

It is important to select or design a system that is tailored to the needs of the client who will use the data. It is especially important to consider compatibility with previous classification systems, if the results are to be compared with data from earlier dates, and to consider compatibility with neighboring jurisdictions, or with higher or lower units (such as state or city systems, in the case of a county survey), as it may be important for the classes to match with these other data sets. A careful selection of classification that meets immediate needs may not be ideal if does not permit comparison with data from these other units.

**TABLE 20.3. New York's Land Use and Natural Resources (LUNR) Inventory Classification**

| | |
|---|---|
| **Area land use** | |
| **A** | Agriculture (10 subclasses) |
| **F** | Forestland (3 subclasses) |
| **W** | Water (3 subclasses) |
| **W** | Wetlands (3 subclasses) |
| **N** | Nonproductive land (2 subclasses) |
| **P** | Public |
| **R** | Residential (7 subclasses) |
| **S** | Shoreline |
| **C** | Commercial (4 subclasses) |
| **I** | Industrial (2 subclasses) |
| **OR** | Outdoor recreation |
| **E** | Extractive (7 subclasses) |
| **T** | Transportation (7 subclasses) |
| **Point data** | |
| | Water (3 classes) |
| | Livestock (3 subclasses) |
| | Water (5 subclasses) |
| | Specialty agriculture (4 subclasses) |
| | Residential public land |
| | Airport |
| | Outdoor recreation (10 subclasses) |
| | Communication (5 subclasses) |
| | Underground extractive |
| **Linear data civil divisions** | |
| | Civil division boundaries (7 subclasses) |
| | Streams, rivers, lakes, reservoirs, shorelines, highways (7 subclasses) |
| | Barge canal |
| | Public land |
| | Railways (5 subclasses) |

*Note.* From Hardy et al. (1971).

## 20.8. Land-Cover Mapping by Image Classification

Land cover can be mapped by applying image classification techniques discussed in Chapter 12 to digital remote sensing images. In principal, the process is straightforward; in practice, many of the most significant factors are concealed among apparently routine considerations:

1. *Selection of images.* Success of classification for land-cover analysis depends upon the astute selection of images with respect to season and date. Therefore, the earlier discussion (Section 6.9) of the design and interpretation of searches of image archives, although ostensibly mundane in nature, assumes vital significance for the success of a project. What season will provide the optimum contrasts between the classes to be mapped? Possibly two or more dates might be required to separate all the classes of significance.

2. *Preprocessing.* Accurate registration of images and correction for atmospheric and system errors (Chapter 11) are required preliminary steps for successful classification. Subsetting of the region to be examined requires careful thought, as discussed in Section 11.3.

**TABLE 20.4. Wetland Classification**

**Systems**
1. Marine
2. Estuarine
3. Riverine
4. Lacustrine
5. Palustrine

**Subsystems**
1. Subtidal
2. Intertidal
3. Tidal
4. Lower perennial
5. Upper perennial
6. Intermittent
7. Limnetic
8. Littoral

**Classes**
1. Rock bottom
2. Unconsolidated bottom
3. Aquatic bed
4. Reef
5. Rocky shore
6. Unconsolidated shore
7. Streambed
8. Emergent wetland
9. Scrub–shrub wetland
10. Forested wetland
11. Moss–lichen wetland

*Note.* From Cowardin et al. (1979).

3. *Selection of classification algorithm.* The discussion in Chapter 12 reviewed many of the classification algorithms available for land-cover analysis. Selection of the classification procedure should also be made on the basis of local experience. AMOEBA, for example, tends to be accurate in landscapes dominated by large homogeneous patches, such as the agricultural landscapes of the midwestern United States, and less satisfactory in landscape composed of many smaller heterogeneous parcels, such as those found in mountainous regions. Local experience and expertise are likely to be more reliable guides for selection of classification procedures than are universal declarations about their performance. Even when there is comparative information on classification effectiveness, it is difficult to anticipate the balance between effects of the choice of classifier, selection of image date, characteristics of the landscape, and other factors.

4. *Selection of training data.* Accurate selection of training data is one process that is universally significant for image classification, as we emphasized in Chapter 12. Training data for each class must be carefully examined to be sure that it is represented by an appropriate selection of spectral subclasses, to account for variations in spectral appearance due to shadowing, composition, and the like. Many individual laboratories and image analysis software packages have applied unsupervised classification in various forms to define homogeneous regions from which to select training fields for supervised classification (e.g., Chuvieco and Congalton, 1988). Another approach to the same issue is an algorithm that permits the analyst to select a

pixel or group of pixels that forms the focal point for a region that grows outward until a sharp discontinuity is encountered. This process identifies a region of homogenous pixels from which the analyst may select training fields for that class.

5. *Assignment of spectral classes to informational classes.* Because of the many subclasses that must be defined to accurately map an area by digital classification, a key process is the aggregation of spectral classes and their assignment to informational classes. For example, accurate classification of the informational class *deciduous forest* may require several spectral subclasses, such as *north-facing forest, south-facing forest, shadowed forest,* and the like. When the classification is complete, these subclasses should be assigned a common symbol to represent a single informational class.

6. *Display and symbolization.* The wide range of colors that can be presented on color displays, and the flexibility in their assignment, provides unprecedented opportunity for effective display of land-cover information. Although unconventional choices of colors can sometimes be effective, it is probably sensible to seek some consistency in symbolization of land-cover information to permit users to quickly grasp the meaning of a specific map or image without detailed examination of the legend. Therefore, the color symbols recommended by Anderson et al. (1976) (Table 20.4) may be useful guides. Another strategy for assignment of colors to classes is to mimic the colors used for USGS 7.5-minute quadrangles (Table 20.4).

Within such general strategies, it is usually effective to assign related colors to related classes to symbolize Level II and Level III categories. For example, subclasses of agricultural land can be represented in shades of brown (using Anderson et al.'s strategy as a starting point), water and wetlands in shades of blue, subclasses of forest in shades of green, and so forth.

## 20.9. Digital Compilation of Land-Use Change

The issue of compatibility retains its significance in the digital domain. Images must register and have comparable levels of spatial, spectral, and radiometric resolution. Unless the objective is to compare images from one season to another, the two images typically represent the same season but different years. Jensen (1981) lists a selection of preprocessing operations effective in preparing data for use in change detection comparisons. In any specific situation, the analyst must employ an intimate knowledge of the area to be studied, then apply a selection of methods tailored to the specifics of the study area. It seems unlikely that there is any single procedure that can be equally effective in all situations.

Jensen's (1996) list of digital change detection procedures includes the following:

1. *Image algebra* applies arithmetic operations to pixels in each image, then forms the change image from the resulting values. *Image differencing,* for example, simply subtracts one digital image from another digital image of the same area acquired at a different date. After registration, the two images are compared pixel by pixel to generate a third image composed of numerical differences between paired pixels from the two images. Values at or near zero identify pixels that have similar spectral values, and therefore presumably have experienced no change between the two dates. This procedure is typically applied to a single band of a multispectral data set; usually a constant value is added to eliminate negative values. The analyst must select (sometimes by trial and error) a threshold level to separate those pixels that have experienced change in land cover from those that have not changed land cover, but may exhibit

small spectral variations caused by other factors. Jensen (1981, 1982) reports that image differencing is among the most accurate change detection algorithms, but that it is not equally effective in detecting all forms of land-use change. Image ratios (Chapters 17 and 18), another form of image algebra, can be important in change detection because they may assist the analyst in standardizing for variations in illumination, thereby isolating spectral changes caused by land-cover changes.

2. *Postclassification comparison* requires two or more independent classifications of each individual scene, using comparable classification strategies. The two classifications are then compared, pixel by pixel, to generate an image that shows pixels placed in different classes on the two scenes. Successful application of this method requires accurate classifications of both scenes so that differences between the two scenes portray true differences in land use rather than differences in classification accuracy. In urban and suburban landscapes, the high percentages of mixed pixels (at Landsat MSS resolution) have tended to decrease classification accuracy, and therefore to generate inflated estimates of the numbers of pixels that have experienced change. Because postclassification comparison permits compilation of a matrix of *from–to changes,* it provides more useful results than some other methods (e.g., image algebra that simply identifies pixels that have changed, without specifying the classes involved).

3. *Multidate composites* are formed by assembling all image bands from two or more dates together into a single data set. The composite data set can then be examined using methods described in earlier chapters, including principal components analysis (Chapter 11) and image classification (Chapter 12). Thus, for two dates of Landsat MSS data, the data would consist of eight bands of spectral values. A classification of the composite identifies not only the usual land-use categories, but also classes composed of pixels that have experienced change from one date to the next. Due to the large size of the data set and its unwieldy character, this approach is usually inefficient and has been reported to be less accurate.

4. *Spectral change vector analysis* examines each pixel's position in mutispectral data space. If the corresponding pixel in an image from another data set occupies a nearby position, then the ground area represented by the pixel has not changed much during the interval between the two dates. If the two pixels occupy different positions, then the ground area has experienced changes. Usually the procedure is applied by preparing a pair of images representing the magnitude of the multispectral changes and the detection of the changes; these images can then be inspected by the analyst to set thresholds separating substantive changes from incidental changes caused by atmospheric effects, shadowing, or other ephemeral causes.

5. Use of a *binary change mask* requires classification of the image from the first date. Image algebra is then applied to the original image data for both dates to generate an image of changes, as described in (1) above. The image is used to prepare a binary mask, representing only changed and unchanged pixels. The binary change mask is then superimposed over the image data for the second date, and the classification of this image examines only those pixels that have been identified as changed. This method is often effective, provided that the binary change mask is accurate.

6. *On-screen digitization* requires specialized image analysis software that permits the analyst to view both images side by side on the same screen and to outline changed areas manually using on-screen digitization, based upon visual interpretation. This methods has been used primarily, if not exclusively, with large-scale digital images.

7. *Change detection by image display* is essentially a digital version of the photographic change detection methods mentioned above. Corresponding bands from different dates are used as separate overlays in a red–green–blue color display (Chapter 4). Pixels that have ex-

perienced change appear in distinctive colors (depending upon the assignment of images to colors), thereby flagging those pixels that have changes during the interval between the two dates. This technique does not reveal the from–to change information provided by some other techniques.

Jensen (1996) provides additional detail, examples, and discussion of additional methods.

## 20.10. Broad-Scale Land-Cover Studies

The availability of multispectral AVHRR data (and data from similar meteorological satellites) on a regular basis has provided the capability to directly compile broad-scale land-cover maps and data. In this context, broad scale refers to images that represent entire continents, or even entire hemispheres, based on data collected over a short period of time, perhaps about 10 days to 2 weeks. Previously, data for such large regions could be acquired only by generalizing more detailed information—a task that was difficult and inaccurate because of the incompleteness of coverage and the inconsistencies of the many detailed maps required to prepare small-scale maps. The finer resolution data from the Landsat and SPOT systems provide information of local and regional interest, but are not really suitable for compilation of data at continental scales because of the effects of cloud cover, differences in sun angle, and other factors that prevent convenient comparisons and mosaicking of many scenes into a single data set representing a large region.

AVHRR data, described in Section 17.9, provide coverage of entire continents over relatively short time periods. Accumulation of data over a period of a week to 10 days usually permits each pixel to be observed at least once under cloud-free conditions. Although the scan angle varies greatly, data are acquired at such frequent intervals that it is often possible to select coverage of the region of interest from the central section of each scene to reduce the effects of the extreme perspective at the edges of each scene.

Tucker et al. (1985) examined AVHRR data for Africa, using images acquired over a 19-month time interval. They examined changes in the vegetation index (the "normalized difference" described in Chapter 17) at intervals of 21 days; their results clearly illustrate the climatic and ecological differences between the major biomes, as observed using the vegetation index and seasonal variations in the vegetation index. They conducted a second analysis using eight of the 21-day segments selected from the 19-month interval mentioned above. They calculated principal components (Chapter 11) as a means of condensing the many variables into a concise yet potent data set that describes both seasonal and place-to-place variations in the vegetation index. Their land-cover map (based upon the first three principal components) is an extraordinary representation of key environmental conditions over an entire continent. It shows what are clearly major climate and ecological zones, as defined by conventional criteria, although they have been derived from data that are completely independent of the usual climatic data. The map has a uniformity, a level of detail, and (apparently) an accuracy not obtainable using conventional vegetation and climatic analysis. Although the authors have not fully evaluated their map by systematic comparison with field observations, it seems clear that products of this sort will provide an opportunity to examine broad ecological zones at continental scales, and furthermore, to examine seasonal and year-to-year changes in such zones.

More recently, AVHRR data have been used to compile other kinds of broad-scale land-cover maps, to be discussed in Chapter 21.

## 20.11. Sources of Compiled Land-Use Data

During the past decades, remote sensing has contributed greatly to the ability of organizations to prepare comprehensive land-cover surveys of large regions. Satellite imagery in particular has provided near-simultaneous acquisition, broad areal coverage, uniform detail, and other qualities not available from alternative sources. In the United States, several governmental initiatives have exploited this capability to prepare broad-scale land-cover surveys.

### *Land Use and Land Cover*

In the 1970s the USGS developed a national land-cover mapping program (described by Anderson et al., 1976, and Campbell, 1983) based on manual interpretation of high-altitude aerial photography at scales of 1:60,000 and smaller. The resulting USGS Land Use and Land Cover (LULC) data files describe the vegetation, water, natural surfaces, and cultural features for large regions of the United States. The USGS National Mapping Program provides these data sets and associated maps using map bases at 1:250,000. (In some cases, such as Hawaii, 1:100,000 scale maps are also used as the base.) For more information, see *http://edc.usgs.gov/glis/hyper/guide/1_250_lulc#lulc10*

### *Multiresolution Land Characteristics Consortium*

To economize in the acquisition of image data, in 1992 several agencies of the U.S. government formed a consortium designed to acquire satellite-based remote sensor data for their separate environmental monitoring programs. The MultiResolution Land Characteristics (MRLC) Consortium was sponsored originally by the USGS, the U.S. Environmental Protection Agency (EPA), the National Oceanographic and Atmospheric Administration (NOAA), and the U.S. Forest Service (USFS). MRLC was designed as an effort to pool financial resources to acquire satellite remote sensing imagery data in a cost-effective manner. Mapping efforts are based upon TM data; for more information, see *http://www.epa.gov/mrlc/nlcd.html*

During the 1990s, MRLC supported land-cover mapping programs such as the GAP analysis project and the National Land-Cover Dataset (NLCD) project, both mentioned below.

### *Gap Analysis*

In the United States, a national program to assess biodiversity and threats to biodiversity has been based on assessment of land cover derived from multispectral Landsat TM data. The objective of *gap analysis* is to identify "gaps" in biodiversity by surveying land-cover patterns to assess cover types. For example, regions with high species diversity are often associated with high ranges of elevations, contrasts in soil characteristics, and the presence of varied vegetation classes. Gap analysis permits identification of high priorities for conservation and for development and land-use policies (Scott and Jennings, 1998).

Each state conducts its own survey, using its own classification, mapping, and accuracy assessment strategy, developed in collaboration with neighboring states. The national project is coordinated by the National Gap Analysis Program within the USGS, Biological Resources Division (USGS-BRD). For more information, see *http://www.gap.uidaho.edu/*.

### *National Land-Cover Datasets (NLCD 1992 and NLCD 2001)*

One of the projects sponsored by the MRLC Consortium was production of land-cover data for the coterminous United States the National Land-Cover Dataset 1992 (NLCD 1992). Land cover was mapped using general land-cover classes, based on the Landsat 5 TM data archive and a host of ancillary sources. The classification is loosely based upon Anderson (1976) Level II classes. NLCD provides the first national land-cover data set produced since the early 1970s (see Table 20.5).

In 1999 MRCL began compilation of a new land cover data set known as (NLCD 2001, based upon Landset 7 TM imagery.

NLCD 2001 uses multitemporal imagery from three dates to prepare a national land-cover data set; it is intended to be flexible enough to meet the needs of a wide spectrum of users. The TM imagery forms the core of a database designed to provide intermediate products tailored for local applications (users can access materials to prepare their own classifications). NCLD 2001 is designed to apply simple, efficient, and consistent methods, and to maintain compatibility with earlier products.

NLCD 2001 is based upon normalized TM imagery for three seasons (spring, summer, and fall), and ancillary data, especially DEMs (used to prepare slope, aspect, and slope position). Imagery and ancillary DEM products were used to prepare intermediate products for each pixel, including (1) normalized tasseled cap (TC) transformations for each season, (2) DEM data and related slope and aspect, (3) per-pixel estimates of percent imperviousness and percent tree canopy (Figure 20.15), (4) 29 classes of land-cover data based upon the imagery, and (5) metadata and classification rules.

These intermediate products were then used to form a database that could be employed to derive land-use and land-cover products tailored to meet the specific needs of individual institutions. This approach differs from the classical classification approach strategies described in Chapter 12 in which a single classification was intended to serve the needs of diverse users. The database approach permits each institution to tailor its classification to meet its specific needs.

This approach is applied to the U.S. states and Puerto Rico by defining 66 mapping zones (Figure 20.16) according to physiography and overall landscape characteristics. Studies have shown that land-cover mapping accuracies can be improved by stratifying by physiographic provinces. Therefore, the 66 regions are designed to provide degree of consistency in implementing the classification system.

Further details are presented by Homer et al. (2004), and at *www.mrlc.gov*

## 20.12. Summary

Study of land use and land cover reveals the overall pattern of mankind's occupation of the Earth's surface and the geographic organization of its activities. At broad scales, the land-cover map provides a delineation of the broad patterns of climate and vegetation that form the environmental context for our activities. At local and regional scales, knowledge of land use and

**TABLE 20.5. NLCD 2001 Land-Cover Class Descriptions**

11. *Open water*—All areas of open water, generally with less than 25% cover of vegetation or soil.
12. *Perennial ice/snow*—All areas characterized by a perennial cover of ice and/or snow, generally greater than 25% of total cover.
21. *Developed, open space*—Includes areas with a mixture of some constructed materials, but mostly vegetation in the form of lawn grasses. Impervious surfaces account for less than 20% of total cover. These areas most commonly include large-lot single-family housing units, parks, golf cources, and vegetation planted in developed settings for recreation, erosion control, or aesthetic purposes.
22. *Developed, low intensity*—Includes areas with a mixture of constructed materials and vegetation. Impervious surfaces account for 20–49% of total cover. These areas most commonly include single-family housing units.
23. *Developed, medium intensity*—Includes areas with a mixture of constructed materials and vegetation. Impervious surfaces account for 50–79% of the total cover. These areas most commonly include single-family housing units.
24. *Developed, high intensity*—Includes highly developed areas where people reside or work in high numbers. Examples include apartment complexes, row houses, and commercial/industrial. Impervious surfaces account for 80–100% of the total cover.
31. *Barren land (rock/sand/clay)*—Barren areas of bedrock, desert pavement, scarps, talus, slides, volcanic material, glacial debris, strip mines, gravel pits, and other accumulations of earthen material. Generally, vegetation accounts for less than 15% of total cover.
32. *Unconsolidated shore*—Unconsolidated materials such as silt, sand, or gravel that is subject to inundation and redistribution due to the action of water. Characterized by substrates lacking vegetation except for pioneering plants that become established during brief periods when growing conditions are favorable. Erosion and deposition by waves and currents produce a number of landforms representing this class.
41. *Deciduous forest*—Areas dominated by trees generally greater than 5 m tall, and greater than 20% of total vegetation cover. More than 75% of the tree species shed foliage simultaneously in response to seasonal change.
42. *Evergreen forest*—Areas dominated by trees generally greater than 5 m tall, and greater than 20% of total vegetation cover. More than 75% of the tree species maintain their leaves all year. Canopy is never without green foliage.
43. *Mixed forest*—Areas dominated by trees generally greater than 5 m tall, and greater than 20% of total vegetation cover. Neither deciduous nor evergreen species are greater than 75% of total tree cover.
51. *Dwarf scrub*—Alaska-only areas dominated by shrubs less than 20 cm tall with shrub canopy typically greater than 20% of total vegetation. This type is often coassociated with grasses, sedges, herbs, and nonvascular vegetation.
52. *Shrub scrub*—Areas dominated by shrubs less than 5 m tall with shrub canopy typically greater than 20% of total vegetation. This class includes true shrubs, young trees in an early successional stage, or trees stunted from environmental conditions.
71. *Grassland/herbaceous*—Areas dominated by grammanoid or herbaceous vegetation, generally greater than 80% of total vegetation. These areas are not subject to intensive management such as tilling, but can be utilized for grazing.
72. *Sedge/herbaceous*—Alaska-only areas dominated by sedges and forbs, generally greater than 80% of total vegetation. This type can occur with significant other grasses or other grass-like plants, and includes sedge tundra and sedge tussock tundra.

land cover forms a basic dimension of resources available to any political unit; both the citizens and the leaders of any community must understand the land resources available to them and the constraints that limit uses of land and environmental resources.

Although the formal study of land use and land cover dates from the early 1800s, systematic mapping at large scale was not attempted until the 1920s, and aerial photography and remote sensing were not routinely applied until the 1960s. Thus effective land-use mapping is a rela-

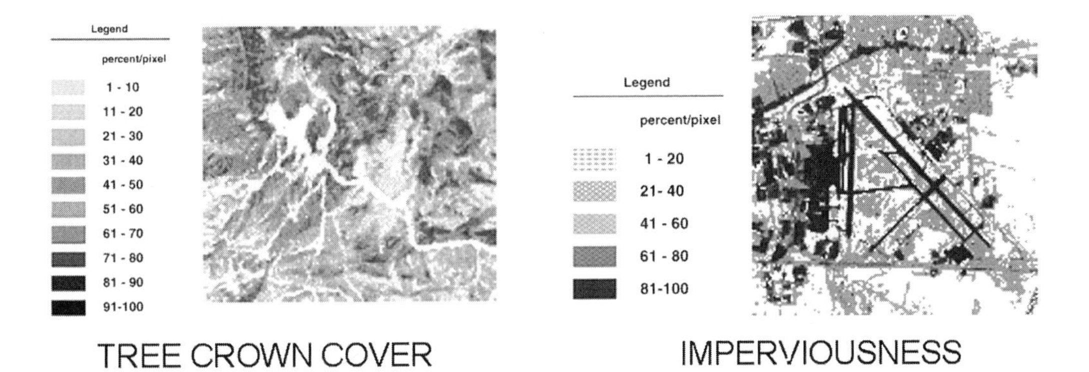


**FIGURE 20.15.** Sample images illustrating tree crown and imperviousness data used for NLCD 2001. Left: percent tree crown cover, for each pixel (for a mountainous area in Utah). Right: imperviousness, derived from TM data (for a built-up region in Virginia). From USGS, Homer et al. (2004).

tively recent capability. We have yet to fully assemble and evaluate all of the data that is available and to develop the techniques for acquiring and interpreting imagery.

Without the aerial images acquired by remote sensing, there can be no really practical method of observing the pattern of land cover or of monitoring changes. Systems such as MSS, TM, SPOT, and AVHRR have provided a capability for observing land cover at broad scales and at intervals that previously were not practical. Images from such systems have not only provided vital information, but have also presented data at new scales that have changed the intellectual perspective with which we consider the environment by recording broad-scale patterns and relationships that otherwise could not be accurately perceived or analyzed.

**FIGURE 20.16.** The 66 regions defined to provide consistent terrain and land-use conditions for the NLCD 2001. From USGS, Homer et al (2004).

## Review Questions

1. Using aerial photographs and other information provided by your instructor, design Level III categories (compatible with the USGS classification) for a nearby region.
2. Review Chapters 11 and 12 to refresh your memory concerning resolutions of satellite sensors and the effects of mixed pixels. A typical city block is said to be about 300 ft. × 800 ft. in size. Make rough assessments of the effectiveness of the MSS, TM, and SPOT systems for depicting land use and land cover within urban regions. List factors other than the sizes of objects that would be important in making such assessments.
3. Outline some of the difficulties that would be encountered in compiling land-use and land-cover maps if aerial photography and remotely sensed data were not available.
4. Compare the relative advantages and disadvantages of alternative kinds of imagery, including photography, thermal imagery, and radar imagery (at comparable scales and resolutions), for compiling land-use and land-cover maps. List the advantages and some problems that might be encountered in using all three kinds of images in combination.
5. The following issues all require, directly or indirectly, accurate land-use and land-cover information. For each, identify, in a few sentences or in short paragraphs, the role of accurate land-use maps and data.
   a. Solid waste disposal
   b. Selecting a location for a new electrical power plant
   c. Establishing boundaries of a state park or wildlife preserve
   d. Zoning decisions in a suburban region near a large city
   e. Abandoned toxic waste dumps
6. Some scientists have advocated development of a classification system with categories based on the appearance of features on specific kinds of remotely sensed images. In contrast, the approach used by Anderson et al. (Table 20.1) is based on the idea that remotely sensed data should be categorized using classes that remain the same for all forms of remotely sensed images and that match those used by planners. Compare the advantages and disadvantages of the two strategies, considering both the ease of application to the imagery and the ease of use by those who must actually apply the information.
7. About 18% of urban land is devoted to streets. About 20% of the land area of large cities is said to be undeveloped. Assess the ability of remotely sensed images to contribute to assessing the amount and pattern of these two kinds of land use; consider TM and SPOT data, as well as aerial photography at 1:10,000.

## References

### *General*

Anderson, J. R. 1977. Land Use and Land Cover Changes: A Framework for Monitoring. *Journal of Research, U.S. Geological Survey,* Vol. 5, No. 3, pp. 143–153.

Anderson, J. R., E. E. Hardy, J. T. Roach, and R. E. Witmer. 1976. *A Land Use and Land-Cover Classifi-*

*cation for Use with Remote Sensor Data* (USGS Professional Paper 964). Washington, DC: U.S. Government Printing Office, 18 pp.

Baker, R. D., J. E. deSteiger, D. E. Grant, and M. J. Newton. 1979. Land-Use/Land-Cover Mapping from Aerial Photographs. *Photogrammetric Engineering and Remote Sensing,* Vol. 45, pp. 661–668.

Campbell, J. B. 1983. *Mapping the Land: Aerial Imagery for Land Use Information* (Resource Publications in Geography). Washington, DC: Association of American Geographers, 96 pp.

Christian, C. S. 1959. The Eco-Complex and Its Importance for Agricultural Assessment. Chapter 36 in *Biogeography and Ecology in Australia. Monographiae Biologicae,* Vol. 8, pp. 587–605.

Chuvieco, E. and R. Congalton, 1988. Using Cluster Analysis to Improve the Selection of Training Statistics in Classifying Remotely Sensed Data. *Photogrammetric Engineering and Remote Sensing,* Vol. 54, pp. 1275–1281.

Cicone, R. C., and M. D. Metzler. 1984. Comparison of Landsat MSS, Nimbus–7 CZCS, and NOAA–7 AVHRR Features for Land-Cover Analysis. *Remote Sensing of Environment,* Vol. 15 pp. 257–265.

Drake, B. 1977. Necessity to Adapt Land Use and Land Cover Classification Systems to Readily Accept Radar. In *Proceedings, 11th International Symposium on Remote Sensing of Environment.* Ann Arbor: Environmental Research Institute of Michigan, pp. 993–1000.

Hoeschele, W. 2000. Geographic Information Systems and Social Ground Truth in Attappadi, Kerala State, India. *Annals, Association of American Geographers,* Vol. 90, pp. 293–321.

Jensen, J. R. 1978. Digital Land Cover Mapping Using Layered Classification Logic and Physical Composition Attributes. *American Cartographer,* Vol. 5, pp. 121–132.

Liverman, D., E. F. Moran, R. R. Rindfuss, and P. C. Stren. 1998. *People and Pixels: Linking Remote Sensing and Social Science.* Washington, DC: National Academy Press, 244 pp.

Loelkes, G. L. 1977. *Specifications for Land Cover and Associated Maps* (U.S. Geological Survey Open File Report 77–555). Reston, VA: U.S. Geological Survey.

Meyer, W. B., and B. L. Turner (eds.). 1994. *Changes in Land Use and Land Cover: A Global Perspective.* New York: Cambridge University Press, 537 pp.

Nunnally, N. R., and R. E. Witmer. 1970. Remote Sensing for Land Use Studies. *Photogrammetric Engineering and Remote Sensing,* Vol. 36, pp. 449–453.

Robinove, C. J. 1981. The Logic of Multispectral Classification and Mapping of Land. *Remote Sensing of Environment,* Vol. 11, pp. 231–244.

Scott, J. M., and M. D. Jennings. 1998. Large Area Mapping of Biodiversity. *Annals of the Missouri Botanical Garden,* Vol. 85, pp. 34–47.

Turner, B. L., W. C. Clark, R. W. Kates, J. F. Richards, J. T. Mathews, and W. B. Meyer (eds.). 1990. *The Earth as Transformed by Human Action: Global and Regional Changes in the Biosphere over the Past 300 Years.* New York: Cambridge University Press.

Vogelmann, J. E., T. Sohl, and S. M. Howard. 1998. Regional Characterization of Land Cover Using Multiple Sources of Data. *Photogrammetric Engineering and Remote Sensing,* Vol. 64, pp. 45–57.

## *Change Detection*

Congalton, R. G., M. Balough, C. Bell, K. Green, J. Milliken, and R. Ottman. 1998. Mapping and Monitoring Agricultural Crops and Other Land Cover in the Lower Colorado River Basin. *Photogrammetric Engineering and Remote Sensing,* Vol. 64, pp. 1107–1113.

Estes, J. E., and L. Senger. 1972. Remote Sensing in the Detection of Regional Change. In *Proceedings, 8th International Symposium on Remote Sensing of Environment.* Ann Arbor: Environmental Research Institute of Michigan, pp. 317–324.

Jensen, J. R. 1979. Spectral and Textural Features to Classify Elusive Land Cover at the Urban Fringe. *Professional Geographer,* Vol. 31, pp. 400–409.

Jensen, J. R. 1981. Urban Change Detection Mapping Using Landsat Digital Data. *American Cartographer,* Vol. 8, pp. 127–147.

Jensen, J. R. 1996. *Introductory Digital Image Processing: A Remote Sensing Perspective.* Upper Saddle River, NJ: Prentice-Hall, 316 pp.

Jensen, J. R., and D. L. Toll. 1982. Detecting Residential Land-Use Development at the Urban Fringe. *Photogrammetric Engineering and Remote Sensing,* Vol. 48, pp. 629–643.

Lunetta, R. S., and S. D. Elvidge. 1998. *Remote Sensing Change Detection: Environmental Monitoring Methods and Applications,* Ann Arbor MI: Ann Arbor Press, 318 pp.

Merteng, B., and E. F. Lambin. 2000. Land-Cover-Change Trajectories in Southern Cameroon. *Annals, Association of American Geographers,* Vol. 90, pp. 467–494.

## *AVHRR for Land-Cover Studies*

Defries, R. S., and J. R. G. Townshend. 1994. NDVI-Derived Land Cover Classifications at a Global Scale. *International Journal of Remote Sensing,* Vol. 15, pp. 3567–3586.

Erb, T. L., et al. 1981. Analysis of Landfills with Historic Airphotos. *Photogrammetric Engineering and Remote Sensing,* Vol. 47, pp. 1363–1369.

Livingston, G. P., C. A. Clark, and M. H. Trenchard. 1984. *An Investigation of NOAA/AVHRR Data in the Classification and Mapping of Landcover* (L. B. Johnson Space Center Technical Report 18918). Houston, TX: Johnson Space Center, 40 pp.

Lyon, J. G. 1987. Use of Maps, Aerial Photographs, and Other Remote Sensor Data for Practical Evaluations of Hazardous Waste Sites. *Photogrammetric Engineering and Remote Sensing,* Vol. 53, pp. 515–519.

Stamp, L. D. 1951. Land Use Surveys with Special Reference to Britain. In T. G. Taylor (ed.) *Geography in the Twentieth Century.* London: Methuen, pp. 372–393.

Townshend, J. R. G. 1994. Global Data Sets for Land Applications from the Advanced Very High Resolution Radiometer: An Introduction. *International Journal of Remote Sensing,* Vol. 15, pp. 3319–3332.

Tucker, C. J., J. R. G. Townshend, and T. E. Goff. 1985. African Land Cover Classification Using Satellite Data. *Science,* Vol. 227, pp. 369–375.

Vogelman, J. E., S. M. Howard, L. Yang, C. R. Larson, B.K. Wylie, and N. Van Driel. 2001. Compilation of the 1990s National Land Cover Data Set for the Coterminous United States from the Landsat Thematic Data and Ancillary Data Sources. *Photogrammetric Engineering and Remote Sensing,* Vol. 70, pp. 650–662.

Homer, C., C. Huang, L. Yang, B. Wylie, and M. Coan. 2004. Development of a 2001 National Land-Cover Database for the United States. *Photogrammetric Engineering and Remote Sensing,* Vol. 70, pp. 829–840.

Stehman, S. V., and R. L. Czaplewski. 1998. Design and Analysis for Thematic Map Accuracy Assessment: Fundamental Principals. *Remote Sensing of Environment,* Vol. 64, pp. 331–344.

Loveland, T. R., T. L. Shol, S. V. Stehman, A. L. Gallant, K. L. Sayler, and D. E. Napton, 2002. A Strategy for Estimating the Rates of Recent United States Land Cover Changes. *Photogrammetric Engineering and Remote Sensing,* Vol. 68, pp. 1091–1100.

Vogelmann, J. E., D. Helder, R. Morfit, M. J. Choate, J. W. Merchant, and H. Bulley. 2001. Effects of Landsat 5 Thematic Mapper and Landsat 7 Enhanced Thematic Mapper Plus Radiometric Calibrations and Corrections on Landscape Characterization. *Remote Sensing of Environment,* Vol. 78, pp. 55–70.

Cowardin, L. M., V. Carter, F. C. Golet, and E. T. LaRoe, 1979. *Classification of Wetlands and Deepwater Habitat of the United States.* Washington, DC: Fish and Wildlife Service, U.S. Department of the Interior.

CHAPTER TWENTY-ONE

# Global Remote Sensing

## 21.1. Introduction

The history of remote sensing can be visualized as one of increasing scope of vision, with respect both to space and time (Figure 21.1). Before the availability of aerial photography, human observation of the landscape was limited in scope to the area visible from a high hill or building. As aerial photography became available, it became possible to view larger areas over longer intervals. Later, as the routine availability of infrared emulsions permitted high-altitude aerial photography, larger areas could be represented within a single frame. Still later, the larger areas viewed by satellite images permitted analysts to consider even broader perspectives. Finally, by merging images from several dates to remove effects of cloud cover, it became possible to compile images representing continents, hemispheres, and the entire globe (Plate 28). Such images have made it possible to directly examine broad-scale environmental patterns and their changes over years, decades, and even longer intervals. This capability has had profound impact on perspectives that scientists and the general public apply to their consideration of broad-scale environmental issues. Issues that once could be only matters of speculation now can be considered in the light of data and information that are based on systematic observation.

## 21.2. Biogeochemical Cycles

Such capabilities open new opportunities for understanding basic processes that underlie the Earth's climatic and biologic patterns. The elements and nutrients that make life on Earth possible reside differentially in biologic, geologic, and atmospheric components of the Earth. Over years, decades, and centuries, atmospheric and geologic processes cycle these nutrients and chemicals through the varied components of the biosphere. The Earth's atmosphere and oceans move at rates measured in minutes, hours, or days. At the other extreme, minerals of the Earth's crust can be considered mobile only when viewed over intervals of many millions of years.

Understanding the Earth's environment requires examination of movements and interchanges between its varied components. For example, minerals weather from geologic material, enter the soil, are absorbed by plant roots, then are incorporated into plant tissues as plants

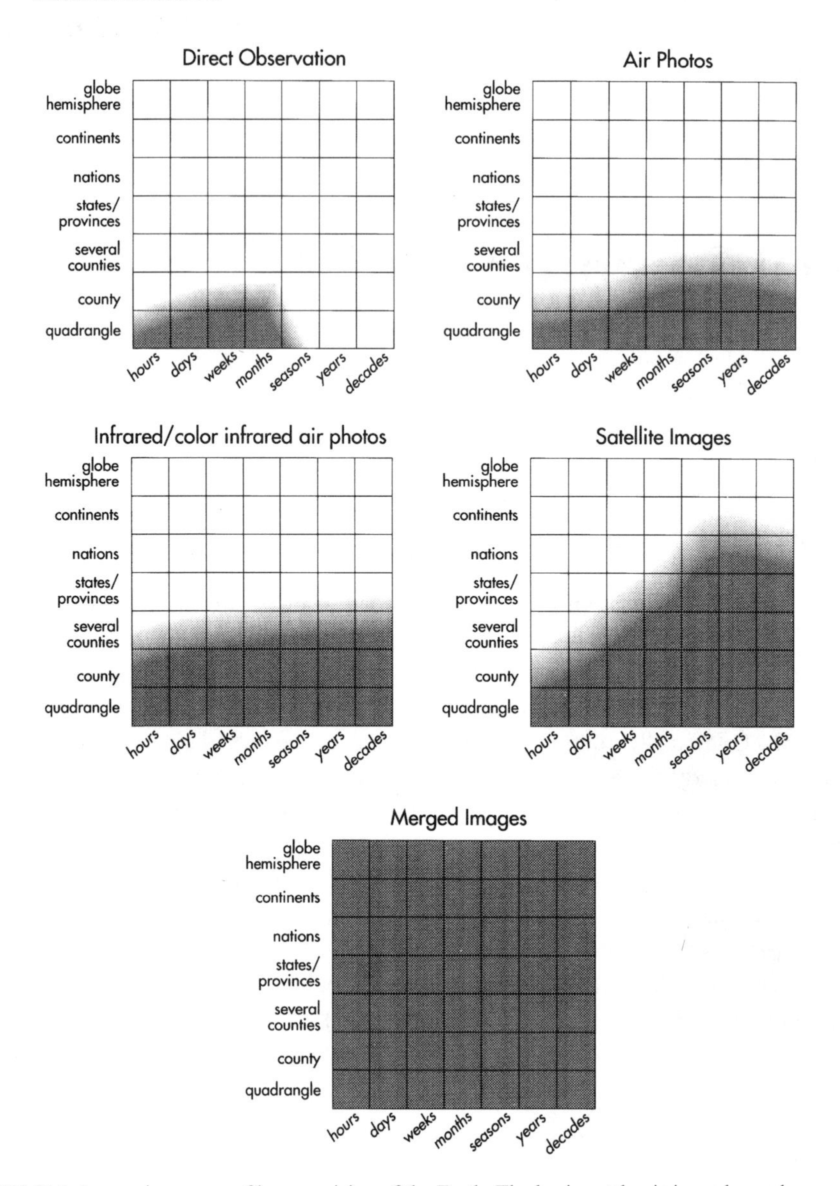


**FIGURE 21.1.** Increasing scope of human vision of the Earth. The horizontal axis in each graph represents time (in an arbitrary scale of increasing intervals from hours and days at one extreme to years and decades at the other). The vertical axis represents space (in arbitrary units, from quadrangle-sized areas at one extreme to continents and hemispheres at the other). From MacEachren et al. (1992). Copyright 1992 by Rutgers University Press. Reproduced by permission.

grow. As these plants die and decompose, nutrients are released into groundwater or surface runoff, then again assume geologic form as sedimentological and lithological processes concentrate and mineralize materials. Such are the biogeochemical processes that govern the movement of carbon, sulfur, nitrogen, oxygen, and other elements.

The following sections outline major components of the most important biogeochemical cycles; these are not intended to provide complete descriptions, but to illustrate the kinds of issues that require global perspectives.

### *Hydrologic Cycle*

The hydrologic cycle is the best known and most central of the many processes that cycle materials through the Earth's biosphere (Figure 21.2). Because of the central role of moisture in biologic, climatic, and geologic processes, the hydrologic cycle (discussed earlier in Chapter 19) has obvious significance for all other processes that cycle materials through the Earth's environment. Hydrologic processes at all scales are intimately connected with land cover (Chapters 17 and 20), so the broad-scale land-cover data provided by AVHRR data form a central component of efforts to examine the hydrologic cycle at continental and global scales. Further, aerial photography and the satellite data outlined in previous chapters provide, at least in some regions, a record of land-cover changes extending several decades into the past, which offer the opportunity to develop broad-scale analyses of land-cover changes and their impact upon key components of the hydrologic cycle.

### *Carbon Cycle*

Carbon (C), in its varied forms, is one of the essential components of living organisms and is of key interest in any consideration of global cycling within the biosphere (Figure 21.3). Carbon is available for living organisms as carbon dioxide ($CO_2$) and as organic carbon (plant and animal tissues), sometimes known as fixed carbon, because it is much less mobile than $CO_2$. Photosynthesis reduces $CO_2$ to organic carbon, also much less mobile than $CO_2$; within organisms respiration can convert organic carbon to its gaseous form.

Chapter 2 described the significance of atmospheric $CO_2$ in the Earth's energy balance; although the atmosphere holds only a very small proportion of the Earth's carbon, atmospheric $CO_2$ is very effective in absorbing long-wave radiation emitted by the Earth. Larger amounts of carbon are stored in oceans.

The largest reservoir of carbon is formed by carbonate rocks (such as limestone). Very slow-

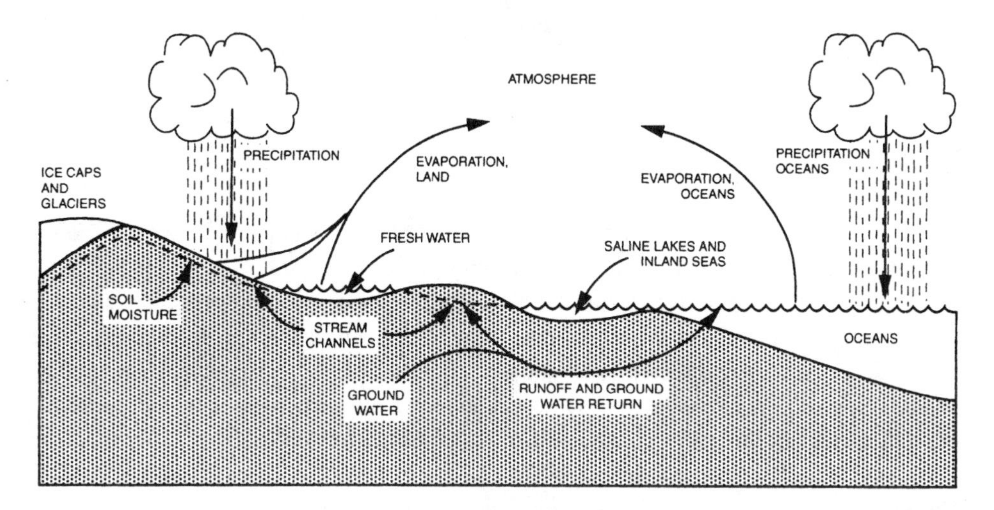


**FIGURE 21.2.** Hydrologic cycle. Adopted from Strahler and Strahler (1973). Copyright 1973 by John Wiley & Sons, Inc. Reproduced by permission.

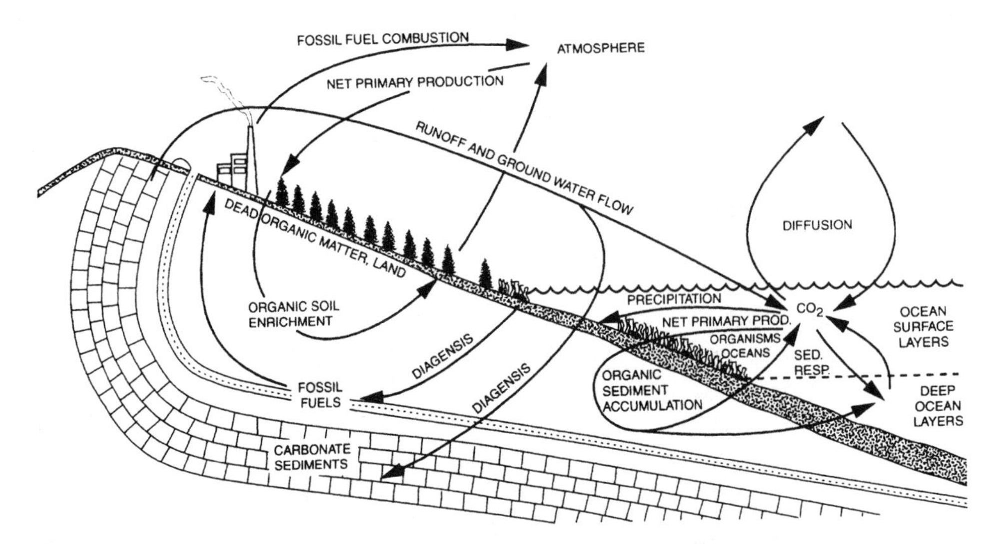


**FIGURE 21.3.** Carbon cycle. Adopted from Strahler and Strahler (1973). Copyright 1973 by John Wiley & Sons, Inc. Reproduced by permission.

ly, carbon from the atmosphere, oceans, and living organisms enters this reservoir as sedimentary rocks are formed. And equally slowly, carbon is released through weathering of these rocks. An exception, of course, is carbon residing as fossil fuels, which is rapidly released as it is burned for heat, to generate electrical power, and to propel automobiles and aircraft. Estimation of the amounts of carbon released by these activities, and of their effects upon the Earth's climate, is a high priority for those who hope to gauge effects of human activities upon Earth's climates.

Key components in the carbon cycle are illustrated in Figure 21.3. Atmospheric $CO_2$ is incorporated into the structure of living organisms by photosynthesis. The decay of organic matter releases $CO_2$ into the atmosphere. $CO_2$ is also released by the burning of fossil fuels.

Remote sensing instruments assist scientists in understanding the carbon cycle by providing estimates of the areas covered by plants, identifying the kinds of plants, and estimating the time period for which they are photosynthetically active.

### *Nitrogen Cycle*

Nitrogen (N) occurs in large amounts on Earth and is of great significance in the food chain because it occurs as a principal element in the atmosphere and in tissues of plants and animals (Figure 21.4). It occurs in several forms, both organic and inorganic, and participates in important biological processes. Very large amounts of the Earth's nitrogen are considered inactive; it is held in sedimentary and crustal rocks and does not participate in the short-term cycling of nitrogen. Within the active pool, the largest amount of nitrogen resides in the atmosphere, where it forms one of the major constituents of the atmosphere. Atmospheric nitrogen, however, is not available for biological use.

Nitrogen becomes available for biological use through *fixation,* the chemical reaction in which nitrogen combines with hydrogen (H) to create ammonia ($NH_3$). Although nitrogen fixa-

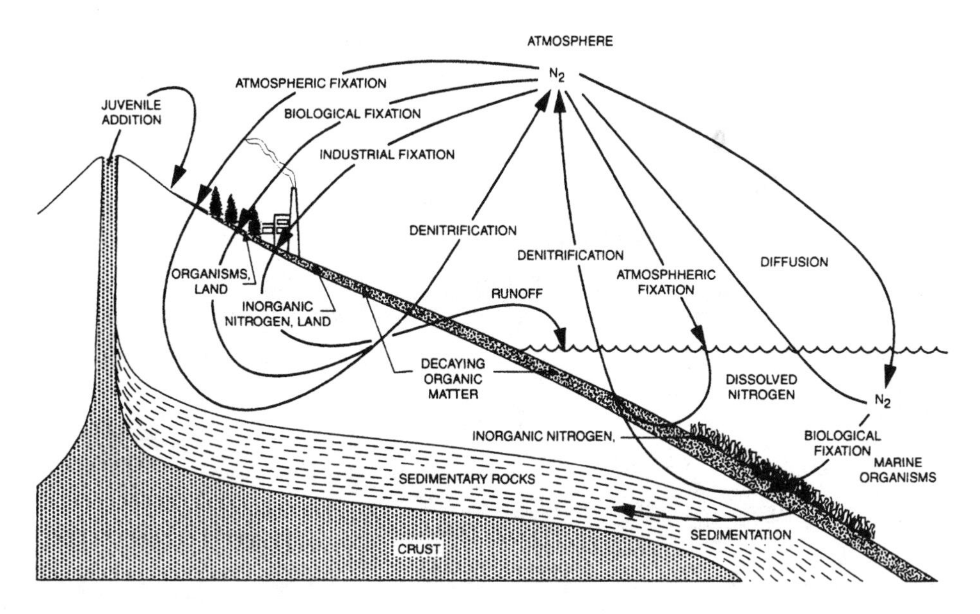


**FIGURE 21.4.** Nitrogen cycle. Adopted from Strahler and Strahler (1973). Copyright 1973 by John Wiley & Sons, Inc. Reproduced by permission.

tion can occur within the atmosphere (through lightning strikes), in the burning of fossil fuels, and industrially (in the manufacture of artificial fertilizers), the most significant source by far is biologically based. *Biological fixation* occurs by the action of living plants that change atmospheric nitrogen ($N_2$) to a fixed form ($NH_3$) that can be used by plants and animals. Microorganisms, both free-living and those symbiotic with specific plants (legumes), are the only animals that can accomplish this process. Amino acids from plant or animal tissue can be broken down by the process of *ammonification,* which releases energy, $H_2O$, and $CO_2$. Ammonification is accomplished by decomposing microorganisms.

By this process, nitrogen enters the food chain and the tissues of plants and animals. *Denitrification,* the process by which nitrites ($NO_{2-}$) or nitrates ($NO_{-3}$) are converted to molecular nitrogen or nitrous oxide, is accomplished by soil bacteria under anaerobic conditions. This process occurs in a two-step process during which ammonia is oxidized to nitrate in two steps. The first, completed by the bacteria *Nitrosommonas,* yields nitrate ions and water. The second step, accomplished by the bacteria *Nitrobacter,* consumes nitrite ions to yield nitrate. Industrial fixation accomplishes the same process using energy to transform atmospheric nitrogen to ammonia, which can be used in other industrial processes to manufacture fertilizers and explosives.

One of the largest reservoirs of nitrogen is decaying organic matter. Remote sensing assists in understanding the nitrogen cycle by providing accurate estimates of areas where organic matter accumulates for decay, such as swamps and marshes.

### *Sulfur Cycle*

In comparison with other biogeochemcial cycles discussed here, the sulfur (S) cycle (Figure 21.5) is somewhat less significant with respect to total amounts in active circulation and with

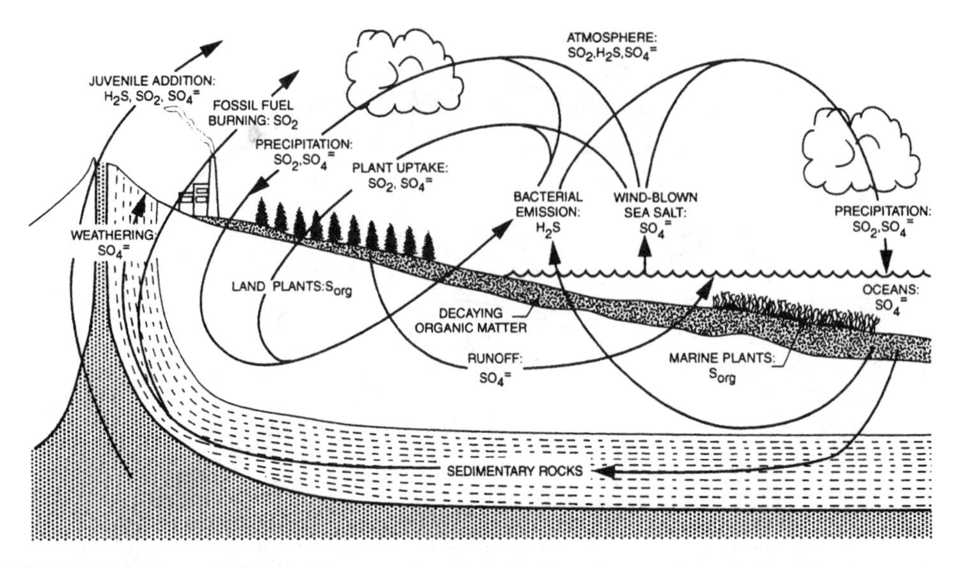


**FIGURE 21.5.** Sulfur cycle. Adopted from Strahler and Strahler (1973). Copyright 1973 by John Wiley & Sons, Inc. Reproduced by permission.

respect to its role in the food chain. But it has great significance if we examine ways in which humans influence the sulfur cycle. Sulfur resembles nitrogen in the sense that it occurs in several forms within the environment and participates in several organic and inorganic reactions that alter its forms and behavior. By far the largest amount of sulfur is in geologic form, in sedimentary deposits that are removed from short-term participation in the sulfur cycle. Weathering releases small amounts of geologic sulfur to the atmosphere. Of greater interest, however, is the release of sulfur into the atmosphere by the removal and burning of fossil fuels.

Oceans form the main reservoir for sulfur in active circulation. Only small amounts reside in the atmosphere. On land, some is incorporated in the tissues of plants and animals, and some is retained as decaying organic matter (e.g., in swamps and marshes). Sulfur enters the atmosphere from four sources: emissions from bacteria as organic matter decays, combustion of fossil fuels, airborne dust from evaporating sea spray, and volcanic sources.

Accelerated release of geologic sulfur by burning of fossil fuels has contributed to reactions within the atmosphere that have increased the acidity of rainfall near industrialized regions of northern latitudes, which in turn is believed to have decreased the pH of lakes, rivers, and even soils in some regions. Such changes in water and soil acidity, if sustained, would have profound impact on wildlife, vegetation patterns, and soil fertility. Because such impacts have already been observed, governments have instituted efforts to reduce sulfur emissions and to encourage industries to switch to low-sulfur fuels.

## *Significance of Remote Sensing*

Remote sensing is the primary means by which we can observe the dynamic character of the Earth's biosphere. Although basic outlines of biogeochemical cycles are well known, specific quantities, rates of movement, and regional variations are not well understood, nor is the true impact of human activities upon their functioning. Such detail is necessary for scientists to assess the significance of human influences on biogeochemical processes and their significance in relation to natural variations. Therefore, scientists seek not only to understand the general structure of such cycles but also to develop a more precise knowledge of quantities and rates of movement within these cycles. Preparation of such estimates is very difficult because the figures commonly given are necessarily very rough estimates and conventional data are not adequate to provide the required information. For example, the existing network of climate stations consists of a series of point observations at locations not representative of specific environments. This network can provide only the very crudest estimates of temperature patterns and is clearly inadequate for answering the critical questions concerning the possible warming of the Earth's atmosphere during recent history.

Mather and Sdasyuk (1991, pp. 96–113) review existing sources of environmental data suitable for assessing broad-scale environmental trends and identify the many defects in the present sources of data. They, like many others, conclude that broad-scale remote sensing provides the only practical means to acquire data of sufficient scope for addressing these issues.

## 21.3. Advanced Very-High-Resolution Radiometer

Previously introduced in Chapters 6 and 17, the advanced very-high-resolution radiometer (AVHRR) is a multispectral radiometer carried by a series of meteorological satellites operated by NOAA in near-polar, sun-synchronous orbits. They can acquire imagery over a swath width of approximately 2,400 km. AVHRR acquires global coverage on a daily basis; the frequent global coverage is possible because of the wide angular field of view, but areas recorded near the edges of images suffer from severe geometric and angular effects. As a result, AVHRR data selected from the regions near the nadir provide the most accurate information. Although designed initially for much narrower purposes, AVHRR is the first sensor to provide the basis for collection of worldwide data sets that permit assessment of environmental issues.

AVHRR pixels are about 1.1 km on each side at nadir, with fine radiometric resolution at 10 bits. Details of spectral coverage vary with specific mission, but in general AVHRR sensors have recorded radiation in the visible and infrared regions of the spectrum. AVHRR spectral regions have been selected to attain meteorological objectives, including spectral data to distinguish clouds, snow, ice, and open water. Nonetheless, AVHRR has been employed on an ad hoc basis for land resource studies. The AVHRR sensor includes one channel in the visible spectrum, one in the near infrared, and three in the thermal infrared. It is possible to use the data to calculate vegetation indices of the type described in Chapter 17.

From these AVHRR data, the *global vegetation index* (GVI) data set has been prepared for the period 1982–1985. GVI data are presented using an arbitrary grid system, with cells that vary in size with latitude, from about 13 km at the equator to 26 km at the poles. Data collected within a 7-day period are examined for the brightest ("greenest") NDVI value in each cell for the reporting interval. These data form the AVHRR composites described in Chapter 17 and referred to indirectly in Chapter 6. GVI images remove effects of clouds, except for instances in which clouds are present in all seven images. For more information, see *http://edcdaac.usgs.gov/1KM/1kmhomepage.html*

Such data provided one of the first tools that permitted study of the Earth's major biomes. Previously, the areas and extents of major regions of forest, savanna, tundra, deserts, and the like could only estimated roughly from climate and vegetation patterns as observed at ground level. Although satellite data show that these earlier approximations were quite accurate, the use of satellite data permits closer studies of detail of the patterns and examination of seasonal variations and long-term trends with accuracy and detail that were not previously possible. Specifically, AVHRR permits closer examination of effects of droughts and other broad-scale climatic changes, as well as changes related to human activities.

### *Pathfinder Data*

In 1990 NASA and NOAA began a program to compile *Pathfinder data sets,* designed to provide data for analysis of long-term environmental processes at continental and global scales. Pathfinder data are compiled from systems placed in service prior to the Earth Observing System (EOS), to provide experience with global-scale data systems and to extend the period of analysis to as early a date as possible. These objectives require use of systems with accurate calibration and consistent qualities to permit analysis over time and from one sensor system to another. The first Pathfinder data sets have been prepared from AVHRR data, other meteorological satellite data, and selected Landsat data. Pathfinder data are sampled over time (in part to select cloud-free data and to reduce other atmospheric effects) and over space (to create consistent levels of detail and data sets of manageable sizes), and then processed to permit convenient use by a wide variety of users. Separate Pathfinder initiatives are underway to compile data describing ocean surface temperature, atmospheric conditions, and land surfaces.

Pathfinder AVHRR land data are global in scope, but sampled over time and space. The *daily data set* consists of AVHRR terrestrial data, mapped to an equal area projection on an 8-km grid. Pixels near the edges of the AVHRR swath are excluded. The daily data can be used directly for analysis or as source data for composite data sets. The *composite data set* is derived from the daily data set; the compositing process filters out effects of clouds and selects data for each date from the scene with the highest NDVI. The *climate data set* is compiled to a 1° × 1° resolution, for 8- to 11-day composite periods. The result forms a set of cloud-free, averaged, NDVI data. Thirty-six such images are produced each year, to provide data for examination by scientists studying broad-scale climatic and biosphere changes. *Browse images* are prepared from the daily data set and the composite data set to provide a reduced data set to assist in data selection and evaluation. Browse images are not designed for analysis. For more information, see AVHRR Pathfinder Program: *http://xtreme.gsfc.nasa.gov/pathfinder/*

## 21.4. Earth Observing System

Although existing satellite data such as AVHRR and those described in previous chapters might seem at first consideration to provide a means to study global environmental problems, in fact these images provide piecemeal rather than systematic long-term coverage. Likewise, much of the existing archive of imagery records arbitrary regions of the spectrum (rather than regions targeted on those required for specific tasks) and are not coordinated with related data-collection efforts. Attacking these broad-scale, long-term data-collection requirements requires instruments tailored for specific tasks at hand. This is the context in which NASA, in the late

1980s and early 1990s, developed a program to acquire the environmental data required to address questions posed by concerns over global environmental change.

Earth Science Enterprise (formerly Mission to Planet Earth [MTPE]) is one of NASA's four strategic enterprises designed to establish a long-term, coordinated research effort to study the total Earth system and the effects of natural and human-induced changes on the global environment. Earth Science Enterprise is a long-term effort to observe the Earth's environments and how they change over time, as an essential component to developing an understanding of the nature, rates, and consequences of broad-scale environmental change. The early activities of Earth Science Enterprise (in the 1990s, when it was still named MTPE) focused on the design of numerous satellites that require specialized orbits and instruments.

The second, long-term, component of the Earth Science Enterprise is a system of satellites that form an integrated system of Earth-orbiting satellites designed to provide a continuous stream of data for a period of 15 years or more. The Earth Observing System (EOS) is to consist of several satellites carrying numerous sensors designed to monitor environmental variables on a worldwide basis. EOS efforts have been coordinated nationally through various agencies of the U.S. government and internationally through agreements with other nations. The design of EOS has changed in response to scientific, operational, and budgetary considerations, and will no doubt continue to change. Thus the following sections outline some of the instruments likely to be of interest to readers of this text, but do not attempt an exhaustive description of the complete system. For more information, see

NASA Earth Observing System: *http://eospso.gsfc.nasa.gov/*
NASA Science Mission Directorate: *http://science.hq.nasa.gov/directorate/index.html*
Destination Earth: *http://www.earth.nasa.gov*

## 21.5. Earth Observing System Instruments

EOS is designed to collect data to provide a comprehensive overview of the dynamic components of the Earth's atmosphere and land and water surfaces (Figure 21.6). To accomplish this objective, the EOS plan has included 30 or more instruments designed to monitor physical and biological components of the biosphere. However, as budget priorities have changed, so has the program design evolved. No doubt, it will continue to evolve. Therefore, individual sensors have been redesigned or dropped from the plan. Here we briefly describe only a few of the many instruments proposed for EOS that are most closely related to those already discussed in this volume. The reader may find that further program changes will alter even the brief list of instruments presented here.

### *Moderate-Resolution Imaging Spectroradiometer*

The moderate-resolution imaging spectroradiometer (MODIS) (Figure 21.7) was launched on the Terra satellite (Section 21.6) in December 1999. MODIS is designed to acquire imagery that will permit compilation of global data sets at frequent intervals. It uses a cross-track scanning mirror and CCD detectors in 36 spectral bands, covering spectral regions from the visible to the

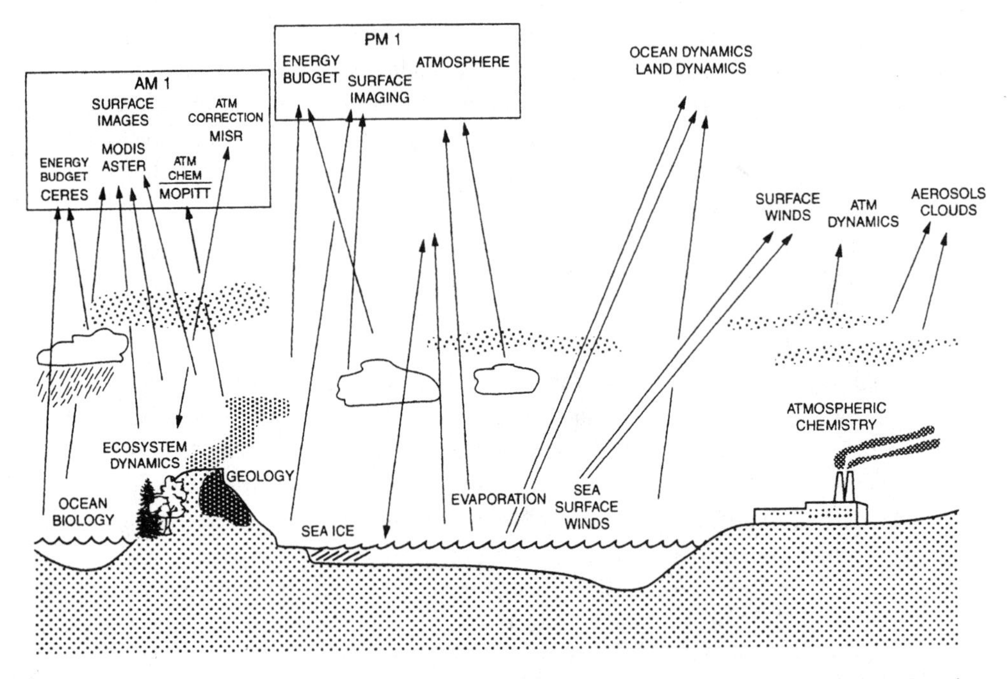


**FIGURE 21.6.** EOS overview. This diagram represents the planned configuration for EOS missions. Based on a NASA diagram given in Assrar and Dokken (1993).

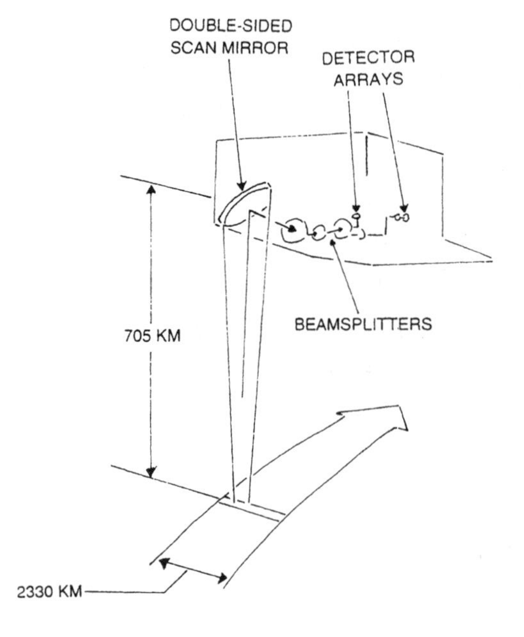


**FIGURE 21.7.** Schematic sketch of MODIS. Based on Running et al. (1994).

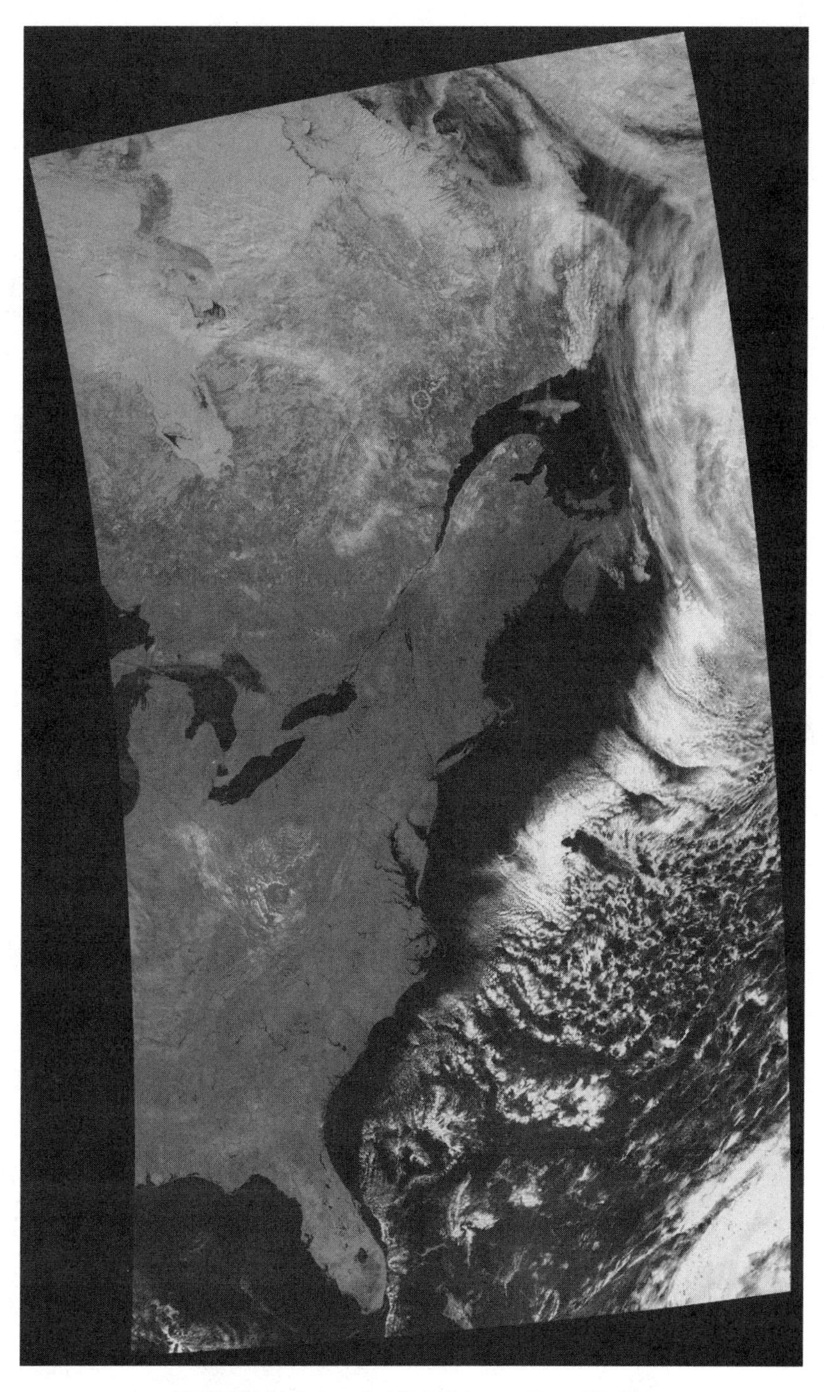

**FIGURE 21.8.** Sample MODIS image. From NASA.

thermal infrared. Spatial resolutions vary from 500 m to 1 km; the satellite provides coverage every 2 days. MODIS instruments are designed to extend the kinds of data collected by AVHRR, TM, and some of the meteorological satellites (Figure 21.8).

MODIS is designed for monitoring the Earth's land areas; in essence, it is an instrument similar to AVHRR, but tailored to monitoring land resources (rather than the meteorological mission of AVHRR). *Moderate resolution* refers to its 500-m pixels, which may seem coarse relative to SPOT or Landsat, but are fine relative to the 1-km pixels of AVHRR, and achieve global coverage to meet EOS objectives. MODIS data permit analysts to monitor biological and physical processes at the Earth's land surfaces and water bodies and to collect data describing cloud cover and atmospheric qualities. MODIS collects data describing range and forest fires, snow cover, chlorophyll concentrations, surface temperature, and other variables. For more information, see *http://modis.gsfc.nasa.gov/*

### *Earth Observing System Ocean Color Experiment*

The EOS Ocean Color Experiment (EOS-Color) is an optical-mechanical imaging radiometer provided for EOS service by Japan's Ministry of International Trade and Industry. It is designed to operate from a sun-synchronous orbit, with a noon equatorial crossing. It collects data in eight spectral channels over a swath width of 2,800 km. Spatial resolution at nadir is 1.1 km, and as much as 4.5 km off-nadir. The instrument is designed to acquire data to study marine phytoplankton, the oceanic role in the carbon cycle, mixing of oceanic and coastal waters, and the nature of heat exchange within the upper layers of the ocean.

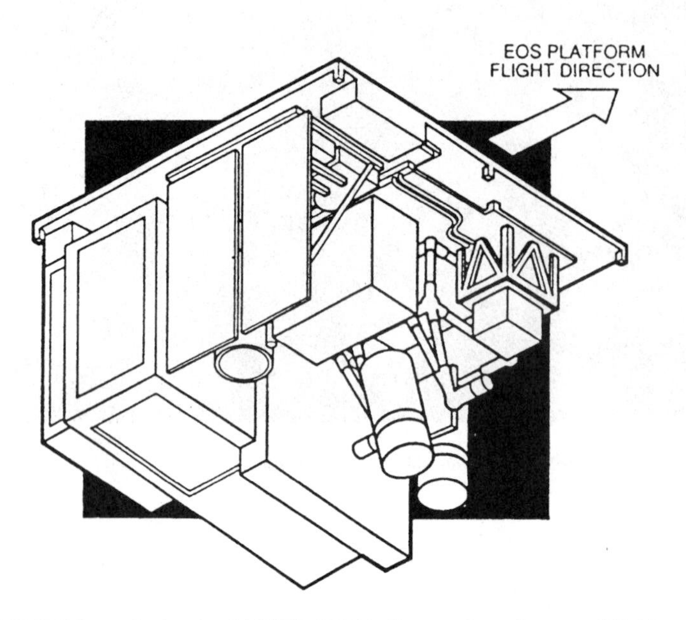


**FIGURE 21.9.** Schematic sketch of ASTER. NASA diagram from Assrar and Dokken (1993).

### *Advanced Spaceborne Thermal Emission and Reflection Radiometer*

The advanced spaceborne thermal emission and reflection radiometer (ASTER) is an imaging radiometer designed to collect data in 14 spectral channels (Figure 21.9). Three channels in the visible and near infrared provide 15-m resolution in the region 0.5 m–0.9 m; six channels in the short-wave infrared provide 30-m resolution in the interval 1.6 m–2.5 m; and five channels in the thermal infrared provide 90-m resolution in the interval 8 μm–12 μm. Imagery can be acquired in a 60-km swath centered on a pointable field of view ±8.5° in the mid-infrared and thermal infrared and ±2.4° in the visible and near infrared (Figure 21.10). The instrument acquires a stereo capability through use of an aft-pointing telescope that acquires imagery in channel 3. The ASTER can provide 16-day repeat coverage in all 14 bands, and 5-day coverage in the visible and near infrared channels. The instrument is designed to provide data to describe land use, cloud coverage and kind and heights of clouds, land and ocean temperature, topographic data, and information on glacial activity and snow cover.

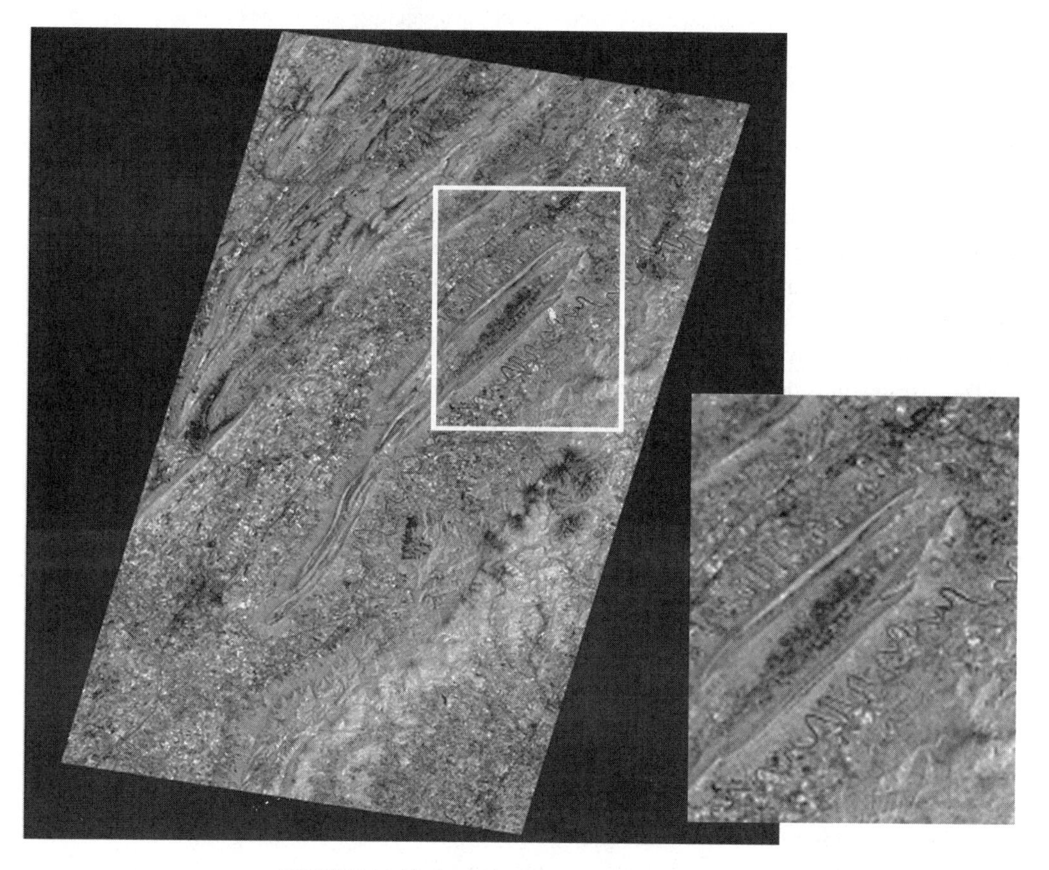

**FIGURE 21.10.** Sample ASTER image. From NASA.

**FIGURE 21.11.** Schematic sketch of MIMR. NASA diagram from Assrar and Dokken (1993).

### *Multi-Angle Imaging Spectroradiometer*

The multi-angle imaging spectroradiometer (MISR) employs nine CCDs to observe the Earth's surface from nine angles (Figure 21.11). One looks to the nadir, and four each look to the fore and aft along the ground track of the satellite in each of four spectral regions, centered on 0.443 m, 0.555 m, 0.670 m, and 0.865 m. These channels have resolutions of 240 m, 480 m, 960 m, and 1.92 km, respectively.

The instrument is designed to observe the atmosphere at different angles and to observe features through differing path lengths and angles of observation. Data from MISR will permit the study of atmospheric properties, including effects of pollutants, and the derivation of information to correct data from other sensors for atmospheric effects. The instrument will also contribute to the development of models for the optical behavior of surfaces (e.g., forest canopies) by developing accurate BRDFs. The configuration of the sensors provides the ability to acquire stereo imagery of both ground surfaces and clouds.

### *Multifrequency Imaging Microwave Radiometer*

The multifrequency imaging microwave radiometer (MIMR) is a passive microwave radiometer designed to acquire data at six frequencies (6.8, 10.65, 18.7, 23.8, 36.5, and 90 GHz) at dual polarization. It observes a 1,400-km swath at resolutions varying from 60 km to 5 km. It is designed to observe surface temperature, sea ice, sea surface, atmospheric water vapor, and soil moisture patterns.

### *Earth Orbiting System Synthetic Aperture Radar*

The EOS program includes a synthetic aperture radar (EOS-SAR) (Chapter 7) designed to monitor sea ice, glaciers, snow, soil moisture, patterns of deforestation, and moisture in forest canopies. The EOS-SAR is planned to acquire imagery in the L-, C-, and X-bands, using multi-polarization. The SAR will use look angles from 15° to 40°, to either side of the ground track. The instrument is designed to operate in any of three modes, which provide different trade-offs between coverage and spatial resolution. At one extreme (local-high-resolution mode), the SAR uses a swath width of 30 km to 50 km to collect data with resolution of 20 m to 30 m. In the regional mapping mode, the SAR acquires data for a 100-km to 200-km swath width at 50-m to 100-m resolution. In global mapping mode, images can show a swath width of as much as 500 km, at resolutions of 250 m to 500 m.

## 21.6. Earth Observing System Bus

EOS sensors are to be placed on a series of satellites designed to provide standard support functions, including communications, thermal control, power supply, and related functions. The standardized components are known as the *EOS bus* (Figure 21.12).

The first of the EOS busses, EOS-AM, known as Terra, was launched in December 1999 and began service in February 2000, carrying CERES, MISR, MODIS, ASTER, and MOPITT sensors. Terra was placed in an orbit designed for morning observation (local sun time), which optimizes observation of terrestrial surfaces by minimizing land areas obscured by cloud cover. A second satellite, EOS-PM (known as Aqua), launched in May 2002, is in an orbit designed to observe the Earth at a later time of day, designed for optimum observation of ocean and atmos-

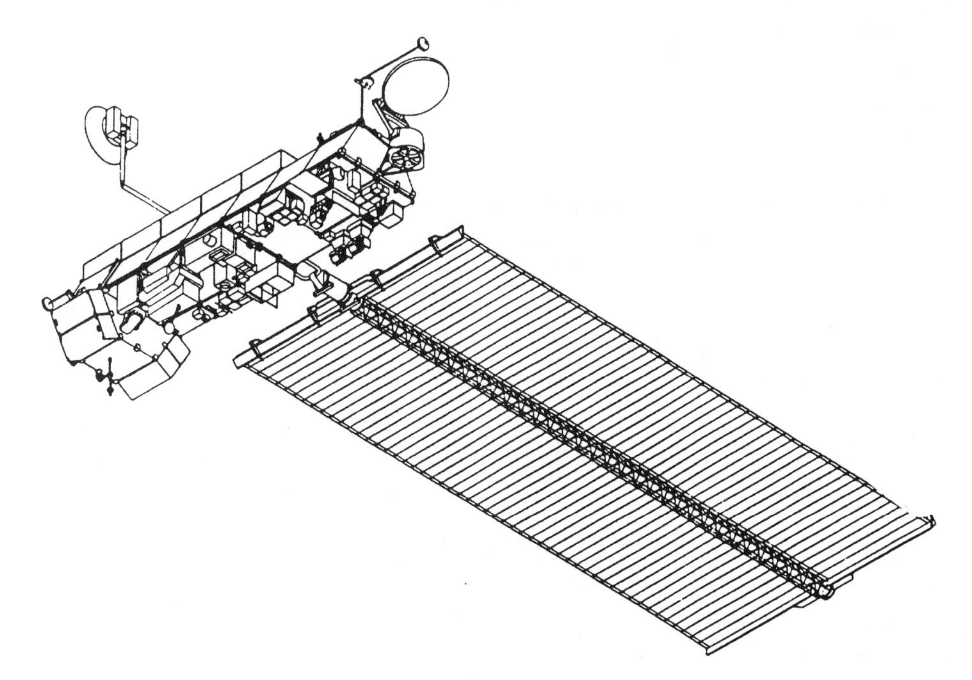

**FIGURE 21.12.** EOS bus. From NASA.

pheric features, as well as selected terrestrial phenomena. Aqua carries some of the same sensors (e.g., MODIS) carried on Terra to permit features to be imaged by the same instrument at different times of day, providing an opportunity to examine diurnal changes such as those suggested in Chapter 9. For more information, see *http://terra.nasa.gov/* and *http://aqua.nasa.gov/about/*

## 21.7. Earth Observing System Data and Information System

The EOS Data and Information System (EOSDIS) was established to handle the unprecedented volume of data to be acquired by the system of satellite sensors. Operation of EOS will require a system for acquiring, processing, and distributing data to scientists for interpretation and analysis. Key components within EOSDIS are the eight distributed active archive centers (DAACs), established to process and distribute data. Each DAAC specializes in handling specific kinds of data (e.g., "land processes," "sea ice and polar processes," "hydrology"), with the objective of providing sustained access to the data for a broad community of users (Figure 21.13 and Table 21.1). Components of EOSDIS are connected electronically, so users can quickly access archives, indices, and data from their own institutions, as outlined in Chapter 4. For more information, see

*http://nasadaacs.eos.nasa.gov/*
*http://edcdaac.usgs.gov/main.asp*

Related global-scale environmental data sets were described in Section 16.6, including

- Hydrology: *http://edcdaac.usgs.gov/gtopo30/hydro/index.html*
- Elevation: *http://edcdaac.usgs.gov/gtopo30/gtopo30.html*

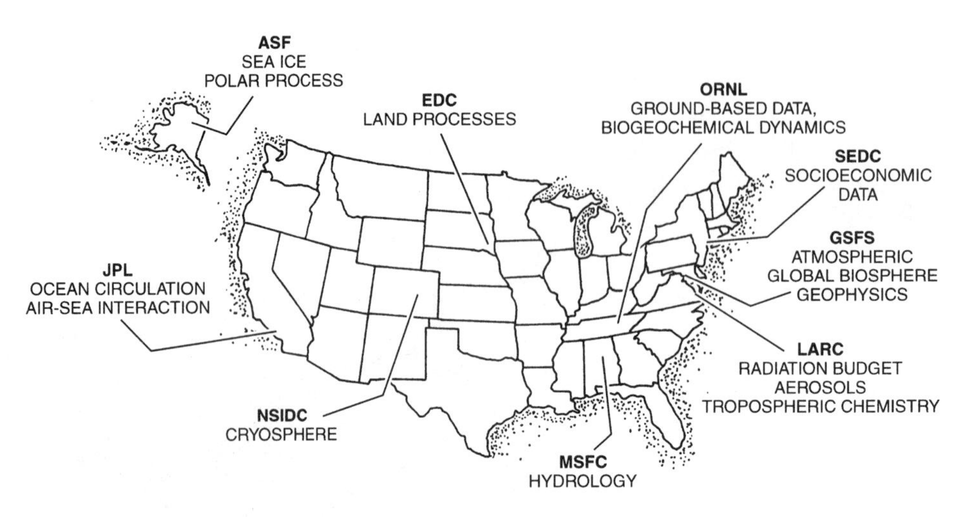


**FIGURE 21.13.** EOS distributed active archive centers (see Table 21.1). Based on NASA diagram in Assrar and Dokken (1993).

TABLE 21.1. EOS Distributed Active Archive Centers

| | |
|---|---|
| **ASF:** Alaska SAR Facility: *http://www.asf.alaska.edu/* | Acquire, process, archive, and distribute satellite synthetic aperture radar (SAR)<br>University of Alaska, Fairbanks |
| **EDC:** EROS Data Center: *http://lpdaac.usgs.gov/main.asp* | Process, archive, and distribute land-related data collected by EOS sensors<br>Sioux Falls, SD |
| **GSFC:** Goddard Space Flight Center: *http://daac.gsfc.nasa.gov/* | Atmospheric chemistry; atmospheric dynamics; field experiments; hydrology; global land, ocean, and atmospheric data; land biosphere; ocean color<br>Greenbelt, MD |
| **JPL:** Jet Propulsion Laboratory: *http://podaac-www.jpl.nasa.gov/* | Physical oceanography<br>Pasadena, CA |
| **LARC:** Langley Research Center: *http://eosweb.larc.nasa.gov/* | Radiation budget, clouds, aerosols, tropospheric chemistry<br>Langley, VA |
| **NSIDC:** National Snow and Ice Data Center: *http://nsidc.org/NASA/GUIDE/index.html* | Snow and ice data<br>Boulder, CO |
| **ORL:** Oak Ridge National Laboratories: *http://www-daac.ornl.gov/* | Biogeochemical and ecological data<br>Oak Ridge, TN |
| **SEDAC:** Socioeconomic Data and Applications Center: *http://sedac.ciesin.columbia.edu/* | Socioeconomic data<br>Columbia University, New York, NY |

## 21.8. Long-Term Environmental Research Sites

In this context, information provided by field data assumes a new meaning. The usual field observations, collected to match with specific images on a one-time basis, at isolated times and places, have less significance when the objectives extend over very large areas and very long time intervals. To support the search for long-term, broad-scale patterns of change, field data must exhibit continuity over time and be positioned to represent ecosystems at continental scales. To address these needs, a network of long-term ecological research sites has been established within many of the Earth's major ecological zones. At each site long-term studies are underway to document ecological behavior and changes over time, and to understand relationships between natural changes and those that might be caused or accelerated by human activities.

Research conducted at some sites records local patterns of nutrient cycling as a component in understanding changes in the global cycles. Some long-term ecological research sites (LTER) (Table 21.2) were already established as reserves or research sites, so some have 30–40 years of records recording climate and ecological conditions. Figure 21.14 and Table 21.2 show a network of LTER stations supported in part by the U.S. National Science Foundation. Most of these sites represent principal ecological zones of the western hemisphere, and a broad sample of global environments. Similar efforts are underway throughout the world to establish a network of international sites representing a full range of biomes. For more information, see *http://www.lternet.edu/*

TABLE 21.2. Long-Term Ecological Research Sites

1. **Andrews LTER:** Temperate coniferous forest: J. J. Andrews Experimental Forest, Blue River, OR
2. **Arctic Tundra LTER:** Arctic lakes and tundra, Brooks Range, AK
3. **Baltimore Ecosystem Study:** Eastern deciduous forest/suburban agriculture fringe, Baltimore, MD
4. **Bonanza Creek LTER:** Tiaga, Fairbanks, AK
5. **Central Arizona-Phoenix:** Sonoran Desert scrub, Phoenix, AZ
6. **California Current Ecosystem:** Upwelling current biome, San Diego, CA
7. **Cedar Creek LTER:** Eastern deciduous forest and tall grass prairie, Cedar Creek Natural History Area, Minneapolis, MN
8. **Coweeta LTER:** Eastern deciduous forest, Coweeta Hydrological Laboratory, Otto, NC
9. **Florida Coastal Everglades:** Marsh ecosystems, Miami, FL
10. **Georgia Coastal Ecosystems LTER:** Coastal barrier island/marsh complex, Sapelo Island, GA
11. **Harvard Forest:** Eastern deciduous forest, Petersham, MA
12. **Hubbard Brook LTER:** Eastern deciduous forest, West Thornton, NH
13. **Jordana Basin:** Hot desert, Jordana Experimental Range, Las Cruces, NM
14. **Kellogg Biological Station:** Row-crop agriculture, Hickory Corners, MI
15. **Konza LTER:** Tall-grass prairie, Manhattan, KS
16. **Luquillo LTER:** Tropical rain forest, San Juan, PR
17. **McMurdo Dry Valleys:** Antarctic valleys, Antarctica
18. **Moorea Coral Reef:** Tahiti, French Polynesia
19. **Niwot Ridge LTER:** Alpine tundra, Boulder, CO
20. **North Temperate Lakes:** Madison, WI
21. **Palmer Station:** Polar marine, Antarctica
22. **Plum Island Ecosystem:** Coastal estuary, Plum Island, MA
23. **Santa Barbara Coastal LTER:** Semi-arid coastal zone, Santa Barbara, CA
24. **Sevilleta LTER:** Arid mountains/hot desert/cold desert, Albuquerque, NM
25. **Short-grass Steppe:** Nunn, CO
26. **Virginia Coastal Reserve:** Coastal barrier island, Oyster, VA
27. **LTER Network Office:** Albuquerque, NM: *http://lternet.edu/sites/*

## 21.9. Earth Explorer

Earth Explorer, developed by the USGS EROS Data Center (Sioux Falls, SD, 57198-0001; 800–252–4547), is an interactive computer system designed to permit scientists to identify information pertaining to the Earth's land surfaces. Earth Explorer consists of metadata (Chapter 6) describing dates, coverages, cloud cover, quality ratings, and other characteristics of broad-scale data sets of the Earth's land areas. Included are summaries of land cover, soil, geologic, cultural, topographic, and remotely sensed data. Users can connect to Earth Explorer computers remotely to review data for their specific regions of interest. Earth Explorer permits examination of the suitability of data for specific projects and provides a vehicle for placing orders to acquire data.

Earth Explorer descriptions are organized to permit users to search for coverage of specific regions by date to ascertain availability of data from specific years or by class of data (aerial photography, digital line graphs, maps, elevation data, or satellite imagery). For more information, see *http://edcsns17.cr.usgs.gov/Earth Explorer/*. Another data distribution effort, NASA's Visible Earth, is intended to provide imagery to the press and public, chiefly though a website at: *http://visibleearth.nasa.gov/about.html*

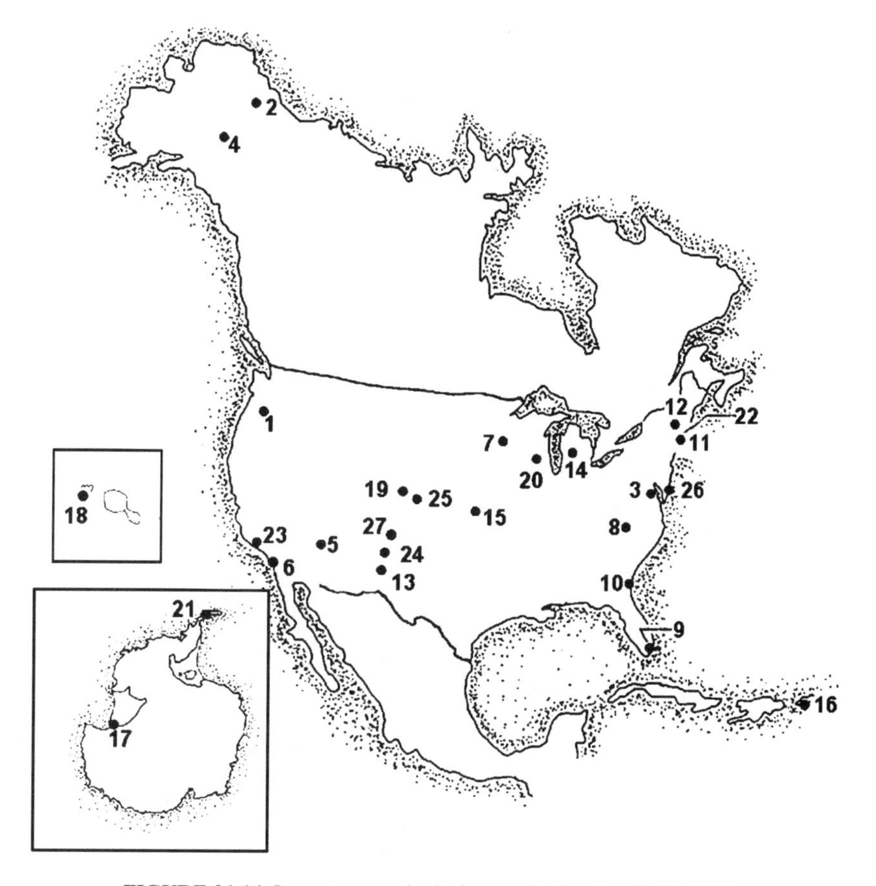


**FIGURE 21.14.** Long-term ecological research sites (see Table 21.2).

## 21.10. Summary

The instruments and programs outlined in this chapter are designed to gather data that provide the geographic coverage and consistency required to compile global data sets and to examine fundamental questions concerning place-to-place variations in climate, vegetation patterns, and biophysical processes and their changes over time. Although existing data have permitted humans to define the general framework for understanding these phenomena, it is only by collection of consistent data over large areas, over long intervals of time, that we will be able to develop an intimate understanding that will permit us to assess effects of human actions in the context of inherent geographic and historical variations.

## Review Questions

1. Enumerate some of the factors important in preventing, until recently, the practice of global remote sensing.

2. How might the practice of global remote sensing differ from the practice of other, more conventional, remote sensing?
3. Discuss problems in collecting and evaluating field data to support global remote sensing.
4. How might global remote sensing assist in global change studies?
5. In what ways might hyperspectral data assist in global change studies?
6. Global remote sensing developed initially from applications of data collected by meteorologic satellites for purposes other than their intended function. Can you identify problems (either practical or conceptual) that arise from these origins?
7. What are some of the special practical and theoretical problems inherent in the practice of global remote sensing?
8. Enumerate some of the special problems or research questions that can be addressed by global remote sensing, but not by more conventional remote sensing.
9. The question of continuity of data (i.e., maintaining a continuous and consistent data archive) is a crucial issue for global change research. Discuss some of the problems this presents. For example, what is a sensible response when new, improved technology becomes available? Is it better to improve instruments or to maintain continuity of the older instruments?
10. Discuss some of the problems you might expect to encounter in maintaining an archive for global change data and in distribution of data to scientific analysts throughout the world.

## References

Assrar, G., and D. J. Dokken. 1993. *EOS Reference Handbook.* Washington, DC: NASA, 140 pp.

Crowley, T. J., and G. R. North. 1991. *Paleoclimatology.* New York: Oxford University Press, 339 pp.

Defries, R. S., and J. R. G. Townshend. 1994. NDVI-Derived Land Cover Classifications at Global Scale. *International Journal of Remote Sensing,* Vol. 15, pp. 3567–3586.

Delwiche, C. C. 1970. The Nitrogen Cycle. *Scientific American,* Vol. 223, pp. 136–146.

Eidenshink, J. C. 1992. The 1990 Conterminous AVHRR Data Set. *Photogrammetric Engineering and Remote Sensing,* Vol. 58, pp. 809–813.

EOS Science Steering Committee. (n.d.). *From Pattern to Process: The Strategy of the Earth Observing System* (EOS Science Steering Committee Report, Vol. 3). Washington, DC: NASA, 140 pp.

Foody, G., and P. Curran. 1994. *Environmental Remote Sensing from Global to Regional Scales.* New York: Wiley, 238 pp.

Gabrynowicz, J. I. 1995. The Global Land 1-km AVHRR Project: An Emerging Model for Earth Observations Institutions. *Photogrammetric Engineering and Remote Sensing,* Vol. 61, pp. 153–160.

Gallo, K. P., and J. C. Eidenshink. 1988. Differences in Visible and Near-IR Responses, and Derived Vegetation Indices, for the NOAA–9 and NOAA–10 AVHRRs: A Case Study. *Photogrammetric Engineering and Remote Sensing,* Vol. 54, pp. 485–490.

Goward, S. N., D. Dye, A. Kerber, and V. Kalb. 1987. Comparison of North and South American Biomes from AVHRR Observations. *Geocarto International,* Vol. 1, pp. 27–39.

Goward, S. N., C. J. Tucker, and D. Dye. 1985. North American Vegetation Patterns Observed with the NOAA–7 Advanced Very High Resolution Radiometer. *Vegetatio,* Vol. 64, pp. 3–14.

Harries, J. E., D. T. Llewellyn-Jones, C. T. Mutlow, M. J. Murray, I. J. Barton, and A. J. Prata. 1995. The

ASTER Programme: Instruments, Data and Science. Chapter 1 in *TERRA 2: Understanding the Terrestrial Environment: Remote Sensing Data Systems and Networks* (P. M. Mather, ed.). New York: Wiley, pp. 19–28.

Hastings, D. A., and W. J. Emery. 1992. The Advanced Very High Resolution Radiometer (AVHRR): A Brief Reference Guide. *Photogrammetric Engineering and Remote Sensing,* Vol. 58, pp. 1183–1188.

Iverson, L. R., E. A. Cook, and R. L. Graham. 1989. A Technique for Extrapolating and Validating Forest Cover across Large Regions: Calibrating AVHRR Data with TM Data. *International Journal of Remote Sensing,* Vol. 10, pp. 1085–1812.

Jenkinson, D. S. 1990. An Introduction to the Global Nitrogen Cycle. *Soil Use and Management,* Vol. 6, No. 2, pp. 56–61.

Justice, C. O., and J. R. Townshend. 1994. Data Sets for Global Remote Sensing: Lessons Learnt. *International Journal of Remote Sensing,* Vol. 17, pp. 3621–3639.

Justice, C. O., J. R. G. Townshend, B. N. Holben, and C. J. Tucker. 1985. Analysis of the Phenology of Global Vegetation Using Meteorological Satellite Data. *International Journal of Remote Sensing,* Vol. 6, pp. 1271–1318.

Loveland, T. R., J. W. Merchant, J. F. Brown, D. O. Ohlen, B. K. Reed, P. Olson, and J. Hutchinson. 1995. Seasonal Land Cover Regions of the United States. *Annals of the Association of American Geographers,* Vol. 85, pp. 339–335 (with map at 1;7,500,000).

Loveland, T., B. Reed, J. Brown, D. Ohlen, Z. Zhu, L. Yang, and J. Merchant. 2000. Development of a Global Land Cover Characteristics Database and IGBP DISCover from 1-km AVHRR Data. *International Journal of Remote Sensing,* Vol. 21, pp. 1303–1330.

MacEachren, A. M., et al. 1992. Visualization. Chapter 6 in *Geography's Inner Worlds* (R. F. Abler, M. G. Marcus, and J. M. Olson, eds.). New Brunswick, NJ: Rutgers University Press, pp. 99–137.

Maiden, M. F., and S. Grieco. 1994. NASA's Pathfinder Data Set Programme: Land Surface Parameters. *International Journal of Remote Sensing,* Vol. 15, pp. 3333–3345.

Mather, J. R., and G. V. Sdasyuk. 1991. *Global Change: Geographical Approaches.* Tucson: University of Arizona Press, 289 pp.

Mather, P. M. (ed.). 1995. *TERRA 2: Understanding the Terrestrial Environment: Remote Sensing Data Systems and Networks.* New York: Wiley, 236 pp.

Mounsey, H. (ed.), and R. Tomlinson (gen. ed.). 1988. *Building Databases for Global Science.* Philadelphia: Taylor & Francis, 419 pp.

Powlson, D. S. 1993. Understanding the Soil Nitrogen Cycle. *Soil Use and Management.* Vol. 9, No. 3, pp. 86–94.

Roller, N. E. G., and J. E. Colwell. 1986. Coarse-Resolution Satellite Data for Ecological Surveys. *BioScience,* Vol. 36, pp. 468–475.

Running, S. W., et al. 1994. Terrestrial Remote Sensing Science and Algorithms Planned for EOS/MODIS. *International Journal of Remote Sensing,* Vol. 15, pp. 3587–3620.

Skekielda, K.-H. 1988. *Satellite Monitoring of the Earth.* New York: Wiley, 326 pp.

Steinwand, D. R., J. R. Hutchinson, and J. P. Snydor. 1995. Map Projections for Global and Continental Data Sets and an Analysis of Pixel Distortion. *Photogrammetric Engineering and Remote Sensing,* Vol. 61, pp. 1487–1497.

Strahler, A. N., and A. H. Strahler. 1973. *Environmental Geoscience: Interaction between Natural Systems and Man.* New York: Wiley, 511 pp.

Tatum, A. J., S. J. Goetz, and S. I. Hay. 2004. Terra and Aqua: New Data for Epidemiology and Public Health. *Journal of Applied Earth Observation and Geoinfromation,* Vol. 6, pp. 33–46.

Townshend, J. R. G. 1994. Global Data Sets for Land Applications from the Advanced Very High Resolution Radiometer: An Introduction. *International Journal of Remote Sensing,* Vol. 15, pp. 3319–3332.

Tucker, C. J., W. W. Newcomb, and H. E. Dregne. 1994. AVHRR Data Sets for Determination of Desert Spatial Extent. *International Journal of Remote Sensing,* Vol. 15, pp. 3547–3565.

# The Outlook for the Field of Remote Sensing

This section presents a concluding look at the prospects for the field of remote sensing during the next decade or so. It identifies processes that will shape the character of the field for the immediate future. Each item discussed here identifies a continuation of a process now already well underway, so it requires no acute insight to identify them, but presenting them as a list may highlight their significance. Here they are presented in no specific order and are not intended to suggest that they form a comprehensive inventory of significant developments.

## Society and Institutions

1. At the time of this writing, present developments provide strong indicate of continued decline in the role of the U.S. government in supporting flagship remote sensing enterprises, such as Landsat (Chapter 6) and EOS (Chapter 21). Although U.S. federal support for high-visibility remote sensing programs may well continue, we can anticipate funding gaps and conflicts, and greater difficulty in marshalling political and public support in the face of competing priorities for public financing. In the United States, consolidation of federal governmental infrastructures supporting remote sensing across agencies and across application areas will provide a more focused, less diverse, spectrum of activities.

2. We can expect continued internationalization of broad-scale remote sensing enterprises, in part because of the inherent significance of such activities, and in part to fill the gap created by the contraction of U.S. governmental programs. Participants will likely include the European Union, China, Japan, India, Indonesia, Malaysia, and Canada. The U.S. government will continue to lead and to participate, but will less often assume its former pioneer role in developing new civil remote sensing enterprises.

3. Corporate remote sensing enterprises will continue to grow and play a significant role in the practice of remote sensing, often with a strong international component. Although the most visible organizations may focus on collection of imagery, those enterprises that process and distribute value-added products will increase with respect to the number and diversity of their specializations. Many will specialize less in the traditional remote sensing-based products than in unconventional products that will combine a remote sensing focus with unexpected application areas—for example, real estate, banking, public health, and law enforcement. Sophistication of

remote sensing applications for disaster management at the local, state, and federal levels will increase.

4. Availability of imagery to the public, and the availability of tools to use it, will increase. Society will accelerate its rate of establishing new applications that directly influence the day-to-day concerns of individual citizens who are not specialists in the field. The number and range of applications will increase, extending to include not only institutions with specialized expertise in remote sensing, but also individuals, households, businesses, local governments, and community enterprises that previously made little use of imagery, or required products produced by specialists.

Such developments will be beneficial to society by expanding economies and increasing efficiency, but will also create controversy, including disputes about issues such as privacy, use by law enforcement, ownership of imagery (and the information derived from it), and issues that arise from merging aerial imagery with other forms of data. Such developments will create a context for the discipline that differs greatly from that of the recent past, in which the field was seen as the province of specialists largely detached from such issues.

## Technology and Analysis

1. The next decade will see the continued displacement of traditional film-based aerial photography by digital imagery, lidar, and fine-resolution satellite imagery. Likewise, routine remote sensing activities will continue their migration from photographic to digital formats. There may well be a continuing role for conventional aerial photography but one much reduced from its status in the present and the recent past.

2. The significance of photointerpretation skills will continue to grow, especially in context-reading of imagery in digital formats. This development will be based upon the increase in the significance of fine-resolution digital satellite imagery, and the recognition that machine-assisted interpretation is not yet adequate for the needs of many fields that require spatially detailed, discipline-specific information. This development will require a cadre of image analysts—skilled interpreters, with discipline-specific knowledge, proficient both in image interpretation and manipulation of digital imagery.

3. Accelerated interest in analysis of sets of data, and assemblages and sequences of data (per the discussion of HDFs and NetCDFs in Chapter 4) that integrate very large arrays of multispectral or hyperspectral data, in combination with other large data sets such as meteorological or environmental data. Remote sensing will play a key, but supporting, role for such analyses. Image fusion may well assume greater significance, with the development of more sophisticated algorithms for combining information in many spectral channels.

4. We can expect the convergence of remote sensing with allied technologies, especially GPS and GIS, and with related systems such as wireless communications and miniaturization of imaging systems. Remote sensing will be influence by progress in advanced technologies, such as "formation flying" of remote sensing satellites and innovations in UAVs (Chapter 13).

5. We can expect expansion of applications of remote sensing into fields with more immediate significance for the public, including security, plant pathology, epidemiology, support of disaster relief, natural hazards, and homeland security. Although remote sensing will often contribute an essential component, other geospatial technologies will have significant roles in such systems.

6. One consequence of the combined effects of such developments will be recognition of the significance of human–computer interaction (HCI) with respect to remote sensing imagery and its interpretation. Advanced display systems and related software, dedicated to the visualization of complex data, will be required to achieve many of the advances listed above. Some of these advances will be based upon display software that will permit presentation of complex data in ways that enhance the ability of humans to understand the meaning of data when it forms part of a dynamic series or must be evaluated with respect to related data sets.

Although some of these likely developments are compatible with one another and many of the technological dimensions are synergistic, there are societal effects that will limit their effectiveness. Even effective technologies must compete with one another, and with other societal needs, for funding and investment. It seems likely that such competition will continue to create obstacles for the deployment of flagship civil remote sensing systems. As a result, sharing of costs with defense and national security projects will continue as a means of supporting costly systems.

The accelerated pace of the development of digital sensors, processing, and storage will continue the decline in traditional photographic technology. Although many in the field of remote sensing recognize the scientific and historical significance of the immense archive of photographic aerial images, marshaling the resources to preserve, organize, and index these resources (probably by digitization) remains one of the principal challenges of the next decade.

# Index

(*i* indicates an illustration; *t* indicates a table)

# About the Author

**James B. Campbell, PhD,** is Professor of Geography at Virginia Tech, where he teaches remote sensing, quantitative methods, and geomorphology. He has worked closely with students and faculty in related fields such as forestry, geology, agronomy, environmental sciences, and planning. Since 1997 he has served as codirector of Virginia Tech's Center for Environmental Analysis of Remote Sensing (CEARS). The author of numerous technical articles and several books, Dr. Campbell received the Outstanding Service Award of the American Society for Photogrammetry and Remote Sensing in 1994 and its Fellow Award in 1996. In 1997 he received the Outstanding Service Medal awarded by the Remote Sensing Specialty Group of the Association of American Geographers. Dr. Campbell's research has been sponsored by numerous academic, governmental, and private organizations, including NASA, the National Science Foundation, and the United States Geological Survey.